T0180792

Smart Innovation, Systems and Technologies

21

Editors-in-Chief

Prof. Robert J. Howlett
KES International
PO Box 2115
Shoreham-by-sea
BN43 9AF
UK
E-mail: rjhowlett@kesinternational.org

Dr. Lakhmi C. Jain
Adjunct Professor
University of Canberra
ACT 2601
Australia
and
University of South Australia
Adelaide
South Australia SA 5095
Australia
E-mail: Lakhmi.jain@unisa.edu.au

For further volumes:
http://www.springer.com/series/8767

Smart Innovation, Systems and Technologies

21

Editors-in-Chief

Prof. Robert J. Howlett
KES International
PO Box 2115
Shoreham-by-sea
BN43 9AP
UK
E-mail: rjhowlett@kesinternational.org

Dr. Lakhmi C. Jain
Adjunct Professor
University of Canberra
ACT 2601
Australia

and

University of South Australia
Adelaide
South Australia SA 5095
Australia
E-mail: Lakhmi.jain@unisa.edu.au

For further volumes:
http://www.springer.com/series/8767

Jeng-Shyang Pan, Ching-Nung Yang,
and Chia-Chen Lin (Eds.)

Advances in Intelligent Systems and Applications – Volume 2

Proceedings of the International Computer
Symposium ICS 2012 Held at Hualien, Taiwan,
December 12–14, 2012

Springer

Editors

Prof. Jeng-Shyang Pan
Department of Electronic Engineering
National Kaohsiung University
of Applied Sciences
Taiwan
Republic of China

Prof. Chia-Chen Lin
Department of Computer Science
and Information Management
Providence University
Taiwan
Republic of China

Prof. Ching-Nung Yang
Department of Computer Science
and Information Engineering
National Dong Hwa University
Taiwan
Republic of China

ISSN 2190-3018   e-ISSN 2190-3026
ISBN 978-3-662-52331-5   ISBN 978-3-642-35473-1 (eBook)
DOI 10.1007/978-3-642-35473-1
Springer Heidelberg New York Dordrecht London

Printed on acid-free paper

Springer is part of Springer Science+Business Media (www.springer.com)

Preface

The field of Intelligent Systems and Applications has expanded enormously during the last two decades. Theoretical and practical results in this area are growing rapidly due to many successful applications and new theories derived from many diverse problems. This book is dedicated to the proceedings of International Computer Symposium (ICS). ICS is a biennial event and is one of the largest joint international IT symposiums held in Taiwan. Founded in 1973, its aim was to provide a forum for researchers, educators, and professionals to exchange their discoveries and practices, and to explore future trends and applications in computer technologies. ICS 2012 consists of twelve workshops. Totally, we received 257 submissions. The Program Committee finally selected 150 papers for presentation at the symposium. This volume contains papers from the following workshops. We would like to express our gratitude to all of the authors, the reviewers, and the attendees for their contributions and participation.

- Workshop on Computer Architecture, Embedded Systems, SoC, and VLSI/EDA
- Workshop on Cryptography and Information Security
- Workshop on Digital Content, Digital Life, and Human Computer Interaction
- Workshop on Image Processing, Computer Graphics, and Multimedia Technologies
- Workshop on Parallel, Peer-to-Peer, Distributed, and Cloud Computing
- Workshop on Software Engineering and Programming Languages

In ICS 2012, we are very pleased to have the following four distinguished invited speakers, who delivered state-of-the-art information on the conference topics:

- Professor Fedor V. Fomin from University of Bergen, Norway
- Professor L. Harn from University of Missouri-Kansas City, USA
- Professor C.-C. Jay Kuo from University of Southern California, USA
- Mr. Michael Wang, an Enterprise Architect, from Oracle, USA

ICS 2012 would not have been possible without the support of many people and organizations that helped in various ways to make it a success. In particular, we would like to thank the Ministry of Education of ROC (especially, the Computer Center of the MOE), National Science Council of ROC, Computer Audit Association of ROC, and Taiwan Association of Cloud Computing for their assistance and financial supports.

December 2012   Jeng-Shyang Pan
Ching-Nung Yang
Chia-Chen Lin

Local Arrangement Chairs

Wen-Kai Tai	National Dong Hwa University, Taiwan
Shi-Jim Yen	National Dong Hwa University, Taiwan
Min-Xiou Chen	National Dong Hwa University, Taiwan
Shou-Chih Lo	National Dong Hwa University, Taiwan
Chih-Hung Lai	National Dong Hwa University, Taiwan

Publication Chairs

Chia-Chen Lin	Providence University, Taiwan
Chang-Hsiung Tsai	National Dong Hwa University, Taiwan
I-Cheng Chang	National Dong Hwa University, Taiwan
Pao-Lien Lai	National Dong Hwa University, Taiwan

Registration Chairs

Guanling Lee	National Dong Hwa University, Taiwan
Mau-Tsuen Yang	National Dong Hwa University, Taiwan

Web Chairs

Chenn-Jung Huang	National Dong Hwa University, Taiwan
Hsin-Chou Chi	National Dong Hwa University, Taiwan
Tao-Ku Chang	National Taitung University, Taiwan

Organization

Honorary Chairs

Wei-ning Chiang   Ministry of Education, Taiwan
Cyrus Chin-yi Chu   National Science Council, Taiwan
Han-Chieh Chao   Computer Audit Association, Taiwan
Maw-Kuen Wu   National Dong Hwa University, Taiwan

Conference Chairs

Rong-Guey Ho   Computer Center of Ministry of Education, Taiwan
Wei-Pang Yang   National Dong Hwa University, Taiwan
Ruay-Shiung Chang   National Dong Hwa University, Taiwan

Conference Co-chairs

Shin-Feng Lin   National Dong Hwa University, Taiwan
Chenn-Jung Huang   National Dong Hwa University, Taiwan

Program Chairs

Ching-Nung Yang   National Dong Hwa University, Taiwan
Shiow-Yang Wu   National Dong Hwa University, Taiwan
Sheng-Lung Peng   National Dong Hwa University, Taiwan

Publicity Chairs

Cheng-Chin Chiang   National Dong Hwa University, Taiwan
Han-Ying Kao   National Dong Hwa University, Taiwan
Chung Yung   National Dong Hwa University, Taiwan

Workshop on Cryptography and Information Security

Chairs

Chin-Chen Chang   Feng Chia University, Taiwan
Jeng-Shyang Pan   National Kaohsiung University of Applied
                                 Sciences, Taiwan

Co-chairs

Ching-Nung Yang   National Dong Hwa University, Taiwan
Chia-Chen Lin   Providence University, Taiwan

Program Committee

Lakhmi C. Jain   University of South Australia, Austraila
Stelvio Cimato   University of Milan, Italy
Lein Harn   University of Missouri-Kansas City, USA
Asifullah Khan   Pakistan Institute of Engineering and Applied
                                 Sciences, Nilore, Pakistan
Cheonshik Kim   Sejong University, Korea
Lina Wang   Wuhan University, China
Daoshun Wang   Tsinghua University, China
Chu-Sing Yang   National Cheng Kung University, Taiwan
Hung-Min Sun   National Tsing Hwa University, Taiwan
Sung-Ming Yen   National Central University, Taiwan
Chun-I Fan   National Sun Yat-sen University, Taiwan
Jung-Hui Chiu   Chang-Gung University, Taiwan
Bo-Chao Cheng   National Chung Cheng University, Taiwan
Chih-Hung Wang   National Chiayi University, Taiwan
Chia-Mei Chen   National Sun Yat-sen University, Taiwan
Der-Chyuan Lou   Chang-Gung University, Taiwan
Jinn-Ke Jan   National Chung Hsing University, Taiwan
Shyong-Jian Shyu   Ming Chuan University, Taiwan
Wen-Chung Kuo   National Yunlin University of Science
                                 and Technology, Taiwan
Wu-Chuan Yang   I-Shou University, Taiwan
Yung-Cheng Lee   WuFeng University, Taiwan

Workshop on Image Processing, Computer Graphics, and Multimedia Technologies

Chairs

Chung-Lin Huang   National Tsing Hua University, Taiwan
Hsi-Jian Lee   Tzu Chi University, Taiwan

Co-chairs

I-Cheng Chang   National Dong Hwa University, Taiwan
Wen-Kai Tai   National Dong Hwa University, Taiwan
Chia-Hung Yeh   National Sun Yat-sen University, Taiwan

Program Committee

Chuan-Yu Chang   National Yunlin University of Science Technology, Taiwan
Bing-Yu Chen   Taiwan University, Taiwan
Chu-Song Chen   Academia Sinica, Taiwan
Jiann-Jone Chen   National Taiwan University of Science and Technology, Taiwan
Mei-Juan Chen   National Dong Hwa University, Taiwan
Wei-Ming Chen   National Ilan University, Taiwan
Fang-Hsuan Cheng   Chung Hua University, Taiwan
Cheng-Chin Chiang   National Dong Hwa University, Taiwan
Chin-Chuan Han   National United University, Taiwan
Shang-Hong Lai   National Tsing Hua University, Taiwan
Shin-Feng Lin   National Dong Hwa University, Taiwan
Chung-Ming Wang   National Chung Hsing University, Taiwan
Mau-Tsuen Yang   National Dong Hwa University, Taiwan

Workshop on Digital Content, Digital Life, and Human Computer Interaction

Chair

Yi-Ping Hung   National Taiwan University, Taiwan

Co-chairs

Cheng-Chin Chiang   National Dong Hwa University, Taiwan
Mau-Tsuen Yang   National Dong Hwa University, Taiwan

Program Committee

Chu-Song Chen   Academia Sinica, Taiwan
Mei-Juan Chen   National Dong Hwa University, Taiwan
Jen-Hui Chuang   National Chiao Tung University, Taiwan
Kuo-Liang Chung   National Taiwan University of Science
                                        and Technology, Taiwan
Pau-Choo Chung   National Cheng Kung University
Kuo-Chin Fan   National Central University, Taiwan
Chiou-Shann Fuh   National Taiwan University, Taiwan
Jun-Wei Hsieh   National Taiwan Ocean University, Taiwan
Chung-Lin Huang   National Tsing Hua University, Taiwan
Shang-Hong Lai   National Tsing Hua University, Taiwan
Chung-Nan Lee   National Sun Yat-sen University, Taiwan
Tong-Yee Lee   National Cheng Kung University, Taiwan
Hong-Yuan Mark Liao   Academia Sinica, Taiwan
Chia-Wen Lin   National Tsing Hua University, Taiwan
Chin-Teng Lin   National Chiao Tung University, Taiwan
Tyng-Luh Liu   Academia Sinica, Taiwan
Chia-Hung Yeh   National Sun Yat-sen University, Taiwan
Shiaw-Shian Yu   Industrial Technology Research Institute, Taiwan

Workshop on Parallel, Peer-to-Peer, Distributed, and Cloud Computing

Chair

Wang-Chien Lee   Pennsylvania State University, USA

Co-chair

Shiow-Yang Wu   National Dong Hwa University, Taiwan

Program Committee

Jiann-Liang Chen   National Taiwan University of Science
                           and Technology, Taiwan
Ge-Ming Chiu   National Taiwan University of Science
                           and Technology, Taiwan
Yeh-Ching Chung   National Tsing Hua University, Taiwan
Michael J. Franklin   UC Berkeley, USA
Kuen-Fang Jack Jea   National Chung Hsing University, Taiwan
Chung-Ta King   National Tsing Hua University, Taiwan
Chiang Lee   National Cheng Kung University, Taiwan
Chung-Nan Lee   National Sun Yat-sen University, Taiwan
Deron Liang   National Central University, Taiwan
Yao-Nan Lien   National Chengchi University, Taiwan
Chuan-Ming Liu   National Taipei University of Technology, Taiwan
Pangfeng Liu   National Taiwan University, Taiwan
Dusit (Tao) Niyato   Nanyang Technological University, Singapore
Wen-Chih Peng   National Chiao Tung University, Taiwan
Zheng Yan   Aalto University, Finland

Workshop on Software Engineering and Programming Languages

Chair

Wuu Yang   National Chiao Tung University, Taiwan

Co-chairs

Shih-Chien Chou   National Dong Hwa University, Taiwan
Chung Yung   National Dong Hwa University, Taiwan

Program Committee

Barrett Bryant   University of North Texas, USA
Wei-Ngan Chin   National University of Singapore, Singapore
Kung Chen   National Chengchi University, Taiwan
Tyng-Ruey Chuang   Academica Sinica, Taiwan
Pao-Ann Hsiung   National Chung Cheng University, Taiwan
Gwan-Hwan Hwang   National Taiwan Normal University, Taiwan
Wen-Hsiang Lu   National Cheng Kung University, Taiwan
Marjan Mernik   University of Maribor, Slovenia
Shin-Cheng Mu   Academica Sinica, Taiwa
Jiann-I Pan   Tzu Chi University, Taiwan
Yih-Kuen Tsay   National Taiwan University, Taiwan
Hsin-Chang Yang   National University of Kaohsiung, Taiwan

Workshop on Computer Architecture, Embedded Systems, SoC, and VLSI/EDA

Chair

Cheng-Wen Wu   Industrial Technology Research Institute, Taiwan

Co-chair

Hsin-Chou Chi   National Dong Hwa University, Taiwan

Program Committee

Robert Chen-Hao Chang   National Chung Hsing University, Taiwan
Yao-Wen Chang   National Taiwan University, Taiwan
Chung-Ho Chen   National Cheng Kung University, Taiwan
Tien-Fu Chen   National Chiao Tung University, Taiwan
Chung-Ping Chung   National Chiao Tung University, Taiwan
Shen-Fu Hsiao   National Sun Yat-sen University, Taiwan
Chun-Lung Hsu   Industrial Technology Research Institute, Taiwan
Ying-Jer Huang   National Sun Yat-sen University, Taiwan
Yin-Tsung Hwang   National Chung Hsing University, Taiwan
Gene-Eu Jan   National Taipei University, Taiwan
Yeong-Kang Lai   National Chung Hsing University, Taiwan
Gwo-Giun Lee   National Cheng Kung University, Taiwan
Tay-Jyi Lin   National Chung Cheng University, Taiwan
Tsung-Ying Sun   National Dong Hwa University, Taiwan
Chua-Ching Wang   National Sun Yat-sen University, Taiwan
Ro-Min Weng   National Dong Hwa University, Taiwan
An-Yeu Wu   National Taiwan University, Taiwan

Additional Reviewers

Bao Rong Chang
Jiann-Jone Chen
Cheng-Chieh Chiang
Shin-Yan Chiou
Chi-Hung Chuang
Jerry Chou
D.J. Guan
Wen-Zhong Guo
Kai-Lung Hua
JS Jang Jiang
Jong Yih Kuo
Yuchi Lai
Harn Lein
Wen-Hung Liao

Tzong-Jye Liu
Huang-Chia Shih
Raylin Tso
Chih-Hung Wang
Hsin-Min Wang
L.N. Wang
Shiuh-Jeng Wang
Wei-Jen Wang
Der-Lor Way
Tin-Yu Wu
Chao-Tung Yang
Martin Yang
C.Y. Yao

Contents

Track 1: Authentication, Identification, and Signature

Track 2: Intrusion Detection

Track 3: Steganography, Data Hiding, and Watermarking

Track 4: Database, System, and Communication Security

Track 5: Computer Vision, Object Tracking, and Pattern Recognition

Track 6: Image Processing, Medical Image Processing, and Video Coding

Track 7: Digital Content, Digital Life, and Human Computer Interaction

Track 8: Parallel, Peer-to-Peer, Distributed, and Cloud Computing

Track 9: Software Engineering and Programming Language

Track 10: Computer Architecture, Embedded Systems, SoC, and VLSI/EDA

A Secure ECC-Based RFID Authentication Scheme Using Hybrid Protocols

Yi-Pin Liao and Chih-Ming Hsiao

Department of Computer Science and Information Engineering, University of St. John,
Taipei, Taiwan
{newsun87,cm}@mail.sju.edu.tw

Abstract. Radio Frequency Identification (RFID) has grown tremendously and has been widely applied in various applications. RFID tags are becoming very attractive devices installed a small microchip for identification of products. This chip functionality makes it possible to verify the authenticity of a product. It is well known that elliptic curve cryptosystem (ECC) receive much attention due to their small key sizes and efficient computations. Recently, some ECC-based authentication schemes are proposed to apply well to the limited resources of the tags. Unfortunately, these schemes ignore some security and operational issues. In this paper, we proposed a secure ECC-based RFID authentication scheme to achieve mutual authentication using both secure ID-verifier transfer and challenge-response protocols. Moreover, the proposed scheme can satisfy the security requirements of RFID. Performance analysis and function comparisons demonstrate that the proposed scheme is well suited for RFID tags with the scarceness of resources.

Keywords: Radio Frequency Identification, Elliptic curve cryptosystem, ID-verifier transfer.

1 Introduction

Recently, RFID has grown tremendously and has been widely applied in various applications such as inventory tracking, supply chain management, theft-prevention, and the like. Radio Frequency Identification (RFID) systems can identify hundreds of objects in a contactless manner at one time. Although RFID technology has potentials to improve our lives, it also presents a privacy risk. Privacy for RFID system is challenging problems due to tags response to nearby readers without discretion. In addition, other security issues make RFID tags an easy target for malicious attacks. Hence, it is essential to design authentication protocol that make RFID system more secure before it is viable for mass deployment. That is, privacy and authentication are the two main security issues that need to be addressed for the RFID technology. The required cryptographic primitives range from symmetric and asymmetric algorithms to hash functions and random number generators. We simply classify the RFID authentication schemes published in the literatures [1-18] into non-public key cryptosystem (NPKC) based schemes and public key cryptosystem (PKC) based schemes.

J.-S. Pan et al. (Eds.): *Advances in Intelligent Systems & Applications*, SIST 21, pp. 1–13.
DOI: 10.1007/978-3-642-35473-1_1   © Springer-Verlag Berlin Heidelberg 2013

The suitability of PKC for RFID is an open research problem due to the limitation in tag cost, gate area and power consumption. Moreover, it was previously proven that PKC algorithms are necessary to solve the requirements of RFID system [19]. That is, it is not possible to satisfy the requirements only with symmetric cryptographic algorithms such as hash algorithms and symmetric key encryption algorithms. To achieve significant consumer market penetration, RF tags will need to be priced in the US$0.05-US$0.10 range and contains only 500 to 5K gates. This causes many researchers deem the PKC based RFID systems to be infeasible at present. Fortunately, the CMOS technologies steadily advance and the fabrication costs decrease, which allows stronger security solutions on tags. Recently, a few papers [20-21] try to discuss the feasibility of PKC primitive cheap implementations on RFID tags; for example, Gaubatz et. al implements Rabin's encryption with cost about 17K gates [20], and Kaya and Savaş design NTRU public encryption which costs only about 3K gates [21].

Among PKC algorithms, elliptic curve cryptosystem (ECC) based algorithms would be best choice for RFID systems due to their small key sizes and efficient computations. However, ECC is still considered to be impracticable for very low-end constrained devices like sensor networks and RFID tags. Very recently, Lee et al. (2008) [22] presents the proposed RFID processor is composed of a microcontroller, an EC processor (ECP), and a bus manager, where the ECP is over $GF(2^{163})$. For an efficient computation with restrictions on the gate area and the number of cycles, several techniques are introduced in the algorithms and the architecture level. As a result, the overall architecture takes 12.5K gates. Lee et al.'s scheme shows the plausibility of meeting both security and efficiency requirements even in a passive RFID tag. That is, an ECC based solution would be one of the best candidates for the RFID system.

In this paper, we will adopt ECC primitives [23] to design an efficient RFID mutual authentication scheme. Compared with the related works based on ECC, the proposed authentication scheme has remarkable features as follows. (1) It integrates both secure ID-verifier transfer and challenge-response protocols to achieve mutual authentication; (2) It solves the security risks neglected by previous ECC-based works; (3) Our work can be applied well to other authentication applications which are similar to RFID environment. The remainder of this paper is organized as follows. In section 2, we discuss all possible vulnerabilities and requirements in RFID system. In section 3, we review the recent PKC based authentication schemes. Next, we propose a secure ECC-based authentication scheme for RFID system in section 4. Then, we make security analysis in section 5, and then performance and functionality comparisons are shown in section 6. Finally, the conclusion is given in section 7.

2   Essential System Requirements

To enhance the security strength of RFID system to be suitable for various applications, we define the system requirements that need to be considered when designing an authentication protocol to solve some security issues. The system requirements are defined in terms of mutual authentication, confidentiality, anonymity, availability, forward security and scalability.

(1) Mutual authentication: It is essential that authentication should occur between the objects of the RFID system. In cases when communication between only the tag and reader is insecure, the authentication process is performed between the tag and the database of the back-end server.

(2) Confidentiality: Confidentiality requires that all of the secret information is securely transmitted during all communications. Therefore, to ensure confidentiality, the tag transmit the encrypt information so that only the server can recognize it.

(3) Anonymity: Anonymity is the most important security requirement for privacy [2]. Anonymity is the property that adversary cannot trace tag by using interactions with tag. If the transmitted tag information cannot satisfy anonymity, an attacker with the same reader can continuously trace the owner of a specific tag or detect the real-time location of the tag owner by using readers dispersed over several locations.

(4) Availability: Authentication process should be run all the time between the server and the tag. To provide privacy protection, after a successful protocol run, most RFID authentication schemes update the secret information between the back-end database and the tag. Hence, the de-synchronization attack causing the secret information to refresh out of phase must be prevented.

(5) Forward security: It is essential that the previously transmitted information cannot be traced using the present transmission tag information. If the past location of the specific tag owner can be traced using the compromised information, it constitutes a serious privacy.

(6) Scalability: Scalability is a desirable property in almost any system, enabling it to handle growing amounts of work in a graceful manner. In RFID system, the server must find the matching record from the database to identify the tag, and a scalable RFID protocol should therefore avoid any requirement for work proportional to the number of tags. Hence, the computational workload must be sustained by the server with the growth for the amount of the tags.

3   Related ECC-Based Works

Some features are especially attractive for security applications where computational power and integrated circuit space is limited, such as smart cards, PC cards, and wireless devices. Such is the case with elliptic curve groups, which were first proposed for cryptographic use independently by Neal Koblitz and Victor Miller in 1985 [29]. For introducing ECC-based RFID schemes in this subsection, we should describe the concepts of ECC and related logarithms. In view of simplification, the details refer to [29]. Next, we will discuss some published schemes based on ECC in RFID system [30-32] as follows.

3.1   Tuyls et al.'s Scheme Using Schnorr Protocol [30]

Tuyls et al. (2006) [30] proposed an ECC-based RFID identification scheme using Schnorr identification protocol [33]. They claimed their scheme can resist against tag counterfeiting, but Lee et al. (2008) [31] pointed their protocol suffers some weakness. The attacker can eavesdrops and collects the exchange messages aiming at a target tag. Hence, he/she can analyze the exchange messages to find the ID-verifier of the target tag. In other words, Tuyls et al.'s scheme is vulnerable to location tracking attack. Moreover, the attacker collects the exchange messages and the ID-verifier of the specific tag. Hence, he can identify the unknown tag as the specific tag using an active attack. Hence, the attacker can then use the ID-verifier to distinguish the tag from the past conversations easily. In other word, their protocol does not achieve forward security. Especially, their protocol only considers tag-to-reader authentication, excluding reader-to-tag authentication. This makes tags easy to suffer malicious queries, because they are not capable of confirming whom they are talking to. In other hand, a scalability problem also exists in it. This means that the server requires linear search to identity each tag and thus increases considerable computational cost. Hence, their protocol lacks scalability.

3.2   Batina et al.'s Scheme Using Okamoto Protocol [31]

Batina et al. (2007) [32] proposed an ECC-based RFID identification protocol using Okamoto's identification protocol [34]. Although they claimed their protocol can avoid active attacks, Lee et al. (2008) [31] pointed their protocol is vulnerable to location tracking attack. Similarly, a scalability problem and forward secrecy also exists in Batina et al.'s scheme.

3.3   Lee et al.'s Scheme Based on Random Access Control [32]

To solve all the requirements for RFID systems, Lee et al. (2008) [31] designed a new RFID protocol based on ECDLP. However, the works in [35-36] showed Lee et al.'s vulnerability against tracking attacks and forgery attacks. The failure of the security proof is caused by neglecting the possibility that an attacker can use multiple sets of authentic communication history [35]. The result shows that a tag can be traced by an attacker. Besides, Bringer et al. [36] show how tags can be tracked if the attacker has intercepted the same tag twice and that a tag can be impersonated if it has been passively eavesdropped three times. Similarly, their protocol only considers tag-to-reader authentication, excluding reader-to-tag authentication. This makes tags easy to suffer malicious queries.

4   The Proposed Scheme

This paper proposes an ECC-based mutual authentication schemes that satisfies all the requirements in RFID system. To assure the security of the ID-verifier transmitted from the tag over radio frequency, a secure ID-verifier transfer protocol need to be design. Moreover, a challenge-response protocol is involved to refresh the communication messages. The proposed scheme is secure against various types of attacks and completely solves the existing research problems. Our scheme consists of two phases: the setup phase and the authentication phase. In the proposed scheme, communication between the reader and back-end server is secure, while communication between each tag and reader is insecure.

4.1 Setup Phase

In the setup phase, the server generates system parameters. The server chooses a random number $x_S \in Z_n$ as its private key and sets $P_S (= x_S P)$ as its public key. It also chooses $x_T \in Z_n$ as the private key of each tag and sets public key $Z_T (=x_T P)$ as the tag's ID-verifier. Hence, the server inserts the entry $\{Z_T, x_T\}$ of each tag into its database. Moreover, each tag stores $\{Z_T, x_T\}$ and system parameters in the memory. The system parameters and the storage of each entity are summarized in Table 1.

Table 1. The system parameters and the storage of each entity

System parameters	$P_S (= x_S P)$: Server's public key.
	P : Base point in $E(Z_p)$, whose order is n .
Server storage	Each tag's entry $[Z_T, x_T]$, server private key x_S and common parameters(P , n)
Tag storage	The tag's public key Z_T as ID-verifier, private key x_T and common parameters (P_S , P , n)

4.2 Authentication Phase

The authentication phase is depicted in Fig. 1. The interactions between the tag and the server are described as follows.

Step 1. The server generates a random number $r_2 \in Z_n$ and computes $R_2 = r_2 P$. Then it sends R_2 along with query message to the tag.

Step 2. After receiving the query message $< Query, R_2 >$, the tag chooses a random number $r_1 \in Z_n$ and computes $R_1 = r_1 P$. And then the tag computes two temporary secret keys $TK_{T1} = r_1 R_2$ and $TK_{T2} = r_1 P_S$. Next, the tag computes $Auth_T = Z_T + TK_{T1} + TK_{T2}$ to encrypt the ID-verifier Z_T , and sends $< Auth_T, R_1 >$ to the server.

Step 3. After receiving $< Auth_T, R_1 >$, the server recovers two temporary secret keys by way of computing $TK_{S1} = r_2 R_1$ and $TK_{S2} = x_S R_1$. Next, the server utilizes the following equation to retrieve the ID-verifier Z_T of the tag:

$$\text{Auth}_T - TK_{S1} - TK_{S2} = (Z_T + TK_{T1} + TK_{T2}) - TK_{S1} - TK_{S2}$$
$$= (Z_T + r_1 R_2 + r_1 x_S P) - r_2 R_1 - x_S r_1 P = (Z_T + r_1 r_2 P + r_1 x_S P) - r_2 r_1 P - x_S r_1 P \quad (1)$$
$$= Z_T$$

Then, the reader searches tag's ID-verifier in the database. If it is found, the reader confirms the tag to be legitimate and obtains the corresponding private key x_T. Next, the server calculates $\text{Auth}_S = x_T R_1 + r_2 Z_T$ and sends back $< \text{Auth}_S >$ to be authenticated by the tag.

Step 4. Next, the tag computes $r_1 Z_T + x_T R_2$ and checks if the value is equal to the received Auth_S. If it is equal, the tag conforms that the server is authentic.

5   Security Analysis

In this section, we will analyze the security of the proposed scheme to verify whether the system requirements have been satisfied. For correctness analysis, an efficient and convincing formal methodology is needed to evaluate the proposed scheme. Before that, we make some reasonable assumptions to sustain the security analysis.

A1: The tag believes r_1 is fresh in every session.

A2: The reader believes r_2 is fresh in every session.

A3: x_S is unknown for anyone except the reader.

A4: Z_T and x_T are unknown for anyone except the tag and the server.

5.1   System Requirements Analysis

In the following, we give an in-depth analysis of the proposed scheme in terms of system requirements. Before that, we draw some inferences to prove our authentication protocol as follows:

I1: The tag believes that the ID-verifier Z_T is securely transmitted to the server. As step 2 of the authentication phase, the tag sends response message $< \text{Auth}_T, R_1 >$ to the server. The message $\text{Auth}_T (= Z_T + TK_{T1} + TK_{T2})$ can be interpreted as an encryption of Z_T with the temporary secret keys (TK_{T1}, TK_{T2}). The attacker cannot decrypt Z_T from Auth_T since the security of both TK_1 and TK_{T2} is based on ECDHP.

Hence, Z_T is embedded in Auth_T and securely transmitted to the server.

I2: The server believes that the ID-verifier Z_T is securely transmitted to the tag. As step 3 of the authentication phase, the server sends $< \text{Auth}_S >$ to the tag. The message $\text{Auth}_S (= x_T R_1 + r_2 Z_T)$ can be interpreted as an encryption of $r_2 Z_T$ with the secret key of $x_T R_1$. In other hand, as step 4 of the authentication phase, the message $\text{Auth}_S (= r_1 Z_T + x_T R_2)$ can be regarded as an encryption of $r_1 Z_T$ with the secret key of $x_T R_2$. Since neither (r_2, x_T) nor (r_1, x_T) is known by the attacker, the ID-verifier Z_T cannot be extracted from Auth_S.

By I1 and I2, a secure ID-verifier transfer protocol can be achieved.

I3: The freshness of exchange messages $< \text{Auth}_S, \text{Auth}_T >$ is assured in every session. By I1 and I2, the messages Auth_T and Auth_S are controlled using two random numbers (r_1, r_2). According to A1 and A2, two random numbers (r_1, r_2) is unpredictable and different in every session. That is, the attacker cannot reuse the previous messages to cheat the tag or the server.

SR1: Mutual Authentication between the Tag and the Server

Proof: In general, the main goal of the authentication protocol shows that the communication entities can achieve mutual authentication. The server believes the tag is authentic by checking the correctness of ID-verifier (i.e. Z_T) embedded in the received Auth_T. As step 3 of the authentication phase, the server receives message $< \text{Auth}_T, R_1 >$. According to I1, only the server can decrypt Z_T by way of calculating $\text{Auth}_T - TK_{S1} - TK_{S2}$. If the result matches the entry listed in database, the identity of the tag is authenticated by the server. In other hand, the tag believes the server

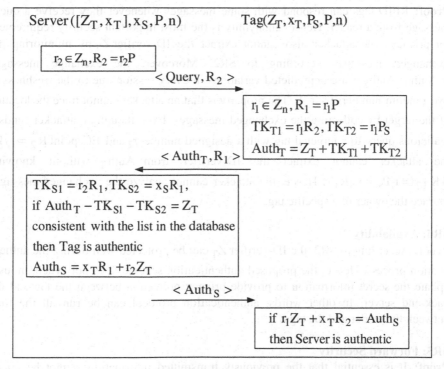

Fig. 1. The proposed scheme

is authentic by checking the correctness of ID-verifier (i.e. Z_T) embedded in the received $Auth_S$. As step 4 of the authentication phase, the tag receives message $Auth_S$ can be deduced as follows:

$$Auth_S = x_T R_1 + r_2 Z_T = x_T (r_1 P) + r_2 (x_T P) = r_1 (x_T P) + x_T (r_2 P) = r_1 Z_T + x_T R_2 \quad (2)$$

After receiving $Auth_S$, only the tag with $\{Z_T, x_T\}$ can compute $r_1 Z_T + x_T R_2$ using (r_1, R_2). If the computed result matches the received $Auth_S$, the tag believes the corresponding party owns the secret information $\{Z_T, x_T\}$. According to A4, the identity of the server is authenticated by the tag. Hence, we prove that the server and the tag authenticate each other. Moreover, the protocol can satisfy the system requirements discussed below.

SR2: ID-Verifier Confidentiality
Proof: During authentication process, the ID-verifier Z_T of the tag should be protected well over unsecure channel. According to I1 and I2, the attacker cannot extract Z_T from the collected messages $< Auth_T, Auth_S >$. Hence, the proposed protocol can achieve ID-verifier confidentiality.

SR3: Anonymity
Proof: RFID tags can respond with some messages whenever they receive a query message from a reader. Hence, anonymity is the most important security requirement for privacy. The attacker also cannot extract the ID-verifier Z_T by monitoring the exchanged messages according to SR2. Moreover, the exchange messages $< Auth_T, Auth_S >$ are unpredicted variations in every session due to the freshness of two random numbers (r_1, r_2). The property is that an attacker cannot trace the location of the target by collecting the exchanged messages. Even though an attacker sends a malicious query to a targeted tag with a designed number r_2^* and EC point $R_2^* = r_2^* P$, the attacker cannot extract the ID-verifier from $Auth_T$ without knowing $TK_{T2} (= r_1 P_S = x_S R_1)$. Hence, the attacker cannot analyze the exchanged messages to trace the owner of a specific tag.

SR4: Availability
Proof: According to SR2, the ID-verifier Z_T can be protected well during the authentication process. Hence, the proposed authentication scheme does not synchronously update the secret information to provide privacy protection between the tag and the back-end server. In other words, authentication protocol can be run all the time between the reader and the tags.

SR5: Forward Security
Proof: It is essential that the previously transmitted information cannot be traced using the present transmission tag information. We assume an attacker knows the

secret keys of a tag, i.e. Z_T and x_T, by way of physical attack on a corrupted tag. However, an attacker still does not know random numbers temporarily generated and used inside of a tag and the server. Hence, the proposed scheme still provides on unpredictable variations in the past communication messages.

SR6: Scalability
Proof: According to step3 in the authentication phase, the server extracts the ID-verifier Z_T from the received $Auth_T$, and then search the matched entry in database. This means the server does not requires linear search to identity each tag and thus save considerable computation cost while the number of the tags increases.

5.2 Attack Analysis

Next, we will prove that the proposed scheme can resist the following attacks

AKR1 Replay Attack Resisting
Proof: Having intercepted previous communication, the attacker can replay the same message of the receiver or the sender to pass the verification of the system. Hence, the attacker may masquerade as the reader or the tag to launch replay attack by reusing previous $Auth_S$ or $Auth_T$. By I3, the action will fail because the freshness of the messages transmitted in the authentication phase is controlled by two random numbers, i.e. (r_1, r_2).

AKR2 Tag Masquerade Attack Resisting
Proof: The attacker may intercept and modify the previous message of the legal tag to pass the authentication of the server. If the attacker may construct a valid authentication message $< Auth_T, R_1 >$ to pass the server's examination, he/she need to extract the ID-verifier Z_T from the previous $Auth_T$. By SR2, the ID-verifier Z_T is securely embedded in transmitted message over unsecure channel. Hence, the attacker cannot construct a valid authentication message without knowing the ID-verifier Z_T. That is, the tag masquerade attack will fail.

AKR3 Server Spoofing Attack Resisting
Proof: Server spoofing attack means the attacker may masquerade as the server to gain the benefits. The attacker constructs a valid message $Auth_S$, where the ID-verifier Z_T is also embedded. By SR2, the attacker cannot succeed without knowing the ID-verifier Z_T.

AKR4 DoS Attack Resisting
Proof: According to SR4, the proposed authentication scheme does not synchronously update the secret information to provide privacy protection between the back-end databases. Hence, our scheme can eliminate the risk against DoS attack.

AKR5 Location Tracking Attack Resisting
Proof: According to SR2, the data transmitted between the server and the tag is well protected so that the tag's ID-verifier Z_T could not be retrieved from the message flow. Moreover, the message flow is provided on unpredictable variations in every session. Hence, the location tracking fail will fail.

AKR6 Cloning Attack Resisting
Proof: If a group of tags share the same secret key and use it for the authentication, it is vulnerable to cloning attacks. In the proposed scheme, there is no shared secret key in all of the tags. That is, the attacker cannot use the revealed secret to clone some other tags.

6   Performance and System Requirements Comparisons

It is well-known that most of RFID tags have limited resources. Hence, it is very important issue for performance analysis in the real applications. In general, performance analysis includes the estimation of computation cost and communication cost. We focus the performance analysis in tag since the server is regarded as a powerful device. In this section, we analyze the efficiency of the proposed scheme. In general, all ECC protocols include a few point scalar multiplications and additions. Besides EC point scalar multiplication and addition, general modular operations are also needed for the computation of the authentication protocols. Recently, Lee et al. [22] proposed a compact architecture of an EC-based security processor for RFID. It is composed of a microcontroller, an EC processor (ECP), and a bus manager, where the ECP is over $GF(2^{163})$. ECP, which computes EC point scalar multiplications, is composed of a controller, MALU (Modular Arithmetic Logic Unit) and a register file. Since the modular operations can be performed in parallel with the EC point scalar multiplication, the former operations do not contribute to the latency. In the proposed scheme, the tag performs five point scalar multiplication computations and three point addition computations. Moreover, the server performs five point multiplication computations and three point addition computations.

Table 2 shows that the comparison among the existing ECC-based schemes in computation cost and communication cost. Seemingly, other ECC-based schemes [30, 32-33] are more efficient than the proposed scheme in tag's performance. However, they do not only provide mutual authentication but also suffer from some attacks discussed above. Hence, the performance of the proposed scheme is reasonable and acceptable. Moreover, we summarize the comparisons of system requirements among the existing ECC-based schemes in Table 3. The result concludes that our scheme is more secure and practical in real applications.

Table 2. Performance comparisons among ECC-based authentication schemes for RFID system

		Ours	Tuyls et al. [30]	Batina et al. [32]	Lee et al. [33]
Computation cost	Tag	(5,3)	(1,0)	(2,1)	(2,0)
(ECm, ECa)*	Server	(5,4)	(2,1)	(3,2)	(4,2)

ECm: ECC point scalar multiplication. ECa: ECC point addition.

Table 3. System requirements comparisons among ECC-based authentication schemes for RFID system

	Ours	Tuyls et al. [30]	Batina et al. [32]	Lee et al. [33]
Mutual authentication	Yes	No	No	No
Confidentiality	Yes	No	No	Yes
Anonymity	Yes	No	No	No
Availability	Yes	Yes	Yes	Yes
Forward security	Yes	No	No	Yes
Scalability	Yes	No	No	Yes

7   Conclusion

We present an ECC-based authentication scheme for RFID combined with hybrid protocols, including secure ID-verifier transfer and challenge-response protocols. Previously proposed schemes based on ECC cannot satisfy the requirements of RFID systems, including mutual authentication, confidentiality, anonymity, forward security and scalability. In this paper, the proposed scheme can be proven to satisfy all essential system requirements through security analysis. Performance analysis of the proposed scheme is well suited for RFID tags embedded a compact architecture of an EC-based security processor. In addition, we conclude that the proposed scheme can be applied well to other authentication applications which are similar to RFID environment.

References

1. EPCglobal: Specification for RFID Air Interface, http://www.epcglobalinc.org
2. Chien, H.Y., Chen, C.H.: Mutual Authentication Protocol for RFID Conforming to EPC Class 1 Generation 2 Standards. Computers Standards & Interfaces 29(2), 254–259 (2007)
3. Duc, D.N., Park, J., Lee, H., Kim, K.: Enhancing Security of EPCglobal Gen-2 RFID Tag against Traceability and Cloning. In: Proc. 2006 Symp. Cryptography and Information Security (2006)

4. Juels, A.: Strengthening EPC Tag against Cloning. In: Proc. ACM Workshop Wireless Security, WiSe 2005, pp. 67–76 (2005)
5. Yeh, T., Wang, Y., Kuo, T., Wang, S.: Securing RFID systems conforming to EPC Class 1 Generation 2 standard. Expert Systems with Applications 37, 7678–7683 (2010)
6. Chien, H.Y., Huang, C.W.: Security of Ultra-Lightweight RFID Authentication Protocols and Its Improvements. ACM Operating System Rev. 41(2), 83–86 (2007)
7. Li, T., Wang, G.: Security Analysis of Two Ultra-Lightweight RFID Authentication Protocols. In: Venter, H., Eloff, M., Labuschagne, L., Eloff, J., von Solms, R. (eds.) New Approaches for Security, Privacy and Trust in Complex Environments. IFIP, vol. 232, pp. 109–120. Springer, Boston (2007)
8. Peris-Lopez, P., Hernandez-Castro, J.C., Estevez-Tapiador, J.M., Ribagorda, A.: LMAP: A Real Lightweight Mutual Authentication Protocol for Low-Cost RFID Tags. In: Proc. Second Workshop RFID Security (July 2006)
9. Peris-Lopez, P., Hernandez-Castro, J.C., Estevez-Tapiador, J.M., Ribagorda, A.: EMAP: An Efficient Mutual-Authentication Protocol for Low-Cost RFID Tags. In: Meersman, R., Tari, Z., Herrero, P. (eds.) OTM Workshops 2006. LNCS, vol. 4277, pp. 352–361. Springer, Heidelberg (2006)
10. Peris-Lopez, P., Hernandez-Castro, J.C., Estevez-Tapiador, J.M., Ribagorda, A.: M^2AP: A Minimalist Mutual-Authentication Protocol for Low-Cost RFID Tags. In: Ma, J., Jin, H., Yang, L.T., Tsai, J.J.-P. (eds.) UIC 2006. LNCS, vol. 4159, pp. 912–923. Springer, Heidelberg (2006)
11. Chien, H.Y.: SASI: A New Ultralightweight RFID Authentication Protocol Providing Strong Authentication and Strong Integrity. IEEE Trans. Dependable and Secure Computing 4(4), 337–340 (2007)
12. Weis, S.A., Sarma, S.E., Rivest, R.L., Engels, D.W.: Security and Privacy Aspects of Low-Cost Radio Frequency Identification Systems. In: Hutter, D., Müller, G., Stephan, W., Ullmann, M. (eds.) Security in Pervasive Computing 2003. LNCS, vol. 2802, pp. 201–212. Springer, Heidelberg (2004)
13. Chien, H.Y.: Secure Access Control Schemes for RFID Systems with Anonymity. In: Proc. 2006 Int'l Workshop Future Mobile and Ubiquitous Information Technologies, FMUIT 2006 (2006)
14. Lim, J., Oh, H., Kim, S.-J.: A New Hash-Based RFID Mutual Authentication Protocol Providing Enhanced User Privacy Protection. In: Chen, L., Mu, Y., Susilo, W. (eds.) ISPEC 2008. LNCS, vol. 4991, pp. 278–289. Springer, Heidelberg (2008)
15. Liu, A.X., Bailey, L.R.A.: A privacy and authentication protocol for passive RFID tags. Computer Communications 32, 1194–1199 (2009)
16. Kang, S.Y., Lee, D.G., Lee, I.Y.: A study on secure RFID mutual authentication scheme in pervasive. Computer Communications 31, 4248–4254 (2008)
17. Cho, J.S., Yeo, S.S., Kim, S.K.: Securing against brute-force attack: A hash-based RFID mutual authentication protocol using a secret value. Computer Communications 34, 391–397 (2011)
18. Juels, A., Molner, D., Wagner, D.: Security and Privacy Issues in E-Passports. In: Proc. First Int'l Conf. Security and Privacy for Emerging Areas in Comm. Networks, SecureComm 2005 (2005)
19. Burmester, M., Medeiros, B., Motta, R.: Robust, anonymous RFID authentication with constant key-lookup. Cryptology ePrint Archive: listing for 2007 (2007/402) (2007)

20. Gaubatz, G., Kaps, J.P., Ozturk, E., Sunar, B.: State of the Art in Ultra-Low Power Public Key Cryptography for Wireless Sensor Networks. In: Proc. in the Third IEEE International Conference on Pervasive Computing and Communications Workshops, PERCOMW 2005 (2005)
21. Kaya, S.V., Savas, E., Levi, A., Erçetin, Ö.: Public key cryptography based privacy preserving multi-context RFID infrastructure. Ad Hoc Networks 7, 136–152 (2009)
22. Lee, Y.K., Sakiyama, K., Verbauwhede, I.: Elliptic-Curve-Based Security Processor for RFID. IEEE Trans. on Computers 11(57) (November 2008)
23. Koblitz, N.: Elliptic curve cryptosystems. Mathematics of Computation 48, 203–209 (1987)
24. Bono, S., Green, M., Stubblefield, A., Juels, A., Rubin, A., Szydlo, M.: Security analysis of a cryptographically enabled RFID device. Pre-print, http://www.rfidanalysis.org (May 4, 2006)
25. RFID Journal: EPC Tags Subject to Phone Attacks. News Article (February 24, 2006), http://www.rfidjournal.com/article/articleview/2167/1/1/ (May 4, 2006)
26. Chen, Y., Chou, J.S., Sun, H.M.: A novel mutual authentication scheme based on quadratic residues. Computer Networks 52, 2373–2380 (2008)
27. Cao, T., Shen, P.: Cryptanalysis of some RFID authentication protocols. Journal of Communications 3(7) (December 2008)
28. Yeh, T.C., Wua, C.H., Tseng, Y.M.: Improvement of the RFID authentication scheme based on quadratic residues. Computer Communications 34, 337–341 (2011)
29. Koblitz, N.: Elliptic curve cryptosystems. Mathematics of Computation 48, 203–209 (1987)
30. Tuyls, P., Batina, L.: RFID-Tags for Anti-counterfeiting. In: Pointcheval, D. (ed.) CT-RSA 2006. LNCS, vol. 3860, pp. 115–131. Springer, Heidelberg (2006)
31. Lee, Y.K., Batina, L., Verbauwhede, I.: EC-RAC (ECDLP Based Randomized Access Control): Provably Secure RFID authentication protocol. In: IEEE International Conference on RFID, pp. 97–104 (2008)
32. Batina, L., Guajardo, J., Kerins, T., Mentens, N., Tuyls, P., Verbauwhede, I.: Public-Key Cryptography for RFID-tags. In: Fifth IEEE International Conference on Pervasive Computing and Communications Workshops, pp. 217–222 (2007)
33. Schnorr, C.-P.: Efficient Identification and Signatures for Smart Cards. In: Brassard, G. (ed.) CRYPTO 1989. LNCS, vol. 435, pp. 239–252. Springer, Heidelberg (1990)
34. Okamoto, T.: Provably Secure and Practical Identification Schemes and Corresponding Signature Schemes. In: Brickell, E.F. (ed.) CRYPTO 1992. LNCS, vol. 740, pp. 31–53. Springer, Heidelberg (1993)
35. Deursen, T., Radomirović, S.: Attacks on RFID Protocols. Cryptology ePrint Archive: listing for 2008 (2008)
36. Bringer, J., Chabanne, H., Icart, T.: Cryptanalysis of EC-RAC, a RFID Identification Protocol. In: Franklin, M.K., Hui, L.C.K., Wong, D.S. (eds.) CANS 2008. LNCS, vol. 5339, pp. 149–161. Springer, Heidelberg (2008)

A Dynamic Approach to Hash-Based Privacy-Preserving RFID Protocols

Chih-Yuan Lee[1], Hsin-Lung Wu[1], and Jen-Chun Chang[1]

Department of Computer Science and Information Engineering,
National Taipei University,
New Taipei City, Taiwan
s710083108@webmail.ntpu.edu.tw, {hsinlung,jcchang}@mail.ntpu.edu.tw

Abstract. We study how to design a hash-based identification protocol in a RFID system which obtains security and privacy against active
adversaries. Here, an active adversary can not only track a tag via successful or unsuccessful identifications with legal or illegal readers but
also perform a compromised attack. In *SPC 2003*, Weis et al. used the
technique of the randomized hash lock to design a privacy-preserving
protocol against such active adversaries. However, in their protocol, the
time complexity of identifying a requested tag is linear in the number of
legal tags. It is still an open problem to design a protocol which obtains
privacy against active adversaries and has a sublinear time complexity
of tag identification.

In this work, we revisit this open problem. We modify the protocol of
Weis et al. by using a dynamic key management scheme to manage tag
identities stored in the back-end database instead of a static approach.
For privacy, our protocol obtains the same privacy level as the protocol
of Weis et al.. For performance, the amortized cost of tag identification of
our protocol is almost twice the optimal amortized cost by a competitive
analysis. For practical implementation, our protocol is very suitable to
be realized in RFID systems due to its online property.

1 Introduction and Related Works

RFID (Radio-Frequency Identification) is a technology in which one can identify objects or people by embedding tags, a small microchip capable of wireless
data transmission. By tagging wares in shops, one can speed up the process of
registration with wireless scanning. RFID tags have several characteristics. First
of all, each tag has an identifier to represent itself. Moreover, such identifiers are
long enough so that it has a unique code. When a tiny tag is implanted within an
object, finding such a tag means discovering the corresponding object. Second,
tag identification via radio frequency allows tagged objects to be read multiple
times at a distance. These characteristics introduce security and privacy issues.
Objects embedded with insecure tags may reveal private information as they
are queried by legal or illegal readers. For the privacy issue, objects embedded
with tags that do not reveal any sensitive information may also be tracked by

J.-S. Pan et al. (Eds.): *Advances in Intelligent Systems & Applications*, SIST 21, pp. 15–23.
DOI: 10.1007/978-3-642-35473-1_2   © Springer-Verlag Berlin Heidelberg 2013

the implanted tags. This is because the tag responses to the requesting readers are possible to help locate the tagged objects by analyzing information from the protocol view between the embedded tag and the reader. This may cause objects to reveal their private data such as their identifications in the future. We refer the readers to Juels' excellent survey [4] on the privacy issue.

In [3,2], formal definitions of privacy are given. Privacy of tags is defined by the ability of adversaries to trace tags by using their responses to readers' interrogations. The authors define two degrees of privacy for RFID tags. Adversaries who try to distinguish two given tags only from their successful identifications with a legitimate reader are called passive adversaries. On the other hand, adversaries who try to differentiate two given tags from their successful or unsuccessful identifications with any reader (legal or illegal) are called active adversaries. In [2], privacy against passive adversaries is called universal traceability whereas privacy against active adversaries is called existential traceability.

On one hand, for privacy-preserving protocols against passive adversaries, Alomair et al. propose a nice protocol in which tag identification can be obtained with constant time [3]. On the other hand, Weis et al. give an identification protocol called the randomized hash lock [7] which obtains privacy against active adversaries. However, in this protocol, the time complexity of tag identification is linear in n where n is the number of legitimate tags. To improve the time efficiency of tag identification, Molnar and Wagner [5] propose a tree-based protocol in which tag identification can be done within $O(\log n)$. However, in [1], Avoine et al. propose new attacks on RFID privacy called compromised attacks in which adversaries may know secrets of some tags. Avoine et al. show that one can obtain compromised attacks for the tree-based protocols with high successful probability. Note that the compromised attacks threat not only tree-based protocols but also those protocols in which the tag identities have high correlation. In this paper, we allow active adversaries perform compromised attacks. It is still an open problem whether there is a privacy-preserving protocol against active adversaries which has identification complexity in sublinear in n.

1.1 Our Contributions

In this work, we construct a privacy-preserving protocol whose privacy level is the same as the protocol of Weis et al [7]. In fact, our proposed protocol obtains privacy against active adversaries and against compromised attacks. In order to improve the efficiency of tag identification, we use a dynamic key management scheme called the move-to-front scheme to store tag identities in the back-end database. By a competitive analysis, the amortized cost of our proposed protocol is almost twice the amortized cost of the optimal key management scheme. We also show that, in some cases, our proposed protocol has amortized constant time to obtain tag identification while the original randomized-hash-lock protocol may require linear time to do it. For practical implementation, the proposed move-to-front key management scheme is easy to implemented by using a data structure such as linked lists.

The remaining part of the paper is organized as follows. In Section 2, we give necessary privacy definitions and some notations. We also introduce the randomized-hash-lock protocol of Weis et al. there. In Section 3, we propose our move-to-front protocol and its efficiency analysis. Finally we conclude in Section 5.

2   Preliminaries

In our proposed protocol, we assume that communication channel between the reader and the back-end database is secure while communication channel between the reader and each tag is insecure. For convenience, we use the following notation in the rest of the paper.

Notation	Corresponding Meaning
n	Number of tags participating in the RFID system
ID_k	Identity of the k-th tag
$h()$	Hash operation

2.1   Privacy Definitions

Here, we give two definitions for the RFID privacy.

Definition 1. *[3](Universal Untraceability) An RFID protocol is universally untraceable if an adversary cannot track a tag based on information obtained from the protocol view between the tag and a legal reader.*

Definition 2. *[3](Existential Untraceability) An RFID protocol is existentially untraceable if an active adversary cannot track a tag based on its responses to multiple interrogation even if the tag has not been able to accomplish mutual authentication with an authorized reader.*

For more formal definitions of the above two definitions, we refer the readers to [3].

2.2   The Randomized-Hash-Lock Protocol

In [7], Weis et al. propose a hash-based RFID identification protocol which obtains existential untraceability. Usually, because of using hash functions and randomness, their protocol is called randomized-hash-lock protocol. Their protocol is describes as follows.

Setup. There are n identities $ID_1, \ldots, ID_n$ which are stored in a fixed array in the back-end database. The i-th tag has ID_i as its identity. Each Tag and each reader have random number generators and share a hash function h.

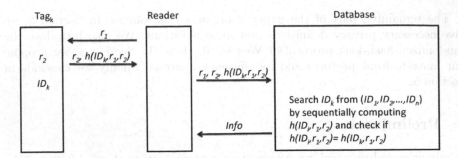

Fig. 1. One single round of the protocol of Weis et al. in [7]

Identification Process. The identification process goes as follows. The implemented version is illustrated in Fig. 1.

1. The reader requests the tag and sends a random string r_1 to it.
2. The k-th tag generates a random string r_2 and computes $h(ID_k, r_1, r_2)$. Next the tag sends them as well as r_1 and r_2 to the reader which passes them to the database.
3. Assume that the tag identities are stored in the linked list whose order is $(ID_1, \ldots, ID_n)$. To identify the tag, the database sequentially computes $h(ID_i, r_1, r_2)$ and check if it is equal to $h(ID_k, r_1, r_2)$ for i from 1 to n. The above protocol of Weis et al. obtains existential untraceability.

3   MTF Protocol

In this section, we propose a hash-based RFID protocol which uses a dynamic key management scheme in the back-end database. For convenience, we call the proposed protocol $\mathcal{MTF}$.

Setup. There are n identities $ID_1, \ldots, ID_n$ which are stored by using a linked list in the database. We illustrate such a linked list in Figure 2. The i-th tag has ID_i as its identity. Each Tag and each reader have random number generators and share a hash function h.

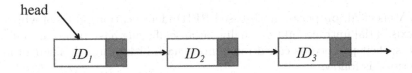

Fig. 2. A linked list of Tag identities

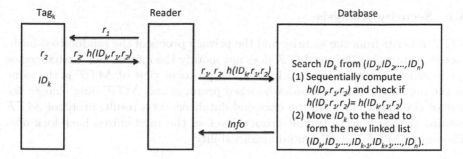

Fig. 3. One single round of protocol $\mathcal{MTF}$

Identification Process. Now the identification process goes as follows. The implemented version is illustrated in Fig. 3.

1. The reader requests the tag and sends a random string r_1 to it.
2. The k-th tag generates a random string r_2 and computes $h(ID_k, r_1, r_2)$. Next the tag sends them as well as r_1 and r_2 to the reader which passes them to the database.
3. Assume that the tag identities are stored in the linked list whose order is $(ID_1, \ldots, ID_n)$. To identify the tag, the database sequentially computes $h(ID_i, r_1, r_2)$ and check if it is equal to $h(ID_k, r_1, r_2)$ for i from 1 to n. After finding ID_k, the database updates the linked list by moving ID_k to the first position of the linked list. The order of the resulting linked list is $(ID_k, ID_1, \ldots, ID_{k-1}, ID_{k+1}, \ldots, ID_n)$.

The difference between our protocol and protocol of Weis et al. is that the database of our protocol updates the order of the lists of tags while the database of the protocol of Weis et al. does not. Here we give an example to illustrate the modification of the linked list after identifying a specific tag in Figure 4. In this example, after Tag 5 is found, it is moved to the head of the linked list.

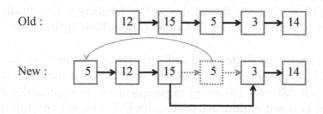

Fig. 4. The updated linked-list after one round of protocol $\mathcal{MTF}$

3.1   Security Analysis

$\mathcal{MTF}$ inherits from the security and the privacy proofs of the randomized-hash-locked protocol in [7] since $\mathcal{MTF}$ does not modify the information exchanged or the internal content of the tag. In fact, the protocol view of $\mathcal{MTF}$ is the same as the one of the randomized-hash-locked protocol and $\mathcal{MTF}$ only change the way of key management in the back-end database. As a result, protocol $\mathcal{MTF}$ obtains the same security and privacy level as the randomized-hash-lock one. Therefore, $\mathcal{MTF}$ has existential untraceability.

4   A Competitive Analysis on Efficiency

Let n be the number of items in the linked-list. Given a protocol $\mathcal{P}$, we define the cost $C_{\mathcal{P}}(i)$ of the protocol $\mathcal{P}$ to identify tag i by the number of cryptographic hash operations used by the back-end database. Let σ be the requested tag sequence of length m, that is $\sigma = (i_1, i_2, \ldots, i_m)$ where i_k means that Tag i_k is requested by the reader in the k-th order. In addition, let $C_{\mathcal{A}}(\sigma)$ be the cost of the protocol $\mathcal{A}$ on the requested tag sequence σ, that is

$$C_{\mathcal{A}}(\sigma) = \sum_{k=1}^{m} C_{\mathcal{A}}(i_k).$$

Static Offline Optimal Protocol: Let f_i be the frequency of accessing the i-th tag on a requested tag sequence σ. Suppose we know f_i for each $1 \le i \le n$. An obvious way to reduce the cost of searching tags is to arrange the tag list in a decreasing order of the frequencies. For convenience, we assume that f_i is decreasing with respect to i. We call such a list a static offline optimal tag list. Let $\mathcal{SOOP}$ be the randomized-hash-lock protocol which uses such a static offline optimal tag list in the back-end database. It is easy to see that

$$C_{\mathcal{SOOP}}(\sigma) = \sum_{i=1}^{n} i f_i.$$

We call $C_{\mathcal{SOOP}}(\sigma)$ the static offline optimal cost on σ. The main drawback of protocol $\mathcal{SOOP}$ is that we do not know the frequency f_i initially. Thus one cannot expect to arrange the tag list in a static offline optimal tag list.

Optimal Self-organizing Protocol: A self-organizing protocol can move the identity of the requested tag into any position of the linked list after the database finds or inserts it. Given a requested tag sequence σ, the optimal self-organizing protocol on σ is a self-organizing protocol which obtains the minimal cost on σ. Note that the optimal self-organizing protocol knows the whole requested sequence σ as the static offline optimal protocol. Furthermore, it can perform exchanges of tag identities after identifying a tag whereas the static offline optimal protocol cannot. Note that the complexity of exchanges of tag identities can be easily obtained in a linked list. Let $C_{OPT}(\sigma)$ be the cost of the optimal self-organizing protocol on σ. Clearly we have

$$C_{OPT}(\sigma) \leq C_{SOOP}(\sigma)$$

for any sequence σ. Similar to the static offline optimal protocol, the drawback of the optimal protocol is that it should be implemented by an offline key management scheme.

Move-to-Front Protocol: This is just our proposed protocol $\mathcal{MTF}$. In this protocol, after identifying or inserting a tag, the algorithm moves the identity of the requested tag to the head of the linked list while preserving the relative order of the other tags. Obviously its key management scheme on tag identities is an online scheme. Hence $\mathcal{MTF}$ is suitable to be used in the back-end database of the RFID system. Given a requested sequence σ, let $C_{\mathcal{MTF}}(\sigma)$ be the cost of the Move-to-Front protocol on σ. By the same argument in the seminar result of [6], it can be proved that

$$C_{\mathcal{MTF}}(\sigma) \leq 2C_{OPT}(\sigma)$$

for any sequence σ if the protocol starts from the empty linked list. On the other hand, if the protocol starts from a nonempty list, then we have

$$C_{\mathcal{MTF}}(\sigma) \leq 2C_{OPT}(\sigma) + O(n^2)$$

for any sequence σ. As a corollary, we have

$$C_{\mathcal{MTF}}(\sigma) \leq 2C_{SOOP}(\sigma) + O(n^2)$$

for any sequence σ.

4.1   Some Examples

Here we consider some distributions on tag-accessing frequency.

Example 1. Define $f_i \doteq 2^{n-i}$ for $1 \leq i \leq n$. Suppose σ is any requested sequence in which the frequency of the i-th tag is f_i for $1 \leq i \leq n$. Then, the cost of the static offline optimal protocol is as follows:

$$C_{SOOP}(\sigma) = \sum_{i=1}^{n} i f_i$$

$$= \sum_{i=1}^{n} i 2^{n-i}$$

$$= 2^{n+1} - n - 2.$$

Note that the length of σ is

$$m = \sum_{i=1}^{n} f_i = \sum_{i=1}^{n} 2^{n-i} = 2^n - 1.$$

Since $C_{\mathcal{MTF}}(\sigma) \leq 2C_{\mathcal{SOOP}}(\sigma) + O(n^2)$, we have

$$C_{\mathcal{MTF}}(\sigma) \leq 2C_{\mathcal{SOOP}}(\sigma) + O(n^2) = 2(2^{n+1} - n - 2) + O(n^2).$$

The amortized cost of $\mathcal{MTF}$ protocol for the requested sequence σ is

$$\frac{C_{\mathcal{MTF}}(\sigma)}{m} \leq \frac{2(2^{n+1} - n - 2) + O(n^2)}{2^n - 1} \leq 4$$

if n is large enough.

On the other hand, the worst static case occurs when tags are listed in an increasing order according to frequencies f_i's. The cost is $\sum_{i=1}^{n} i2^{i-1} = (n - 1)2^n - 3$. The amortized cost is at least $\frac{(n-1)2^n - 3}{2^n - 1} \geq n - 1$.

Example 2. In this example, we show that $\mathcal{MTF}$ has a better performance than $\mathcal{SOOP}$. Suppose σ is a requested sequence such that

$$\sigma = (n, \underbrace{n-1,, \ldots, n-1}_{2}, \ldots, \underbrace{i, \ldots, i}_{n-i+1}, \ldots, \underbrace{1, \ldots, 1}_{n}).$$

Clearly the frequency $f_i = n - i + 1$ for $1 \leq i \leq n$ in the sequence σ. The cost of the static offline optimal protocol is as follows:

$$
\begin{aligned}
C_{\mathcal{SOOP}}(\sigma) &= \sum_{i=1}^{n} if_i \\
&= \sum_{i=1}^{n} i(n - i + 1) \\
&= \frac{n(n+1)(2n+1)}{6}.
\end{aligned}
$$

Next, the length of σ is

$$m = \sum_{i=1}^{n} f_i = \sum_{i=1}^{n} n - i + 1 = \frac{n(n+1)}{2}.$$

So the amortized cost of $\mathcal{SOOP}$ for σ is at least $\frac{2n}{3}$. Let us see the cost $C_{\mathcal{MTF}}(\sigma)$. We have

$$C_{\mathcal{MTF}}(\sigma) \leq \sum_{i=1}^{n} n + (i - 1) = \frac{3n^2 - n}{2}.$$

Thus the amortized cost of $\mathcal{MTF}$ protocol for the requested sequence σ is

$$\frac{C_{\mathcal{MTF}}(\sigma)}{m} \leq 3.$$

5   Conclusion

In this paper, we construct a hash-based RFID identification protocol called $\mathcal{MTF}$ which obtains existential untraceability and can be against compromised attacks. In addition, via a competitive analysis, $\mathcal{MTF}$ has almost twice optimal amortized cost on the time efficiency of tag identification. Moreover, in a practical sense, the proposed $\mathcal{MTF}$ protocol is suitable to be implemented in RFID systems due to its online property.

References

1. Avoine, G., Dysli, E., Oechslin, P.: Reducing Time Complexity in RFID Systems. In: Preneel, B., Tavares, S. (eds.) SAC 2005. LNCS, vol. 3897, pp. 291–306. Springer, Heidelberg (2006)
2. Alomair, B., Poovendran, R.: Privacy versus scalability in radio frequency identification systems. Computer Communications 33(18), 2155–2163 (2010)
3. Alomair, B., Clark, A., Cuellar, J., Poovendran, R.: Scalable RFID systems: a privacy-preserving protocol with constant-time identification. In: Procedings of the 40th Annual IEEE/IFIP International Conference on Dependable Systems and Networks V, DSN 2010, Chicago, Illinois, USA. IEEE (2010)
4. Juels, A.: RFID security and privacy: a research survey. IEEE Journal on Selected Areas in Communications 24(2), 381–394 (2006)
5. Molnar, D., Wagner, D.: Privacy and security in library RFID: issues, practices, and architectures. In: Proceedings of the 11th ACM Conference on Computer and Communications Security, pp. 210–219 (2004)
6. Sleator, D., Tarjan, R.: Amortized efficiency of list update and paging rules. Communications of the ACM 28(2), 202–208 (1985)
7. Weis, S.A., Sarma, S.E., Rivest, R.L., Engels, D.W.: Security and Privacy Aspects of Low-Cost Radio Frequency Identification Systems. In: Hutter, D., Müller, G., Stephan, W., Ullmann, M. (eds.) Security in Pervasive Computing 2003. LNCS, vol. 2802, pp. 201–212. Springer, Heidelberg (2004)

An Extension of Harn-Lin's Cheater Detection and Identification

Lein Harn[1] and Changlu Lin[2]

[1] Department of Computer Science and Electrical Engineering
University of Missouri-Kansas City, MO 64110-2499, USA
harnl@umkc.edu
[2] Key Laboratory of Network Security and Cryptology
Fujian Normal University, Fujian, 350007, P.R. China
cllin@fjnu.edu.cn

Abstract. Cheater detection and identification are important issues in the process of secret reconstruction. Most algorithms to detect and identify cheaters need the dealer to generate and distribute additional information to shareholders. In a recent paper, algorithms have been proposed to detect and identify cheaters based on shares only without needing any additional information. However, more than t (i.e. the threshold) shares are needed in the secret reconstruction. In this paper, we extend the algorithms to the situation when there are exact t shares in the secret reconstruction. We adopt the threshold changeable secret sharing which shareholders work together to change the threshold t into a new threshold t' (i.e., $t' < t$) and generate new shares of a (t', n) secret sharing; while at the same time, maintain the original secret. Since $t' < t$, there are redundant shares. We also include discussion on how to select the new threshold t' in order to detect and identify cheaters successfully.

Keywords: Secret sharing, threshold changeable secret sharing, cheaters, redundant share.

1 Introduction

In a (t, n) secret sharing scheme, a dealer divides the secret into shares in such a way that any t (i.e., the threshold) or more than t shares can reconstruct the secret; while any fewer than t shares cannot obtain any information about the secret. Shamir's (t, n) secret sharing scheme [16] is based on the linear polynomial. Secret reconstruction uses Lagrange interpolating polynomial.

When shareholders present their shares in the secret reconstruction, dishonest shareholders (i.e. cheaters) can always exclusively derive the secret by presenting fake shares and thus the other honest shareholders get nothing but a fake secret. It is easy to see that Shamir's (t, n) secret sharing scheme does not prevent dishonest shareholders in the secret reconstruction. Cheater detection and identification are important features in order to provide fair reconstruction of a secret.

There are many research papers in the literature to propose algorithms for cheater detection and identification. Most of these algorithms

J.-S. Pan et al. (Eds.): *Advances in Intelligent Systems & Applications*, SIST 21, pp. 25–32.
DOI: 10.1007/978-3-642-35473-1_3   © Springer-Verlag Berlin Heidelberg 2013

[17,4,15,6,5,11,9,14,13,1] assume that there are exact t shareholders participated in the secret reconstruction. The dealer needs to provide additional information to enable shareholders to detect and identify cheaters. Some algorithms [12,3] use error-correcting codes to detect and identify fake shares.

In a recent paper, Harn and Lin [7] proposed a new approach to detect and identify cheaters. The algorithm uses shares to detect and identify cheaters. When there are more than t (i.e., the threshold) shares in the secret reconstruction, the redundant shares can be used to detect and identify cheaters. In this approach, shares in a secret sharing scheme serve for two purposes; that are, (a) reconstructing the secret and (b) detecting and identifying cheaters. Since Harn and Lin's algorithm requires more than t shares in the secret reconstruction, the algorithm does not work if there are exact t shares. In this paper, we generalize Harn and Lin's algorithm to the situation when they are exact t shares in the (t, n) secret reconstruction. We adopt the threshold changeable secret sharing (TCSS) which shareholders work together to change the threshold t into a new threshold t' and generate new shares of a (t', n) secret sharing; while at the same time, maintain the original secret. Since $t' < t$, there are redundant shares. The new shares can be verified without revealing the secret and new shares. We also include discussion on how to select the new threshold t' in cheater detection and identification.

The Rest of This Paper Is Organized as Follows. In the next section, we briefly review Shamir's (t, n) secret sharing scheme [16] and Harn and Lin's algorithm [7]. In Section 3, we propose our generalized scheme. We conclude in Section 4.

2   Preliminaries

2.1   Review of Shamir's Secret Sharing Scheme [16]

In Shamir's (t, n) secret sharing scheme based on the polynomial, there are n shareholders and a mutually trusted dealer. The scheme consists of two algorithms:

Scheme 1: Shamir's (t, n) secret sharing scheme

---

1. Share generation algorithm: the dealer first picks a random polynomial of degree $t - 1$, $f_i(x) = a_{t-1}x^{t-1} + \cdots + a_1 x + a_0 \pmod{p}$, such that the secret s satisfies $f(0) = a_0 = s$ and all coefficients, $a_0, a_1, \ldots, a_{t-1} \in \mathbb{Z}_p$, p is a prime with $p > s$. The dealer computes shares as, $f(x_i)$, for $i = 1, 2, \ldots, n$, and distributes each share $f(x_i)$ to shareholder U_i secretly.
2. Secret reconstruction algorithm: it takes any t or more than t shares, for example, with following t shares, $\{(x_1, f(x_1)), (x_2, f(x_2)), \ldots, (x_t, f(x_t))\}$, as inputs, and outputs the secret s using the Lagrange interpolating formula as

$$s = \sum_{i=1}^{t} f(x_i) \prod_{r=1, j \neq i}^{t} \frac{-x_j}{x_i - x_j} \pmod{p}.$$

We note that the above algorithms satisfy the basic requirements of the secret sharing scheme, that are, (1) with the knowledge of any t or more than t shares, shareholders can reconstruct the secret s; and (2) with the knowledge of any $t - 1$ or fewer than $t - 1$ shares, shareholders cannot obtain the secret s. Shamir's secret sharing scheme is unconditionally secure since the scheme satisfies these two requirements without making any computational assumption. For more information on this scheme, please refer to the original paper [16].

2.2 Review of Harn and Lin's Algorithm [7]

We briefly review the algorithm [7] to detect and identify cheaters using the property of strong t-consistency and majority voting mechanism. The algorithm assumes that there are more than t shareholders participated in the secret reconstruction.

Benaloh [2] presented a notion of t-consistency to determine whether a set of n (i.e., $n > t$) shares are generated from a polynomial of degree $t - 1$ at most. Recently, Harn and Lin [8] proposed a new definition of strong t-consistency which is the extension of Benaloh's definition.

Definition 1 (Strong t-consistency [8]). *A set of n shares (i.e., $t < n$) is said to be strong t-consistent if (a) any subset of t or more than t shares can reconstruct the same secret, and (b) any subset of fewer than t shares cannot reconstruct the same secret.* □

It is obvious that if shares in Shamir's (t, n) secret sharing scheme are generated by a polynomial with degree $t - 1$ exactly, then shares are strong t-consistent. Checking strong t-consistency of n shares can be executed very efficiently by using the Lagrange interpolating formula. In fact, to check whether n shares are strong t-consistent or not, it only needs to check whether the interpolation of n shares yields a polynomial with degree $t - 1$ exactly. If this condition is satisfied, we can conclude that all shares are strong t-consistent. However, if there are some invalid shares, the degree of the interpolating polynomial of these n shares is more than $t - 1$ with very high probability. In other words, these n shares are most likely to be not strong t-consistent.

- **Method for Detecting Cheaters:** If there are more than t shares in Shamir's (t, n) secret sharing scheme and all shares are valid, all shares must be strong t-consistent. Cheater detection is determined by checking the property of strong t-consistency of all shares.
- **Method for Identifying Cheaters:** If there are n (i.e., $n > t$, the threshold) shares in the secret reconstruction and there have some invalid shares, the reconstructed secrets must be inconsistent. This is because any t shares can construct a secret and there are $\binom{n}{t}$ different combinations. Any t shares including some invalid shares is very likely to reconstruct a different secret from the true secret reconstruct based on all valid shares. After cheaters being detected, if the true secret is the majority of reconstructed secrets, we can use

the majority voting mechanism to identify fake shares. The cheater identification method needs to figure out the majority of the reconstructed secrets first. A set, A, consisting of t valid shares is identified. Then, cheaters (i.e., having fake shares) can be identified one at a time by computing the reconstructed secret using shares in A and the testing share.

The primary advantage of Harn and Lin's algorithm is its simplicity. Shamir's (t, n) secret sharing scheme is capable to detect and identify cheaters without any modification. In [7], it also investigates the bounds of detection and identification which are functions of the threshold, the number of cheaters, and the number of redundant shares in the secret reconstruction. Interest readers can refer to the original paper.

Remark 1. As pointed out in [7], the computational complexity of method to detect cheaters is $O(1)$ and the complexity to identify cheaters is $O(j!)$, where j is the number of shares in the secret reconstruction. The method of cheater identification only works properly when there is small number of shares in the secret reconstruction.

3   Proposed Algorithm

From now on, we assume that there are t, where $t \leq n$, shareholders with their shares $\{(x_1, f(x_1)), (x_2, f(x_2)), \ldots, (x_t, f(x_t))\}$, obtained from a trusted dealer in Shamir's (t, n) secret sharing scheme want to reconstruct the secret.

The basic idea of our approach is to adopt the threshold changeable secret sharing (TCSS) which shareholders work together to change the threshold t into a new threshold t' and generate new shares of a (t', n) secret sharing; while at the same time, maintain the original secret. Since $t' < t$, there has enough redundant shares in the secret reconstruction to detect and identify cheaters; while at the same time, keep the same secret. The new shares of the (t', t') secret sharing scheme are generated and are used to reconstruct the secret. In our proposed algorithm, each shareholder M_i acts like a dealer to select a random $(t' - 1)$-th degree polynomial $f_i(x)$ with the constant term $f_i(0) = f(x_i) \prod_{j=1, j \neq i}^{t} \frac{-x_j}{x_i - x_j}$ (mod p). Then, each shareholder M_i computes sub-shares $f_i(x)$ for other shareholders. After receiving all shares from other shareholders, each shareholder releases the sum of all sub-shares which is the share of sum of polynomials as $F(x) = \sum_{r=1}^{t} f_r(x)$ (mod p). The interpolation of all released sums can construct the polynomial $F(x)$ with constant term $F(0) = s$. The TCSS scheme in this algorithm is similar to the strong (n, t, n) verifiable secret sharing scheme proposed in [8]. However, in current application, there are t shareholders working together to change the threshold t into a new threshold t' and generate new shares. Thus, it is a (t, t', t) verifiable secret sharing scheme. In addition, these new shares can be verified without revealing the secret and new shares. We will give detail discussions in the extended version of this paper.

Scheme 2: Secret reconstruction algorithm

Step 1. For each shareholder M_i, uses his share $f(x_i)$ obtained from the dealer to compute $y_i' = f(x_i) \prod_{j=1, j \neq i}^{i} \frac{-x_i}{x_i - x_j} \pmod{p}$ and selects a random polynomial $f_i(x)$ with $(t'-1)$-th degree satisfying $f_i(0) = y_i'$. Then, shareholder M_i computes sub-shares, $f_i(x_j)$, for all other shareholders, M_j, for $j = 1, 2, \ldots, t$, $j \neq i$, and sends each sub-share $f_i(x_j)$ to shareholder M_j secretly. Shareholder M_i computes and keeps a self-generated sub-share $f_i(x_i)$. By the end of this step, every shareholder receives $t - 1$ sub-shares from other shareholders.

Step 2. For each shareholder M_i, after receiving all sub-shares, $f_r(x_i)$, for $r = 1, 2, \ldots, t$, computes $z_i = \sum_{j=1}^{t} f_j(x_i) \pmod{p}$. z_i is the new share. In Theorem 1, we will prove that the threshold of z_i, for $i = 1, 2, \ldots, j$, is t'. z_i is t'. In the extended version of this paper, we will describe complete procedures to verify these new shares.

Step 3. With knowledge of z_i, for $i = 1, 2, \ldots, t$, shareholders can follow Harn and Lin's algorithm [7] to detect and identify cheaters. If there is no cheater, the secret s can be computed following Lagrange interpolating formula.

Theorem 1. *If shareholders act honestly and present valid shares in above algorithm, the threshold of z_i is t', and the secret s can be recovered successfully following Lagrange interpolating formula.*

Proof. If shareholders act honestly in the algorithm, each new share z_i is the additive sum of sub-shares of random polynomials $f_i(x)$, for $i = 1, 2, \ldots, t$, selected by shareholders. According to the property of secret sharing homomorphisms, z_i is the share of polynomial $F(x) = \sum_{r=1}^{t} f_r(x) \pmod{p}$. It is obvious that the degree of polynomial $F(x)$ is $t' - 1$. Thus, the threshold of z_i, for $i = 1, 2, \ldots, t$, is t'. In addition, if each shareholder owns a valid share in Step 1, the random polynomial $f_i(x)$ selected by shareholder M_i satisfies $f_i(0) = y_i' = f(x_i) \prod_{j=1, j \neq i}^{i} \frac{-x_i}{x_i - x_j} \pmod{p}$. Knowing z_i, for $i = 1, 2, \ldots, t$, the secret s can be recovered since the polynomial $F(x)$ satisfies $F(0) = \sum_{i=1}^{t} f_j(0)$ $\pmod{p} = \sum_{i=1}^{t} f(x_i) \prod_{j=1, j \neq i}^{t} \frac{-x_i}{x_i - x_j} \pmod{p} = s$. However, if there are some invalid shares, the secret s cannot be computed from the released new shares. $\square$

Remark 2. Since the threshold of the new shares z_i is t', there are $t - t'$ redundant shares in above algorithm. In the following, we will discuss how to choose the new threshold t' in order to detect and identify cheaters in our proposed secret reconstruction algorithm.

3.1 Selecting t' in Our Design

Harn and Lin [7] have classified three types of attack according to the behavior of attackers; that are, (a) Type 1 attack - attackers present fake shares without any collaboration; (b) Type 2 attack - shares are released synchronously and colluded attackers modify their shares to fool honest shareholders; and (c) Type 3 attack - shares are released asynchronously and colluded attackers modify their shares to fool honest shareholders. The bounds of detection and identification of cheaters are functions of the threshold, the number of cheaters, and the number of shares in the secret reconstruction. In a recent paper, Ghosting [10] has proposed a *wise cheating attack* on the cheater detection method based on the property of strong t-consistency. New bounds of detection of cheaters can be found. In the following, we list the bounds of detection and identification of cheaters incorporating the attack proposed by Ghosting [10].

Theorem 2. *Under* Type 1 *attack, Harn-Lin's scheme can successfully detect cheaters if $j \geq t + 1$, and identify cheaters if $j - c > t$, where j is the number of shares, t is the threshold and c is the number of cheaters in the secret reconstruction.*

Theorem 3. *Under* Type 2 *attack, Harn-Lin's scheme can successfully detect cheaters if $j - c \geq t$, and identify cheaters if $\{(c < t) \cap (j - c \geq t + 1)\} \cup \{(c \geq t) \cap (j - c > c + t - 1)\}$, where j is the number of shares, t is the threshold and c is the number of cheaters in the secret reconstruction.*

Theorem 4. *Under* Type 3 *attack, Harn-Lin's scheme can successfully detect cheaters if $j - c \geq t$, and identify cheaters if $\{j \geq t + 1\} \cap \{j - c > c + t - 1\}$, where j is the number of shares, t is the threshold and c is the number of cheaters in the secret reconstruction.*

In this paper, we consider the situation when there are exact t shares in the secret reconstruction. In order to create redundant shares to detect and identify cheaters, the proposed secret reconstruction algorithm enables shareholders to work together to change the threshold from its original value t to a new value t' such that there are $t - t'$ redundant shares in the secret reconstruction. New shares of the (t', t') secret sharing scheme are generated and are used in the secret reconstruction.

   Let us re-evaluate the upper and lower bounds in terms of the new threshold t'. In above theorems, the symbols, j is the number of participated shares, t is the threshold, and c is the number of cheaters in the secret reconstruction. In our proposed algorithm, the number of participated shares is t and the threshold is t'. From Theorems 2, 3 and 4, we can obtain the following results: (1) Under Type 1 attack, the proposed algorithm can successfully detect cheaters if $t' \leq t - 1$, and identify cheaters if $t' \leq t - c - 1$; (2) Under Type 2 attack, the proposed algorithm can successfully detect cheaters if $t' \leq t - c$, and identify cheaters if $\{c + 1 \leq t' \leq t - c - 1\} \cup \{t' \leq \min\{c, t - 2c\}\}$; (3) Under Type 3 attack, the proposed algorithm can successfully detect cheaters if $t' \leq t - c$, and identify cheaters if $t' \leq \min\{t - 1, t - 2c\}$. We summarize this result in Table 1.

   We use the following example to explain how to choose the new threshold t' in our proposed algorithm to meet the requirements of cheater detection and identification. Assume that in Shamir's $(7, 15)$ secret sharing scheme, our proposed secret reconstruction algorithm needs to detect and identify at most two

Table 1. Bounds of the threshold t' when t and c are given

	Detectability	Identifiability
Type 1	$t' \leq t - 1$	$t' \leq t - c - 1$
Type 2	$t' \leq t - c$	$\{c + 1 \leq t' \leq t - c - 1\} \cup \{t' \leq \min\{t - 1, t - 2c\}\}$
Type 3	$t' \leq t - c$	$t' \leq \min\{t - 1, t - 2c\}$

Table 2. Maximum values of t' for $t = 7, n = 15$ and $c = 2$

	t'_{max} for detectability	t'_{max} for identifiability
Type 1	6	4
Type 2	5	4
Type 3	5	3

cheaters. From Table 1, we can compute the maximal values of the new threshold t'. We list the threshold values in Table 2.

4 Conclusion

We propose a generalized cheater detection and identification algorithm for Shamir's (t, n) secret sharing scheme. Our scheme allows shareholders to detect and identify cheaters using their shares only without needing any additional information. When t shareholders need to reconstruct the secret, shareholders work together to change the threshold to a new threshold so redundant shares can be used to detect and identify cheaters. New shares are generated and used in the secret reconstruction. We include discussion on how to choose the new threshold to meet the requirements of cheater detection and identification.

Acknowledgment. This research is in part supported by the National Natural Science Foundation of China under Grant No. 61103247, the Natural Science Foundation of Fujian Province under Grant No. 2011J05147, and the Foundation for Excellent Young Teachers of Fujian Normal University under Grant No. fjsdjk2012049.

References

1. Araki, T.: Efficient (k,n) Threshold Secret Sharing Schemes Secure Against Cheating from $n - 1$ Cheaters. In: Pieprzyk, J., Ghodosi, H., Dawson, E. (eds.) ACISP 2007. LNCS, vol. 4586, pp. 133–142. Springer, Heidelberg (2007)
2. Benaloh, J.C.: Secret Sharing Homomorphisms: Keeping Shares of a Secret Secret. In: Odlyzko, A.M. (ed.) CRYPTO 1986. LNCS, vol. 263, pp. 251–260. Springer, Heidelberg (1987)
3. Blundo, C., De Santis, A., Gargano, L., Vaccaro, U.: Secret Sharing Schemes with Veto Capabilities. In: Cohen, G., Lobstein, A., Zémor, G., Litsyn, S.N. (eds.) Algebraic Coding 1993. LNCS, vol. 781, pp. 82–89. Springer, Heidelberg (1994)

4. Brickell, E.F., Stinson, D.R.: The Detection of Cheaters in Threshold Schemes. In: Goldwasser, S. (ed.) CRYPTO 1988. LNCS, vol. 403, pp. 564–577. Springer, Heidelberg (1990)
5. Carpentieri, M.: A perfect threshold secret sharing scheme to identify cheaters. Designs, Codes and Cryptography 5(3), 183–187 (1995)
6. Carpentieri, M., De Santis, A., Vaccaro, U.: Size of Shares and Probability of Cheating in Threshold Schemes. In: Helleseth, T. (ed.) EUROCRYPT 1993. LNCS, vol. 765, pp. 118–125. Springer, Heidelberg (1994)
7. Harn, L., Lin, C.: Detection and identification of cheaters in (t, n) secret sharing scheme. Designs, Codes and Cryptography 52(1), 15–24 (2009)
8. Harn, L., Lin, C.: Strong (n, t, n) verifiable secret sharing scheme. Information Sciences 180(16), 3059–3064 (2010)
9. He, J., Dawson, E.: Shared secret reconstruction. Designs, Codes and Cryptography 14(3), 221–237 (1998)
10. Ghosting, H.: Comments on Harn-Lin's cheating detection scheme. Designs, Codes and Cryptography 60(1), 63–66 (2011)
11. Kurosawa, K., Obana, S., Ogata, W.: t-Cheater Identifiable (k, n) Threshold Secret Sharing Schemes. In: Coppersmith, D. (ed.) CRYPTO 1995. LNCS, vol. 963, pp. 410–423. Springer, Heidelberg (1995)
12. McEliece, R.J., Sarwate, D.V.: On sharing secrets and Reed-Solomon codes. Communications of the ACM 24(9), 583–584 (1981)
13. Ogata, W., Kurosawa, K., Stinson, D.R.: Optimum secret sharing scheme secure against cheating. SIAM Journal on Discrete Mathematics 20(1), 79–95 (2006)
14. Pieprzyk, J., Zhang, X.-M.: Cheating Prevention in Linear Secret Sharing. In: Batten, L.M., Seberry, J. (eds.) ACISP 2002. LNCS, vol. 2384, pp. 121–135. Springer, Heidelberg (2002)
15. Rabin, T., Ben-Or, M.: Verifiable secret sharing and multiparty protocols with honest majority. In: Proceedings of the 21st Annual ACM Symposium on the Theory of Computing, pp. 73–85 (1989)
16. Shamir, A.: How to share a secret. Communications of the ACM 22(11), 612–613 (1979)
17. Tompa, M., Woll, H.: How to share a secret with cheaters. Journal of Cryptology 1(3), 133–138 (1989)

Cryptanalysis on the User Authentication Scheme
with Anonymity

Yung-Cheng Lee

Department of Security Technology and Management, WuFeng University
Minhsiung, Chiayi 62153, Taiwan
yclee@wfu.edu.tw

Abstract. Nowadays, people obtain a variety of services through networks. Many systems provide services without verifying users, but in many applications, the users obtain services only after they are authenticated. Remote user authentication scheme provides the server a convenient way to authenticate users before they are allowed to access database and obtain services. For the sake of security, anonymity is an important requirement for some user authentication schemes. In 2012, Shin et al. proposed a smart card based remote user authentication scheme. Their scheme has merits of providing user anonymity, key agreement, freely updating password and mutual authentication. They also declared that their scheme provides resilience to potential attacks of smart card based authentication schemes. In this article, we show that their scheme cannot resist impersonation attack, denial-of-service attack and guessing attack. Furthermore, the scheme suffers high hash computation overhead and validations steps redundancy.

Keywords: Authentication, Anonymity, Smart Cards.

1   Introduction

Remote user authentication scheme is a widely used mechanism to allow users and servers communications via insecure channel, it is the most common method used to check the validity of the login message and authenticate the users. For security and efficiency consideration, many schemes authenticate users by using the smart cards [2, 4, 6, 7, 9-12, 14, 15, 17].

In 1981, Lamport [11] proposed the first remote password authentication scheme by using smart cards. However, Lamport's scheme has the drawbacks such as high hash overhead and vulnerable to stolen-verifier attack. Many schemes use one-way hash functions and exclusive-or operations to reduce the computing complexity in smart cards [3, 13, 16]. Hwang et al. [6] proposed a smart card based user authentication scheme in 2000. However, their scheme can not withstand masquerade attack. In 2002, Chien et al. [4] presented a scheme with merits of mutual authentication and freely updating password. But Ku et al. [9] showed that Chien et al.'s scheme is vulnerable to the reflection attack and insider attack. Ku et al. also proposed an improved scheme to fix the flaws. However, Yoon et al. [17] indicated that the improved scheme was also susceptible to parallel session attack and presented an improvement scheme.

J.-S. Pan et al. (Eds.): *Advances in Intelligent Systems & Applications*, SIST 21, pp. 33–39.
DOI: 10.1007/978-3-642-35473-1_4   © Springer-Verlag Berlin Heidelberg 2013

Chien et al. [3] proposed an improved scheme to preserve user anonymity, however, Bindu et al. [1] showed that the scheme is vulnerable to the insider attack and man-in- the-middle attack. Lin et al. [13] presented a strong password authentication protocol with one-way hash function. But the scheme is insufficient of mutual authentication and user anonymity. Juang [7] presents a simple authentication scheme in 2004, but the users cannot change passwords freely and the scheme does not provide mutual authentication. Das et al. [5] and Liao et al. [12] introduced dynamic ID to achieve user's anonymity, but both schemes are vulnerable to insider attacks and neither scheme really provides user anonymity [14]. Khan et al. [8] and Tseng et al. [16] proposed remote authentication schemes to provide user anonymity. However, both schemes require time synchronization to resist replay attack [14].

In 2012, Shin et al. [14] proposed a remote user authentication scheme with merits of mutual authentication and user anonymity. The scheme overcomes the weaknesses of Das et al.'s scheme [5] and Liao et al.'s scheme [12]. However, in this article, we show that Shin et al.'s scheme is vulnerable to impersonation attack, denial-of-service attack and guessing attack.

The remainder of the article is organized as follows. Shin et al.'s smart card based remote user authentication scheme is briefly described in next Section. The security analysis of their scheme is analyzed in Section 3. Finally, we make conclusions.

2   Shin et al.'s Remote User Authentication Scheme

The scheme comprises four phases: registration phase, login phase, key agreement phase and password updating phases as follows.

2.1   Registration Phase

If the legitimate user U_i wants to join the system, U_i performs the following steps.

Step R-1. $U_i \Rightarrow S : \{ID_i, h(PW_i)\}$.

The user U_i chooses his/her identity ID_i and password PW_i and submits $\{ID_i, h(PW_i)\}$ to the server S via a secure channel.

Step R-2. The server computes user's TID_i, A_i and B_i.

After receiving $\{ID_i, h(PW_i)\}$, the server obtains the user's transform identity TID_i by $TID_i = h(ID_i \| h(PW_i))$ and computes A_i and B_i by:

$$A_i = h(K_U) \oplus K_S \tag{1}$$

$$B_i = (g^{A_i} \bmod p) \oplus h(PW_i) \tag{2}$$

Where g is a primitive element in Galois field $GF(p)$, p is a large prime number, K_S is the server's secret key, and K_U is the common key of user for S.

Step R-3. $S \Rightarrow U_i$: Smart card.

The server stores $\{TID_i, B_i, h(\cdot), K_U\}$ in a smart card and sends it to the user.

2.2   Login Phase

If the user wants to log into the system, the login steps are as follows.

Step L-1. The user attaches smart card to a card reader and then keys in ID_i and PW_i .

Step L-2. $U_i \rightarrow S : \{DID_i, CTID_i, C_i, k_i\}$.

The smart card generates two nonces n_i and k_i , and computes:

$$CTID_i = TID_i \oplus n_i \qquad (3)$$

$$C_i = h(B_i \oplus h((PW_i)) \oplus n_i \qquad (4)$$

$$M_i = K_U \bmod k_i \qquad (5)$$

$$DID_i = h^{M_i}(TID_i \oplus h(B_i \oplus h(PW_i))) \qquad (6)$$

The user sends $\{DID_i, CTID_i, C_i, k_i\}$ along with the login request message to the server. Note that there has typo in Shin et al.'s scheme, DID_i should be computed by $h^{M_i}(TID_i \oplus h(B_i \oplus h(PW_i)))$ rather than by $h^{M_i}(TID_i \oplus B_i \oplus h(PW_i))$.

Step L-3. After receiving $\{DID_i, CTID_i, C_i, k_i\}$, the server computes A_i by:

$$A_i = h(K_U) \oplus K_S \qquad (7)$$

Since $C_i = h(B_i \oplus h((PW_i)) \oplus n_i = h(g^{A_i}) \oplus n_i$, the nonce n_i can be recovered by:

$$n_i = C_i \oplus h(g^{A_i}) \qquad (8)$$

With n_i and $CTID_i$, the user's transform identity TID_i is obtained by:

$$TID_i = CTID_i \oplus n_i \qquad (9)$$

Then S checks whether the transform identity TID_i is in the database. If it isn't, the server terminates the connection; otherwise, continue the next steps.

Step L-4. The server authenticates the legitimate user.

The server computes $M_i = K_U \bmod k_i$. Then S obtains DID_i' by:

$$DID_i' = h^{M_i}(TID_i \oplus h(g^{A_i})) \qquad (10)$$

If $DID_i' = DID_i$, S authenticates the user U_i . Otherwise, S stops the connection.

Step L-5. $S \rightarrow U_i : \{DID_S, CTID_S\}$.

The server generates a nonce n_S and computes $\{DID_S, CTID_S\}$ by:

$$DID_S = h(DID_i \oplus n_i \oplus n_S) \qquad (11)$$

$$CTID_S = CTID_i \oplus n_S \qquad (12)$$

The server forwards $\{DID_S, CTID_S\}$ to U_i.

Step L-6. The user U_i authenticates the server S.

On receiving $\{DID_S, CTID_S\}$, the user obtains $n_S{'}$ by:

$$n_S{'} = CTID_S \oplus CTID_i \qquad (13)$$

Thereby, U_i computes $DID_S{'}$ with:

$$DID_S{'} = h(DID_i \oplus n_i \oplus n_S{'}) \qquad (14)$$

If $DID_S{'} = DID_S$, the user authenticates the remote server. Otherwise, U_i terminates the login steps.

Step L-7. $U_i \rightarrow S : \{DID_{iS}\}$.

After S is authenticated, U_i computes DIS_{iS} and sends it to S. Where

$$DID_{iS} = DID_S \oplus n_i \oplus (n_S + 1) \qquad (15)$$

Step L-8. The server S authenticates the user U_i.

After receiving DIS_{iS}, the server obtains $(n_S + 1)'$ by:

$$(n_S + 1)' = DID_{iS} \oplus DIS_S \oplus n_i \qquad (16)$$

The server S computes $(n_S + 1)$ and compares it with $(n_S + 1)'$. If $(n_S + 1)' = (n_S + 1)$, mutual authentication is obtained. Otherwise, S terminates connection with U_i.

2.3   Key Agreement Phase

After mutual authentication is obtained, the user and the server compute common session key SK_i and SK_S, respectively, by:

$$SK_i = h(B_i \oplus h(PW_i) \oplus n_i \oplus n_S) \qquad (17)$$

$$SK_S = h((g^{A_i} \bmod p) \oplus n_i \oplus n_S) \qquad (18)$$

The generated common session keys of SK_i and SK_S are the same since $B_i \oplus h(PW_i) = g^{A_i}$.

2.4   Password Updating Phase

When the user wants to change password, the steps are as follows.

Step U-1. $U_i \rightarrow S : \{DID_i, CTID_i, C_i, k_i, \text{Password updating Request}\}$.

Similar to the login steps, the user attaches the smart card to a reader and forwards $\{DID_i, CTID_i, C_i, k_i, \text{Password updating Request}\}$ to the server.

Step U-2. The user and the server obtain mutual authentication.

Similar to the steps in the login phase, U_i and S obtain mutual authentication.

Step U-3. $U_i \rightarrow S : \{(E_{SK_i}(TID_i^*)\}$

U_i chooses a new password PW_i^* and the smart card computes new transform identity TID_i^* by $TID_i^* = h(ID_i \| h(PW_i^*))$. Then the smart card encrypts TID_i^* by using the session key SK_i and sends $(E_{SK_i}(TID_i^*))$ to the server.

Step U-4. The server replaces TID_i with TID_i^* in the database.

After receiving $(E_{SK_i}(TID_i^*))$, S decrypts it by using SK_S and replaces TID_i with TID_i^*. Next, S sends response message to U_i.

Step U-5. U_i replaces TID_i and B_i with TID_i^* and B_i^*, respectively.

After receiving the response message from S, U_i computes $B_i^* = B_i \oplus h(PW_i) \oplus h(PW_i^*)$. Then the user replaces the old values TID_i and B_i with TID_i^* and B_i^*, respectively.

3   Security Analysis on Shin et al.'s Scheme

In Shin et al.'s scheme, the smart card computes $DID_i = h^{M_i}(TID_i \oplus h(B_i \oplus h(PW_i)))$ at the login session, where $M_i = K_U \mod k_i$. Thus their scheme suffers high hash overhead if k_i is very large. Moreover, their scheme is vulnerable to the following attacks:

(1) *Impersonation Attack*

Suppose that an adversary Eve (E, for short) wants to impersonate as the legitimate user U_i to login the system. Firstly, Eve intercepts $CTID_i$ from Step L-2 and $CTID_S$ from Step L-5. Then, with Eq.(12), n_S can be obtain by $n_S = CTID_S \oplus CTID_i$. Next, Eve intercepts DID_S from Step L-5 and DID_{iS} from Step L-7. Then, with Eq.(15), n_i also can be obtain by $n_i = DID_{iS} \oplus DID_S \oplus (n_S + 1)$. By n_i, the user's $h(g^{A_i})$ and TID_i also be obtained with Eq.(8) and Eq.(9). With TID_i and $h(g^{A_i})$, Eve impersonate as the legitimate user U_i with the steps as follows.

Step I-1. $E \rightarrow S : \{DID_i, CTID_i, C_i, k_i\}$

Eve selects two integers for nonces n_i and k_i, and chooses a small integer for M_i. Thereby she computes $\{CTID_i, C_i, DID_i\}$ by $CTID_i = TID_i \oplus n_i$, $C_i = h(g^{A_i}) \oplus n_i$ and $DID_i = h^{M_i}(TID_i \oplus h(g^{A_i}))$. Next, Eve sends $\{DID_i, CTID_i, C_i, k_i\}$ along with the login request message to S.

Step I-2. $S \rightarrow E : \{DID_S, CTID_S\}$.

After receiving $\{DID_i, CTID_i, C_i, k_i\}$, the server computes A_i and obtains $\{n_i, TID_i, M_i\}$ as the steps in login phase. Thereby the server authenticates the legitimate user. Note that $M_i = K_U \bmod k_i$ and Eve doesn't know K_U, so M_i is also unknown by Eve. The probability for Eve to pass the verification is $P = 1/2^{|k_i|}$.

If Eve chooses a very small k_i such that M_i is small enough, then the forwarded DID_i will pass the verification with very high probability. That is, in Step I-1, Eve should choose a very small k_i and selects an integer for M_i, where $M_i < k_i$. If Eve is authenticated, the server generates a nonce n_S, computes $\{DID_S, CTID_S\}$ and sends it to Eve.

Step I-3. $E \rightarrow S : \{DID_{iS}\}$.

After receiving $\{DID_S, CTID_S\}$, Eve obtains n_S by Eq.(13). Then DIS_{iS} can be obtained by Eq.(15). Next, Eve sends DIS_{iS} to the server.

Step I-4. S and Eve obtain a common session key.

After receiving $\{DID_S, CTID_S\}$, S and U_i obtain mutual authentication and a common session key.

Hereafter, the adversary can successfully to impersonate as a legitimate user to communicate with the server by using the common session key. Thus the Shin et al.'s scheme is vulnerable to the impersonation attack.

(2) ***Guessing Attack***

Similar to the cryptanalysis steps in the impersonation attack, Eve obtains n_S by $n_S = CTID_S \oplus CTID_i$ and knows n_i by $n_i = DID_{iS} \oplus DID_S \oplus (n_S + 1)$. With n_i, the user's transform identity TID_i will be obtained by $TID_i = CTID_i \oplus n_i$. Since $TID_i = h(ID_i \| h(PW_i))$ and user's identity ID_i is public, password PW_i can be easily guessed. Thus Shin et al.'s scheme cannot resist the guessing attack.

(3) ***Denial-of-Service Attack***

In password updating phase, Eve intercepts $(E_{SK_i}(TID_i^*)$ and sends a random message X to the server. After receiving X, the server will decrypted it to Y and replace the old transform identity TID_i with Y, where $Y = D_{SK_S}(X)$. Hereafter, the legitimate cannot login the system for services since $Y \neq TID_i^*$. Thus Shin et al.'s scheme cannot withstand denial-of-service attack.

4   Conclusions

Recently, Shin et al. proposed a remote authentication scheme. In this article, we show that their scheme is vulnerable to impersonation attack, denial-of-service attack and guessing attack. Furthermore, the scheme has drawbacks such as high hash overhead and validations steps redundancy.

Acknowledgment. This work was partial supported by the National Science Council of the Republic of China under the contract number NSC 101-2632-E-274-001-MY3.

References

1. Bindu, C.S., Reddy, P.C.S., Satyanarayana, B.: Improved Remote User Authentication Scheme Preserving User Anonymity. Int. J. of Computer Science and Network Security 8(3), 62–65 (2008)
2. Chang, C.C., Wu, T.C.: Remote Password Authentication with Smart Cards. IEE Proceedings-E138 3, 65–168 (1993)
3. Chien, H.Y., Chen, C.H.: A Remote Authentication Scheme Preserving User Anonymity. In: Proc. of the 19th In. Conference on Advanced Information Networking and Applications, AINA 2005, vol. 2, pp. 245–248 (2005)
4. Chien, H.Y., Jan, J.K., Tseng, Y.M.: An Efficient and Practical Solution to Remote Authentication: Smart Card. Computer Security 4(21), 372–375 (2002)
5. Das, M.L., Saxena, A., Gulati, V.P.: A Dynamic ID-Based Remote User Authentication Scheme. IEEE Trans. Consumer Electronics 50(2), 28–30 (2004)
6. Hwang, M.S., Li, L.H.: A New Remote User Authentication Scheme Using Smart Cards. IEEE Trans. on Consumer Electronics 1(46), 28–30 (2000)
7. Juang, W.S.: Efficient Password Authentication Key Agreement Using Smart Cards. Computer & Security 23, 167–173 (2004)
8. Khan, M.K., Kim, S.K., Alghathbar, K.: Cryptanalysis and Security Enhancement of A More Efficient & Secure Dynamic ID-Based Remote User Authentication Schemes. Computer Communications 34(3), 306–309 (2011)
9. Ku, W.C., Chen, S.M.: Weaknesses and Improvements of an Efficient Password Based Remote User Authentication Scheme Using Smart Cards. IEEE Trans. on Consumer Electronics 50(1), 204–207 (2004)
10. Kumar, M.: A New Secure Remote User Authentication Scheme with Smart Cards. Int. J. of Network Security 11(2), 88–93 (2010)
11. Lamport, L.: Password Authentication with Insecure Communication. Communications of the ACM 24(11), 770–772 (1981)
12. Liao, C.H., Chen, H.C., Wang, C.T.: An Exquisite Mutual Authentication Schemes with Key Agreement Using Smart Card. Informatica 33, 125–132 (2009)
13. Lin, C.W., Tsai, C.S., Hwang, M.S.: A New Strong Password Authentication Scheme Using One-Way Hash Functions. J. of Computer and Systems Sciences International 45(4), 623–626 (2006)
14. Shin, S., Kim, K., Kim, K.-H., Yeh, H.: A Remote User Authentication Scheme with Anonymity for Mobile Devices. International Journal of Advanced Robotic Systems 9, 1–7 (2012)
15. Sun, H.M.: An Efficient Remote User Authentication Scheme Using Smart Cards. IEEE Trans. on Consumer Electronics 4(46), 958–961 (2000)
16. Tseng, H.R., Jan, R.H., Yang, W.: A Bilateral Remote User Authentication Scheme That Preserves User Anonymity. J. of Security and Communication Networks 1(4), 301–308 (2008)
17. Yoon, E.J., Ryu, E.K., Yoo, K.Y.: Further Improvement of an Efficient Password Based Remote User Authentication Scheme Using Smart Cards. IEEE Trans. on Consumer Electronics 50(2), 612–614 (2004)

Deniable Authentication Protocols with Confidentiality and Anonymous Fair Protections

Shin-Jia Hwang, Yun-Hao Sung, and Jen-Fu Chi

Department of Computer Science and Information Engineering,
Tamkang University, Tamsui, New Taipei City, 251, Taiwan, R.O.C.
sjhwang@mail.tku.edu.tw, 697411758@s97.tku.edu.tw,
698410528@s98.tku.edu.tw

Abstract. Hwang and Chao proposed interactive deniable authentication protocols providing anonymity and fair protection both for senders and receivers. However, no non-interactive deniable authentication protocols are proposed to provide anonymity and fair protection both for senders and receivers. A non-interactive deniable authentication protocol with anonymity and fair protection is proposed to improve performance. Moreover, our protocol provides confidentiality but Hwang and Chao's protocol does not.

Keywords: Deniable authentication protocols, promise of digital signatures, signcryption, anonymity, intended receivers, confidentiality.

1 Introduction

Deniability and intended receiver properties are two security requirements of deniable authentication protocols (DAP). Even if the intended receiver reveals some secret information about the received data m, no one, expect the designated receiver, can be convinced that the sender sent the data m. After the first DAP[1], various DAPs are proposed. Those DAPs are classified into two classes. One is interactive DAPs [1-4] and one is non-interactive DAPs [5-6]. In general, non-interactive DAPs are more efficient than interactive DAPs, by reducing the communication cost [5].

Deniability of DAPs is provided since receivers have the same ability to produce the same authenticator as senders. However, a malicious receiver is able to prejudice sender's benefit by forging a valid authenticator. To remove this injurious problem, Hwang and Ma [7] proposed the first non-interactive DAP with sender protection. The sender protection property means that the sender can convince anyone that the authenticator is sent from him/her. Moreover, Hwang and Ma [8] also proposed a non-interactive DAP with anonymous sender protection to protect senders' identity privacy. To improve the efficiency of Hwang and Ma's DAP, Hwang and Chao [9] proposed a new DAP with anonymous sender protection.

However, only the sender protection [7-9] is not fair to the receivers. To protect both senders' and receivers' benefit and privacy, Hwang and Chao [10] proposed an interactive deniable authentication protocol with anonymous fair protection. To protect the receivers' benefit, in Hwang and Chao's protocol, the sender must also send

J.-S. Pan et al. (Eds.): *Advances in Intelligent Systems & Applications*, SIST 21, pp. 41–51.
DOI: 10.1007/978-3-642-35473-1_5   © Springer-Verlag Berlin Heidelberg 2013

receiver some evidence that is validated only with the help of the sender. After receiving evidences, the receiver should interact with the sender to validate senders' evidences. To reduce the communication load caused by senders' evidences, our non-interactive protocol are proposed by adopting the concept in Kudla's non-interactive designated verifier (NIDV) proof scheme [11].

However, those protocols [7-10] provide (fair) protection for senders or receivers, without confidentiality. Without confidentiality, these non-interactive DAPs may reveal some sensitive information about senders or receivers since the messages are sent in plaintext. These revelations may damage sender's and receiver's benefit, and even destroy some announced security properties. Thus confidentiality is important for DAPs with senders'/receivers' protection.

To provide confidentiality may use symmetric cryptosystems. Two additional costs should be paid. One is the cost to construct the session keys between senders and receivers. One is the encryption/decryption cost for message. To efficiently provide signing and encryption at the same time, Zheng [12] first proposed the signcryption schemes. However, the signcryption scheme has the non-repudiation property resulting in that signcryption schemes cannot be used directly in DAPs. Thus Hwang and Sung [13] first proposed the promised signcryption scheme to design their non-interactive DAP with confidentiality and anonymous sender protection. However, there is no non-interactive DAP with confidentiality and anonymous sender protection is proposed. Being inspired of the promised signcryption scheme, our non-interactive DAP with confidentiality, anonymity, and fair protection is proposed.

Based on Schnorr signature scheme [14] and its promise [15], and NIDV proof scheme, Section 2 describes our DAP with confidentiality, anonymity, and fair protection. Session 3 is the brief security proof of our DAP. Session 4 gives the security and performance comparison between our protocol and Hwang and Chao's DAP. The last session is our conclusion.

2   Our Deniable Authentication Protocol with Confidentiality and Anonymous Fair Protection (DAP-CAFP)

Our DAP-CAFP has three parties: Sender A, Receiver B, and a trustworthy Judge J. Our DAP-CAFP consists of three phases: Setup, authentication, and clarification phases. A produces the promised signcryptext and the transferring evidence for B in the authentication phase. B proves the message is transferred by A for Judge J in the clarification phase.

Setup Phase
Some system parameters and functions are published in this phase. Two large public primes p and q are first chosen to satisfy $p= 2q-1$. The element g in Z_p^* with order q and the multiplicative cyclic-subgroup $G= <g>$ of Z_p^* of order q are published. The public symmetric-key encryption $E_k(m)$ and decryption function $D_k(m)$ are also published, where m is the message and k is the session key. Four one-way hash functions $H_{q_1}(.)$, $H_{q_2}(.)$, $H_G(.)$, and $H_l(.)$ are published for all legal users. $H_{q_1}(.)$ and $H_{q_2}(.)$ map from $\{0,1\}^*$ to Z_q^*, $H_G(.)$ maps from $\{0,1\}^*$ to $G= <g>$, and $H_l(.)$ maps from $\{0,1\}^*$ to

$\{0,1\}^l$, where l is the length of a bit string. Assume each user i has a randomly-chosen private key x_i from Z_q^* and a computed public key $y_i = g^{x_i} \bmod p$.

Authentication Phase

This phase consists of Signcrypt_PGen and Designcrypt_PVerify algorithms. By Signcrypt_PGen, Sender A generates the sender's promised signcryptext (C, V, S), the transferring evidence σ, and the proof (w, r, h, d) of the transferring evidence for B. By Designcrypt_PVerify, the intended receiver decrypts and verifies the promised signcryptext and validates the evidence. The validation of the evidence is only performed by the intended receiver.

Signcrypt_PGen Algorithm

Signcrypt_PGen consists of three major steps. The promised signcryptext is generated in Step 1. The transferring evidence is generated in Step 2. The proof of the transferring evidence is generated in Step 3.

Step 1: Sender A generates of the promised signcryptext (C, V, S) on the message m.

   Step 1.1: Choose two random integers R and $k \in Z_q^*$.

   Step 1.2: Compute $V = H_{q_1}(g^k \bmod p \| H_G(m) \| R \| y_B)$, $s = k + V x_A \bmod p$, and $S = g^s \bmod p$, and $K = H_l((y_B)^s \bmod p)$.

   Step 1.3: Encrypt m by the symmetric encryption function $C = E_K(m \| R)$.

Step 2: A computes the transferring evidence $\sigma = H_G(m)^{x_A} \bmod p$ for the intended Receiver B.

Step 3: A generates the proof of the transferring evidence σ for Receiver B.

   Step 3.1: Choose three random integers w, r, and $t \in Z_q^*$.

   Step 3.2: Compute $c = g^w y_B^r \bmod p$, $T = g^t \bmod p$, $M = H_G(m)^t \bmod p$, $h = H_{q_2}(c \| T \| M \| m \| R \| \sigma \| S)$, and $d = t - x_A(h + w) \bmod q$.

Finally, A transmits the promised signcryptext (C, V, S) and the transferring evidence σ with its proof (w, r, h, d) to B.

Designcrypt_PGen

Designcrypt_PGen consists of two steps. To decrypt and verify the promised signcryptext is in Step 1. Step 2 is the confirmation of the evidence σ.

Step 1: B designcrypts and verifies the promised signcryptext.

   Step 1.1: Compute $K = H_l(S^{x_B} \bmod p)$.

   Step 1.2: Performing the symmetric decryption function $m \| R = D_K(C)$ to gain the message m and the random number R.

   Sept 1.3: Verify $V = H_{q_1}(S \times y_A^{-V} \bmod p \| H_G(m) \| R \| y_B)$. If $V = H_{q_1}(S \times y_A^{-V} \bmod p \| H_G(m) \| R \| y_B)$, Receiver B is convinced that the message m is sent by Sender A; otherwise, B rejects the promised signcryptext (C, V, S).

Step 2: B validates the proof of evidences.

   Sept 2.1: Compute $c = g^w y_B^r \bmod p$, $T = g^d y_A^{(h+w)} \bmod p$, and $M = g^d y_A^{(h+w)} \bmod p$.

   Sept 2.2: Verify $h = H_{q_2}(c \| T \| M \| m \| R \| \sigma \| S)$. If $h = H_{q_2}(c \| T \| M \| m \| R \| \sigma \| S)$, B is convinced that the sender knows the same discrete logarithm of σ and S; otherwise, B rejects the evidence.

Clarification Phase

If A declares that he/she did not transmitted the message m to B, the receiver's benefit is damaged. To protect receiver's benefit, B transmits Judge J the promise of signcryptext (V, S), the hash value $H_G(m)$, and the evidence σ. The clarification procedure between Sender A and Judge J are described below.

Sept 1: J validates (V, S) by $V = H_{q_1}(S \times y_A^{-V} \bmod p \| H_G(m) \| R \| y_B)$.

Step 2: J chooses two random numbers $a,\ b \in Z_q^*$, computes and sends $t = \sigma^a y_A^b$ $\bmod\ p$ to A.

Step 3: A computes and returns $d_1 = t^{x_A^{-1}} \bmod p$ to J.

Step 4: After receiving d_1 form A, J computes $d_2 = H_G(m)^a g^b \bmod p$. If $d_2 \equiv d_1$ (mod p), J stops and confirms that A is the real sender.

Step 5: J chooses two random numbers $a',\ b' \in Z_q^*$, computes and sends $t' = \sigma^{a'} y_A^{b'}$ $\bmod\ p$ to A.

Step 6: A computes $d_1' = t'^{x_A^{-1}} \bmod p$ and returns d_1' to J.

Step 7: After receiving d_1' form A, J computes $d_2' = H_G(m)^{a'} g^{b'} \bmod p$ and compares d_1' and d_2'. If $d_2' \equiv d_1'$ (mod p), J confirms that A is the real sender and stops to clarify; otherwise, J continues performing the following steps.

Step 8: If $(d_1 g^{-b})^{a'} \equiv (d_1' g^{-b'})^a$ (mod p), J confirms A is not the real sender, otherwise, J confirms A is.

Only $H_G(m)$ is sent to Judge, so the message confidentiality is still satisfied.

Sender A proves that he/she is the real sender of the promise of signcryptext by publishing s. Then anyone performs the following steps to decrypt the signcryptext to obtain m and transfer signcryptext (C, V, S) to a Schnorr signature (V, s) on m.

Step 1: Compute the session key $K = H_l((y_B)^s \bmod p)$.

Step 2: Perform the decryption $m \| R = D_K(C)$ with the session key K to gain $m \| R$.

Step 3: Check whether or not $V = H_{q_1}(g^s y_A^{-V} \bmod p \| H_G(m) \| R \| y_B)$ holds. If $V = H_{q_1}(g^s y_A^{-V} \bmod p \| H_G(m) \| R \| y_B)$, anyone is convinced that m is really sent by A. Otherwise, A is not the real sender.

3   Security Proof and Analysis

The underlying hard problem assumption is given below.

DDHP[16]: Let G be a group of order q, where q is a prime. Let g be a generator of G. Given g and the elements g^a, g^b, and g^t in G, determine whether $g^t = g^{ab}$.

DDHP Assumption: No polynomial-time algorithm solving DDHP with non-negligible probability exists.

Our protocol satisfies five properties: Message confidentiality, deniability, intended receiver, anonymity, and fair protection. The indistinguishable game for confidentiality is defined first. Some proofs are skipped in this conference version.

Definition 1 (Indistinguishable Game for Message Confidentiality)

Our DAP-CAFP satisfies indistinguishable security against chosen message attacks, if no polynomial-time adversary T winning the indistinguishable game with a non-negligible probability exists.

This game has two participators: Challenger U and Adversary T. U controls some oracles, Signcrypt_PGen oracle and four hash oracles.

Signcrypt_PGen Oracle S_p

Adversary T chooses a message m to query S_p. Then S_p returns U the corresponding promised signcryptext (C, V, S) and the evidence σ with its proof (w, r, h, d).

Hash Oracle H_V

Oracle S_p queries H_V by giving $(g^k \bmod p \| H_p(m) \| R \| y_B)$ and a digest value V'. H_V first check whether V' is null or not. If V' is null, H_V searches its local record. If $((g^k \bmod p \| H_p(m) \| R \| y_B), V)$ exists, H_V returns the same digest V; otherwise, H_V returns a random value $V \in Z_q^*$ and saves $((g^k \bmod p \| H_p(m) \| R \| y_B), V)$ into its local record. If V' is not null, H_V searches the local record first. If $((g^k \bmod p \| H_p(m) \| R \| y_B), V)$ exists, H_V returns an error message; otherwise, H_V returns the inputted digest value V' and stores $((g^k \bmod p \| H_p(m) \| R \| y_B), V')$ into its local record.

Hash Oracle H_{key}

Oracle S_p inputs the receiver's public key y_B and s to query H_{key}. H_{key} searches its local record. If a record $((y_B, s), K)$ exists, H_{key} returns the same value K; otherwise, H_{key} returns a random value $K \in \{0, 1\}^l$ and stores $((y_B, s), K)$ into its local record.

Hash Oracle H_p

Oracle S_p queries H_p for a message m. For the queried m, H_p returns the same digest X by searching its local record to find (m, X). Otherwise, H_p returns a random value $X \in G$ and saves (m, X) into its local record.

Hash Oracle H_h

Oracle S_p queries H_h by giving (c, T, M), an evidence σ, a message with a random number $m \| R$, and a promise S. H_h searches its local record first. If $((c \| T \| M \| m \| \sigma \| R \| S), h)$ is found, H_h returns the same h; otherwise, H_h returns a random value $h \in Z_q^*$ and stores $((c \| T \| M \| m \| \sigma \| R \| S), h)$ into its local record.

This game consists of setup, probing, challenging and guessing phases.

Setup Phase

U generates all system parameters and the public/private key pairs of Sender A and Receiver B. Then Adversary T is given A's and B's public keys y_A and y_B, and the system parameters.

Probing Phase

T collects some promised signcryptexts (C, V, S) and its evidence σ with proof (w, r, h, d) by choosing a message m to query U. U utilizes the Signcrypt_PGen oracle $S_p(m)$ to return T the promised signcryptext (C, V, S) and its corresponding evidence σ with proof (w, r, h, d).

Challenging and Guessing Phase

T randomly sends two legal messages m_0 and m_1 with the same length to U. U chooses a random bit e and produces the provable promised signcryptext $((C', V', S'),$ $\sigma',$ $(w',$ $r',$ $h',$ $d'))= S_p(m_e)$ with the help of the oracle S_p. U sends the provable promised signcryptext to T as a challenge.

Finally, T outputs a guessing bit e'. If $e'= e$, U returns '1' to show that Adversary T wins the game; otherwise, returns '0'. If T gives the correct e' with probability $1/2+\varepsilon$ and the winning advantage ε is non-negligible, he/she attacks successfully.

Theorem 1 (Message Confidentiality)

Let the symmetric encryption cryptosystem satisfy indistinguishable security against chosen ciphertext attacks (IND-CCA). Our protocol satisfies IND-CCA, if there is no probabilistic polynomial-time algorithm solving the DDHP with a non-negligible probability.

Proof

Let T be a polynomial-time adversary whose goal is to distinguish a message among two message candidates from a signcryptext under chosen message attack in our protocol. Suppose that T wins the indistinguishable game with probability $(1/2)+\varepsilon$, where ε is a non-negligible advantage. By using Adversary T as subroutines, a probabilistic polynomial-time algorithm U exists to solve the DDHP. Suppose that the DDHP instance is $(g^a \bmod p, g^b \bmod p, g^c \bmod p)$.

Setup Phase

U generates Sender A's and Receiver B's public/private key pairs and the system parameters. Then Adversary T is given the system parameters and the public keys $y_A= g^{x_A} \bmod p$ and $y_{B.}= g^b \bmod p$, where x_A is a randomly chosen integer by U.

Probing Phase

T collects some provable promised signcryptext $((C, V, S),$ $\sigma,$ $(w, r, h, d))$ on the legal chosen message m, and querying oracles S_p, H_V, H_{key}, H_p, and H_h. The producing procedure of the provable promised signcryptext by Oracle S_p is described below.

Step 1: Choose five random integers w, r, t, R, and k in Z_q^*.

Step 2: Get $V= H_V(g^k \bmod p||H_p(m)||R||y_B)$ by using the oracle H_p on the input $(m,$ null digest) and then the hash oracle H_V on the input consisting of $(g^k \bmod p||H_p(m)||R||y_B)$ and null digest.

Step 3: Compute $s= k+ Vx_A \bmod q$ and $S= g^s \bmod p$.

Step 4: Gain $K= H_{key}((y_B)^s \bmod p)$ by using the hash oracle H_{key}.

Step 5: Perform $C= E_K(m||R)$.

Step 6: Compute $c = g^w y_B{}^r \bmod p$, $\sigma = H_p(m)^{x_A} \bmod p$, $T = g^t \bmod p$, $M = H_p(m)^t \bmod p$, $h = H_h(c||T||M||m||R||\sigma||S)$, and $d = t - x_A(h + w) \bmod q$.

Step 7: Return $((C, V, S),$ $\sigma,$ $(w, r, h, d))$ to T.

Challenging and Guessing Phase

Adversary T sends U two legal messages m_0 and m_1 with the same length. After randomly choosing a bit e, U produces the provable promised signcryptext $((C', V', S'),$ $\sigma',$ $(w', r', h', d'))= S_p(m_e)$ by the following procedure.

Step 1: Let $S' = g^a \bmod p$ and choose a random value $V' \in Z_q^*$.

Step 2: Compute $g^k = S' \times y_A^{-V'} \bmod p$.

Step 3: Choose three random integers w', r', and t' in Z_q^*. Then compute $c' = g^{w'} y_B^{r'}$ mod p, $\sigma' = H_p(m_e)^{x_A} \bmod p$, $T' = g^{t'} \bmod p$, $M' = H_p(m)^{t'} \bmod p$, $h' = H_h(c' \| T' \| M' \| m_e \| R' \| \sigma' \| S')$, and $d' = t' - x_A(h' + w') \bmod q$.

Step 4: Obtain $V' = H_V(g^k \bmod p \| H_p(m_e) \| R' \| y_B)$ by using the hash oracle H_V on the input consisting $(g^k \bmod p \| H_p(m_e) \| R' \| y_B)$ and the digest value V'.

Step 5: Compute $K' = H_{key}(g^c \bmod p)$ with the help of the hash oracle H_{key}.

Step 6: Perform $C' = E_{K'}(m_e \| R')$.

Step 7: Send $((C', V', S'), \sigma', (w', r', h', d'))$ to the adversary T as a challenge.

On the challenge, T outputs the guessing bit e'. Finally, U returns '1', if $e' = e$ or T outputs nothing after its polynomial-time bound. Otherwise, U returns '0'.

Probability Analysis of U Solving DDHP

Notation $\Pr[U\_Fail]$ denotes the failure probability of U solving DDHP. The analysis of the failure probability of U consists two cases. The message sent from sender to receiver consists of two parts. One is the promised signcryptext (C, V, S) and another is the proof $(\sigma, (w, r, h, d))$. The analysis of the promised signcryptext (C, V, S) is given first.

Case 1: $g^c \equiv g^{ab} \pmod p$ and $e' \neq e$.

In this case, U returns the incorrect answer of the yes-instance (g^a, g^b, g^c), where $g^c \equiv g^{ab} \pmod p$. Since T's losing probability $\Pr[e' \neq e]$ is $(1/2)$-ε, the failure probability is $\Pr[e' \neq e \text{ and } g^c \equiv g^{ab} \pmod p] = ((1/2)\text{-}\varepsilon)/q$.

Case 2: $g^c \bmod p \neq g^{ab} \bmod p$ and $e' = e$.

Only when the collisions of H_{key} occurs, the encryption key K is correct. Assume H_{key} is an ideal hash function, so the collision probability of H_{key} is $1/2^l$. This case means that U returns the incorrect answer of no-instance (g^a, g^b, g^c) because $g^c \bmod p \neq g^{ab}$ mod p. T's winning probability $\Pr[e' = e]$ is $(1/2)$+ε, so U's failure probability given g^c mod $p \neq g^{ab} \bmod p$ and the correct K is $\Pr[e' = e, g^c \bmod p \neq g^{ab} \bmod p, \text{ and } K \text{ is correct}] = (1/2 + \varepsilon)(1/2^l) \times (1 - 1/q)$.

So U's failure probability is $\Pr[U\_Fail] = \dfrac{(\frac{1}{2}\text{-}\varepsilon)}{q} + (1/2 + \varepsilon) \times (1/2^l) \times (1 - 1/q) \leq \dfrac{1}{2q}$

$+ (1/2^l)$. $\dfrac{1}{2q} + (1/2^l)$ is negligible, since both q and 2^l are large. So $\Pr[U\_Fail]$ is negligible. Based on DDHP assumption, the confidentiality of our protocol is IND-CCA.

The digest $h' = H_h(c' \| T' \| M' \| m_e \| R' \| \sigma' \| S')$ provides negligible information about the message because h' is also randomized by the secret random number R'. Only the one knowing R' is able to adopt $h' = H_h(c' \| T' \| M' \| m_e \| R' \| \sigma' \| S')$ to distinguish the messages. On the other hand, the ones without the secret random value R' adopt the value h' to distinguish the messages with negligible probability. For the promised signcryptext (C, V, S), five cases are considered one by one without using the value h'.

Lemma 1 shows the promise property of our DAP-CAFP.

Lemma **1 (Promise Property):** The promise of Schnorr signature (V, S) on message m provides promise property in our protocol, where $S = g^s \pmod{p}$.

Proof: (This proof is skipped in this version.)
By Lemma 1, the deniability of our protocol is proved in Theorem 2.

Theorem **2 (Deniability)**
Our DAP-CAFP satisfies deniability property because both the intended receiver B and the sender A can generate the promised signcryptext (C, V, S) and the evidence σ with the proof (w, r, h, d).

Proof: (This proof is skipped in this version.)

Sender Anonymity
A DAP satisfies the sender anonymity against adaptively chosen message attacks if no probabilistic polynomial-time algorithm wins the sender anonymity game with non-negligible advantage more than 1/2.

Sender Anonymity Game
This game has two participators, an adversary and a challenger, and consists of three phases, Setup, probing, and challenging and guessing phases. In Setup phase, the challenger constructs the system parameters, the hash oracles, the encryption oracle, and two senders' and one receiver's public keys. Adversary knows the public keys and public parameters and functions. In the probing phase, by querying the encryption oracle, the adversary is allowed to choose some legal messages to obtain the promised signcryptexts and the evidences with the proofs that are from someone between the two senders. Finally, the adversary sends the challenger one un-queried message. In the challenging and guessing phase, the challenger first randomly selects one between the two senders. Then the challenger generates the challenge such that the promised signcryptext and the evidence with the proof on the received message are generated on behalf of the selected one. After receiving the challenge, the adversary guesses about who the chosen sender is.

Theorem **3 (Sender Anonymity):** Except Sender A and the intended receiver B, no one wins the sender anonymity game against adaptively chosen message attacks with non-negligible advantage over 1/2 based on the hardness of DDHP in the random oracle model.

Proof: (The proof is skipped in this version.)

Receiver Anonymity
A DAP satisfies receiver anonymity against adaptively chosen message attacks if no probabilistic polynomial-time algorithm wins the following receiver anonymity game with non-negligible advantage more than 1/2.

Receiver Anonymity Game
This game has two participators, one adversary and one challenger. The game consists of setup, probing, and challenging and guessing phases. In Setup phase, the challenger constructs the system parameters, the hash and encryption oracles, and one sender's and two receivers' public keys. The adversary knows those public keys and public

parameters and functions. In the probing phase, the adversary chooses some legal random messages and queries the encryption oracle to obtain the promised signcryptexts and the evidences with the proofs that are from the sender sending to anyone between two receivers. Finally, the adversary sends the challenger one chosen message. In the challenging and guessing phase, the challenger first randomly chooses one receiver and generates the challenge that is the promised signcryptext and the evidence with the proof on the received message for the chosen receiver. Finally, the adversary guesses about who the receiver is.

Theorem 4 (Receiver Anonymity): Except Sender A and the intended receiver B, no one can win the receiver anonymity game with non-negligible advantage against adaptively chosen message attacks based on the hardness of DDHP in the random oracle model.

Proof(The proof is skipped in this version.)

Theorem 5 (Intended Receiver): The $((C, V, S), \sigma, (w, r, h, d))$ generated by the sender A can be verified only by the intended receiver B based on the hardness of DDHP in the random oracle model.

Proof: (The proof is skipped in this version.)

Fair Protections

Fair protections contain the sender and receiver protection. By the sender protection, Sender A can convince anyone that the message is actually sent by him/her. If A denies that the message is sent from him/her, B owns some evidence to prove that the message is actually from A, with the help of the trusted Judge J.

Theorem 6 (Sender Protections): The sender protection of our protocol is based on the unforgeability of Schnorr signature scheme.

Proof: (The proof is skipped in this version. The proof is based on the results in [17])

Theorem 7 (Receiver Protections): The receiver protection of our protocol is guaranteed by the undeniability of the Chaum and van Antwerpen's undeniable signature scheme.

Proof: (The proof based on the results in [18] is skipped in this version)

4   Comparison and Discussions

Table 1 shows the security property comparison between Hwang and Chao's DAP-AFP and our protocol. Both two protocols satisfy intended receiver, deniability, unforgeability, sender anonymity properties. Furthermore, the two protocols both provide sender and receiver protection. Only our protocol provides confidentiality property to prevent the sensitive information revelation. Moreover, our protocol is no-interactive for receiver protection, so our protocol efficiently provides the receiver protection by reducing communication cost.

Table 1. Security Property Comparison Two DAPs

	Hwang and Chao's DAP-AFP	Our Protocol
Intended Receiver	Yes	Yes
Deniability	Yes	Yes
Unforgeability	Yes	Yes
Sender Anonymity	Yes	Yes
Sender Protection	Yes	Yes
Receiver Protection	Yes(Interactive)	Yes(Non-interactive)
Confidentiality	No	Yes

5   Conclusions

Our non-interactive DAP not only satisfies the basic properties of deniable authentication protocols, but also provides some other useful properties: Confidentiality, sender anonymity, and fair protection. Beside the sender protection part, our protocol always keeps the confidentiality of transmitted message to prevent revealing the sensitive information.

References

1. Dwork, C., Naor, M., Sahai, A.: Concurrent Zero-Knowledge. In: Proc. of 30th ACM STOC 1998, Dallas, TX, USA, pp. 409–418 (1998)
2. Aumann, Y., Rabin, M.: Efficient Deniable Authentication of Long Messages. Presented at International Conference on Theoretical Computer Science in Honor of Professor Manuel Blum's 60th Birthday (1998),
   http://www.cs.cityu.edu.hk/dept/video.html
3. Deng, X., Lee, C.H., Zhu, H.: Deniable Authentication Protocols. IEE Proceeding-Computers and Digital Techniques 148(2), 101–104 (2001)
4. Fan, L., Xu, C.X., Li, J.H.: Deniable Authentication Protocol Based on Diffie-Hellman Algorithm. Electronics Letters 38(4), 705–706 (2002)
5. Shao, Z.: Efficient Deniable Authentication Protocol Based on Generalized ElGamal Signature Scheme. Computer Standards and Interfaces 26, 449–454 (2004)
6. Wang, B., Song, Z.X.: A Non-Interactive Deniable Authentication Scheme Based on Designated Verifier Proofs. Information Sciences 179, 858–865 (2009)
7. Hwang, S.J., Ma, J.C.: Deniable Authentication Protocols with Sender Protection. In: 2007 National Computer Symposium, NCS 2007, Wufeng, Taiwan, pp. 762–767 (2007)
8. Hwang, S.J., Ma, J.C.: Deniable Authentication Protocols with (Anonymous) Sender Protection. In: 2008 International Computer Symposium, ICS 2008, Tamsui, Taiwan, pp. 412–419 (November 2008)
9. Hwang, S.J., Chao, C.H.: An Efficient Non-Interactive Deniable Authentication Protocol with Anonymous Sender Protection. In: Cryptology and Information Security Conference, Taipei City, Taiwan, R.O.C. (2009)

10. Chao, C.H.: Deniable Authentication Protocols with Anonymous Fair Protections. M.S. thesis, University of Tamkang, Taipei country, Taiwan, R.O.C. (2009)
11. Kudla, C.J.: Special Signature Scheme and Key Agreement Protocols. Ph.D. dissertation, Royal Holloway, University of London, Egham, Surrey, England (2006)
12. Zheng, Y.: Digital Signcryption or How to Achieve Cost (Signature & Encryption) << Cost(Signature) + Cost(Encryption). In: Kaliski Jr., B.S. (ed.) CRYPTO 1997. LNCS, vol. 1294, pp. 165–179. Springer, Heidelberg (1997)
13. Hwang, S.-J., Sung, Y.-H.: Confidential Deniable Authentication Using Promised Signcryption. Journal of Systems and Software 84(10), 1652–1659 (2011)
14. Schnorr, C.-P.: Efficient Identification and Signatures for Smart Cards. In: Brassard, G. (ed.) CRYPTO 1989. LNCS, vol. 435, pp. 239–252. Springer, Heidelberg (1990)
15. Nguyen, K.: Asymmetric Concurrent Signatures. In: Qing, S., Mao, W., López, J., Wang, G. (eds.) ICICS 2005. LNCS, vol. 3783, pp. 181–193. Springer, Heidelberg (2005)
16. Diffie, W., Hellman, M.E.: New Directions in Cryptography. IEEE Transactions on Information Theory 22(6), 644–654 (1976)
17. Pointcheval, D., Stern, J.: Security Arguments for Digital Signatures and Blind Signatures*. Journal of Cryptology 13(3), 361–396 (2000)
18. Chaum, D., van Antwerpen, H.: Undeniable Signatures. In: Brassard, G. (ed.) CRYPTO 1989. LNCS, vol. 435, pp. 212–216. Springer, Heidelberg (1990)

A Novel Authentication Scheme Based on Torus Automorphism for Smart Card

Chin-Chen Chang[1,2], Qian Mao[2,3,*], and Hsiao-Ling Wu[1]

[1] Department of Information Engineering and Computer Science, Feng Chia University,
No. 100, Wenhwa Rd., Seatwen, Taichung, 40724, Taiwan
{alan3c,wuhsiaoling590}@gmail.com
[2] Department of Computer Science and Information Engineering, Asia University,
No. 500, Lioufeng Rd., Wufeng, Taichung, 41354, Taiwan
maoqiansh@gmail.com
[3] Department of Optical-Electrical and Computer Engineering,
University of Shanghai for Science and Technology, No. 516, Jungong Rd.,
Yangpu, Shanghai, 200093, P. R. China

Abstract. A novel authentication scheme for smart card is proposed in this paper. In this scheme, the cardholder's photograph is printed on the card. Meanwhile, the compressed image of the same photograph is encrypted by the torus automorphism. The encrypted image is stored in the smart card. The secret keys for decryption are shared by a trusted third party and the user. Only when all the secret keys are presented can the original image be recovered. The recovered image should be the same as the photograph printed on the card. The combination of the image encryption using torus automorphism and secret sharing provides high security for the proposed authentication scheme.

Keywords: smart card, authentication, image encryption, torus automorphism, secret sharing.

1 Introduction

Smart cards are used extensively, but the threat exists for the card to be stolen and used by unauthorized people. Many user authentication schemes have been proposed to prevent the illegal use of smart cards [1-3]. In many authentication schemes, the cardholder's photograph is printed on the card, and it can be authenticated by other people when the card is used. However, this procedure cannot completely protect the cardholder, because people's faces change with time, the illegal user may resemble the cardholder, or the illegal user may replace the photo by her/his own photograph. All these issues make the authentication unreliable. Zhao and Hsieh proposed an authentication scheme based on image morphing. The photograph of the cardholder was morphed and hid in some cover image [4]. Thongkor and Amornraksa proposed another authentication scheme that hid the cardholder's ID in her/his photograph as a watermark [5]. In addition, many biometric methods also have been used to

---

* Corresponding author.

J.-S. Pan et al. (Eds.): *Advances in Intelligent Systems & Applications*, SIST 21, pp. 53–60.
DOI: 10.1007/978-3-642-35473-1_6   © Springer-Verlag Berlin Heidelberg 2013

authenticate smart cards [6, 7]. All these authentication schemes improved the security of smart card.

Image encryption encrypts images by making them chaotic [8-10], which can also provide high security for smart cards. Torus automorphism is a dynamic system that can be used in image encryption, providing a high level of chaos [11]. Images scrambled by torus automorphism have high levels of chaos, and a certain number of the permutations from the original image can recover it [12, 13]. An image copyright protection scheme that uses torus automorphism has been proposed, and it provided good protection and a high-quality cover image [14].

In this paper, we proposed a novel authentication scheme that uses image encryption and secret sharing. The combination of these two methods achieved high security for the authentication. Section 2 presents some important theorems for our scheme, and the novel authentication scheme is proposed in Section 3. The security of the proposed scheme is analyzed in Section 4, and our conclusions are presented in Section 5.

2   Torus Automorphism for Image Encryption

The two-dimensional automorphism $\mathbb{F}_N$ of group G_N, denoted as $\mathbb{F}_N : G_N \rightarrow G_N$, where $G_N = \{0, 1, \cdots, N-1\} \times \{0, 1, \cdots, N-1\}$, is defined by the following map:

$$\mathbb{F}_N : X^{(n)} = A \cdot X^{(n-1)} = A^n \cdot X^{(0)} \Rightarrow \begin{bmatrix} x^{(n)} \\ y^{(n)} \end{bmatrix} = \begin{bmatrix} a_{11} & a_{12} \\ a_{21} & a_{22} \end{bmatrix}^n \cdot \begin{bmatrix} x^{(0)} \\ y^{(0)} \end{bmatrix} \mod N, (1)$$

where $X^{(0)} = [x^{(0)} \quad y^{(0)}]^T$ is the initial state, $X^{(n)} = [x^{(n)} \quad y^{(n)}]^T$ is the nth state, and $X^{(0)}, X^{(n)} \in G_N$. In (1), $a_{ij} \in Z$ $(i, j = 1, 2)$, $\det(A) = 1$, and the eigenvalues λ_1 and λ_2 of matrix A satisfy $\lambda_{1,2} \notin \{-1, 0, 1\}$. The parameter $r = a_{11} + a_{22}$ is defined as the trace of matrix A. Percival and Vivaldi proved that when $r^2 > 4$, the torus automorphism $\mathbb{F}_N$ has strong chaos [11].

For the torus automorphism $\mathbb{F}_N$, the iterations from the initial state form a set of orbits $\mathcal{O}(X) = \{X^{(0)}, X^{(1)}, X^{(2)}, \ldots\}$ that is periodic. That is to say, there exists an integer R such that $X^{(0)} = X^{(R)}$. The period R is defined as the recurrence time of the torus automorphism.

Since the matrix A in (1) is restricted by the conditions $\det(A) = 1$ and its trace t, the torus automorphism $\mathbb{F}_N$ is actually a two-parameter map. Therefore, (1) can be generalized in the following form [12]:

$$\mathbb{F}_N: X^{(n)} = A^n \cdot X^{(0)} \Rightarrow \begin{bmatrix} x^{(n)} \\ y^{(n)} \end{bmatrix} = \begin{bmatrix} 1 & a \\ b & ab+1 \end{bmatrix}^n \cdot \begin{bmatrix} x^{(0)} \\ y^{(0)} \end{bmatrix} \bmod N, \quad (2)$$

where $a, b \in Z$. The recurrence time R of $\mathbb{F}_N$ depends on the values of a, b, and N. We can prove that the two different scrambling matrices,

$$A_1 = \begin{bmatrix} 1 & a \\ b & ab+1 \end{bmatrix} \text{ and } A_2 = \begin{bmatrix} 1 & a+N \\ b+N & (a+N)(b+N)+1 \end{bmatrix}$$

lead to the same scrambling result, where N is the modulus in (1).
Prove:

$$\begin{bmatrix} x^{(n)} \\ y^{(n)} \end{bmatrix} = A_1 \cdot \begin{bmatrix} x^{(n-1)} \\ y^{(n-1)} \end{bmatrix} = \begin{bmatrix} 1 & a \\ b & ab+1 \end{bmatrix} \cdot \begin{bmatrix} x^{(n-1)} \\ y^{(n-1)} \end{bmatrix} = \begin{bmatrix} x^{(n-1)} + ay^{(n-1)} \\ bx^{(n-1)} + aby^{(n-1)} + y^{(n-1)} \end{bmatrix} \bmod N$$

$$\begin{bmatrix} x^{(n)} \\ y^{(n)} \end{bmatrix} = A_2 \cdot \begin{bmatrix} x^{(n-1)} \\ y^{(n-1)} \end{bmatrix} = \begin{bmatrix} 1 & a+N \\ b+N & (a+N)(b+N)+1 \end{bmatrix} \begin{bmatrix} x^{(n-1)} \\ y^{(n-1)} \end{bmatrix}$$

$$= \begin{bmatrix} x^{(n-1)} + ay^{(n-1)} + Ny^{(n-1)} \\ bx^{(n-1)} + Nx^{(n-1)} + aby^{(n-1)} + N(a+b)y^{(n-1)} + N^2 y^{(n-1)} + y^{(n-1)} \end{bmatrix}$$

$$= \begin{bmatrix} x^{(n-1)} + ay^{(n-1)} \\ bx^{(n-1)} + aby^{(n-1)} + y^{(n-1)} \end{bmatrix} \bmod N$$

The recurrence time R of $\mathbb{F}_N$ can be found by simulations. The following table gives some examples. We see that the values of a and b in A_1 and A_2 are periodic with 128, therefore, when $N = 128$, A_1 and A_2 lead to the same recurrence time, as do A_3 and A_4.

Table 1. Examples of Torus Automorphism

Matrix A	Modulus N	Recurrence Time R
$A_1 = \begin{bmatrix} 1 & 7 \\ 10 & 71 \end{bmatrix}$	128	32
$A_2 = \begin{bmatrix} 1 & 135 \\ 138 & 18631 \end{bmatrix}$	128	32

Table 1. (*continued*)

$A_3 = \begin{bmatrix} 1 & 3 \\ 4 & 13 \end{bmatrix}$	64	64
$A_4 = \begin{bmatrix} 1 & 67 \\ 68 & 4557 \end{bmatrix}$	64	64

3   Proposed Card-User Authentication

The structure of our proposed authentication scheme is shown in Fig. 1. First, a trusted third party, such as a bank, accepts a user's registration and then builds a smart card for the user. A photograph of the cardholder is printed on the card, while an encrypted image of the same photograph is stored in the card. When the smart card is used, the terminal, which may be in a shop or another place of business, reads the encrypted information and decrypts it. The secret key is shared by the bank and the cardholder. If the right keys are provided, the secret image will be decrypted correctly, which should look the same as the photograph printed on the card. By this means, the legality of the user is authenticated.

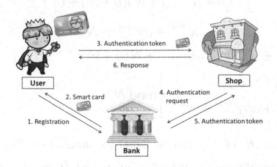

Fig. 1. Framework of the Proposed Authentication System

We assume that the original photograph is a gray image with a size of $N \times N$ and assume that the pixels' gray values vary from 0 to 255, then each pixel can be denoted as an eight-bit byte. Take the limited storage capacity of a smart card into consideration. There are only m ($m \le 8$) bits to represent each pixel's gray value in secret image $I^{(0)}$.

After that, the two-dimensional torus automorphism is used to encrypt the image. Assume that the map of the torus automorphism $\mathbb{F}_N$ is shown as (2) and that the recurrence time is R. Then, after n ($n < R$) iterations from the original image $I^{(0)}$, an arbitrary pixel with coordinates $[x^{(0)} \quad y^{(0)}]$ in $I^{(0)}$ is permutated to a new location with coordinates $[x^{(n)} \quad y^{(n)}]$ by (2). This permutation leads to an image, $I^{(n)}$, that is strongly chaotic. Since $\{I^{(0)}, I^{(1)}, ..., I^{(n)}, ..., I^{(R)}\}$ is a torus automorphism system with recurrence time R, another $n' = R - n$ iterations from $I^{(n)}$ will recover the original image. That is to say $I^{(0)} = I^{(R)}$. Therefore, the decryption algorithm is shown as the following:

$$\begin{bmatrix} x^{(R)} \\ y^{(R)} \end{bmatrix} = A^{n'} \cdot \begin{bmatrix} x^{(n)} \\ y^{(n)} \end{bmatrix} \bmod N, \quad x^{(R)}, y^{(R)}, x^{(n)}, y^{(n)} \in [1, N], \tag{3}$$

where $\begin{bmatrix} x^{(R)} & y^{(R)} \end{bmatrix} = \begin{bmatrix} x^{(0)} & y^{(0)} \end{bmatrix}$ if $n' = R - n$ and if A is same as that in (2). By this means, the original secret image is recovered.

Therefore, we use (2) to encrypt the original image $I^{(0)}$, and we use (3) to get the decrypted image $I^{(R)}$. To make $I^{(0)} = I^{(R)}$, both of the following requirements must be satisfied:

- The same matrix A must be used in encryption and decryption.
- The sum of the iterations in encryption and decryption should be the recurrence time R, i. e., $n + n' = R$.

If either one of the requirements is not satisfied, the decrypted image is chaotic. An example is given in the following. The size of the original photo is 128×128 pixels. The scrambling matrix is A_1, and the recurrence time is 32, as shown in Table 1. Fig. 2 shows the experimental results. In this figure, (a) is the original image, (b) is same as (a) except that there are only four highest bits to represent each pixel, and (c), (d), (e), and (f) are scrambled images with 1, 9, 19, and 32 iterations from (b), respectively. The Peak Signal-to-Noise Ratio (PSNR) between (a) and (b) is 31.7 dB. The loss of image quality is due to the lack of the four lowest bits of each pixel. The PSNRs of (c), (d), (e), and (f) between (b) are 9.2 dB, 9.2 dB, 9.2 dB, and 90.3 dB, respectively. In fact, (f) is totally the same as (b), since 32 iterations from (b) can recover itself. And the intermediate images, which are (c), (d), and (e), are totally chaotic images.

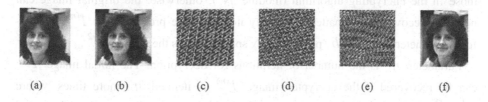

(a)   (b)   (c)   (d)   (e)   (f)

Fig. 2. Experimental Results of Image Encryption Using Torus Automorphism

Therefore, parameters a, b, and n' are crucial for successful decryption. In order to provide high security for the smart card, the three parameters must be kept safely. Therefore, a secret sharing algorithm is used in our scheme.

The t-out-of-k secret sharing scheme was proposed by Shamir [13]. If a secret message s is to be shared by k users, at least t ($2 \leq t \leq k$) of whom can recover s cooperatively, then the sharing function $f(z)$ is:

$$f(z) = s + c_1 z + c_2 z^2 + \dots + c_{t-1} z^{t-1}, \text{ where } z \in Z, \tag{4}$$

and c_1, c_2, $\cdots$, c_{t-1} are random numbers excepting 0. This function provides a point $(x_i, f(x_i))$ ($1 \leq i \leq k$) for the ith user. And t of such points can solve (4), by which the message s can be recovered.

In order to protect a, b, and n', our secret sharing function is designed as follows:

$$f(z) = a + bz + n'z^2, \tag{5}$$

where a and b are parameters in matrix A in (2), and n' is the number of iterations for decryption. If three random numbers z_1, z_2, and z_3 ($z_1, z_2, z_3 \in Z$ and $z_1 \neq z_2 \neq z_3$) are chosen, then three points, $(z_1, f(z_1))$, $(z_2, f(z_2))$, and $(z_3, f(z_3))$ can be computed by (5). The first point is kept secretly by the bank, and the second and the third points are kept secretly by the cardholder. When authentication occurs, the terminal gets the three points, one from the bank and two from the user, and then it can solve (5) and decrypt the secret image using a, b, and n'.

4   Security of the Proposed Scheme

The security of our proposed scheme depends on a, b, and n'. Parameters a and b construct the scrambling matrix A. In order to recover the original image successfully, the values of a and b in the decrypting algorithm must be the same as those in the encrypting algorithm (module N); otherwise, the original image can never be recovered, no matter how many iterations are processed on $I^{(n)}$. Therefore, parameters a and b provide a key space that has the size of N^2.

Parameter n' is the number of iterations for decryption. The initial image $I^{(0)}$ can be recovered if the encrypted image $I^{(n)}$ is iterated n' more times, where $n' = R - n$. Only when the correct number of iterations is executed can the initial image be recovered successfully. Since $1 \leq n' < R$, the size of the key space provided by n' is $R - 1$.

Therefore, the key space S of the proposed authentication scheme is:

$$S = (R-1) \cdot N^2. \tag{6}$$

5   Conclusions

The proposed authentication scheme provides high security for smart cards by the combination of image encryption and secret sharing. The encryption method using torus automorphism provides high chaos, and it is difficult for illegal users to pass the authentication process. Our future work may focus on methods to improve the key space of the torus automorphism.

References

1. Sonwanshi, S.S., Ahirwal, R.R., Jain, Y.K.: An Efficient Smart Card Based Remote User Authentication Scheme Using Hash Function. In: Proc. 2012 IEEE Students' Conference on Electrical, Electronics and Computer Science, SCEECS, pp. 1–4 (2012)
2. Sood, S.K., Sarje, A.K., Singh, K.: Smart Card Based Secure Authentication and Key Agreement Protocol. In: Proc. 2010 International Conference on Computer and Communication Technology, ICCCT, pp. 7–14 (2010)
3. Matanovic, G., Mikuc, M.: Implementing Certificate-Based Authentication Protocol on Smart Cards. In: Proc. 2012 Proceedings of the 35th International Convention, MIPRO, pp. 1514–1519 (2012)
4. Zhao, Q.F., Hsieh, C.H.: Card User Authentication Based on Generalized Image Morphing. In: Proc. 2011 3rd International Conference on Awareness Science and Technology, iCAST, pp. 117–122 (2011)
5. Thongkor, K., Amornraksa, T.: Digital Image Watermarking for Photo Authentication in Thai National ID Card. In: Proc. 2012 9th International Conference on Electrical Engineering/Electronics, Computer, Telecommunications and Information Technology, ECTI-CON, pp. 1–4 (2012)
6. Yahaya, Y.H., Isa, M., Aziz, M.I.: Fingerprint Biometrics Authentication on Smart Card. In: Proc. Second International Conference on Computer and Electrical Engineering, ICCEE, pp. 671–673 (2009)
7. Das, A.K.: Analysis and Improvement on an Efficient Biometric-Based Remote User Authentication Scheme Using Smart Cards. IET Information Security 5(3), 145–151 (2011)
8. Luo, X.Z., Fan, J.H., Wu, J.H.: Single-Channel Color Image Encryption Based on the Multiple-Order Discrete Fractional Fourier Transform and Chaotic Scrambling. In: Proc. 2012 International Conference on Information Science and Technology, ICIST, pp. 780–784 (2012)
9. Bhatnagar, G., Wu, Q.M.J.: Chaos-Based Security Solution for Fingerprint Data During Communication and Transmission. IEEE Transactions on Instrumentation and Measurement 61(4), 876–887 (2012)
10. Tao, R., Meng, X.Y., Wang, Y.: Image Encryption with Multiorders of Fractional Fourier Transforms. IEEE Transactions on Information Forensics and Security 5(4), 734–738 (2010)

11. Percival, I., Vivaldi, F.: Arithmetical Properties of Strongly Chaotic Motions. Physica D: Nonlinear Phenomena 25(1-3), 105–130 (1987)
12. Voyatzis, G., Pitas, I.: Chaotic Mixing of Digital Images and Applications to Watermarking. In: Proc. ECMAST 1996, vol. 2, pp. 687–694 (1996)
13. Chen, G., Mao, Y., Chui, C.: A Symmetric Image Encryption Scheme Based on 3D Chaotic Cat Maps. Chaos, Solitons, Fractals 21(3), 749–761 (2004)
14. Chang, C.C., Hsiao, J.Y., Chiang, C.L.: An Image Copyright Protection Scheme Based on Torus Automorphism. In: Proc. the First International Symposium on Cyber Worlds, pp. 217–224 (2002)
15. Shamir, A.: How to share a secret. Communications of the ACM 22(11), 612–613 (1979)

Cryptanalysis of a Provably Secure Certificateless Short Signature Scheme

Yu-Chi Chen[1], Raylin Tso[2], and Gwoboa Horng[1]

[1] Department of Computer Science and Engineering,
National Chung Hsing University, Taiwan
[2] Department of Computer Science, National Chengchi University, Taipei, Taiwan
{s9756034,raylin,gbhorng}@cs.nchu.edu.tw

Abstract. Certificateless public key cryptography, introduced by Al-Riyami and Paterson, simplifies the complex certificate management in PKI-based public key cryptography and solves the key escrow problem of identity-based cryptography. Huang et al. in 2007 showed security models of certificateless signature to simulate possible adversaries according to their attack abilities. Recently, Choi et al. proposed a certificateless short signature scheme. They claimed their scheme to be the only certificateless short signature scheme achieving the strongest security level presented by Huang et al.. They also give their security proofs to support their claim. However, we find that their scheme is not as secure as the authors claimed. In this paper, we give comments on the paper of Choi et al. including the cryptanalysis of their scheme and the weakness of the security proof.

Keywords: certificateless cryptography, certificateless signature, cryptanalysis, security models, short signature.

1 Introduction

Certificates of public keys must be fully managed and maintained by a trusted certificate authority (CA) in conventional public key infrastructure. CA plays an important role of authenticating the public keys. However, with development of wireless networks such as ad hoc networks, communication cost is required to decrease between users and CA. A straightforward solution is a cryptsystem which does not adopt CA. Therefore, both of identity-based public key cryptography (ID-PKC) [11] and certificateless public key cryptography (CL-PKC) [1] are developed without the trusted CA to manage certificates. Simultaneously, lower communication costs comparing with those of traditional cyptsystems are also achieved since a certificate is not required to be send along with a public key. Technically, ID-PCK and CL-PKC only depend on a trusted entity to generate keys. One of the security issues of ID-PKC is the *key escrow* problem in which the private key generator, the trusted entity in ID-PKC, has every user's secret key. However, the core of CL-PKC is the key generation center (KGC) which cannot have the user's actual secret key. The KGC only owns user's *partial secret*

J.-S. Pan et al. (Eds.): *Advances in Intelligent Systems & Applications*, SIST 21, pp. 61–68.
DOI: 10.1007/978-3-642-35473-1_7 　　　 © Springer-Verlag Berlin Heidelberg 2013

key, which is the most different property from ID-PKC. As a result, CL-PKC is one of the most dependable methods to avoid the key escrow problem in practice.

Certificateless public key cryptography has attracted significant research attention, since it was first introduced by Al-Riyami and Paterson in 2003. Certificateless signature (CLS) therefore becomes popular for a decade [3,4,5,8,15]. Existential unforgeability is an important issue when designing a provably secure CLS scheme. As well-known, there are two types of adversaries in CLS: the first one is referred to as the Type I adversary acting as an outside attacker, and the second one is referred to as the Type II adversary acting as the curious KGC. Type I adversary can replace any user's public key, but it cannot access the system master key which is generated and held by the KGC. Type II adversary holds the system master key, but it cannot replace public keys of users.

Taking the security of CLS into consideration, the paper of Huang et al. [9] (the full version [10]) discuss the security models of CLS schemes in details. Adversaries are classified into Normal, Strong, and Super adversaries which are ordered by their attack abilities. Among them, the super Type I and II adversaries are more powerful than others respectively.

On the other hand, Boneh et al. [2] introduced the concept of short signatures in 2001, which are useful for systems with low bandwidth and/or low computation power. Inheriting the advantages of both certificateless cryptography and short signatures, certificateless short signatures are introduced and have come into limelight in recent years [5,6,7,13,14]. However, Shim[12] presented an attack which is performed by the Strong or Super Type I adversary and claimed that to design a secure short CLS schemes withstand the attack is an open problem. Recently, Choi et al. [5] proposed a CLS scheme and proved their scheme to be secure against both of the super Type I and II adversaries as the strongest security level.

In this paper, we find Choi et al.'s CLS scheme is not as secure as they proved. We thus cryptanalysis this scheme and indicate the weakness of the security proof. Choi et al.'s scheme is insecure against the Super or Strong Type I adversary in our analysis. Actually, there are some loopholes in the security proof, which causes that the proof seems correct but actually not.

The rest of this paper is organized as follows. We briefly describe the definition and security model of CLS in Section 2. We then review an efficient certificateless short signature, proposed by Choi et al. [5], in Section 3. We show the cryptanalysis of this scheme and point out the weakness of the security analysis in Section 4. Finally, the conclusions of this paper are given in Section 5.

2   Certificateless Signature (CLS)

2.1   Definition of CLS

A certificateless signature scheme involves three entities, the KGC, a user/signer, and a verifier. Generally, it consists of the following algorithms: Setup, Partial-Secret-Key-Extract, Set-Secret-Value, Set-Secret-Key, Set-Public-Key, CL-Sign, and CL-Verify:

- Setup: This algorithm, run by the KGC, takes a security parameter as an input, and then returns master-key and system parameter, params.
- Partial-Secret-Key-Extract: This algorithm, run by the KGC, takes params, master-key and a user's identity ID as inputs. It generates a partial-secret-key D_{ID}, and sends it to the user via a secure channel.
- Set-Secret-Value: This algorithm, run by a user, returns a secret value, r_{ID}.
- Set-Secret-Key: This algorithm, run by a user, takes the user's partial-secret-key D_{ID} and the secret value r_{ID} as inputs, then returns the user's full secret key.
- Set-Public-Key: This algorithm, run by a user, takes params and the user's full secret key as inputs, and returns a public key pk_{ID} for the user.
- CL-Sign: This algorithm, run by a signer/user, takes params, a message m, and the user's full secret key as inputs. It then generates σ as the signature for the message m.
- CL-Verify: This algorithm, run by a verifier, takes params, a public key pk_{ID}, a message m, a user's identity ID, and a signature σ as inputs. It returns 1 as the verifier accepts σ if σ is the signature of the message m, the public key pk_{ID}, and the user with identity ID. It returns 0 if not.

2.2   Security Model of CLS

For security of CLS, there are several adversaries which act as different roles. We usually assume that Type I adversary is an outsider and Type II adversary is the curious KGC. Both of their goals are to generate a forged signature existentially. Nevertheless, Huang et al. [10] classified the Type I and II adversaries into three levels based on their different abilities: Normal, Strong, and Super adversaries respectively. Since we want to show the security flaw of Choi et al. 's scheme[5] against Strong and Super Type I adversaries, in what follows, we only present Game Strong I which modelling the Strong Type I adversary and Game Super I which modelling the Super Type I adversary.[1]

Game Strong I. An adversary $\mathcal{A}$ interacts with a challenger $\mathcal{C}$. $\mathcal{A}$ acts as an outsider and it can replace any public key.

Setup: The challenger $\mathcal{C}$ runs Setup to generate the system parameters and sends them to $\mathcal{A}$.

Attack: $\mathcal{A}$ can query (1) the public key of identity ID, (2) the secret value of ID, (3) the partial-secret-key of ID, and (4) the signature of (m, ID, r_{ID}) where r_{ID} is a secret value. Moreover, $\mathcal{A}$ also can replace a public key with a new one, pk'_{ID}.

Forgery: $\mathcal{A}$ outputs a forged signature σ^* of (m^*, ID^*, r_{ID^*}).

$\mathcal{A}$ wins this game if and only if the following conditions hold.

---

[1] We will not present Game Normal I or any Game II modelling the normal Type I adversary or the Type II adversary, since these are not the major point discussed by this paper. However, readers can refer to the paper by Huang et al. [10] for more details.

1. The signature σ^* is valid, which means CL-Verify$(\sigma^*, m^*, ID^*, r_{ID^*}) = 1$.
2. The partial-secret-key of ID^* has never been queried before.
3. Signatures of (m^*, ID^*, r_{ID^*}) has never been queried before.

Definition 1. *A certificateless signature scheme is existentially unforgeable against the Strong Type I adversary if no probabilistic polynomial time adversary wins Game Strong I with non-negligible probability.*

Game Super I. An adversary $\mathcal{A}$ interacts with a challenger $\mathcal{C}$. $\mathcal{A}$ acts as an outsider but it can replace any public key.

Setup: The challenger $\mathcal{C}$ runs Setup to generate the system parameters and sends them to $\mathcal{A}$.

Attack: $\mathcal{A}$ can query (1) the public key of identity ID, (2) the secret value of ID, (3) the partial-secret-key of ID, and (4) the signature of (m, ID).[2] Moreover, $\mathcal{A}$ also can replace a public key with a new one, pk'_{ID}.

Forgery: $\mathcal{A}$ outputs a forged signature σ^* of (ID^*, m^*).

$\mathcal{A}$ wins this game if and only if the following conditions hold.

1. The signature σ^* is valid, which means CL-Verify$(\sigma^*, m^*, ID^*) = 1$.
2. The partial-secret-key of ID^* has never been queried before.
3. Signatures of (m^*, ID^*) have never been queried before.

Definition 2. *A certificateless signature scheme is existentially unforgeable against the Strong Type I adversary if no probabilistic polynomial time adversary wins Game Super I with non-negligible probability.*

3   Choi et al.'s Certificateless Short Signature Scheme

We first briefly describe the definition of the bilinear map before reviewing Choi et al.'s scheme [5]. A bilinear map is a mapping $\hat{e} : \mathbb{G}_1 \times \mathbb{G}_2 \to \mathbb{G}_3$. $\mathbb{G}_1$ and $\mathbb{G}_2$ are an additive cyclic group of prime order q, and $\mathbb{G}_3$ is a multiplicative cyclic group of the same order q. A bilinear map concerns the following properties:

(1) Computable: given any $P \in \mathbb{G}_1$ and $Q \in \mathbb{G}_2$, there exists a polynomial time algorithm to compute $\hat{e}(P, Q) \in \mathbb{G}_3$.
(2) Bilinear: for any $x, y \in \mathbb{Z}_q^*$, we have $\hat{e}(xP, yQ) = \hat{e}(P, Q)^{xy}$ for any $P \in \mathbb{G}_1, Q \in \mathbb{G}_2$.
(3) Non-degenerate: $\hat{e}(P_1, P_2) \neq 1$ if P_1 is a generator of $\mathbb{G}_1$ and P_2 is a generator of $\mathbb{G}_2$.

The above is the normal form; however, if $\mathbb{G}_1 = \mathbb{G}_2$, the bilinear map will be denoted by $\hat{e} : \mathbb{G} \times \mathbb{G} \to \mathbb{G}_3$ sometimes, where $\mathbb{G}$ is an additive cyclic group of

---

[2] The input of the sign queries of Game Super I is different from that of Game Strong I. Huang et al. [10] had proven the super Type I adversary is more powerful than the Strong Type I adversary.

prime order q, and $\mathbb{G}_3$ is a multiplicative cyclic group of the same order q. In what follows, we only consider $\hat{e} : \mathbb{G} \times \mathbb{G} \to \mathbb{G}_3$.

The scheme of Choi et al. [5] is claimed to be secure against the Super Type adversaries. Now we describe this scheme as follows.

Setup: This algorithm, run by the KGC, takes a security parameter k as input. It determines a bilinear map $\hat{e} : \mathbb{G} \times \mathbb{G} \to \mathbb{G}_3$ where $\mathbb{G}$ is a cyclic additive group of prime order q with a generator P, $\mathbb{G}_3$ is a cyclic multiplicative group of the same order, and several hash functions, $H_1, H_1', H_2 : \{0,1\}^* \to \mathbb{G}$ and $\mathcal{H}, \mathcal{H}' : \{0,1\}^* \to \mathbb{Z}_q^*$. It randomly chooses $s \in \mathbb{Z}_q^*$ as master-key and accordingly sets master-public-key $P_{pub} = sP$. Finally, it returns master-key $= s$ and system parameter params $= \langle \mathbb{G}, \mathbb{G}_3, \hat{e}, q, P, P_{pub}, H_1, H_1', H_2, \mathcal{H}, \mathcal{H}' \rangle$.

Set-Secret-Value: This algorithm, run by a user with identity ID, returns a random value $r_{ID} \in \mathbb{Z}_q^*$. The user sets r_{ID} as his secret value.

Set-Public-Key: This algorithm, run by a user, takes params, the user's secret value r_{ID} and identity ID as inputs, and returns $pk_{ID} = r_{ID}P$ as his public key.

Partial-Secret-Key-Extract: This algorithm, run by the KGC, first sets $Q_{ID} = H_1(ID)$ and $\tilde{Q}_{ID} = H_1'(ID)$. It then takes params, master-key, and a user's identity ID as inputs, and computes $D_{ID} = sQ_{ID}$ and $\tilde{D}_{ID} = s\tilde{Q}_{ID}$. Finally, it generates a partial-secret-key psk_{ID} to the user, where $psk_{ID} = (D_{ID}, \tilde{D}_{ID})$.

Set-Secret-Key: This algorithm, run by a user, takes the user's partial-secret-key psk_{ID} and secret value r_{ID} as inputs and returns the full secret key $sk_{ID} = (psk_{ID}, r_{ID})$.

CL-Sign: This algorithm, run by a signer (a legitimate user), takes params, a message m, and the signer's secret key sk_{ID} as inputs. It sets $h = \mathcal{H}(m, ID, pk_{ID})$ and $\tilde{h} = \mathcal{H}'(m, ID, pk_{ID})$. Then, it returns $\sigma = hD_{ID} + \tilde{h}\tilde{D}_{ID} + r_{ID}H_2(m, ID, pk_{ID})$ as the signature of (m, ID, pk_{ID}).

CL-Verify: To verify a signature σ of $(m, ID, , pk_{ID})$, a verifier takes params, the public key pk_{ID}, the message m, the user's identity ID, and the signature σ as inputs. The verifier first obtains $h = \mathcal{H}(m, ID, pk_{ID})$ and $\tilde{h} = \mathcal{H}'(m, ID, pk_{ID})$, then checks whether the equation, $\hat{e}(\sigma, P) = e(hH_1(ID) + \tilde{h}H_1'(ID), P_{pub}) \cdot \hat{e}(pk_{ID}, H_2(m, ID, pk_{ID}))$, holds or not. Return 1 if it holds, and return 0 if not.

For the security analysis, Choi et al. proved this scheme is secure (existentially unforgeable) against Super Type I adversary. Straightly, it should be secure against Strong Type I adversary as well.

4   Cryptanalysis

4.1   Breaking Choi et al.'s Scheme

We find that Choi et al.'s scheme is insecure against the Strong or Super Type I adversary. The cryptanalysis is showed as follows, where the Strong Type I adversary can forge a user's signature on any message existentially.

1. Let $\mathcal{A}$ be the Strong Type I adversary and $\mathcal{C}$ be the challenger in the Game Strong I.
2. $\mathcal{A}$ chooses a secret value, $r'_{ID} \in \mathbb{Z}_q^*$, at random.

3. $\mathcal{A}$ thus queries two signatures of (m_1, ID, r'_{ID}) and (m_2, ID, r'_{ID}). $\mathcal{C}$ will return two valid signatures, σ'_1 of (m_1, ID, pk'_{ID}) and σ'_2 of (m_2, ID, pk'_{ID}), hence

$$\sigma'_1 = \mathcal{H}(m_1, ID, pk'_{ID})D_{ID} + \mathcal{H}'(m_1, ID, pk'_{ID})\tilde{D}_{ID} + r'_{ID}H_2(m_1, ID, pk'_{ID}), \quad (1)$$

$$\sigma'_2 = \mathcal{H}(m_2, ID, pk'_{ID})D_{ID} + \mathcal{H}'(m_2, ID, pk'_{ID})\tilde{D}_{ID} + r'_{ID}H_2(m_2, ID, pk'_{ID}). \quad (2)$$

4. Then, $\mathcal{A}$ queries $\mathcal{H}(m_1, ID, pk'_{ID})$, $\mathcal{H}'(m_1, ID, pk'_{ID})$, $\mathcal{H}(m_2, ID, pk'_{ID})$, $\mathcal{H}'(m_2, ID, pk'_{ID})$, $H_2(m_1, ID, pk'_{ID})$ and $H_2(m_2, ID, pk'_{ID})$. $\mathcal{C}$ returns the correct values, $(h_1, h_2, \tilde{h}_1, \tilde{h}_2, T_1, T_2)$, where

$$h_1 = \mathcal{H}(m_1, ID, pk'_{ID}), \quad h_2 = \mathcal{H}(m_2, ID, pk'_{ID}),$$

$$\tilde{h}_1 = \mathcal{H}(m_1, ID, pk'_{ID}), \quad \tilde{h}_2 = \mathcal{H}'(m_2, ID, pk'_{ID}),$$

$$T_1 = H_2(m_1, ID, pk'_{ID}), T_2 = H_2(m_2, ID, pk'_{ID}).$$

5. $\mathcal{A}$ can obtain the following equations, Eq 3 and 4, due to Eq 1 and 2.

$$\sigma'_1 - r'_{ID}T_1 = h_1 D_{ID} + \tilde{h}_1 \tilde{D}_{ID} \quad (3)$$

$$\sigma'_2 - r'_{ID}T_2 = h_2 D_{ID} + \tilde{h}_2 \tilde{D}_{ID} \quad (4)$$

Since σ'_1, σ'_2, T_1, T_2 and r'_{ID} are known, let $S_1 = \sigma'_1 - r'_{ID}T_1$ and $S_2 = \sigma'_2 - r'_{ID}T_2$. $\mathcal{A}$ can straightly have Eq 5 and 6.

$$S_1 = h_1 D_{ID} + \tilde{h}_1 \tilde{D}_{ID} \quad (5)$$

$$S_2 = h_2 D_{ID} + \tilde{h}_2 \tilde{D}_{ID} \quad (6)$$

$\mathcal{A}$ thus has Eq 7 and 8, then gets Eq 9 as a result.

$$h_1^{-1}(S_1) = D_{ID} + h_1^{-1}\tilde{h}_1 \tilde{D}_{ID} \quad (7)$$

$$h_2^{-1}(S_2) = D_{ID} + h_2^{-1}\tilde{h}_2 \tilde{D}_{ID} \quad (8)$$

$$h_1^{-1}(S_1) - h_2^{-1}(S_2) = (h_1^{-1}\tilde{h}_1 - h_2^{-1}\tilde{h}_2)\tilde{D}_{ID} \quad (9)$$

Finally, $\mathcal{A}$ infers $\tilde{D}_{ID}$ by computing $\tilde{D}_{ID} = (h_1^{-1}\tilde{h}_1 - h_2^{-1}\tilde{h}_2)^{-1}(h_1^{-1}S_1 - h_2^{-1}S_2)$, and $\mathcal{A}$ also can infers D_{ID} by using $\tilde{D}_{ID}$. Upon getting the partial-secret-key $(D_{ID}, \tilde{D}_{ID})$, $\mathcal{A}$ can forge valid signatures existentially.

We conclude that Choi et al.'s scheme suffers from the above attack in Game Strong I, which means the scheme is insecure against the Super Type I adversary. We thus transform the above attack to Game Super I as follows. Step 4 and 5 are similar to the Strong Type I adversary.

1. Let $\mathcal{A}$ be the Super Type I adversary and $\mathcal{C}$ be the challenger in the Game Super I.
2. $\mathcal{A}$ queries a secret value of ID. $\mathcal{C}$ returns $r_{ID} \in \mathbb{Z}_q^*$ to $\mathcal{A}$.
3. $\mathcal{A}$ thus queries two signatures of (m_1, ID) and (m_2, ID) respectively. $\mathcal{C}$ will return two valid signatures, σ'_1 of (m_1, ID, pk_{ID}) and σ'_2 of (m_2, ID, pk_{ID}), hence

$$\sigma'_1 = \mathcal{H}(m_1, ID, pk_{ID})D_{ID} + \mathcal{H}'(m_1, ID, pk_{ID})\tilde{D}_{ID} + r_{ID}H_2(m_1, ID, pk_{ID}),$$

$$\sigma'_2 = \mathcal{H}(m_2, ID, pk_{ID})D_{ID} + \mathcal{H}'(m_2, ID, pk_{ID})\tilde{D}_{ID} + r_{ID}H_2(m_2, ID, pk_{ID}).$$

4.2   Discussions for the Security Proof of Choi et al.'s Scheme

A CLS scheme is provably secure under the security model, which implies that the adversary, modelled by the security game, has no polynomial time algorithm to win the game. Therefore, a scheme is proven but insecure under the security model if and only if the security proof is incorrect without doubt. There are several weaknesses listed with respect to the security proof of Choi et al. [5] as follows.[3] Here we assume the forged signature σ^* is of (m^*, ID^*).

(1) $\mathcal{H}$ and $\mathcal{H}'$ are not random oracles. The outputs of $\mathcal{H}$ and $\mathcal{H}'$ queries are not truly random.
(2) According to the conventional form for proving security, a signature scheme is (t, ϵ, q_S)-secure if and only if it is existentially unforgeable, where t is the running time, ϵ is the probability of winning the security game, and q_S is the most queried times of signatures and usually sufficiently large. However, Choi et al.'s scheme is $(t, \epsilon, 1)$-secure. This means that any Strong Type I or Super Type I adversary of their scheme is only allowed to query one signature, otherwise, the adversary can break the scheme and existentially forge a valid signature.
(3) The probability of $\mathcal{A}$ is not correct, since the author only take the success of partial-private-key queries of $\mathcal{A}$ into consideration. The successes of signature queries and forgery are necessary.

At present, no appropriate solution overcomes the open problem where the certificateless short signature schemes are not provably secure against the Strong and Super Type I adversaries, since Choi et al.'s scheme is insecure.

There exists another Type of adversary presented by Tso et al. [14], which is exactly the same as the Strong or the Super Type adversary, and is a little weaker than the Super Type adversary. Based on the security model defined in [14], we find that Choi et al.'s scheme is also insecure against this kind of Type I adversaries. Consequently, we conclude that Choi et al.'s scheme is only secure against the Normal Type I adversary, not against higher ones.

5   Conclusions

There are many certificateless signature schemes in the literature which apparently can work but they have been showed to be insecure under other different realizable security models. The very recent certificateless short signature scheme of Choi et al. is reviewed and the security is broken in this paper. We cryptanalysis the scheme under different security models of the Strong and Super Type adversaries. Eventually, to propose a certificateless short signature scheme secure against the Strong and Super Type adversaries is still an open problem.

---

[3] Readers can refer to the security proof of this scheme in the paper [5] for more details.

Acknowledgment. This work was partially supported by the National Science Council of Taiwan, under contracts NSC100-2221-E-005-062, NSC101-2221-E-005-083 and NSC101-2628-E-004-001-MY2.

References

1. Al-Riyami, S.S., Paterson, K.G.: Certificateless Public Key Cryptography. In: Laih, C.-S. (ed.) ASIACRYPT 2003. LNCS, vol. 2894, pp. 452–473. Springer, Heidelberg (2003)
2. Boneh, D., Lynn, B., Shacham, H.: Short Signatures from the Weil Pairing. In: Boyd, C. (ed.) ASIACRYPT 2001. LNCS, vol. 2248, pp. 514–532. Springer, Heidelberg (2001)
3. Chen, Y.C., Liu, C.L., Horng, G., Chen, K.C.: A provably secure certificateless proxy signature scheme. International Journal of Innovative Computing, Information and Control 7(9), 5557–5569 (2011)
4. Choi, K.Y., Park, J.H., Hwang, J.Y., Lee, D.H.: Efficient Certificateless Signature Schemes. In: Katz, J., Yung, M. (eds.) ACNS 2007. LNCS, vol. 4521, pp. 443–458. Springer, Heidelberg (2007)
5. Choi, K., Park, J., Lee, D.: A new provably secure certificateless short signature scheme. Computers and Mathematics with Applications 61, 1760–1768 (2011)
6. Du, H., Wen, Q.: Efficient and provably-secure certificateless short signature scheme from bilinear pairings. Computer Standards & Interfaces 31, 390–394 (2009)
7. Fan, C.I., Hsu, R.H., Ho, P.H.: Truly non-repudiation certificateless short signature scheme from bilinear pairings. Journal of Information Science and Engineering 24, 969–982 (2011)
8. Hu, B.C., Wong, D.S., Zhang, Z., Deng, X.: Certificateless signature: a new security model and an improved generic construction. Designs, Codes and Cryptography 42(2), 109–126 (2007)
9. Huang, X., Mu, Y., Susilo, W., Wong, D.S., Wu, W.: Certificateless Signature Revisited. In: Pieprzyk, J., Ghodosi, H., Dawson, E. (eds.) ACISP 2007. LNCS, vol. 4586, pp. 308–322. Springer, Heidelberg (2007)
10. Huang, X., Mu, Y., Susilo, W., Wong, D.S., Wu, W.: Certificateless signatures: new schemes and security models. Computer Journal (2011), doi:10.1093/comjnl/bxr097
11. Shamir, A.: Identity-Based Cryptosystems and Signature Schemes. In: Blakely, G.R., Chaum, D. (eds.) CRYPTO 1984. LNCS, vol. 196, pp. 47–53. Springer, Heidelberg (1985)
12. Shim, K.: Breaking the short certificateless signature scheme. Information Sciences 179, 303–306 (2009)
13. Tso, R., Yi, X., Huang, X.: Efficient and short certificateless signatures secure against realistic adversaries. Journal of Supercomputing 55, 173–191 (2011)
14. Tso, R., Huang, X., Susilo, W.: Strongly secure certificateless short signatures. Journal of Systems and Software 85, 1409–1417 (2012)
15. Yum, D.H., Lee, P.J.: Generic Construction of Certificateless Signature. In: Wang, H., Pieprzyk, J., Varadharajan, V. (eds.) ACISP 2004. LNCS, vol. 3108, pp. 200–211. Springer, Heidelberg (2004)

Impact of Identifier-Locator Split Mechanism on DDoS Attacks

Ying Liu, Jianqiang Tang, and Hongke Zhang

School of Electronic and Information Engineering, Beijing Jiaotong University,
Beijing 100044, P.R. China
yliu@bjtu.edu.cn

Abstract. The dual semantics of IP address, representing not only the identifiers of nodes but also the locators of nodes, is one of the fundamental reasons for hindering the development of current Internet. Therefore, the identifier-locator split mapping network which separates the identifier role and the locator role of an IP address has become one of the most-watched techniques in the field of future Internet architecture. However, DDoS attacks are still in existence in this network. In this paper, by using the attack traffic we discuss and compare the effects of DDoS attacks between the current Internet and the identifier-locator split mapping network. In particular, the numerical analysis and simulation show that the identifier-locator split mapping network alleviates DDoS attacks more effectively, compared with the current Internet.

Keywords: Network security, identifier-locator split, DDoS attacks, LISP.

1   Introduction

In recent years, the problems in routing scalability, security, and mobility of the current Internet have become notable and prominent [1-5]. The research shows that the main reasons for the problems mentioned above include the open, trustworthy, and self-governing design of the Internet and the ambiguity of the IP address [6,7]. For example, in the aspect of security, since a traditional IP address contains the terminal's identity information and location information, the correspondent node and the malicious eavesdroppers can obtain the terminal's identity information and topology location information from the terminal's IP address, resulting in the exposure of users' privacy. The attackers can easily use the dual attributes of an IP address to probe into the network topology, counterfeit the identity information, and make ARP cheats, IP cheats, man-in-the-middle attacks, and Distributed Denial of Service (DDoS) attacks.

Therefore, in the research of the next generation of Internet's system structure, the design idea to split the identifier information from the locator information is widely concerned and accepted. The main achievements are as follows: the network-based separation mechanism includes LISP (Locator/ID Separation Protocol) [8], IP^2 (IP-Based IMT Network Platform) [9,10], Ivip (Internet Vastly Improved Plumbing) [11],

J.-S. Pan et al. (Eds.): *Advances in Intelligent Systems & Applications*, SIST 21, pp. 69–78.
DOI: 10.1007/978-3-642-35473-1_8   © Springer-Verlag Berlin Heidelberg 2013

TIDR (Tunneled Inter-domain Routing) [12] and the identifier-based universal network [13,14]; the host-based separation mechanism, includes HIP (Host Identity Protocol) [15], Hi3 (Host Identity Indirection Infrastructure) [16], SHIM6 [17], ILNP (Identifier Locator Network Protocol) [18,19] and so on. Although these proposals are different, they are the solutions all based on the separation of node identity information and location information to meet the needs of the future network. Compared with the host-based separation mechanism, the network-based separation mechanism has better routing scalability and users' location privacy, etc.

This paper makes an in-depth study on the characteristics of the identifier-locator split mapping network (hereinafter referred to as split mapping network) and the theory of DDoS attacks. Furthermore, it makes a quantitative and comparative analysis of the threats caused by DDoS attacks in the current Internet and the split mapping network. It proves that, compared with the current Internet, the identifier-locator split mapping network can effectively mitigate the DDoS attacks. The framework of this paper is as follows: the second part talks about the model of split mapping network; the third part makes a theoretical analysis of the DDoS attacks in the current Internet and the split mapping network; the fourth part describes the calculation analysis; the fifth part presents the simulation result; the final part draws a conclusion.

2   The Identifier-Locator Split Mapping Network Model

After studying and analyzing the architectures of the split mapping network which are proposed in the literature from [8] to [14], their common features are summarized as follows:

1. According to the topology position, the network is divided into two parts: the access network and the core network. The access network is a collection of various types of terminals or subnets, such as fixed, mobile, sensor networks, allowing access to the Internet. The core network is responsible for location management and global routing. Therefore, no matter when and where, an identity can never perform the dual functions of "identifier" and "locator".

2. For the data forwarding plane, IP addresses can be separated into two numbering spaces: the Endpoint Identifier (ID) and the Routing Locator (LOC). The ID used in the access network represents the identity of a terminal. The LOCs used in the core network are responsible for global routing. Additionally, the IDs and the LOCs can't directly communicate with each other. They should be mapped to each other when the packets traverse the different space.

3. The mapping services between the ID and the LOC are realized by some specific devices. Usually, the router in the joint between the access network and the core network, known as Access Router (AR), is responsible for realizing the services. The mapping information of the whole network is stored in a specific device in the core network. In this paper, except the access routers, all the devices in the core network maintaining the mapping information are referred to Mapping Servers (MS).

Therefore, without loss of generality and without considering the specific realization mechanism, the identifier-locator split mapping network model is shown in Figure 1.

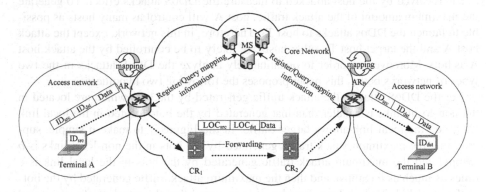

Fig. 1. The Identifier-Locator Split Mapping Network Model

3   DDoS Attacks Analysis

As shown in Figure 1, this paper uses the network topology in Figure 2 to analyze the DDoS attacks in the current network and the split mapping network. In Figure 2, the access routers are not distinguished between the current network and the split mapping network, represented uniformly with AR.

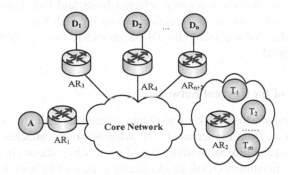

Fig. 2. Network Topology

As shown in Figure 2, the network is a fully connected network, in which any two hosts have access to each other. In this network, host A is a hostile attack host. The hosts represented by $D_1, D_2 \ldots D_n$ and $T_1, T_2 \ldots T_m$ are all common hosts. The hosts $T_1, T_2 \ldots T_m$ are connected to the same sub-network through AR_2. In this topology, it is assumed that the attack target of host A is host T_2, so the access network to which AR_2 is connected is defined as the local link in this paper and the number of hosts is m. Other access networks are non-local links and the number of hosts is n.

The reference [20] generally classifies DDoS attacks into two types, the utilization of software vulnerability and the flow attack, and this paper will analyze the DDoS flow attack. According to the analytical procedure in [21], this paper uses the attack traffic received by the host attacked to measure the DDoS attacks effect. To generate the maximum amount of the attack traffic, host A will control as many hosts as possible to launch the DDos attacks to host T_2. Therefore, in this network, except the attack host A and the target host T_2, other hosts are likely to be controlled by the attack host A as bots. Moreover, in order to quantitatively analyze the DDoS attacks in the two types of networks better, this paper proposes the following two hypotheses.

For the DDoS attacks, the attack traffic generated by the bots which are located in the non-local links is smaller than that generated by the bots located in the local link [22], i.e., the local link allows larger traffic. In the Figure 2, to make it simple, suppose that the maximum attack traffic generated by the bots in the non-local links is b (Mbps) and the maximum attack traffic generated by the bots in the local link is c times as much as b (Mbps), and then the maximum attack traffic generated by the bots in the local link is cb (Mbps). Besides, it is assumed that the attack traffic generated by the attacker is not more than the maximum bandwidth of the local link.

The attacker's ability is limited, so it can only control some of the hosts in the network. Suppose that the number of bots is k, and, for the attack host A, it is of the same difficulty to control any of the hosts in the network.

In the current network, the attacker can obtain a terminal's identity information and location information from the terminal's IP address. To generate more attack traffic, the attacker will first choose the hosts in the local link as bots. However, in the split mapping network, because of the location privacy, the attacker can only obtain the terminal's identity information rather than the location information; therefore, the attacker can only control the randomly selected hosts and then launch attack. This feature can effectively mitigate the threat caused by the DDoS attacks. The following part will make a detailed analysis of the DDoS attacks in the current network and the split mapping network.

3.1   Analysis of the Current Network

In the current network, the attacker can obtain a terminal's identity information and location information from the terminal's IP address, so the attacker will choose as many hosts in the local link as possible. Based on this hypothesis, in the current network, suppose the maximum DDoS attacks traffic generated by host A is F_t, then:

$$F_t = \begin{cases} k * c * b & k < m - 1 \\ (k - m + 1) * b + (m - 1) * c * b & k \geq m - 1 \end{cases}$$

(1)

3.2   Analysis of the Identifier-Locator Split Mapping Network

In the split mapping network, since the attacker is unable to obtain a terminal's location information, the attacker can only control the randomly selected hosts instead of finding the hosts correctly which can do the maximum damage to the target host.

Therefore, according to this hypothesis, when $k < m-1$, the maximum attack traffic that the DDoS attackers can generate has $k+1$ possible situations, which are shown as follows.

Possible situation 1. All the randomly selected hosts are in the non-local links of target host T_2, and now the maximum attack traffic is $F_{s0} = k*b$.

The probability of this situation is $P_{s0} = \dfrac{C_n^k}{C_{n+m-1}^k}$.

Possible situation 2. Only one of the randomly selected hosts is in the local link of target host T_2, and now the maximum attack traffic is $F_{s1} = (k-1)*b + c*b$.

The probability of this situation is: $P_{s1} = \left\{ \dfrac{C_n^{k-1}C_{m-1}^1}{C_{n+m-1}^k} \right.$.

Possible situation $k+1$. k of the randomly selected hosts are in the local link of target host T_2, and now the maximum attack traffic is $F_{sk} = k*c*b$.

The probability of this situation is $P_{sk} = \dfrac{C_n^0 C_{m-1}^k}{C_{n+m-1}^k}$.

At this moment, the average maximum attack traffic that the DDoS attackers can generate is:

$$F_{sa} = \sum_{i=0}^{k} \frac{C_n^{k-i}C_{m-1}^i}{C_{n+m-1}^k}\left[(k-i)*b + i*c*b\right]$$

(2)

Similarly, when $k \geq m-1$, the situation is:

$$F_{sb} = \sum_{i=0}^{m-1} \frac{C_n^{k-i}C_{m-1}^i}{C_{n+m-1}^k}\left[(k-i)*b + i*c*b\right]$$

(3)

To sum up, in the split mapping network, the average maximum attack traffic that the DDoS attackers can generate is:

$$F_s = \sum_{i=0}^{\min\{k,m-1\}} \frac{C_n^{k-i}C_{m-1}^i}{C_{n+m-1}^k}\left[(k-i)b + i*c*b\right]$$

(4)

4   Numerical Analysis

This part presents the concrete numerical analysis and the calculation of formula (1) and formula (4). Suppose that the number of hosts in the non-local links is $n=100$, the attack traffic generated by each bot in the non-local links is $b=1$(Mbps), and the attack traffic generated by each bot in the local link is c times as much as that generated by each bot in the non-local links, with $c=5$. F_t, the attack traffic generated by the current network, is shown in Figure 3 and F_s, the attack traffic generated by the split mapping network, is shown in Figure 4 where m, representing the number of hosts in the local link, varies from 2 to 50 and k, representing the number of attack hosts, varies from 1 to 100. In the two figures, the different colors in the fill area represent different amounts of attack traffic and the specific numerical values are corresponding to the color bar at the right hand of the Figures.

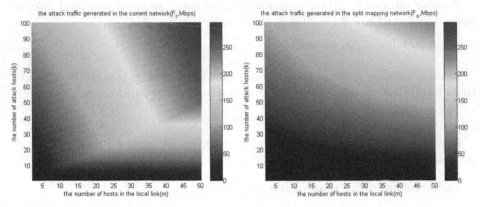

Fig. 3. Variation of the attack traffic generated in the current network results from the changes in the values of *m* and *k*

Fig. 4. Variation of the attack traffic generated in the split mapping network results from the changes in the values of *m* and *k*

From Figure 3 and Figure 4, it can be found that in the same network topology environment, as *m* and *k* increase, the attack traffic in the two networks increase as well. But, when m is equal to *k*, the attack traffic in the split mapping network is less than that in the current network. The calculation result shows that compared with the attack traffic in the current network, the attack traffic in the split mapping network can be reduced by 79% at most and 37% in average. Only when *m* is small, and *k* is much larger than *m*, can the attack traffic in the split mapping network be close to that in the current network. It can be found from Figure 3 and Figure 4 that under this condition, the attack traffic in both networks remains relatively small. Thus, it can be seen that the split mapping network can better mitigate the bad influences caused by the DDoS attacks.

The following part analyzes the attack traffic in both networks where the numerical value of *m* is fixed and the numerical values of *n* and *k* are changing. F_t and F_s are shown in Figure 5 and Figure 6 respectively. In the two figures, *m*, *b* and *c* are set to 20, 1 and 5, respectively, while *n* ranges from 60 to 150, and *k* ranges from 1 to 40.

From Figure 5, it can be found that in the current network, when the number of hosts in the local link is fixed (*m*=20), the attack traffic increases with the growth of the number of bots and when the number of hosts in the non-local link is big enough, the attack traffic is not influenced by the change of the number of hosts in the non-local links. From Figure 6, it can be seen that in the split mapping network, as *k* increases, the attack traffic increases as well, but obviously it is less than the attack traffic generated in the current network under the same condition. In addition, with the further increase of n, the attack traffic will become smaller and smaller, gradually approaching to the situation where all the bots are the hosts in the non-local links.

The calculation result shows that compared with the attack traffic in the current network, the attack traffic in the split mapping network can be reduced by 71% at most, 43% at least, and 60% in average. Thus, it can be seen that the split mapping network can better mitigate the bad influences caused by the DDoS attacks.

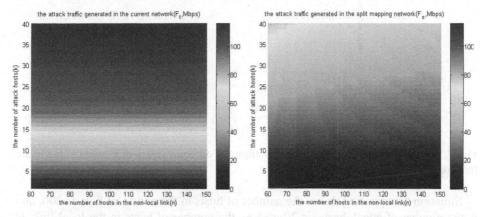

Fig. 5. Variation of the attack traffic generated in the current network results from the changes in the values of n and k

Fig. 6. Variation of the attack traffic generated in the split mapping network results from the changes in the values of n and k

5   Simulation Analysis

This part makes the simulation for the network topology shown in Figure 2 by using OMNET++ [23]. It mainly makes the simulation for the attack traffic received by the target host in 10 seconds in the following two situations.

Situation 1: supposing n, the number of hosts in the non-local link, is 200, and m, the number of hosts in the local link, is 20, and k, the number of attack hosts, is respectively 5, 20, and 35, the simulation of the attack traffic received by the target host in the current network and in the split mapping network is shown in Figure 7.

From Figure 7, it can be seen that when the number of hosts in the non-local links and in the local link is fixed, with the increase of the number of bots, the attack traffic received by the target host in the current network and in the split mapping network will increase as well. But, as shown in Figure 7, when k is 5, 20 and 35 respectively, compared with the average DDoS attacks traffic generated in the current network, the average DDoS attacks traffic generated in the split mapping network is reduced by 63%, 62%, and 44% respectively, which suggests that the split mapping network reduces the influences of DDoS attacks. At the same time, it can be also found that the smaller the number of bots is, the more obvious the mitigation effect is. In the actual network environment, the number of bots is much smaller than the total number of hosts in the network, so in the actual environment, the split mapping network can mitigate the DDoS attacks better.

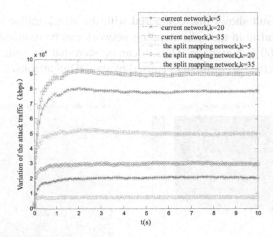

Fig. 7. Variation of the attack traffic results from the different values of k, when n and m are fixed to 200 and 20 respectively

Situation 2: supposing that n, the number of hosts in the non-local link, is 200, and k, the number of attack hosts, is 20, and m, the number of hosts in the local link, is respectively 10, 20, and 30, the simulation of the attack traffic received by the target host in the current network and in the split mapping network is shown in Figure 8.

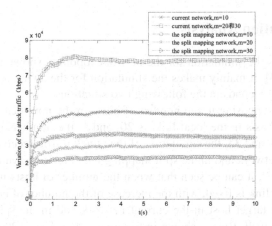

Fig. 8. Variation of the attack traffic results from the different values of m, when n and k are fixed to 200 and 20, respectively

In Figure 8, when m is respectively 10, 20, and 30, compared with the average DDoS attacks traffic generated in the current network, the average DDoS attacks traffic generated in the split mapping network is reduced by 51%, 62%, and 54% respectively. When $m=k=20$, the attack traffic generated in the current network will reach the maximum (when all the bots are the hosts in the local link); later, with the increase of m, the attack traffic generated in the current network no longer increases, but the attack traffic generated in the split mapping network keeps increasing all the time, so the percentage of the reduced attack traffic will decrease. In general, in the actual network environment, the number of hosts in the local link is smaller than the number of hosts in the non-local links, so the split mapping network's role in mitigating the DDoS attacks is still obvious.

The simulation results of the two situations mentioned above show that in the actual network environment, the split mapping network can better mitigate the influences caused by the DDoS attacks, which is consistent with the results of the theoretical analysis and numerical analysis.

6   Conclusion

As a new type of network architecture, the identifier-locator split mapping network successfully separates a user's identity information from the location information and protects the user's location privacy. This paper analyzes the split mapping network's role in mitigating the DDoS attacks. The numerical analysis and the simulation analysis indicate that compared with the current network, the identifier-locator split mapping network can mitigate the DDoS attacks better.

Acknowledgments. This work was partially supported by the Fundamental Research Funds for the Central Universities under grant NO.2011JBM016, 2012YJS019, supported by the National Natural Science Foundation of China under grant NO.61202428 and 60903150.

Reference

1. BGP Report, http://bgp.potaroo.net/
2. http://www.cert.org.cn/articles/docs/common/
   2011042225342.shtml
3. Nakajima, N., Dutta, A., Das, S., et al.: Handoff delay analysis and measurement for SIP based mobility in IPv6. In: IEEE International Conference on Communications, vol. 2, pp. 1085–1089. IEEE Press, New York (2003)
4. Koodli, R.: IP Address Location Privacy and Mobile IPv6: Problem Statement. RFC 4882 (2007)
5. Feng, Q., Xiaoqian, L., Wei, S., et al.: A Novel Location Management Scheme based on DNS in Proxy Mobile IPv6. China Communications 7, 43–52 (2010)
6. Saltzer, J.: On the Naming and Binding of Network Destinations. RFC 1498 (1993)
7. Clark, D., Braden, R., Falk, A., et al.: FARA: Reorganizing the Addressing Architecture. In: Proc. ACM SIGCOMM Workshop on Future Directions in Network Architecture, FDNA, pp. 313–321. ACM Press, New York (2003)
8. Farinacci, D., Fuller, V., Meyer, D., et al.: Locator/ID Separation Protocol (LISP). Internet Draft, draft-farinacci-lisp-15.txt (2011)
9. Yumiba, H., Imai, K., Yabusaki, M.: IP-Based IMT Network Platform. IEEE Personal Communications Magazine 8(5), 18–23 (2001)

10. Okagawa, T., Nishida, K., Miura, A.: A Proposed Routing Procedure in IP^2. In: Vehicular Technology Conference, vol. 3, pp. 2083–2087 (2003)
11. Whittle, R.: Ivip (Internet Vastly Improved Plumbing) Architecture. Internet Draft, draft-whittle-ivip-arch-04 (2010)
12. Adan, J.J.: Tunneled Inter-domain Routing (TIDR). Internet Draft, draft-adan-idr-tidr-01 (2006)
13. Hongke, Z., Wei, S.: Fundamental Research on the Architecture of New Network—Universal Network and Pervasive Services. Acta Electronica Sinica 35(4), 593–598 (2007) (in Chinese)
14. Ping, D., Yajuan, Q., Hongke, Z.: Research on Universal Network Supporting Pervasive Services. Acta Electronica Sinica 35(4), 599–606 (2007) (in Chinese)
15. Moskowitz, R.: Host Identity Protocol Architecture (HIP). Internet Draft, draft-ietf-hip-rfc4423-bis-03 (2011)
16. Nikander, P., Arkko, J., Ohlman, B.: Host Identity Indirection Infrastructure. In: 2nd Swedish National Computer Networking Workshop, pp. 1–4. SNCNW Press, Swedish (2004)
17. Nordmark, E., Bagnulo, M.: Shim6: Level 3 Multihoming Shim Protocol for IPv6. RFC 5533 (2009)
18. Atkinson, R., Bhatti, S.: An Introduction to the Identifier-Locator Network Protocol (ILNP). In: London Communications Symposium (2006)
19. Atkinson, R.J.: ILNP Concept of Operations. Internet Draft, draft-rja-ilnp-intro-11 (2011)
20. Hussain, A., Heidemann, J., Papadppoulos, C.: A Framework for Classifying Denial of Service Attack. In: Proceedings of the ACM SIGCOMM Workshop on Applications, Technologies, Architectures, and Protocols for Computer Communications, pp. 99–110. ACM Press, New York (2003)
21. Jiejun, K., Mirza, M., Shu, J., et al.: Random Flow Network Modeling and Simulations for DDoS Attack Mitigation. In: IEEE International Conference on Communications, vol. 1, pp. 487–491. IEEE Press, New York (2003)
22. Chunfeng, W., Hurwitz, J.G., Newman, H., et al.: Optimizing 10-Gigabit Ethernet for Networks of Workstations, Clusters, and Grids: A Case Study. In: Proceedings of the 2003 ACM/IEEE Conference on Supercomputing, pp. 50–62. IEEE Press, New York (2003)
23. OMNET++3.3, http://www.omnetpp.org

Detecting Web-Based Botnets
with Fast-Flux Domains

Chia-Mei Chen, Ming-Zong Huang, and Ya-Hui Ou

Department of Information Management
National Sun Yat-sen University
Kaohsiung 80424
Taiwan
cchen@mail.nsysu.edu.tw

Abstract. Botnet is one of the most threatening attacks recently. Web-based botnet attacks are serious, as hacker takes advantage of the HTTP connections hiding malicious transmissions in a vast amount of normal traffic that is not easily detectable. In addition, integrating with fast-flux domain technology, botnet may use a web server to issue attack commands and fast-flux technology to extend the lifespan of the malicious website. This study conducts anomalous flow analysis on web-based botnets and explores the effect of fast-flux domains. The proposed detection mechanism examines flow traffic and web domains to identify a botnet either using HTTP as control and command channel or using fast-flux domain for cloaking. Based on the experiments on both testbed and real network environments, the results prove that the proposed method can effectively identify these botnets.

Keywords: Fast-flux domain, web-based botnet, malware.

1   Introduction

The Internet brought great convenience to the masses. In contrast, the pervasiveness of the Internet also makes hackers envious of the economic effects brought together by the status quo. This led to network security problems that are severe and difficult to prevent. Botnet is a combination of various forms of malware, such as Trojan horse, virus, worm, and spyware. Traditional firewalls, anti-virus software, and IDS (intrusion detection systems) are ineffective against a botnet attack. Infections can occur through system vulnerabilities or through social engineering that can induce a user to click on a malicious image or website, which can automatically trigger the execution of a pre-set function. This could then result in the system being remotely controlled by the pre-defined set of commands. Once partial control of the bot machine is achieved, spam or other attacks are then launched to the detriment of the user. Methods for detecting botnets are mostly based on analysis of the behavioral characteristics of the attack. The methodology used in the attack must first be fully understood before any attempt at prevention can be made.

By using an HTTP connection as a communication channel, a web-based botnet attack can avoid detection by a firewall and increase the threat of the attack. One of

J.-S. Pan et al. (Eds.): *Advances in Intelligent Systems & Applications*, SIST 21, pp. 79–89.
DOI: 10.1007/978-3-642-35473-1_9   © Springer-Verlag Berlin Heidelberg 2013

the attack characteristics is its small traffic signature, which also fits perfectly well within the normal traffic flow. Since most firewalls do not filter HTTP traffic, it is therefore not easy to detect any abnormal behavior.

According to security-related information provided by various research institutions, web-based botnets not only present serious threats to today's network security, they are also found to be using fast-flux domain technology for seeking routes of transmission. From the hacker's perspective, the advantage of fast-flux domain is that one domain name can have multiple IP addresses that can be used for rapid switching, which can keep the malicious website from being detected, and thereby extend its lifespan. This technique can not only block blacklist detection methods, but also benefits from load balance.

Attacks based on web-based botnet and fast-flux domain offer the benefit of nearly undetectability and rapid spreading. These attacks raise attack capability but also increase the difficulty of anti-hacking defense. In this study, a malicious web-based botnet detection mechanism is proposed with the ability of fast-flux domain detection.

2   Related Studies

Since most malicious attacks use the fast flux domain technique to evade detection, many studies have focused on analyzing the characteristics of fast flux server domain (FFSN) to formulate a method for determining the presence of fast flux domains. Along with the ever increasing threat of web-based botnets, a plethora of information security issues have arisen from web-based botnets and FFSN. The purpose of this study is to detect web-based botnets while considering fast flux domain technology, by exploring the changing relationship between the two; thereby helping users achieve a more accurate detection of this type of attack.

The authentication mechanism proposed by Holz et al. [6] for detecting a fast-flux domain in a network is the use of double DNS queries to obtain the following three features of DNS response: non-duplicated IP addresses, the number of name servers (NS) and the number of autonomous system numbers (ASN). Holz et al. [10] did not use the TTL feature, because CDNs also exhibit the characteristic of possessing a short term TTL. Such a characteristic cannot distinguish clearly between a fast-flux domain and a CDN.

Based on the above detection method, Zhou et al. [7] proposed two additional methods to improve detection. One method involves the use of simultaneous queries to check multiple DNS hosts to observe the number of non-duplicated IPs to reduce the time required to detect a fast-flux domain. The other is through cross-comparison of fast-flux domain detection results to accelerate query performance. Because many fast-flux domains share the same IPs, if the query results from a FQDN to be tested are similar to those of the known fast-flux domains, then the probability that the FQDN to be tested is a fast-flux domain is high.

Although the detection methods proposed by these studies already have a good detection rate, some web sites that legitimately use fast-flux domain technology—such as pool.ntp.org and database.clamav.net—are still classified as malicious websites, thereby increasing the rate of false alarm. Passerini et al. [8] used even more features divided into three categories, as shown in Table 1.

Table 1. Fast-flux domain feature classification

Category	#	Description
Domain name	F_1	Domain age
	F_2	Domain registrar
Availability of the network	F_3	Number of distinct DNS records of type "A"
	F_4	Time-to live of DNS resource records
Heterogeneity of the agents	F_5	Number of distinct networks
	F_6	Number of distinct autonomous systems
	F_7	Number of distinct resolved qualified domain names
	F_8	Number of distinct assigned network names
	F_9	Number of distinct organisations

Passerini et al. believed that TTL is important in determining a fast-flux domain, in that a short TTL can quickly match to various different IPs. The classification of domain information by Passerini et al. is not the same as those of other studies. Because hackers often use the personal information of victims or randomly generated names to register malicious domains, such classification can find other malicious domains based on the personal registration information that has already been detected from a fast-flux domain.

Fast-flux domains are not limited to web applications. Any applications using DNS can also use fast-flux domains. However, presently, most fast-flux domains facilitate web services. Regardless of which service a fast-flux domain uses, its detection method and characteristics are identical [9]. Yu et al. [9] developed a system and proposed two detection methods to validate the effectiveness of a service. They are average online rate (AOR) and minimum availability rate (MAR). Once the existence of a fast-flux domain agent is discovered, its activities are monitored every hour using calculations based on AOR and MAR.In a legitimate fast-flux domain, these activities should be under complete control for around the clock service, and its AOR value should be close to 100 %. If a fast-flux domain is malicious, then its AOR value is significantly less than that of a legitimate domain. Similarly, its MAR value is also smaller than that of a legitimate domain. Thus, finding MAR can help identify whether a fast-flux domain is malicious. Yu et al. also developed an agent monitoring system that builds on an IP database. Each time a new IP is found, the system immediately determines if the IP address is suspicious, and its findings are documented.

These findings showed that a fast-flux domain service network has some fixed characteristics that can be used for detection. This study uses the number of ASNs and the time of registration as characteristics for measuring fast-flux domains. The number of ASNs can be used to determine whether a website is using a fast-flux domain. Because a malicious website using a fast-flux domain exists for a very short period, the registration time can be used to narrow down the selection range. Furthermore, through observation, this study discovered a correlation between the A records, the IP reverse lookup of its domain name, and the original FQDN as features of a malicious website. A hacker may be able to decide the FQDN, but their use of fast-flux technology is subject to the following restrictions: Unlike a CDN, which can choose its own hardware device and a specific IP, a fast-flux domain cannot guarantee service up-time.

In addition, the domain name of a reverse IP lookup is determined by the network administrator of that IP address. Therefore, the FQDN of a malicious website is not usually relevant to the domain name of a reverse IP lookup.

3   Research Methodology

The primary purpose of fast-flux domain is load balance and such domain usually has multiple IP addresses to serve user requests. A botnet can use the same fast flux technology serving for the purpose of cloaking and intrusion evasion. This study demonstrates that a fast-flux domain not only shields malicious websites, but also masks the C & C server from being detected. Hackers can then use the fast-flux domain to send commands, taking control of the entire botnet. Additionally, a fast-flux domain agent is simply a relay station that possesses the traffic-redirection features of port 80 or port 53. Therefore, a web-based botnet using HTTP as communication channel can more easily redirect users to malicious websites.

3.1   System Architecture

The architecture in this study aims to develop a detection system which can identify web-based botnets or malicious website using fast-flux domain technology to evade the detection. The data sources include URL traffic and spam archive. Observed from spam archives, the majority of the malicious activities relate to fast-flux domains. Therefore, spam is often used to discover fast-flux domains. Traffic data with URL information provides the website address of each HTTP connection for distinguishing visiting pattern of a machine which is an important attribute for web-based botnet as well as fast-flux domain. The proposed system architecture is described in Figure 1. In the following sections, the botnet detection mechanism is explained first, followed by fast-flux domain detection.

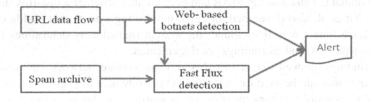

Fig. 1. System architecture diagram

3.2   Web-Based Botnet Detection

The web-based botnet attack model revealed that bot herder communicates with the server, the command & control server (C&C server), that issues commands and controls bots through HTTP connections. Based on the preliminary study conducted on

Testbed@TWISC, the HTTP connections exhibit the characteristics of periodic repeatability and identical webpage access. This study conducted preliminary experiments choosing web-based botnets from BlackEnergy and Zeus families. The results show that BlackEnergy does possess the characteristics of periodic repeatability and identical webpage accesses and can be identified through layer 4 flow traffic. The first sight of Zeus bots however seems no bot behaviors found, as its download time indicates a lack of regularity and the accessed web sites are not the same ones. When the URL data is taken into consideration, each web page (i.e., URL) visited by Zeus bots reveals regular browsing pattern, a characteristic of periodic repeatability.

Based on the above preliminary experiments, the results demonstrate that using URL information as the basis for detection allows for more accurate web-based botnet detection. In addition, it can also improve the false alarms generated by a hybrid botnet, integration of web-based and fast-flux domain technology. As shown in Figure 2, initially the bot herder connects to the C&C server which facilitates fast-flux domain technology, turns it into a fast-flux domain agent, and establishes a connection between the two sides. Unfortunately, the detection based on IP address may fail to identify despite its fixed connection. It is also the case that since the server loses regularity of connectivity due to frequent changes of the IP addresses, which can cause the connectivity of the same original group to lose its grouping functionality as a result of dissimilar IPs. If URL information is used for detection based on groupings of FQDN in the accessed web pages, the detection can overcome the problem of dissimilar IPs and it can identify the connection regularity.

The proposed web-based botnet detection first screens out successful web connections which usually are issued by real clients, not bots. The connections to the same web server are grouped together for further examination. A normal web server often provides dynamic contents for users, while a C&C server's webpage remains the same for a period of time and bots repeatedly and periodically visit the same contents. Webpage attributes of each grouped traffic are extracted to discover the anomalous web connections. The attribute of webpage dynamicity indicates the degree of variance of a webpage and that of regularity rate represents the degree of periodic repeatability.

A normal website usually provides rich and diversified contents for surfers. Therefore, the web links in a domain will be visited variously by different users. Due to the diversified contents and visitors, the first attribute, webpage dynamicity, is expressed by entropy, an information-theoretic statistic which measures the variability of the data. Let $f_1, f_2, \cdots, f_m$ be a group of flow traffic on a domain X . $Sim(f_i, f_j)$ is the similarity of the flows with the selected fields.

The second attribute, regularity rate, is defined as $\dfrac{\sum d^x i}{\sum \max((d^x i + 1 - d^x i), 1)}$, where

$d^x i$ be the timespan of two consecutive flows, f_{i+1} and f_i . To avoid the two timespans are equal and the divider zero, a minimum value 1 is used in case they are the same. The reciprocal gives higher degree of regularity to small period discrepancy.

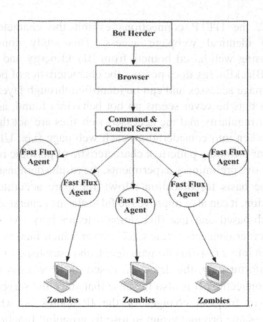

Fig. 2. Web-based botnet and fast-flux domain integrated system architecture

Let G^X be a set of web connections accessing website X , $g^x i$ be the i th connection to website X , $d^x i$ be the timespan of two consecutive connections, $g^x i$ and $g^x i+1$, without loss of generosity, the i th connection is earlier than $(i+1)$ th's. Assume a flow traffic data contains m fields of information, $f_1, f_2, \cdots, f_m$. The flow data of connection $g^x i$ can be represented as $f_1(g^x i), f_2(g^x i), \cdots, f_m(g^x i)$. Let $f_t(g^x i)$ be the timestamp of connection $g^x i$ and $d^x i$ be the timespan of two consecutive connections, $g^x i$ and $g^x i+1$, where $p^x i = f_t(g^x i+1) - f_t(g^x i)$. The degree of periodic repeatability of a grouped traffic G^X , $R(G^X)$, is defined as $\dfrac{\sum p^x i}{\sum (p^x i+1 - p^x i)}$. The attribute is normalized by its period and the reciprocal gives higher degree to small period discrepancy. The definition of webpage dynamicity is similar, but the rest of flow information is taken into consideration. The flow data remains the same, if the accessed webpage replies the same web contents.

3.3   Fast Flux Domain Detection

This study adopted command *dig* to get information about the associated IP and domain and to determine whether a domain name, FQDN, uses fast-flux. Furthermore, the A and NS records recorded in DNS are used to determine if the FQDN is legitimate or malicious. This study used dissimilar ANSs, the reverse lookup of the DNS, and the time of the domain registration as characteristics for distinguishing benign and illegal domains.

Dissimilar ASNs (autonomous system number): Each IP has an ASN; IPs that are geographically close have the same ASN. The appearance of different ASNs indicates the presence of a potential fast-flux domain. As fast-flux domain agents, mostly victim's machines, spread all over the world, one IP may correspond to several dissimilar ASNs. Though a legitimate site may also have multiple IPs, the regional distribution of their ASNs remains the same. The reason for having multiple IPs is to maintain a balanced load, thereby reducing traffic to the host.

www.ava**.com	2000	IN	A	67.228.*.*	(36351)
www.ava**.com	2000	IN	A	74.55.*.*	(21844)
www.ava**.com	2000	IN	A	74.55.*.*	(21844)
www.ava**.com	2000	IN	A	74.55.*.*	(21844)
www.ava**.com	2000	IN	A	74.55.*.*	(21844)
www.ava**.com	2000	IN	A	74.86.*.*	(36351)
www.ava**.com	2000	IN	A	174.36.*.*	(36351)
www.ava**.com	2000	IN	A	174.36.*.*	(36351)
www.ava**.com	2000	IN	A	174.37.*.*	(36351)
www.ava**.com	2000	IN	A	174.37.*.*	(36351)
www.ava**.com	2000	IN	A	174.123.*.*	(21844)

Fig. 3. ASN in A record of www.ava**.com

Most malicious websites use fast-flux domain to make multiple IPs appear in the same DNS resource records. However, this does not mean that all websites using fast-flux domain are malicious. In Figure 3, for example, www.ava**.com is a legitimate website. Its two groups of ASN numbers are 36351 and 21844, indicating that these host machines are divided geographically into two groups, and representing that fast-flux domain technology is used to place multiple IPs in the same DNS. Therefore, more features are required to determine malicious websites, rather than simply judging a website to be malicious based on the use of fast-flux domain.

Reverse lookup of DNS server: DNS server records the DNS resource records of the domain. The reverse lookup of a FQDN in that domain should match with that of the DNS server. The domain name can be used to determine if the reversed domain name has any correlation to the original domain name. If it is correlated, then it is judged to be a legitimate site. Conversely, illegal websites often have uncorrelated domain names, indicating that they do not belong to the same domain system. As shown in Figure 4, for example, the reverse IP lookup in the A record of a legitimate FQDN www.ava**.com generates the same domain name, www.ava**.com. Conversely, wavecable.com, which is the reverse IP lookup in the A records of an illegal website www.tax.state-ca.net is fundamentally unrelated to the original domain name.

www.ava**.com	2000	IN	A	67.228.*.*	a***sl.ava***.com
www.tax.sta**-ca.net	180	IN	A	24.113.*.*	24.113.*.*.wave***.com

Fig. 4. Reverse IP lookup corresponding to legitimate and malicious websites

Registration Time: The time for a fast-flux domain to stay active is 18.5 days on average. Domain names with a long survival time can usually be judged as legitimate sites. Figure 5, for example, shows that legitimate sites have a relatively long registration time, while most malicious websites have short registration times.

taipei****.com	benign	Creation Date: 06-dec-1998
ava****.com	benign	Creation Date: 06-oct-1997
jx2d****.com	malicious	Creation Date: 19-may-2012
lit****.com	malicious	Creation Date: 20-jul-2012
roko****.com	malicious	Creation Date: 12-jun-2012

Fig. 5. Registration time comparison between legitimate and malicious websites

The above attributes are used for detection. The attribute values from benign and malicious websites were studied and evaluated to find good threshold for better detection rate. Therefore, Bayesian probability theory is chosen as the analysis method to explore the impact of parameters on the result predictions.

Number of dissimilar ASNs: Based on our study, a legitimate site usually has only one ASN, but there are some legitimate sites with one to three dissimilar ASNs. However, illegal sites often have multiple dissimilar ASNs. As the discreteness of the attribute, the definition of ASN parameter, w1, use a step-wise function: If a website has more than three dissimilar ASNs, it is judged to be illegal. One with one to three ASNs is given a figure between 0 and 1. Based on our experiments, 0.2 exhibits a better performance.

$$w_1 = \begin{cases} 0 & if\,|ASN| = 1 \\ 0.2 & if\,1 < |ASN| \le 3 \\ 1 & if\,|ASN| > 4 \end{cases}$$

Reverse IP Lookup: This study used the domain name obtained from reverse lookup to match the domain name of FQDN to calculate the degree of similarity as the basis of web link similarity. This study using $D_Y = \{d_{yj} | 1 \le j \le n_2 - r\}$ represents the domain name of a FQDN. $D_Y = \{d_{yj} | 1 \le j \le n_2 - r\}$ represents the domain name after the reverse IP lookup; $M_Y \rightarrow X = \{d_{yj} | d_{yj} \in D_X\}$ was used to calculate the difference between the domain name of the reverse IP lookup and the domain name of FQDN based on the calculation of each level domain (LD). The LD must match exactly to be listed into the numerical calculation $\left| M_Y \to X \right|$. The application for a top level domain (TLD) must follow stringent qualifications; thus, the number is small. Therefore, similarities can be generated more easily. When $\left| M_Y \to X \right| = 1$, it does not mean that the domain name is legitimate. Typically, it must be $\left| M_Y \to X \right| \geq 2$. Regarding the degree of domain name similarity between FQDN and reverse IP lookup, the larger the $\left| M_Y \to X \right|$, the closer $Sim(X,Y)$ is to 1.

$$w_2 = Sim(X,Y) = \frac{\left| M_Y \to X \right|}{n_1}$$

Registration Time: Malicious website typically has been registered for less than one month. Nevertheless, some malicious websites use fast-flux domain technology to evade detection and may have longer time. Based on our observation, one year is considered to be a trustful domain and less than one month is not. The registration time in between does not have significant difference and hence a step-wise function, w3, is defined to represent the weight of the registration time to the legitimacy of a domain.

$$w_3 = \begin{cases} a = \dfrac{registrar\ age \leq 1\ mouth}{malicious\ number} \\[2mm] b = \dfrac{registrar\ age \leq 1 year}{malicious\ number} \\[2mm] c = \dfrac{registrar\ age > 1 year}{malicious\ number} \end{cases}$$

The summation of the above three attribute values gives an anomaly score of the inspected domain name.

4   Experimental Results and Analysis

This study obtained a list of legitimate websites from the 1,000 most renowned enterprises, and that of malicious websites derived from spam for performance evaluation. The legitimacy of the websites is inspected by McAfee [11] and other monitoring services, including a Blocklist Removal Center [12], a WOT [13], and Free PC Security [14], as each monitoring service checks for different security threats, spyware, phishing, malware, and other security threats. McAfee applied reputation analysis detecting the above mentioned threats actively. Its results are labeled as 'mark1' and other monitoring as 'mark2'[1].

The discussion focuses on the spam archive of malicious websites as the basis for evaluating system performance. The monitoring report from McAfee website was

---

[1] Note that a website is considered as malicious if one of the monitoring services returns malicious.

considered as a reference and labeled as test result mark1, while that from Blocklist Removal Center, WOT, and Free PC Security as mark2. The malicious websites from spam using fast-flux shown in Table 2 were classified as malicious by the proposed system as well as other monitoring sites mostly.

Table 2. System-determined malicious websites in the spam archive

ID	URL	source	Mark1	Mark2
1	sigo****.com	spam	Malicious	Malicious
2	wvisit****osr29.com	spam	Malicious	Malicious
3	goph****good13.com	spam	Malicious	Malicious
4	hot****line10.com	spam	Malicious	Malicious
5	visit****sr23.com	spam	Malicious	Malicious
6	vph****daa.tk	spam	Malicious	benign
7	ruy****.com	spam	Malicious	Malicious
8	medphar*****count9.com	spam	Malicious	Malicious
9	medphar*****count2.com	spam	Malicious	Malicious
10	bestgood****24.com	spam	Malicious	Malicious

5   Conclusion

Apart from using the bot program to identify web-based botnet characteristics, this study also used it to find malicious websites that use fast-flux domain technology by identifying its characteristics. In web-based botnet detection, this study used previous literature as the basis for modifying IP-based detection methods to help improve its accuracy. Apart from solving the botnet problem from a programming perspective, the role that botnet played in benefiting hackers also provided us with the direction for solving the botnet problem. Fast-flux domain requiring numerous IPs is a characteristic that makes it closely related to botnet, so that detection of fast-flux domain techniques can indirectly solve the botnet problem. Previous botnet detection methods mostly started from the bot program or malicious traffic, and did not solve the botnet problem from a different angle. Therefore, this study proposed to use connection regularity as the basis for web-based botnet detection, and combined it with fast-flux domain detection. In addition to enhancing the accuracy of detection, it can also detect different types of botnet.

The longer-term goal of this study is for the system to integrate with IPS or Layer 7 firewalls. The output results of this system include IP and FQDN, which writes problem IPs into IPS (intrusion prevention system) rules, or writes FQDN into Layer 7 firewalls so that it can block the connection to suspected botnet. IPS or Layer 7 firewalls have also been widely used in local area networks. If they were integrated with the system of this study, the host in the local area network could be protected from botnet threats. Additionally, IPS or Layer 7 firewalls are scalable, which can be integrated with various malicious software detection methods to achieve the most rigorous protection.

References

1. Gu, G., Zhang, J., Lee, W.: BotSniffer: Detecting Botnet Command and Control Channels in Network Traffic. In: Proc. 15th Annual Network and Distributed System Security Symposium (2008)
2. Polychronakis, M., Mavrommatis, P., Provos, N.: Ghost turns Zombie: Exploring the Life Cycle of Web-based Malware. In: Proc. 1st Usenix Workshop on Large-Scale Exploits and Emergent Threats (2008)
3. Lee, J.S., Jeong, H.C., Park, J.H., Kim, M., Noh, B.N.: The Activity Analysis of Malicious HTTP-based Botnets using Degree of Periodic Repeatability. In: International Conference on Security Technology, SECTECH 2008, pp. 13–15 (2008)
4. Lakhina, A., Crovella, M., Diot, C.: Mining Anomalies Using Traffic Feature Distribution. In: Proc. 2005 Conference on Applications, Technologies, Architectures, and Protocols for Computer Communications, vol. 11(12), pp. 217–228 (2005)
5. Wang, K.M.: A Netflow Based Internet-worm Detecting System in Large Network. In: Third International Conference on Digital Information Management, ICDIM 2008, pp. 581–586 (2008)
6. Holz, T., Gorecki, C., Freiling, F., Rieck, K.: Measuring and Detecting of Fast-Flux Service Networks. In: Proc. 15th Annual Network & Distributed System Security Symposium (2008)
7. Zhou, C.A., Leckie, C., Karunasekera, S.: Collaborative Detection of Fast Flux Phishing Domains. Journal of Networks 4(1), 75–84 (2009)
8. Passerini, E., Paleari, R., Martignoni, L., Bruschi, D.: FluXOR: Detecting and Monitoring Fast-Flux Service Networks. In: Zamboni, D. (ed.) DIMVA 2008. LNCS, vol. 5137, pp. 186–206. Springer, Heidelberg (2008)
9. Yu, S., Zhou, S., Wang, S.: Fast Flux Attack Network Identification Based on Agent Lifespan. In: IEEE International Conference on Wireless Communications, Networking and Information Security, WCNIS 2010, pp. 658–662 (2010)
10. McAfee (2003), http://www.siteadvisor.com/
11. SPAMHAUS (1998), http://www.spamhaus.org/lookup.lasso
12. WOT (2010), http://www.mywot.com/
13. Free PC Security (2007), http://www.freepcsecurity.co.uk/
14. Testbed @ NCKU (2007), https://testbed.ncku.edu.tw

References

1. Gu, G., Zhang, J., Lee, W.: BotSniffer: Detecting Botnet Command and Control Channels in Network Traffic. In: Proc. 16th Annual Network and Distributed System Security Symposium (2008)

2. Paparrizos, M., Mavrommatis, P., Provos, N.: Ghost turns Zombie: Exploring the Life Cycle of Web-based Malware. In: Proc. 1st Usenix Workshop on Large-Scale Exploits and Emergent Threats (2008)

3. Lee, S., Jeong, H.C., Park, J.H., Kim, M., Noh, B.N.: The Activity Analysis of Malicious HTTP-based Botnets using Degree of Periodic Repeatability. In: International Conference on Security Technology, SECTECH 2008, pp. 83–86 (2008)

4. Lakhina, A., Crovella, M., Diot, C.: Mining Anomalies Using Traffic Feature Distributions. In: Proc. 2005 Conference on Applications, Technologies, Architectures, and Protocols for Computer Communications, vol. 11(12), pp. 217–228 (2005)

5. Wang, K.H., Aziz, A.: Online Fast-fluxing worm Detecting System in Large Networks. In: Third International Conference on Digital Information Management, ICDIM 2008, pp. 541–546 (2008)

6. Nazario, J., Gobel, J., Giffin, J., Rieck, K.: Measuring and Detecting of Fast Flux Service Networks. In: Proc. 15th Annual Network & Distributed System Security Symposium (2008)

7. Zhou, C.V., Leckie, C., Karunasekera, S.: Collaborative Detection of Fast Flux Phishing Domains. Journal of Networks 4(1), 75–84 (2009)

8. Passerini, E., Paleari, R., Martignoni, L., Bruschi, D.: FluXOR: Detecting and Monitoring Fast-Flux Service Networks. In: Zamboni, D. (ed.) DIMVA 2008. LNCS, vol. 5137, pp. 186–206. Springer, Heidelberg (2008)

9. Yu, S., Zhou, W., Wang, S.: Fast Flux Attack Network Identification Based on Agent Lifespan. In: IEEE International Conference on Wireless Communications, Networking and Information Security, WCNIS 2010, pp. 658–662 (2010)

10. Malware 2005, http://www.malware.com

11. SPAMHAUS (1998), http://www.spamhaus.org/rokso/index.lasso

12. WOT COOL, http://www.mywot.com

13. Phishtank (Summer 2005), http://www.phishtank.com/phish_search.php

14. Techdo, T., NCKU (2007), http://apsimon.cs.nthu.edu.tw/cisc

Improvements of Attack-Defense Trees for Threat Analysis

Ping Wang and Jia-Chi Liu

Department of Information Management, Kun Shan University, Taiwan
Tainan, Taiwan
pingwang@mail.ksu.edu.tw, momolu22@gmail.com

Abstract. Attack trees technique is an effective method to investigate the threat analysis (TA) problem to known cyber-attacks on the Internet for risk assessment. Therefore, Protection Trees (PT) have been developed to migrate the system weaknesses against attacks. However, existing protection trees scheme provided a converse approach to counter against attacks, ignored the interactions between threats and defenses. Accordingly, the present study proposes a new method for solving threat analysis problem by means of an improved ADT (iADTree) scheme considering the best defense policy to select the countermeasures associated with each of attack path. Defense evaluation metrics for each node for probabilistic analysis is used to assisting defender simulate the attack results. Finally, a case of threat analysis of typical cyber security attack is given to demonstrate our approach.

Keywords: Threat analysis, iADTree, Attack–Defense Trees, probabilistic analysis.

1 Introduction

The problem of identifying the attack profile of possible hackers over the Internet is referred to as the Threat Analysis (TA) Problem. Generally, the TA focuses on evaluating the competition between *attack and defense actions* to determine the feasible defense strategies based on attack profile. Compared to TA, risk assessment more emphases collecting sufficient system vulnerabilities information to evaluate the risk level of asset, given a constraint on both the probability of attack occurrence and the potential impact loss.

From 1990s, tree structure have been applied for exploring the attack profile based on FTA (Fault tree analysis) thru discovering all possible vulnerabilities associated with attack action (namely attack paths of threat list). Available Threat Analysis schemes with Attack Trees (AT) [3], such as Defense Trees (DT) [10], Protection trees (PT) [4] and Attack Response Tree (ART) [9]are capable of identifying the risk level and threats of an information asset via accumulating the system vulnerabilities, the corresponding impacts and estimating the attack costs and defense costs.

Existing threat risk analysis schemes, such as DT, PT and ART provide a means of stating the theoretical defense costs and lowering the risk, but do not reasonably answer the critical questions regarding: (i) cost-effective solution of defense mechanisms (ii) suitable attack paths to put safeguards in place.

J.-S. Pan et al. (Eds.): *Advances in Intelligent Systems & Applications*, SIST 21, pp. 91–100.
DOI: 10.1007/978-3-642-35473-1_10   © Springer-Verlag Berlin Heidelberg 2013

To mitigate this problem, Kordy *et al.*(2010) proposed a new tree structure, namely Attack–Defense Trees (ADTree) [2] to model the interactions between attacks and defenses using game theory for arbitrary alternation between these two types of actions. Practically, ADTree suffers from two facts: (i) only two notations are used to specify the complex attack defense scenarios. (ii) absence of the defense metrics for probabilistic analysis to real cyber-attack cases.

Accordingly, an improved ADT scheme (iADTree) is proposed to investigate the threat analysis for Advance Persistent Threat (APT) considering the best defense policy to select the countermeasures associated with each of attack path. In the proposed approach, probabilistic analysis with defense evaluation metrics for each nodes is used to assisting defender analyze the attack sequence taking into account the proponent and opponent attitude. The effectiveness of the proposed approach was evaluated by a set of metrics for mitigating new cyber threats.

In developing the model proposed, there are three important aspects of focusing on our work: (i) explore the *ROA*(Return Of Attack) of targeted goal, (ii) examine the effect of *ROI*(Return Of Investment) with countermeasures, (iii) evaluate the possible countermeasure in accordance with the overall defense cost of the responding to these specified attacks.

The remainder of the paper is organized as follows. The proposed model is introduced in Section 2. Section 3 takes an example to illustrate the method. Section 4 discusses how to select the optimal countermeasure to defense practically. Section 5 draws the conclusions.

2   An Analysis Model For Attack Profiles and Countermeasures

2.1   Basic Attack Modeling

Our model is designed to describe the attack profile, estimate the metrics of each node for appropriately selecting the proper safeguard under the circumstance of interleaving attacks. In ADTree, there are two basic types of events: attack node and defense node. It is too simple for two notations to specify the complex attack scenarios. Thus, the present study redefines the notation: attack event is break into two sub-events: detection (e.g., network exploits) and attack; defense event is separated into deception (e.g., honeypot) and countermeasure (fix vulnerabilities of host) as depicted in Fig.1 and Table 1.

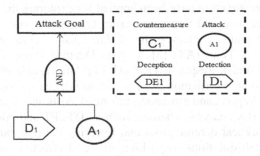

Fig. 1. Notation of an iADTree

Table 1. Meaning of notation

Action	Examples
detection	DNS query, port scan
attack	Registry modification, open ports
deception	Honeypot deployment
countermeasure	Vulnerability fix, safeguards put in place

In real attack and defense scenarios, an iADTree can be consists of (i) a detection event and an attack event, (ii) multiple detection events and an attack event, (iii) multiple detection events, a deception, an attack and multiple countermeasures as shown in Fig.2.

Definition 1. iADTree The universe of the iADTree structure (see Fig.1) is denoted by T, where $T=(N,\rightarrow, n_0)$ is a 3-tuple tree structure, where N is a set of attack nodes or defense nodes; $\rightarrow$ is a set of acyclic relations of type $\rightarrow \subseteq N \times M$ (where N, M represents arbitrary sets of two connected nodes); and n_0 is the root node defined such that every node in N is reachable from n_0.

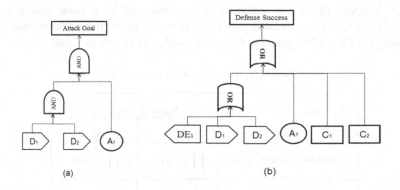

(a)   (b)

Fig. 2. Attack and defense scenarios depicted by iADTree

2.2 Estimating the Success Probability

For probabilistic analysis, defender need estimate the probability of attack success for each node in iADTree. In Fig.1 and Fig.2(a), the probability of attack success at the goal can be derived by

$$p(t) = p_{A_1}(t)(1-p_{D_1}(t)) \tag{1}$$

$$p(t) = p_{A_1}(t)(1-p_{D_1}(t))(1-p_{D_2}(t)) \tag{2}$$

In solving the TA problem, a fundamental difficulty exists in assessing the success probability of basic attack actions, such as $p_{A_1}(t)$. Inspired by attack analysis concept on Intrusion Detection System (IDS), the success probability of attack occurrence is solved using 'episode frequency rules' [5] thru accumulating and associating the alert events as follows.Generally, episodes are partially ordered sets of events. The frequent episode rule is used to discover the specific event sequences as a means of appropriately estimating the probability of attack occurrence.

Given an event sequence $s = (s; T_s; T_e)$ and a window width win. Let time window of an episode $w=(w; T_s; T_e)$. The *support degree* of an episode is defined as the fraction of windows where the episode occurs. In other words, given an event sequence s and a window width win, the support degree of an episode (α). in s is

$$\sup(\alpha) = p_i(\alpha, s, win) = \frac{|\{\alpha \text{ occurs in } \omega\}|}{|\{W(s, win)\}|} \tag{3}$$

Once $\sup(\alpha)$ has obtained, it can be used to predict $p_{A_i}(t)$ that describe connections between attack events in the given event sequence (i.e., signature).

2.3   Attack and Defense Actions for Threat Analysis

The analysis of iADTree is constructed by the start from attacker's actions (leafs) thru recursively occupied sub-goals until attacker's goal (root node), as illustrated in Fig.3(a). Suppose threat i is assumed to be composed of q basic attack actions $(k=1,\dots,q)$, the metrics associated with the leaf nodes in the tree structure are calculated using the FTA AND-gate and Or-gate formulae as shown in Table 2.

Table 2. Rule set for iADTree metrics

Root node	Non-leaf node k	
	AND	OR
Probability of success $p_o(t)$	$\prod_{k=1}^{q} p_k(t)$	$1-\prod_{k=1}^{q}(1-p_k(t))$
Attack cost $C_o(t)$	$\sum_{k=1}^{q} c_k(t)$	$\forall\, Min(c)_k$
Impact $l_o(t)$	$\sum_{k=1}^{q} l_k(t)$	$\forall\, Max(l)_k$

In Table 2, the probability of success of threat i (p_i) represents the chances of threat i ($i=1,\dots,m$) successfully hacking into the system, and has a value in the interval [0,1]. Meanwhile, the attack cost (c_i) represents the manpower cost required to carry out the attack, and is stated in terms of U.S dollars. The impact associated with a specific threat is measured on the scale of [1~10], where a higher value indicates a more severe loss. Finally, the defense cost (d_i) represents the cost of defending against specific threat i and comprises both the security hardware cost and the defense manpower. Note that for simplicity, the man-hours used in evaluating the attack cost and defense cost, respectively, are converted to dollars at the rate of $100 per man-hour.

In evaluating the performance of iADTree, two important metrics, i.e., *ROA* and *ROI* modified from [1] associated with each of the nodes is evaluated as follows: (two formulas in [1] cannot be evaluated rationally due to the consistence problems of metric units)   Return On Attack (*ROA*) of a non-leaf node for specific threat *i* at time *t* can be evaluated by aggregating the attack cost (c_i), the success probability (p_i) and the impact loss (l_i) as (see Fig.3)

$$ROA_k(t) = \frac{p_k(t) \cdot l_k(t)}{c_k(t)} \tag{4}$$

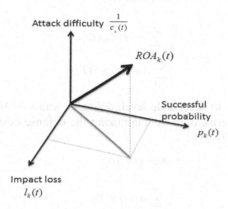

Fig. 3. Affecting factors of *ROA*

For the root node, the overall *ROA* of the entire system at time *t* can be evaluated by either AND-gate or Or-gate computation

$$ROA_o(t) = \sum_{k=1}^{q} ROA_i(t) \tag{5}$$

$$ROA_o(t) = \forall \max_k ROA_k(t) \tag{6}$$

After analyzed *ROA* associated with a specific threat *i*, defender adopts the safeguards during the countermeasure stage (i.e., *t+1*) to decrease the *ROA* of attacker. Thus, *ROI* of defense actions is given by

$$ROI_k(t+1) = ROA_k(t+1) - ROA_k(t), \tag{7}$$

In order for defender to easily compare the *ROA* values between nodes, the raw numerical value of defense cost is normalized for the complete tree using logarithmic scale as

$$d_{kn}(t+1) = \log_{10}(\frac{10 * d_k(t+1)}{\min(d_k(t+1))}) \tag{8}$$

The Normalized Return On Investment (*NROI*) of countermeasure at each non-leaf node is calculated as

$$ROI_{kn}(t+1) = \frac{ROA_k(t+1) - ROA_k(t)}{d_{kn}(t+1)},$$ (9)

where d_i represents the defense cost.

Having assigning values to the leaf nodes, the metrics are propagated up the tree until the goal node metrics are determined thru the link of AND-gate and OR-gate logic. Finally, the ROA and ROI value is obtained in each node.

Attacker's goal is to obtain the best result of ROA in terms of minimizing the attack cost AC_k within the attack time constraint AT_k, i.e.

$$\forall_i \ Max \ ROA_o(t),$$ (10)

$$s.t. \quad \begin{cases} \sum_{k=1}^{q} c_k(t) \leq AC_k, \\ \sum_{k=1}^{q} \Delta at_k(t) \leq AT_k, \end{cases}$$

To eliminate the ROA to acceptable level, defender wants to Max ROI to ensure the effectiveness of countermeasure by minimizing the defense cost DC_k within the time constraint of defense DT_k, i.e.,

$$\forall_i \ Max \ ROI_o(t),$$ (11)

$$s.t. \quad \begin{cases} \sum_{k=1}^{q} d_k(t) \leq DC_k, \\ \sum_{k=1}^{q} \Delta dt_k(t) \leq DT_i, \end{cases}$$

The detailed algorithm for finding the possible set of countermeasures is described by PDL as Fig.4.

Input: Parameters of attack actions of cyber threats and a pool of possible safeguards
Output: Suggested defense mechanisms given in a budget constraint

Algorithm FSCA: Finding the Suitable Countermeasures Algorithm
1: initial an iACTtree(ID);
2: Input the mincuts of the iACTtree(from
ISOGRAHP attack tree+) with lowest cost and highest impact;
3: Assign metric values to iACTtree(ID);
4: **loop**
5:   **for** each node () in iACTtree **do**
6:   Select the safeguards which cover the maximum no of attack events;
7:   **if** (the safeguard_cost<total defense);
8:       Select the safeguards;
9:       Total defense= total defense + safeguard_cost;
10:      Output_defense_list←safeguards_id;
11:      **endif**
12:       return (output_ defense_list);
13:   **end for**
14: **end loop**

Fig. 4. Algorithm FSCA

3   Cyber Security Application

In the present study, a new zero day PDF exploit the attack profiles of hackers on Cloud Computing services, Zeus attacks [11] will be analyzed. First identified in July 2007 when it was used to steal information from the United States Department of Transportation, it became more widespread in March 2009. In 2010, there were reports of various attacks by Zeus such that the credit cards of more than 15 unnamed US banks were compromised. The threat analysis is constructed using the following four-step procedure.

Step 1: Understand of the System Vulnerability
Generally, the recognized security vulnerabilities of computer have been investigated, examined and reported. For example, Mitre Corporation maintains a list of disclosed vulnerabilities in a system called Common Vulnerabilities and Exposures (CVE), where vulnerabilities are scored using Common Vulnerability Scoring System (CVSS). Vulnerability issued by US-CERT for Zeus botnet is listed in Table 3.

Table 3. Vulnerabilities for Zeus botnet

ID	Name	Last revised
CVE-2010-0359	DoS Exec Code Overflow	2010-01-21
CVE-2010-0362	spoof DNS responses	2011-05-06
CVE-2010-0363	XSS(cross site scripting)	2010-02-02

Step.2: Collect the Information of Recognized Attacks
Once the system vulnerabilities are identified, defender may focus on the issues of understanding possible network attacks that hackers maliciously attempt to compromise network security, as well as discovering attack profile with the probability of an event occurrence and its impact.

Step 2.1 Collect the Malware
Deploy honeypot Dionaea at switching edge node in the camp networks, log the alerts, and capture the payloads.

Step 2.2 Signature Analyses
In the present study, attack profile is validated by CWSandBox and SandNets [8] in a dynamic malware analysis environment supported by Testbed@NCKU project [12] as shown in Fig.7. Defender can examine the details of attack sequence to discover the possible attack profile. After collected the information from the aforementioned three sources of cyber-attacks, defender constructs the iADTree and predict the success probability of malware infection and hacker attack.

Step 3: Perform iADTree Analysis
The metrics for intermediate and goal nodes shown in Table 2 operates on lower level nodes beginning with the leaf nodes. The partial iADTree in Figure 5 (see Fig.6) shows how an attacker might intrude into the servers for gain the root privilege thru exploiting IE vulnerability. After assigning values to the leaf nodes, the metrics are propagated up the tree until the goal node metrics are determined using Attack

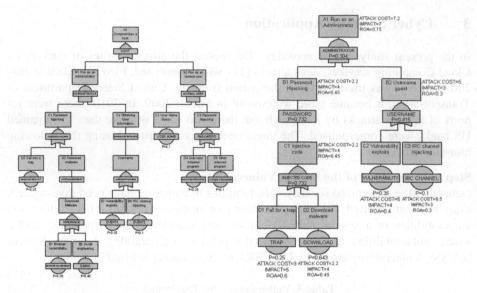

Fig. 5. Compromised a host   **Fig 6.** Attack path with attack cost

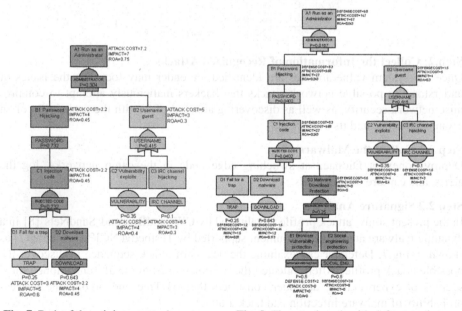

Fig. 7. Path of the minimum attack cost   **Fig. 8.** The attack path with defense actions

Tree+ tool [8], as shown in Fig. 7. Fig.8 illustrates that the possible way is given as a malware with the minimum attack cost is to gain root access in a host (see orange nodes). This attack path is the first priority of security controls to be deployed.

Step.4: Countermeasure Analysis

In practice, implement each security control can lower the distinct ROA values to attacker and increase the corresponding attack cost. Two crucial parameters α and β, are defined to specify the protection capability of safeguards as shown in Table 4.

Table 4. Parameters for protection capability of safeguards

Safeguards ID	α	β	ROA	
S_1	0.325	0.40	Before:0.45	
			After: 0.243	
S_2	0.25	0.75	Before:0.51	
			After:0.340	
S_3	0.275	0.65	Before:0.56	
			After:0.220	
S_4	0.35	0.80	Before:0.35	
			After:0.175	

where α represents the capability ratio of lowering the impact, β means the increasing ratio of the attack cost. Obviously, the higher α and β is the better choice. For this example, S_1 is selected as an illustration case with setting(α =0.325, β =0.40). After implemented with safeguard S_1 against infected code (see green blocks D_3) in Fig.8, the metrics are analyzed and filled in Table 5. Table 5 illustrates that assigning the safeguards will cost 5.6k indirectly converted to increase 40% defense cost to attack cost and eliminate 25% impact effect of attack to infected code(C_1). Consequently, the final ROA (A_1) will decrease from 0.75 to 0.543, impact loss declined from 7.0 to 6.7, success probability falls from 0.304 to 0.0167 and impose attack cost increase from 7.2k to 14.7k. The *ROI* of countermeasure on node D_3 is 0.207.

Table 5. Metrics of attack path for IE vulnerability

Metric	Value
ROA_o (root)	0.75→0.543
Success probability (root)	0.304→0.0167
Attack cost (root)	7.2k→14.7k $dollars
Impact (root)	7.0→6.7
Defense cost$_k$	5.6k $dollars
ROI_k	0.207

4   Discussion

Generally, the defender desires to implement the security controls in a cost-effective way to cover all the attack and detection events, i.e., mincuts of iACTtree. Fig.8 shows the defender constantly chooses the minimum attack cost associated with countermeasure implemented. At some cases, the highest ROA or impact loss will be chosen to maximizing the security controls. In contrast to [3], iADTree approach take advantages on analyzing the interactions of attack and defenses, holding better flexibility by incorporating countermeasures.

5   Conclusions

This paper has presented an improved ADT scheme [2] by revising the formula in [1] as an enhancement method for threat analysis of cloud security, allowing defender to convert defense cost (security controls) with attack cost, and estimating the impact losses for the evolution of a system's security concerns. Additionally, our scheme can consider not only the interactive scenarios of attacks and defenses, but also estimate the required costs, in order to practically analyze the risk regarding a specified threat. Consequently, the proposed method improves the precision of the risk solution by enabling defenders to make an appropriate decision when tracing possible attack actions. The qualitative analysis of iADTree for different scenarios in accordance with mini-cuts analysis will be tackled in future studies. Especially, sensitivity analysis for ROA and ROI computation may be considered in the selection the best strategy of countermeasures to bring defenses in place against APT attacks.

Acknowledgments. This work was supported partly by TWISC@ NCKU, National Science Council under the Grants No: NSC101- 2221-E-168-034 and NSC 101-2219-H-168-001.

References

[1] Roy, A., Kim, D., Trivedi, K.S.: Cyber Security Analysis using Attack Countermeasure Trees. In: Proc. of Cyber Security and Information Intelligence Research Workshop, CSIIRW 2010, Oak Ridge, TN, USA. ACM (2010)

[2] Kordy, B., Mauw, S., Radomirović, S., Schweitzer, P.: Foundations of Attack–Defense Trees. In: Degano, P., Etalle, S., Guttman, J. (eds.) FAST 2010. LNCS, vol. 6561, pp. 80–95. Springer, Heidelberg (2011)

[3] Schneier, B.: Attack Trees: Modeling Security Threats. Dr. Dobbs' Journal (December 1999)

[4] Edge, K.S., Dalton II, G.C., Raines, R.A., Mills, R.F.: Using Attack and Protection Trees to Analyze Threats and Defenses to Homeland Security. In: MILCOM 2007, pp. 1–7 (2007)

[5] Mannila, H., Toivonen, H., Verkamo, I.A.: Discovery of Frequent Episodes in Event Sequences. Data Mining and Knowledge Discovery 1(3), 259–289 (1997)

[6] Honeynet Project, honeypot Dionaea, http://dionaea.carnivore.it/

[7] ISOGraph, attack tree+, http://www.isograph-software.com/2011

[8] Stewart, J.: Behavioral malware analysis using Sandnets. Computer Fraud & Security 2006, 4–6 (2006)

[9] Zonouz, S.A., Khurana, H., Sanders, W.H., Yardley, T.M.: RRE: A Game-Theoretic Intrusion Response and Recovery Engine. In: Proc. DSN, pp. 439–448 (2009)

[10] Bistarelli, S., Dall'Aglio, M., Peretti, P.: Strategic Games on Defense Trees. In: Dimitrakos, T., Martinelli, F., Ryan, P.Y.A., Schneider, S. (eds.) FAST 2006. LNCS, vol. 4691, pp. 1–15. Springer, Heidelberg (2007)

[11] Symantec, Zeus: King of the Bots (PDF),
http://www.symantec.com/content/en/us/enterprise/media/
security_response/whitepapers/zeus_king_of_bots.pdf

[12] Testbed @TWISC, http://testbed.ncku.edu.tw/index.php3

Design and Implementation of a Linux Kernel Based Intrusion Prevention System in Gigabit Network Using Commodity Hardware

Li-Chi Feng, Chao-Wei Huang, and Jian-Kai Wang

Department of Computer Science and Information Engineering
Chang Gung University
lcfeng@mail.cgu.edu.tw

Abstract. Due to the development of the Internet, much valuable information is stored in the networked computer or transmitted on the network. System and network security is more and more important than before. Intrusion detection system (IDS) is developed to monitor network and/or system activities for malicious or unwanted behavior. Intrusion Prevention System offer stronger protection. When an attack is detected, IPS can drop the offending packets while still allowing all other traffic to pass. Recently, the speed of backbone network has already reached Gbit-scale, the intrusion detection or prevention is more difficult than before. The price of the related products in the market is above two million new Taiwan dollars. In this paper, we design and implement an in-kernel Intrusion Prevention System in Gigabit network using commodity hardware and Linux operating systems. Preliminary experiment results show that, our system outperforms traditional intrusion prevention system (snort inline) substantially. Besides, our system can reach the wire speed under a typical set of detection rules.

Keywords: Intrusion Detection System, Intrusion Prevention System, Gigabit Network, Linux Kernel.

1 Introduction

With the progress of computer and network technology, much valuable information is stored in the networked computer or transmitted on the network. System and network security is becoming more and more important than before. In order to defense diverse network attacks, various network security products and tools, such as Intrusion Detection Systems (IDS) and firewalls, have been development and have became the essential components in Internet.

An Intrusion detection system (IDS) is a network security device that monitors network and/or system activities for malicious or unwanted behavior. In a further, Intrusion Prevention System (IPS) can react, in real-time, to block or prevent those activities. When an attack is detected, it can drop the offending packets while still allowing all other traffic to pass.

J.-S. Pan et al. (Eds.): *Advances in Intelligent Systems & Applications*, SIST 21, pp. 101–109.
DOI: 10.1007/978-3-642-35473-1_11   © Springer-Verlag Berlin Heidelberg 2013

There are two main categories of IDS, one is host-based IDS and the other is network-based IDS. Host-based IDSs analyze operation systems' files, processes and audit trails to recognize invasions in system; Network-based IDSs analyze network traffic to filter out network attacks over the network.

However traditional NIDS has been criticized with the following drawbacks. First of all, NIDS need to implement part of communication protocols in user level, it may cause the incorrect analysis just because of the implementation is insufficient. Invaders may use some evasive tricks such as IP Fragmentations or IP overlap to dodge from the detection of NIDS. For example, hackers can take advantage of ip fragmentation by dividing attack packets into smaller portions to dodge NIDS examinations [6]. If the IDS and the host reassemble the packets differently the IDS will not see the packets, but the reassembling host will [2].

Secondly, most NIDSs just only monitor the network traffic. It informs firewalls or system administrators to adjust the network security policies as suspicious behaviors intercepted. Such passive protection can't resist intrusions immediately. Invaders can still attack the target network until firewalls or system administrators enforce the security policy to block the suspicious connections [3].

More critically, NIDS's ability to process traffic at the maximum rate offered by the network is insufficient [1, 10, 11]. Under mass network traffic or huge rule database, some packets may bypass the examinations of NIDS. Besides, invaders may generate mass normal network traffic to cover few suspicious packets. NIDS would consume lots of system resource to inspect these unharmful packets. Performance limitation becomes the most significant problem in traditional NIDS.

Several researches focus on boosting the throughput of Snort under Linux [17, 18]. In [18], the performance of Snort was improved by tuning key configurable parameters of the Linux kernel networking subsystem related to packet reception mechanism. But performance limitation is still a significant problem due to large memory copy across user and kernel space.

Recently, the speed of backbone network has already reached Gigabit scale, intrusion detection or prevention is more difficult to accomplish than before. High-end intrusion detection or prevention systems usually apply special design hardware/software components to accelerate network packets capture and processing. The price of these products in the market is too high for many nonprofit organizations. For example, Juniper IDP series and IBM Proventia cost more than two million new Taiwan dollars. How to provide more cost effective solution becomes an important issue.

The performance of PC server has increased rapidly in recent years and there are many excellent open source security tools. If we can design and build a Giga-bit level Intrusion Detection (Prevention) System using PC server and open source software, it can provide a cost-effective solution.

To address these problems described above, we design and implement an in-kernel Intrusion Prevention System (KIPS) using commodity hardware and Linux operating systems. With the advantages of kernel support, such as direct handling of network packets, we can improve the flaw of traditional NIDS. We also set our system as an in-line system to provide an active prevention service. Moreover, we reuse part of kernel IP protocol stack codes to avoid re-implement packet analyzer by our self. Preliminary experiment results show that, our system outperforms traditional intrusion prevention system (snort inline) substantially.

2   Traditional Network Intrusion Detection Systems

Fig. 1 illustrates the traditional NIDS architecture implemented on Linux platform. NIDS monitors all the communications between the network gateway and intranet. As suspicious behaviors are detected, NIDS records all alert messages and informs system administrators or network security software such as firewalls to block the suspicious connections.

In order to monitor network behaviors, NIDS must to intercept all packets over the network from Network Interface Cards (NICs). Most Ethernet Network Interface Cards support two different kinds of modes. Promiscuous mode allows a NIC to intercept and read all network packets in the same collision domain; In non-promiscuous mode, NIC would only accept packets belongs to the host and drops all the other hosts' packets. As we put the NIC into promiscuous mode, packets will be captured by NIC and passed to Linux kernel. However, most of packets captured by the promiscuous mode do not belong to the local host and will bypass the processing of communication protocol stacks.

Most NIDSs such as Snort [16] use Pcap library [19] or raw socket system calls to receive packets which captured by promiscuous mode in kernel-space. Large amount of packets duplicated from kernel space to user space always cause the performance downcast rapidly. As NIDS gets these packets from kernel space, it needs to analyze these packets to find out network attacks by comparing these data with signature database. Due to performance limitation, some packets will bypass the examinations of NIDS and cause network damage.

As described in Section 1 and previous paragraph, there are many defects in traditional NIDS architecture. In order to address these defects, we implement NIDS in kernel space to avoid heavy packet duplications from kernel to user space. Instead of implementing the communication protocols by ourselves, we reuse kernel communication protocol directly. Direct handling network packets in kernel space also results in outstanding performance. In next session, we will describe the design and implementation of our system named KIPS in detail.

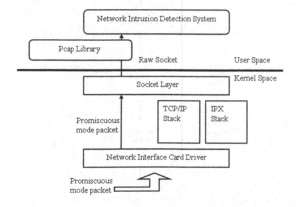

Fig. 1. Architecture of traditional NIDS (Linux platform)

3   System Design and Implementation

3.1   Design Issues

In order to design and implement a robust in-kernel IDS, there are some problems must be addressed. Firstly, stability is an important factor to network IDS. Any crash of NIDS will lead the network to be unprotected. For this reason, we will modify the Linux kernel as less as possible.

Secondly, in order to protect the network more positively we design KIDS as an in-line IDS system. Any suspicious packet will be dropped by our KIDS system to avoid any damage to the network. To support the in-line IDS architecture, we use the Netfilter [10] mechanism to redirect the incoming packets into KIDS and transfer the passing packets into the other NIC interface.

Thirdly, to provide high packet handling performance, we set the network IDS into kernel level and reuse part of communication protocols to handle network packets.

NetFilter is a network framework of Linux kernel 2.4.x and 2.6.x which provides five different hook points for kernel modules to make packet filtering, network address translation or other packet processing. Netfilter allows kernel module to register callback functions. All the packets received by NIC would redirect to the register callback function through these five hook points.

3.2   System Architecture

Our system was implemented between network interface card device driver and TCP/IP protocol stacks. We set NIC into promiscuous mode to receive packets from the network and reuse some part of TCP/IP protocols to support our Intrusion Detection System.

We can divide our system into some functional modules to deal with different works. The system architecture is shown in Figure 2.

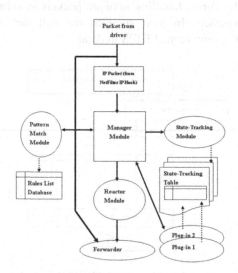

Fig. 2. System architecture

As packets redirected from PREROUTING hook functions, the Manager Module starts to redirect the packets to other modules in KIDS. The Manager Module also controls the main function flows and all the communications between these modules. It is the core of our system.

State-Tracking module maintains the network communication constructions and terminations. Pattern-Match module compares the rule database with packet data. Plug-in module analyzes application level protocol such as HTTP or SMTP. Reactor Module makes reactions according to the result of signatures comparison.

Pattern Match Module compares the data with the signature in rule database. If suspicious packet is found, Pattern Match Module will pass "Packet Drop" message to Reaction Module. Reaction Module will drop the packet. If the result of Pattern Match is "not match", Reaction Module will allow the packet to pass.

In order to speed up the operation of packet inspection, We adopt ACBM algorithm to implement the Pattern Match Module. ACBM combines the advantage of Aho-Corassick algorithm and the speed of Boyer-Moore algorithm, it can be used to improve the performance of NIDS [6, 8].

Snort [16] is a popular open source Network IDS in user space. It develops a complete rule database and keeps these rules updated all the time. In order to develop an usable system, we decide to adopt snort rule database in our system. We develop a tool to translate Snort's rule files and import the rule database into our system.

3.3   Operating Flow

We use kernel module mechanism to implement our system. When we insert kernel module into Linux kernel, its initial function create a proc file system entry, search NIC device, initial connection track table and so on. Then, we can start to use the Kernel Intrusion Prevention System (KIPS) to monitor the packets in the network. The operating flow of our system is illustrated in Figure 3.

Manager Module is the main module of the system. It controls the communications between these modules. Conceptually, Manager Module would give the control authority to other modules at the right moment.

First, Manager Module receives network packet from the Netfilter hook point. Then, it checks if the packet is a TCP packet or not. If the packet belongs to TCP packets, Manager Module would connect to State-Tracking module to maintain TCP connection tracing; if the packet belongs to UDP packet, Manager Module would send UDP packet to Pattern-Match Module to inspect the packet with rule database.

If Manager Module observes that the packet have any data, it would search Plug-in Modules List to determine if there are any Plug-in modules can be used to analyze the packet. If there are any Plug-in modules need to be executed, Manager Module would pass the control authority to the Plug-In module.

Then, Manager Module would call Pattern-Match Module to compare the signature strings in rule database with skb data.

Finally, Manager Module would decide how to deal with the packet. Reactor Module would select appropriate actions such as redirect or drop packet according to the state of the system.

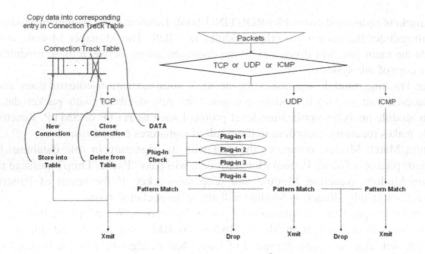

Fig. 3. Operating flow of our system

4   Experimental Results

Our experimental environment is shown in Figure 4. A is the front-end source host. C is the back-end destination host. In order to generate enough network traffic, we use two PCs as the source hosts. All of these PCs are running CentOS 5.5, the kernel version 2.6.18. Each PC has an Intel Core 2 Duo E8500 CPU, 4GB RAM and one Gigabit Ethernet card.

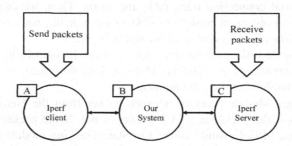

Fig. 4. Experimental environmental organization chart

B is the central server. It has dual Intel® Xeon® E5335 2.00 GHz 4 core processor, 4GB Memory, 500 GB Disk, dual Gigabit Ethernet network cards. The server is running CentOS 5.5, kernel version 2.6.18. Our intrusion prevention system (KIPS) and Snort-inline 2.6.1 (snort version of IPS) will be running on the central server.

Iperf [12] is an open source tool that can be used to measure network performance. We adopt Iperf 2.05 to test our system.

In the experiment, 2 front-end PC(A) connect to the back-end PC(C) with iperf tool at the same time. A large number of packets are sent by iperf. All of test packets send from A to C will be inspected by our system or snort-inline running on the central server B. At the end of test, maximum bandwidth of our system or snort-inline would be reported.

In order to evaluate the ability of our system under various system loads, we divide the snort rule set into 4 groups illustrated in Table 1. In the experiment, the performance of Linux bridge is used as a baseline. The experimental results are shown in Figure5.As shown in Figure 5, the performance of our system is almost equivalent to the Linux bridge (the upper limit) under 485 rules. As the number of rules increased, our system also performs well. When the number of rules arrived in 1899, the maximum bandwidth is still more than 850Mb/s.

Consider the performance of snort-inline, the maximum bandwidth only achieves 357 Mb/s under 485 rules. When the number of rules arrived in 1899, the maximum bandwidth reduces to 229 Mb/s. As shown in Table 2, our system performs much better than snort-inline. The speed-up of our system ranges from 163% to 272%.

In addition, we use sar utility to measure the CPU utilization of KIPS and snort-inline. The results are shown in Figure 6. CPU utilization of our system is always lower than snort-inline.

Table 1. Classification of snort rule set

Number of Rules	Snort rule set
485	chat、pop3、dns、rpc、finger、snmp、icmp、multimedia、icmp-info、rservices、imap、smtp、mysql、sql、netbios、telnet、nntp、tftp、oracle、x11、p2p、ftp、pop2、web-attacks、web-client、web-coldfusion、web-frontpage
1000	web-iis、web-misc、web-php
1500	web-cgi、attack-responses、backdoor、ddos、dos、exploit
1899	bad-traffic、experimental、info、local、misc、other-ids、policy、virus、eleted、scan、shellcode

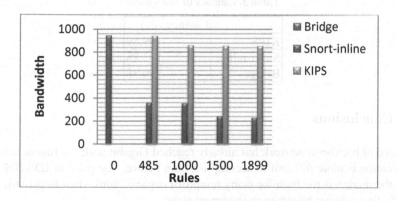

Fig. 5. Performance comparison of our system and snort-inline under different number of snort rules

Table 2. Speedup up of our system

Number of rules \ System	Snort-inline	KIPS	Speed up
485	357 Mb/s	941 Mb/s	163.59%
1000	353 Mb/s	860 Mb/s	143.63%
1500	242 Mb/s	856 Mb/s	253.72%
1899	229 Mb/s	853 Mb/s	272.49%
Average Value	295.25 Mb/s	877.5 Mb/s	208.35%

Besides bandwidth, we use LMbench [13] to measure the latency of our system. The result is shown in Table 3. The latency of our system is as small as Linux bridge and 25% smaller than snort-inline.

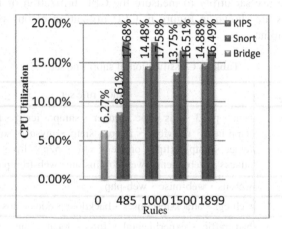

Fig. 6. The curve of the comparison of the CPU usage rate in the different number of rules

Table 3. Lanency of our system

	Latency (μs)
KIPS	99.89
Snort-inline	133.88
Bridge	99.88

5   Conclusions

The speed of backbone network has already reached Gigabit scale, intrusion detection or prevention is more difficult to accomplish than before. The price of IDS/IPS products in the market is too high for many nonprofit organizations. How to provide more cost effective solution becomes an important issue.

In this paper, we design and implement an in-kernel Intrusion Prevention System using commodity hardware and Linux operating systems. With the advantages of kernel support, such as direct handling of network packets, we can improve the flaw of traditional NIDS. Preliminary performance evaluation shows that our system can achieve up to 2.7x improvement in maximum bandwidth; 25% improvement in latency.

References

1. Alserhani, F., Akhlaq, M., Awan, I.U., Cullen, A.J., Mellor, J., Mirchandani, P.: Snort Performance Evaluation. Informatics Research Institute, University of Brad-ford, Bradford, BD7 1DP, United Kingdom (2009)
2. Baggett, M.: IP Fragment Reassembly with scapy, SANS Institute InfoSec Reading Room (2012)
3. Brown Jr., B.J.: IDS, the Silver Bullet!? A conversation with your CEO. SANS GIAC Security Essentials Certification Practical V.1.4b (2004)
4. Charitakis, I., Anagnostakis, K., Markatos, E.P.: A Network-Processor-Based Traffic Splitter for Intrusion Detection, ICS-FORTH Technical Report, vol. 342 (2004)
5. Coit, J., Staniford, S., McAlerney, J.: Towards Faster String Matching for Intrusion Detection or Exceeding the Speed of Snort. In: DARPA Information Survivability Conference and Exposition, DISCEX II 2001, pp. 367–373 (2001)
6. Daniel, N., Kristina, M., Ed, T.: Intrusion Detection Overview – Intrusion Detection Evasive Technologies (2004)
7. Deri, L.: Passively Monitoring Networks at Gigabit Speeds Using Commodity Hardware and Open Source Software. In: Passive and Active Measurement Workshop (2003)
8. Desai, N.: Increasing Performance in High Speed NIDS,
   http://www.snort.org/docs/Increasing-Performance-in
   -High-Speed-NIDS.pdf
9. Dorothy, E.D.: An Intrusion Detection Model. IEEE Transactions on Software Engineering SE-13(2), 222–232 (1987)
10. Fu, T., Chou, T.S.: An Analysis of Packet Fragmentation Attacks vs. Snort Intrusion Detection System. International Journal of Computer Engineering Science, IJCES 2(5) (2012)
11. Schaelicke, L., Slabach, T., Moore, B., Freeland, C.: Characterizing the Performance of Network Intrusion Detection Sensors. In: Vigna, G., Kruegel, C., Jonsson, E. (eds.) RAID 2003. LNCS, vol. 2820, pp. 155–172. Springer, Heidelberg (2003)
12. Iperf, http://processors.wiki.ti.com/index.php/Iperf
13. LMbench, http://www.bitmover.com/lmbench/
14. NetFilter/IPTable, http://www.netfilter.org
15. NFR Network Intrusion Detection System,
    http://www.nfr.com/solutions/system.php
16. Snort, http://www.snort.org
17. Salahh, K., Kahtanti, A.: Boosting throughput of Snort NIDS under Linux. In: Proc. Fifth IEEE Int. Conf. Innovations in Information Technology, Innovations 2008, December 16-18 (2008)
18. Salah, K., Kahtani, A.: Improving Snort performance under Linux. IET Communications 3(12), 1883–1895, 13p. 5 diagrams, 4 graphs (2009)
19. TCPDump/Libpcap, http://www.tcpdump.org
20. Zhou, Z., Chen, Z., Zhou, T., Guan, X.: The study on network intrusion detection system of Snort. In: 2nd International Conference on Networking and Digital Society, ICNDS, vol. 2, pp. 194–196 (2010)

Performance Evaluation on Permission-Based Detection for Android Malware

Chun-Ying Huang, Yi-Ting Tsai, and Chung-Han Hsu

Department of Computer Science and Engineering
National Taiwan Ocean University, Keelung, Taiwan 20224
chuang@ntou.edu.tw, {yttsai,chhsu}@snsl.cs.ntou.edu.tw

Abstract. It is a straightforward idea to detect a harmful mobile application based on the permissions it requests. This study attempts to explore the possibility of detecting malicious applications in Android operating system based on permissions. Compare against previous researches, we collect a relative large number of benign and malicious applications (124,769 and 480, respectively) and conduct experiments based on the collected samples. In addition to the requested and the required permissions, we also extract several easy-to-retrieve features from application packages to help the detection of malicious applications. Four commonly used machine learning algorithms including *AdaBoost, Naïve Bayes, Decision Tree (C4.5)*, and *Support Vector Machine* are used to evaluate the performance. Experimental results show that a permission-based detector can detect more than 81% of malicious samples. However, due to its precision, we conclude that a permission-based mechanism can be used as a quick filter to identify malicious applications. It still requires a second pass to make complete analysis to a reported malicious application.

Keywords: Android, classification, malware, mobile security, permission.

1 Introduction

An Android application requires several permissions to work. Consequently, an essential step to install an Android application into a mobile device is to allow all permissions requested by the application. Android users must have ever seen a similar screen shot to Figure 1. Before an application is being installed, the system prompts a list of permissions requested by the application and asks the user to confirm the installation. Although Google announced that a security check mechanism is applied to each application uploaded to their market [1], the open design of the Android operating system still allows a user to install any applications downloaded from an untrusted source. Nevertheless, the permission list is still the minimal defense for a user to detect whether an application could be harmful.

Google classifies built-in Android permissions into four categories: normal, dangerous, signature, and signatureOrSystem [2]. Therefore, a straightforward

J.-S. Pan et al. (Eds.): *Advances in Intelligent Systems & Applications*, SIST 21, pp. 111–120.
DOI: 10.1007/978-3-642-35473-1_12   © Springer-Verlag Berlin Heidelberg 2013

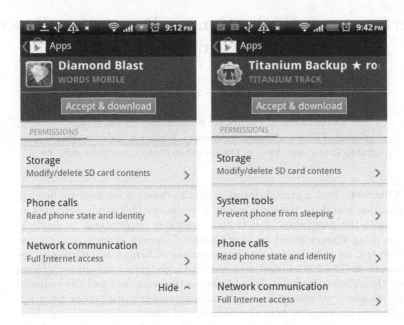

Fig. 1. Example screenshots of asking a user to confirm the installation of applications

idea to determine a harmful application is to check whether it requires a *dangerous permission*. Access permissions to several common activities are classified as dangerous. For example, permissions to read the location of a user (ACCESS_COARSE_LOCATION and ACCESS_FINE_LOCATION), access bluetooth devices (BLUETOOTH), and access Internet (INTERNET) are all classified as dangerous. However, an application requesting one or more dangerous permissions does not indicate that it is a harmful application. A simple application such as a location-based real-time weather forecast application would need some dangerous permissions such as INTERNET and ACCESS_COARSE_LOCATION. Although Android adopts a coarse-grained permission model to control access to its built-in components, it is not known how good (or bad) it is to detect a malicious application based on permissions or combinations of permissions. It should be noticed that the permissions shown to a user during an installation process are *requested permissions* instead of *required permissions*. The requested permissions are declared by an application developer *manually*. However, not all declared permissions are required by the application. Researchers [3,4] have shown that many developers often declare much more permissions than they actually required. It thereby increases the difficulty on detecting malicious applications based on the permissions.

This study attempts to explore the possibility of detecting malicious applications based on permissions, including both requested and required permissions. Compare against previous researches, a relative large number of benign and malicious applications (124,769 and 480, respectively) were collected and used to conduct the experiments. In addition to the requested and the required permissions, several easy-to-retrieve features from application packages were extracted as well to help the detection of malicious applications. Four commonly used machine learning algorithms including *AdaBoost, Naïve Bayes, Decision Tree (C4.5)*, and *Support Vector Machine* are used to explore the possibility.

The remaining of this paper is organized as follows. Section 2 provides a review on several interesting works that analyze permissions of Android applications. Section 3 explains how the features (permissions and other features included in an Android package) are obtained and how the features are labeled (as benign or malicious). Section 4 analyzes the permission requirements of applications and discusses the performance of detectors. Finally, a conclusion and future works are discussed in Section 5.

2   Related Work

A number of researches have introduced and discussed Android permissions. Enck et al. [5] wrote a good introduction on Android's security design in 2009. Basically the Android operating system provides a coarse-grained mandatory access control (MAC). It is able to enforce how applications access components based on granted permissions. Consequently, each Android application must have a list of requested permissions and all these permissions must be granted at the time of installation. The requested permission list is often declared by an application developer manually. Hence, a number of interesting researches are devoted to review how permissions are declared in applications. Barrera et al. [6] analyzed how developers of Android applications use the permissions. They explored and analyzed 1,100 applications using the Self-Organizing Map (SOM) algorithm. They found that although Android has a rich set of permissions, only a small number of these permissions are actively used by developers. Felt et al. [3] studied Android applications to determine whether Android developers follow least privilege with their permission requests. They built a tool and applied it to 940 applications and found that about one-third of evaluated applications are over privileged. They also concluded that developers are trying to follow least privilege but failed due to insufficient API documentation. Johnson et al. [4] developed an architecture that automatically searches for and downloads Android applications from Android Market. With the application, they created a detailed mapping of Android API calls to the required permissions. The idea is similar to [3] but they collected a large number (141,372) of applications to conduct the experiments. They found that the majority of developers are not using the correct permission set. The applications are either over-specify or under-specify their security requirements. Zhou and Jiang [7] systematically characterized 1,260 Android malicious applications from various aspects, including their installation methods, activation mechanisms, and the carried malicious payloads. In addition, they also compared the permission requests of the 1,260 malicious applications against another top free 1,260 benign applications on Android market. The comparison shows that the top 20 frequently requested permissions are similar for both benign and malicious applications.

In addition to analyze permissions, a number of researches tried to detect malicious application using static analysis or dynamic analysis techniques. These techniques are similar to those used to detect traditional malware on desktop personal computers. Besides many well-known signature-based virus scanners, androguard [8] is an open source project that dedicated to detect Android malware. Androguard detect a malicious application or an injected malicious code based on *control flow graph*. A given application package is first disassembled and each identified method in assembly source codes is converted into a formatted string that represents the control flow graph [9] of the method. A number of predefined malware's control flow graphs are then compared against the obtained control flow graph strings to check if they are similar [10] to malware. Schmidt et al. [11] proposed a static analysis solution to detect malicious application based on the output of the *readelf* tool, which contains a list of symbols that involved with an executable. They then differentiate malicious applications from benign ones based on the combinations of system calls used in the executable. Burguera et al. [12] proposed to detect malware using dynamic analysis techniques. They developed a client named *Crowdroid* that is able to monitor Linux kernel system call and report them to a centralized server. Based on the collected dataset, they cluster each dataset using a partition clustering algorithm and hence differentiate between benign and malicious applications. Due to the lack of malware samples, most existing works conduct experiments using self-made malware or a limited number of real malware. It still requires more evidence to prove the effectiveness of these solutions.

3   Feature

For each Android application, we retrieved several selected features from the corresponding application package (APK) file. In addition, we analyzed the source codes of an application, identified real permissions required by the application, and adopted the features for malware detection. The values of selected features are stored as a feature vector, which is represented as a sequence of comma separated values. We enumerate all selected features in the following items. Each item includes the name of a feature, the data type of the feature, and a detailed description about how the features are retrieved.

1. ext_so (*integer*): We list all files found from an APK file and count the number of files with a ".so" extension filename.
2. file_elf (*integer*): We use the UNIX *file* utility to determine the type of each file in an APK file and counts the number of executable and linking format (ELF) files.
3. file_exe (*integer*): Similar to Item 2, but this feature counts only executables.
4. file_so (*integer*): Similar to Item 2, but this feature counts only shared objects.
5. dex.all (*integer*): This feature counts the total number of *required permissions*. As introduced in Section 1, requested permissions and required permissions are different. There is not a file that describes the actual

permissions required by an application. Therefore, it is a must to retrieve the required permissions by analyzing the application from the source-code level. Although Android applications are often written in the Java programming language, here "source codes" are the assembly source codes represented in Jasmin's (dedexer's) syntax. We disassemble byte codes of each Java class file in an APK file into assembly codes using the *baksmali* [13] disassembler. We then identify invoked Android system functions from the assembly codes and look up the required permissions from the permission map table provided by [3]. It should be noticed that currently we only map from function calls to permissions. Although the obtained required permission would be less than all the required permissions, it still improves the performance.

6. dex.normal (*integer*): Google classifies all permissions into four categories, i.e., normal, signature, dangerous, and signatureOrSystem. Among all the 139 built-in permissions[1], 21 permissions are classified as normal, 27 permissions are classified as signatureOrSystem, 35 permissions are classified as signature, and 56 permissions (approximately 40% of all permissions) are classified as dangerous. Similar to Item 5, but this feature counts only the number of permissions that are classified as "normal."

7. dex.sign (*integer*): Similar to Item 5, but this feature counts only the number of permissions that are classified as "signature."

8. dex.dangerous (*integer*): Similar to Item 5, but this feature counts only the number of permissions that are classified as "dangerous."

9. dex.signOrSys (*integer*): Similar to Item 5, but this feature counts only the number of permissions that are classified as "signatureOrSystem."

10. List of all *required permissions* (*boolean*): In addition to count the number of required permissions, we also list the permissions required by an analyzed application. With the retrieved required permission, we convert the permissions into a boolean vector. Suppose the 139 built-in permissions are labeled from 1 to 139 (P_1, P_2, ..., P_{139}), an application that requests P_2 and P_3 would have a boolean vector of values (0, 1, 1, 0, ..., 0). This feature contains 139 boolean values.

11. xml.all (*integer*): This feature counts the number of permissions requested by an application. The requested permissions are retrieved directly from the *AndroidManifest.xml* file that is placed at the root of an APK file. Reading requested permissions from an *AndroidManifest.xml* file is simple. This file can be extracted from an APK file by using the *unzip* tool, convert to a human-readable format using a tool such as *AXMLPrinter2*, and then parsed using the *libxml* library.

12. xml.normal (*integer*): Similar to Item 11, but this feature counts only the number of permissions that are classified as "normal."

13. xml.sign (*integer*): Similar to Item 11, but this feature counts only the number of permissions that are classified as "signature."

14. xml.dangerous (*integer*): Similar to Item 11, but this feature counts only the number of permissions that are classified as "dangerous."

---

[1] The number is retrieved from the Android 2.3 version (codename: Gingerbread) source tree. Readers can refer to the `frameworks/base/core/res/AndroidManifest.xml`.

15. xml.signOrSys (*integer*): Similar to Item 11, but this feature counts only the number of permissions that are classified as "signatureOrSystem."

16. List of all *requested permissions*: In addition to count the number of requested permissions, we also list the exact permissions requested by an analyzed application. The format is the same as Item 10. This feature contains 139 boolean values as well.

17. under (*boolean*): This feature is a boolean value to indicate that an application is *under-privileged*. Since the requested permissions listed in an AndroidManifest.xml are declared by the application developer, there are often inconsistencies between the requested permissions and the required permissions. Although a developer should be able to determine which permissions are required by reading the official developer's API reference document, researchers [3] found the documented permission requirements are somewhat different from the actual requirements. Therefore, an application may be *under-privileged* or *over-privileged* depending on how its permission request is declared.

    An under-privileged application means that an application developer requests less permissions than actually the application needs. It could be malfunctioned because of security exceptions raised by the Android operating system when accessing unprivileged system functions. In contrast, an over-privileged application means that an application developer requests more permissions than actually it needs. Although an over-privileged application breaks the ideal least privilege scenario, it does not have any side-effect. Therefore, to prevent an application from being blocked by the Android operating system due to insufficient permissions, a developer often chooses to request more permissions than actually the application needs.

18. ucount (*integer*): This feature counts the number of under-privileged permissions by comparing required permissions against requested permissions. For example, if an application requires INTERNET permission but it does not request the permission, the counter increases by one.

19. over (*boolean*): In contrast to Item 17, this feature is a boolean value to indicate that an application is *over-privileged*.

20. ocount (*integer*): This feature counts the number of over-privileged permissions by comparing required permissions against requested permissions. For example, if an application does not require BLUETOOTH permission but it request the permission, the counter increases by one.

In addition to the selected features, a label BoM is appended at the end of a feature vector to show that the vector belongs to a benign or a malicious application. The value of the BoM contains only malicious and benign. Labeling an application correctly is an important task. We label the obtained feature vectors using three different strategies—*site-based labeling*, *scanner-based labeling*, and *mixed labeling*. Site-based labeling labels an application based on the source we obtain the corresponding APK file. If an APK file is downloaded from Google Play or third party markets, it is labeled as benign. If an APK file is downloaded from a malicious repository, it is labeled as malicious. Scanner-based labeling

```
com.rovio.angrybirds: 2,2,0,2,9,3,6,0,0,0,0,0,1,1,0,0,1,0,1,0,0,0,0,0,0,0,0,
0,0,0,0,0,0,0,0,0,0,0,0,0,0,0,0,0,0,0,0,0,0,0,0,0,0,0,0,0,0,0,0,0,0,0,0,0,0,
0,0,0,0,0,0,0,0,0,0,0,0,0,1,0,0,0,0,0,0,0,0,0,0,0,0,0,0,0,0,0,1,0,0,0,1,0,0,0,
0,0,0,0,0,0,0,0,0,0,0,0,0,0,0,0,0,0,0,0,0,0,0,0,0,0,0,0,0,0,0,0,0,0,1,1,0,0,
0,0,0,0,0,0,0,0,0,0,0,6,2,4,0,0,0,0,0,1,0,0,0,1,0,1,0,0,0,0,0,0,0,0,0,0,0,0,
0,0,0,0,0,0,0,0,0,0,0,0,0,0,0,0,0,0,0,0,0,0,0,0,0,0,0,0,0,0,0,0,0,0,0,0,0,0,0,
0,0,0,0,0,0,0,0,1,0,0,0,0,0,0,0,0,0,0,0,0,0,0,0,0,0,0,0,0,1,0,0,0,0,0,0,0,0,
0,0,0,0,0,0,0,0,0,0,0,0,0,0,0,0,0,0,0,0,0,0,0,0,0,0,0,0,0,0,0,0,0,0,0,0,1,0,0,
0,0,0,0,0,0,0,1,4,1,1,benign
```

Fig. 2. An example feature vector for the *AngryBird* application retrieved from its Android application package file

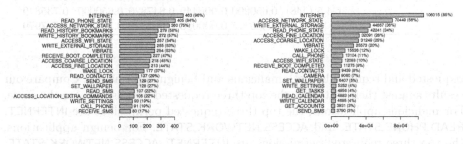

(a) Requested Permission (Malicious)   **(b)** Requested Permission (Benign)

Fig. 3. Count of requested permissions for both benign and malicious Android applications. The malicious applications are labeled using mixed labeling strategy.

labels an application based on the decision of an anti-virus scanner. Currently we use the open source ClamAV anti-virus software to make the decision. If an APK file is reported to be malicious, it is labeled as malicious. Otherwise it is labeled as benign. Mixed labeling is the union of site-based labeling and scanner-based labeling. If an APK is downloaded from a malicious repository or it is reported as malicious by an anti-virus scanner, it is labeled as malicious. Otherwise, it is labeled as benign. We obtain feature vectors for all the collected 125,249 applications. A complete example of a feature vector for the *AngryBird* game is shown in Figure 2. All the features are retrieved from the *AngryBird*'s APK file. Since we have three different strategies to label the feature vectors, there are three corresponding datasets. The datasets are named by its labeling strategy. We then feed the datasets to machine learning algorithms and evaluate the performance of permission-based detection for malicious Android applications.

4   Result

Before digging into the results of classification, we have a quick look on the most frequently used permissions in Android applications. We plot a similar statistics on top 20 permissions to [7]. Figure 3 shows the top 20 permissions

Table 1. The Performance of Classifiers on Detection of Malicious Applications

	Classifier	TP Rate	FP Rate	Precision	Recall	F-Measure
Dataset #1	AdaBoost	0	0	n/a	0	n/a
Site-based label	Naïve Bayes	0.720000	0.057769	0.019544	0.720000	0.038055
	C4.5 (J48)	0.460000	0.000080	0.901961	0.460000	0.609272
	SVM	0.445000	0.000048	0.936842	0.445000	0.603390
Dataset #2	AdaBoost	0	0.000008	0	0	n/a
Scanner-based label	Naïve Bayes	0.811905	0.076625	0.034424	0.811905	0.066047
	C4.5 (J48)	0.714286	0.000401	0.857143	0.714286	0.779221
	SVM	0.616667	0.000080	0.962825	0.616667	0.751814
Dataset #3	AdaBoost	0	0.000024	0	0	n/a
Mixed label	Naïve Bayes	0.762500	0.086536	0.032787	0.762500	0.062870
	C4.5 (J48)	0.650000	0.000449	0.847826	0.650000	0.735849
	SVM	0.585417	0.000088	0.962329	0.585417	0.727979

requested and required by both malicious and benign applications. Compare our results against their statistics, the top three requested permissions are the same. For malicious applications, the top three requested permissions are INTERNET, READ_PHONE_STATE, and ACCESS_NETWORK_STATE. For benign applications, the top three requested permissions are INTERNET, ACCESS_NETWORK_STATE, and WRITE_EXTERNAL_STORAGE. Although the number of malicious application we evaluated is less than [7], the ranks of requested permissions are similar.

We then use the *Weka* data mining software [14] to classify benign and malicious applications based on permissions. We feed the permission datasets retrieved from the 125,249 applications to four commonly used classifiers. They are *AdaBoost, Naïve Bayes, C4.5 (J48)*, and *support vector machine (SVM)*. Each classifier builds classification models from the three datasets, distinguishes malicious applications from benign ones based on the models, and then evaluates how good (or bad) it performs. Readers should notice that it is a difficult problem for classifiers because the datasets are *extremely imbalanced datasets*. The ratio of the number of malicious applications and benign applications, in the best case, is 480:124769 (less than 0.004). Finding a malicious application is just like finding a needle in a haystack.

Table 1 shows the performance of each classifier on detection of malicious Android applications. The performance of a classifier is measured using the following metrics: the true positive (TP) rate, the false positive (FP) rate, the precision, the recall rate, and the F-measure. All values range from 0.0 to 1.0. Note that the precision and the F-measure fields of some classifiers are labeled *n/a* in the table. This is because precision is evaluated by *true-positives/(true-positives + false-positives)* but the classifier does not classify any instance into the malicious class. Hence, both *true-positives* and *false-positives* are zero. Similarly, the F-measure is evaluated by $2 \cdot (precision \cdot recall)/(precision + recall)$ and therefore it cannot be evaluated if either precision or recall rate cannot be obtained.

From the table, we also find that the *AdaBoost* classifier does not perform well. It classifies all applications as benign applications. The *Naïve Bayes* classifier does not also perform well because it has a very low precision. The C4.5 (J48) and the SVM would be better choices. They have a much higher precision and the recall rate for the default C4.5 classifier ranges from 0.46 to 0.71. This means that the default C4.5 classifier is possible to identify more than 70% of evaluated malicious applications. For the support vector machine (SVM) classifier, we have tried to optimize it by tuning its *cost* and *gamma* parameter via *cross-validation and grid-search* [15]. Although the optimized performance shows that the recall rate is lower than the C4.5 classifier, the SVM has a very high precision. Based on the result, we are able to choose a classifier to fit different usage different scenarios. If precision is the concern, the C4.5 and the SVM would be good choices. In contrast, if recall rate is the concern, the Naïve Bayes would be a good choice. It is of course that we can combine results from multiple classifiers to get the maximal set of malicious applications. However, we would need a second phase to further examine a detected malicious application.

5 Conclusion and Future Work

Application requested permissions are currently the minimal defense for an Android user to decide whether or not to install an application. This paper explores the possibility of detection malicious Android applications based on permissions and several easy-to-retrieve features from Android application packages. Our large scale experiments show that a single classifier is able to detect about 81% of malicious applications. By combining results from various classifiers, it can be a quick filter to identify more suspicious applications. Although the performance numbers are not perfect, permission-based classifications can be further improved in two directions. First, the retrieval of the required permissions can be further improved by considering permissions relevant to event handling and content accessing. Second, more features can be retrieved by statically analyzing assembly source codes of an application. We believe that permission-based classifications can be a good auxiliary to detect malicious applications.

Acknowledgement. This research was supported in part by National Science Council under the Grants NSC 100-2221-E-019-045 and NSC 101-2219-E-019-001. We would like to thank the anonymous reviewers for their valuable and helpful comments.

References

1. Lockheimer, H.: Android and security. Official Google Mobile Blog (February 2012), http://googlemobile.blogspot.tw/2012/02/android-and-security.html
2. <permission>. Android Developer - API Guides - Android Manifest, http://developer.android.com/guide/topics/manifest/permission-element.html

3. Felt, A.P., Chin, E., Hanna, S., Song, D., Wagner, D.: Android permissions demystified. In: Proceedings of the 18th ACM Conference on Computer and Communications Security, pp. 627–638 (2011)
4. Johnson, R., Wang, Z., Gagnon, C., Stavrou, A.: Analysis android applications' permissions. In: Proceedings of the 6th International Conference on Software Security and Reliability (2012)
5. Enck, W., Ongtang, M., McDaniel, P.: Understanding android security. IEEE Security and Privacy 7(1), 50–57 (2009)
6. Barrera, D., Kayacik, H.G., van Oorschot, P.C., Somayaji, A.: A methodology for empirical analysis of permission-based security models and its application to android. In: Proceedings of the 17th ACM Conference on Computer and Communications Security, pp. 73–84 (2010)
7. Zhou, Y., Jiang, X.: Dissecting android malware: Characterization and evolution. In: Proceedings of the 33rd IEEE Symposium on Security and Privacy, pp. 95–109 (2012)
8. Desnos, A.: androguard - reverse engineering, malware and goodware analysis of android applications ... and more (ninja !). Googld Project Hosting, http://code.google.com/p/androguard/
9. Cesare, S., Xiang, Y.: Classification of malware using structured control flow. In: Proceedings of the 8th Australasian Symposium on Parallel and Distributed Computing (2010)
10. Pouik, G0rfi3ld: Similarities for fun & profit. Phrack #68 (April 2012), http://www.phrack.org/issues.html?issue=68&id=15#article
11. Schmidt, A.D., Bye, R., Schmidt, H.G., Clausen, J., Kiraz, O., Yuksel, K.A., Camtepe, S.A., Albayrak, S.: Static analysis of executables for collaborative malware detection on android. In: Proceedings of IEEE International Conference on Communications (2009)
12. Burguera, I., Zurutuza, U., Nadjm-Tehrani, S.: Crowdroid: behavior-based malware detection system for android. In: Proceedings of the 1st ACM Workshop on Security and Privacy in Smartphones and Mobile Devices (2011)
13. Freke, J.: smali - an assembler/disassembler for android's dex format. Google Project Hosting, http://code.google.com/p/smali/
14. Hall, M., Frank, E., Holmes, G., Pfahringer, B., Reutemann, P., Witten, I.H.: The WEKA data mining software: an update. SIGKDD Explorations Newsletter 11(1), 10–18 (2009)
15. Hsu, C.W., Chang, C.C., Lin, C.J.: A practical guide to support vector classification. Technical report, Department of Computer Science and Information Engineering, National Taiwan University (2009), http://www.csie.ntu.edu.tw/%7ecjlin/libsvm/

Image Steganography Using Gradient Adjacent Prediction in Side-Match Vector Quantization

Shiau-Rung Tsui[1], Cheng-Ta Huang[1], and Wei-Jen Wang[1,2,*]

[1] Department of Computer Science and Information Engineering,
National Central University, Taiwan
[2] Software Research Center, National Central University, Taiwan
wjwang@csie.ncu.edu.tw

Abstract. This study presents a new steganographic method that embeds secret data into a cover digital image using VQ encoding. The core concept of the proposed method uses the gradient adjacent prediction (GAP) algorithm, which enhances prediction accuracy of neighboring blocks in SMVQ encoding. To embed secret data into the cover image, the proposed method utilizes the features of GAP to decide the capacity of the secret data per pixel in a block. It then embeds the secret data accordingly. It also embeds an index value in each block to ensure that the secret data can be recovered back. The index value points to the closest codeword of a state codebook to the encoding block, where the state codebook is generated by GAP-based SMVQ. The result shows that the proposed method has better performance than a recent similar work proposed by Chen and Lin in 2010.

Keywords: Steganography, Gradient adjacent prediction (GAP), VQ, SMVQ.

1   Introduction

Steganographic methods for digital images have been widely studied in the last decade. They employ various encoding techniques to embed secret data into the cover image, and to produce the stego-image carrying secret data. Those methods have been used in many types of applications, such as watermarking and secret communication. Encoding techniques used in those methods can be assorted to three kinds of technique domains — the spatial domain [1–5], the frequency domain [6–8], and the compression (substitution) domain [6], [7], [9–13]. Each technique domain has different features. For example, the spatial domain usually offers large embedding capacity for secret data and excellent visual quality for stego-images, but may not pass statistical steganalysis. On the other hand, the compression domain performs better in statistical steganalysis, but may produce less embedding capacity for secret data and lower visual quality for stego-images. As for the frequency domain, a method of the domain is usually used in watermarking applications since it is more robust against image distortion attacks.

---

[*] Corresponding author.

J.-S. Pan et al. (Eds.): *Advances in Intelligent Systems & Applications*, SIST 21, pp. 121–129.
DOI: 10.1007/978-3-642-35473-1_13   © Springer-Verlag Berlin Heidelberg 2013

This study focuses on developing a new steganographic method in the compression (substitution) domain. The most popular technique in this domain is to use an algorithm based on vector quantization (VQ) [14], to encode the cover image along with the secret data. The VQ uses a codebook to quantize an image to non-overlapping blocks of 4×4 pixels. In VQ, each block in the image is replaced by a short index value, which points to a codeword of the codebook and the codeword can best represents the block. Side match vector quantization (SMVQ) [15] is a VQ-based encoding approach that is also popular in image steganography. It is based on the observation that neighboring pixels are usually similar. The property can be utilized to construct state codebooks using smaller index size to improve compression rate, which can create larger capacity for embedding secret data. The gradient adjacent prediction (GAP) algorithm [17] can be used to improve SMVQ compression. It uses seven neighboring pixels to predict what the current encoding/decoding block should look like. Each successful prediction can potentially improve the compression rate of the SMVQ encoding method. This algorithm has the advantage of being able to detect the feature of the vertical side and the horizontal side precisely. This advantage can be used to improve the compression rate of SMVQ.

Most VQ-based steganographic methods produce VQ-based stego-images as the output; some VQ-based steganographic methods produce raw image as the output. This study focuses on the latter since the types of steganographic methods have much room for performance improvement. In 2006, Chang et al. proposed an SMVQ-based method that produces raw images as the output stego-images [11]. To encode a block, the method uses the two closest codewords to the block in the codebook, and constructs a substitute block by applying an adjustable weight ratio to the two closest codewords. A secret bit can be embedded in each block by adjusting different ratios. In 2010, Chen and Lin proposed two similar steganographic methods [13] that produce raw images as the output. It adopts the least-significant-bit embedding approach [16] and the concept of codeword radius from Chang et al.'s method. The methods, compared with prior similar studies, provide significantly larger embedding capacity for secret data. This is because the methods can embed multiple secret bits in each pixel of an image block.

Section 2 introduces Chen and Lin's methods and their limitations. Section 3 describes the proposed method. Section 4 summarizes the experimental results and makes comparisons to other methods. Section 5 draws the conclusions.

2   Chen and Lin's Steganographic Methods

In the section, we will briefly introduce the concepts of the two methods proposed by Chen and Lin [13]. Both the methods are very similar and employ the same concept, the distance radius concept. That is, their methods need to define a distance radius for each used codeword in the codebook. To encode a block R, the method retrieves the closest codeword X to R and the closest codeword Y to X from the codebook. The half of the distance between X and Y is the distance radius of X, namely t. Any block value within the distance radius can be mapped to X, which represents the same result of using SMVQ to map block R into a codeword. Given the integer representation of the secret bits I, the encoded value $X+I$ can be extracted if I is smaller than t. This is because $X+I$ is always mapped to X under SMVQ encoding. The first steganographic method proposed by Chen and Lin is shown below:

Input: Cover image G, Codebook C, Secret data S
Output: Stego-image G'
Step 1: Divide the image G into non-overlapping blocks of 4×4 pixels.
Step 2: For each block R, do Steps 3 to 6.
Step 3: If R is on the first row/ column, do VQ encoding. If not, go to Step 4.
Step 4: Do SMVQ encoding to establish state codebook.
Step 5: Find the closest codeword X to R and the closest codeword Y to X in the state codebook. Set $v = \lfloor \log_2 (\|X - Y\|/2) \rfloor$ to be the largest embedding capacity for each pixel of R.
Step 6: Retrieve v bits of secret data, namely I, and then replace the least significant v bits of X by the secret bits I. The block is saved in Z. Go back to Step 2 until all blocks are processed.
Step 7: Output Stego-image Z.

Chen and Lin proposed another aggressive encoding method based on the above steganographic method. They pointed out that, for a 4×4 block, the constraint for embedding v bits per pixel is:

$$16(2^v-1)^2 \le t^2 \qquad (1)$$

Since the embedding is pixel-based, each pixel can embed different number of bits. As a result, they proposed an aggressive encoding method, which finds the largest v and then embeds one more bit ($v+1$ bits) for the first k pixels of the block and v bits for the remaining pixels of the block, such that the following equation holds:

$$k(2^{v+1}-1)^2 + (16-k)(2^v-1)^2 \le t^2 \qquad (2)$$

3   Proposed Method

Traditional VQ encoding uses a codebook to encode an image for the purpose of data compression. The SMVQ and GAP can be used to compress the VQ-encoded image to a smaller representation. The proposed method uses all of the techniques to encode the cover image along with the secret data. Since the proposed method produces a raw image as the output, it does not need to follow the steps of SMVQ and GAP encoding strictly. Our strategy is to leave the blocks in the first row/column as they are, and then to perform SMVQ and GAP on other blocks. Based on SMVQ and GAP, the proposed steganographic method can achieve higher PSNR than what VQ/SMVQ encoding can achieve. In the embedding procedure, the proposed method partitions the input image into blocks of 4×4 pixels. Then, it uses the LSB method [19] to embed an index value, which points to the most similar codeword to the block, into the first n pixels of each image block, where $n=1\sim3$. The remaining pixels of the block are used to embed secret bits. To achieve this purpose effectively, it uses GAP to decide the number of bits to be embedded. The following steps show how the embedding procedure works:

> **Input:** Cover image Z, codebook C, state codebook size 2^k and secret data S
> **Output:** Stego-image Z'
> **Step 1:** Divide the image Z into non-overlapping blocks of 4×4 pixels.
> **Step 2:** For each block R, do Steps 3 to 7.
> **Step 3:** If R is in the first row/column, save R in Z' and go back to Step 2. Otherwise, go to Step 4.
> **Step 4:** Use GAP to predict what R should be and get R'. Generate a state codebook SC by R' and codebook C, and then use the state codebook SC to get an index value pointing to the most similar codeword g.
> **Step 5:** Partition the index value (*i.e.* the size is k bits) into $1\sim3$ three-bit segments. Use LSB to embed the segments into the first n pixels except the last segment. The last segment may not contain 3 bits because k is not dividable by 3. Assume the size of the last segment is m. Embed m bits by LSB into the corresponding pixel.
> **Step 6:** The GAP algorithm labels the uppermost pixels and the leftmost pixels with their property: "sharp", "weak", or "normal." Count the total number of each label appeared in the block. If "sharp" dominates, retrieve four bits from the secret data and embed them into each of the rest of the pixels. If "normal" dominates, retrieve and embed three bits. Otherwise, retrieve and embed one bit. For each remaining pixel that corresponds to the position p in the block, the embedding approach adds or subtracts the value of the secret bits to the g_p, depending on which operation results in better visual quality.
> **Step 7:** Combine the pixels generated in Steps 5 and 6 to create a new stego-image block. Save the block in Z'. Go back to Step 2 until all blocks are processed.
> **Step 8:** Output stego-image Z'.

The secret data extraction procedure uses GAP to generate a state codebook for each image block, and uses LSB to restores an index value from the first to the third pixels. Then we use GAP to calculate the number of secret bits we had embedded in each of the remaining pixels. The secret data extraction procedure is shown below:

> **Input:** Stego-image Z', Codebook C and state codebook size 2^k
> **Output:** Secret data S
> **Step 1:** Divide the stego-image Z' into non-overlapping blocks of 4×4 pixels.
> **Step 2:** For each block R, do Steps 3 to 7.
> **Step 3:** If R is in the first row/column, do nothing and return to Step 2. Otherwise, go to Step 4.

Step 4: Use GAP to predict what R should be and get R'. Use R' and codebook C to generate a state codebook SC.

Step 5: Retrieve $\lceil k/3 \rceil$ three-bit segments from the least significant three bits of the first $\lceil k/3 \rceil$ pixels except the last segment. Read only $k \bmod 3$ bits from the corresponding pixel that embeds the last segment. Recover the index value and obtain the codeword g of the block.

Step 6: Use GAP to see which property ("sharp", "weak", or "normal") of block R' dominates. If "sharp" dominates, retrieve three secret bits s from each of the rest of the pixels. If "normal" dominates, retrieve two bits. Otherwise, retrieve one bit. For each remaining pixel h that corresponds to the position p in the block, the retrieval procedure calculates the absolute difference between the g_p and h, and then gets the secret bits s.

Step 7: Append s to S. Go back to Step 2 until all blocks are processed.

Step 8: Output secret data S.

4   Experimental Results

We have conducted some experiments to evaluate the performance of the proposed method, and show the results in this section. We used six 512×512 gray-level images for performance evaluation —"Lena," "Baboon," "Sailboat" "Pepper," "F16," and "Boat," as shown in Fig. 2. In the experiments, we used two different codebooks, one with 256 codewords and the other with 512 codewords, respectively. The codebooks are generated by the LBG algorithm [18], given many images for training.

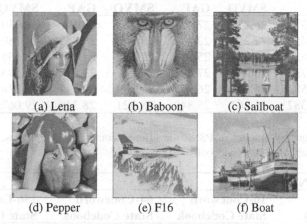

(a) Lena   (b) Baboon   (c) Sailboat

(d) Pepper   (e) F16   (f) Boat

Fig. 1. The six testing cover images, each of which has a size of 512×512 pixels

Table 1 and Table 2 show the visual quality of the testing images, encoded by VQ, SMVQ, and GAP-based SMVQ respectively. The results of Table 1 and Table 2 are generated by using a codebook of 256 codewords and a codebook of 512 codewords, respectively. Note that the visual quality of an SMVQ-encoded image is always worse than the visual quality of a VQ-encoded image. However, a GAP-based-SMVQ-encoded image can potentially achieve better visual quality than a VQ-encoded image because we use raw image blocks in the first row/column.. Generally, a GAP-based-SMVQ-encoded image has better visual quality than an SMVQ-encoded image in terms of PSNR. The improvement comes from the accurate prediction scheme of GAP. From Table 1 and Table 2, we found that GAP-based SMVQ could enhance SMVQ image quality.

In Table 3 and Table 4, we can see that all the embedding capacity values are close, and so are the visual quality values, given different sizes of state codebook and different codebooks. Although the difference is small, we still can find that the visual quality goes up as the size of the state codebook increases if we are given the same codebook. On the other hand, the embedding capacity decreases as the size of the state codebook increases. Note that the embedding capacity is the same for state codebook sizes 16 and 32. This is because our method decides the number of pixels for embedding an index value by $\lceil k/3 \rceil$, given the state codebook size 2^k. For example, a state codebook with 2^8 codewords should be partitioned into $\lceil (\log_2 8)/3 \rceil$ segments, implying the use of the first pixel to embed the index value. When the size of the state codebook doubles, we need $\lceil (\log_2 16)/3 \rceil$ segments, implying the use of the first two pixels to embed the index value.

Table 1. Image quality in terms of PSNR for different compression strategies, given different sizes of state codebooks and a codebook of 256 codewords

Image	PSNR VQ	16-Codeword State Codebook		64-Codeword State Codebook		128-Codeword State Codebook	
		SMVQ	GAP	SMVQ	GAP	SMVQ	GAP
Lena	31.373	26.731	27.278	30.088	30.169	31.049	31.107
Baboon	24.452	22.267	22.375	23.642	23.842	24.180	24.213
Sailboat	28.622	25.359	25.585	27.792	27.956	28.435	28.684
Pepper	30.728	26.714	27.291	29.787	29.878	30.543	30.562
F16	30.582	26.227	26.957	29.389	29.511	30.228	30.411
Boat	29.387	25.863	25.840	28.321	28.348	29.067	29.146

Table 2. Image quality in terms of PSNR for different compression strategies, given different sizes of state codebooks and a codebook of 512 codewords

Image	PSNR VQ	16-Codeword State Codebook		64-Codeword State Codebook		256-Codeword State Codebook	
		SMVQ	GAP	SMVQ	GAP	SMVQ	GAP
Lena	32.248	25.898	26.836	29.776	29.910	32.115	32.079
Baboon	24.703	21.947	22.176	23.349	23.398	24.602	24.648
Sailboat	29.251	24.569	25.072	27.433	27.635	29.141	29.222
Pepper	31.408	26.060	26.662	29.565	29.736	31.306	31.342
F16	31.578	25.705	26.166	29.330	29.350	31.412	31.561
Boat	30.163	25.044	25.639	28.072	28.106	30.063	30.657

Table 3. The visual quality in PSNR and the embedding capacity (EC) in bits of the stego-images produced by the proposed method, given different sizes of state codebooks and a codebook of 256 codewords

Image	state codebook size							
	8		**16**		**32**		**128**	
	PSNR	EC	PSNR	EC	PSNR	EC	PSNR	EC
Lena	26.339	518910	28.044	529018	29.781	527240	31.877	489099
Baboon	22.561	614813	23.111	596904	23.644	599844	24.953	554320
Sailboat	25.268	551855	26.191	547064	28.126	546070	29.825	504556
Pepper	26.274	528911	28.048	542434	29.778	532574	31.641	492089
F16	25.840	469413	27.693	477172	29.025	486612	31.345	449644
Boat	24.923	508001	26.597	508550	28.293	507640	30.162	469313

Table 4. The visual quality in PSNR and the embedding capacity (EC) in bits of the stego-images produced by the proposed method, given different sizes of state codebooks and a codebook of 512 codewords

Image	state codebook size							
	8		**16**		**32**		**256**	
	PSNR	EC	PSNR	EC	PSNR	EC	PSNR	EC
Lena	26.254	510894	27.642	523334	28.937	521136	32.906	482404
Baboon	22.219	622877	23.147	608146	23.426	612220	25.266	566761
Sailboat	24.523	537813	25.835	534562	27.174	532798	30.311	490555
Pepper	25.634	525712	27.400	529256	28.802	526764	32.293	484757
F16	25.261	441601	26.882	460292	28.587	459480	32.311	424541
Boat	24.536	505854	25.982	518490	27.578	518112	31.155	480025

Table 5. Comparison of the proposed method and Chen and Lin 's method [13], given the state codebook size of 16 codewords and a codebook of 512 codewords

Image	Codebook of 512 Codewords and State Codebook of 16 Codewords			
	Proposed method		**Chen and Lin**	
	PSNR	EC(bit)	PSNR	EC(bit)
Lena	27.642	523334	26.512	355952
Baboon	23.147	608146	22.155	456286
Sailboat	25.835	534562	25.274	424587
Pepper	27.400	529256	26.804	374148
F16	26.882	460292	26.386	296154
Boat	25.982	518490	25.786	375021
average	26.148	528897	25.486	380358

To show how good the proposed steganographic method is, we compare the proposed method with Chen and Lin's first steganographic method, which is proposed in [13]. We do not compare ours with Chen and Lin's aggressive embedding method (the second method) in [13] because it is not applicable to all images, as explained in Section 2. The comparison results are shown in Table 5. The proposed method achieves better visual quality for all testing images. The average improvement rate is 2.5% better. As for the embedding capacity, the proposed method achieves significantly larger embedding capacity. The embedding capacity provided by the proposed method is 28% better. The results indicate that the proposed method is a much better steganographic method in the compression domain.

5   Conclusions

In this paper, we proposed a steganographic method based on SMVQ and GAP. The proposed method can embed huge amount of secret data, and yet maintain good visual quality for the stego-images. The key idea of the proposed method is to utilize the first to the third pixels of an image block, depending on the state codebook size, to embed a GAP-adjusted SMVQ index. The index is used to calculate and to recover the embedded secret bits in the remaining pixels of the block. The experimental results show that the proposed method has significantly larger embedding capacity for secret data and better image quality than a prior method proposed by Chen and Lin [13] in 2010.

Acknowledgments. This work was partially supported by the National Science Council, Taiwan, under Grant No. 101-2218-E-008-003-.

References

1. Xu, H., Wang, J., Kim, H.J.: Near-optimal solution to pair-wise LSB matching via an immune programming strategy. Inf. Sci. 180(8), 1201–1217 (2010)
2. Peng, F., Li, X., Yang, B.: Adaptive reversible data hiding scheme based on integer transform. Signal Processing 92(1), 54–62 (2012)
3. Lin, C.C., Tai, W.L., Chang, C.C.: Multilevel reversible data hiding based on histogram modification of difference images. Pattern Recogn. 41(12), 3582–3591 (2008)
4. Liao, X., Wen, Q.Y., Zhang, J.: A steganographic method for digital images with four-pixel differencing and modified LSB substitution. Journal of Visual Communication and Image Representation 22(1), 1–8 (2011)
5. Lou, D.C., Hu, C.H.: LSB steganographic method based on reversible histogram transformation function for resisting statistical steganalysis. Information Sciences 188(1), 346–358 (2012)
6. Chung, K.L., Shen, C.H., Chang, L.C.: A novel SVD- and VQ-based image hiding scheme. Pattern Recognition Letters 22(9), 1051–1058 (2001)
7. Chang, C.C., Lin, C.C., Tseng, C.S., Tai, W.L.: Reversible hiding in DCT-based compressed images. Information Sciences 177(13), 2768–2786 (2007)

8. Noda, H., Niimi, M., Kawaguchi, E.: High-performance JPEG steganography using quantization index modulation in DCT domain. Pattern Recognition Letters 27(5), 455–461 (2006)
9. Chang, C.C., Nguyen, T.S., Lin, C.C.: A reversible data hiding scheme for VQ indices using locally adaptive coding. Journal of Visual Communication and Image Representation 22(7), 664–672 (2011)
10. Chang, C.C., Wu, W.C., Hu, Y.C.: Lossless recovery of a VQ index table with embedded secret data. Journal of Visual Communication and Image Representation 18(3), 207–216 (2007)
11. Chang, C.C., Tai, W.L., Lin, C.C.: A reversible data hiding scheme based on side match vector quantization. IEEE Transactions on Circuits and Systems for Video Technology 16(10), 1301–1308 (2006)
12. Yang, C.H., Wang, W.J., Huang, C.T., Wang, S.J.: Reversible steganography based on side match and hit pattern for VQ-compressed images. Information Sciences 181(11), 2218–2230 (2011)
13. Chen, L.S.T., Lin, J.C.: Steganography scheme based on side match vector quantization. Optical Engineering 49(3), 0370080–0370087 (2010)
14. Gray, R.: Vector quantization. IEEE ASSP Magazine 1(2), 4–29 (1984)
15. Kim, T.: Side match and overlap match vector quantizers for images. IEEE Transactions on Image Processing 1(2), 170–185 (1992)
16. Bender, W., Gruhl, D., Morimoto, N., Lu, A.: Techniques for data hiding. IBM Systems Journal 35(384), 313–336 (1996)
17. Fallahpour, M., Megias, D., Ghanbari, M.: Subjectively adapted high capacity lossless image data hiding based on prediction errors. Multimedia Tools Appl. 52(2-3), 513–527 (2011)
18. Linde, Y., Buzo, A., Gray, R.: An algorithm for vector quantizer design. IEEE Transactions on Communications 28(1), 84–95 (1980)

A Data Hiding Scheme Based on Square Formula Fully Exploiting Modification Directions

Wen-Chung Kuo

Department of Computer Science and Information Engineering,
National Yunlin University of Science & Technology,
Taiwan, R.O.C.
simonkuo@yuntech.edu.tw

Abstract. Recently, Kieu and Chang modified the extract function and proposed a new data hiding scheme to improve the data hiding capacity from 1 bpp to 4.5 bpp. However, they must use the search matrix method to embed the secret data, i.e., they do not give the close form to embedding secret data. In order to quick embedding secret and improve the above shortcoming, a new data hiding scheme based on square formula fully exploiting modification directions method will be proposed in this paper. From the experiment results, we can prove that proposed scheme not only to enhance the embedding rate and good embedding capacity but also to keep stego-image quality.

Keywords: Data hiding, Extracting function, Stego-image, Modulus operation.

1   Introduction

As the rapid growth of network and smart phone technology, a lot of private image information such as digital photos or videos are communicated in Internet. As a result, people can be shared the happy or unhappy things each other, immediately. However, there are many attacks such as illegal duplication, forgery and spoofing when digital multimedia is transmitted through a public channel. Therefore, how to protect the digital data security has become very important. A common way to solve this problem is to hide personal data behind a meaningful image such that an unintended observer will not be aware of the existence of the hidden secret message.

Until now, many data hiding schemes based on different embedding methods (such as replacing or extraction function) have been proposed [1-8]. From replacing view, the most common data hiding technique is the least significant bit replacement method (LSB-R) proposed. Its major embedding formula is that the secret data will be embedded into the kth bit ($1 \leq k \leq 8$) of each pixel of the cover image. Generally, the stego-image's quality will be acceptable when k>3. In other ways, from the extraction function, a data hiding scheme based on the Exploiting Modification Direction (EMD) method to achieve the data hiding goal is proposed by Zhang and Wang [8] in 2006. The EMD-scheme characteristically uses the relationship of n adjacent pixels to embed the secret data. That is to say, the binary secret data stream will be separated into blocks and transformed into a (2n+1)-ary. Then, this secret will be embedded into

J.-S. Pan et al. (Eds.): *Advances in Intelligent Systems & Applications*, SIST 21, pp. 131–139.
DOI: 10.1007/978-3-642-35473-1_14   © Springer-Verlag Berlin Heidelberg 2013

n adjacent pixels where n > 1. For instance, if the secret data is embedded in two adjacent pixels, i.e., it only modifies only one of two adjacent pixels in the EMD scheme – add one, subtract one, or stay the same. From the experimental results, they claimed that EMD-scheme can enhance secret data embedding capacity and maintain good stego-image quality. Since then, there are many EMD-type embedding methods have been proposed[2-4]. Recently, Kieu and Chang [1] proposed robust data hiding scheme based on the fully exploiting modification direction(FEMD) to improve the original data hiding capacity from 1 bpp to 4.5 bpp and maintain good stego-image quality. In the Kieu-Chang scheme, they use the search matrix method to finish embedding the secret data. In order to quick embedding secret and improve on these shortcomings, a data hiding scheme based on square fully exploiting modification directions method will be proposed in this paper. So, we will propose a new extraction function and a formula to get the shifting distance when the secret data will be embedded. According to our experiment results, we can prove this proposed scheme not only enhances the embedding rate and provides good embedding capacity but also keeps stego-image quality.

This paper is organized as follows: Section 2 will introduce the EMD-scheme and Kieu-Chang scheme. Then, we will propose a data hiding scheme based on square fully exploiting modification direction in Section 3 and give the experimental results in Section4. Finally, concluding remarks will be given in Section 5.

2   Review Two Data Hiding Schemes

2.1   EMD Data Hiding Scheme

In 2006, a novel data hiding scheme based on the exploiting modification direction method was proposed by Zhang and Wang [8]. The characteristic of the EMD-scheme is that propose a weighing extraction function to embed secret data for a cover image. Therefore, Zhang and Wang propose the following extraction function shown as Eq.(1):

$$f(g_1, g_2, \dots, g_n) = [\textstyle\sum_{i=1}^{n}(g_i \times i)]mod(2n + 1) \tag{1}$$

where g_i is the ith pixel value, *n* as the number of pixels. For example, the 5-ary secret data stream will be embedded in two adjacent pixels, i.e., it only modifies one of two adjacent pixels in the EMD scheme – add one, subtract one, or stay the same. According their analysis, from theoretical view, the embedding capacity of EMD at most is $(\log_2^{(2n+1)})/n$ bpp and PSNR is 51.9dB. As a result, the best hiding bit rate for a pixel (1 bpp) exists when it is 5-ary, i.e., n=2, in Zhang-Wang scheme. When *n* increases, the number of pixels in a group increases, and the hiding bit rate will be decreased each time [8].

2.2   Kieu and Chang Scheme[1]

In order to improve the data hiding capacity, Kieu and Chang modified the extract function and proposed a new data hiding scheme to improve the data hiding capacity from 1 bpp to 4.5 bpp. In other words, the main idea of Kieu-Chang scheme is that the value of s^2 can be hidden into 2 adjacent pixels in the cover image. Therefore, Kieu and Chang [1] proposed another extraction function $F(x_i, x_{i+1})$ shown as Eq.(2):

$$F(x_i, x_{i+1}) = [x_i \times (s - 1) + x_{i+1} \times s] \bmod s^2 \tag{2}$$

where x_i is the ith pixel value, s is the weighting coefficient. Then, Kieu and Chang use a 256×256 S -matrix to represent $F(x_i, x_{i+1})$ where the value of the x_i^{th} row and the x_{i+1}^{th} column in S -matrix is $F(x_i, x_{i+1})$. The symbol $S[x_i][x_{i+1}]$ is used to represent the matrix, i.e. $S[x_i][x_{i+1}] = F(x_i, x_{i+1})$. The result of the extraction function $F(x_i, x_{i+1})$ with different s ($s \in \{2,3,4\}$) is shown in Fig.1.

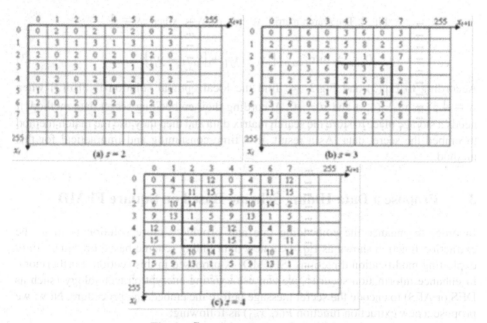

Fig. 1. S -matrix when s = 2, 3 and 4

Subsequently, Kieu and Chang use the search matrix structure $W_{(2r+1) \times (2r+1)}(s, (x_i, x_{i+1}), r)$ to embed the secret data. In other words, the k-bit secret data can be embedded into pair (x_i, x_{i+1}) of cover image by using the S-matrix structure with the search range r, where $k = \lfloor \log_2 s^2 \rfloor$ and $r = \lfloor s/2 \rfloor$. For example, given the pair pixel $(x_i, x_{i+1}) = (4, 3)$ and $s = 4$, the resultant search matrix $W_{5 \times 5}(4, (4,3), 2)$ is shown as Fig.2.

In order to reduce stego-image distortion, Kieu and Chang use the minimum distortion strategy method shown as Eq.(3) to select local optimal solution $D_{\min}$ in $S(x_j, y_j)$.

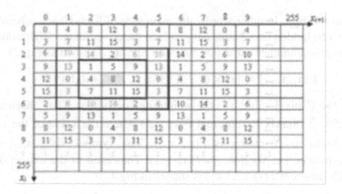

Fig. 2. The search matrix $W_{5\times5}(4,(4,3),2)$ when $s = 4$

$$D_{\min} = \min_{j=a,b,c}\{|x_i - x_j| + |x_{i+1} - y_j|\} \quad\quad (3)$$

According to Eq.(3) and Fig.2, we find the local optimal solution is $S[6][2]$ when $s = 4$, $(x_i, x_{i+1})=(4,3)$ and $d = 10$. According their embedding secret data method, it needs a storage place to store the search matrix data and then they use the match method to embed the secret data. As a result, it is time-consuming and impractical for this method.

3   Propose a Data Hiding Scheme Based on Square FEMD

In order to enhance the embedding rate and provide a close solution form to the extraction function shown as Eq.(4), a new data hiding scheme based on square fully exploiting modification directions (SFEMD) is proposed in this section. Furthermore, to enhance information security, we can use existing encryption technology (such as DES or AES) to encode the secret message before the embedding procedure. Now, we propose a new extraction function $F(x_i, x_{i+1})$ as following:

$$F(x_i, x_{i+1}) = [x_i \times (s^2 - 1) + x_{i+1} \times s^2]mod\ s^4 \quad\quad (4)$$

where x_i is the ith pixel value, s is the weighting coefficient. In order to solve (x_i, x_{i+1}) quickly, we will propose a theorem to satisfy this requirement.

Theorem 1: If $F(x_i, x_{i+1})$ and modulus s are given, then we can find out (x_i, x_{i+1}) directly, $x_i = (s^2 - 1) \times F(x_i, x_{i+1})\ mod\ s^2$ and
$x_{i+1} = (\frac{F(x_i,x_{i+1})-(s^2-1)\times x_i}{s^2})\ mod\ s^2$, such that $F(x_i, x_{i+1}) =$ $[x_i \times (s^2 - 1) + x_{i+1} \times s^2]mod\ s^4$.

In order to understand the theorem 1, we give an example to explain it.

Example 1: If $F(x_i, x_{i+1})=13$ and $s=2$, then the pair $(x_i, x_{i+1})=(3,1)$ by the following steps.

    Step 1. From $s=4$, we can get the extraction function $F(x_1, x_2)=(x_1)\times(3)+$ $(x_2)\times(4)mod\ 16$.

Step 2. Compute $x_1 = 3 \times 13 \bmod 4 = 3$.

Step 3. Calculate $x_2 = \left(\frac{13 - 3 \times 3}{4}\right) \bmod 4 = 1$.

So, there are some notations are defined before we introduce the square FEMD scheme.

IC-New-FEMD:The grayscale cover image $X = \{1 \leq i \leq H \times W, x_i \in [0,255]\}$.

IS-New-FEMD: The grayscale cover image $Y = \{1 \leq i \leq H \times W, y_i \in [0,255]\}$.

3.1   Embedding Algorithm

Input : Cover image $I_{C\text{-}New\text{-}FEMD}$ and the secret data $(S)_{s^4} = \{s_1, s_2, \ldots, s_{(H \times W)/2}\}$.

Output: the stego-image $I_{S\text{-}New\text{-}FEMD}$.

Step 1. For $i = 1$

Step 2. Select two pixels pair (x_i, x_{i+1}) and s_i,

Step 3. Compute $d = F(x_i, x_{i+1}) = (x_i) \times (s^2 - 1) + (x_{i+1}) \times (s^2) \bmod s^4$,

Step 4. If $(d = s_i)$ then $(x'_i, x'_{i+1}) = (x_i, x_{i+1})$,

Otherwise, compute {

1. $t_i = (s^2 - 1)^{-1} \times s_i \bmod s^2$,

2. $t_{1,i} = t_i - (x_i \bmod s^2)$,

3. $x'_{1,i} = x_i + t'_{1,i}$,

4. $t_{1,j+1} = \left(\frac{s_i - (s^2 - 1) \times x_{1,j}}{s^2}\right) \bmod s^2$;

5. Compute $t'_{1,i+1} = t_{1,i+1} - (x_{i+1} \bmod s^2)$,

6. Compute $x'_{1,i+1} = x_{i+1} + t'_{1,i+1}.$}

Step 5. $i = i + 2$

Step 6. Repeat Steps 2–6 until all secret bits are embedded.

Here, we give an example to explain the embedding algorithm.

Example 2: If the cover's pixels pair is $(x_1, x_2) = (163,167)$ and the secret data $s_1 = 13$ when $s = 4$, then the stego-image's pixels pair $(x'_1, x'_2) = (163,168)$ by the following steps.

Step 1. Select two pixels pair $(x_1, x_2) = (163,167)$ and $s_1 = 13$,

Step 2. Compute $d = ((4^2 - 1) \times 163 + 4^2 \times 167 \bmod 4^4 = 61$;

Step 3. Since $d = 61 \neq 13$, compute {

1. $t_i = 15 \times 13 \bmod 16 = 3$,

2. $t'_{1,i} = 3 - (163 \bmod 16) = 0$,

3. $x'_{1,i} = 163 - 0 = 163$,

4. $t_{1,j+1} = \left(\frac{13 - 15 \times 163}{16}\right) \bmod 16 = 8$;

5. $t'_{1,i+1} = 8 - (167 \bmod 16) = 1$,

6. $x'_{1,i+1} = 167 + 1 = 168$}.

So, the stego-image's pixel pair is $(x'_1, x'_2) = (163,168)$.

3.2   Extraction Procedure

The designated receiver can recover the secret data when receiving stego-image $I_{S\text{-}New\text{-}FEMD}$. The extraction algorithm is detailed as following:

Extraction Algorithm
Input: Stego-image $I_{S\text{-}New\text{-}IFEMD}$.
Output: The secret data $(S)_{s^4} = \{s_1, s_2, \ldots, s_{(H \times W)/2}\}$.

Step 1.   For i=1
Step 2.   Select two pixels pair (x_i, x_{i+1}),
Step 3.   Compute $s_{\lfloor i/2 \rfloor + 1} = F(x_i, x_{i+1}) = [x_i \times (s^2 - 1) + x_{i+1} \times s^2] \bmod s^4$,
Step 4.   Convert $s_{\lfloor i/2 \rfloor + 1}$ into s,
Step 5. $i = i + 2$
Step 6.   Repeat Steps 2–5 until all secret bits are extracted.

Example 3: If the stego-image's pixels pair is (x_1, x_2)=(163,168) then we can get
$s_1 = 13$ by   using   the   extraction   function $F(163,168)$=
163×15+168×16mod 64=13.

4   Simulation and Discussion

The proposed scheme was tested on ten 512×512 gray images (Lena, Baboon, F16, Barbara, Boat, Goldhill, Elaine, Tiffany, Pepper and Bridge) as shown in Fig.3. The corresponding stego images when s=2 and s=3 are shown in Fig.4 and Fig.5, respectively. There is no perceivable difference in appearance between cover images and stego images when s=2. However, there are significant difference in visual between cover images and stego images when s=3.

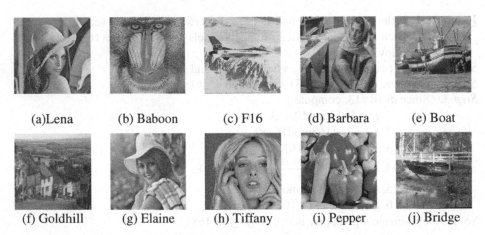

(a)Lena   (b) Baboon   (c) F16   (d) Barbara   (e) Boat

(f) Goldhill   (g) Elaine   (h) Tiffany   (i) Pepper   (j) Bridge

Fig. 3. Ten 512x512 gray test images

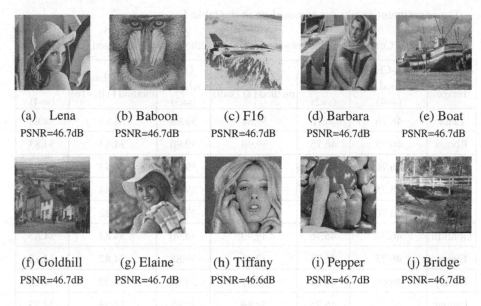

(a) Lena	(b) Baboon	(c) F16	(d) Barbara	(e) Boat
PSNR=46.7dB	PSNR=46.7dB	PSNR=46.7dB	PSNR=46.7dB	PSNR=46.7dB

(f) Goldhill	(g) Elaine	(h) Tiffany	(i) Pepper	(j) Bridge
PSNR=46.7dB	PSNR=46.7dB	PSNR=46.6dB	PSNR=46.7dB	PSNR=46.7dB

Fig. 4. Ten 512x512 gray stego images (s=2)

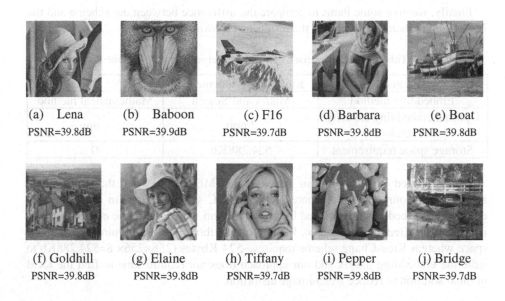

(a) Lena	(b) Baboon	(c) F16	(d) Barbara	(e) Boat
PSNR=39.8dB	PSNR=39.9dB	PSNR=39.7dB	PSNR=39.8dB	PSNR=39.8dB

(f) Goldhill	(g) Elaine	(h) Tiffany	(i) Pepper	(j) Bridge
PSNR=39.8dB	PSNR=39.8dB	PSNR=39.7dB	PSNR=39.8dB	PSNR=39.7dB

Fig. 5. Ten 512x512 gray stego images (s=3)

The experiment results are summed up in the table 1.

Table 1. Comparsion table between Kuo-Kao method and our proposed scheme

Cover Image	Kieu-Chang method [1] (s=4)	Our scheme (s=2)	Kieu-Chang method [1] (s=9)	Our scheme (s=3)	Kieu-Chang method [1](s=16)	Our scheme (s=4)
Lena	46.76	46.76	39.88	39.88	34.83	34.83
Bboon	46.75	46.75	39.90	39.90	34.83	34.83
F16	46.76	46.76	39.89	39.89	34.83	34.83
Barbara	46.75	46.75	39.89	39.89	34.84	34.84
Boat	46.76	46.76	39.89	39.89	34.82	34.82
Goldhill	46.76	46.76	39.90	39.90	34.83	34.83
Elaine	46.75	46.75	39.88	39.88	34.82	34.82
Tiffany	46.69	46.69	39.81	39.81	34.72	34.72
Pepper	46.75	46.75	39.89	39.89	34.83	34.83
Birdge	46.70	46.70	39.83	39.83	34.75	34.75

Finally, we give some items to compare the difference between our scheme and the Kieu-Chang scheme [1] and then the results are shown as table 2.

Table 2. The comparison between our scheme and KC scheme

Items	Kieu-Chang scheme [1]	Our scheme
Embedding method	Matrix and Search	Mathematical method
Optimal embedding capacity When s=2	2 bpp	4 bpp
Storage space requirement	524.288Kb	0

Our proposed scheme is also based on FEMD model, but there are many advantages in our scheme compared with the KC scheme. First, in our model all embedding procedures are finished by using theorem 1. Secondly, the embedding rate is better than Kieu-Chang scheme. Finally, our method does not require any memory space whereas Kieu-Chang scheme requires ~524 Kbytes (256×256× 8=524.288Kb to store the embedding matrix and our approach does not require time to find the local optimal solution to reduce stego-image distortion.

5   Conclusions

In order to enhance embedding capacity, an efficient data hiding scheme based on the FEMD method is proposed by Kieu and Chang in 2011. According to their experimental results, the embedding capacity of their scheme is better than the other EMD-type data hiding scheme. However, they use an extraction matrix and a search method to achieve the data embedding goal. In this paper, we proposed a new extraction function to setup the data hiding scheme by using a mathematical approach. According to our simulation results, our scheme can enhance the embedding rate and maintain the same embedding capacity but also keep good stego-image quality.

Acknowledgment. This work was supported by NSC 101-2221-E-224-100.

References

1. Kieu, T.D., Chang, C.C.: A steganographic scheme by fully exploiting modification directions. Expert Systems with Applications 38, 10648–10657 (2011)
2. Kuo, W.C., Wang, C.C.: Data Hiding Based on Generalized Exploiting Modification Direction Method. Imaging Science Journal, http://dx.doi.org/10.1179/1743131X12Y.0000000011
3. Kuo, W.C., Wuu, L.C., Shyi, C.N., Kuo, S.H.: A Data Hiding Scheme with High Embedding Based on General Improving Exploiting Modification Direction Method. In: 2009 Conference on Hybrid Intelligent Systems, HIS 2009, pp. 69–72 (August 2009)
4. Lee, C.F., Wang, Y.R., Chang, C.C.: A Steganographic Method with High Embedding Capacity by Improving Exploiting Modification Direction. In: Third International Conference on IIHMSP 2007, November 26-28, vol. 1, pp. 497–500 (2007)
5. Mielikainen, J.: LSB Matching Revisited. IEEE Signal Processing Letters 13(5), 285–287 (2006)
6. Wang, R.Z., Lin, C.F., Lin, J.C.: Image Hiding by Optimal LSB Substitution and Genetic Algorithm. Pattern Recognition 34(3), 671–683 (2001)
7. Wu, H.C., Wu, N.I., Tsai, C.S., Hwang, M.S.: Image Steganographic Scheme Based on Pixel-Value Differencing and LSB Replacement Methods. IEE Proceedings-Vision, Image and Signal Processing 152(5), 611–615 (2005)
8. Zhang, X., Wang, S.: Efficient Steganographic Embedding by Exploiting Modification Direction. IEEE Comm. Letters 10(11), 1–3 (2006)

Digital Watermarking Based on JND Model and QR Code Features

Hsi-Chieh Lee[1,*], Chang-Ru Dong[2], and Tzu-Miao Lin[2]

[1] National Quemoy University
Department of Computer Science and Information Engineering
No.1 University Rd, Jinning, Kinmen, 89250, Taiwan, R.O.C.
[2] Yuan Ze University
Department of Information Management,
135 Yuan-Tung Rd, Chung-Li, 32003, Taiwan, R.O.C.
imhlee@gmail.com

Abstract. In this study, we proposed a method to hide QR Codes into images; the method combines JND model and digital watermark techniques and consists of three parts. First, Using JND Model to find the JND map of images and using Sobel operation to find contour of images. Second, find the area there QR Code will embed by scanning the JND map and contour of images with the mask. Finally, in order to hide QR Codes in Images , we adjust the pixel value in the area QR Code embedded and add in invisible watermark and verify information for security propose.

Experimental results showed that the proposed method performs better than the way that puts QR Code in the image without processing, we can not only hide QR Code into images but also the QR Code can be detected clearly.

Keywords: QR Code, JND, Sobel operation.

1 Introduction

The human visual system (HVS) [1], [2] and [3] is quite complex. Just Noticeable Difference (JND) [4] and [5] is the maximum difference that the human visual system can not to detect. JND value is mainly determined by two factors. The first factor is the average background brightness (it also called Weber's Law).The contents of this law are that the visible grayscale threshold and background brightness are in direct proportion. The second factor is the variation of brightness of the background. It is also referred to as spatial masking effect, such as the gradient of the pixels surrounding point in the image. If it has a strong change at the edge of the images, then the change can be detected easily by the human eye. Conversely, if a noise in the background where is not smooth, such boundary would be more difficult to detect. Since the gradient can be a measurement, thus it can be manipulated through digital image processing.

---

* Corresponding author.

J.-S. Pan et al. (Eds.): *Advances in Intelligent Systems & Applications*, SIST 21, pp. 141–148.
DOI: 10.1007/978-3-642-35473-1_15   © Springer-Verlag Berlin Heidelberg 2013

QR Code [6], [7] and [8], is the two-dimensional barcode, invented by the Japanese Denso-Wave in 1994. It was originally used to track car parts. After more than 10 years of development, QR Code became efficient marketing tools. The related applications were also developed such as virtual reality and personal business cards and other presents QR Code. QR Code can carry more information than the general 2D-barcode. There were more devices support decode for QR Code, it has three patterns to assist decoding software locate the pattern, users can scan QR Code with any angle, the content can be read correctly. QR Code has a higher capacity and supports multiple encoding formats; read speed is faster and resistant to damage of digital image processing. The performance of the QR Code error correction ability up to 30% of the code area still can be restored. QR Codes were classified forty versions, from version 1 to version 40, and adjust the barcode size with the amount of information content. QR Code has four error correction levels L / M / Q / H, are available for users on demand, select the higher capability of correcting means QR Code needs more capacity for extra information.

2   JND Model

JND model [9] and [10] is a function which increase or decrease depends on inputs. The ratio of JND value and original stimulus is roughly a constant. If variable I is original stimulus then ΔI is the minimum change which can detected by human eyes for intensity I, and k is a constant. The law was originated when Ernst Heinrich Weber took the experiment of the weightlifting perceived boundaries, so it called Weber Law.

$$\frac{\Delta I}{I} = k \tag{1}$$

When we use the formula, generally assumed there are not relations between two factors of Weber Law and Spatial masking, JND value is decided by the factor which has stronger influence, the formula signs are as follows:

$$JND(x, y) = \max \{f_1(bg(x, y), mg(x, y)), f_2(bg(x, y)) \} \tag{2}$$

f_1 means the second factor and f_2 means the first factor, bg(x, y) is average background brightness and mg(x, y) is the variation of brightness of the background. Hsieh used mask with size 5x5 as a filter to calculate the average background brightness. The point far away from the center of the mask the weight is lower. The point close the center of the mask the weight is higher. By this way we can get more accurate average brightness value.

$$f_1(bg(x, y), mg(x, y)) = mg(x, y)\alpha(bg(x, y)) + \beta(bg(x, y))$$
$$\alpha(bg(x, y)) = bg(x, y)0.0001 + 0.115 \tag{3}$$
$$\beta(bg(x, y)) = \lambda - bg(x, y)0.01$$

$$f_2(bg(x,y)) = \begin{cases} T_0(1 - (bg(x,y)/127)^{\frac{1}{2}}) + 3 \\ \qquad\qquad \text{if } bg(x,y) \leq 127 \\ \gamma(bg(x,y) - 127) + 3 \\ \qquad\qquad \text{if } bg(x,y) > 127 \end{cases} \qquad (4)$$

Gradient value was determined by the change of background brightness. Hsien et al. use four 5x5 gradient filter, respectively to calculate the gradient value of the four directions, and then take the maximum as computing result, $\alpha(bg(x, y))$ and $\beta(bg(x, y))$ are the slope and intercept of the equation of the background brightness. T_o, γ and λ are the experimental data obtained from the experiment which premise the distance from experimenters and observed image is six times of a image, they get 17, 3/128 and 1/2, but all these values will increase when we increase the viewing distance.

The above formula is the practice to obtain the JND value for the grayscale image, while Chou and Liu proposed a method to calculate the JND for color image. They get the JND value by calculating with CIE Lab color space, it is a absolute color space, that means the selected color will present the same in all devices, so we have to convert RGB color space to CIE Lab color space, but before the step, it is essential to get the JND map of Y plane from YCbCr color space and use the JND map as the benchmark for CbCr, then we can calculate the JND value in the three-dimensional space.

3   Proposed Method

This section describes the watermark embedding algorithm that QR Code can be read although as a watermark (the algorithm architecture shows in Fig.1). The first calculate the JND map with the JND formula for color image and get the contour of the image by using Sobel operator. The contour will assist to find the position where QR Code will embed. If the area contains a portion of apparent contour then the QR Code is easy to be detected visually. So it is important to embed QR Code with the condition that embed QR Code in the area where exist no apparent contour. When scan the image with the mask and find the area where the sum of JND value is highest means the area can accommodates maximum variation. If the position to embed QR Code was found, then using the computers and mobile devices for testing data. This data is the maximum value changed by QR Code and QR Code still can be detected, after embedding the QR Code into the image. Using the image as a carrier and adding the invisible watermark by frequency domain techniques, the watermark can verify the image has been tampered with or not.

Step1 Computing JND Map
In order to find the area suitable to embed QR Code, first we must obtain the JND value of the image. It was known that the different devices may have different corresponding color in RGB and YCbCr color space, which are not an absolute color space. The CIE Lab is an absolute color space. So it is essential to convert color space from

RGB to CIE Lab. First get JND value of the Y plane in YCbCr color space, this value is used to be a benchmark for calculating the JND value of Cb and Cr planes. Converting the color space of benchmark to CIE Lab, new benchmark was used to calculate the JND value, and then convert the color space from CIE Lab to YcbCr. We will obtain the JND map of Y. Cb and Cr planes, again.

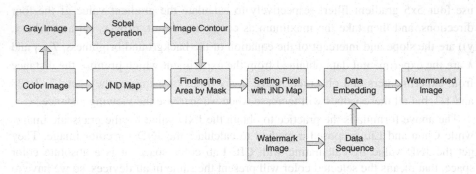

Fig. 1. Algorithm flow chart

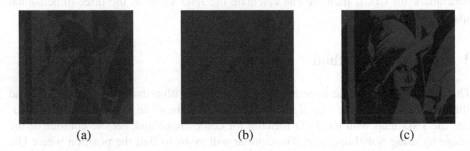

(a)　　　　　　　　　　　(b)　　　　　　　　　　　(c)

Fig. 2. JND Map of Y, Cb and Cr planes: (a) is Y, (b) is Cb, (c) is Cr

Step2 Contour Detection

The edge of image usually has high contrast. So it could be identified by visual easily, in order to exclude the area where is strong edges, this study use Sobel edge detection to find the edges, calculated to strengthen the surveillance of the edge of the x-component and y-component respectively. Then superimposed two images and obtained an apparent contour image. For the different characteristics of the image must set the appropriate threshold in order to remove the required edge. For example, the threshold setting for Lena image is 70, by this value getting a required contour map.

Step3 Find the Most Suitable Region

The effect of the condition that embed QR Code in the area does not cover the edge is greater than the condition that change of JND value. First scan the contour map with mask to find the area meet the conditions that is the area contains least contour pixels.

Then calculate the sum of JND value at JND map in the same position and find the area has maximum sum of JND value. When finding the area where conform the two conditions, it will be determined as the location of the QR Code embedded.

Step4 Adjusting the JND Value

JND value is based on the observation of the experimental by human eyes, but the decoding of QR Code can not rely on human eyes. It needs some devices for decoding.

The value must be determined and adjust the value with the basis of JND. So that it can be ensured the changes are still keeping in minimum, by using a value device the QR Codes can be read clearly and make minimal impact to the image. Fig.3 shows the measurement test by handle-device HTC Desire, and test two size for QR Code.

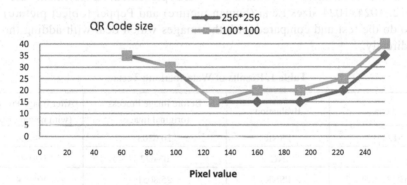

Fig. 3. The required pixel values to produce contrast imaging

Step5 Watermark Embedding

To embed a watermark into the image is for the purpose of enhancing the security of image. Using a binary image as the watermark and after the processing with Torus Automorphism [11] and [12], the binary image will become meaningless information, this way prevent the destruction to watermark and also help verify the image integrity

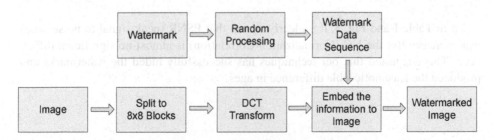

Fig. 4. Watermark embedding flowchart

Operating procedures:

1. Image pre-processing with Torus Automorphism.
2. Transform the binary information into the sequence.
3. Split the carrier image to 8x8 blocks and deal with DCT.
4. Embed watermark sequence into image.
5. Take reverse DCT for image and get the watermarked image.

4   Experiment and Result

The experimental results of implement the algorithm embedded in the QR Code image and test if the code can be read or not, and test the ability to detect tampering and the robustness of watermark were presented. Experiment use two 512x512, 1024x1024 sizes Lena (human picture) and Pepper (object picture) images to do the test and compared with the images which deal with adding the QR Code directly.

Table 1. Results of Watermarking Tests

Lena	Evaluation	Before image Process (original image)	After image Process (with new process)
512x512	PSNR	20.6203	34.89
—	NC	0.9672	0.9897
1024x1024	PSNR	25.8051	40.0206
—	NC	0.9883	0.99166
Pepper	—	—	—
512x512	PSNR	19.8894	30.9393
—	NC	0.9576	0.9874
1024x1024	PSNR	26.5551	40.6212
—	NC	0.9914	0.9912

From Table 1 and Fig. 5, it is clearly shown that PSNR (peak signal to noise ratio) was increase. But the NC (Normalization Correlation) is almost no significant difference. This suggested that our techniques has successfully hided the watermarks and produced the least noticeable difference in ages.

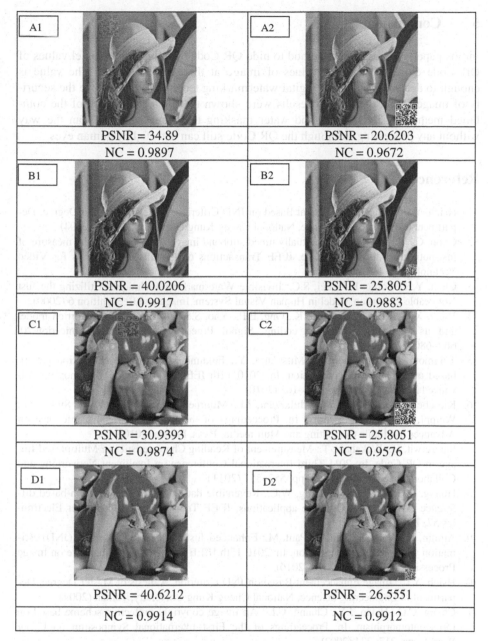

A1 PSNR = 34.89	**A2** PSNR = 20.6203
NC = 0.9897	NC = 0.9672
B1 PSNR = 40.0206	**B2** PSNR = 25.8051
NC = 0.9917	NC = 0.9883
C1 PSNR = 30.9393	**C2** PSNR = 25.8051
NC = 0.9874	NC = 0.9576
D1 PSNR = 40.6212	**D2** PSNR = 26.5551
NC = 0.9914	NC = 0.9912

Fig. 5. Results of embedding QR Code into the image A1, B1, C1, D1 images after processed and A2, B2, C2, D2 before processing, respectively

5   Conclusion

In this paper we presented a method to hide QR Code by adjusting the pixel values of QR Code close to the pixel values of image at the embedded area. The value is enough to detected and add in digital watermarking techniques to improve the security of image. The experimental results were shown that the performance of the combined method of JND model and water masking technique is better than the way without any processing. Although the QR Code still can detected by human eyes

References

1. Hsich, Y.C.: Image Enhancement Based on JND Criterion. A thesis of Master Degree, Department of Engineering Science, National Cheng Kung University, Tainan (2004)
2. Chou, C.H., Li, Y.C.: A perceptually tuned subband image coder based on the measure of just-noticeable-distortion profile. IEEE Transactions on Circuits and Systems for Video Technology 5, 467–476 (1995)
3. Chen, Y.H., Torng, R.F., Pei, S.C.: Invisible Watermarking Techniques Utilizing the Just Noticeable Distortion Model in Human Visual System. Image and Recognition 6 (2000)
4. Yang, X.K., Ling, W.S., Lu, Z.K., Ong, E.P., Yao, S.S.: Just noticeable distortion model and its applications in video coding. Signal Processing: Image Communication 20, 662–680 (2005)
5. Takimoto, H., Yoshimori, S., Mitsukura, Y., Fukumi, M.: Invisible calibration pattern based on human visual perception. In: 2010 11th IEEE International Workshop on Advanced Motion Control, pp. 159–163 (2010)
6. Kieseberg, P., Leithner, M., Mulazzani, M., Munroe, L., Schrittwieser, S., Sinha, M., Weippl, E.: QR code security. In: Proceedings of the 8th International Conference on Advances in Mobile Computing and Multimedia, Paris, France (2010)
7. Samretwit, D., Wakahara, T.: Measurement of Reading Characteristics of Multiplexed Image in QR Code. In: 2011 Third International Conference on Intelligent Networking and Collaborative Systems, INCoS, pp. 552–557 (2011)
8. Huang, H.C., Chang, F.C., Fang, W.C.: Reversible data hiding with histogram-based difference expansion for QR code applications. IEEE Transactions on Consumer Electronics 57, 779–787 (2011)
9. Anmin, L., Weisi, L., Fan, Z., Paul, M.: Enhanced Just Noticeable Difference (JND) estimation with image decomposition. In: 2010 17th IEEE International Conference on Image Processing, ICIP, pp. 317–320 (2010)
10. Hsieh, Y.C.: Image Enhancement Based on JND Criterion. A thesis of Master Degree, Department of Engineering Science, National Cheng Kung University, Tainan (2004)
11. Chang, C.C., Hsiao, J.Y., Chiang, C.L.: An image copyright protection scheme based on torus automorphism. In: Proceedings of the First International Symposium on Cyber Worlds, pp. 217–224 (2002)
12. Chen, Y.S.: Application of Digital Watermarking to Real-Time Surveillance System. A thesis of Master Degree, Department of Information Management, Yuan Ze University, Chungli (2006)

Multi-dimensional and Multi-level Histogram-Shifting-Imitated Reversible Data Hiding Scheme[*]

Zhi-Hui Wang[1], Chin-Chen Chang[2,3,**], Ming-Li Li[1], and Shi-Yu Cui[1]

[1] School of Software, Dalian University of Technology, Dalian, China
wangzhihui1017@gmail.com, 513657245@qq.com
[2] Department of Information Engineering and Computer Science,
Feng Chia University, Taichung, Taiwan
[3] Department of Computer Science and Information Engineering, Asia University,
Taichung, Taiwan
alan3c@gmail.com

Abstract. Wang et al. proposed a histogram-shifting-imitated reversible data hiding scheme. Because the pixel value after transformation for each peak point is limited to a fixed range, the visual quality of the stego-image is very good and is independent of the volume of the embedded data. In this paper, we have proposed an extra information-free multi-dimensional multi-layer data embedding scheme, which improves two aspects of Wang et al.'s scheme. First, our scheme expands the hiding capacity by extending the frame of Wang et al.'s scheme from a one-dimensional one-layer scheme to a multi-dimensional multi-layer scheme. Second, our scheme treats the location map as a part of the data to be embedded into the image rather than treating it as extra information to be transmitted to the receiver through the secure channel. The experimental results demonstrated that the proposed scheme can achieve a higher hiding capacity and a lower distortion than Wang et al.'s scheme without transmitting any extra information.

Keywords: information hiding, steganography, run-length encoding, histogram shifting.

1   Introduction

Image steganography (also called image information hiding) is a novel technique that can be used to transmit information securely by embedding information into digital images. The embedding procedure is accomplished by changing the content of the digital images, and one important requirement of information hiding is to make the changes undetectable by the human eye so that it would be difficult for attackers to be aware of the existence of the embedded message. In an information hiding scheme,

[*] This work was supported by the National Natural Science Foundation of China (No. 61201385).
[**] Correspondence author.

J.-S. Pan et al. (Eds.): *Advances in Intelligent Systems &Applications*, SIST 21, pp. 149–158.
DOI: 10.1007/978-3-642-35473-1_16   © Springer-Verlag Berlin Heidelberg 2013

the image in which the information is to be embedded is called the cover image; after the information has been embedded, the image is called the stego image. The reversible information hiding method is an information hiding method in which the cover image can be restored losslessly after the embedded message has been extracted. The reversible information hiding method is very useful for applications that cannot tolerate any distortion of the cover image at the receiver end. For example, assume that the cover image is an image of a medical X-ray that has the patient's personal information embedded in it. It is important to be able to recover the cover image losslessly for the doctor's use in diagnosising the disease. Thus, in this paper, our focus was on reversible information hiding techniques. There are three types of reversible information hiding techniques, i.e., 1) the spatial domain reversible data hiding technique, 2) the transformed domain reversible data hiding technique, and 3) the compression domain reversible data hiding technique. Our proposed scheme is based on the histogram-shifting technique (HS-based technique), which belongs to the spatial domain reversible data-hiding technique.

The histogram-shifting technique was proposed by Ni et al. [1], and it uses a histogram of the cover image to embed the message into the pixels with the peak point value and decrease degradation by shifting a segment of the histogram slightly. The HS-based technique has a short, but significant, development history. In 2007, Fallahpour and Sedaaghi used the peak point and zero pixel point pairs of the segments of the cover image to embed the secret message rather than using the entire image, as was done in Ni et al.'s scheme [2]. Although this method improved the hiding capacity by utilizing more peak points, the image quality was diminished due to the modification of more pixels. In 2008, Lin et al. [3] presented a data-hiding scheme based on the difference image, which is generated by the difference between every two adjacent pixels in an image. As a result, the hiding capacity can be improved through the property that it is significantly probable that adjacent pixels have the same or similar values, thus the peak point value in the difference image is approximately zero, and its appearance number is much greater than the corresponding number in the cover image. In 2009, Tai et al. proposed a method that used a binary tree to pre-determine multi-peak points to hide a secret message [4]. They also used the difference image to improve the hiding capacity. Tsai et al. presented another data hiding scheme [5] that applied the modification of the prediction error to acquire larger data hiding capacity while embedding information, because the height of the peak point in the prediction error histogram is higher than the peak point in the histogram of the image. In 2010, Hong et al. proposed an enhanced version of Tsai et al.'s scheme [6]. In order to increase the prediction accuracy and data hiding capacity, they used the orthogonal projection method to evaluate the optimal weights of a linear predictor. In 2011, Luo et al. further enhanced the HS-based, reversible data hiding technique by using prediction methods in which the histogram was formed by block differences [7]. Both the block differences and their integer medians were used to restore the cover image. In 2012, Wang et al. proposed a histogram-shifting-imitated reversible data hiding scheme [8]. They separated the range of the pixel values into several segments and utilized the peak point in each segment to hide information. The advantage of Wang et al.'s scheme is high hiding capacity with limited distortion, which means the distortion is limited to a small interval, since the modification of the peak point was limited to each segment. In this paper, we further improved Wang et al.'s scheme in the following two aspects. The first improvement was expanding the hiding capacity via extending the frame of Wang et al.'s scheme from a one-dimensional one-layer scheme to a multi-dimensional multi-layer scheme. Since the proposed scheme generates new mapping between the image and points in multi-dimensional space, the embedding capacity can be improved by increasing the number of points in the subspaces. The second improvement resulted from treating the location map as a part of the data to be embedded into the image, rather than treating it as extra information to be transmitted to the receiver via a secure channel. The proposed scheme hides the location map in front of the cover image by using the modified run-length encoding (RLE) method and the least significant bit (LSB) substitution method. The receiver could recover the location map losslessly without using any extra information by utilizing the invariability of the most significant bits while embedding data.

The rest of this paper is organized as follows. In Section 2, some previous work related to our proposed scheme is introduced. The proposed scheme is illustrated in Section 3. In Section 4, the experimental results are shown to verify the validity of the proposed scheme.

2   Related Work

This section briefly introduces two techniques that are relevant to our proposed scheme. The first technique is run-length encoding method, which could be used to process the location map; the second technique is a data hiding method proposed by Wang et al., which was used to process the data embedding procedure, secret extraction and image recovery.

2.1   Run-Length Encoding Method

The idea of the run-length encoding (RLE) method is substituting continuous repeated symbols with the repeated times of the symbol and the symbol itself. As a result, the stream obtained after using the RLE compression method is composed by several segments and each segment of the stream consists of three parts, i.e., indicator number, count number, and token. Indicator number indicates whether repeated times of one symbol is greater than 1. If the repeated times of one symbol is greater than 1, indicator number is set to 1; otherwise, it is set to 0. Count number shows the number of times that one symbol is repeated. Token means repeated symbol 0/1, which can be coded by using 1 bit. Assume that three bits are used to express the number of times the symbol is repeated and assume that the count number is larger than 1. Then, the form of RLE is: 1‖XXX ‖ token; otherwise, if the count number equals 1, the form of RLE is: 0 ‖ token. For example, assume that the data stream is to be encoded by RLE is 1111 1111 1000 and that three bits are used to condense repeated symbols. Then the compressed result is 1 111 1 1 010 1 1 011 0.

2.2   Wang et al.'s Data Hiding Scheme

Wang et al.'s method uses the peak points of the segments' histograms rather than the peak point of the entire image's histogram to hide the message. It uses a location map to mark whether a pixel is embeddable or not, in which bit "1" represents the segment-peak pixel, which is embeddable except for the first one that appears in each segment, and bit "0" indicates the unembeddable pixels. We briefly describe Wang et al.'s scheme in the following subsections.

The Embedding Phase

Assume a secret message SM is to be embedded into a grayscale cover image, where the pixel value is between 0 and 255. The detailed presentation of the embedding procedure is provided below.

First, divide the pixel value set, $PV = \{0, 1, ..., 255\}$, of the grayscale cover image I into mutually-exclusive, equal-sized pixel segments. Assume the size of a segment is 2^k, where $0 \le k \le 7$. Second, identify the pixel value that occurs most often in each segment, and call it a segment peak. Third, process the cover image I in a zig-zag scanning order and extract the first k bits data S_d from the secret message SM. To increase the security of the system, Wang et al. used a private key $Key^{(i)}$ to generate a mapping list between cover pixel values in the ith segment and 2^k k-bit binary codes. Then, hide S_d in an embeddable pixel, which is one of the peak-point pixels except for the first one that appeared, by transforming the peak-point pixel to another pixel in the same segment according to the mapping list. Fourth, mark an indicator bit at the location map to record whether the pixel that is currently being processed is the peak-point pixel. Afterwards, continue the third step and the fourth step until all secret messages in SM are embedded or the entire cover image has been processed. Then, output the stego image I'. Next, use efficient and lossless compression method, such as the JBIG technique, to compress the location map. The compressed code and $Key^{(i)}$ are considered as extra information that can be sent to the receiver by using a secure channel.

The Secret Extraction and Image Recovery Phase

The secret extraction and image recovery procedure of Wang et al.'s scheme decompresses the location map first and then utilizes it to recover the original pixels in each segment losslessly and completely extract the embedded message. This process is described in detail below.

First, retrieve $Key^{(i)}$ and decompress the location map L, which is the combination of sub-location maps L_i for t segments, where $0 \le i \le t$ and $t = 256/2^k$. Second, scan location map L_i to find the first encountered bit "1", which can be used to recover the peak value for the ith segment. Third, check other locations of bit "1" of the decompressed sub-location map L_i and extract the pixel values in every bit "1" location in each segment, except the first one. According to the mapping list generated by $Key^{(i)}$, the message embedded in the extracted pixel values could be extracted correctly, and the original pixel values can be recovered by substituting the peak point pixel value for the extracted pixel values in the same segment. After processing all the segments, the stego-image can be restored completely to the original image, and the secret message SM can be extracted completely.

3 Proposed Scheme

Wang et al.'s scheme divided the pixel-value space into several segments and embedded the message by modifying the segment's peak point pixel to another pixel value in the same segment, so it is a one-dimensional method. If the number of dimensions of Wang et al.'s scheme is increased to two, i.e., every two pixels are considered as a point in two-dimensional space, it is obvious that there is a chance to improve the embedding capacity while maintaining the same distortion. For example, if the segment size of Wang et al.'s scheme is 8, every peak point represented by one pixel, except the first one that appears, can be used to embed 3 bits. However, in the two-dimensional scenario, every peak point represented by two pixels, except for the first that appears, can be used to embed 6 bits. Even though the height of the peak point may decrease by adding one dimension, the hiding capacity of Wang et al.'s scheme still can be improved when the height of the new peak point is greater than the half of the old peak piont. In addition, the hiding capacity can be improved further without creating a large distortion of the stego image by processing multi-layer embedding, since Wang et al.'s scheme is a reversible scheme and limits the extent to which the pixel value can be modified. In this section, we describe the data embedding, data extraction, and cover image recovery procedures for the proposed, reversible, n-dimensional and N-layer data hiding method.

3.1 The Data Embedding Procedure

The detailed description of the data embedding procedure is as follows:

Input: A cover image I_1 and a series of messages SM.

Step 1: Compress the first most significant bit of every pixel in the cover image by using the run-length encoding technique. The compressed data are $cmsb$, and the length of $cmsb$ is len_cmsb, which is used to extract the location map. Reserve the first len_cmsb/k pixels in the cover image I_1 for embedding the location map, where k is the number of bits that are substituted in each pixel ($1 \leq k \leq 8$) by using the k bits LSB substitution method. The pixel-reserved image is denoted as I'.

Step 2: Divide the n-dimensional pixel value space P into M equal-sized sub-spaces, and make sure that the pixel values 127 and 128 are separated into different sub-spaces, so that the content of the first most significant bit of every pixel in the stego image is same as the content of the first most significant bit of every pixel in the cover image. The sub-space size 2^{e_n} is determined by segment size D_i in every dimension, i.e., $2^{e_n} = D_1 \times D_2 \times ...D_i \times ...\times D_n$, where D_i is the ith dimensional segment size, $i \in [1, n]$.

Step 3: A point in the sub-space is described by a vector V_n which is formed by non-overlapping and adjacent n pixel values in I'. Scan every vector V_n in cover image I' in zig-zag order, and find the peak value vector in every sub-space, which is the point that appears most often in each sub-space, and then denote them as $P_n(v_{max})$, $P_n(v_{max}) = \{P_i(v_{max}) \mid 1 \leq i \leq M\}$.

Step 4: Scan the image I' again in zig-zag order from the $(len\_cmsb/k+1)^{th}$ pixel. Assume that the current processing vector is C_{Vn}, and, if it belongs to $P_n(v_{max})$, put '1' into the location map. If C_{Vn} does not belong to $P_n(v_{max})$, put '0' into the location map. After processing the entire cover image I', get the location map of I'. Afterwards, compress the location map by using JBIG, and get the length of the compressed stream len_Jloc. In order to extract the location map correctly, we modified the run-length encoding method to further compress the compressed JBIG stream. We changed the one-bit indicator bit to 2 bits, so we can use "00" and "01" to indicate the continuous mode and discontinuous mode, respectively. Also, we can use "10" as the stop sign of the run-length encoding procedure. The compressed result of run-length encoding is $crlc$, the length of which is len_crlc. Next, we compare len_crlc with $(len\_cmsb-1) \times k$, and, if $len\_crlc \leq (len\_cmsb-1) \times k$, we can use this cover image to embed the secret data. Otherwise, the current image cannot be used as the cover image, since the location map cannot be embedded and extracted correctly in this case. Assume that $len\_crlc \leq (len\_cmsb-1) \times k$, use the 8-bit LSB substitution method to embed k into the first pixel in the cover image and embed $crlc$ into the next $(len\_cmsb-1)/k$ pixels in the cover image by using the k-bit LSB substitution method. Attach the first eight substituted bits of the cover image to the front of the information SM to be embedded to get the new message S. After that, add the other substituted bits on the front of S to get a new message S'. $S' = \{s_c \mid s_c \in \{0,1\}$ for $c = 1, 2, ..., j \times l\}$, where $j \times l$ is the length of S'.

Step 5: Transform the binary stream S' into decimal values, changing each value from j-bit data, so the range of the decimal values is $[0, 2^j-1]$. Transform the decimal information into 2^{e_n}-system $S^{(j)}$, where 2^{e_n} is the size of the sub-space. The binary stream S' is transformed as $S^{(j)}$ and indicated as: $S^{(j)} = \{s_d \mid s_d \in \{0,1,2,...,2^j-1\}$ for $d = 1, 2, ..., l\}$.

Step 6: From the $(len\_cmsb/k+1)^{th}$ pixel, extract the j-bit data s_d from $S^{(j)}$ and embed s_d into an embeddable pixel according to the mapping list generated by the private key.

Step 7: Continue Step 6 until all secret data have been embedded or the scanning process of the cover image is completed.

Step 8 : Output stego-image I_1''.

Step 9: Treat stego-image I_1'' as the cover image I_2 in the embedding procedure of the second layer, and, since the proposed method is reversible, we can embed data into the cover image again. Following this rule, we can embed data into the Nth layer of the cover image by using the $(N-1)^{th}$ layer's output stego-image. In each layer, only the last k bits of the first len_crlc/k pixels are substituted bits, and the embeddable pixels can only be changed in fixed sub-space, so the quality of the stego image is not decreased noticeably after data have been embedded in the N layers.

3.2   The Secret Extraction and Cover Image Recovery Procedure

The detailed description of the secret extraction and cover image recovery procedure is provided below:

Step1: Compress the first most significant bit of every pixel in the stego image to get compressed data $Jmsb'$, and, because two pixels 127 and 128 are separated into two different sub-spaces, we can ensure that the first most significant bit of every pixel does not change during the data embedding procedure described above. Assume that the length of $Jmsb'$ is len_Jmsb'.

Step 2: Decode the first 8 bits of the image to get k, which is the number of bits having been substituted in each pixel. Extract every k least significant bits from the next $(len\_Jmsb'-1)/k$ pixels and decode the code stream using run-length decoding. If the decoding process shows that the current indicator is "10", then stop decoding and keep the processed length of the stego image as len_indi, and the decoded code stream is the JBIG compressed result of the location map. Afterwards, further decompress the JBIG compressed result to get the location map.

Step 3: By using the location map, extract the embedded information S in the stego image from the len_Jmsb'/k +1 pixel by using Wang et al.'s data extracting procedure. Divide S into two parts. The length of the first part is $(len\_indi-1) \times k+8$, which is used to recover the original first pixel and the least significant bits of other few pixels in the stego image of $(N-1)^{th}$ layer, which were used to embed the location map. The second part is the secret information denoted as SM_N in the Nth layer.

Step 4: Continue Step 3 to get the other parts of the secret information embedded in layer N-1, layer N-2, ..., until layer 1 and attach them to the front of SM_N. Finally, we can get all of the secret data and the original cover image.

The pseudo codes of the secret extraction and cover image recovery procedure of one layer are shown below:

4   Experimental Results

In this section, the experimental results of the proposed scheme are shown to verify the validity of our proposed scheme. All of the experiments were performed on an Intel Core i5 machine at 2.40 GHz with 4GB of main memory. Programs of experiments were implemented using MATLAB software. Fig. 1 shows the cover images used in the experiment, which are all 512×512 grayscale images.

In the image steganography field, peak signal-to-noise ratio ($PSNR$) [8] is used to evaluate the quality of stego-images. Embedding capacity illustrates the number of secret bits that can be hidden in a cover image.

In the proposed scheme, the number of dimensions, the segment size of each dimension, the sub-space size, and the number of embedding layers influence the experimental results, including $PSNR$ and embedding capacity.

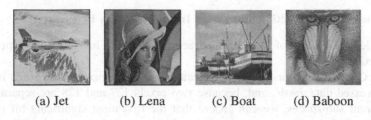

(a) Jet   (b) Lena   (c) Boat   (d) Baboon

Fig. 1. Cover images

Table 1. Influence of dimension number for a sub-space size of 4

Dimension number (segment size) / Images		2			3	
		(1, 4)	(2, 2)	(4, 1)	(1, 4, 1)	(1, 2, 2)
Jet	capacity (bpp)	0.35	0.38	0.39	0.22	0.20
	PSNR	58.48	54.10	54.19	56.80	61.21
Lena	capacity (bpp)	0.32	0.34	0.35	0.21	0.18
	PSNR	59.3	55.05	54.93	57.65	61.96
Boat	capacity (bpp)	0.31	0.33	0.33	0.16	0.15
	PSNR	59.18	54.7	54.66	58.24	62.48
Baboon	capacity (bpp)	0.30	0.31	0.31	0.09	0.09
	PSNR	59.67	56.15	56.02	61.49	64.93

Table 1 shows the influence of setting different dimensions on the proposed scheme by setting the subspace-size as 4. The second row of Table 1 indicates the different segment size of each dimension, e.g., (1, 4) means that the segment size of the first dimension is 1, and the size of the second dimension is 4. We conducted experiments with two dimensional and three dimensional space here, respectively. As Table 1 shows, as the number of dimensions increased, embedding capacity decreased and *PSNR* increased. This occurs because, as the number of dimensions used is increased, the modificaitons of the pixel values become smaller if we set constant subspace size. However, the number of embeddable pixels decreased when more pixels are used to form a vector, which represents a point in a sub-space, so the embedding capacity desreased when the number of dimensions increased.

The influences of the segment size of each dimension and the sub-space size are shown in Table 2. It can be observed that the embedding capacity increased for Jet, Lena and Boat as the sub-space size increased. For the Baboon image, the embedding capacity initially increased, but it decreased later. This occurred because Baboon is more complicate than the other images. Obviously, for the same sub-space size, the segment size does not have a sigificant influence on the performance of the proposed scheme.

Fig. 2 shows the influence of the number of embedding layers and the comparison between the proposed scheme and other recently published HS-based information hiding methods, such as Kim et al.'s method [9], Ni et al.'s method [1], Tai et al's method [4], Thodi and Rodriguez's method [10], Tian's method [11], and Tsai et al.'s method [5]. Here, the experimental results of the proposed multi-dimensional multi-layer scheme shown in Fig. 2 are obtained by setting the dimension number as 2, the sub-space size as 4, the segment size as 2 for both dimensions and the number of the embedding layers tested from 1 to 10. It can be observed that the embedding capacity was increased dramatically, while the *PSNR* decreased only minimally as the number of embedding layers increased. This occurred because an increase in the number of embedding layers does not affect the range of the modification for each embeddable pixel. Comparing with other schemes, Fig. 2 shows that our proposed scheme achieved excellent quality, exceeding all the other schemes, when the embedding capacity was limited by choosing the multi-dimensional, one-layer mode. Here, three-dimensional one-layer mode by setting sub-space sizes as 2, 4, 8, 16 and 32 are tested to get the experimental results shown in Fig. 2. Also, Fig. 2 shows that our proposed scheme, using the multi-dimensional multi-layer mode achieved a greater embedding capacity and better quality stego images than other methods.

Table 2. Influence of segment size for the two-dimensional one-layer proposed scheme

Subspace size		2		4			8			
(Segment size) Images		(1, 2)	(2, 1)	(1, 4)	(2, 2)	(4, 1)	(1, 8)	(2, 4)	(4, 2)	(8, 1)
Jet	capacity (bpp)	0.29	0.28	0.35	0.38	0.39	0.42	0.34	0.34	0.42
	PSNR	59.51	59.51	58.48	54.1	54.19	48.76	55.35	55.39	48.86
Lena	capacity (bpp)	0.27	0.27	0.32	0.34	0.35	0.36	0.30	0.30	0.36
	PSNR	60.03	60.13	59.3	55.05	54.93	49.94	56.48	56.52	49.92
Boat	capacity (bpp)	0.26	0.26	0.31	0.33	0.33	0.34	0.29	0.29	0.34
	PSNR	59.98	59.84	59.18	54.7	54.66	49.96	56.12	56.24	49.89
Baboon	capacity (bpp)	0.26	0.26	0.30	0.31	0.31	0.28	0.26	0.26	0.28
	PSNR	60.35	60.26	59.67	56.15	56.02	51.46	57.59	57.55	51.42

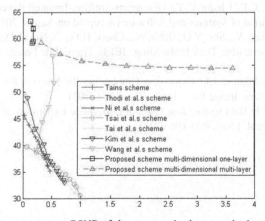

Fig. 2. Embedding rate versus *PSNR* of the proposed scheme and other schemes for Lena

5   Conclusions

We proposed a multi-dimensional multi-layer extra information free data embedding scheme with high visual quality stego image and large embedding capacity. One advantage of the proposed scheme is that it can hide the location map in the front of the cover image without using any other auxiliary information, which is innovative and effective for dealing with the problem of transmitting the location map to the receiver end. The experimental results proved that another advantage of the proposed scheme is that it increases embedding capacity and decreases the distortion in the stego image compared to other methods. In the future, we will focus on improving the performance of the proposed scheme by modifying the prediction errors in the multi-dimensional multi-layer space of digital images.

References

1. Ni, Z., Shi, Y.Q., Ansari, N., Su, W.: Reversible Data Hiding. IEEE Trans. Circuits Syst. Video Technol. 16(3), 354–362 (2006)
2. Fallahpour, M., Sedaaghi, M.H.: High Capacity Lossless Data Hiding based on Histogram Modification. IEICE Electron. Express 4(7), 205–210 (2007)
3. Lin, C.C., Tai, W.L., Chang, C.C.: Multilevel Reversible Data Hiding based on Histogram Modification of Difference Images. Pattern Recognit. 41(12), 3582–3591 (2008)
4. Tai, W.L., Yeh, C.M., Chang, C.C.: Reversible Data Hiding based on Histogram Modification of Pixel Differences. IEEE Trans. Circuits Syst. Video Technol. 19(6), 906–910 (2009)
5. Tsai, P.Y., Hu, Y.C., Yeh, H.L.: Reversible Image Hiding Scheme Using Predictive Coding and Histogram Shifting. Signal Process. 89(6), 1129–1143 (2009)
6. Hong, W., Chen, T.S., Chang, Y.P., Shiu, C.W.: A High Capacity Reversible Data Hiding Scheme Using Orthogonal Projection and Prediction Error Modification. Signal Process. 90(11), 2911–2922 (2010)
7. Luo, H., Yu, F.X., Chen, H., Huang, Z.L., Li, H., Wang, P.H.: Reversible Data Hiding based on Block Median Preservation. Inf. Sci. 181(2), 308–328 (2011)
8. Wang, Z.H., Lee, C.F., Chang, C.Y.: Histogram-Shifting-Imitated Reversible Data Hiding. To Appear in Journal of Systems and Software (accepted on August 2012)
9. Kim, H.J., Sachnev, V., Shi, Y.Q., Jeho, N., Choo, H.G.: A Novel Difference Expansion Transform for Reversible Data Embedding. IEEE Trans. Inf. Foren. Sec. 3(3), 456–465 (2008)
10. Thodi, D.M., Rodríguez, J.J.: Expansion Embedding Techniques for Reversible Watermarking. IEEE Trans. Image Proc. 16(3), 721–730 (2007)
11. Tian, J.: Reversible Data Embedding Using a Difference Expansion. IEEE Trans. Circuits Syst. Video Technol. 13(8), 890–896 (2003)

A Threshold Secret Image Sharing with Essential Shadow Images

Ching-Nung Yang and Chih-Cheng Wu

Department of Computer Science and Information Engineering,
National Dong Hwa University, Taiwan
cnyang@mail.ndhu.edu.tw

Abstract. In (k, n) threshold secret image sharing (TSIS), a secret image is shared into n shadow images. Any k or more shadow images can be collaborated together to reconstruct the secret image, while less than k shadow images cannot reveal any secret. All shadow images have the same importance in reconstruction process. In some applications, some participants are accorded special privileges due to their status or importance. In this paper, we consider the (t, s, k, n) essential TSIS (ETSIS) scheme. All n shadows in the proposed (t, s, k, n)-ETSIS scheme are classified into s essential shadows and $(n-s)$ non-essential shadows. In reconstruction, we needs k shadow images with at least t essential shadow images.

Keywords: Threshold secret sharing, Threshold secret image sharing, Essential shadow image.

1   Introduction

Threshold secret sharing (TSS) is one of main research topics in modern cryptography and has been studied extensively in the literatures. In 1979, Blakley [1] and Shamir [2] independently proposed TSS solutions for safeguarding cryptographic keys. In Shamir's (k, n)-TSS scheme, the secret value is embedded into the constant coefficient of a random $(k-1)$-degree polynomial. Based on Shamir's scheme, (k, n) threshold secret image sharing (TSIS) schemes, where $k \le n$, were accordingly proposed. In a (k, n)-TSIS scheme, a secret image is shared into n shadow images (referred to as shadows) in such a way that any k shadows can be used to reconstruct the secret image exactly, but use of any number of shadows less than k will not provide any information about the secret image.

Thien and Lin [3] used all coefficients of the polynomial for embedding secret pixels, and reduced shadow size $1/k$ times to the secret image. Since noise-like shadow images are suspected, it is desirable to design a (k, n)-TSIS scheme using steganography so that shadows are meaningful [4-7]. There is also a novel scalable TSIS scheme [8-11], the information amount of reconstructed image is proportional to the number of shadows engaged in decryption.

In previous (k, n)-TSIS schemes, each participant plays the same role in the revealing process. However, there are many examples that some participants are

J.-S. Pan et al. (Eds.): *Advances in Intelligent Systems & Applications*, SIST 21, pp. 159–166.
DOI: 10.1007/978-3-642-35473-1_17   © Springer-Verlag Berlin Heidelberg 2013

accorded special privileges due to their status or importance, e.g., heads of government, CEO of company, ..., etc. So, we have to give special treatments to some persons for some reasons. In this paper, we consider the (t, s, k, n) essential TSIS (ETSIS) scheme, which has s essential shadows and $(n-s)$ non-essential shadows. We need k shadows with at least t essential shadows for reconstruction. The following sections are organized as follows. In Section 2, we review the (k, n)-TSIS scheme and. Motivation is introduced in Section 3. Section 4 describes the proposed (t, s, k, n)-ETSIS scheme. Experiment is given in Section 5, and Section 6 is the conclusion.

2   The (k, n)-TSIS Scheme

In 1979, Shamir [1] introduced the (k, n)-TSS scheme to share a secret to n shares by a $(k-1)$-degree polynomial $f(x) = (a_0 + a_1 x + \cdots + a_{k-1} x^{k-1}) \mod p$, in which p is a prime number and a_0 is the secret value. The dealer randomly selects a polynomial, and generates the n shares $(i, f(i))$, $i=1, 2, \ldots, n$, which are delivered to n participants. In reconstruction, any k shares (say 1, 2, ..., k) can be used to recover the polynomial $f(x)$ by Lagrange's interpolation $f(x) = \sum_{j=1}^{k} f(j) \prod_{i=1,i \neq j}^{k} \frac{(x-i)}{(j-i)} \mod p$, and then the secret is obtained as $f(0)$. However, any $k-1$ or fewer shadows cannot get any information about the secret.

Through Shamir's (k, n)-TSS scheme, we could take every secret pixel as a_0 in a $(k-1)$-degree polynomial $f(x)$ to construct n random grayscale values on n noise-like shadows. To further reduce the share size, Thien and Lin's (k, n)-TSIS scheme [3] reduced the shadow size by using all coefficients in $f(x)$ for sharing secret pixels. We first divide a secret image into non-overlapping blocks. Every block has k pixels is represented as the $(k-1)$-degree polynomial $f(x)$. By substituting the image identification id, the value of polynomial $f(id)$ is generated to form a shadow. Because we embed k pixels each time, the shadow size is $1/k$ of the secret image.

When sharing image, the prime number p is often chosen as 251 such that the coefficients are constrained between 0 and 250 and suitable to represent a conventional 8-bit gray-scale or color images. A possible value of an 8-bit gray pixel is from 0 to 255, and the gray-scale values (>250) need to be modified to 250. Obviously, we can use Galois Field $GF(2^8)$ instead of modulus 251 rather than ordinary arithmetic to achieve a lossless scheme. For simplicity, some schemes adopt $GF(251)$, while some papers use $GF(2^8)$ to achieve a secret image with no distortion. In this work, we use $GF(2^8)$ to avoid distortion.

3   Motivation

In this paper, we consider the case that some shadows are essential. Our ETSIS scheme has not only the threshold property (i.e., a threshold value is necessary to reveal the secret) but also the essentiality (i.e., need essential shadows involved in the recovery process). All previous (k, n)-TSIS schemes do not have the essentiality so that the reconstruction may not need essential shadows.

All shadows of the conventional (k, n)-TSIS scheme have the same importance. However, in some application environments, some shadows may be more important than others. These shadows are essential and necessary for reconstruction. In this paper, we discuss a general (t, s, k, n)-ETSIS scheme, where $t \leq s \leq n$, and $t < k \leq n$. Since $t=k$ implies that we do not need non-essential shadows for reconstruction. Thus, we do not consider the case $t=k$. Obviously, our (t, s, k, n)-ETSIS scheme is reduced to a (k, n)-TSIS scheme for $t=k$ and $s=n$.

Here, we describe the reasons why we study ETSIS scheme. An application scenario using ETSIS scheme is shown below. In United Nations Security Council (UNSC), there are fifteen members consisting of five permanent members (China, France, Russia, the United Kingdom, and the United States) and ten elected non-permanent members. Under Article 27 of the UN Charter V, the decisions of UNSC on all non-procedure matters require the affirmative votes of nine members (i.e., the threshold value is $k=9$). According to the rule of "great power unanimity", even though receiving nine votes, a veto by a permanent member may prevent adoption of a proposal. This implies that all five permanent members should give affirmative votes (i.e., the number of essential votes is $t=5$). There are total fifteen members ($n=15$) including five permanent members ($s=5$). Therefore, this voting scenario has the similar threshold property and essentiality like a (5, 5, 9, 15)-ETSIS scheme.

Indeed, we may need to deliver special treatments to some persons for some reasons. There are more examples that can be implemented by applying our ETSIS scheme, when some participants are accorded special privileges due to their status or importance, e.g., heads of government, CEO of company, high-level corporate officers, major employers, ..., etc. Therefore, (t, s, k, n)-ETSIS scheme has potential applications and deserves studying.

4   The Proposed (t, s, k, n)-TSIS Scheme

The threshold of reconstructing a secret image in (k, n)-TSIS scheme is k. Let P be the set of all participants. A qualified subset of participants $Q \subseteq P$ should satisfy the threshold condition: $|Q| \geq k$, where $|Q|$ is the cardinality of Q. This condition allows any k participants for reconstructing the secret. Every participant has the same importance. Let EP and NEP be the set of essential participants and the set of non-essential participants in our (t, s, k, n)-ETSIS scheme, where $P = EP \cup NEP$. We then have $|EP|=s$ and $|NEP|=(n-s)$. Let $Q \backslash NEP$ denote the set having elements in Q but not in NEP. A qualified subset of participants Q in the proposed (t, s, k, n)-ETSIS scheme should satisfy the following two conditions.

$$\left\{ \begin{array}{ll} \text{(i) Threshold condition: } |Q| \geq k, & (1\text{-}1) \\ \text{(ii) Essentiality condition: } |Q \backslash NEP| \geq t. & (1\text{-}2) \end{array} \right.$$

Our approach is based on the derivative of $f(x)$. Here, we first describe the concept of using the derivative of polynomial. Let $f^{(t)}(x)$ be the t-th derivative of $f(x)$. Suppose that we have $(k-t)$ outputs (x_1, y_1), (x_2, y_2), ..., (x_{k-t}, y_{k-t}) of $f^{(t)}(x)$. We may reconstruct the polynomial $f^{(t)}(x)$ by using Lagrange's interpolation. By computing the integration of $f^{(t)}(x)$ t times, we can derive the polynomial $f(x)$ in the following equation, where there are t unknowns.

$$f(x) = (\overbrace{u_0 + u_1 x + \cdots + u_{t-1}x}^{t \text{ unknown coefficients}} + a_t x^t \cdots + a_{k-1}x^{k-1}) \bmod p \qquad (2)$$

Combining at least t outputs of $f(x)$ and Eq. (2), we can determine the t unknowns (u_0, u_1, ..., u_{t-1}) in $f(x)$. Finally, we can reconstruct the polynomial $f(x)$ and gain the secret pixel $f(0)=a_0$.

Fig. 1 shows the proposed (t, s, k, n)-ETSIS scheme. The formal encryption and decryption of our (t, s, k, n)-ETSIS scheme are shown in Algorithm 1 and Algorithm 2, respectively. Notations used in our algorithms are defined in Table 1.

Table 1. Notation used in the proposed (t, s, k, n)-ETSIS scheme

Notation	Description
I	The secret image with the size l/l.
$f(x)$	The $(k-1)$-degree polynomial $f(x) = (a_0 + a_1 x + \cdots + a_{k-1}x^{k-1}) \bmod p$.
$f^{(t)}(x)$	The t-th derivative of $f(x)$ with $(t-k+1)$-degree, and $f^{(t)}(x) = b_0 + b_1 x$ $+ ... + b_{k-t+1}x^{k-t+1}$, where $b_0=(a_t \times t!),..., b_{k-t+1}=(a_{k-1}\times((k-1)!/t!))$.
$E_{k,s}(\cdot)$	Encryption function of (k, s)-TSIS scheme based on $f(x)$, where the secret pixel is embedded into t coefficients $a_0, a_1, ..., a_{t-1}$.
$E_{k-t,n}(\cdot)$	Encryption function of $(k-t, n)$-TSIS scheme based on $f^{(t)}(x)$, and its reverse function is $D_{k-t,n}(\cdot)$.
I_i	The n intermediate shadows of $(k-t, n)$-TSIS scheme generated by $E_{k-t,n}(I) = I_i$, $1 \leq i \leq n$.
J_i	The s intermediate shadows of (k, k)-TSIS scheme generated by $E_{s,k}(I) = J_i$, $1 \leq i \leq s$.
O_i	The n shadows of the proposed (t, s, k, n)-ETSIS scheme, $1 \leq i \leq n$, where $O_1 - O_{n-s}$ are non-essential shadows and $O_{n-s+1} - O_n$ are essential shadows

As shown in Fig. 1, we embed the secret pixels into t coefficients $a_0, a_1, ...,$ and a_{t-1}. in $f(x)$. By using (k, s)-TSIS scheme, we generate s intermediate shadows $J_1 - J_k$ and deliver to essential participants. Then, we obtain a $(k-t+1)$-degree $f^{(t)}(x)$ from $f(x)$, and apply $(k-t, n)$-TSIS scheme based on this polynomial $f^{(t)}(x)$ to generate n intermediate shadows $I_1 - I_n$. Finally, $(n-s)$ non-essential participants have the shadow

$O_i=I_i$, $1\leq i\leq(n-s)$, and s essential participants have the shadow $O_i=(I_i\|J_{i-n+s})$, $(n-s+1)\leq i\leq n$. Theorem 1 theoretically proves that our (t, s, k, n)-ETSIS scheme satisfies the threshold property (Condition (1-1)) and the essentiality property (Condition (1-2)).

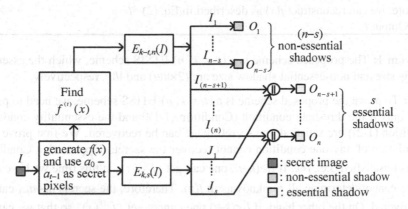

Fig. 1. Block diagram of the proposed (t, s, k, n)-ETSIS scheme

---

Algorithm 1. Encryption of the Proposed (t, s, k, n)-ETSIS Scheme

Input: a secret image I; the values of t, s, k and n.

Output: n shadows O_i, $1\leq i\leq n$.

(1) Obtain $I_i = E_{k-t,n}(I)$, $1\leq i\leq n$;

(2) Obtain $J_i = E_{k,s}(I)$, $k\leq i\leq s$;

(3) for $i=1$ to $n-s$ $O_i=I_i$; /* non-essential shadows */

(4) for $i=n-s+1$ to n $O_i=(I_i\|J_{i-n+s})$; /* essential shadows */

(5) Output n shadows $O_1, O_2, \ldots, O_n$.

Algorithm 2. Decryption of the Proposed (t, s, k, n)-ETSIS scheme

Input: l ($\geq k$) shadows including t or more essential shadows.

/* Suppose that there are l_1 non-essential shadows (say $O_1 - O_{l_1}$) and l_2 ($\geq t$)

essential shadows (say $O_{n-s+1} - O_{n-s+l_2}$), where $l_1+l_2=l$. */

Output: the secret image I.

(1) From $(O_1,\ldots,O_{l_1},O_{n-s+1}\cdots,O_{n-s+l_2})$, we have $(I_1,\ldots,I_{l_1},I_{n-s+1}\cdots,I_{n-s+l_2})$;

/* Since $I_i = O_i$ for $1\leq i\leq(n-s)$ and $I_i \subset O_i$ for $(n-s+1)\leq i\leq n$. */

(2) Obtain $D_{k-t,n}(I_1,\ldots,I_{l_1},I_{n-s+1}\cdots,I_{n-s+l_2})$;

/* note: $l \geq k$ achieves the threshold of $D_{k-t,n}(\cdot)$, but there are t unknown pixels in

$f(x)$ */

(3) From $(O_{n-s+1}\cdots,O_{n-s+l_2})$, we have $(J_{n-s+1}\cdots,J_{n-s+l_2})$;

/* Since $J_i \subset O_i$ for $(n-s+1)\leq i\leq n$. */

---

(4) Obtain all unknown pixels by using the result
$D_{k-t,n}(I_1,\ldots,I_{l_1},I_{n-s+1}\ldots,I_{n-s+l_2})$ and $(J_{n-s+1}\ldots,J_{n-s+l_2})$;

(5) The secret pixels can obtained from $f(0)=a_0$.

/* Note: we can reconstruct $f(x)$ as described in Eq. (2) */

(6) Output I.

Theorem 1: The proposed scheme is a (t, s, k, n)-ETSIS scheme, which the essential shadow size and non-essential shadow size are $(2\times|I|/t)$ and $|I|/t$, respectively.

Proof: To prove the proposed scheme is a (t, s, k, n)-ETISS scheme, we need to prove if and only if the threshold condition (Condition (1-1)0 and the essentiality condition (Condition (1-2)) are satisfied, the secret image can be recovered. We first prove that the violation of any one condition cannot recover the secret. Suppose that Condition (1-1) is unsatisfied, i.e., $l<k$. If $l_1\geq(k-t)$, one can get $f^{(t)}(x)$, at this time we have $l_2<t$, and we cannot determine all t unknowns in $f(x)$. Therefore, the secrets a_0-a_{t-1} cannot be recovered. On the other hand, if $l_1<(k-t)$ one cannot get $f^{(t)}(x)$, so that we cannot use the integration to get $f(x)$. Meantime $l_2<l<k$, so the $f(x)$ cannot be obtained from $(J_{n-s+1}\ldots,J_{n-s+l_2})$. Next, we prove that when Conditions (1-1) and (1-2) are satisfied we can reconstruct the secret. Since $l\geq k$, we can get $f^{(t)}(x)$. Also, $l_2\geq t$, so we can use $(J_{n-s+1}\ldots,J_{n-s+l_2})$ to determine all t unknowns in $f(x)$ and get the secrets a_0-a_{t-1}.

Afterwards, we determine the sizes of essential shadow and non-essential shadow. In Algorithm 1, non-essential shadows are O_1-O_{n-s} and essential shadows are $O_{n-s+1}-O_n$. Since $O_i=I_i$, $1\leq i\leq(n-s)$, so non-essential shadow sizes are $|O_i|=|I_i|=|I|/t$ (note: we embed t secret pixels each time). Essential shadows are $O_i=(I_i\|J_{i-n+s})$, $n-s+1\leq i\leq n$, and thus their shadow sizes are $|O_i|=(|I_i|+|J_{i-n+s}|)=(2\times|I|/t)$.   □

5   Experiment and Discussion

We conduct an experiment in an attempt to test the effectiveness of the proposed (t, s, k, n)-ETSIS scheme

Example 1: Test the proposed $(2, 4, 4, 8)$-ETSIS scheme by using the secret image Lena.

A secret image 300×300-pixel Lena (Fig. 2(a)) is shared by $(4, 4)$-TSIS scheme into four intermediate shadows (J_1-J_4). Then, find $f^{(2)}(x)$ of the corresponding $f(x)$ in $(4, 4)$-TSIS scheme. Generate six intermediate shadows (I_1-I_8) by $(2, 8)$-TSIS scheme based on $f^{(2)}(x)$. Then four non-essential shadows O_1-O_4 are generated by $O_1=I_1$, $O_2=I_2$, $O_3=I_3$, and $O_4=I_4$, and the other four essential shadows O_5-O_8 are generated by $O_5=(I_5\|J_1)$, $O_6=(I_6\|J_2)$, $O_7=(I_7\|J_3)$, $O_8=(I_8\|J_4)$. Finally, we have four non-essential shadows with the size 300×150 pixels (see Fig. 2(b)), and four essential shadows with the size 300×300 pixels (Fig. 3(c)).   □

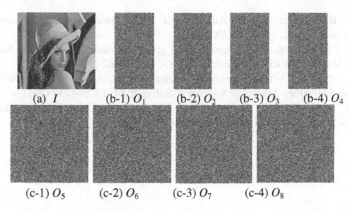

(a) I (b-1) O_1 (b-2) O_2 (b-3) O_3 (b-4) O_4

(c-1) O_5 (c-2) O_6 (c-3) O_7 (c-4) O_8

Fig. 2. The proposed (2, 4, 4, 8)-ETSIS scheme: (a) the secret image of 300×300 pixels (b) four non-essential shadows of 300×150 pixels (c) four essential shadows of 300×300 pixels

Actually, some (k, n)-TSIS schemes reduced the shadow size by using all coefficients in $f(x)$ for sharing secret pixels. For example, Thien and Lin's (k, n)-TSIS scheme [2] adopted all coefficients in $f(x)$ for embedding. A secret image is divided into non-overlapping k-pixel blocks. Because we embed k pixels each time, the shadow size is $1/k$ of the secret image. However, in the proposed (t, s, k, n)-ETSIS scheme, we only embed the secret pixels into t coefficients in $f(x)$. How to embed the secret pixels into all coefficients for reduction of shadow size requires further study.

6   Conclusion

In this work, we propose a new (t, s, k, n)-ETSIS scheme, where essential shadows are more important than non-essential shadows. A qualified subset of shadows Q should satisfy the threshold condition ($|Q| \geq k$) and the essentiality condition ($|Q \backslash NEP| \geq t$).

References

1. Blakley, G.R.: Safeguarding Cryptographic Keys. In: Proceedings of the AFIPS 1979 National Computer Conference, vol. 48, pp. 313–317. AFIPS Press (1979)
2. Shamir, A.: How to Share A Secret. Communications of the Association for Computing Machinery 22, 612–613 (1979)
3. Thien, C.C., Lin, J.C.: Secret Image Sharing. Computer & Graphics 26, 765–770 (2002)
4. Lin, C.C., Tsai, W.H.: Secret Image Sharing with Steganography and Authentication. Journal of Systems & Software 73, 405–414 (2004)
5. Yang, C.N., Chen, T.S., Yu, K.H., Wang, C.C.: Improvements of Image Sharing with Steganography and Authentication. Journal of Systems & Software 80, 1070–1076 (2007)
6. Chang, C.C., Hsieh, Y.P., Lin, C.H.: Sharing Secrets in Stego Images with Authentication. Pattern Recognition 41, 3130–3137 (2008)
7. Yang, C.N., Ouyang, J.F., Harn, L.: Steganography and Authentication in Image Sharing without Parity Bits. Optics Communications 285, 1725–1735 (2012)

8. Wang, R.Z., Shyu, S.J.: Scalable Secret Image Sharing. Signal Processing: Image Communication 22, 363–373 (2007)
9. Yang, C.N., Huang, S.M.: Constructions and Properties of k out of n Scalable Secret Image Sharing. Optics Communications 283, 1750–1762 (2010)
10. Yang, C.N., Chu, Y.Y.: A General (k, n) Scalable Secret Image Sharing Scheme with the Smooth Scalability. Journal of Systems & Software 84, 1726–1733 (2011)
11. Lin, Y.Y., Wang, R.Z.: Scalable Secret Image Sharing With Smaller Shadow Images. IEEE Signal Processing Letters 17, 316–319 (2010)

Theoretical Analysis and Realistic Implementation of Secure Servers Switching System

Yu-Hong Chen, Kuang-Tse Chen, and Lei Wang

Department of Electrical Engineering, Feng Chia University, Taiwan
tomo0502@gmail.com, lookingthestar@livemail.tw,
leiwang@fcu.edu.tw

Abstract. Protective measures for server invasions should not solely focus on events before an invasion occur. Recording and monitoring successful server invasions with endless streams of security mechanisms should be employed attentively to reduce the loss of data due to successful intrusion attacks on any system. This paper focus on the implementation of an embedded system technology developed in a host control module group in which the entire server achieves coordinated group allocation of resources combined with a larger number of group hosts designed to meet demand. The security server switching system uses server load balancing to prevent system failures, errors, and interruptions, accompanied with the ever so important theory of fault-tolerance for grace degradation purposes.

Keywords: Redundancy, Information Security, Virtual Machine, Intrusion Tolerance, Intrusion Elimination.

1   Introduction

There have been many marketed tools designed to protect and prevent data leakage from entering the wrong hands. Intrusion Management System (IMS)[1][2] was assumed that it is able to detect all intrusion and the system will be activated when it detects a deliberate intrusion or suffered damage. However, computer hackers (intruders) have not only found a way around most firewall protection schemes, they are accustomed to creating numerous ways to breach critical systems for political and/or personal gain. With that said, any person or organization who is absolutely sure that their on-line network is totally secure is in for a rude awakening.

Client-server architecture (Client-Server Model) is and has been the Internet's main mode of application for some time now. All Web Services (WWW), Domain Name Services (DNS), File Transfer Services (FTP) along with other services are molded to fit this framework. There is plenty at stake within many social civilizations because of everyday educational, cultural, and consumer transactional activity over the Internet. System failures and human vandalism of these systems will most definitely paralyze the functionality of many societies. In turn, studies have proven that these kinds of mishaps could lead to chaotic situations throughout the entire world. Therefore, it is necessary to attack the process of invasion deeply and determine ways to reduce the amount of data loss coupled with reducing the amount of time an attacker has on the system. By doing this one could possibly save the reliability/credibility of their organization.

J.-S. Pan et al. (Eds.): *Advances in Intelligent Systems & Applications*, SIST 21, pp. 167–176.
DOI: 10.1007/978-3-642-35473-1_18   © Springer-Verlag Berlin Heidelberg 2013

Although there are many security architectures developed to prevent and detect intrusions, it is an undeniable fact that servers have never been totally secure from invasions, so protective measures for server invasions should not solely focus on events before invasions occur. How to reduce the loss caused by attacking events that arise from failures of protective mechanism is also important. In 1985, Fraga and Powell first presented intrusion tolerance concepts, they pointed that security issues are inevitable, and separation and exclusion are no longer the primary means when face to security problem. How to provide normal and correct services in the case of system under attack is also taken into account.

The Secure Server Switching System (4S)[3][4] takes a practical point of view against aggressive behavior geared toward all internet invasions, as well as, focuses on how to reduce the invasion time of successful invasions. System design incorporates embedded system technology in to achieve coordination amongst the entire server group (Server Pool) via server host control module. Distribution of host resources to meet a large number of server demands has been implemented accordingly. In addition, 4S uses server load balancing effectively, enabling the system to function positively without interrupts due to system and/or service errors. 4S server load balancing techniques also tackles all processing power and starvation issues that can be caused by redundant server and host switching. Using virtual technology[5][6] to achieve fast server engine replacement services in hosts, 4S allows intrusion, eliminates the intruders invasion, then activates the system normality in a very short time.

The detailed structure of this paper is as follows: Section 2 will introduce the framework and algorithm of secure server switch system; A brief description about an analytical model for the reliability achieved by 4S with the variance in switching interval is illustrated in Section 3; Section 4 shows the implementation details of 4S and some observations about the implementation; Finally, the conclusion is made in Section 5.

2   Secure Server Switching System

The Secure Server Switching System (4S) takes a practical point of view against aggressive behavior geared toward all internet invasions, as well as, focuses on how to reduce the invasion time of successful invasions. Its redundant structure achieves fault tolerance. Services periodically switch to a different server restricting the intruder's residence time. Fig. 1 shows the 4S system architecture.

The request/response operations in the 4S can be described briefly as the following three steps:

- Clients send their requests to pool controller directly.
- Pool controller transfer the requests to the corresponding server by means of the mapping records maintained in the pool controller.
- While server completing the service process, the reply packet is sent to the client directly via internet connection.

4S architecture contains a Pool Controller and several hosts. The Pool Controller connects with several entities composed of several computers as follows:

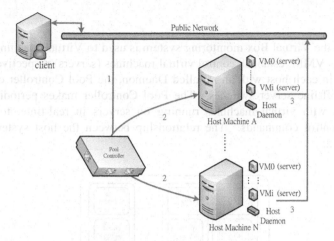

Fig. 1. 4S architecture

2.1   Pool Controller

The Pool Controller is a customized embedded system device responsible for monitoring and collecting all information associated with all host and virtual service entities. It incorporates a predefined rule for running a dynamic control server switch. It is composed of the three main modules: Workload Monitor, Schedule Component and Control Kernel.

Workload Monitor
It sends a request periodically to the Host daemon for the information of active servers. When the messages that include the rate of CPU utilization, memory usage and IO usage of the host server are received, Workload Monitor will check the information to find out any abnormal server and notice the Schedule Component to switch services among virtual servers.

Schedule Component
Follows administrator's pre-defined parameter rules and makes good decisions for online services based on workload monitor's real-time information. The Schedule Component sorts its hosts based on workloads of each host and the time quantum defined for each active server. It consults the sorting result to decide which server has to go On-Line at the right time. The switching time is decided mainly by the switching period window size determined by a theoretical model described in next section, but other factors such as the utilization rate of each CPU, the memory usage, and IO usage are also takes into account.

Control Kernel
Based on scheduled information, installs management server daemon; scheduling component. Based on the scheduled decision of the Schedule Component, the Control Kernel manages all communications among pool controller and Hosts. In the current design, Control Kernel performs the following tasks: switching commands for servers, receiving host/server states from hosts, and control commands for the creation/removing of virtual machines.

2.2   Host

In this paper, the Virtual Box monitoring system is used to Virtual Machine Monitor, abbreviated as VMM, establish control virtual machines (servers) respectively to their host servers. In each host with an installed Daemon, the Pool Controller controls all on-line and off-line server switches. The Pool Controller makes periodic machine transmissions with virtual machines running on servers in real-time, followed by transfer of control commands. The relationship between the host system stack is shown in Fig.2.

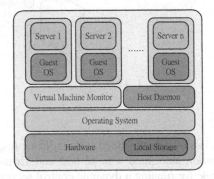

Fig. 2. Software stack of the Host

The use of Full-virtualization technology constructs a single host to several virtual servers (Server 1, Server 2, ... Server n). Virtual machine snapshots with the systems architecture of the switching mechanism allows access to server off-line data based on a snapshot of the server. This quickly restores the original/initial state to ensure that services are completely removed. Even if the previous cycle was invaded by tampering through the back-door, the line will be reset and erroneous information will be erased.

3   Analysis for the Switching Interval of 4S

Intrusion tolerance system (ITS) is a branch of the solutions for securing computer information systems. As distinct from the intrusion avoidance of current systems, ITS systems focus on containing the losses while system has been intruded. From the view of containing the losses, 4S is a time-based ITS since 4S will limit the time for intruders to break into the system and acquire illegal profits in the intrusion. In this section we present a theoretical model for the behavior of intrusion and derive a mathematical expression for assessing the intrusion tolerance of 4S, then the relation between expected reliability and the exposure time of a server is then be created.

A server system should set up several lines of defense to prevent intrusions. It means that a successful intrusion needs to conquer several security mechanisms successively. At first, the intruder should avoid the filtering of firewall. The IP address and the type of message may need to be faked as another legal operation to avoid the checking of firewall and other intrusion detection systems. The intrusion also needs to conquer the protection schemes of the operation systems to get enough privilege to intrude in the system to get rights for passwords or access privilege. In the whole process of intrusion, intruders must also act very carefully to hide themselves from the detection of intrusion detection systems. Besides, applications always equipped with some protection schemes to find malicious operations out and eliminate the intrusion. So intruders should be carefully avoiding the protection logic by issuing several continuous operations to achieve the goal of the intrusion.

By assuming there are N defense lines need to be conquered for a successful intrusion. We can define N as the number of defense lines that a intruder should be conquered, and $N_S(t)$ as the number of defense lines that are not conquered at time t. On the contrary, $N_F(t)$ stands for the number of defense lines that have been conquered at time t. According to the definition, we can define the function of secure confidence under attacking as:

$$C(t) = \frac{N_S(t)}{N} = \frac{N - N_F(t)}{N} = 1 - \frac{N_F(t)}{N} \qquad (1)$$

Since the N lines are not intruded at the beginning, so $C(0) = 1$. For a success intrusion, we can further indicate that $C(\infty) = 0$.

By differentiating equation (1), we can find:

$$\frac{dC(t)}{dt} = \frac{-1}{N} \times \frac{dN_F(t)}{dt} \Rightarrow \frac{dN_F(t)}{dt} = -N \frac{dC(t)}{dt} \qquad (2)$$

Let us consider about the probability of a server system been attacked by intruders, it is reasonable that the probability, or say risk, that a server system been intruded will increased with its exposed time. Let $R_F(t)$ means the risk growth rate at time t, we can derivate the feature of $R_F(t)$ as follow:

$$R_F(t) \times N_S(t) = \frac{dN_F(t)}{dt} \Rightarrow R_F(t) = \frac{1}{N_S(t)} \times \frac{dN_F(t)}{dt} \qquad (3)$$

The character function of $R_F(t)$ can be further derived from equation (3) by equation (2) as:

$$R_F(t) = \frac{1}{N_S(t)} \times \frac{dN_F(t)}{dt} = -\frac{N}{N_S(t)} \times \frac{dC(t)}{dt} = -\frac{1}{C(t)} \times \frac{dC(t)}{dt} \qquad (4)$$

It is well-known that the risk growth rate, $R_F(t)$, can be modeled as a Poisson cumulative distribution with parameter k shown in equation (5), by replacing the definition of $R_F(t)$ into equation (4), we can conclude the relation as illustrated in equation (6):

$$R_F(t) = 1 - e^{-kt} \qquad (5)$$

$$R_F(t) = 1 - e^{-kt} = -\frac{1}{C(t)} \times \frac{dC(t)}{dt} \Rightarrow -\frac{dC(t)}{C(t)} = (1 - e^{-kt})dt \qquad (6)$$

Based on the relation of equation (6), the distribution of the confidence function can be further estimated as follow:

$$\int -\frac{dC(t)}{C(t)} = \int (1 - e^{-kt})dt \Rightarrow \ln(C(t)) = -(t - \frac{1}{k}e^{-kt}) \Rightarrow C(t) = ce^{-(t-\frac{1}{k}e^{-kt})} \qquad (7)$$

The distribution of the confidence function is mainly determined by the parameter k and constant c. Although the values of k and c can not be resolved by the analysis; the values can be estimated by a comprehensive simulation for the attacker behavior. Erland Johnsson and Tomas Olovsson had ever proposed a study about the behavior of intrusion process. They performed a practical intrusion test on a distributed computer system and collected data related to the difficulty of making these intrusions. Time-related data have been found to be especially valuable for the modeling of the intrusion process. Based on empirical data collected from intrusion experiments, the study have worked out a hypothesis on typical attacker behavior. The collected data indicates that the times between breaches are exponentially distributed. The mean time to breach was found to be as low as four hours. This would actually imply that traditional methods for reliability modeling could be applicable.

We select the test that held by 11 groups of attackers who attempted to make 12 breaches. There are totally 51 successful breaches are made in the test. By analyzing the working time and successful time of these 51 successful breaches, the accumulative successful rate to working time is shown as the solid curve in Fig. 3. Because the character of intrusion behavior this research derived is the same as the study made by Johnsson and Olovsson[7], the values of k appeared in equation (5) can be estimated by means of the simulation results produced by their study. Matlab is used to analysis the value of k by substituting the simulation results into the model and then the value of k is estimated by means of least-squares error method. According to the analysis, we found the value of k is about 0.06154. The dot line appeared in Fig. 3 shows the curve of equation (5) with k equal to 0.06154. We can find from the figure that the behaviors proposed by the two researches are matched well.

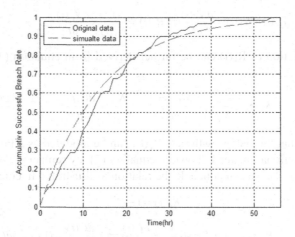

Fig. 3. The curves produce by analysis model and simulation results

When the value of k is determined, the formula of confidence function can be further clarified. As described before, $C(0) = 1$. It means that:

$$C(0) = c \times e^{-(t - \frac{1}{k}e^{-kt})} = 1, when \ t = 0 \tag{8}$$

$$\Rightarrow c = e^{\frac{1}{k}}$$

Fig. 4. is the curve of confidence function calculated by substituting the values of k and c as described above. According to the confidence curve, we can find that all defense lines will be conquered by three hours. It is noted that the analysis result does not mean that the server switching interval should be set on three hours. This result can be viewed as a upper bound of switching interval since the time is deduced by ideal theoretical model, and the data provided to form the realistic time estimation are collected some simulation tests. There are many factors can influence the final result. For example, an intruder may try to attack a server indefatigably even though the server always cleansing itself before the attack success. The unremitting attacks make the intruder more skillful to develop more innovative attack at the beginning of a new service on-line. This consideration means the confidence curve should be drop more and more quickly by the sophisticate attacker.

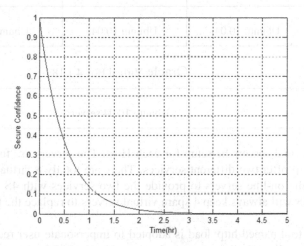

Fig. 4. The curve of confidence function

For the implementation of 4S, the switching interval upper bound is used to judge whether the system is out of control. When the interval is extended to the limit, the system will turn off some service to reduce system overhead and alarm administrators to resolve the situation.

4   Implementation and Simulation Results

For the simulation, we use a laptop equipped with an Intel i5-2450m CPU and 8GB memory as Pool Controller and three PCs as Hosts. Pool Controller will route the client requests to the dedicated servers by direct routing and collect information about all hosts to manage the system. The detail specifications of Hosts are different as shown in Table 1. It is evident that the execution potential of Host2 is best and Host1 is worst. Every Host has been setup four virtual machines as servers. Each virtual server was set 512MB memory and 8GB disk.

Table 1. Host specifications

Host	Host0	Host1	Host2
CPU	Intel i5-2400	Intel 6320	Intel i5-2400
Memory	8Gib	4Gib	16Gib
Host OS	Ubuntu 10.04	Ubuntu 10.04	Ubuntu 12.04
VMM	Oracle Virtual Box 4.1.18		
Guest OS	Ubuntu 10.04 server		

There are 4 virtual servers be setup for each Host. The servers are further divided into two groups to offer two different services. There are totally 8 virtual servers will be activated as the on-line servers to provide the two services with 4S operations at the same time. 4S will always keep 4 spare virtual servers to replace the failed on-line servers.

A simulation tool named http_load is adopted to impersonate user requirements in the simulation. It will send client requests by a workload of 50 requests per second to each server in the first 109 seconds; the workload is then periodically increased by 10 requests per 30 second. The whole simulation is held for 10 minutes to test the functions of 4S and observe the reflections by arranging some bombshells during the simulation.

Fig. 5 shows the detail of the simulation, there are three subgraphics are shown in the figure: Service Distribution, CPU Usage and Number of Running VM.

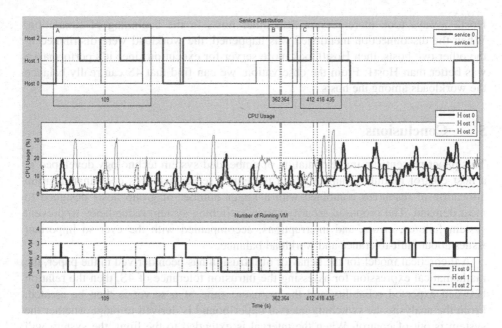

Fig. 5. Simulation Results

The first subgraph shows the service distribution during the whole simulation, it shows that which host is working for corresponding service. From the block A in the subgraph, we can observe that the two services were switched in different frequency since the switching intervals of the two services are set to be different value. The switching period of service 0 was set to be 30 seconds and service 1 was set to be 50 seconds. The block B lined in the subgraph shows an emergency condition happened and the succeeding reflection made by the system: The service1 was damaged caused by an internal bombshell at 362 second. In this case, host daemon detected the error and reported the condition to pool controller immediately. Pool controller then redirected this service to backup server right away. The whole process was completed in 4 seconds. Block C is another type of error happened during the simulation: The cable between pool controller and Host2 was pulled out to cut off the connection. The condition was treated as a hardware failure happened on Host2. In this case the Pool Controller detected the cessation of services in this host. Service1 was immediately resumed in 6 seconds, but unfortunately, the service0's backup server was also at host2. Pool controller has to wake up a server in another host for service0. At last, it took 23 seconds to recover the service for this worst case.

From the subgraph about the number of running VM, we can observe how many servers was running in the hosts, including working servers and backup servers. By comparing the lines with the subgraph about cpu usage, we can find that when a host wake up a server or close a server, cpu usage will always rise up to a peak. It means that cleansing will consume a certain amount of host resources.

Another interesting condition can be found that the running servers were distributed on hosts evenly base on the performance of hardware resources. For example,

Host2 always took more servers than other hosts in the beginning of simulation. When the disconnection mentioned was happened, the workload were distributed to the other two hosts, then Host0 took more server for execution since it's performance was better than Host1. From the observation, we can find that 4S can really balance the workloads among the hosts.

5   Conclusions

This paper presents an implementation of embedded system technology developed in a host control module group in which the entire server achieves coordinated group allocation of resources combined with a larger number of group hosts designed to meet demand. The proposed security server switching system redundancy to prevent system failures, errors, and interruptions, accompanied with the ever so important issues about fault-tolerance and workload balance.

A theoretical model for the behavior of intrusion has been derived in the paper, the mathematical expression for assessing the intrusion tolerance of 4S, then the relation between expected reliability and the exposure time of a server is then be created. From the analysis, the switching interval upper bound is derived to judge whether the system is out of control. When the interval is extended to the limit, the system will turn off some service to reduce system overhead and alarm administrators to resolve the situation.

References

1. Lunt, T.F.: A survey of intrusion detection techniques. Computers and Security 12(4), 405–418 (1993)
2. Smith, J.E., Nair, R.: The Architecture of Virtual Machines. IEEE Computer 38(5), 32–38 (2005)
3. Wang, Y.-S., Wang, L.: Secure Server Switching System. In: Cryptology and Information Security Conference (2009)
4. Wang, Y.-S., Wang, L.: Secure Server Switching System. In: 2010 Second International Conference on Computer Engineering and Applications (ICCEA), Bali Island, pp. 224–228 (2010)
5. Singh, A.: An Introduction To Virtualization (2004), http://research.ihost.com/osihpa/osihpa-hensbergen.pdf
6. Oracle Virtual Box, http://www.virtualbox.org/
7. Jonsson, E., Olovsson, T.: A Quantitative Model of the Security Intrusion. IEEE Transactions on Software Engineering 23(4) (1997)

Design and Implementation of a Self-growth Security Baseline Database for Automatic Security Auditing

Chien-Ting Kuo[1], He-Ming Ruan[2], Shih-Jen Chen[3], and Chin-Laung Lei[4]

[1,2,4] Department of Electrical Engineering, National Taiwan University,
No. 1, Sec. 4, Roosevelt Rd., 106 Taipei, Taiwan
[1,3] CyberTrust Technology Institute, Institute for Information Industry,
105 Taipei, Taiwan
{protools,tannhauser}@fractal.ee.ntu.edu.tw,
{ctkuo,sjchen}@iii.org.tw, lei@cc.ee.ntu.edu.tw

Abstract. As the security consciousness rising, information security audit has become an important issue nowadays. This circumstance makes the security audit baseline database a crucial research domain. In this paper, we proposed a security baseline database to assist information security auditors to maintain the security update patch baseline automatically with the help of the Microsoft knowledge base and automatic audit process. A practical implementation demonstrates that the proposed structure is both useful and effective.

Keywords: vulnerability, baseline database, security auditing.

1 Introduction

In recent years, security defenses of plenty of enormous enterprises and organizations have been breached and it causes tremendous loss of profit or other assets. Traditional information attacks such as DDoS attacks or official site modifications are replaced by the advanced persistent threat (APT) attacks. The aims of the attacks are not only to damage the target system but also to steal high value business secrets or sensitive customer data such as credit card numbers. To avoid the damage from all kinds of attacks, more and more organizations begin to pay attention on internal security audit. Accordingly, information security audit, personal data protection, and personal data inventory have become three of the most popular research issues on information security and management recently.

There are already some researches on developing risk assessment process [1][2]. In 2002, Aagedal et al. [2] gave us a clear view of the methodology of Consultative Objective Risk Analysis System (CORAS), which is a model-driven risk analysis approach. Later in 2008, Fu et al. [1] provided us an example about how to use CORAS to establish an Information Sharing and Analysis Center (ISAC).

Besides the risk assessment process, analyzing and scoring known vulnerabilities is also an important research area. The most famous vulnerability scoring system might be the "Common Vulnerability Scoring System (CVSS)". The CVSS uses six parameters

J.-S. Pan et al. (Eds.): Advances in Intelligent Systems & Applications, SIST 21, pp. 177–184.
DOI: 10.1007/978-3-642-35473-1_19   © Springer-Verlag Berlin Heidelberg 2013

and simple equations to calculate the basic scores for each of the vulnerabilities. With the help of CVSS, it becomes possible to develop automatic auditing software, which can scan devices inside an organization and evaluate the risk level according to the discovered vulnerabilities automatically. Therefore, the baseline information, which records all the security patch information, plays a very crucial role in automatic auditing software development.

In this paper, we aim to provide a security baseline database to grant automatic auditing software the capability to evaluate the risk level without the intervention of security experts. The proposed security baseline database integrates security information from the Microsoft Knowledge Base, the common vulnerabilities and exposures (CVE), and the common vulnerabilities scoring system (CVSS) to provide a global and clear view of security risk to benefit the security audit. In section 2, we will introduce the Microsoft knowledge base, CVE, and CVSS. The proposed baseline database structure will be detailed in section 3. The implementation of the security baseline database will be presented in section 4. Finally, we will have some conclusions in section 5.

2   Background

In this section, we will introduce the Microsoft knowledge base in section 2.1, the common vulnerability and exposures (CVE) in section 2.2, and the common vulnerability scoring system (CVSS) in section 2.3. Those technology and theory are the cornerstones of the proposed Microsoft security update patch database in this paper.

2.1   The Microsoft Knowledge Base

The Microsoft Knowledge Base is sponsored and maintained by the Microsoft Corporation [3]. The Knowledge Base offers the knowledge for products of the Microsoft. Each knowledge record has a unique ID number, such as "KB2722913". Every Microsoft patch is associated with a knowledge base. If someone want to know the detail information of the update patch, he/she could find the knowledge base number in his/her own computer and query the Microsoft online support center. There also exists a security bulletin in the Microsoft online support center. Fig 1 shows the general information from KB2722913.

Affected Software

Operating System	Component	Maximum Security Impact	Aggregate Severity Rating	Updates Replaced
Internet Explorer 6				
Windows XP Service Pack 3	Internet Explorer 6 (KB2722913)	Remote Code Execution	Critical	KB2699988 in MS12-037 replaced by KB2722913
Windows XP Professional x64 Edition Service Pack 2	Internet Explorer 6 (KB2722913)	Remote Code Execution	Critical	KB2699988 in MS12-037 replaced by KB2722913
Windows Server 2003 Service Pack 2	Internet Explorer 6 (KB2722913)	Remote Code Execution	Moderate	KB2699988 in MS12-037 replaced by KB2722913
Windows Server 2003 x64 Edition Service Pack 2	Internet Explorer 6 (KB2722913)	Remote Code Execution	Moderate	KB2699988 in MS12-037 replaced by KB2722913
Windows Server 2003 with SP2 for Itanium-based Systems	Internet Explorer 6 (KB2722913)	Remote Code Execution	Moderate	KB2699988 in MS12-037 replaced by KB2722913

Fig. 1. The snapshot of general information from KB2722913 (source from [3])

2.2   The Common Vulnerability and Exposures (CVE)

The Common vulnerability and exposures (CVE) [4] is a vulnerability list and each of the vulnerabilities is identified by a unique identity. For example, "CVE-2012-2521" means the 2521th vulnerability discovered in 2012. CVE is co-sponsored by National Cyber Security Division of the U.S. Department of Homeland Security and the MITRE Corporation and has been used as a public dictionary. Security software developers or researchers can download the digital copy with XML format for free or look up the information on the National Vulnerability Database (NVD) website. Fig.2 shows the vulnerability information of KB2722913. The KB2722913 security update patch fixes the following four vulnerabilities: CVE-2012-1526, CVE-2012-2521, CVE-2012-2522, and CVE-2012-2523.

Vulnerability Information
  Severity Ratings and Vulnerability Identifiers
  Layout Memory Corruption Vulnerability - CVE-2012-1526
  Asynchronous NULL Object Access Remote Code Execution Vulnerability - CVE-2012-2521
  Virtual Function Table Corruption Remote Code Execution Vulnerability - CVE-2012-2522
  JavaScript Integer Overflow Remote Code Execution Vulnerability - CVE-2012-2523

Fig. 2. The snapshot of vulnerability information from KB2722913 (source from [3])

2.3   The Common Vulnerability Scoring System (CVSS)

The Common Vulnerability Scoring System (CVSS) was proposed by National Infrastructure Advisory Council (NIAC) in 2004 [5]. It became an international telecommunication union standard in April 2011. Different from traditional security risk analysis mechanisms, which focus on defining the important resources in the enterprise and assess the security risk using assure metrics [6][7][8][9][10], the CVSS scoring mechanism provides a set of comprehensible risk formulas for all the discovered vulnerabilities. According to the scoring mechanism, researchers and managers can understand how severe the vulnerability is. Table 1 shows all the six parameters, possible items for each of the parameters, and the values of the selected items. Experts can select items for every parameter and use the equations in Fig. 3 to calculate the risk scores, which are so called *BaseScore*. The *BaseScore* is normalized to a specific range such as zero to ten. The higher the risk score is, the greater the risk will be.

$$BaseScore = (0.6 * Impact + 0.4 * Exploitability$$
$$- 1.5) * f(Impact)$$
$$Impact = 10.41 * (1 - (1 - ConfImpact) *$$
$$(1 - IntegImpact) * (1 - AvailImpact))$$
$$Exploitability = 20 * AccessComplexity$$
$$* Authentication * AccessVector$$
$$f(Impact) = 0 \ if \ Impact = 0; 1.176 \ otherwise$$

Fig. 3. The *BaseScore* equations of CVSS vulnerability scoring system

Table 1. CVSS Parameters and associate values

Parameters	Selection Items	Associate value
AccessVector	Requires local access	0.395
	Adjacent network accessible	0.646
	Network accessible	1.0
AccessComplexity	High	0.35
	Medium	0.61
	Low	0.71
Authentication	Requires multiple instances	0.45
	Requires single instance	0.56
	No authentication	0.704
ConfImpact	None	0.0
	Partial	0.275
	Complete	0.660
IntegImpact	None	0.0
	Partial	0.275
	Complete	0.660
AvailImpact	None	0.0
	Partial	0.275
	Complete	0.660

3   Security Baseline Database

The main idea of the security baseline database is to use the security update patches from different operating system versions on the internal devices along with the security information from the Microsoft knowledge base to build a baseline database for automatic information security auditing software. Besides, we construct a new database to store the ID numbers of each of the security patches, the CVE ID of the related vulnerabilities of each patch, the CVSS scores of the CVE vulnerabilities, and the dependency between the security patches. With the proposed security baseline database, we can periodically obtain up to date data from the Microsoft knowledge base and each of the devices to be audited, and then update the implemented database flexibly. With the scheduled update, it becomes possible for the proposed security baseline database to self-growth as the security update patches be announced.

In Fig. 5 is our first structure, which is constructed under network environments, to automatically sense the version of Microsoft operating system. The advantage of this structure is that it can automatically collect the security update patches with each kind version of the operating systems. However, it could collect the security update patch list only and no corresponding vulnerability information such as records from CVE or CVSS will be available in this structure. Therefore, we designed another information colleting method to enhance the first structure in Fig. 6. We designed a web crawler, which parses the corresponding CVE information of each security update patch that we sensed on the proposed automatic sensing environment. In this structure, we also query the CVSS scores of the corresponding CVE vulnerabilities. Combine the two structures in Fig. 5 and Fig. 6, we could automatically compute the latest security update patch information including related vulnerabilities and security risk score as a baseline for the information security audit.

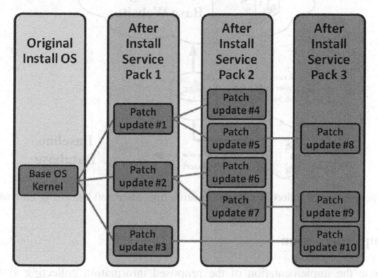

Fig. 4. The relation of different patches on different service packs

The relations between different security update patches on different service packs are shown in Fig. 4.

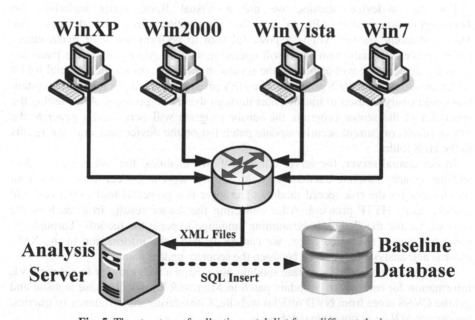

Fig. 5. The structure of collecting patch list from different devices

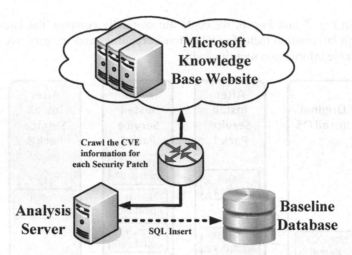

Fig. 6. The structure of collecting CVE information from Microsoft knowledge base website

4 Implementation

We separate the implementation of the proposed information collecting system in Section 3 into two parts: the first part consists of the on-device sensors and the other one is the central server. We will detail the tools and technologies for each part of the implementations in the remaining part of this section. To clarify our idea, Fig. 7 illustrates the overview of our implementation.

For the on-device sensors, we use a Visual Basic script including the *Microsoft.update.session* API to get the information from the devices. The *Microsoft.update.session* API is a powerful tool to perform query or management patch update information on Microsoft operating system. After parsing the patch list from the devices, we will also store the results in the XML format in a special folder which is setting from *HFS* program. The *HFS* program is a simple HTTP file system that could easily be used to transfer files through the HTTP protocol. After setting the scheduler of the sensor program, the sensor program will periodically generate the sensor results of current security update patch list on the device and store the results in the HFS folder.

In our central server, the server consists of two modules: the risk record module and the security baseline database module. We use *wget* to collect the records from the devices for the risk record module. The *wget* is a powerful tool to perform file transfer using HTTP protocol. After collecting the sensor results from each of the devices, we use the *Python* programming language to parse the records. Through the powerful toolkit, *lxml*, in *Python*, we can easily parse the information in the XML records and analysis the relation between the security update patches.

In our security baseline database module, we design a web crawler to get the CVE information for each security update patch in Microsoft knowledge base website and get the CVSS score from NVD official website. Considering the efficiency of queries, we choose *SQLite* as our store database.

Moreover, we integrate the risk record module and the security baseline database module with the web interface which is consisted of *PHP* and other web development tools such as *AJAX* and *JQuery* to provide a cloud service in a SaaS form. By using this system, one can learn about not only the current situation of the internal devices, but also the vulnerabilities and the corresponding risk level for each of the security update patches.

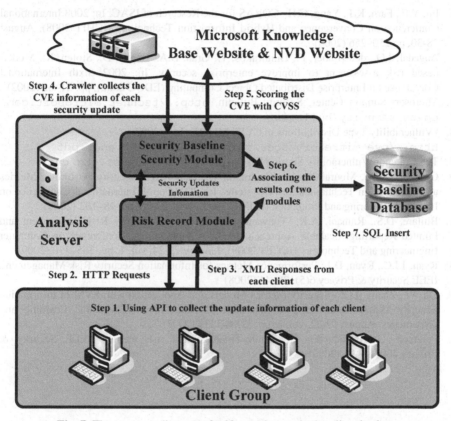

Fig. 7. The structure diagram of self-growth security baseline database

5   Conclusion

Nowadays, most organizations audit security update patches and vulnerabilities of their internal devices, and use some baseline database to evaluate the audit result. Therefore, we propose a security baseline database, which integrates security information from the Microsoft Knowledge Base, the common vulnerabilities and exposures (CVE), and the common vulnerabilities scoring system (CVSS) and grant automatic auditing software the capability to evaluate the risk level without the intervention of security experts. Also, we use the lightweight database to implement an automatic audit system. The implemented automatic audit system shows that the proposed security baseline database is useful and effective.

Acknowledgments. This study is conducted under the "ICT Security and Data Protection Technology Development Project" of the Institute for Information Industry which is subsidized by the Ministry of Economic Affairs of the Republic of China.

References

1. Fu, Y.P., Farn, K.J., Yang, C.H.: CORAS for the Research of ISAC. In: 2008 International Conference on Convergence and Hybrid Information Technology (ICHIT 2008), August 28-30, pp.250–256 (2008)
2. Aagedal, J.O., den Braber, F., Dimitrakos, T., Gran, B.A., Raptis, D., Stolen, K.: Model-based risk assessment to improve enterprise security. In: 2002 Sixth International Conference on Enterprise Distributed Object Computing (EDOC 2002), pp. 51–62 (2002)
3. Microsort Support Center, Security Bulletin, http://technet.microsoft.com/en-us/security/bulletin
4. "Vulnerability Type Distributions in CVE" MITRE (May 2007), http://cve.mitre.org/docs/vuln-trends/vuln-trends.pdf
5. The Common Vulnerability Scoring System, http://www.first.org/cvss/
6. Ouedraogo, M., Mouratidis, H., Khadraoui, D., Dubois, E.: Security Assurance Metrics and Aggregation Techniques for IT Systems. In: 2009 Fourth International Conference on Internet Monitoring and Protection (ICIMC 2009), May 24-28, pp. 98–102 (2009)
7. Bhilare, D.S., Ramani, A.K., Tanwani, S.: Information Security Risk Assessment and Pointed Reporting: Scalable Approach. In: 2009 International Conference on Computer Engineering and Technology (ICCET 2009), January 22-24, vol. 1, pp. 365–370 (2009)
8. Ryan, J.J.C., Ryan, D.J.: Performance Metrics for Information Security Risk Management. IEEE Security & Privacy 6(5), 38–44 (2008)
9. Qu, W., Zhang, D.Z.: Security Metrics Models and Application with SVM in Information Security Management. In: 2007 International Conference on Machine Learning and Cybernetics, August 19-22, vol. 6, pp. 3234–3238 (2007)
10. Peterson, G.: Introduction to identity management risk metrics. IEEE Security & Privacy 4(4), 88–91 (2006)

Enhancing Cloud-Based Servers by GPU/CPU Virtualization Management

Tin-Yu Wu[1], Wei-Tsong Lee[2], Chien-Yu Duan[2], and Tain-Wen Suen[3]

[1] Institute of Computer Science and Information Engineering, National Ilan University,
Taiwan, R.O.C
[2] Department of Electrical Engineering, Tamkang University, Taiwan, R.O.C
[3] Chung-Shan Institute of Science and Technology, Taiwan, R.O.C
tyw@niu.edu.tw, wtlee@mail.tku.edu.tw,
{jason84195,qoosuntw}@hotmail.com

Abstract. This paper proposes to add the multithreaded Graphic Processing Units (GPUs) to some virtual machines (VMs) in the existing cloud-based VM groups. To handle the multidimensional or multithreaded computing that a CPU cannot process quickly by a GPU that has hundreds of Arithmetic Logic Units (ALUs), and to regulate the time for initiating physical servers by real-time thermal migration, our proposed scheme can enhance the system performance and reduce the energy consumption of long-term computing. Four major techniques in this paper include: (1) GPU virtualization, (2) Hypervisor for GPU, (3) Thermal migration implementation, and (4) Estimation of multithreaded tasks. In no matter quantum mechanics, astronomy, fluid mechanics, or atmospheric simulation and prediction, a GPU suits not only parallel multithreaded computing for its tens of times performance than a CPU, but also multidimensional array operations for its excellent efficiency. Therefore, how to distribute the computing performance of CPUs and GPUs appropriately becomes a significant issue. In general cloud computing applications, it is rarely seen that GPUs can outperform CPUs. Furthermore, for groups of virtual servers, many tasks actually can be completed by CPUs without the support of GPUs. Thus, it is a waste of resources to implement GPUs to all physical servers. For this reason, by integrating with the migration characteristic of VMs, our proposed scheme can estimate whether to compute tasks by physical machines with GPUs or not. In estimating tasks, we use Amdahl's law to estimate the overall performance include communication delays, Synchronization overhead and me possible additional burden.

Keywords: Virtual Machine (VM), Multithreading, GPU, CUDA, Mapreduce.

1 Introduction

In traditional computer science, computers processed tasks mainly by Central Processing Units (CPUs). However, the development of CPU has recently encountered bottlenecks because the computation speed-up of single-core processors may result in overheating and power consumption problems. Therefore, in place of

J.-S. Pan et al. (Eds.): *Advances in Intelligent Systems & Applications*, SIST 21, pp. 185–194.
DOI: 10.1007/978-3-642-35473-1_20   © Springer-Verlag Berlin Heidelberg 2013

single-core processors, multi-core processors are gradually used for parallel computing to enhance computer performance.

Parallel computing in the early days was usually executed by several computers and processed by traditional CPUs. Thus, organizations those needed to process large amounts of data established large-scale multicomputer systems and exchanged data through Message Passing Interface (MPI). In such a kind of multicomputer environment, every computer is a computational node, which has its own CPU, memory and networking interface. Thus, a multicomputer system usually transforms a parallel program into single program multiple data (SPMD) for every computer in the system to operate the same program but process different data.

In the past, display cards were defined as the auxiliary to CPUs to process image and graph related tasks. Later, GPU was presented to reduce display cards' dependence and occupancy of CPUs. Although the number of computing units on GPUs was not large previously, the computing ability of display cards has been enhanced recently: not only the improvements of computing clock, but also the increasing number of GPUs on display cards, which enhances the floating-point operations per second. Instead of being designed to finish heavy computing tasks within a limited number, GPUs are expected to process large amounts of data and tasks by parallel computing to improve the system performance. Because a large number of GPUs are suitable for parallel computing, many supercomputers in the word have started to use GPUs to support CPUs for a great deal of complicated computing tasks.

Since the future of computers keep stepping into cloud computing, we propose to add the multithreaded GPUs to some VMs in the existing cloud-based VM groups. To handle the multidimensional array operations or multithreaded computing that a CPU cannot process quickly by a GPU that has hundreds of ALUs, and to regulate the time for initiating physical servers by real-time thermal migration, our proposed scheme can estimate whether to compute tasks by GPUs or not, enhance the system performance, and reduce the energy consumption of long-term computing

2   Related Works

2.1   CPU/GPU Collaborative Computing

In modern computer science, traditional CPU computing has reached a bottleneck while high performance computing systems are experiencing a revolution, in which novel architectures are presented one after another and the combination of multi-core microprocessor and GPU is one of the highest potential and prospective method.

GPU (Graphics Processing Unit) was first presented by NVidia in 1999[1]. With the evolution of semiconductor industry, the growth of GPU has been exceeding Moore's Law and reached more than 500 gigaflops of double-precision floating point operations. As for researches about GPGPU (General-Purpose computation on GPU), papers [2-5] have specified the history, architecture, software environment and several cases of GPU.

Because of its powerful computational capabilities, high cost performance and high performance but low power consumption, GPU has received great attention in such an eco-friendly era. In addition to traditional graphical computing, GPU has been greatly applied to general-purpose computing and thus formed GPGPU or General-purpose computing on graphics processing units (GP²U). Due to its excellent general-purpose computational capabilities, GPU has been regarded as the future of computer science since 2003 [2].

CPU and GPU are designed with absolutely different goals. The design concept of CPU is to execute instructions and operations quickly with low delay and to use a great deal of IC for control and temporary storage. On the other hand, GPU is designed for graphical computing, in which great amounts of IC are used as ALUs for high intensity computing. Therefore, by utilizing CPU/GPU collaborative computing, we can use CPU for control and buffer and use GPU for processing a great deal of computing tasks.

In the scope of CPU/GPU collaborative computing, CUDA (Compute Unified Device Architecture)[5] presented by NVidia is currently the leading technique of GPGPU. The CUDA is a C-language development environment, in which tasks are computed by GPUs after NVidia GeForce 8 together with Quadro GPU. Commands in either CUDA C-language or OpenCL will be compiled into PTX code by driver programs for the display core to compute.

The latest CUDA-x86 compilers can support traditional multi-core CPU architecture and execute all parallel programs written by CUDA. Instead of outperforming CPU in all computing aspects, GPU only surpasses CPU in matrix computing and parallel computing, which are still rarely seen in the present program structure. Thus, when a VM of a physical server without GPU executes parallel computing, the system will estimate the computing cost of GPU servers. Supposing the computing amount is not large, we use CPU for parallel computing. On the contrary, we will use GPU for a great deal of computing amount.

2.2   GPU Virtualization

When cloud computing becomes the future of computers, how to virtualizes all computer interfaces and optimize the computer performance is the goal for all cloud service providers. While GPU computing has been integrated into new computer structure, traditional virtual structures will be challenged. VMware, the leading company in the virtualization, has presented some concepts about GPU virtualization in [6]. Like traditional virtualization, physical GPU is cut into several virtual GPUs and GPU resource is managed by a resource manager, which is similar to a VM monitor. However, to distribute GPU resource to all VMs equally is not exactly the best method, especially when the VMs are greatly different. For the diversification of server client mode after virtualization, we propose a novel distribution method of GPU virtualization together with thermal migration to achieve reasonable application of GPU.

2.3   MapReduce

[7] proposes to use GPU based on a computing concept similar to Hadoop. According to MapReduce, the program is first sent to a master node. In the Map phase, the proposed scheme divides the program into suitable sizes, distributes the tasks equally to worker nodes, and tracks the tasks. After the nodes complete the tasks, the worker nodes collect the results by Reduce, which can greatly decrease the computing time. But, one great restriction of this scheme is that only when the computers at all ends belong to the same specification and the type specification of CPU and GPU are the same, can the scheme find out α, the performance ratio of a GPU map task execution to a CPU map task execution. In addition, this scheme does not consider the loading conditions of all worker nodes in the virtual environment and the time difference may occur to parallel computing under different loading conditions, which causes further delay in Reduce.

3   Enhancing Cloud-Based Virtual Servers by GPU Parallel Computing

3.1   Integrating GPU into Cloud Server Virtualization

In the virtualization architecture, because GPU is one of the necessities for future computers, GPU virtualization is inevitable. According to the basic architecture of VMs and the initial ideas about GPU virtualization presented by VMware, GPU is virtualized, just like CPU and other computer devices, for resource management. As shown in Figure 1, each VM has a pass-through GPU to form a channel to stride the resource manager for GPU utilization and to establish GPU driver for Apps on all kinds of VMs. Moreover, there is another channel from the resource manager to Emulation for GPU management.

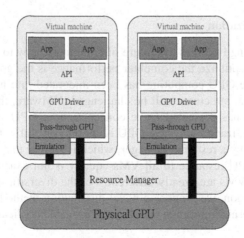

Fig. 1. GPU virtualization of VMware (Source: VMware)

Nevertheless, GPU is not suitable for public sharing because large number of data transmissions will influence the efficiency of GPU computing. Furthermore, in general cloud computing applications, only high-performance and multithreaded computing, including real-time image processing, atmospheric simulation and prediction, astrophysics, quantum mechanics, fluid mechanics, etc., can make good use of GPU computing.

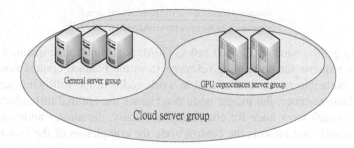

Fig. 2. Cloud Server Group

Therefore, we propose to add GPUs to a cloud server group and classify the servers into two subgroups: general server group that occupy the great majority of the group, and GPU coprocessors server group, as displayed in Figure 2. As for GPU coprocessors server group, we suggest that a VM occupies a GPU at one time to complete one single task within the minimum time. Figure 3 shows that only one VM controls one GPU at one time. When a VM needs a GPU, the VM monitor gives the GPU usage right to the VM and completes the computing task by MapReduce within the minimum time. Finally, the GPU usage right will be returned to the VM hypervisor.

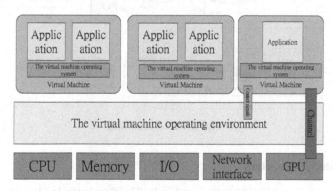

Fig. 3. Architecture of GPU Virtualization

By making an improvement of the method presented in [8], our proposed scheme aims to achieve balance control but the control node does not take charge of all data transmissions for fear of causing heavy burdens. In our opinion, the balance control node is only responsible for resource management while all work nodes have to return CPU loading, GPU loading and the current user list to the balance control node.

Our task scheduling is displayed in Figure 4. First, the client sends the task to the cloud server and the master node estimates the task. Second, when the program asks for more CPU/GPU resource, the master node cuts the task into small tasks of the same size. By referring to [7], we can find out the performance ratio of a GPU map task execution to a CPU map task execution, α. Let

$$\alpha = \frac{mean\,map\,task\,execution\,time\,on\,CPU\,cores}{mean\,map\,task\,execution\,time\,on\,GPU\,cores} \tag{1}$$

Then we can get the number of small tasks. Third, according to the quantity of tasks, the program requests the computing resource from the balance control node, which distributes obtainable resource to available work nodes in the following step. Based on the available resource, the master node determines the optimal task allocation and maps them to each work node for computation. Finally, the master node reduces the calculation results and informs the control node the completion of the task to release computing resource.

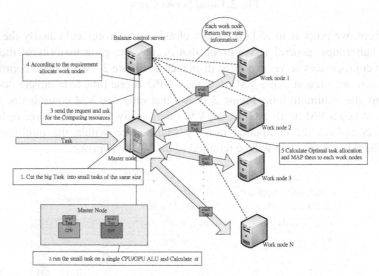

Fig. 4. Task Scheduling

3.2   Related Parameters

Related parameters are divided into two kinds. The first is the loading condition of the server, which is sent to the control node as the index for the master node to request computing resource. We define CPU load as:

$$C_L = \frac{Computing\,the\,amount\,of\,use}{The\,maximum\,computation} \tag{2}$$

and GPU load as:

$$G_L = \frac{Number\ of\ busy\ GPU\ devices}{max\ number of GPU devices} \tag{3}$$

The resource for the control node to distribute is C_L <90% and 0%<G_L<100%.

The second kind of parameters are related to small task blocks and total calculation time. Let

- α be the CPU/GPU calculate time rate.
- C_n be the number of Assigned CPU cores.
- G_n be the number of Assigned CPU cores.
- t be the time for 1 GPU core calculate a task.
- $T = max\ \{\frac{x}{C_n}\alpha t, \frac{y}{G_n}\}$ be the task computation time.

According to Amdahl's law, we know that

$$Speedup = \frac{1}{s+\frac{p}{N}} \tag{4}$$

where s denotes the part that is not improved in the system, p refers to the improved part in the system, and N means the enhance ratio. Next, we will integrate our defined parameters with the formula. Because our assumed scenario is a cloud-based parallel architecture, in which extra overhead must be computed, the formula is revised into:

- T_s = serial code Execution time
- T_p= parallelizable codeExecution time
- T_o = overhead expend time

$$Speedup' = \frac{T_s + T_p}{T_s + \frac{T_p}{N} + T_o} \tag{5}$$

4   Performance Simulation and Analysis

According to the architecture presented in the previous section, we made the following simulation and analysis: Assume that the specifications of the cloud servers are the same and the operational speed of a single CPU core is five times faster than a single GPU core. Each cloud server has four CPU cores and two CPU devices (2*128cores). The considered overhead include the time for data transmissions and MapReduce. Five kinds of program types are taken into consideration:

1. 1% serial code and 99% parallelizable code
2. 5% serial code and 95% parallelizable code
3. 10% serial code and 90% parallelizable code
4. 25% serial code and 75% parallelizable code
5. 50% serial code and 50% parallelizable code

Suppose that all servers are free.

Figure 5 reveals the parallel system gain according to traditional computer based concept. In our simulation, the system gain reaches the maximum when the number of nodes is unit digit. Also, the more proportion the parallelizable code occupies the program, the better the efficiency will be. When the parallelizable code occupies 99% of the program, the system gain reaches 89 times when there are 3 nodes. However, when the parallelizable code occupies 90% of the program, the system gain reduces quickly and the maximal system gain is 19.5 times when the number of nodes is 3 or 4. When the parallelizable code occupies only 50% of the program, the gain is only 2 times at most.

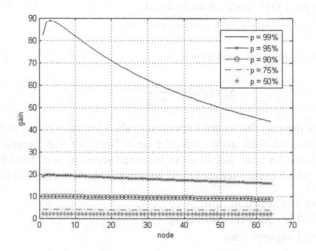

Fig. 5. Theoretical gain of servers

Next, without considering the execution time for collaborative computing, we analyzed the transmission time and the execution time for small chunks. Supposing the execution time for a 10-megabyte serial code on a CPU is 15 seconds, 1% of the program, and 990-megabyte parallelizable code occupies the rest of the program. It takes 1485 seconds to compute the task by CPUs only and 29 seconds by two GPUs. But, it takes only 26.9 seconds to complete the task by 4 CPUs and 2 CPUs.

Table 1. Total Execution Time (1% serial code)

	serial code	parallelizable code	Total time
1 CPU	15s	1485s	1500s
2 GPU	75s	29.01s	104.01s
1 node	15s	26.9s	41.9s
2 nodes	15s	13.45s	28.45s
3 nodes	15s	8.967s	23.967s

In the same way, by using one single CPU to using 1-3 nodes, we estimate the execution time of different program types, as displayed in Figure 6.

Figure 6 shows that to use CPUs only for computing, all kinds of program types can be completed within 1500s. If the program code occupies 99%, the task can be completed in 44 seconds and even 20 seconds with the support of more GPUs. However, the more proportion un-parallelized computing occupies, the less improvements parallel computing can make. Theoretically speaking, with the support of more GPUs, the execution time of the parallelizable code should be reduced to 0. But, more GPUs and more nodes in fact will result in more overhead.

In our proposed scheme, because the server's loading condition is considered, few nodes cannot offer enough GPUs for parallel support (compared with the above-mentioned situation that all servers are free). The maximal gain is obtained when there are 15 nodes. But, more nodes will decrease the performance due to synchronization and transmission. Moreover, when the server is near end, our simulation occupies approximately 2 seconds for transmission.

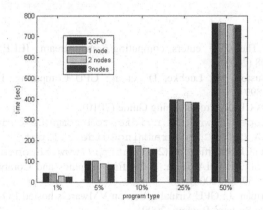

Fig. 6. Execution time of different program types in GPU parallel computing

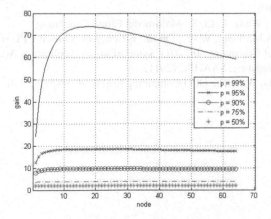

Fig. 7. Server Gain under Load Condition

5   Conclusion and Future Objective

This paper proposed a scheme to integrate GPU with cloud computing for users to utilize high-performance but low-cost GPU resource without building devices by themselves. However, the biggest limitation of GPU computing is the market share of parallel computing, which needs to be popularized by future parallel computing service providers and developed/adopted by numerous application developers. Moreover, the transmission amount of the network and the internal bus in the computer is another bottleneck of parallel computing. Supposing the transmission amount can be further enhanced, a great deal of overhead will be reduced and the maximal gain will appear when much more nodes are utilized for collaborative computing.

Acknowledgement. This study was supported by the National Science Council, Taiwan, under grant no. NSC 100-2219-E-032-001.

References

1. Macedonia, M.: The GPU enters computing's mainstream. IEEE Computer Society 36(10), 106–108 (2003)
2. Owens, J.D., Houston, M., Luebke, D., et al.: GPU Computing. Proceedings of the IEEE 96(5), 879–899
3. NVIDIA, NVIDIA CUDA Programming Guide (2010),
   http://developer.download.nvidia.com/compute/cuda/3_1/toolki t/docs/NVIDIA_CUDA_C_ProgrammingGuide_3.1.pdf
4. Khronos, The OpenCL Specification (2011), http://www.khronos.org/opencl/
5. Lindholm, E., et al.: NVIDIA Tesla: A Unified Graphics and Computing Architecture. IEEE Micro 28(2), 39–55 (2008)
6. Dowty, M., Sugerman, J.: GPU virtualization on VMware's hosted I/O architecture. ACM SIGOPS Operating Systems Review (2009)
7. Shirahata, K., Sato, H., Matsuoka, S.: Hybrid Map Task Scheduling for GPU-based Heterogeneous Clusters. In: IEEE International Conference on Cloud Computing Technology and Science (2010)
8. Zhu, W., Luo, C., Wang, J., Li, S.: Multimedia Cloud Computing
9. Daga, M., Aji, A.M., Feng, W.-C.: On the Efficacy of a Fused CPU+GPU Processor (or APU) for Parallel Computing. In: 2011 Symposium on Application Accelerators in High-Performance Computing, SAAHPC (2011)
10. Nickolls, J., Dally, W.J.: The GPU Computing Era. IEEE Micro (2010)

Controlled Quantum Secure Direct Communication Based on Single Photons

Wei-Lin Chang, Fang-Jhu Lin, Guo-Jyun Zeng, and Yao-Hsin Chou

Department of Computer Science and Information Engineering
National Chi-Nan University,
No. 1, University Rd., Puli 54561, Taiwan
{s98321045,s100321529,s100321542}@mail1.ncnu.edu.tw, yhchou@ncnu.edu.tw
http://www.csie.ncnu.edu.tw

Abstract. We propose a quantum secure direct communication protocol with a controller, who assists the agent gets the secret messages sent from the dealer. Single photons are used to carry dealer's message, so the cost of our protocol is less than others use entangled qubits. If any eavesdropper tries to steal dealer's messages, the lawful participants will perceive it and abort their transmission.

Keywords: quantum secure direct communication, single photons.

1   Introduction

The development of quantum mechanics gets more and more attention nowadays, because many algorithms, which base on quantum mechanics, have been proved efficiently than classical algorithms, likes fast Z buffers, instant radiosity, double-speed transmission, factorization of an integer, etc. In the quantum mechanics, factorization of an integer can be completed in polynomial time by using the Shors algorithm [1]. RSA is one kind of application of factorization of an integer, which bases its security on mathematics complexity. It means RSA is insecure, if quantum mechanics can be implemented on computers. Because of this reason, we consider that the security of algorithms should be based on physical complexity not mathematics complexity. That is also why quantum cryptography attracts much attention now.

Quantum communication becomes one of the most important applications of quantum mechanics nowadays. Quantum key distribution (QKD) is one of the most mature techniques of branch of quantum information theory [2, 3], which enables two remote legitimate users establish a shared secret key through the transmission of photons, and use this key to encrypt (decrypt) the secret messages. Since the first QKD scheme was proposed in 1984 [4], many QKD schemes have been presented and improved [5–10].

Recently two novel quantum communications were proposed [11–29]: quantum secure direct communication (QSDC) [11–23] and quantum secret sharing (QSS) [24–29]. In QSDC, secret messages can be transmitted directly from the sender to the receiver without the classical communication of ciphertext, in other words,

J.-S. Pan et al. (Eds.): *Advances in Intelligent Systems & Applications*, SIST 21, pp. 195–204.
DOI: 10.1007/978-3-642-35473-1_21   © Springer-Verlag Berlin Heidelberg 2013

the plaintext doesnt have to be encrypted and decrypted with the shared key, and the messages just transmit during the quantum communication. QSDC has a great potential in the future because it relies on quantum mechanics. QSS is the combination of classical secret sharing and quantum mechanics, it can share both classical and quantum messages among sharers. The methods of message transmission can be classified into two types: entanglement swapping [8, 12, 13, 16–22] and single photons transmission [11, 14, 15, 23], and the message of transmission can be classified into quantum and classical information.

In this paper, we present controlled quantum secure direct communication (CQSDC) scheme by using single photons. In the present scheme, the sender's secret message is transmitted directly to the receiver and can only be reconstructed by the receiver with the help of the controller. Different from QSS, the sender transmits his/her secret messages to the receiver directly, and the information of the receiver is asymmetric to that of the controller. Our scheme employs single photons and the transformation which bases on Cai et al. [23], but improves its security, all single photons are used to transmit the secret message except those chosen for eavesdropping check. We also discuss the security of the scheme and compare performance with Gao et al. [30] and Wang et al. [31].

2   Related Work

In this section, we will briefly introduce the concept of Cai et al. [23], Gao et al. [30] and Wang et al. [31]. Before we introduce these schemes, we need to define some operations as follows:

$$|0\rangle = \begin{bmatrix} 1 \\ 0 \end{bmatrix}, |1\rangle = \begin{bmatrix} 0 \\ 1 \end{bmatrix}, |+\rangle = \frac{1}{\sqrt{2}}(|0\rangle + |1\rangle), |-\rangle = \frac{1}{\sqrt{2}}(|0\rangle - |1\rangle)) \qquad (1)$$

$$I = |0\rangle\langle 0| + |1\rangle\langle 1|, \sigma_z = |0\rangle\langle 0| - |1\rangle\langle 1|,$$

$$\sigma_x = |0\rangle\langle 1| + |1\rangle\langle 0|, i\sigma_y = |0\rangle\langle 1| - |1\rangle\langle 0| \qquad (2)$$

$$U_1 = \sigma_z \otimes \sigma_z, U_2 = I \otimes \sigma_z, U_3 = i\sigma_y \otimes \sigma_z, U_4 = \sigma_x \otimes \sigma_z,$$

$$U_5 = I \otimes \sigma_x, U_6 = \sigma_z \otimes \sigma_x, U_7 = \sigma_x \otimes \sigma_x, U_8 = i\sigma_y \otimes \sigma_x \qquad (3)$$

Besides, we refer Alice as sender, Bob as receiver and Charlie as controller in these protocols.

2.1   Deterministic Secure Communication without Using Entanglement

In 2004, Cai and Li [23] proposed a deterministic secure direct communication protocol using single photon in a mixed state, following some ideas from the ping-pong protocol [12]. Now we give a brief description of this protocol.

Step1. Bob prepares a single photon randomly in one of the two states $\{|0\rangle, |1\rangle\}$, and then sends this photon to Alice.

Step2. Alice chooses message mode or control mode randomly to deal with this photon, which is similar to the ping-pong protocol. If Alice chooses control mode, she will replace the photon with a new one which is in one of the four state $\{|0\rangle, |1\rangle, |+\rangle, |-\rangle\}$, otherwise she encodes the photon with I or σ_y according to the bit value of the secret message is 0 or 1, respectively, and then sends this qubit back to Bob.

Step3. Bob performs a measurement on the photon with the same measurement basis as he originally chooses for preparing it, and informs Alice that he had received the qubit. If the photon is in control mode, Alice will tell Bob which state she prepared, so Bob can check whether there exists eavesdropper or not. If the photon is in message mode, then Bob can decode Alice's messages successfully.

The method of communication is insecure in the lossy quantum channel, because eavesdropper Eve has high probability to know which basis Alice prepared. For example, Eve intercepts the photon from Alice and using Z-basis (X-basis) to measure it, if the outcome is $|1\rangle$ ($|-\rangle$), she sends $|+\rangle$ ($|0\rangle$) to Bob, otherwise, she sends nothing. Because the quantum channel is lossy, both Alice and Bob couldn't detect the existence of Eve. For this reason, our scheme chooses randomly in one of four state $\{|0\rangle, |1\rangle, |+\rangle, |-\rangle\}$ in prepare phase to prevent this kind of attack.

2.2 Controlled Quantum Teleportation and Secure Direct Communication

In 2005, Gao et al. [30] proposed a controlled quantum teleportation and secure direct communication, this scheme takes three steps to accomplish the communication.

Step1. At first, Alice, Bob, and Charlie share a set of triplets of qubits in an entangled state $|\xi\rangle_{ABC} = \frac{1}{2}(|000\rangle + |110\rangle + |011\rangle + |101\rangle)_{ABC}$, sequence of particles A is in Alice's hand, sequence of particles B is in Bob's hand and the others belong to Charlie. Alice prepares her message particles in $|+\rangle$ or $|-\rangle$ according to her message, which corresponds to 1 and 0 respectively. The quantum state of the whole system is in

$$|\varphi\rangle_M |\xi\rangle_{ABC} = \frac{1}{\sqrt{2}}(|0\rangle + |1\rangle)_M \otimes \frac{1}{2}(|000\rangle + |110\rangle + |011\rangle + |101\rangle)_{ABC},$$

where $b = 0$ and $b = -1$ correspond to $|+\rangle$ and $|-\rangle$, respectively, and M represents the message. If Charlie allows the communications between the two users, he performs measurements on his qubit C in Z-basis and announces publicly the measurement result on classical channel.

Step2. After Charlie announces his measurement result, Alice performs Bell measurement on her particle M and A. That leads particle C, M and A to collapse into one of eight states $\{ |0\rangle_C|\Phi^+\rangle_{MA}, |0\rangle_C|\Psi^+\rangle_{MA}, |0\rangle_C|\Phi^-\rangle_{MA}, |0\rangle_C|\Psi^-\rangle_{MA}, |1\rangle_C|\Phi^+\rangle_{MA}, |1\rangle_C|\Psi^+\rangle_{MA}, |1\rangle_C|\Phi^-\rangle_{MA}, |1\rangle_C|\Psi^-\rangle_{MA} \}$. Here $|\Phi^\pm\rangle_{MA} = \frac{1}{\sqrt{2}}(|00\rangle) \pm |11\rangle)_{MA}, |\Psi^\pm\rangle_{MA} = \frac{1}{\sqrt{2}}(|01\rangle) \pm |10\rangle)_{MA}\}$

Step3. Bob can 'fix up'his particle B to recover the signal state $|\varphi\rangle_B = \frac{1}{\sqrt{2}}(|0\rangle + b|1\rangle)_B$ by applying appropriate quantum gate like quantum teleportation. Then Bob measures these particles in X-basis and reads out the message that Alice wants to transmit to him.

2.3   Multiparty CQSDC Using Greenberger-Horne-Zeilinger State

In 2006, Wang et al. [31] proposed a multiparty controlled quantum secure direct communication using Greenberger-Horne-Zeilinger state, this scheme consists of six steps as following:

Step1. Charlie prepares ordered N three-photon in $|\psi\rangle = \frac{1}{\sqrt{2}}((|000\rangle + |111\rangle_{ABC})$. We denote the ordered N three-photon with $\{ [P_1(A), P_1(B), P_1(C)], [P_2(A), P_2(B), P_2(C)], \ldots, [P_n(A), P_n(B), P_n(C)] \}$, we call $[P_1(A), P_2(A), \ldots, P_1(A)]$ as A sequence, and so on. Charlie selects one of the four unitary operations $\{ I, \sigma_z, \sigma_x, i\sigma_y \}$ randomly, and performs it on each of the photons in B sequence, then sends A, B sequences to Alice and keeps C sequence.

Step2. Alice selects randomly a sufficiently large subset from those received photons for eavesdropping check as follows: (a) Alice announces publicly the positions of the selected photons. (b) Charlie publishes his operations performed on these photons in B sequence. (c) Alice chooses measuring basis Z or X randomly to measure each of the selected photons in A, B sequences and announces those basis publicly. (d) Alice and Charlie both measure the corresponding photons according to the basis Alice announced, and then Charlie publishes his measurement outcomes. (e) Now Alice has Charlie's operations and his measurement results on selected photons B, so she can check the security by comparing their measurement outcomes.

Step3. Charlie chooses randomly one of the four unitary operations $\{I, \sigma_z, \sigma_x, i\sigma_y\}$ and performs these operations on each photon in C sequence. He then sends C sequence to Bob.

Step4. Alice and Bob analyze the error rate of the transmission of C sequence similar to step 2.

Step5. Alice first selects randomly two sufficiently large subsets from A, B sequences for eavesdropping checks and then performs randomly one of the eight operations $U_k(k = 1, \ldots, 8)$ on each of single sample photons, and do the same thing on the remaining photons according to secret messages, and then sends B sequence to Bob. After Bob receives B sequence, Alice performs randomly Z or X basis measurement on the sampling photons of one subset in A sequence. She then publishes the positions of the sampling photons of this subset and which measurement basis she uses for each of the sampling photons, and then Bob measures with the same basis on each photon in B, C sequences. Before Bob publishes his measurement results, Alice let Charlie announce his operation information on the sampling photons in B and C sequences, so Alice and Bob can check eavesdropping by comparing their measurement results. If the channel is safe, Alice sends A sequence to Bob. After Bob receives A sequence, Alice publishes the positions of the sampling photons of the other subset and lets Bob make GHZ measurement on the sampling photons in A, B and C sequences. Then Bob announces his measurement results and let Charlie publish his operations on the sampling photons in B and C sequences for eavesdropping check.

Step6. Now Bob owns A, B and C sequences, but without Charlie's operation information Bob cant obtain Alices secret message. After Charlie publishes his operations on the photons in B and C sequences, Bob can perform GHZ measurements to get Alice secret message.

3   Description of Our Protocol

In this section, we present a controlled quantum secure direct communication (CQSDC) by using N single qubits. In Cais protocol [23], each time Alice sends one single qubit to Bob and repeats the transmission until Bob gets all secret messages, but in our scheme, Alice transmits her message at one single round. Suppose Alice wants to transmit her secret messages of N bits directly to Bob. The point is that Alice hopes that Bob must recover his secret under Charlies permission. We show the schematic demonstration in Fig. 1, and our protocol works with the following steps:

Step1: Controller Prepares for the Communication Qubits
Charlie is the controller and prepares N single qubits in one of following four states randomly $\{I, \sigma_z, \sigma_x, i\sigma_y\}$. Then Charlie sends these N single qubits to Alice.

Step2: Security Checking of the First Time Qubits Transmission
After receiving these N single qubits sent by Charlie, Alice selects some single qubits (the number of qubits is X) from them and chooses randomly a measuring basis Z-basis or X-basis, measuring the selected qubits for eavesdropping check. Alice must tell Charlie the information including which basis she uses, the positions of selected qubits and her measurement results. Comparing with the qubits states in the initial, Charlie could know whether the eavesdropper Eve exists or not. If the Eve is detected, the communication stops. If the channel is safe, Alice throws away the qubits used for checking and the communication continues.

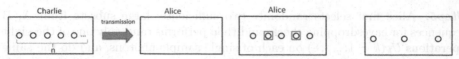

(a) Step1. Controller prepares the commu-   (b) Step2. Alice checks channel security.
nication qubits and sends to Alice.

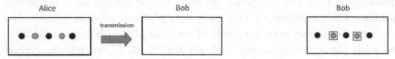

(c) Step3. Alice encodes the messages and   (d) Step4. Bob checks channel security.
prepares the second time security checking
bits then sends to Bob.

(e) Step5. Bob extracts the message with
the controller's help.

Fig. 1. Schematic demonstration of this present protocol. The hollow circles represent unencoded qubits, the solid circles represent encoded qubits, and the blue circles denote additional single qubits which used for security checking. The red squares in step 2 and step 4 denote the projective measurements. In step 5, after Charlie publishes his unitary operations, Bob could get Alices messages.

Step3: Sender Encodes the Messages and Prepares the Second Time Security Checking

Now Alice could encode her secret messages on the remaining photons by doing these following things. (1) According to what secret messages she wants to send to Bob, Alice performs operation I or $i\sigma_y$ on each remaining quit (the number is N-X). If Alice wants to send the bit 0, then she performs operation I,if Alice wants to send the bit 1, then she will perform operation $i\sigma_y$.(2) Next, Alice prepares additional single qubits (the number is X), which states is one of four states $\{I,\sigma_z,\sigma_x,i\sigma_y\}$ randomly, and inserts these qubits into coding qubits in random positions. Then Alice sends both these additional qubits and the coding qubits (the number of all qubits is N) to Bob. The purpose of preparing these additional qubits is to check the existence of eavesdropper.

Step4: Security Checking of the Second Time Qubits Transmission

After Bob receives qubits from Alice, Alice must tell Bob the positions of additional qubits she prepares. Bob chooses randomly a measuring basis Z-basis or X-basis and measures on the additional qubits Alice prepares. Bob must tell Alice which basis he uses and his measurement results. Comparing with the qubits states Alice prepares, Alice could know whether there is an eavesdropper or not. If Eve is detected, the communication stops.

Step5: Receiver Extracts the Message with the Controllers Help
Bob throws away the qubits used for checking and now owns (N-X) single qubits from Alice. Without Charlies permission, Bob cant acquire Alices secret message. Only after Charlie publishes the initial states of the single qubits, could Bob recover Alices secret message by comparing with the initial state.

4   Security Analysis

So far, we have described our protocol in detail, now we will discuss the security against possible eavesdroppers attacks. There two types of attacks, one is external eavesdropper and the other is internal. In this analysis, we assume that Bob and Charlie all could be the internal eavesdropper. We first consider the external eavesdropper attacks.

Suppose Eve is an external eavesdropper who tries to get the Alices secret messages without permission. That means she needs to know the initial states Charlie prepared, and intercepts the photons after Alice encodes so she could extract the message. There are two ways for Eve to get Charlies initial states, she can capture the photons in Alice-Charlie quantum channel or waits until Charlie declares it. The following strategies are the attacks that Eve may take.

(1)Intercept-Measure-Resent Attack. Since Eve wants to know initial states, she may intercepts those photons sent from Charlie to Alice, and then measures it before returns to Alice. Because she doesnt know which basis Charlie prepares for each photon, she may measure each photon in Z-basis or X-basis randomly. The probability that Eve chooses correct measurement basis for each photon is $(\frac{1}{2})^N$,which is extremely small. So when Alice and Charlie check whether there is an eavesdropper or not, it is almost impossible for Eve to escape from it. Similarly, if Eve captures the photons sent from Alice to Bob, she will face the same situation like we describe above.

(2)Control-Not Attack. After Eve intercepts photons in Alice-Charlie quantum channel, she may use control-not gates to entangle a qubit in $|0\rangle$ on each photon and sends the original qubits back to Alice. By doing this, the qubits she entangles now have the same states like the original ones. But like the first attack, when Alice and Charlie check the security of quantum channel, they will find the existence of Eve. For example, suppose there has one qubit in $|+\rangle$ and Eve uses control-not gates to entangle $|0\rangle$ on the qubit, which makes the entangled qubits become

$$|\Phi^+\rangle = \frac{1}{\sqrt{2}}(|00\rangle + |11\rangle) = \frac{1}{\sqrt{2}}(|++\rangle + |--\rangle)$$

That means even Alice measures with X-basis, she still may find the outcome which is opposite to the initial state, so the participants will know there has an eavesdropper between them.

Now we discuss the internal eavesdropper attacks. If Charlie, the controller, is an eavesdropper and he tries to steal the photons after Alice encodes to get the final states. He may use the two kinds of attacks we described above, but it only makes the transmission abort or even worse, he may obtain the wrong secret messages. It is because before Alice sends the encoded photons to Bob, she inserts some additional qubits in random positions, and both of the attacks would not only change the states of additional qubits but also the original qubits. Charlie cannot gain any profit of being the eavesdropper.

For Bob, if he is the eavesdropper, he wants to secretly get the initial states Charlie prepared without his help. Like the case of external eavesdropper attacks, Alice and Charlie would perceive it by security checking and terminate their transmission. It only makes Bob get nothing about Alices message.

From the above discussion, we can see if any eavesdropper tries to gain knowledge about those qubits, it will inevitably disturb the qubit states, so the lawful participants know there has an eavesdropper between them. Therefore, our protocol is secure against eavesdroppers attacks.

5   Performance

We found the research papers of CQSDC are still rare, so we choose two one-way transmission protocols to compare with. They are Gao et al. [30] and Wang et al [31]. We assume Alice wants to transmit M classical bits to Bob and calculate the cost of transmission that each of protocols takes. Here, the cost of security checking is not considered. From the Table 1, we can see the most costs of our protocol are equal or less than the others, expect the cost of classical bits and the number of projective measurement when it compares to Wangs. As we know, the cost of classical bits is common and the single qubit projection is also cheaper than GHZ measurement. Besides, single photons are much easier to produce than entangled qubits, so our entire cost is less than Wangs. Finally, this present protocol is more efficient than the others, when it comes to achieve the same goal.

Table 1. The performance comparison between Gao[30], Wang[31] and ours. In the column of the number of Bell measurement /GHZ measurement, the m in Gaos row means it takes m times Bell measurement, and the m/3 in Wangs row denotes it takes m/3 times GHZ measurement.

protocols	The cost of each protocol and the performance comparison				
	cost of qubits	cost of bits	the number of qubit transmission	the number of Bell measurement /GHZ measurement	the number of projective measurement
Gao[30]	4M	3M	2M	M	2M
Wang[31]	M	4M/3	5M	M/3	0
ours	M	2M	2M	0	M

6   Conclusion

In general, the works on quantum secure direct communication (QSDC) attracted a great deal of attention and can be divided into two kinds, one utilizes single photon, and the other bases on entangled qubits. However, in QSDC, without using single photons but entangled qubits may need more complicated experimental setup for transmission and the cost is higher. Under this consideration, we choose single qubits as our transmission media, and it turns out to be a more efficient communication protocol as we show in the performance section. In some special circumstances, we may need a third party to authenticate the communication, so our scheme has a controller, Charlie. He is in charge of preparing the initial state of N single qubits and without his permission Bob cant recover the secret message Alice wants to send to him. In our protocol, CQSDC with single-photon is simple, useful and efficient. It will be more convenient for being implemented and we still can guarantee its security of the communication transmission.

References

1. Shor, P.W.: Proc. 35th Annual Symposium on Foundations of Computer Science, p. 124 (1994)
2. Nielsen, M.A., Chuang, I.L.: Quantum computation and quantum information (2000)
3. Vernam, G.S.: J. Amer. Inst. Elec. Eng. 45, 109 (1926)
4. Bennett, C.H., Brassard, G.: Proceedings of IEEE international Conference on Computer, Systems, and Signal Processing, Bangalore, p. 175 (1984)
5. Ekert, A.K.: Phys. Rev. Lett. 67, 661 (1991)
6. Bennett, C.H., Brassard, G., Mermin, N.D.: Phys. Rev. Lett. 68, 557 (1992)
7. Bruß, D.: Phys. Rev. Lett. 81, 3018 (1998)
8. Long, G.L., Liu, X.S.: Phys. Rev. A 65, 032302 (2002)
9. Xue, P., Li, C.F., Guo, G.C.: Phys. Rev. A 65, 022317 (2002)
10. Acin, A., Gisin, N., Masanes, L.: Phys. Rev. Lett. 89, 187902 (2002)
11. Beige, A., Englert, B.G., Kurtsiefer, C., Weinfurter, H.: Acta Phys. Pol. A 101, 357 (2002)
12. Boström, K., Felbinger, T.: Phys. Rev. Lett. 89, 187902 (2002)
13. Deng, F., Long, G., Liu, X.: Phys. Rev. A 68, 042317 (2003)
14. Deng, F., Long, G.: Phys. Rev. A 69, 052319 (2004)
15. Lucamarini, M., Mancini, S.: Phys. Rev. Lett. 94, 140501 (2005)
16. Shimizu, K., Imoto, N.: Phys. Rev. A 60, 157 (1999)
17. Deng, F.G., Li, C.Y., Zhou, P., Zhou, H.Y.: Phys Lett. A 359, 359 (2006)
18. Li, X.H., Zhou, P., Liang, Y.J., Li, C.Y., Zhou, H.Y., Deng, F.G.: Chin. Phys. Lett. 23, 1080 (2006)
19. Wang, C., Deng, F.G., Li, Y.S., Liu, X.S., Long, G.L.: Phys. Rev. A 71, 044305 (2005)
20. Wang, C., Deng, F.G., Long, G.L.: Opt. Commun. 253, 15 (2005)
21. Zhou, P., Li, X.H., Liang, Y.J., Deng, F.G., Zhou, H.Y.: Physica A: Statistical Mechanics and its Applications 381, 164 (2007)
22. Cai, Q.Y., Li, B.W.: Phys. Rev. A 69, 054301 (2004)
23. Cai, Q.Y., Li, B.W.: Chin. Phys. Lett. 21, 601 (2004)

24. Hillery, M., Buzek, V., Berthiaume, A.: Phys. Rev. A 59, 1892 (1999)
25. Karlsson, A., Koashi, M., Lmoto, N.: Phys. Rev. A 59, 162 (1999)
26. Zhang, Z.J.: Phys. Lett. A 342, 60 (2005)
27. Guo, G.P., Guo, G.C.: Phys. Lett. A 310, 247 (2003)
28. Zhang, Z.J., Li, Y., Man, Z.X.: Phys. Rev. A 71, 044301 (2005)
29. Xiao, L., Long, G.L., Deng, F.G., Pan, J.W.: Phys. Rev. A 69, 052307 (2004)
30. Gao, T., Yan, F.L., Wang, Z.X.: Chin. Phys. 14(5), 1009 (2005)
31. Wang, J., Zhang, Q., Tang, C.J.: Opt. Commun. 266, 732 (2006)

Construction of a Machine Guide Dog Using a Two-Mirror Omni-camera and an Autonomous Vehicle*

Chih-Wei Huang[1] and Wen-Hsiang Tsai[1,2,**]

[1] Institute of Computer Science and Engineering, National Chiao Tung University, Taiwan
sakuraai1986.cs99g@nctu.edu.tw
[2] Department of Information Communication, Asia University, Taiwan
whtsai@cis.nctu.edu.tw

Abstract. A system for use as a machine guide dog composed of an autonomous vehicle and a two-mirror omni-camera for navigations on sidewalks to guide blind people is proposed. Methods for extracting 3D information from acquired omni-images to localize the vehicle using landmarks of curb lines, tree trunks, stop lines on roads, lawn corners, signboards, and traffic cones are proposed. The methods are based on a space-mapping scheme and three new space line detection techniques. Each space line detection technique can be applied directly on omni-images to compute the 3D locations of a specific type of space line in the landmark shapes. Good experimental results show the feasibility of the proposed system.

Keywords: machine guide dog, autonomous vehicle, landmark detection, vehicle localization, omni-images.

1 Introduction

There are millions of blind people in the world. Some of them use blind canes to walk on the road. However, blind canes can only detect obstacles at short distances. A better choice is to use guide dogs as shown in Fig. 1(a). However, guide dogs are very few; e.g., there are about 60,000 blind people but just about 30 guide dogs in Taiwan in 2012. So it is of great advantages if *machine guide dogs* can be designed for use by the blind. To implement a machine guide dog, one way is to use a vision-based autonomous vehicle which can navigate automatically in outdoor environments and keep watch over the camera's field of view (FOV) to avoid collisions with obstacles. In this study, we use a specially-designed omni-camera with two mirrors as the vision system on an autonomous vehicle for this purpose.

Localization is a critical issue in implementing a navigation system, by which a vehicle can move on correct paths. Willis and Helal [1] provided a navigation system for the blind using the RFID technology to identify building and room locations.

* This work was supported financially by the Ministry of Economic Affairs under Project No. MOEA 100-EC-17-A-02-S1-032 in Technology Development Program for Academia.
** Corresponding author.

J.-S. Pan et al. (Eds.): *Advances in Intelligent Systems & Applications*, SIST 21, pp. 205–219.
DOI: 10.1007/978-3-642-35473-1_22   © Springer-Verlag Berlin Heidelberg 2013

Chen and Tsai [2] proposed an indoor autonomous vehicle system using ultrasonic sensors. In outdoor spaces, the GPS can be used as a localization system for the vehicle [3]. Also, Atiya and Hager [4] proposed a vision-based system which can localize a mobile robot in real time. To enhance localization accuracy, Lui and Jarvis [5] constructed an outdoor robot with a GPU-based omni-vision system possessing an automatic baseline selection capability. To detect landmarks in environments, Fu et al. [6] proposed a navigation system with embedded omni-vision for multi-object tracking. A vehicle which achieves self-localization by matching omni- images was proposed by Ishizuka et al. [7].

  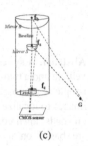

      (a)   (b)   (c)

Fig. 1. Guide dogs and system configuration. (a) A guide dog. (b) View of system. (b) Architecture of camera system (upright).

The goal of this study is to navigate a vision-based machine guide dog on outdoor sidewalks automatically. To achieve this goal, the major task is *vehicle localization*. The strategy for vehicle localization proposed in this study is to detect as many types of landmarks as possible along paths on sidewalks. The operation of the proposed system includes two stages: *learning* and *navigation*. The learning stage includes primarily the task of training the vehicle to acquire the along-path information useful for later vehicle guidance in the navigation stage. A scheme for training a vehicle for outdoor navigation along sidewalks is proposed first. Then, new space line detection and localization techniques based on the space-mapping method [9] are proposed next. These techniques then are applied to detect natural landmarks of lawn corner and tree trunks as well as artificial landmarks of signboards, stop lines on roads, and traffic cones for vehicle localization. Also proposed are methods for dynamically adjusting the vehicle guidance scheme to overcome varying outdoor lighting conditions.

In the remainder of this paper, the proposed path learning process and navigation strategies will be presented in Sections 3 and 4, respectively. The proposed new space line detection techniques and their applications for landmark detection and localization will be described in Sections 4 and 5, respectively. Finally, some experimental results and conclusions will be given in Section 6.

2   Learning Stage

The purpose of the proposed learning process is to create a path consisting of nodes on a sidewalk to be visited by the vehicle toward a destination. At first, some landmarks are selected for vehicle localization. Then, the used camera system is calibrated. At last, some parameter information of each landmark is extracted and recorded in the path.

2.1   Learning of Selected Landmarks

When the vehicle navigates for a time period, mechanic errors will accumulate to cause imprecise odometer readings of the vehicle location and orientation. To solve this problem, Chou and Tsai [8] proposed methods for detecting light poles and hydrants as landmarks to localize the vehicle. In this study, we select additionally two types of *natural* landmarks, *tree trunk* and *lawn corner*, and three types of *artificial* landmarks, *signboard*, *traffic cone*, and *stop line on the road*. With these additional types of landmarks, more information along the path can be used for vehicle localization, and so the vehicle can be guided more reliably to the destination. We "learn" these landmarks by driving the vehicle to get close to each of them and recording the *vehicle direction* with respect to the nearby curb line as well as the *vehicle location* with respect to the landmark.

2.2   Construction of Pano-mapping Table as Camera Calibration

The used camera system as illustrated in Figs. 1(b) and 1(c) consists of a perspective camera, a lens, and two reflective mirrors of different sizes, all integrated into a single structure. We call the big mirror *Mirror B*, and the small *Mirror S*, respectively. The camera system is slanted for an angle of γ to enlarge the imaged frontal scene portion.

To "calibrate" the camera system, we use the space-mapping method proposed by Jeng and Tsai [9] by creating a *pano-mapping table* to record the relations between the locations of image points and those of the corresponding world-space points. A light ray is assumed to go through each world-space point P with an elevation angle α and an azimuth angle θ, be reflected by the mirror of the camera system, and be projected onto the omni-image plane as a point p at coordinates (u, v), as illustrated in Fig. 2. The pano-mapping table, like the one shown in Table 1, specifies the relation between the coordinates (u, v) of the image point p and the azimuth-elevation angle pair (θ, α) of the corresponding world-space point P.

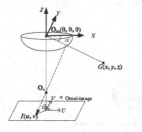

Fig. 2. Omni-imaging principle

Table 1. A pano-mapping table used for the omni-camera

	θ_1	θ_2	θ_3	...	θ_S
β_1	(u_{11}, v_{11})	(u_{21}, v_{21})	(u_{31}, v_{31})	...	(u_{S1}, v_{S1})
β_2	(u_{12}, v_{12})	(u_{22}, v_{22})	(u_{32}, v_{32})	...	(u_{S2}, v_{S2})
β_3	(u_{13}, v_{13})	(u_{23}, v_{23})	(u_{33}, v_{33})	...	(u_{S3}, v_{S3})
β_4	(u_{14}, v_{14})	(u_{24}, v_{24})	(u_{34}, v_{34})	...	(u_{S4}, v_{S4})
...	...	...	...	...	...
β_T	(u_{1T}, v_{1T})	(u_{2T}, v_{2T})	(u_{3T}, v_{3T})	...	(u_{ST}, v_{ST})

3   Navigation Stage

After learning the navigation path like that shown in Fig. 3, we use it to guide the vehicle. The path includes a series of nodes, through which the vehicle can move to the destination. More details are described in the following sections.

3.1   Navigation Strategy Adopted in This Study

The vehicle presumably could localize itself by the on-board odometer readings to conduct *node-based navigation*; that is, the three readings (P_x, P_y, P_θ) provided by the odometer might be used as the *vehicle pose* to identify the vehicle position and orientation in a global coordinate system (GCS) at each path node, and so can be utilized to navigate the vehicle correctly on the path. However, these odometer values are in general imprecise because of the mechanic errors accumulated during the navigation. Therefore, a *vehicle localization* process should be conducted at each node. The proposed strategy for this purpose includes two major steps: (1) adjust the erroneous vehicle orientation by the use of the detected curb line orientation; and (2) correct the vehicle position by the use of the estimated pre-selected landmark location. Note that the GCS is defined on the sidewalk for each navigation session with the start vehicle position as the origin, and the forward moving direction of the vehicle as the vertical axis.

3.2   Vehicle Localization by Selected Landmarks

To reduce the influence of accumulated mechanic errors, we conduct vehicle localization by the use of curb lines and landmarks as mentioned. Specifically, when the vehicle arrives at a node with inaccurate odometer readings $(P_x'', P_y'', P_\theta'')$, according to the first step of the above-mentioned vehicle localization strategy, we detect the straight curb line in the acquired omni-image, and compute its orientation θ' with respect to the moving direction of the vehicle [8], as illustrated in Fig. 4. Comparing θ' with the recorded curb line orientation θ in the learned path data, we compute the deviation of the vehicle orientation as $\Delta\theta = \theta' - \theta$ and adjust the vehicle

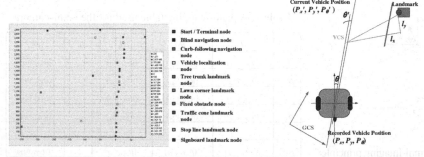

Fig. 3. Illustration of learned navigation path   **Fig. 4.** Computing current vehicle location

orientation for the amount of $\Delta\theta$ to obtain a *calibrated* vehicle orientation P_θ' computed as $P_\theta' = P_\theta'' - \Delta\theta$.

Afterward, we start to detect the landmark of the current node, if recorded, and obtain its position (l_x, l_y) with respect to the vehicle coordinate system (VCS) (the details described later). According to the learned landmark position (L_x, L_y) in the GCS recorded in the path data and the calibrated vehicle orientation P_θ', we compute the current vehicle position (P_x', P_y') by the following equations (see Fig. 4 for an illustration):

$$\begin{bmatrix} P_x' \\ P_y' \end{bmatrix} = \begin{bmatrix} L_x \\ L_y \end{bmatrix} + \begin{bmatrix} \cos P_\theta' & \sin P_\theta' \\ -\sin P_\theta' & \cos P_\theta' \end{bmatrix} \begin{bmatrix} l_x \\ l_y \end{bmatrix}. \tag{1}$$

Finally, we replace the odometer readings, $(P_x'', P_y'', P_\theta'')$, with the above corrected vehicle pose parameters (P_x', P_y', P_θ') and navigate the vehicle forward to the next node.

4   Natural Landmark Detection for Vehicle Localization

It is found in this study that the uses of space lines are sufficient to localize many types of landmarks. However, compared with the result of using images acquired by the traditional projective camera, the projection of a space line onto an *omni-image* taken by an omni-camera is not a line but a *conic-section curve* [10]. Wu and Tsai [10] proposed a method for detecting *directly* such curves in omni-images of an H-shaped landmark used in automatic helicopter landing. In this study, we propose instead new space line detection techniques based on the space-mapping method [9] using *pano-mapping tables*. An essence of the proposed techniques is to utilize the *space plane* which goes through the space line and the center of the mirror of the omni-camera, instead of trying to obtain directly the conic section curve in the omni-image. More details are described next.

4.1   Line Detection Using Pano-mapping Table

Assume that a pano-mapping table has been set up for the omni-camera, and that a space line L to be detected is projected by *Mirror B* onto the omni-image with G being a point on L. A light ray going through G is projected by *Mirror B* onto the omni-image to become an image point I as shown in Fig. 5. The mirror center O_B and G together form a vector $V_G' = [G_x', G_y', G_z']^T$ where T means "transpose." The components of V_G' can be described in terms of the azimuth and elevation angles θ and α of the light ray as: $G_x' = \cos\alpha \times \cos\theta$, $G_y' = \cos\alpha \times \sin\theta$, $G_z' = \sin\alpha$. Also, as mentioned previously, to increase the frontal FOV, we have slanted the camera system up for the angle of γ. So, there is a transformation between the coordinates (X', Y', Z') of the original camera coordinate system (CCS) and the new coordinates (X, Y, Z) of the slanted CCS, described by:

$$\begin{bmatrix} X \\ Y \\ Z \end{bmatrix} = \begin{bmatrix} 1 & 0 & 0 \\ 0 & \cos(-\gamma) & -\sin(-\gamma) \\ 0 & \sin(-\gamma) & \cos(-\gamma) \end{bmatrix} \begin{bmatrix} X' \\ Y' \\ Z' \end{bmatrix} \tag{2}$$

so that $V_G' = [G_x', G_y', G_z']^T$, after being slanted with the above transformation, becomes $V_G = [G_x, G_y, G_z]$ with $G_x = \cos\alpha\times\cos\theta$; $G_y = \cos\alpha\times\sin\theta\times\cos\gamma$ and $G_z = -\cos\alpha\times\sin\alpha + \sin\alpha\times\cos\gamma$. Then, as shown in Fig. 6, let I_L be the conic section curve resulting from projecting the space line L onto the omni-image, and Q be the space plane going through L and the mirror center O_B. Also assume that the coordinates (X, Y, Z) describe a point on Q, and that $N_Q = (l, m, n)$ describe the normal vector of Q. Because N_Q and V_G are perpendicular to each other, we have:

$$N_Q{\cdot}V_G = (l, m, n){\cdot}(G_x, G_y, G_z) = l\times G_x + m\times G_y + n\times G_z = 0. \tag{3}$$

where "$\cdot$" denotes the inner product operation. By Eq. (2), we can rewrite Eq. (3) as:

$$l + m\times\frac{(\cos\alpha\times\sin\theta\times\cos\gamma + \sin\alpha\times\sin\gamma)}{(\cos\alpha\times\cos\theta)} + n\times\frac{(-\cos\alpha\times\sin\theta + \sin\alpha\times\cos\gamma)}{(\cos\alpha\times\cos\theta)} = 0. \tag{4}$$

However, Eq. (4) consists of three unknown parameters l, m, and n which describe the normal vector of Q. Assuming $n \neq 0$, we may divide Eq. (4) by n to get:

$$B + Aa_0 + a_1 = 0 \tag{5}$$

where

$$A = m/n; B = l/n;$$

$$a_0 = \frac{(\cos\alpha\times\sin\theta\times\cos\gamma + \sin\alpha\times\sin\gamma)}{(\cos\alpha\times\cos\theta)}; \quad a_1 = \frac{(-\cos\alpha\times\sin\theta + \sin\alpha\times\cos\gamma)}{(\cos\alpha\times\cos\theta)}. \tag{6}$$

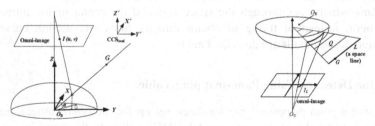

Fig. 5. A space point with elevation α & azimuth θ   **Fig. 6.** A space line projected in omni-image

Chou and Tsai [8] proposed a method to localize a vertical line with respect to the vehicle (actually, with respect to the CCS on the vehicle). The vertical line, called the L_Y line hereafter, is parallel to the Y-axis line in the GCS. See Fig. 7 for an illustration. In this study, we extend their method to localize two more specific lines: one parallel to the X-axis, called the L_X line; and the other parallel to the Z-axis, called the L_Z line. We now derive equations for use to localize these three types of lines.

The direction vector of L_Y is $D_Y = [0, 1, 0]^T$. Therefore, Eq. (3) leads to $0\times l + 1\times m + 0\times n = 0$, ;or equivalently, $m = 0$, and so Eq. (5) can be reduced to be:

$$B = -a_1. \qquad (7)$$

For our cases here, the direction vector of the L_X line is $D_X = [1, 0, 0]^T$. So, Eq. (3) leads to $1{\times}l + 0{\times}m + 0{\times}n = 0$, or equivalently, $l = 0$, and so Eq. (5) can be reduced to be:

$$A = -a_1/a_0. \qquad (8)$$

About L_Z line detection, the direction vector of the L_Z line is $D_Z = [0, 0, 1]^T$. So, Eq. (3) leads to $0{\times}l + 0{\times}m + 1{\times}n = 0$, or equivalently, $n = 0$, and so Eq. (4) can be reduced to be:

$$l + m \times a_0 = 0. \qquad (9)$$

or equivalently,

$$-a_0 = K \qquad (10)$$

where

$$K = l/m. \qquad (11)$$

For each case above, to detect the L_X, L_Y, or L_Z line in the omni-image M directly with the tilt angle γ of the camera system known in advance, the following steps are performed: 1) apply binarization and edge detection operations to M to obtain edge points in M; 2) set up a 1D Hough space H of the parameter A, B, or K; 3) for each edge point p with coordinates (u, v), look up the pano-mapping table of the omni-camera to obtain the azimuth-elevation angle pair (θ, α) of the world-space point P corresponding to p; 4) compute a_0 and a_1 according to Eqs. (6); 5) for each cell c in H with value A, B, or K, if Eq. (8), (7), or (10) is satisfied, then increment the count of c by one; 6) find the maximum cell count in H with value A_{max}, B_{max}, or K_{max} which, according to Eqs. (6) and (11), are equal to m/n, l/n, or l/m, respectively.

Note that in the above algorithm, we do not *really* detect the line L_X, L_Y, or L_Z (in the form of a conic section) but just its related parameter A, B, or K described by Eq. (8), (7), or (11), respectively.

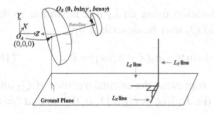

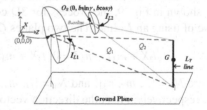

Fig. 7. Three specific space lines   **Fig. 8.** An L_Y line projected onto two mirrors

4.2   3D Data Computation Using Detected Space Lines

Based on the detected parameters of the three types of space lines described above, we can derive the 3D locations of each type of space line, as described subsequently.

(A) 3D Data Computation Using an L_Y Line

As shown in Fig. 8, an L_Y line is projected by *Mirrors B* and *S* onto the image plane to form lines I_{L1} and I_{L2}, respectively. The center O_B of *Mirror B* is assumed to be located at coordinates $(0, 0, 0)$ in the CCS. The position of the center O_S of *Mirror S* may be described in terms of the slant angle γ of the camera system and the *baseline* value b between the two mirrors as $(0, b\sin\gamma, b\cos\gamma)$ in the CCS. Let the two space planes going through L_Y and the centers of the two mirrors, O_B and O_S, respectively, be denoted as Q_1 and Q_2. Also, let G be a point on L_Y with coordinates (X, Y, Z), and the normal vector of Q_1 be described by $N_1 = [l_1, m_1, n_1]^T$. Denote the vector from the origin O_B at coordinates $(0, 0, 0)$ to G at coordinates (X, Y, Z) as V_{G1} which is just $V_{G1} = [X, Y, Z]^T$. Then, since N_B and V_{G1} are perpendicular, we have $V_{G1} \cdot N_1 = 0$, leading the following equation:

$$l_1X + m_1Y + n_1Z = 0. \tag{12}$$

Furthermore, we know that the mirror center O_S is at coordinates $(0, b\sin\gamma, b\cos\gamma)$. Also, suppose that the normal vector of Q_2 is denoted by $N_2 = [l_2, m_2, n_2]^T$. Since point G is on the L_Y line, it is also on Q_2, meaning that the vector V_{G2} from O_S to G is just $V_{G2} = [X, Y - b\sin\gamma, Z - b\cos\gamma]^T$. So, by a similar reasoning using the fact $V_{G2} \cdot N_2 = 0$, we get:

$$l_2X + m_2(Y - b\sin\gamma) + n_2(Z - b\cos\gamma) = 0. \tag{13}$$

Since the direction vector of L_Y is $D_Y = [0, 1, 0]^T$, meaning that $m_1 = m_2 = 0$, the above two space planes described by (12) and (13) can be reduced to be

$$B_1X + Z = 0;\; B_2X + (Z - b\cos\gamma) = 0, \tag{14}$$

with $B_1 = l_1/n_1$ and $B_2 = l_2/n_2$ which can be obtained by the 1D Hough transform process mentioned previously. Solving (14), we can obtain the following desired solutions for X and Z to specify the location of a point P on the L_Y line, or simply, just that of L_Y on the X-Z plane in the CCS (or equivalently, on the ground):

$$X = b\cos\gamma/(B_2 - B_1);\; Z = -B_1b\cos\gamma/(B_2 - B_1). \tag{15}$$

(B) 3D Data Computation Using an L_X Line

As shown in Fig. 9, the process for 3D computation using an L_X line is similar to the case of using an L_Y line. The two planes Q_3 and Q_4 may be described by:

$$l_3X + m_3Y + n_3Z = 0;\; l_4X + m_4(Y - b\sin\gamma) + n_4(Z - b\cos\gamma) = 0 \tag{16}$$

where $N_3 = [l_3, m_3, n_3]^T$ and $N_4 = [l_4, m_4, n_4]^T$ represent the normal vectors of Q_3 and Q_4, respectively. Also, the direction vector of the L_X line is $D_X = [1, 0, 0]^T$ so that $l_3 = l_4 = 0$, and so Eqs. (16) can be reduced to be

$$A_1Y + Z = 0;\; A_2(Y - b\sin\gamma) + (Z - b\cos\gamma) = 0 \tag{17}$$

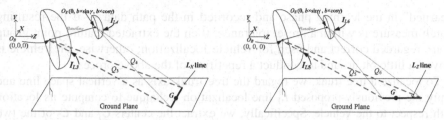

Fig. 9. An L_X line projected onto two mirrors   **Fig. 10.** An L_Z line projected onto two mirrors

where $A_1 = m_3/n_3$ and $A_2 = m_4/n_4$ which can be obtained by the 1D Hough transform process mentioned previously. Eqs. (17) may be solved to get the values of Y and Z as follows to specify the location of a point G on the L_X line in the Y-Z plane of the CCS:

$$Y = (A_2 b\sin\gamma + b\cos\gamma)/(A_2 - A_1); \; Z = -A_1(A_2 b\sin\gamma + b\cos\gamma)/(A_2 - A_1). \quad (18)$$

(C) 3D Data Computation Using an LZ Line

Similarly, as shown in Fig. 10, Q_5 and Q_6 may be described by:

$$l_5 X + m_5 Y + n_5 Z = 0; \; l_6 X + m_6(Y - b\sin\gamma) + n_6(Z - b\cos\gamma) = 0 \quad (19)$$

where $N_5 = [l_5, m_5, n_5]^T$ and $N_6 = [l_6, m_6, n_6]^T$ represent the normal vectors of Q_5 and Q_6, respectively. Eqs. (19) are equivalent to

$$K_1 X + Y = 0; \; K_2 X + (Y - b\sin\gamma) = 0, \quad (20)$$

where $K_1 = l_5/m_5$ and $K_2 = l_6/m_6$. Also, the direction vector of the L_Z line is $D_Z = [0, 0, 1]^T$, meaning that $n_5 = n_6 = 0$, and so Eqs. (2) can be solved to get the values of X and Y as follows to specify the location of a point G on the L_Z line on the X-Y plane in the CCS:

$$X = (b\times\sin\gamma)/(K_2 - K_1); \; Y = -(K_1\times b\times\sin\gamma)/(K_1 - K_2). \quad (21)$$

4.3   Tree Trunk Detection and Localization

Before we can apply the formulas derived previously to conduct localization of a tree trunk, we have to solve the varying lighting problem occurring during image acquisition, which often causes failures of tree trunk detection. For this, we binarize the input image by moment-preserving thresholding [11] and conduct image segmentation to obtain a group G of candidate feature points. To ensure goodness of the result, we apply principal component analysis (PCA) to G. Specifically, as shown in Fig. 11 we compute the following data: the center C of the points in G; the height h of C; the eigenvalue pair (λ_1, λ_1) of the covariance matrix of the points in G; the eigenvectors $e_1 = [u_1, v_1]^T$ and $e_2 = [u_2, v_2]^T$ corresponding to λ_1 and λ_2, respectively; the length ratio η of G in terms of the two eigenvalues: $\eta = \lambda_1/\lambda_2$; and the orientation ω of G: $\omega = \tan^{-1}(v_1/u_1)$. Then, we utilize the three parameters h, ω, and η to describe the shape of the tree trunk, and check the correctness of the extracted tree trunk points by matching these computed parameter values against those "learned" in the learning phase and recorded in the path data — if the resulting match measure is within a preset tolerance, then the extracted feature point group G are regarded correct and used for vehicle localization; otherwise, the vehicle is moved a little bit around to conduct a repetition of the above process.

To localize a tree trunk, we regard the tree trunk axis as a vertical space line and apply the previously-proposed L_Y line localization technique to compute its location with respect to the vehicle. Specifically, we extract the centers C_1 and C_2 of the two groups G_1 and G_2 of tree trunk feature points appearing in the image regions of *Mirrors* S and B, respectively, in the input omni-image, and regard them as projections of a single space point G on the L_Y line, as mentioned in Section 4.1. More detailed steps include: 1) compute the coordinates (u_1, v_1) and (u_2, v_2) of centers C_1 and C_2, respectively; 2) use (u_1, v_1) and (u_2, v_2) to look up the pano-mapping table to get the respective elevation-azimuth angle pairs (α_1, θ_1) and (α_2, θ_2) of C_1 and C_2; 3) use Eqs. (6) and (7) to compute the two parameters B_1 and B_2; 4) use Eqs. (15) to compute the location of the tree trunk axis described by X and Z with respect to the vehicle. An example of tree trunk detection and corresponding vehicle localization results is shown in Fig. 12.

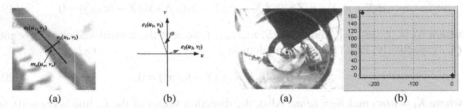

(a)   (b)   (a)   (b)

Fig. 11. Principal component analysis for tree trunk detection. (a) Principal components, e1 and e2. (b) Orientation ω.

Fig. 12. Tree truck detection and localization. (a) Extracted tree axis. (b) Computed tree location (red spot) with respect to vehicle.

4.4 Lawn Corner Detection and Localization

Because the lawn corner is too obscure to be recognized in the omni-image, the proposed lawn corner detection process is divided into two stages. When the vehicle arrives at a proper position for the detection work, a space line forming one side of the corner, appearing as a horizontal space line L_1 on the ground, is detected and localized firstly. Then, the vehicle is guided to turn left, and the other side of the corner, appearing to be another horizontal line L_2 perpendicular to L_1, is detected and localized as well. The two space lines L_1 and L_2 then are drawn in the input omni-image to cross each other to form a corner. Finally, we compute the 3D data of the corner as the localization result.

The details of this process for estimating the coordinates (X_g, Y_g, Z_g) of the lawn corner in the GCS (global coordinate system) include: 1) acquire an omni-image I_1 of the lawn; 2) apply image thresholding and edge detection operations to I_1 to obtain an edge point image I_1' of the lawn boundaries; 2) apply the previously-described 1D Hough transform to I_1' to detect a side boundary L_1 of the lawn as an L_X line and compute its corresponding parameters A_1 and A_2; 3) use Eqs. (18) to compute the values of Y and Z as $Y = -h_1$ and $Z = d_1$ which specify the height of the lawn with respect to the CCS and the distance of the lawn boundary L_1 to the vehicle, respectively; 4) turn the vehicle for an angle of 90° and conduct the above steps to obtain two other values of Y and Z as $Y = -h_2$ and $Z = d_2$ for the other lawn boundary L_2; 5) assign values to the coordinates (X_g, Y_g, Z_g) of the lawn corner by $X_g = d_2$, $Y_g = -(h_1 + h_2)/2$, and $Z_g = d_1$. An example of the results of lawn detection and localization using the above process is shown in Fig.13.

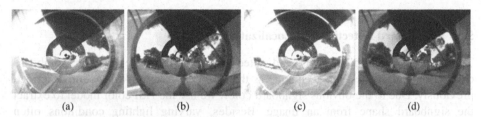

(a)   (b)   (c)   (d)

Fig. 13. Results of lawn corner detection and localization. (a) and (b) Lawn images taken in two perpendicular directions. (c) and (d) Lawn boundary detection results.

5   Artificial Landmark Detection for Vehicle Localization

The artificial landmarks we use in this study include curb line, signboard, stop line on roads, and traffic cones. Their uses for vehicle localization are described now.

5.1   Proposed Technique for Curb Line Following

To conduct vehicle navigations on sidewalks, Chou and Tsai [8] proposed a technique to localize curb lines with respect to the vehicle. In this study, we propose a new method to localize the curb line by the use of the projection of the curb line onto the image region of *Mirror B* in the omni-image. More specifically, we get the feature points of the curb line using its color information. Then, we detect the two boundary lines of the curb which has a certain width. In the resulting edge-point image, we use the previously-proposed L_Z line detection method to find the two boundary lines. Then, we choose the inner boundary line of the curb to compute its location with respect to the vehicle. By the parameter K_1 obtained by the previously-described 1D Hough transform and the height h of the center of *Mirror B*, we can know from Eqs. (20) that $X = -Y/K_1$. Also, it is obvious that $Y = -h$, so we can get the following data of the curb line:

$$X = h/K_1; Y = -h. \tag{22}$$

An example of the experimental results of curb detection and corresponding vehicle location estimation using the proposed localization method are shown in Fig. 14.

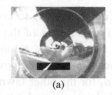
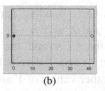

(a)   (b)

Fig. 14. An example of curb line detection and localization results. (a) Curb line segmentation result. (b) Computed curb line location (yellow spot) with respect to vehicle position (blue spot).

5.2   Signboard Detection and Localization

The idea of the proposed signboard detection method is to extract the signboard contour and apply the same technique as that used for tree trunk detection described previously. Due to the obvious signboard color, we use the HSI color model to extract the signboard shape from an image. Besides, varying lighting conditions often influence the hue and saturation features of the HSI colors of the landmark. Based on learned signboard contour information, a dynamic color thresholding scheme is proposed in this study to adjust the saturation threshold value S_{th} to be within a fixed range $[S_0, S_1]$ for the purpose of guaranteeing consistent signboard contour segmentation, where S_0 and S_1 are learned in advance in different lighting conditions in the learning stage. An experimental result of signboard segmentation by dynamic thresholding is shown in Fig. 15.

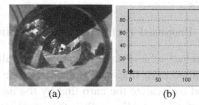

(a)   (b)   (a)   (b)

Fig. 15. Results of signboard segmentation. (a) Result using fixed threshold. (b) Result using proposed dynamic thresholding.

Fig. 16. Signboard localization. (a) Result of extracting L_Y line of signboard (b) Computed signboard position (green spot).

After the signboard is segmented successfully, by regarding the vertical signboard axis as an L_Y line perpendicular to the ground, we apply the method we use for detecting and localizing the tree trunk described previously in Sec. 4 to compute the position of the signboard axis for vehicle localization. The details are omitted due to the page limit. An example of the results of such vehicle localization using a signboard is shown in Fig. 16.

5.3   Stop Line Detection and Localization

Besides landmarks on the *sidewalk*, we may also use those on the *road* for vehicle localization. One of the landmarks commonly seen on the road is the *stop line* which is used in this study as well for vehicle localization. Because the stop line on the road has obvious color information (mostly white), we also utilize the HSI color model to extract it. Then, we detect its boundaries in the input omni-image. Finally, we detect two parallel L_X lines and one perpendicular L_Z line in the edge-point image as illustrated in Fig. 17(a), using the previously-proposed 1D Hough transform, to obtain the parameter information of the entire boundary shape for use in vehicle localization. An experimental result is shown in Figs. 17(b) and 17(c).

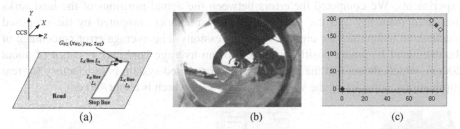

(a)   (b)   (c)

Fig. 17. Result of stop line detection and localization. (a) Result of extracting stop line boundaries as L_X and L_Z lines. (b) Computed positions of stop line (yellow and red spots).

5.4   Traffic Cone Detection and Localization

When engineering works are conducted on sidewalks, the workers usually put traffic cones near the working area to warn people. For this type of situation, we propose to detect traffic cones and use them as landmarks. The proposed method for traffic cone detection is similar to that for stop line detection. But here we detect one L_Z line and one L_X line to carry out the detection of the traffic cone base which is of the shape of a square. After detecting the boundary lines of the traffic cone base, we apply the previously-mentioned 1D Hough transform to compute two parameters, A_1 and K_1, for use in computing its location. Besides, we also utilize traffic cone corner to compute a L_Y line by which we can draw a vertical line to illustrate the position of the traffic cone in the omni-image. An experimental result of detecting the traffic cone using the proposed method is given in Fig. 18.

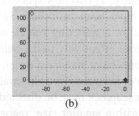

(a)   (b)

Fig. 18. Result of traffic cone localization. (a) Result of extracting L_X, L_Y, and L_Z lines of traffic cone (b) Computed position of traffic cone (yellow spot) with respect to vehicle.

6   Experimental Results

The experimental environment was a sidewalk in National Chiao Tung University as shown in Figs. 19(a) and 19(b). In each navigation session, the vehicle started from an identical spot on the sidewalk just like in the learning process and navigated along the recorded navigation path nodes mainly by the curb line following technique. Then, the vehicle detected the pre-learned landmarks and localized their positions to adjust its odometer readings at each visited path node until reaching the appointed terminal node. Many successful navigation sessions have been conducted. A path map with recorded vehicle positions at the visited nodes of one navigation session is shown in Fig. 19(c).

Furthermore, we have tested the precision of vehicle localization in the experiments. We computed the errors between the actual positions of the landmarks measured manually and the positions of the landmarks computed by the proposed localization techniques for eight navigation sessions. The average error percentage of the estimated landmark position is 7.52% of an average landmark distance of about 200cm, which shows that the precision of the proposed system is satisfactory for real applications, considering the width of the sidewalk which is about 400 cm.

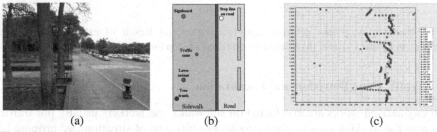

(a)   (b)   (c)

Fig. 19. Experimental environment and a path map. (a) A view of environment. (b) An illustration of environment. (c) A path map resulting from a navigation session.

7   Conclusions

Construction of a machine guide dog using a two-mirror omni-camera and an autonomous vehicle has been proposed, for which several methods have been proposed: 1) by the use of a learning interface designed in this study, a trainer can guide the vehicle to navigate on a sidewalk and construct a navigation path conveniently; 2) two new space line detection techniques based on the space mapping method have been proposed; 3) several landmark detection techniques have been proposed for conducting vehicle navigation; and 4) to conduct the landmark detection works more effectively in the outdoor environment, techniques for dynamic threshold adjustments have also been proposed. Good landmark detection results and successful navigation sessions on a sidewalks show the feasibility of the proposed methods. Future researches may be directed to detecting pedestrians or bike riders; designing a camera with a smaller size; recognizing traffic signals to go through road crossings; etc.

References

1. Willis, S., Helal, S.: RFID information grid for blind navigation and wayfinding. In: Proceedings of the 9th IEEE International Symposium on Wearable Computers, Washington, DC, USA, pp. 34–37 (October 2005)
2. Chen, M.F., Tsai, W.H.: Automatic learning and guidance for indoor autonomous vehicle navigation by ultrasonic signal analysis and fuzzy control techniques. In: Proceedings of 2009 Workshop on Image Processing, Computer Graphics, and Multimedia Technologies, Nat'l Computer Symposium, Taipei, Taiwan, pp. 473–482 (November 2009)
3. Abbot, E., Powell, D.: Land-vehicle navigation using GPS. Proc. of IEEE 87, 145–162 (1999)
4. Atiya, S., Hager, G.D.: Real-time vision-based robot localization. IEEE Transactions on Robotics and Automation 9(6), 785–800 (1993)
5. Lui, W., Jarvis, R.: Eye-Full Tower: A GPU-based variable multibaseline omnidirectional stereovision system with automatic baseline selection for outdoor mobile robot navigation. Robotics and Autonomous Systems 58(6), 747–761 (2010)
6. Fu, H., Cao, Z., Cao, X.: Embedded omni-vision navigator based on multi-object tracking. Machine Vision and Applications 22(2), 349–358 (2011)
7. Ishizuka, D., Yamashita, A., Kawanishi, R., Kaneko, T., Asama, H.: Self-localizaion of mobile robot equipped withomnidirectional camera using image matching and 3D-2D edge matching. In: Proceedings of IEEE Int'l Conference on Computer Vision Workshops, Barcelona, Spain, November 6-13, pp. 272–279 (2011)
8. Chou, Y.H., Tsai, W.H.: Guidance of a vision-based autonomous vehicle on sidewalks for use as a machine guide dog. In: Proceeding of 2011 Conference on Computer Vision, Graphic, and Image Processing, Chiayi, Taiwan (2011)
9. Jeng, S.W., Tsai, W.H.: Using pano-mapping tables to unwarping of omni-images into panoramic and perspective-view Images. Proceeding of IET Image Processing 1(2), 149–155 (2007)
10. Wu, C.J., Tsai, W.H.: An omni-vision based localization method for automatic helicopter landing assistance on standard helipads. In: Proceedings of 2nd Int'l Conference on Computer and Automation Engineering, Singapore, vol. 3, pp. 327–332 (2010)
11. Tsai, W.H.: Moment-preservingthresholding: a new approach. Computer Vision, Graphics, and Image Processing 29(3), 377–393 (1985)

References

1. P. Willis, S.; et al.: RFID information grid for blind navigation and wayfinding. In: Proceedings of the 9th IEEE International Symposium on Wearable Computers, Washington DC, USA, pp. 34–37 (October 2005)

2. Chen, M.F., Tsai, W.H.: Automatic learning and guidance for indoor autonomous vehicle navigation by ultrasonic signal analysis and fuzzy control techniques. In: Proceedings of 2009 Workshop on Image Processing, Computer Graphics, and Multimedia Technologies, Nat'l Computer Symposium, Taipei, Taiwan, pp. 475–482 (November 2009)

3. Abbott, E., Powell, D.: Land vehicle navigation using GPS. Proc. of IEEE 87(1), 145–162 (1999)

4. Atiya, S., Hager, G.D.: Real-time vision-based robot localization. IEEE Transactions on Robotics and Automation 9(6), 785–800 (1993)

5. Lin, W., Jan, C.I.: Five-BRH Tower: A GPS-based scalable omni-directional autonomous navigation system with automatic bus-line detection for bus for a mobile robot navigation. Robotics and Autonomous Systems 58(6), 317–327 (2010)

6. Fu, H., Cao, X., Cao, X.: Embedded active vision navigation based on multi-object tracking. Machine Vision and Applications 22(2), 349–358 (2011)

7. Fukushima, D., Yanishita, A., Kwanishi, R., Kaneko, T., Asano, H.: Self-localization of mobile robot equipped with omnidirectional camera using image matching and 3D–2D edge matching. In: Proceedings of IEEE Int'l Conference on Computer Vision Workshops, Barcelona, Spain, November 6–13, pp. 273–279 (2011)

8. Chiou, Y.H., Tsai, W.H.: Guidance of a vision-based autonomous vehicle on sidewalks for use as a machine guide dog. In: Proceeding of 2011 Conference on Computer Vision, Graphics and Image Processing, Chiayi, Taiwan (2011)

9. Jeng, S.W., Tsai, W.H.: Using pano-mapping tables to unwarping of omni-images into panorama and perspective-view images. IET Image Processing 1(2), 149–155 (2007)

10. Wu, C.J., Tsai, W.H.: An omni-vision based localization method for automatic helicopter landing assistance on standard helipad. In: Proceedings of 2nd Int'l Conference on Computer and Automation Engineering, Singapore, vol. 3, pp. 327–332 (2010)

11. Tsai, W.H.: Moment-preserving thresholding: a new approach. Computer Vision, Graphics, and Image Processing 29(3), 377–393 (1985)

Protection of Privacy-Sensitive Contents in Surveillance Videos Using WebM Video Features[*]

Hsin-Hsiang Tseng[1] and Wen-Hsiang Tsai[1, 2,**]

[1] Institute of Computer Science and Engineering, National Chiao Tung University, Taiwan
itsummervar@gmail.com
[2] Department of Information Communication, Asia University, Taiwan
whtsai@cis.nctu.edu.tw

Abstract. Privacy protection is a critical issue in video surveillance because nowadays video cameras existing everywhere might monitor spaces of individuals and violate protection of personal privacy. Based on the data hiding approach, a new method for protection of privacy-sensitive contents in surveillance videos of the recently-developed WebM format is proposed. With skillful uses of certain special features of the WebM video, techniques for removing privacy-sensitive contents from a given WebM video, embedding the removed contents into the same video imperceptibly, and extracting the hidden contents later to recover the original privacy-sensitive contents are proposed. Experimental results showing the feasibility of the proposed method are also included.

Keywords: data hiding, privacy protection, video surveillance, WebM video.

1 Introduction

Privacy protection is an important issue in video surveillance. Since video cameras exist everywhere in our environments nowadays, which conduct monitoring of public or private spaces for long time periods, it might happen that information of individuals' activities is videotaped, leading to infringement upon personal privacy. Hence, it is necessary occasionally to hide the privacy-violating parts of surveillance video contents to avoid legal disputes or to protect personal privacy from being misused. It is so desirable to propose methods to solve this issue of privacy protection in videos.

Several methods have been proposed for this purpose. Dufaux et al. [1] proposed a method to scramble regions in videos containing personal information; the resulting scene in the video remains visible, but the privacy-sensitive information is not identifiable. A method was proposed by Meuel et al. [2] to protect faces in

---

[*] This work was supported in part by the NSC, Taiwan under Grant No. 100-2631-H-009-001 and in part by the Ministry of Education, Taiwan under the 5-year Project of "Aiming for the Top University" from 2011 through 2015.
[**] Corresponding author.

surveillance videos; any visible information of faces in a video is deleted and embedded in the same video, allowing later reconstructions of the faces when needed. Zhang et al. [3] presented a method to protect authorized persons appearing in videos, which conducts both removal and embedding of the shapes of concerned persons in videos. Also proposed by Yu et al. [4] is a method for privacy protection by controlling the disclosure of individuals' visual information in videos using visual abstraction operations like silhouette generation and transparency enhancement.

About data hiding via videos, many techniques have been proposed in the past decade [5-8]. By data hiding, information can be transmitted covertly or kept securely for various applications. Hu et al. [5] proposed a method for hiding data in H.264/AVC videos based on the idea of modifying 4×4 intra-prediction modes to encode hidden bits. Only intra-coded macroblocks are used to hide data. Hussein [6] proposed a method for embedding data in motion vectors based on their associated prediction errors. Yang and Bourbakis [7] proposed a method for embedding data in the DCT coefficients by means of vector quantization. Kapotas et al. [8] proposed a method for embedding data into encoded video sequences by modulating the partition size; only inter-coded macroblocks are used for embedding information.

In this study, we propose a new method to deal with the problem of privacy protection in WebM videos. By the method, an authorized user is allowed to specify a region R identically-located in each frame of an input *cover video V*. The region R, called the *protection region* hereafter, presumably contains the privacy-sensitive content C_i in each frame F_i of V. The privacy-sensitive contents of all the frames of V are collected sequentially to form a *privacy-sensitive data set E*. Then, a process of replacing the privacy-sensitive content C_i with an identical *background image portion B* in each video frame is conducted automatically. Also, the privacy-sensitive data set E is hidden into the same video to produce a *privacy-protected video* V_p. Thereafter, whenever the privacy-sensitive contents need be recovered, the hidden data in V_p can be extracted to reconstruct the original video. The main contributions of this study are skillful uses of special features of the WebM video format for *removals*, *embeddings*, and *recoveries* of privacy-sensitive contents to achieve the purpose of privacy protection in video surveillance using WebM videos.

In the remainder of this paper, the proposed data hiding technique for embedding privacy-sensitive contents and the corresponding data extraction technique are introduced first in Section 2 after a brief review of the WebM video format is given. Proposed techniques for removing and recovering the privacy-sensitive contents in WebM videos are described in Section 3. In Section 4, some experimental results are presented, followed by conclusions in the last section.

2   Data Hiding in WebM Videos

2.1   Brief Review of WebM Video Format

WebM is an open media file format designed for the web whose openness was offered by Google Inc. in May 2010 [10]. Each WebM file consists of video streams compressed with the VP8 video codec and audio streams compressed with the Vorbis audio codec. The VP8 video codec works exclusively with an 8-bit YUV 4:2:0 image format, each 8-bit chroma pixel in the two chroma color spaces (U and V) corresponds to a 2×2 block of 8-bit luma pixels in the luma color space (Y).

Also, each frame in a WebM video is decomposed into an array of macroblocks. And each macroblock is a square array of pixels whose Y dimensions are 16×16 and whose U and V dimensions are 8×8. The macroblock-level data in a compressed frame are processed in a raster-scan order. Each macroblock is further decomposed into four 4×4 subblocks. So each macroblock has sixteen Y subblocks, four U subblocks, and four V subblocks. These three types of subblocks, when composed together respectively, are called the *Y*, *U*, and *V components* of the macroblock henceforth. So, the *Y* component is a 16×16 array of 8-bit luma pixels, and the U and V components are both 8×8 arrays of chroma pixels. Fig. 1 illustrates one of the 4×4 subblocks of a macroblock.

The VP8 video codec transforms pixels in the spatial domain into coefficients in the frequency domain by the discrete cosine transform (DCT) and the Walsh-Hadamard transform (WHT) at the 4×4 resolution. The DCT is used for the sixteen Y, four U, and four V subblocks and the WHT is used to encode a 4×4 array comprising the average intensities of the sixteen Y subblocks of a macroblock. These average intensities are, up to a constant normalization factor, nothing more than the zeroth DCT coefficients of the Y subblocks. The VP8 video codec considers this 4×4 array as a second-order subblock, called the Y2 subblock.

Furthermore, two frame types are used in the VP8 codec, namely, *intra-frame* and *inter-frame*. Intra-frames, also called *key frames* or *I-frames*, are decoded without reference to other frames in a sequence; and inter-frames, also called *prediction frames* or *P-frames*, are encoded with reference to prior frames, including the recent key frame.

2.2   Ideas of Proposed Data Hiding Method

The pixel values in each 4×4 subblock of a WebM video frame are converted by the DCT into frequency-domain coefficients, and the energy of the coefficient signals is "clumped" at the left-upper corner of the subblock. In addition, after the coefficients are quantized according to an adaptive level and put into a zigzag order as illustrated by Fig. 2, near-zero coefficients will usually appear in the *positive-sloped diagonal* of the subblock as shown by the example illustrated in Fig. 3, where the positive-sloped diagonal means the four red squares in Fig. 3. Furthermore, human eyes have lower sensitivity on high-frequency signals and chrominance than on low-frequency signals and luminance [9]. Accordingly, it is proposed is this study to define 16 data patterns to replace the DCT coefficients in the positive-sloped diagonal of the 4×4 subblock to achieve imperceptible data hiding.

In addition, we pre-compute the PSNR resulting from data hiding for each macroblock, and if the computed value is too large (larger than a pre-selected threshold), then data hiding is abandoned. It is in this way that the proposed method maintains good video quality in the data hiding result. To mark which macroblock has been used for data hiding, we use a feature of the WebM video, namely, *region-of-interest* (ROI) *map*. Each macroblock has its own map index. Such an index is also encoded into the bitstream by tree coding in the compressed video data. Taking advantage of this feature, we use the map index as a *data embedding marker* to label each of those macroblocks whose coefficients have been modified for data hiding.

0	1	2	3
4	5	6	7
8	9	10	11
12	13	14	15

81	20	6	-2
11	4	0	0
1	0	0	0
0	0	0	0

Fig. 1. A subblock with yellow coefficients composing a positive-sloped diagonal line

Fig. 2. Zigzag scan order of coefficients in a subblock of WebM video

Fig. 3. A subblock obtained after DCT and coefficient quantization with red coefficients composing a diagonal line

2.3   Embedding of Message Data into WebM Videos

In this section, we will describe the detailed algorithm of the proposed method for hiding privacy-sensitive contents into cover videos by changing the frequency coefficients into pre-defined data patterns. At first, the aforementioned 16 data patterns for use in the proposed algorithm are defined. For this, let the notations N and 0 denote the meanings "non-zero" and "zero," respectively. Then, a *data pattern* DP_i ($i = 0$ to 15) is defined as a 4×4 block with its positive-sloped diagonal being filled with four symbols $S_3S_2S_1S_0$ of N's and 0's, which correspond to the binary value $b_3b_2b_1b_0$ of i in the following way: if $b_j = 0$, then $S_j = 0$; and if $b_j = 1$, then, $S_j = N$, where $j = 0, 1, 2, 3$. Fig. 4 illustrates all the 16 data patterns DP_0 through DP_{15}. In each of the 4×4 data patterns, the top-rightmost square contains S_0 and the bottom-leftmost one S_3, and so on. For example, when $i = 3$, the corresponding binary value is 0011_2 and the defined data pattern DP_3 is a 4×4 block with its positive-sloped diagonal being filled with the four symbols $S_3S_2S_1S_0 = 00NN$. And when $i = 10$, the corresponding binary value is 1010_2 and the data pattern DP_{10} has its positive-sloped diagonal being filled with $S_3S_2S_1S_0 = N0N0$.

Each of the data patterns DP_i is used in this study to embed a message data item i into a 4×4 frequency-coefficient subblock SB in a video frame by the following *match-and-replace rule* where the four elements of the positive-sloped diagonal of SB are denoted as $B_3B_2B_1B_0$, and those corresponding ones of DP_i as $S_3S_2S_1S_0$, respectively:

if $S_j = 0$, then replace B_j by 0; if $S_j = N$, then

$$\text{if } B_j \neq 0,\text{ keep } B_j \text{ unchanged; otherwise, set } B_j = 1. \tag{1}$$

Reversely, when the message data item i already embedded in SB is to be extracted, the following *match-and-extract rule* is conducted:

$$\text{if } B_j = 0,\text{ then extract } S_j \text{ to be } 0;\text{ if } B_j \neq 0,\text{ then extract } S_j \text{ to be } N. \tag{2}$$

Fig. 4. Sixteen data patterns DP_0 through DP_{15} for use to embed message data 0 through 15, respectively, where the small white squares means 0's and the blue ones mean N's (non-zeros)

In the following algorithm, given a cover WebM video V, we assume that a protection region R has been selected for V and the privacy-sensitive contents of all the video frames in R have been removed (using Algorithm 3 described later in Section 3.2) and collected as a privacy-sensitive data set E, resulting in a *non*-privacy-sensitive cover WebM video which we denote by V_0. We regard E to be associated with V_0. The technique we propose to remove the privacy-sensitive content from the protection region in each prediction frame will be described later in the next section.

Algorithm 1: *Embedding privacy-sensitive contents into a non-privacy-sensitive cover WebM video.*

Input: a non-privacy-sensitive cover WebM video V_0 and its associated privacy-sensitive data set E of V_0, a secret key K, a random-number generating function f, and a threshold value T.

Output: a privacy-protected video V_p with E being embedded in it.

Steps

Step 1. (*Randomizing the privacy-sensitive data set E*) Transform data set E in a character form into a binary string B, use key K and function f to randomize string B, and divide the result into a sequence A of 4-bit segments.

Step 2. (*Generating a random sequence for later uses in randomizing generated data patterns*) Use K and f to generate a sequence Q of random numbers.

Step 3. (*Embedding the random data sequence A*) Take sequentially an *unprocessed* macroblock MB from the prediction frames of V_0, and perform the following steps to embed the data of sequence A into MB.

    3.1 (*Generating data patterns which encode the data to be embedded*) Take sequentially eight *unprocessed* 4-bit elements $A_1, A_2, \ldots, A_8$ from A; and for each A_j with binary value i, generate the corresponding 4×4 data pattern DP_{i_j} (e.g., if $A_j = i_j = 1001_2 = 9_{10}$, then generate DP_9), where $j = 1, 2, \ldots, 8$.

    3.2 (*Randomizing the generated data patterns*) Take sequentially eight *unprocessed* elements, $N_1, N_2, \ldots, N_8$, from sequence Q; and for each N_j with $j = 1, 2, \ldots, 8$, combine it with DP_{i_j} by the exclusive-OR operator $\oplus$ to yield a new data pattern $DP_{i_j}' = DP_{i_j} \oplus N_j$.

    3.3 (*Saving the original macroblock content*) Save the original content of MB into a temporary macroblock MB_{temp}.

    3.4 (*Embedding the generated random data string DP_{i_j}'*) Denote the eight subblocks of the chroma channels, four in the U channel and the other four in the V channel, of macroblock MB by $SB_1, SB_2, \ldots, SB_8$, and for each SB_{i_j} with $j = 1, 2, \ldots, 8$, conduct the match-and-replace rule described previously by (1) to embed DP_{i_j}' into SB_j, resulting in a *modified* macroblock MB'.

3.5 (*Computing the distortion in the modified macroblock MB'*) Denote the U and V components of MB as MB_U and MB_V, respectively, and those of MB' as MB_U' and MB_V', respectively; compute the average peak signal-to-noise ratios (PSNRs), $PSNR_U$ and $PSNR_V$, of MB_U' and MB_V' with respect to MB_U and MB_V, respectively, as well as the average PSNR, $PSNR_{avg}$, of $PSNR_U$ and $PSNR_V$ as $PSNR_{avg} = (PSNR_U + PSNR_V)/2$ for MB' with respect to MB.

3.6 (*Checking the data embeddability of macroblock MB*) If $PSNR_{avg}$ is smaller than the input pre-selected threshold value T, then regard MB as *data-embeddable* by setting the ROI map index mi_{ROI} of MB' to be 1; else, keep the default value, which is 0, of mi_{ROI} unchanged, meaning that MB is non-data-embeddable and resume the original content of MB to be those saved in the temporary macroblock MB_{temp}.

Step 4. (*Ending*) Repeat Step 3 until the entire content of sequence A is embedded, take the final version of video V_o as the desired privacy-protected video V_p, and exit.

2.4   Extraction of Embedded Data from WebM Videos

The proposed process for extracting privacy-sensitive contents from a privacy-protected video is described as an algorithm in the following, which essentially is a reverse version of Algorithm 1 proposed previously for privacy-sensitive content embedding.

Algorithm 2: *Extraction of the privacy-sensitive contents from a privacy-protected WebM video.*

Input: a privacy-protected WebM video V_p; and the secret key K and the random number generating function f used in Algorithm 1.

Output: a privacy-sensitive data set E including the privacy-sensitive contents of all frames in the original video V within a privacy region R.

Steps

Step 1. Use key K and function f to generate a sequence Q of random numbers and take the first eight random numbers from Q, denoted as N_1 through N_8.

Step 2. For each macroblock MB_k in the prediction frames of V_p with its ROI map index $MI_{ROI} = 1$, perform the following steps to extract data embedded in it.

1.1 Take out the eight subblocks $SB_1, SB_2, \ldots, SB_8$ of the chroma channels, four in the U channel and the other four in the V channel, and for each SB_j with $j = 1, 2, \ldots, 8$, conduct the match-and-extract rule described by (2) above to extract a data pattern DP_{i_j}', and combine it with N_j by the exclusive-OR operator $\oplus$ to yield a new data pattern $DP_{i_j} = DP_{i_j}' \oplus N_j$.

1.2 Transform each DP_{i_j} with $j = 1, 2, \ldots, 8$ into a 4-bit segment A_j with its decimal value equal to i_j, and concatenate all A_j sequentially to form a binary string B_k.

Step 3.   Concatenate all the binary strings B_k's obtained above to form a binary string B'.

Step 4.   Use key K and function f to de-randomize B' to get another binary string B, and transform B into a character form as the desired output data set E.

3   Protection of Privacy-Sensitive Contents

3.1   Idea of Proposed Method

Like other codecs, the VP8 video codec has a process to find the best prediction block in blocks. A *motion vector* is used to indicate the location of the best prediction block. The difference between the best prediction block and the currently-processed block is converted by the DCT into a set of *frequency coefficients*. Motion vectors and frequency coefficients are used in the decoding process to decode corresponding blocks, and they together are called the *decoding information* hereafter.

A video can be decoded correctly based on the decoding information generated during the encoding process. The idea we propose to protect the privacy-sensitive content C in a selected protection region R specified by a user is to set the decoding information of C to be some *pre-defined* values. In this way, the privacy-sensitive video content can be removed and replaced by the background image. The decoding information of C is then embedded into the video. If the privacy-sensitive content of C need be recovered, the embedded decoding information of C may be extracted and used to conduct the recovery work.

A critical issue to overcome here is how to replace the privacy-sensitive content C with the background image without causing negative visible effects. In order to remove the privacy-sensitive content C in the user-specified protection region R, we have to replace an *encoded macroblock* in the encoding process, but this will cause a *reference problem* here, which occurs when the encoded macroblock is used as a reference to encode other macroblocks during the encoding process. This problem, if not solved, will cause errors in the decoding result. Fig. 5 shows an example of the reference problem.

In this study, we propose the use of the *golden reference frame*, which is a feature of the WebM video, to solve this reference problem. It is provided for the VP8 video codec to store a video frame from an arbitrary point in the past. The VP8 encoder could use such a type of frame to maintain a copy of the background image when there are objects moving in the foreground part; by using the golden reference frame, the foreground part can be easily and cheaply reconstructed when a foreground object moves away. And this is just what we need for dealing surveillance videos in this study because a surveillance video often comes from monitoring a fixed area for a long time, and the background image is usually still with no moving object included.

Another problem encountered here is caused by the use of the intra-coded macroblock, which is a type of encoded macroblock used by the VP8 video codec. With the intra prediction mode, an intra-coded macroblock does not use any reference frame and appear in the last video frame, a golden reference frame, or an alternative reference frame [10]. Therefore, any modification of the frequency coefficients in an intra-coded macroblock to remove the privacy-sensitive content will result in a grey color macroblock rather than the background image. Fig. 6 shows an example of this problematic phenomenon. To solve this problem, we choose to enforce the VP8 encoder to use the inter prediction mode when displaying the privacy-protected video.

Fig. 5. An example of errors caused by the reference problem

Fig. 6. Modifying coefficients in an intra-coded macroblock yields grey macroblocks

3.2   Process for Removing Privacy-Sensitive Contents

The proposed process for removing privacy-sensitive contents as described in the following is applied to the prediction frames of an input WebM video. After a protection region R in which privacy-sensitive contents, if there is any, should be removed is specified by the user, the motion vectors and frequency coefficients of the currently-processed macroblock within R are all set to be zero. Also, we assume that the first video frame is a background image which is taken to be a golden reference frame. The values of the original motion vectors and frequency coefficients of the macroblocks of all frames within R are grouped sequentially to form a data set E and hidden into the prediction frames in the input video using algorithm 1, as mentioned previously.

Algorithm 3: *Removing the privacy-sensitive contents in a specified protection region.*

Input: a WebM video V and a pre-specified protection region R.

Output: a non-privacy-sensitive WebM video V_o with the privacy-sensitive contents in R of all the frames removed and collected as a privacy-sensitive data set E.

Steps

Step 1.   Take the first frame of video V as the golden reference frame, and set it as the reference frame for each prediction frame F in V.

Step 2.   Restrict the encoder to use inter prediction modes during the prediction step.

Step 3.   For each prediction frame F in V, perform the following steps.

    3.1   Detect motions in region R of F by checking the prediction mode of each macroblock within R; and if there exists any macroblock with its prediction mode other than ZEROMV (meaning "no motion exists") [10], then set a *motion flag* f_m to be 1; otherwise, set f_m to be 0.

    3.2   If $f_m = 1$, then for each macroblock MB within R of F, perform the following steps.

        (1) Record all the sixteen coefficients of the Y2 subblock into E and set all the sixteen coefficients of this subblock to be zero.

(2) Record the DC coefficient of each subblock of the chroma channels (including the U channel and the V channel) into E and set all the coefficients of these subblocks to be zero.

(3) Record, according to the zigzag scan order as shown in Fig. 7, the first seven coefficients of each subblock of the luma color channel (the Y channel) into E and set all the coefficients of these subblocks to be zero.

(4) Record the index of MB and the index of F into E.

(5) Record the motion vector of MB into E and set this vector to be zero.

Step 4. Repeat Step 3 until all frames of V have been processed, take the final versions of V and E as the desired outputs V_o and E, respectively, and exit.

Some considerations involved in designing the above algorithm are reviewed here. Considering the capacities of the proposed hiding data method, we cannot record all the coefficients in a macroblock which will be used to recover privacy-sensitive contents. Therefore, we conduct some tests in order to decide which coefficients of color channels should be recorded. There exists a type of 4×4 second-order subblock in the WebM video called Y2, as mentioned previously, which records the DC coefficients of all the sixteen Y subblocks. If we lose the coefficients of the Y2 subblock, we cannot recover the contents in this macroblock. Fig. 7 shows an example of such cases. Therefore, we record all the coefficients in the Y2 subblock in order to make sure we can recover the privacy-sensitive contents. In addition, because the VP8 video codec uses the zigzag scan order, which is shown in Fig. 2, to encode subblocks, after some experimental tests we decide to record, according to the zigzag scan order, the first seven coefficients of each Y subblock, as described in Step 3.2(3).

(a)   (b)

Fig. 7. Comparison between original image and image with coefficients of Y2 subblocks lost. (a) Original image. (b) Image with coefficients of Y2 subblocks in a region lost.

3.3   Process for Recovering Privacy-Sensitive Contents

Once the privacy-sensitive contents in an input privacy-protected video need be recovered, the recovery information, namely, the privacy-sensitive data set E, extracted by Algorithm 2 may be used to recover the original privacy-sensitive contents. If the extracted data are correct, we can know accordingly the positions of the protected regions in the frames of the input privacy protected video and the original privacy-sensitive contents.

There are two phases in the proposed process for recovery of privacy information in a privacy-protected video. The first is to extract the recovery information of the protected region in the privacy-protected video by Algorithm 2. The second is to replace the contents of the protected regions with the recovery information. A detailed algorithm for the second phase is described in the following.

Algorithm 4: *Recovering the privacy-sensitive contents of a privacy-protected WebM video.*

Input: a privacy-protected WebM video V_p, and a privacy-sensitive data set E of the protected regions in V.

Output: a WebM video V' with the privacy-sensitive contents recovered.

Steps

Step 1.   Set a *recovery flag* f_r initially to be 1.

Step 2.   Take sequentially an unprocessed prediction frame F from V_p and perform the following steps.

    2.1   Take an unprocessed macroblock MB from F and perform the following steps.

        (1) Extract sequentially a set of unextracted recovery information from E, including a frame index i_f, a macroblock index i_{mb}, a motion vector MV, and a set FC of frequency coefficients.

        (2) If the frame index of F is equal to i_f and the macroblock index of MB is equal to i_{mb}, then replace the motion vector in MB by MV, and replace the set of frequency coefficients in MB by the respective ones in FC.

    2.2   Repeat Step 2.1 until the macroblocks in F are exhausted

Step 3.   Repeat Step 2 until the prediction frames in V_p are exhausted (i.e., until reaching the end of V_p).

4   Experimental Results

Four clips of surveillance videos are used in part of our experiments using the previously-proposed algorithms (Algorithms 1 through 4) conducted in a sequence as illustrated in Fig. 8. The first one is acquired by a camera monitoring an aisle of Engineering Building 5 in National Chiao Tung University. Four representative original frames of the video clip are shown in Fig. 9(a). The purpose of surveillance was to monitor activities around the aisle, but it was hoped that the personal information appearing in the window of the second floor would not be revealed. Therefore, we utilized the proposed process for removing privacy-sensitive contents (Algorithm 3) to conceal such personal information. Furthermore, we embedded the information into the resulting video to yield a privacy-protected video by Algorithm 1. The four frames of the privacy-protected video yielded by Algorithm 1 and corresponding to those of Fig. 9(a) are shown in Fig. 9(b). Finally, the four corresponding frames of the recovered video yielded by the proposed processes for recovering the privacy-sensitive contents (Algorithms) are shown in Fig. 9(c).

A comparison between an original frame and the corresponding recovered one is shown in Fig. 10. The average PSNR of the recovered image part in the protection region with respect to the original one is 35.73.

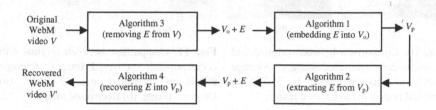

Fig. 8. Illustration of sequence of conducting proposed processes (Algorithms 3, 1, 2, 4)

Another surveillance video used in our experiments came from monitoring the Computer Vision Lab at National Chiao Tung University. A comparison between an original frame and the corresponding recovered one is shown in Fig. 11. The average PSNR of the recovered image part with respect to the original one is 30.372, which together with the previous one (35.73) indicate that the qualities of the recovered video frames are good for practical applications. More PSNR values of other experimental results show the same conclusion.

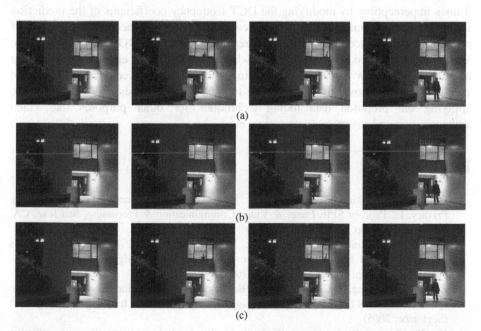

Fig. 9. Representative original and processed video frames. (a) Four original video frames. (b) Four corresponding frames in the privacy-protected video. (c) Four representative frames of recovered video.

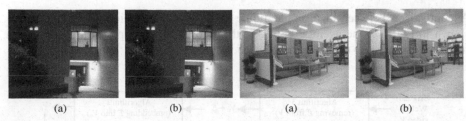

(a)   (b)   (a)   (b)

Fig. 10. Comparison between original and corresponding recovered images. Average PSNR of recovered area is 35.73. (a) Original image. (b) Recovered image.

Fig. 11. Comparison between original and corresponding recovered images. Average PSNR of recovered area is 30.37. (a) Original image. (b) Recovered image.

5   Conclusions

For privacy protection in surveillance videos, a data-hiding method using WebM video features has been proposed. A user can specify a protection region in an input WebM video, and the privacy-sensitive contents in the region in all frames can be removed to protect personal privacy. The problem of yielding unacceptable decoding errors in the resulting video due to such content removals have been solved by assigning a background image as the golden reference frame of the WebM format for use by the video encoder. The removed contents can be embedded into the video frames imperceptibly by modifying the DCT frequency coefficients of the prediction frames of the chroma channels in the compression result, according to a set of predefined data patterns, to encode the removed data. The ROI map index of the WebM format is used to indicate frames where data have been embedded and should be extracted later for recovering the original video. The recovered videos still have good qualities as shown by experimental results. Future studies may be directed to applying the proposed data hiding techniques for other purposes like video authentication and covert communication, etc.

References

1. Dufaux, F., Ebrahimi, T., Emitall, S.A.: Smart video Surveillance System Preserving Privacy. In: Proc. of SPIE Image & Video Communication & Processing, San Jose, CA, USA, vol. 5685, pp. 54–63 (January 2005)
2. Meuel, P., Chaumont, M., Puech, W.: Data Hiding in H.264 Video for Lossless Reconstruction of Region of Interest. In: Proc. of European Signal Processing Conf., Poznań, Poland, pp. 120–124 (September 2007)
3. Zhang, W., Cheung, S.-C.S., Chen, M.: Hiding privacy information in video surveillance system. In: Proc. of IEEE Int'l Conf. on Image Processing, Genova, Italy, vol. 3, pp. 868–871 (September 2005)
4. Yu, X., Chinomi, K., Koshimizu, T., Nitta, N., Ito, Y., Babaguchi, N.: Privacy protecting visual processing for secure video surveillance. In: Proc. of IEEE Int'l Conf. on Image Processing, Los Alamitos, CA, USA, pp. 1672–1675 (October 2008)

5. Hu, Y., et al.: Information hiding based on intra prediction modes for H.264/AVC. In: Proc. of IEEE Int'l Conf. on Multimedia and Expo, Beijing, China, pp. 1231–1234 (July 2007)
6. Aly, H.A.: Data Hiding in Motion Vectors of Compressed Video Based on Their Associated Prediction Error. IEEE Trans. on Information Forensics & Security 6, 14–18 (2011)
7. Yang, M., Bourbakis, N.: A High Bitrate Information Hiding Algorithm for Digital Video Content under H.264/AVC Compression. In: Proc. of IEEE Int'l Conf. on Image Processing Midwest Symp. on Circuits & Systems, Cincinnati, OH, USA, vol. 2, pp. 935–938 (August 2005)
8. Kapotas, S.K., et al.: Data hiding in H.264 encoded video sequences. In: Proc. of Int'l Workshop on Multimedia Signal Processing, Chania, Crete, Greece, pp. 373–376 (October 2007)
9. Winkler, S., van den Branden Lambrecht, C.J., Kunt, M.: Vision and Video: Models and Applications. Springer, USA (2001)
10. John, K.: The WebM project (2010), http://www.webmproject.org/

A Study of Real-Time Hand Gesture Recognition
Using SIFT on Binary Images

Wei-Syun Lin[1], Yi-Leh Wu[1,*], Wei-Chih Hung[1], and Cheng-Yuan Tang[2]

[1] Department of Computer Science and Information Engineering,
National Taiwan University of Science and Technology, Taipei, Taiwan
ywu@csie.ntust.edu.tw
[2] Department of Information Management, Huafan University, Taiwan

Abstract. We present a novel way to use the Scale Invariance Feature Transform (SIFT) on binary images. As far as we know, we proposed employ SIFT on binary images for hand gesture recognition and provide more accurate result comparing to traditional template approaches. There exist many restrictions on template matching approaches, such as the rotation must be less than 15°, and the variation on scale, etc. However, our proposed approach is robust against rotations, scaling, illumination conditions, and can recognize hand gestures in real-time with only off-the-shelf camera such as webcams. The proposed approach employs the SIFT features on binary image, the k-means clustering to map keypoints into a unified dimensional histogram vector (bag-of-words), and the Support Vector Machine (SVM) to classify different hand gestures.

1 Introduction

Hand gesture plays an important role of a social communication bridge. Gestures are the motion of the body or physical action form to convey some meaningful information. The difficulties of hand gestures recognition are to recognize hand gestures in real time with the high degrees of freedom (DOF) of the human hand. The ideal hand gestures recognition system have to meet the requirements in terms of real-time performance, recognition accuracy, robustness against transformations, cluttered background, and with hands from different people.

This work presents a novel way to employ the Scale Invariance Feature Transform (SIFT) on binary images, to the best of our knowledge, this work is the first research to use the SIFT method on binary images for hand gesture recognition. The traditional template matching approaches face many restrictions on the design of templates, such as rotation, scaling, etc. The proposed approach is robust against rotation, scaling, illumination conditions, and with real-time performance for hand gesture recognition.

In [1], Dardas et al. proposed a new technique to detect hand gestures only using face subtraction, skin detection, and hand posture contour comparison algorithm. However, Dardas et al. focused on bare hand gesture recognition without the help of

---

* Corresponding author.

J.-S. Pan et al. (Eds.): *Advances in Intelligent Systems & Applications*, SIST 21, pp. 235–246.
DOI: 10.1007/978-3-642-35473-1_24   © Springer-Verlag Berlin Heidelberg 2013

any markers and gloves but required additional trainings to recognize hands from different people [1]. Dardas et al. employed the Viola–Jones method [11], which is considered the fastest and most accurate learning based method, to detect faces in images [1]. The detected face will be subtracted by replacing the face area with a black circle. After subtracting the face, Dardas et al. detected the skin area using the hue, saturation, value (HSV) color model. The proposed method has real-time performance is robust against rotation, scaling, and lighting condition changes. Then, the contours of skin area were compared with all the predefined hand gesture contours to remove other skin-like objects in the image.

Given an input image, Lowe's method [12] extracts a large collection of feature vectors, each of which is invariant to image translation, scaling, and rotation, partially invariant to illumination changes and robust to local geometric distortion. Therefore, the SIFT is adopted in this work for the bare hand gesture recognition. However, the SIFT features are of too high dimensionality to be used efficiently. The work proposes to alleviate the high dimensionality problem by employing the bag-of-features approach [13-14] to reduce the dimensionality of the feature space.

Many gesture recognition techniques were developed for vision-based hand gesture recognition with different pros and cons. The traditional approaches are the template based hand pose recognition and the appearance based features of hand [2-3]. These approaches have real-time performance because the easier 2-D image features are employed. There are three main steps in the traditional hand gesture recognition approaches: the hand segmentation, the feature extraction, and the posture recognition. However, the traditional hand gesture recognition approaches are very sensitive to illumination conditions.

There are some other approaches for hand gesture recognition which employ additional sensors to collect data and the recognition performance increases with more details of the collected data. But generally speaking, the more details of the collected data increase the processing cost. In [5], Van den Bergh et al. proposed a hand gesture interaction system based on a RGB camera and a Time-of-Flight (ToF) camera for real-time hand gesture interaction with high recognition accuracy. Their proposed system may have many advantages; however, the cost of ToF cameras is extremely high when comparing web-cams, the ones employed in our proposed approach.

Xu et al. proposed a hand gesture recognition system which utilizes both the multi-channel surface electromyogram (EMG) sensors and the 3D accelerometer (ACC) to achieve the average recognition accuracy about 91.7% in real application [6]. However, the time delay between the finished gesture command and the system response as a cube action is about 300ms, which is slower than our proposed approach.

Our proposed method is mainly based on the method proposed in [1] with the main idea of using machine learning algorithms to train and test different hand gesture models. But when the training is imperfect, the machine learning algorithms tend to produce inferior recognition results. For example, if there is a hand gesture that does not exist in the training model or the illumination conditions differ between the training and the testing model, the recognition accuracy will be reduced.

The main overview of [1] is as shown in Fig. 1. The input image captured by the webcam is transformed to gray scale and directly extract the keypoints in the input image using SIFT. The vector quantization (VQ) maps the keypoints of every training image into a unified dimensional histogram vector after the K-means clustering. The histogram is then employed as the input data for the SVM classification. The main problem of this approach is that the SIFT algorithm is a robust algorithm which extracts many local features as keypoints. And too many training keypoints will leads the machine learning process to over-fit the learned models. So this work proposes a new approach using SIFT on images to alleviate the above problem.

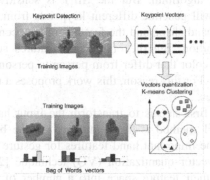

Fig. 1. Dardas el al. method [1] to generate the bag-of-words vectors

The major contributions of this work are as follows:

I.   To our knowledge, this work is the first to employ the SIFT on binary images, which is considered infeasible in the past. The experiment results suggest that the proposed approach increases the robustness to recognize hands from different people.

II.   The proposed approach is also robust against rotation, and scaling, unlike the traditional template matching approaches.

III.   By removing unnecessary features and leaving only the useful features, the proposed system can achieve real-time performance for hand gesture recognition with high accuracy.

2   System Overview

Compared with related works, the main difference in the proposed work is to employ the SIFT on binary images instead of color/grey images. The proposed method removes the redundant information to achieve real-time and high recognition accuracy with different training hands (the hand features do not exist in the bag of features). Our hand gesture recognition system consists of two stages: the offline training stage and the online testing stage. All the images in the training stage and the testing stage are captured from a webcam.

2.1   Training Stage

In the training stage, the hand gesture training images can be represented by sets of keypoint descriptors. However, the numbers of keypoints from individual images are different and the keypoints lack meaningful order. The variant number of features and the lack of feature order create difficulties for machine learning methods such as the multiclass support vector machine (SVM) classifier that require feature vectors of fixed dimension as input. To address this problem, this work proposes to employ the bag-of-features approach, which has several steps.

The first steps is to extract the features (keypoints) from hand gesture training images using the SIFT algorithm. But the SIFT is sensitive to local geometric distortion, so the SIFT will generate different keypoints from images with different people. The features of hand shape and the features of between fingers are important for hand gesture recognition. But the other hand features such as the fingernail, fingerprint, and the skin color that differ from person to person are not important for hand gesture recognition. For this reason, this work proposes a new way that employs the SIFT on binary images.

In our approach, the first step is to transform the input hand gesture images to binary images. The SIFT method is then employed on the binarized hand gesture images to extract only the important hand features for gesture recognition. The next step is to employ the vector quantization (VQ) technique [2], which clusters the keypoint descriptors in their feature space into a number of clusters using the K-means clustering algorithm. Then each keypoint in encoded by the index of the cluster (codevector) to which this keypoint belongs. This VQ process maps keypoints of every training image into a unified dimensional histogram vector as shown in Fig. 2. Finally, each cluster is considered as a visual word (codevector) that stands for a particular local pattern shared by the keypoints in that cluster.

The clustering process constructs a visual word vocabulary (codebook) representing the local patterns in the training images. The size of the vocabulary is

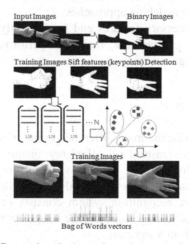

Fig. 2. Generating the bag-of-words vectors for training

determined by the number of clusters (codebook size), which can be varied from hundred to over tens of thousands. Each training image can be described as a "bag-of-words" vector by mapping the many keypoints to one visual word vector. The multiclass SVM models can be trained with the unified dimensional feature representation of visual word vectors. This work assumes that by employing the Dardas et al. method [1] to detect and track the hands and the skin color in images, the input image can be separated into foreground and background. And the training images contain only the hand gestures without any other objects, such as elbow or torso, in the input image. The size of the training images is by default 640×480 pixels in size but any other image sizes will work.

2.1.1   Binary Images

The work assumes that there is only one hand gesture in an input image and the hand gesture image is already divided into foreground (hand) and background (other) by applying method such as [1]. In this case, the foreground (hand gesture) of the input image is set to white and the background (other) is black.

With the above assumption, we effectively constraint some variant conditions such as illumination and skin colors of different test subjects. Thus the resulting SIFT features are inherently more robust. Fig. 3 shows the differences between applying the SIFT on binary images and applying the SIFT on color/grey images. Notice the redundant features vectors on fingernail, fingerprint, knuckle, etc., when applying the SIFT method on color/grey images as shown in Fig. 3(a) and 3(c). So using SIFT on binary image is useful, it can remove most redundant vector. When applying the SIFT method on the binary image as shown in Fig. 3(b) and (d), notice that only the most discriminating SIFT features, between fingers or the shape of hand, are remained. These most discriminating SIFT features are expected to increase the robustness of the proposed hand gesture recognition method.

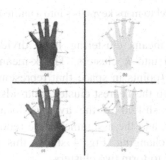

Fig. 3. SIFT extract from hand (a) SIFT keypoints from color image (b) SIFT keypoints from binary image (c) SIFT keypoints from color image (d) SIFT keypoints from binary image

2.1.2   SIFT

Lowe proposed the Scale- Invariant Feature Transform (SIFT) in [12]. The SIFT descriptors describe the local feature in the image geometric variations.

2.1.3   K-means Clustering

In data mining, the k-means clustering is a method of cluster analysis which partition n observations into k clusters in which each observation belongs to the cluster with the nearest mean. The k-means clustering is an approach of unsupervised learning algorithm and an ordinary method for statistical data analysis applied in several fields. In the training stage, when the training images contain only hand gestures in color on a white background, the extracted keypoints represent the hand gesture only. The number of extracted keypoints normally does not exceed 75 for each gesture. But there are some redundant keypoints that have to be removed from the color hand gestures. When applying the SIFT on binary hand gesture images, the maximum number of extracted keypoints decreased to 50.

When using the SIFT keypoints extracted from binary images, if the number of clusters is too small, the classification accuracy will decrease. In the training image set, each gesture has 100 training images and the total number of keypoints for each gesture is about 5000. Because the each keypoints in the training image set is important and unique, we chose the value 800 as the number of clusters (visual vocabularies or codebook) to build our cluster model.

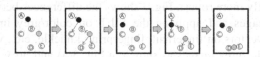

Fig. 4. K-means clustering with two clusters and saving the cluster model

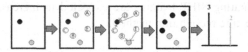

Fig. 5. Using cluster model to maps keypoints into a unified dimensional histogram

The first step in the k-means clustering is to divide the vector space (128-dimensional feature vector) into k clusters. The k-means clustering starts with k randomly located centroids (points in space that represent the center of the cluster) and assigns every keypoint to the nearest cluster centroids. After the assignment, the centroids (codevectors) are shifted to the average location of all the keypoints assigned to the same cluster, and the assignments are redone. This procedure repeats until the assignments stop changing. Fig. 4 shows this process in action for five keypoints: A, B, C, D, and E to form two clusters.

Fig. 5 shows the process that maps the image keypoints to a k dimensional histogram vector. By using the above k-means clustering method, the keypoint vectors for each training image are employed to build the cluster mode. The number of clusters (codebook) will represent the number of centroids in the cluster model. Finally, the cluster model will build codevectors equal to the number of clusters assigned (k) and each codevector will have 128 components, which is equal to the length of each keypoint. Then, the keypoints of each training image are mapped into the k-means clustering model to reduce the dimensionality into one bag-of-words vector with k components, where k is the number of clusters. In this way, each keypoint, extracted from a training image, will be represented by one component in the generated bag-of-words vector with value equal to the index of the centroids in the cluster model with the nearest Euclidean distance. The generated bag-of-words vector, which represents the training image, will be grouped with all the generated vectors of other training images that have the same hand gesture and labeled with the same number, and this label will represent the hand gesture class number. For example, label or class 1 for the hand gesture **C** training images, class 2 for hand gesture **fist**, class 3 for hand gesture **five**, and class 4 for hand gesture **index**.

By analyzing the SIFT keypoints, we discover unique characteristics of the SIFT vectors for each gesture. In Fig. 6, the SIFT keypoints of every gestures have some common regular patterns. But the VQ process maps the keypoints of a training image into one k dimensional histogram vector after K-means clustering. And with the k-means clusters, similar vectors will be mapped into the same dimension as show in Fig. 7. And if there are some minor differences in the SIFT vectors in the hand gesture image taken from different people, the k-means clustering will maps the similar gesture vectors to the same cluster.

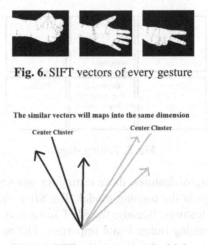

Fig. 6. SIFT vectors of every gesture

Fig. 7. Hand SIFT vectors mapped with k-mean

2.1.4 K-means Clustering

After mapping the keypoints of each training image to one bag-of-words vector, the bag-of-words vector is labeled with the hand gesture class or label number. All the labeled bag-of-words vectors are employed as the training data to build the multi-class SVM classifier model. The SVM is a supervised learning method for classification and regression by creating an n-dimensional hyperplane that optimally divides the data into difference groups. Even though SVMs were initially intended as binary classifiers, other methods that deal with a multiclass problem as a single "all-together" optimization problem exist [7], but are computationally much more costly than solving several binary problems.

A variety of approaches for decomposition of the multiclass problem into several binary problems using the two-class SVM have been proposed. In our implementation, multiclass SVM training and testing are performed using the LIBSVM library [8]. The LIBSVM supports multiclass classification and uses a one-against-one (OAO) approach for multiclass classification in SVM [9]. For the M-class problems (M being greater than 2), the OAO approach creates M(M-1)/2 two-class classifiers, using all the binary pair-wise combinations of the M classes. Each classifier is trained using the samples of the first class as positive examples and the samples of the second class as negative examples. To combine these classifiers, the Max Wins method is used to find the resultant class by selecting the class voted by the majority of the classifiers [10].

2.2   Testing Stage

The testing stage workflow is shown in Fig. 8. The first step is to capture image frames from the webcam or video file. And then apply the Dardas et al. method [1] for face detection and subtraction, hand gesture detection, and hand extraction in each image frame. These extracted hand images are employed for the testing model.

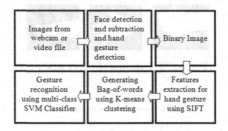

Fig. 8. Testing stage

Before building the bag-of-features, these testing images need to be binarized to be consistent with the image in the training model. The SIFT method is then applied to extract the hand gesture features. Because the SIFT features is invariant to scaling and rotation, the size of the testing image is not important. The next step is to employ the k-means model which to map the keypoints in the testing image into a unified dimensional histogram vector the same as the testing images. Finally, the SVM classifier model built in the training stage is employed to classify the histogram vector as one of the hand gestures trained.

3   Experimental Result

This section discusses the experimental settings and the results in details.

3.1 "Left" and "Right" Hand Gesture Recognition

There are the two experiments. The first case tries to recognize people hands which are already included in the training model. And the second case tries to recognize people hands which are not included in the training model. The training model is built with 100 training images (640×480) for the "left" and the "right" hand gestures. And the testing is performed on 100 images (640×480) to evaluate the accuracy of the multiclass SVM classifier model for each gesture.

In our experiments, we assumed that all the input images had been pre-processed with the hand gesture part of the images extracted. A low-cost Logitech QuickCam web camera provides videos captured with different resolutions such as 640×480, 320×240, and 160×120. The experiments are conducted with 200 testing images to evaluate the performance of the multiclass SVM classifier model for each gesture. In our conjecture, the bag-of-features model proposed in [1] may not perform well with hands from different people. Two experiments are built to verify our conjecture and the experiments model as show in Table 1.

Table 1. Experiment A and B (a) Training and testing both with Subject 1's hands. (b) Training with Subject 1hands and testing with Subject 2's hands

Experiment A		
Training model	Testing	Accuracy of recognition
Subject 1's hands	Subject 1's hands	95%
Experiment B		
Training model	Testing	Accuracy of recognition
Subject 1's hands	Subject 2's hands	76%

The experiment A employs Subject 1's hands to build the training model and to test with Subject 1's hands too. The experiment result achieves high classification accuracy of 95% and suggests that that the bag-of-features with the SIFT method on color/grey images can perform well for hand gesture recognition. The experiment B also employs Subject 1's hands to build the training model but employs Subject 2's hands for testing. The classification accuracy decreases to 76% from 95% as shown in Tables. The recognition accuracy of other hand gestures shows similar results. Based on the above observations, we conjecture that if all the hand gesture images are taken under the same conditions, the recognition accuracy can be improved. We propose to employ the binary images with the removal redundant hand SIFT features by different people. In Fig. 9, the color hand image on the left is binarized as shown in the left. As shown in Fig. 3, the binarized hand image retains only the shape information of the gesture and removes most of the detail information such as shadows.

Fig. 9. Hand image after binarization

And the next experiment is to employ the SIFT method on binary images to build the training model and also in testing. We observe that by employing the SIFT method on binary images, the SIFT features extracted from different subject's hands are still similar, thus the high classification accuracy.

3.2   Five Hand Gestures Recognition

The experiments employ five hand gestures, which are C, Fist, Five, Index (point), and V (two), as shown in Fig. 10. The number of clusters to build the cluster model is an important factor that affects the classification accuracy. In this experiment, the number of clusters to build the cluster model is fixed to 800 clusters for both the proposed method and the comparing method in [1]. The experiment results are as shown in Table 2 and Table 3.

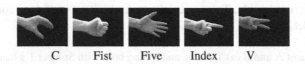

C   Fist   Five   Index   V

Fig. 10. Hand postures used in training images

Table 2. SIFT on color Images [1]

Gesture Name	Number of frames	Correct	Incorrect	Accuracy of recognition
C	200	200	0	100%
Fist	200	77	123	38.50%
Five	200	200	0	100%
Index	200	200	0	100%
V	200	98	102	49%
Average accuracy of recognition = 77.5%				

Table 3. Proposed SIFT on binary Images

Gesture Name	Number of frames	Correct	Incorrect	Accuracy of recognition
C	200	198	2	99%
Fist	200	196	4	98.00%
Five	200	191	9	95.50%
Index	200	193	7	96.50%
V	200	185	15	92.50%
Average accuracy of recognition = 96.3%				

In Table 2, the results show that the method proposed in [1] cannot effectively recognize two of the five gestures, namely Fist and V, but perform well to recognize the rest three gestures. Table 3, however, shows that proposed method to employ the

SIFT on binary images can recognize all five hand gesture accurately. The overall recognition accuracy is decreased when the number of gestures increases from three to five, which is as expected when employing the multi-class SVM as the classifier.

3.3   Variant Numbers of Clusters

The number of clusters to build the cluster model is an important factor that affects the qualification accuracy. The main purpose of this section is to analyze how the number of clusters affects the accuracy when using the SIFT on binary images. Table 4 shows how the number of clusters affects the classification accuracy of using SIFT on binary images. The results suggest that 1600 clusters produce the highest classification accuracy. Fig. 11 shows the comparison of the proposed method and the method in [1] with varying numbers of clusters.

Table 4. Our approach with different numbers of the clusters (codebook size)

Number of the clusters	100	200	400	800	1600	3200
Accuracy of recognition	81%	86%	92.5%	95.6%	96.3%	89.2%

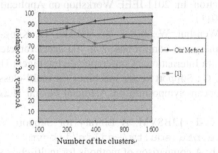

Fig. 11. Comparison with varying numbers of clusters

4   Conclusion

In this paper, we present a novel way to use SIFT on binary images for the first time in gesture recognition research. In our proposed recognition technique is inherently robust against rotations, scaling, and lighting conditions, and even with hands from different people. The proposed approach is low time-consuming approach and can provide real-time hand gesture recognition. Experiment results show that the proposed system can achieve high classification accuracy of 96.3% with hand images of different people. Three important factors affect the accuracy of the system, first is the quality of the webcam in the training and testing stages, second is the number of the training images, and third is the chosen number of clusters to build the cluster model. One of our future research directions is to apply the proposed technique for real-time sign language translation.

Acknowledgments. This work was partially supported by the National Science Council, Taiwan, under the Grants No. NSC101-2221-E-011-141, NSC100-2221-E-011-121, and NSC101-2221-E-211-011.

References

[1] Dardas, N.H., Georganas, N.D.: Real-Time Hand Gesture Detection and Recognition Using Bag-of-Features and Support Vector Machine Techniques. IEEE Transaction on Instrumentation and Measurement (November 2011)

[2] Stenger, B.: Template-Based Hand Pose Recognition Using Multiple Cues. In: Narayanan, P.J., Nayar, S.K., Shum, H.-Y. (eds.) ACCV 2006. LNCS, vol. 3852, pp. 551–560. Springer, Heidelberg (2006)

[3] Tofighi, G., Monadjemi, S.A., Ghasem-Aghaee, N.: Rapid Hand Posture Recognition Using Adaptive Histogram Template of Skin and Hand Edge Contour. In: 2010 6th Iranian Machine Vision and Image Processing (MVIP) (October 2010)

[4] Kanungo, T., Mount, D.M., Netanyahu, N., Piatko, C., Silverman, R., Wu, A.Y.: An efficient k-means clustering algorithm: Analysis and implementation. IEEE Trans. Pattern Analysis and Machine Intelligence (2002)

[5] Van den Bergh, M., Van Gool, L.: Combining RGB and ToF Cameras for Real-time 3D Hand Gesture Interaction. In: 2011 IEEE Workshop on Applications of Computer Vision (WACV) (January 2011)

[6] Xu, Z., Xiang, C., Wen-hui, W., Ji-hai, Y., Vuokko, L., Kong-qiao, W.: Hand gesture recognition and virtual game control based on 3D accelerometer and EMG sensors. In: Proceedings of the 14th International Conference on Intelligent User Interfaces (2009)

[7] Weston, J., Watkins, C.: Support vector machines for multi-class pattern recognition. In: Proceedings of European Symposium on Artificial Neural Networks, Bruges, Belgium (April 1999)

[8] Chang, C.-C., Lin, C.-J.: LIBSVM: A Library for Support Vector Machines (2001), http://www.csie.ntu.edu.tw/~cjlin/libsvm

[9] Hsu, C.-W., Lin, C.-J.: A comparison of methods for multi-class support vector machines. IEEE Transactions on Neural Networks (March 2002)

[10] Friedman, J.H.: Another approach to polychotomous classification. Department of Statistics and Stanford Linear Accelerator Center Stanford University (1997)

[11] Viola, P., Jones, M.: Robust real-time object detection. International Journal of Computer Vision (2004)

[12] Lowe, D.G.: Distinctive image features from scale-invariant keypoints. International Journal of Computer Vision (November 2004)

[13] Lazebnik, S., Schmid, C., Ponce, J.: Beyond bags of features: Spatial pyramid matching for recognizing natural scene categories. In: Proceedings of the 2006 IEEE Computer Society Conference on Computer Vision and Pattern Recognition (2006)

[14] Jiang, Y., Ngo, C., Yang, J.: Towards optimal bag-of-features for object categorization and semantic video retrieval. In: Proceedings of the ACM International Conference on Image and Video (2007)

An Approach for Mouth Localization Using Face Feature Extraction and Projection Technique

Hui-Yu Huang and Yan-Ching Lin

National Formosa University, Yunlin 632, Taiwan
anne.huang@ieee.org

Abstract. In this paper, we propose an efficient approach to perform mouth localization based on face detection, face mask label, and projection. For mouth localization, there are some related techniques which could not exactly locate the correct mouth position on profile-face image. In order to solve this problem, we design a skin-color filter to efficiently segment skin-color region, and edge projection technique to locate optimal mouth area and position. Experimental results verify that this approach can obtain a high accuracy and a fast executed time with face feature extraction and mouth detection.

Keywords: Skin-color filter, face feature, Canny detector, projection.

1 Introduction

In recent years, there are many researches which focus on human attributes, such as face recognition, or face detection, etc. As for applications, one of important schemes is eyes detection which can avoid or alarm a fatigue situation for driving a car. In addition, for mouth detection, it can provide a lip-read recognition for help the blind persons. Based on this interesting scheme, in this paper, we will focus on locate the correct mouth position based on our proposed method. However, based on this topic, firstly, the face region in an image must be clearly labeled out. For face feature detection, there are many approaches which have been presented [1, 2].

Yang *et. al.* [1] put forward an adaptive face detection algorithm. Firstly, face is chosen by the adaptive skin color model and eyebrow and eyes are located by integral projection. Then, the coordinates of eye center are confirmed by the method which makes use of mixed round and elliptic based on Hough transform. Finally, face is detected by the human eye center mapping. Hu *et al.* [2] presented an automatic face recognition method. Authors used the AdaBoost technology to detect face region and wavelet transform and KPCA method to extract face features. These features were fed up into support vector machine for recognition. This method is superior to traditional PCA in the time of features extraction. He and Zhang [3] presented a new real time lip detection based on the discrepancy of skin and lip on chrominance and R/G. Nhan and Bao [4] proposed a mouth detection approach in color image. Firstly, authors segmented image based on skin. After skin process, they added some specific techniques to fit the mouth in color image efficiently in order to determine mouth candidates and then classify those of candidates by neural network.

J.-S. Pan et al. (Eds.): *Advances in Intelligent Systems & Applications*, SIST 21, pp. 247–257.
DOI: 10.1007/978-3-642-35473-1_25   © Springer-Verlag Berlin Heidelberg 2013

In this paper, we propose a mouth detection method based on face extraction, face mask labeling, edge detector and projection technique. The face labeling is to find face mask which can filter out non-skin color regions and raise the detection accuracy before the mouth localization process.

The remainder of the paper is organized as follows. Section 2 presents the proposed method. Experimental results and performance evaluation are presented in Sections 3. Finally, Section 4 concludes this paper.

2   Proposed Method

In the section, we will describe our approach. This approach consists of preprocess, segment the face region, edge detection, and projection. Details of procedures are described in the following subsections.

2.1   Preprocessing

In order to decrease the computational complexity and fast obtain the face features, we take the normalization process for all test data. Here, we use a bi-cubic interpolation to perform the normal size due to the bi-cubic interpolation has better image quality than other interpolation method. The normal size is defined as 500×350, 350×500, and 500×500.

2.2   Viola and Jones Detector

For face detection, one of many researchers, Viola and Jones (V-J) method [5], is usually adopted to work this field that possesses a high efficiency and accuracy to locate the face region in an image. The V-J method consists of three phases. Firstly, the Haar-like features is rectangular type that is obtained by integral image; Secondly, The Adaboost algorithm is a learning process that is a weak classification and then uses the weight value to learn and construct as a strong classification. Finally, we can obtain the non-face region and face region after cascading each of strong classifiers. Details of V-J detector can study Ref. [5]. Here, we don't further describe it.

2.3   Filtering Out Non-faces Area

Although the face detection by used V-J detector has a higher performance, the threshold may affect the location result. Hence, we employ a ratio of skin region filtered candidates to further decide which the correct face region is, so that it can improve the face location accuracy after V-J detector. In other words, the better location of skin-color region implies a better outstanding feature for face.

Owing to the skin characteristic, we will take the different color space and threshold to decide the skin region. As previously researched, many color models, *YCbCr* or *HSV* or *HIS*, provide the advantage information in skin-color detection, in this paper, we use *YCbCr* color space to detect the skin-color area.

In *YCbCr* color space, *Y* is Luma influenced by light factor, we ignore *Y* value and adopt the chrominance components *Cb* and *Cr* to decide the skin range. Using the adaptive threshold assignment to decide the skin region, the related works have been published [6, 7].

Chai and Ngan [6] defined this skin condition as follows.

$$Skin = \begin{cases} 1, if \begin{cases} 77 < Cb < 127 \\ 133 < Cr < 173 \end{cases} \\ 0, otherwise \end{cases} \tag{1}$$

Wu [8] expressed as:

$$Skin = \begin{cases} 1, if \begin{cases} 60 \leq Y \leq 255 \\ 100 \leq Cb \leq 125 \\ 135 \leq Cr \leq 170 \end{cases} \\ 0, otherwise \end{cases} \tag{2}$$

According to Eqs. (1) and (2), the skin range for *Cb* and *Cr* components can be obtained. Because this range is too strict, it will lose some advantage information. Based on this reason, we will modify this range to fit the skin area more feasible. Here, the range is defined as follows.

$$Skin = \begin{cases} 1, if \begin{cases} 90 \leq Cb \leq 124 \\ 136 \leq Cr \leq 180 \end{cases} \\ 0, otherwise \end{cases} \tag{3}$$

After skin segmentation, we will calculate the ratio of number of white pixels (skin color) and number of total pixels. Next, we utilize the threshold to filter out non-face region and to reduce the mistake. The face localization results are presented in Fig. 1.

Fig. 1. Result of filtering out non-faces. (a) Original image, (b) face detection by V-J detector, (c) face detection by skin ratio.

2.4   Face Mask Labeling

In general, the face posture has two types of frontal face and profile face. For mouth location, it existed higher wrong on biased face image because it includes skin color and non-skin color. In order to solve this situation, we propose a face mask to reduce mistake. The face mask is obtained by morphologic processing.

1)   Skin Color Segmentation

This procedure is to transform RGB into skin color before product face mask. The skin color model is to normalize RGB because it can discriminate non-skin color clearly. The normalized RGB function is expressed as Eq. (4) and many related researches have been presented [8-10].

$$r = \frac{R}{R+G+B}, g = \frac{G}{R+G+B}, \tag{4}$$

where the r and g are the normalized pixels.

Here, we present the results of the different skin color segmentation methods compared with Soriano $et\ al.$'s method [8], Huang's method [9], Chen's method [10], and our proposed methods. Figure 3 shows the compared result. From Fig. 2 (d) shows that Chen's method is better than other author's method on facial contour and noise, but the median image in Fig. 2(d) has little defects which non-skin region as skin color, such as hair. Hence, based on Chen's method, in this paper, we further modify his method and redesign a more effective skin constraint to achieve the skin region for a face image, it is expressed as

$$Skin = \begin{cases} 1, & \text{if} \begin{cases} Q_-(r) < g < Q_+(r), \\ (R-G) \geq 25, \\ w > 0.001, \end{cases} \\ 0, & \text{otherwise}, \end{cases} \tag{5}$$

where

$$\begin{cases} Q_+(r) = A_u r^2 + b_u r + c_u, \\ Q_-(r) = A_d r^2 + b_d r + c_d, \\ w = (r-0.33)^2 + (g-0.33)^2. \end{cases} \tag{6}$$

The upper bound quadratic coefficients found are $A_u = -1.3767$, $b_u = 1.0743$, $c_u = 0.1452$, while the lower bound coefficients are $A_d = -0.776$, $b_d = 0.5601$, and $c_d = 0.1766$ [8].

From Fig. 2(e), it is clear that our proposed rule of skin color constrain is superior to Chen's method which let the face out line be good.

2) Design of Ratio Filter

In order to clearly present the skin-color region in a face image, we design a novel morphological process called ratio filter. The filter aims to emphasize skin pixels and to decrease the non-skin pixel. The ratio filter is to compute number of skin pixels and non-skin pixels on 5×5 mask, and then to compare those of skin pixels.

$$\begin{cases} P_{255} = P_{255} + 1, & \text{if } X = 255, \\ P_0 = P_0 + 1, & \text{if } X = 0, \end{cases} \tag{7}$$

where P_{255} is number of skin pixel, P_0 is number of non-skin pixel, and X is a pixel value of the result of skin detection. The current pixel has three conditions on setting as follows.

- If P_{255} is greater than P_0, the value of current pixel is as 255.
- If P_0 is greater than P_{255}, the value of current pixel is as 0.
- If P_0 is equal P_{255}, the value of current pixel is as original pixel.

Figure 3 presents the result using the ratio filter. The facial contour is more compact and the non-skin color is reduced.

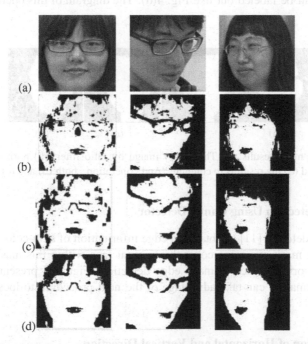

Fig. 2. Skin color detection results. (a) Original images, (b) results of Soriano's method, (c) result of Huang's method, (d) result of Chen's method, and (e) our proposed skin-color rule.

Fig. 2. *(Continued)*

(a)   (b)

Fig. 3. Results. (a) Non using ratio filter, (b) using ratio filter

3) Mask Labeling

The section describes how face mask is produced. The face mask denote white region. The process consists of two phases. One phase is to find first pixel which value is 255 on horizontal (left to right and right to left) and vertical (top to down and down to top). The other phase is do "AND" operation both horizontal result and vertical result, so face mask can be labeled out like Fig. 4(d). The diagram of this operation presents in Fig. 4.

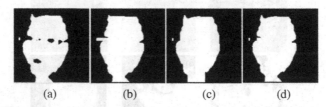

(a)   (b)   (c)   (d)

Fig. 4. The face mask results. (a) The binary image by ratio filter, (b) horizontal result, (c) vertical result, and (d) the result is to compute "And" operation both (b) and (c).

2.5   Edge Detection Using Canny Detector

We use Canny detector [11] to obtain the edge information of skin-color region. After processing face mask, we can recover the original face within this mask region, and then takes this original face to make edge detection. Figure 5 presents the results. This edge information can take advantage of the next procedure to locate the mouth position.

2.6   Projection of Horizontal and Vertical Direction

After processing edge detection, we utilize the projection computation to locate mouth position on horizontal and vertical direction. According to the spatial geometric relationship of facial features, the mouth in general is positioned between one third of face region and one fourth of face region; we define this range as follows. The horizontal/vertical region is expressed as

$$\begin{cases} H_{h1} = H \times 0.7 \\ H_{h2} = H \times 0.1, \\ W_h = W \times 0.1 \end{cases} \begin{cases} H_{v1} = H \times 0.4 \\ H_{v2} = H \times 0.1, \\ W_v = W \times 0.1 \end{cases} \tag{8}$$

where H_{h1}, $H - H_{h2}$, and W_h denote the first position and the last position on image height (H), the rang for horizontal region on image width (W), respectively. H_{v1}, $H - H_{v2}$, and W_v denote the first position, the last position on H, and the rang for vertical region on W, respectively. Figure 6 shows this diagram of horizontal and vertical regions. The results of horizontal projection are shown in Fig. 7.

In order to reduce mistake which doesn't match mouth region, here, we ignore the first edge point computation from left to right and right to left before vertical projection. Figure 9 shows the result of vertical projection.

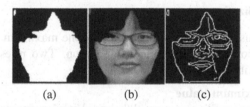

(a)   (b)   (c)

Fig. 5. The Canny results. (a) The binary image by ratio filter, (b) the recovery result, (c) Canny edge detection.

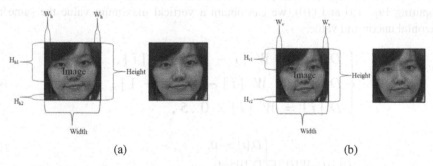

(a)   (b)

Fig. 6. The diagram of mouth region. (a) Horizontal region, (b)vertcal region.

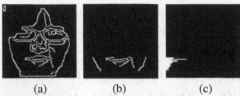

(a)   (b)   (c)

Fig. 7. The horizontal projection. (a) Canny edge detection, (b) the horizontal region of Canny image, (c) the horizontal projection.

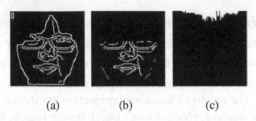

(a)   (b)   (c)

Fig. 8. The vertical projection (delete one pixel). (a) Canny edge detection, (b) vertical region of Canny image, (c) vertical projection.

2.7   Mouth Localization

Based on the previous processing, we will compute the maximum value of horizontal and vertical projection to decide the mouth position. Two rules are to decide the mouth location describe as follows.

1)   Horizontal Maximum Value

If there are two or more positions generated the equal maximum value, we take the average position for these two positions and media position for more than two positions.

2)   Vertical Maximum Value

Owing to the accuracy of mouth location, it usually deviates the vertical position; hence, we design Eqs. (9) and (10) to enhance the correct vertical location. After computing Eqs. (9) and (10), we can obtain a vertical maximum value the same as horizontal maximum value.

$$
\begin{cases}
D_1[i] = W[i-1] - W[i], \\
D_2[i] = W[i] - W[i+1], \\
D[i] = W[i] \times 0.5,
\end{cases}
\tag{9}
$$

$$
\begin{cases}
W[i] = W[i], \text{if} \begin{cases} D_1[i] >= 0, \\ D_2[i] >= 0, \\ D_1[i] < D[i] \text{ or } D_2[i] < D[i], \end{cases} \\[2em]
W[i] = 0, \text{if} \begin{cases} D_1[i] < 0 \text{ or } D_2[i] < 0, \\ D_1[i] >= D[i] \text{ or } D_2[i] >= D[i], \end{cases}
\end{cases}
\tag{10}
$$

where i is position, $W[i]$ is the projection of current position, $D_1[i]$ is the difference between the forward position and the current position, $D_2[i]$ is the difference between the current position and the next position, $D[i]$ is a threshold value.

3 Experimental Results

In this section, we demonstrate our proposed method using Boa database [12] and our database to estimate the performance for mouth localization. The face images have 355 that contain 94 on Boa database and 261 on our database respectively. The computational time of mouth localization is about 0.75 s on image size of 150×150 worked a 2.80 GHz Intel ® Core(TM) is-2300 CPU with 4 GB RAM PC and C# language.

Based on skin-color segmentation rule which will affect the mouth location, in our experiments, the results comparing with Chen's rule and our rule about skin-color segmentation are presented the accuracy percentage of mouth detection. Table 1 presents the results. From Table 1, it is clear that our result is better than Chen's result in Bob database, that is because Chen's condition is lax, it may obtain a big face mask which includes non-skin color region. Hence, it will cause the mistake mouth location. On the whole, the accuracy of mouth localization in our data can achieve 94%.

Figures 9-10 show the results of mouth location. From Figs. 9, it is clear that our result is more precise than Chen's rule because our condition is strict on face mask. However, there are some results that localize the mistake cases, such as Fig. 11.

Table 1. The mouth localization

Database	Samples	Accuracy	
		Chen's	The proposed
Bob	94	90%	92%
Our data	261	94%	94%
Total	355	93%	93%

(a)   (b)

Fig. 9. Results of the mouth location. (a) Chen's condition, (b) our condition.

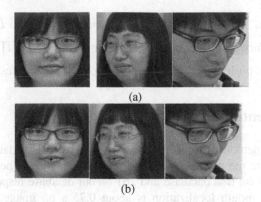

(a)

(b)

Fig. 10. Results of the mouth location include frontal face and profile face. (a) Original image, (b) our proposed condition.

(a)   (b)

Fig. 11. Results of the mouth location. (a) Original image, (b) wrong result.

4   Conclusions

In this paper, we have presented a fast mouth localization method based on face detection, skin-color segmentation, edge detector, and projection. The system purpose is to locate mouth position and to raise the correct accuracy by our designed rule of skin-color segmentation. The results demonstrate that our approach can exactly obtain higher accuracy than Chen's skin-color segmentation condition. Consequently, the accuracy of mouth location can achieve 94% and it is acceptable. In the future, we will further modify the current algorithm to work facial feature extraction and face posture estimation.

Acknowledgement. This work was supported in part by the National Science Council of Republic of China under Grant No. NSC100-2628-E-150-003-MY2.

References

1. Yang, G., Sun, H., Li, H.: Face detection based on adaptive skin model and hough transform. In: Proc. of IEEE Int. Conf. on Electronics, Comm. and Control (ICECC), pp. 258–260 (2011)

2. Hu, T., Liu, R., Zhang, M.J.: Face recognition under complex conditions. In: Proc. of IEEE Int. Conf. on Electronics, Comm. and Control Engineering (ICECE), pp. 960–963 (2010)
3. He, J., Zhang, H.: A real time lip detection method in lipreading. In: Control Conf. on Chinese, pp. 516–520 (2007)
4. Nhan, H.N.D., Bao, P.T.: A new approach to mouth detection using neural network. In: Proc. of IEEE Int. Conf. Control, Automation and System Engineering, pp. 616–619 (2009)
5. Viola, P., Jones, M.: Rapid object detection using a boosted cascade of simple features. In: Proc. of the IEEE Computer Vision and Pattern Recognition, vol. 1, pp. 511–518 (2001)
6. Chai, D., Ngan, K.N.: Face segmentation using skin-color map in videophone applications. IEEE Trans. Circuits and System for Video Technology 9(4), 551–564 (1999)
7. Wu, M.W.: Automatic facial expressions analysis system, Master Thesis, National Cheng King University (2003)
8. Soriano, M., Martinkauppi, B., Huovinen, S., Laaksonen, M.: Using the skin locus to cope whit changing illumination conditions in color-based face tracking. In: Proc. IEEE Nordic Signal Proc. Symp., Kolmarden, Sweden, pp. 383–386 (2000)
9. Huang, T.S.: A smart digital surveillance system with face tracking and recognition capability. Master Thesis, Chung Yuan Christian University (2004)
10. Chen, C.T.: Multiple face recognition based on skin-color regional segmentation and principal component analysis. Master Thesis, National Taiwan Ocean University (2006)
11. Canny, J.: A computational approach to edge detection. IEEE Trans. on Pattern Analysis and Machine Intelligence PAMI-8(6), 679–698 (1986)
12. Boa database,
    http://www.datatang.com/datares/go.aspx?dataid=604374

2. Hu, T., Liu, K., Zhang, M.: Face recognition under complex conditions. In: Proc. of IEEE Int. Conf. on Electronics, Comm. and Control Engineering (ICCCE), pp. 900–903 (2010)

3. He, J., Zhang, H.: A real time lip detection method in threshold. In: Control Conf. on Chinese, pp. 516–520 (2007)

4. Shan, H.N.D., Rao, R.L.: A new approach to pupil detection using neural network. In: Proc. of IEEE Int. Conf. Automation and System Engineering, pp. 616–619 (2009)

5. Viola, P., Jones, M.: Rapid object detection using a boosted cascade of simple features. In: Proc. of the IEEE Computer Vision and Pattern Recognition, vol. 1, pp. 511–518 (2001)

6. Chai, D., Ngan, K.N.: Face segmentation using skin-color map in videophone applications. IEEE Trans. Circuits and Systems for Video Technology 9(4), 551–564 (1999)

7. Wu, M.W.: Automatic facial expressions analysis system. Master Thesis, National Chung Kung University (2001)

8. Sidenbladh, Moeslund, Hu, Bernardin, S.: ... using the skin locus to cope with changing illumination conditions in color-based face tracking. In: Proc. IEEE Nordic Signal Proc. Symp. Kolmarden, Sweden, pp. 383–386 (2000)

9. Hsiao, T.S.: A virtual digital surveillance system with face tracking and recognition capability. Master Thesis, Chung Yuan Christian University (2004)

10. Tren, C.T.: Multiple face recognition based on skin color regional segmentation and principal component analysis system. Master Thesis, National Taiwan Ocean University (2005)

11. Kung, J.: A computational approach to edge detection. IEEE Trans. on Pattern Analysis and Machine Intelligence PAAM 8(6), 679–698 (1986)

12. Borderline...
http://www.datacamp.com/data/ace/Apo_asex?dataset=b04574

Facial Expression Recognition Using Image Processing Techniques and Neural Networks

Hsi-Chieh Lee[1,*], Chia-Ying Wu[2], and Tzu-Miao Lin[2]

[1] National Quemoy University
Department of Computer Science and Information Engineering
No.1 University Rd, Jinning, Kinmen, 89250, Taiwan, R.O.C.
imhlee@gmail.com
[2] Yuan Ze University
Department of Information Management,
135 Yuan-Tung Rd, Chung-Li, 32003, Taiwan, R.O.C.

Abstract. In our daily life, the facial expression contains important information responded to interaction to other people. Human Facial Expression Recognition has been researched in the past years. Thus, this study adds facial muscle streak, for example nasal labial folds and front lines, as another recognition condition.

We used the traditional face detection to extract face area from original image. Then to extract eyes, mouth and eyebrow outlines' position from face area. Afterward, we extracted important contours from different feature areas. Ultimately, we used these features to create a set of feature vector. Then, these vectors were used to process with neural network and to determine user's facial expression.

In summary, this study used TFEID (Taiwanese Facial Expression Image Database) database to determine the expression recognition and face recognition. The experiment result shown, that 96.2% and 92.8% of TFEID database can be recognized in personalizing expression recognition experiment and full member expression recognition, respectively. In face recognition, 97.4% of TFEID sample were recognized.

Keywords: Facial expression recognition, Face detection, Feature extraction, Feature areas.

1   Introduction

Human face researches contain face detection, face recognition and facial expression recognition. There are many scholars and researchers working on these researches. In face detection, Yang, Kriegman and Ahuja [1] had developed single picture face detection method. This method can be divided into four categories: Knowledge-based [2], Feature-based [3], [4], Template-matching [5] and Appearance-based [6], as well as Viola Jones [6] using Rectangle Feature in Integral Image, AdaBoost feature classifier and cascade classifier to find object or human face in pictures quickly.

---

* Corresponding author.

J.-S. Pan et al. (Eds.): *Advances in Intelligent Systems & Applications*, SIST 21, pp. 259–267.
DOI: 10.1007/978-3-642-35473-1_26   © Springer-Verlag Berlin Heidelberg 2013

However, it is more complex in feature extraction. The detail face informations like feature position, size and muscle texture will be needed. Fasel and Luettin [7] had classified the feature extraction methods into two categories as deformation extraction[8], [9]and motion extraction [10]. On the other hand, Contreras had proposed the most important information of human face is contour of features [11]. He applied the edge detectors of Sobel and Canny to achieve completely contour of facial features. Majumder, Behera and Subramanian used the distant of facial features to detect the center of eye and the lip position effectively [9]. Similarly, Bashyal and Venayagamoorthy [12] using applied the Gabor filter in different angles to measure vector value of facial features and locate its position.

In this study, a new approach to including of image preprocessing, face detection, feature extraction, and classification is proposed and in attempting to achieve a more real timed characterization of facial images.

2   The Proposed System

Fig.1 illustrates process of the proposed approach, which includes image preprocessing, face detection, feature extraction and classification. Details of these processes are described in the following subsections.

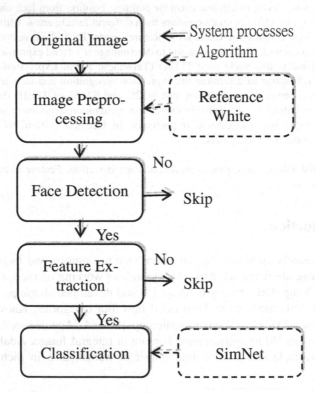

Fig. 1. Flow Diagram

2.1   Image Preprocessing

In image pre-processing, because the different environments, such as the different sources, will cause the final outcome. Our proposed system will adjust the light intensity of the facial expression to be recorded into the image pictures. Higher brightness of the pixel interval to calculated the mean in this image, and used this average value as the reference and calculation in equation (1).

$$\begin{cases} R_{avg} = \frac{\sum_1^n \max(N_R)}{n} \\ G_{avg} = \frac{\sum_1^n \max(N_G)}{n} \\ B_{avg} = \frac{\sum_1^n \max(N_B)}{n} \end{cases} \tag{1}$$

$$\begin{cases} R' = \frac{255}{R_{avg}} * N_R \\ G' = \frac{255}{G_{avg}} * N_G \\ B' = \frac{255}{B_{avg}} * N_B \end{cases} \tag{2}$$

The original image pixel adjustments based on average values. Where N_R, N_G and N_B represent the original image pixel values. R, G and B represent average value of a pixel in the color channel interval of n. n represent the total number of pixels to retrieve a brightness range, and R', G' and B' represent pixel values after the adjustment. This image pre-processing will be used to adjust for the low light images.

2.2   Face Detection

Viola and Jones [6] proposed real-time object detection algorithms to detect expression recognition of human face detection. This method compared with the color detection of face detection that can be excluded from the light source to the formation of the pixel values change in the error detection. During the experiment, the image of a miscarriage of justice will be skipped. Image pre-processing and face detection algorithm is illustrated in the flow chart shown in Fig 2.

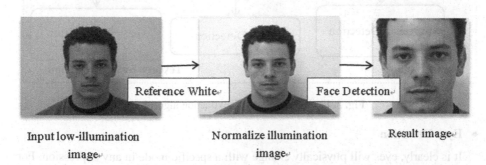

Input low-illumination image ·   Normalize illumination image ·   Result image ·

Fig. 2. Image pre-processing and face detection flowchart

2.3   Feature Extraction

In our proposed method, lots of features will be extracted from human face. We extracted eyebrows, eyes, and mouth all of region. Because the facial features location have relative distance relationship. The eyes will be detected at the first search region from face, and then identify the candidate position of the eyebrows, nostrils and mouth. Fig. 3 showed the feature extraction algorithm of the ROI-based Detection block. In our method, we extract 20 feature points and calculated the distance between points, then using the SimNet classification method to characterize the facial expression recognition.

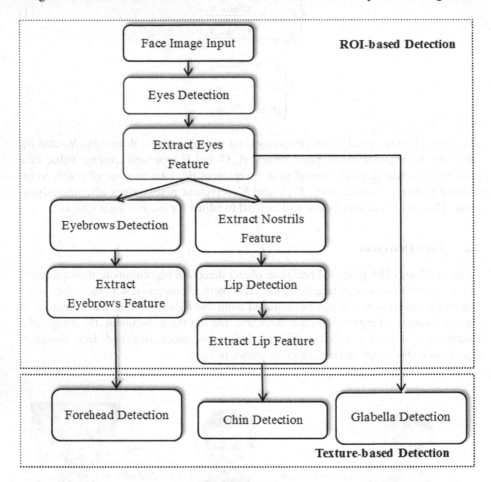

Fig. 3. Flowchart of feature extraction algorithm

- **Eyes Detection**

    It is clearly, eyes will physically change with a specific mode in any expression. For examples, the eyes will enlarge in surprised expression and shrink in disgust expression. In this step, we use the method proposed by Viola and Jones to detect the candidate area surrounding eye in the face image. Then converted to YCbCr space to capture the Cr color channel and the binary image can be captured for determination of the eye contour. Then used as the connected components (see fig. 4).

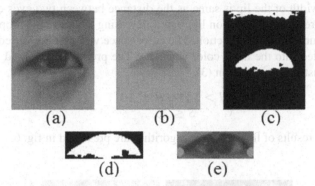

(a)   (b)   (c)

(d)   (e)

Fig. 4. (a) Original image (b) Capture Cr color channel(c) Binarization (d) Connected Component (e) Feature points location

- **Eyebrows Detection**

It is recognized, the above eyes region is the important candidate region of eyebrows. The highest point of the eye feature point coordinates can retrieve the candidate region of the eyebrows. Then eyebrows candidate region converted to grayscale image from the RGB color space image and use Sobel horizontal directional edge detection to find the eyebrows region. In order to preserve the integrity of the eyebrow contour, and use Sobel edge detection will have excessive noise. Therefore, morphological filter to remove noise and then use the connected components and Canny edge detection to capture feature point (see fig. 5).

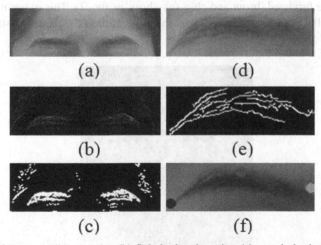

(a)   (d)

(b)   (e)

(c)   (f)

Fig. 5. (a) Eyebrows candidate region (b) Sobel edge detection (c) morphological filtering noise (d) connected components capture feature location of the eyebrow area (e) Canny edge detection (f) feature points

- **Lip Detection**

  Anima Majumder [9] proposed method of mouth detection uses a simple fundamental of facial geometry. From the facial geometry, we can easily observe the approximate width of the lip is same as the distance between two eyes centers. Then use color threshold to detection lip region. By using Chora's [13] proposed method of lip contour uses color detection. The color space was first converted to HSV and then separated into the three color channels. The proposed threshold for detection lips was illustrated in equation (3).

$$\begin{cases} H > 334 \cup H \le 10 \\ \quad\quad S \le 25 \end{cases} \tag{3}$$

The sample results of lips detection algorithm are presented in fig. 6.

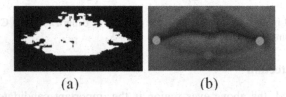

(a)   (b)

Fig. 6. (a)The detection of HSV lip images (b) Lips feature location

- **Texture Based Detection**

  In facial expression recognition, in addition to the characteristics of the facial position changes those different expressions of the facial muscles to form different texture, the forehead, and brow and chin area are important for significantly detail changes. We based on the results of feature point extraction to capture the three regions of the forehead, brow and chin (as show in fig. 7). This experiment uses the mask of Sobel edge detection to detect the edge of the skin characteristics.

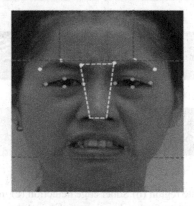

Fig. 7. Texture detection schematic

- **Features Normalization**

Because of the different scales of faces in picture (even the same person), we have to normalize the features. We used the distance between each feature point (as see in fig. 8). At first, we create a personal expression of model and using equation (4).

$$ND_i^{Neutral} = \frac{1}{n}\sum_{j=1}^{n} ND_{i,j}^{Neutral_m} \qquad (4)$$

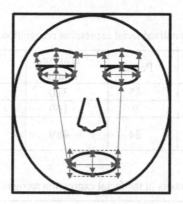

Fig. 8. The distance between feature points

3   Expression Recognition Using SimNet

SimNet is Fuzzy Logic combined with Artificial Neural Network. They have Unsupervised Learning and Supervised Learning. We use the SimNet with two hidden layers and each layer contains twenty neurons. Our proposed system will fed the hidden layer with the twenty normalize features. The output layer outputs eight values and each value represents one kind of expression. The highest value of each image will be indicated to the corresponding facial expression. The network architecture is shown as fig. 9.

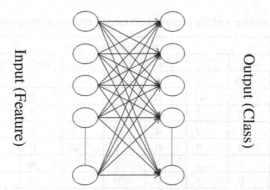

Fig. 9. Proposed network architecture

4   Experiment Results

We used the Taiwanese Facial Expression Image Database (TFEID) [14] to conduct our experiments. In this study, the original image to the right and left offset of 1° to 2°. There are having 960 of images. Then we used two kinds of experiment on database. TABLE 1 show personalized facial expression recognition system and TABLE 2 show all members of the facial expression recognition system.

Table 1. Personalized facial expression recognition system result

Database	People	Training images	Testing images	Accuracy (%)
TFEID (Woman)	15	320	235	92.9%
TFEID (Man)	9	169	128	99.4%
TFEID (Total)	24	489	363	96.2%

Table 2. All members of the facial expression recognition system result

		Recognition expression							Accuracy
		Ang	Con	Dis	Fea	Hap	Neu	Sad	Sur	(%)
Actual expression.	Anger	39	2	2	2	0	0	2	0	83.0%
	Contempt	0	43	0	1	0	1	0	0	95.6%
	Disgust	1	1	44	0	0	0	0	1	93.6%
	Fear	0	0	0	43	0	0	0	0	100.0%
	Happiness	0	0	0	3	47	0	0	0	94.0%
	Neurosis	2	0	0	0	0	48	0	0	96.0%
	Sadness	3	1	0	1	0	2	42	1	84.0%
	Surprise	0	0	0	0	0	0	0	31	100.0%
	Average									92.8%

Table 3. All members of the personalized facial expression recognition system result

		Recognition expression							Accuracy
		Ang	Con	Dis	Fea	Hap	Sad	Sur	(%)
Actual expression.	Anger	40	0	0	0	0	7	0	85.1%
	Contempt	1	41	0	0	1	2	0	91.1%
	Disgust	2	2	41	2	0	0	0	87.2%
	Fear	1	0	4	35	0	2	1	81.4%
	Happiness	0	0	0	1	48	0	0	96.0%
	Sadness	3	3	3	2	2	36	1	72.0%
	Surprise	0	1	0	0	1	0	29	93.5%
	Average								86.3%

5   Conclusions

We presented a system to automatically recognize the facial expressions. Our system extracted and described the features from the contour of eyebrows, eyes and mouth by a scalable rectangle. This is an improvement over the ways for using manual facial characteristic points and complicated face mask model. We defined less features to reduce the recognition time and obtain appropriate recognition accuracy.

References

1. Yang, M.H., Kriegman, D.J., Ahuja, N.: Detecting faces in images: A survey. IEEE Transactions on Pattern Analysis and Machine Intelligence 24(1), 34–58 (2002)
2. Yang, G., Huang, T.S.: Human face detection in a complex background. Pattern Recognition 27(1), 53–63 (1994)
3. Singh, S.K., Chauhan, D.S., Vatsa, M., Singh, R.: A robust skin color based face detection algorithm. Tamkang Journal of Science and Engineering 6(4), 227–234 (2003)
4. Wang, J., Tan, T.: A new face detection method based on shape information. Pattern Recognition Letters 21(6), 463–471 (2000)
5. Brunelli, R., Poggio, T.: Face recognition: Features versus templates. IEEE Transactions on Pattern Analysis and Machine Intelligence 15(10), 1042–1052 (1993)
6. Viola, P., Jones, M.: Rapid object detection using a boosted cascade of simple features. In: Proceedings of the 2001 IEEE Computer Society Conference on Computer Vision and Pattern Recognition, CVPR 2001 (2001)
7. Fasel, B., Luettin, J.: Automatic facial expression analysis: a survey. Pattern Recognition 36(1), 259–275 (2003)
8. Edwards, G.J., Cootes, T.F., Taylor, C.J.: Face Recognition Using Active Appearance Models. In: Burkhardt, H., Neumann, B. (eds.) ECCV 1998. LNCS, vol. 1407, pp. 581–595. Springer, Heidelberg (1998)
9. Majumder, A., Behera, L., Subramanian, V.K.: Automatic and robust detection of facial features in frontal face images. IEEE (2011)
10. Lien, J.J.J.: Automatic Recognition of Facial Expression Using Hidden Markov Models and Estimation of Expression Intensity. Washington University (1998)
11. Contreras, R., Starostenko, O., Pulido, L.F.: An improved method for facial features extraction in images. IEEE (2008)
12. Bashyal, S., Venayagamoorthy, G.K.: Recognition of facial expressions using Gabor wavelets and learning vector quantization. Engineering Applications of Artificial Intelligence 21(7), 1056–1064 (2008)
13. Choraś, M.: Lips Recognition for Biometrics. In: Advances in Biometrics, pp. 1260–1269 (2009)
14. Taiwanese Facial Expression Image Database (TFEID), Library and Information Communication Building Editor, Taiwan (2007)

Search Space Reduction in Pedestrian Detection for Driver Assistance System Based on Projective Geometry*

Karlis Dimza, Te-Feng Su, and Shang-Hong Lai

Department of Computer Science, National Tsing Hua University, Hsinchu, Taiwan
{karlis.dimza@gmail.com,tfsu@cs.nthu.edu.tw,lai@cs.nthu.edu.tw}

Abstract. Vehicles are equipped with smarter and smarter driver assistance systems to improve driving safety year by year. On-board pedestrian detection system is a critical and challenging task for driving safety improvement because driving environment is very dynamic, where humans appear in wide varieties of clothing, illumination, size, speed and distance from the vehicle. Most of existing methods are based on the sliding window search methodology to localize humans in an image. The easiest and also the most popular way is to check the whole image at all possible scales. However, such methods usually produces large number of false positives and are computationally expensive because large number of inappropriate regions were checked. In this paper, we develop a method which reduce the search space in pedestrian detection by using properties of projective geometry in the case when camera parameters are unavailable. The simple user interaction with stochastic optimization is used to estimate projective parameters. We showed the efficiency of our method on public dataset with known camera parameters and self captured dataset without registered camera parameters. Experiment results show that the effectiveness of the proposed method is superior compared to the traditional uniform sliding window selection strategy.

Keywords: Pedestrian Detection, ADAS, Vanishing Points.

1 Introduction

The years when car was a simple tool without safety and comfort equipment, are owned by history. Modern cars are equipped with a large number of electronic devices aiming to improve driving safety and comfort. In recent years, the so-called Advanced Driver Assistance Systems (ADAS) have become prevalent for new automobile market.

For the left image in Figure 1, our intuition says immediately that something is just wrong there, because it is impossible to meet such a giant humans in the real world! But all these humans have exactly the same size in image. How can

* This work was partially supported by National Science Council in Taiwan under the project 101-2220-E-007 -004.

J.-S. Pan et al. (Eds.): *Advances in Intelligent Systems & Applications*, SIST 21, pp. 269–278.
DOI: 10.1007/978-3-642-35473-1_27 © Springer-Verlag Berlin Heidelberg 2013

Fig. 1. Put pedestrian in scene. **Left:** putting pedestrian in image at different y coordinates of image with the same projected size. **Right:** putting pedestrian in image at different y coordinates with the same metric size.

we, as a human beings, tell immediately, that some humans are actually larger than others, since the object size in image is the same? Answer comes from our brain, which can perform projective transformation easily and immediately. Let us take a look at the right part of the Figure 1, where we put humans in the image by taking principle of the projective geometry into account. Obviously, it seems much more natural if the property of projective geometry is used. This simple, but effective principle will be used in our work to reduce the search space in human detection, by avoiding for the search of humans with unreasonable sizes and positions based on projective geometry. Of course, one can argue that this method is limited to humans on the ground plane only, but in real life there are not many situations where the pedestrians are way above the ground level, which are dangerous to the traffic safety.

The rest of the paper is organized as follows: in section 2, we will shortly review the literature for imaging technologies, pedestrian detection and camera calibration. In section 3 we will introduce the imaging hardware installation for driver assistance under limited visibility conditions. Section 4 introduces the proposed method to reduce the search space in pedestrian detection, and section 5 describes the experiment methodology and shows the results of experiments. Finally, we conclude this paper in Section 6.

2   Related Works

During the last decade, pedestrian detection has received significant attention from researchers all over the world. Many researches were funded or co-funded by large automotive manufacturers, such as Daimler[1][2], Mitsubishi[3], VW[4], etc. Research groups sponsored by these large corporations have published a number of papers in the field of automotive safety, automotive vision technologies and object detection from a moving platform. This work can be separated into two main parts: hardware and software. The goal is to detect pedestrians in all possible lighting conditions. An appropriate hardware should be selected to capture images during day and night. It is not considered as a big problem to capture scene at daylight, but imaging at night is a much more challenging task. Passive and active technologies have been used for night vision in general. Additional near infrared(NIR) illumination has been employed to build an active night vision system. In contrast, passive night vision usually makes use of thermal sensor(FIR), which can form images based on thermal intensities in a scene. There are advantages and disadvantages for both systems. Alive objects can be easily segmented from FIR images, but it is difficult to interpret thermal images for human especially for objects with no thermal differences. The structure of NIR image is very similar to the image in visible wavelength, objects appear naturally, but camera can be dazzled easily. The main drawback of thermal imaging is the expensive installation costs. Heat from sun makes thermal sensor difficult to use at daylight, which turns into necessity of two systems in car to be installed. This fact raises the costs to the even higher level. After summarizing all the pros and cons for both systems, we decide to build a cost-effective NIR system. This system can be used in both day and night conditions with minor changes: turning off NIR illumination at day and turning IR-CUT filter for camera on, and for nighttime do the other way round. More detailed comparison of both NIR and FIR systems can be found in [5].

Detection of humans from images is very important step in many vision-related systems. Massive research of human detection has been done during the last decade. Some survey papers [6][7][8] nicely review all the related works on human detection. Dalal and Triggs's HOG [9] method currently is the most important method for human detection. Some later methods [10] are trying to gain some improvement by combining different features together with HOG.

Most of the authors are paying considerable amounts of attention to the feature selection and feature classification, but very few of them focuses on search space reduction. The first work which motivated us to look in the direction of perspective information was the work by Hoiem et. al. [11]. The significance of perspective knowledge was proved there, and notable improvement in detection accuracy for different object categories was shown. Knowledge of camera parameters, which sometimes are not available, is required. The idea of restoring camera parameters with simple user interaction was taken from [12]. The main idea is to build a mutually orthogonal vanishing point triangle, which is the heart of camera calibration. In [12] errorless user input is assumed. However there are many cases when precise user input is just impossible. The method to deal with user interaction errors to estimate three vanishing points was designed. Finally, human size estimation is accomplished by applying the cross ratio principle with the pixel coordinates in the sliding window.

3   Night Vision and Dataset

A high percentage of car-pedestrian accidents happen at nighttime. Sleepy, tired, and careless drivers are the main reasons. Limited visibility is another important

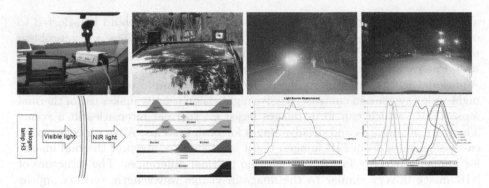

Fig. 2. Row 1: First two images show camera and NIR illuminator from different views. The last two show the WDR images captured from NIR system. **Row 2:** Image 1: The concept of NIR illuminator. Image 2: The concept of IR-PASS filter which was constructed. Image 3: The measurement of pure light source intensity distribution. Image 4: The measured color filters and filter combination.

reason of car accidents. In the case of limited visibility, sometimes pedestrian is noticed simply too late to avoid collision. Active night vision system was built to improve visibility and accuracy of pedestrian detection at night. The main parts of system are WDR image sensor and NIR illuminators. As an image sensor surveillance camera was chosen. NIR illuminators was constructed by combining halogen visible light source with three layer IR-PASS filter. The images of hardware installation and WDR images captured from NIR system are shown in Figure 2.

Several hundreds of kilometers of driving was recorded with the installed system. Data was captured both day and night conditions. The small but challenging parts of data were extracted and pedestrians were annotated. In total 4 sequences were created, calling them Day, Night1, Night2, Night3 respectively. The complete statistics of NTHU CVLAB Day-Night monocular pedestrian dataset is summarized in Table 1.

4   Search Space Reduction for Pedestrian Detection

Pedestrian detection is a complicated task which requires to integrate several methods for successful pedestrian localization in images. Basically there are three main stages in such a system: preprocessing, classification and post processing.

We focus on developing an efficient preprocessing procedure in this paper. Since classification is the core part of the whole process, it will be used as a validation tool to observe the improvement of the preprocessing stage. The same human detector in OpenCV [13], which was trained on Daimler [6] monocular pedestrian training set, will be used as the pedestrian detector in this work.

The method is based on the assumption that the object projected size depends on object distance from camera: the closer the object to the camera, the

Table 1. NTHU CVLAB Day-Night monocular pedestrian dataset

Sequence	#frames	Img. size	Light	#pedestr.
Day	1533	800x600	Day	5821
Night1	1085	800x600	NIR	1721
Night2	2044	800x600	NIR	4839
Night3	1479	800x600	NIR	3469
Total	**6141**			**15877**

larger its projection on the image. Another assumption is based on the fact that pedestrians are located on the ground plane. An expected height map can be built depending on the y coordinate in the image. The regions above the horizon line can just be discarded.

These projective parameters of the camera are different in almost every imaging case. Sometimes these parameters are available and we can just use them to construct a projection matrix and then estimate the height of human through the re-projection. But in the most of the cases camera parameters are unavailable, we have to estimate them. The principle of three mutually orthogonal vanishing points is employed here. Vanishing points can be estimated with some simple user interaction. Reference human also should be inputted.

4.1 Adaptive Scan with Camera Parameters

Most of the object detection methods are based on the sliding window approach to perform object localization. Object in scene might appear in different scales which means different projected size on image. To solve this problem, many detection algorithms adopt the so-called uniform scan methods, where all possible scales and positions are assumed with the same probability of human. In literature principles of adaptive scan are referred as projective geometry with the main assumption that parallel lines in world intersect at a vanishing point in the projective image. In the traffic scenes parallel lines in Z direction usually are the most obvious ones, but in general three directions X, Y and Z are considered. When the projection matrix or three vanishing points with one human reference object are estimated, the generation of the height map can be started. Height map will be generated based on the y coordinate of image. This process should be started from the maximal y coordinate on the image, and continued while the estimated image height is larger than the defined minimal height threshold. When the height map is estimated, the human classifier is applied to the sub regions in the image with size determined from the height map. In case of SVM classification, these sub-images should be resized to the size of detector.

4.2 Adaptive Scan with Unknown Camera Parameter

In general cases, information about imaging conditions is unknown. These cases are the cases when to use camera calibration techniques. Most of the calibration

Fig. 3. Left: Size of vanishing point estimation error, depending on mutual slope of vanishing lines. **Upper:** The case with large mutual slope (Z direction vanishing point). **Lower:** The case with small mutual slope (X and Y direction vanishing points).**Middle:** Vanishing point justification process. **Right:** height estimation using three vanishing points and known reference human height.

methods rely on multi-view calibration objects. Unfortunately, such classical calibration objects are not available on the road environment. The goal is to obtain similar camera calibration result without the calibration objects compared to the method with known camera parameters. Calibration pattern should consist of three mutually orthogonal sets of 3D parallel lines and three vanishing points can be extracted precisely. For the scenarios on the road, usually we can't find objects which are exactly parallel and/or mutually orthogonal. Even if such object exists there, the user interaction can not guarantee estimation to be precise enough. Thus perfection is not expected, but we are trying to estimate height as precise as possible. The subsequent section will describe our method in details.

The trickiest part of the method is to build a mutually orthogonal vanishing point triangle. We should start with the user interaction, by asking user to input three pairs of mutually parallel (parallel in the world) lines and these pairs should be mutually orthogonal to each other (in the real world). By analyzing road scenes, we can easily see that it is very easy to find a frame with parallel lines in Z direction, such as lane markings, boundaries of the road or similar objects. But if we want to find parallel lines in X and Y directions, for example, structure of building, traffic signs or light poles, then this case is much more complicated, because these parallel lines when projected on image, has almost zero mutual slope and a small mistake in user interaction may affect the estimated vanishing point significantly. Example of vanishing point estimation error dependence on mutual slope of vanishing lines is in left image of Figure 3. The user interaction error is assumed to be constant in both cases. To deal with these cases, a method based on stochastic optimization was developed. The method is based on the assumption that the principle point of every modern camera is located at the center of an image, and the orthocenter of vanishing point triangle is located at the principle point of an image. The main iteration of the method is addition of small random numbers to every coordinate of four lines representing user input for the vanishing points along X and Y directions. This process is iterated until the orthocenter of vanishing point triangle is very close to the center of the image. Then we assume that the final vanishing points are estimated. The illustration of vanishing point justification process is showed in left image of Figure 3, where search regions for new endpoints of the parallel lines along X and Y directions, in a schematic way are presented.

When the vanishing points are estimated, we are ready to estimate human heights in the image depending on image coordinates. To do that, we have to ask user to input at least one human reference object with average human projected height. First of all we have to estimate horizon line passing from X vanishing point to Z vanishing point, next form the line (L_1) from the reference human foot point through the foot point of human height of interest and estimate its intersection with horizon line($Point_1$). Next form the line (L_2) from foot point of human height of interest through the Y vanishing point. Next form the line (L_3) from $Point_1$ through reference human head point and compute the intersection point ($HeadOI$) with L_2. Note that $HeadOI$ is the head point of point of interest, which is assumed to be the foot point of expected human. The height of expected human can be computed by simply computing difference in y. The flow of the above process is illustrated in right image of Figure 3.

5   Experiments

Evaluation of the correctness of the proposed method was performed by using a public dataset, with ground truth and camera calibration available, as well as self captured video datasets, which was captured with unknown camera parameters. Daimler[6] monocular pedestrian dataset was chosen as the public dataset for our experiment. The rest of this section is organized as follows: first we will test the human height estimation by comparing result from the vanishing point method with the result from the projection matrix, then the accuracies of the pedestrian detection with the uniform and adaptive scanning schemes are evaluated and compared.

5.1   Human Height Estimtion

Daimler [6] dataset was chosen to evaluate human height estimation, because the camera parameters are available. We assume human height computed from the camera parameters as the ground truth, and the result was compared with result from our method. Position of the reference object affects the accuracy of the estimated object size. As we can imagine, it is better to choose the reference object which is as close as possible to the observer. Results of both cases are shown in Figure 4.

It is easy to observe in Figure 4 that the estimation error increases in case of far reference object, but still error is not very significant in both cases. For classification we use ±20% multi-heightening to deal with different 3D human heights and height estimation errors. The multi-heightening ranges are plotted with dashed lines in Figure 4.

5.2   Pedestrian Detection with Uniform and Adaptive Scan

To test the correctness of adaptive scan, it was compared with uniform scanning strategy for pedestrian detection in on-board sequences. These sequences, used

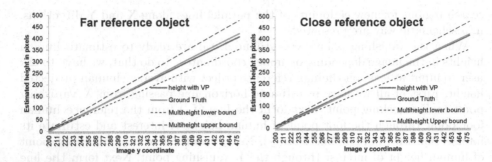

Fig. 4. The results of human height estimation for the conditions with far reference object and close reference object

for tests, can be separated in three main groups: Daimler [6] daytime monocular pedestrian dataset, NTHU CVLAB Day monocular pedestrian dataset and NTHU CVLAB Night monocular pedestrian datasets. For our dataset, user interaction was necessary to estimate camera projection parameters, but for Daimler dataset provided camera parameters was used to estimate camera projection matrix. In all cases as a human detector HOG implementation from OpenCV [13] was used, because the goal is to observe as fair as possible comparison of these different scanning strategies. User input for projective properties estamtion and sample detections are shown in Figure 5. As we can see in sample detection image in Figure 5, the uniform scan spreads the sliding windows equally on whole frame, while adaptive scan concentrates the sliding windows in the regions with the highest probability of pedestrians. Quantitative results of pedestrian

Fig. 5. Left: User input for vanishing point estimation (left) and human reference object (right). **Right:** Sample detections by using uniform(left) and adaptive(right) scans.

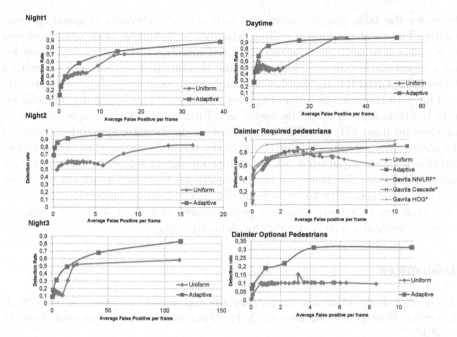

Fig. 6. The results of pedestrian detection on different datasets. The same classifier was used with adaptive and uniform scans.

detection on different datasets proves our intuition that the better region selection can improve pedestrian detection significantly. The methods were tested on 5 datasets, and the results are summarized in Figure 6. Ground Truth in Daimler dataset is classified into two groups so-called optional and required pedestrians. Required pedestrians are pedestrians with no occlusion and projected height 72 or more pixels while optional pedestrians are all others which are not required ones. We measured accuracy on both of the groups, and adaptive scan performs slightly better for required pedestrians, while significant improvement can be observed in the case of optional ones. The comparison with results from [6] was made and the results are similar, in some cases slightly worse, because the post processing is not used for our methods. When the accuracy is compared on our datasets (both day and night), observation is similar: adaptive scan performs better than uniform scan on every tested dataset.

6   Conclusions and Further Works

The algorithm for reducing the search space in the pedestrian detector based on scene geometry was developed. We used simple user interaction with stochastic vanishing point optimization to compute the camera projective information and the results were equivalent to the results of ground truth. Experimental results show that the reduced search space significantly improves the detection rates, decreases the false positive rates and improves the theoretical speed of computing. The total number of sliding windows in Adaptive scan was decreased by a factor 2 in average. The near infrared imaging system was developed for night vision and set up on vehicle. By using this imaging system, four pedestrian datasets captured during the day and night with pedestrian labeling were created for performance assessment of pedestrian detectors. Combining information from multiple sensors might be useful as a future work, for example combining information from FIR and NIR image sensors, to increase spectral coverage of the scene. Another important direction of the research is related to stereo vision, and depth map computation, which enables intelligent foreground segmentation and search space reduction even more. Finally, for real-time application in practical use, it is necessary to implement the method on an embedded multi-core GPU platform to achieve real-time performance, as well as designing rectangle grouping method which is suitable for Adaptive scan.

References

1. Online: Daimler 6D vision, http://www.6d-vision.com/ (accessed April 30, 2012)
2. Online: Daimler Gavrila research, http://www.gavrila.net/ (accessed April 30, 2012)
3. Online: Mitsubishi Electric Research Laboratories, http://www.merl.com/ (accessed April 30, 2012)
4. Online: Volkswagen Electronics Research Laboratory, http://www.vwerl.com/ (accessed April 30, 2012)
5. Luo, Y., Remillard, J., Hoetzer, D.: Pedestrian detection in near-infrared night vision system. In: Intelligent Vehicles Symposium (2010)
6. Enzweiler, M., Gavrila, D.: Monocular pedestrian detection: Survey and experiments. PAMI (2009)
7. Gero Andnimo, D., Lo Andpez, A., Sappa, A., Graf, T.: Survey of pedestrian detection for advanced driver assistance systems. PAMI (2010)
8. Dollar, P., Wojek, C., Schiele, B., Perona, P.: Pedestrian detection: An evaluation of the state of the art. PAMI (2012)
9. Dalal, N., Triggs, B.: Histograms of oriented gradients for human detection. In: CVPR (2005)
10. Wang, X., Han, T.X., Yan, S.: An hog-lbp human detector with partial occlusion handling. In: ICCV (2009)
11. Hoiem, D., Efros, A., Hebert, M.: Putting objects in perspective. In: CVPR (2006)
12. Lee, S.C., Nevatia, R.: Robust camera calibration tool for video surveillance camera in urban environment. In: CVPRW (2011)
13. Online: OpenCV Library, http://opencv.willowgarage.com/wiki/ (accessed April 30, 2012)

Fast Multi-path Motion Estimation Algorithm with Computation Scalability

Kuang-Han Tai[1], Gwo-Long Li[2], Mei-Juan Chen[1], and Haw-Wen Chi[1]

[1] Dept. of Electrical Engineering, National Dong Hwa University, Hualien, Taiwan
{810123002,610023036}@ems.ndhu.edu.tw,
cmj@mail.ndhu.edu.tw
[2] Dept. of Video Coding Core Technology, Industrial Technology Research Institute,
Hsinchu, Taiwan
glli@itri.org.tw

Abstract. Developing computation-efficient motion estimation algorithm is always a significant topic in video coding and becomes more important in green computing. It also helps the reduction of power consumption. This paper proposes a fast multi-path motion estimation algorithm to aim at computational complexity reduction and providing various computation choices. We propose an enhanced intersected diamond search algorithm to take the advantages of low computational complexity. The proposed search patterns will decrease the search points for each step. By searching the best matched block with multiple directions, the drawback of being trapped at the local minimum for distortion measure can be solved. With adaptive thresholding for early termination, the proposed algorithm can provide different computations and performances for various video encoders to adapt different kinds of computation abilities. Experimental results show that the proposed algorithm with computation scalability can give better performance than diamond search and its multi-path algorithm.

Keywords: Motion estimation, block-matching, video coding, diamond search, intersected diamond search, multi-path search, early termination.

1   Introduction

To reduce the temporal redundancy in video data, the technique of motion estimation has been widely adopted in many video coding standards [1-3]. The motion estimation algorithm can significantly eliminate temporal redundancy at the expense of remarkable computational complexity. Therefore, designing efficient and low complexity motion estimation algorithm for video coding is always an important research topic in last decades.

There were many literatures proposed to reduce computational complexity of motion estimation [4-16]. However, the computational requirement of motion estimation still dominates the large portion of computational load in entire video coding system. In addition, these literatures only focused on the computational complexity reduction

J.-S. Pan et al. (Eds.): *Advances in Intelligent Systems & Applications*, SIST 21, pp. 279–288.
DOI: 10.1007/978-3-642-35473-1_28   © Springer-Verlag Berlin Heidelberg 2013

without considering the constraint of computation power availability. As a result, these literatures could not be applicable for battery equipped mobile devices. For the latest video coding standard high efficiency video coding (HEVC) [3] targeting at HDTV application, the problem of high computational complexity of motion estimation further results in noticeable computational power consumption. In addition, for the power limited mobile devices, the high computational complexity of motion estimation leads to the shorter mobile devices operating life. Therefore, if the computational burden of motion estimation can be further reduced, not only the operating life of mobile devices can be extended, but also the goal of green computing can be also achieved.

This paper proposes an enhanced version of intersected diamond search incorporating with multi-path concept for motion estimation. In addition, a criterion for early terminating the search of motion estimation is further combined into our proposed algorithm to reduce the computational requirement. The computation scalability can also be achieved by our proposed method.

The rest of this paper is organized as follows. In Section 2, the algorithm of new intersected diamond search(NIDS) and multi-path motion estimation are briefly described because the two algorithms are the fundamentals of our proposal. Our proposed algorithm is introduced in Section 3 in detail. Section 4 shows some simulation results to demonstrate the efficiency of our proposed algorithm. The conclusion is given in Section 5.

2   Overview of New Intersected Diamond Search and Multi-path Search

2.1   New Intersected Diamond Search (NIDS)

Figure 1 shows the used search patterns of NIDS[13] in which the ISP and SDSP are the abbreviations of intersected search pattern and small diamond search pattern, respectively. The operation of NIDS is briefly described as follows.

Step 1: The ISP is used to evaluate the sum of absolute difference (SAD) of each candidate position. Go to Step 2.

Step 2: If the candidate position with the smallest SAD is located at the center, go to Step 3. Otherwise, go to Step 4.

Step 3: The SDSP is applied to compute the SADs for another four candidate positions and the candidate position with the smallest SAD will be selected as the best result. Search process is terminated.

Step 4: If the candidate position with the smallest SAD doesn't occur at the center position in ISP step, the pair-wised direction determination approach will be used to determine which direction that the following search operation will move to (as shown in Figure 2). Here, the pair-wised direction determination approach is in charge of computing the average SAD of each position pair. Go to Step 5.

Step 5: Once the moving direction of search process has been decided, set up the new search center and go to Step 1.

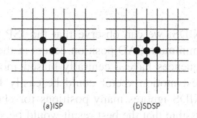

(a)ISP   (b)SDSP

Fig. 1. Illustration of the search patterns in new intersected diamond search(NIDS)

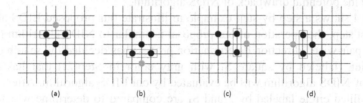

(a)   (b)   (c)   (d)

Fig. 2. Pair-wised direction determination of NIDS

2.2 Multi-path Search

To avoid the search process trapped into local optimal, the multi-path search approach [14] provides a way which allows search process to check multiple paths for finding out the best results. Figure 3 shows an example to illustrate how the multi-path search is applied to diamond search (DS) pattern. For multi-path search, more than one candidate with least SADs will be selected as the search center after evaluating initial search pattern and the search will thus be spread out.

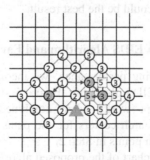

Fig. 3. Illustration of multi-path search [14] applied to DS

3   Proposed Algorithm

The proposed algorithm is mainly composed by two components. One is the multi-path Enhanced Intersected Diamond Search (EIDS). Another one is a dynamic early termination algorithm. The details of each component are introduced as follows.

3.1   Proposed EIDS

From the operation of NIDS, we can observe that the ISP is applied for the entire search process until the minimum SAD position occurs at the center. As described in [9], most real-world sequences have a centrally biased motion vector distribution, which means that the best motion vector could be easily found around the center [15-16]. However, the NIDS ignores many positions for checking during its search operation. It might be possible that the best result would be skipped and thus degrade the prediction accuracy. Our proposed EIDS is the modified search pattern which tries to resolve the potential drawback of NIDS algorithm.

Figure 4 shows a step-by-step example to illustrate our proposed EIDS algorithm. In the first step, the operation of our proposal is similar to NIDS algorithm (Step(1)). Once the pair-wised direction determination process has decided the search direction, the position (labeled by square S_1 with gray color in Step(2)) that has not been checked by NIDS algorithm will be evaluated by our EIDS algorithm. Afterwards, the SADs of filled circle labeled by 2 and S_1 are compared to determine which one has the minimum SAD value. If S_1 is not the position with minimum SAD, the NIDS algorithm will be applied for the rest search processes (Step(3a) to Step(5a)) until the best results have been found. However, if S_1 has the minimum SAD, the SADs of neighboring positions surrounding by S_1 will be compared to determine which position has the minimum SAD among them. If the upper position has minimum SAD (as shown in Step(3b)), this position will be used as the new search center and the two additional positions(labeled by 2 in Step(3b)) will be checked. From now on, our enhanced pattern will be used in the rest search process to find out the best result. The benefit of our proposed EIDS is that our proposal not only takes the advantage of low computation complexity from NIDS but also avoids the miss-checking for the nearby positions, which potentially could be the best result.

3.2   Proposed Multi-path EIDS Algorithm and Early Termination

To further increase the coding performance, we incorporate the multi-path search strategy into our proposed EIDS algorithm. Figure 5 shows an example to illustrate how the multi-path concept works when combined with our proposed EIDS algorithm. The priority of search centers for the paths is according to the order of minimum SADs of the search points in Step 2.

Figure 6 exhibits the flowchart of the proposed algorithm with multi-path EIDS algorithm with early termination. The detailed coding flow is described as follows. First, some variables are initialized for the following motion estimation usage. The initialized variables are introduced as follows.

- *NP*: the number of supported search paths
- *p*: index for indicating current processing path
- $SAD_{Min}[p]$: variable for storing the min SAD of path *p*
- $MV_{Best}[p]$: variable for storing the best MV of path *p*

After the initialization process, the ISP is applied to the first step for deriving initialized search results. If the minimum SAD located at the center position after ISP search, the search pattern is changed to SDSP and the best motion vector (MV) is thus obtained from the SDSP search results. Otherwise, our proposed multi-path EIDS algorithm will be executed. To further increase the motion estimation speed, an early termination mechanism [17] has been adopted in our proposal. From our flowchart, we can find that the every SAD_{Temp} of every search step of every path has to be determined whether the search for current path has to be terminated or not. If SAD_{Temp} is less than a dynamic threshold, *Th*, the rest search for that path will be skipped. In our proposal, the variable of threshold, *Th*, is derived by the following formula.

$$Th = Mean(MinSAD_{Left}, MinSAD_{Top}, MinSAD_{TopRight}) \times \beta \qquad (1)$$

where $MinSAD_{Left}$, $MinSAD_{Top}$, and $MinSAD_{TopRight}$ are the minimum SAD values of left, top, and top right MB, respectively.   The parameter β is adjustable according to the available computation ability. Once all best results of all paths have been found out, the best motion vector will be selected as the MV_{Best} according to the minimum SAD among the paths.

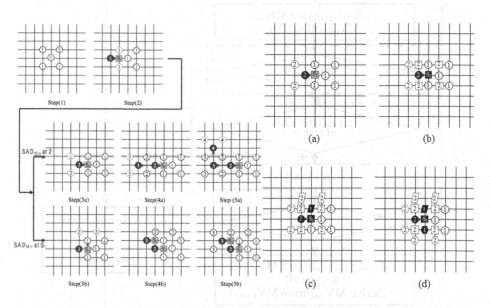

Fig. 4. Illustration of our EIDS algorithm

Fig. 5. Illustration of our proposed multi-path EIDS (a) 1-path (b)2-path (c) 3-path (d) 4-path

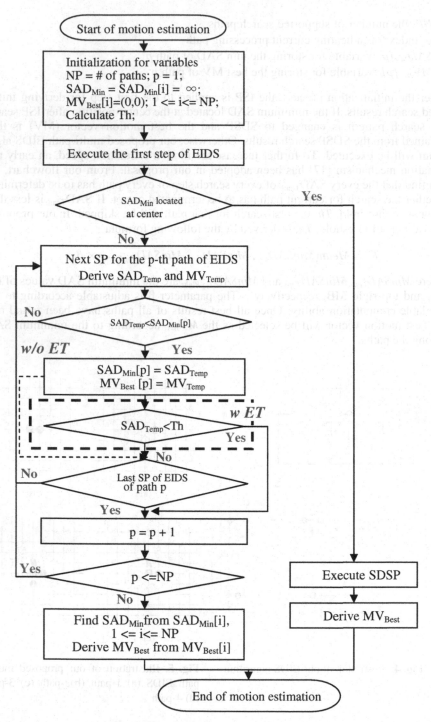

Fig. 6. Flowchart of our proposed algorithm

4   Simulation Results

This section exhibits several simulation results to demonstrate the coding efficiency of our proposed algorithm. In the simulation, full search (FS), new intersected diamond search (NIDS), diamond search (DS), and multi-path diamond search (MPS-DS) are included for comparison. Several test sequences are SIF Football, CIF Table Tennis, 4CIF Soccer, 720p Pedestrian and Rush Hour sequences. The term of ProxP stands for our proposed algorithm incorporating with x-path search. For example, Pro2P means that two paths are used in our proposed EIDS algorithm for search. In addition, the term of ET means that the early termination mechanism is adopted. In our simulation, the search range is set to ±7 and 30 frames are used for encoding. Each macroblock (MB) consists of 16x16 pixels and gets one MV for each algorithm.

Table 1 and Table 2 list the comparisons of PSNR and average search points per MV, respectively. For our proposed algorithm without incorporating multi-path search and early termination (Pro1P), it can individually achieve 1.20dB and 1.89dB PSNR increasing compared to DS and NIDS algorithms on average. For our proposed algorithm incorporating with multi-path concept (Pro4P), 1.78dB and 2.47dB PSNR improvement can be achieved compared to DS and NIDS, respectively. Even through the early termination algorithm has been adopted in our proposed algorithm to further increase motion estimation speed, our proposal can still keep the search performance with significant speedup for motion estimation. On average, 0.38dB and 1.07dB PSNR gain can be derived by our proposed algorithm (Pro4P+ET(β= 1.0)) compared to DS and NIDS, respectively. For the MPS-DS algorithm, since this algorithm searches multi-paths for finding the best result, the higher PSNR performance, said 27.39dB, can be obtained at the expense of high computational complexity as shown in Table 2. From Table 2, it can be seen that the average search points per MV of MPS-DS algorithm is 27.57, which is much higher than our proposed algorithm.

Table 1. PSNR(dB) comparisons for different algorithms

	SIF	CIF	4CIF	720p		Avg.
	Football	Tennis	Soccer	Pedestrian	Rush Hour	
FS	21.77	31.28	24.31	27.94	34.12	**27.88**
DS	20.07	29.19	21.54	24.76	32.32	**25.58**
NIDS	19.47	28.91	20.78	24.08	31.23	**24.89**
MPS-DS	21.20	30.51	23.78	27.46	34.02	**27.39**
Pro1P	20.70	30.00	22.98	26.70	33.50	**26.78**
Pro2P	20.98	30.40	23.47	27.38	33.91	**27.23**
Pro3P	21.09	30.63	23.57	27.43	33.94	**27.33**
Pro4P	21.16	30.63	23.61	27.44	33.95	**27.36**
Pro4P+ET(β= 0.1)	21.15	30.63	23.61	27.44	33.95	**27.36**
Pro4P+ET(β= 0.3)	21.13	30.61	23.58	27.41	33.94	**27.33**
Pro4P+ET(β= 0.5)	21.10	30.40	23.48	27.29	33.89	**27.23**
Pro4P+ET(β= 0.8)	20.75	29.78	23.04	26.73	33.89	**26.83**
Pro4P+ET(β= 1.0)	20.31	28.14	22.60	26.14	32.62	**25.96**
Pro4P+ET(β= 1.5)	19.49	26.28	21.78	25.23	30.89	**24.73**
Pro4P+ET(β= 2.0)	18.98	25.34	21.25	24.19	29.97	**23.95**

From Table 2 , we can observe that our proposed algorithm can result in different PSNR results and computational complexity combinations. For high quality application, we can use Pro4P+ET(β=0.1) for deriving higher PSNR results. Similarly, Pro4P+ET(β=2.0) can aim at much lower computational complexity requirement with acceptable PSNR degradation. By properly selecting the β value, our proposed algorithm can achieve quality as well as computational complexity scalability.

Table 2. Average search points per MV comparisons for different algorithms

	SIF	CIF	4CIF	720p		Avg.
	Football	Tennis	Soccer	Pedestrian	Rush Hour	
FS	203.93	206.19	211.85	212.99	212.99	**209.59**
DS	13.93	12.95	14.89	14.50	14.08	**14.07**
NIDS	9.67	9.67	10.63	10.28	10.63	**10.18**
MPS-DS	23.13	16.75	33.71	34.77	29.48	**27.57**
Pro1P	11.94	10.32	13.66	13.77	12.71	**12.48**
Pro2P	15.03	12.57	17.77	18.30	16.35	**16.00**
Pro3P	18.69	14.84	22.14	23.81	20.02	**19.90**
Pro4P	21.15	16.98	24.98	27.37	22.47	**22.59**
Pro4P+ET(β= 0.1)	21.10	16.97	24.93	25.96	22.44	**22.28**
Pro4P+ET(β= 0.3)	20.34	15.98	24.26	24.89	22.13	**21.52**
Pro4P+ET(β= 0.5)	18.52	12.65	22.82	22.58	21.16	**19.55**
Pro4P+ET(β= 0.8)	15.21	12.65	18.97	17.57	17.43	**16.37**
Pro4P+ET(β= 1.0)	12.69	8.01	16.17	15.58	13.77	**13.24**
Pro4P+ET(β= 1.5)	9.99	6.68	12.78	13.32	10.40	**10.63**
Pro4P+ET(β= 2.0)	8.80	6.06	11.02	11.43	9.19	**9.30**

Figure 7 and Figure 8 exhibit some performance comparison curves of different algorithms for Soccer and Pedestrian sequences. From these figures, we can observe that our proposed algorithm can achieve much better PSNR performance than DS algorithm at the same search points. For the highest PSNR performance case, our proposed algorithm requires much less search points than MPS-DS algorithm. In addition, compared to all other algorithms, our proposal can provide the high flexibility of adjusting the PSNR performance and computational complexity.

5   Conclusion

This paper proposes an enhanced intersected diamond search algorithm to solve the drawback of NIDS algorithm. To avoid to be trapped into local optimal, our proposed algorithm further incorporates the multi-path concept to increase the coding performance. In addition, an early termination mechanism is also adopted in our proposal to speed up the motion estimation process. Simulation results demonstrate that our proposed algorithm can achieve higher PSNR improvement than well-known DS and NIDS algorithms. In addition, compared to MPS-DS, our proposed algorithm requires much less computational complexity than MPS-DS algorithm under the same PSNR condition. Furthermore, through the proper selection for the early termination criteria, our proposed algorithm can aim at coding performance as well as computational complexity scalability.

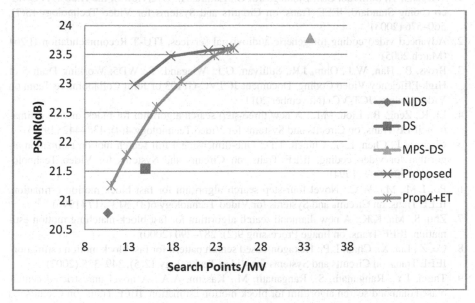

Fig. 7. Performance Comparison for various algorithms for 4CIF Soccer sequence

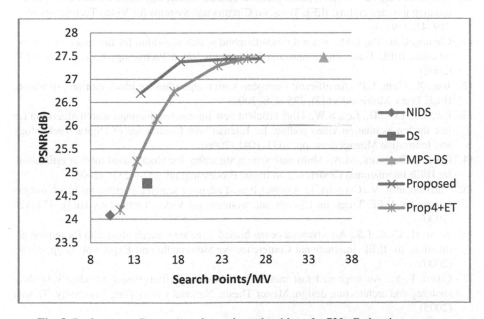

Fig. 8. Performance Comparison for various algorithms for 720p Pedestrian sequence

References

1. Wiegand, T., Sullivan, G.J., Bjontegaard, G., Luthra, A.: Overview of the H.264/AVC video coding Standard. IEEE Trans. on Circuits and Systems for Video Technology 13(7), 560–576 (2003)
2. Advanced video coding for generic audiovisual services, ITU-T Recommendation H.264 (March 2005)
3. Bross, B., Han, W.J., Ohm, J.R., Sullivan, G.J., Wiegand, T.: WD5: Working Draft 5 of High-Efficiency Video Coding. Document JCTVC-G1103 of Joint Collaborative Team on Video Coding (JCT-VC) (November 2011)
4. Li, R., Zeng, B., Liou, M.L.: A new three-step search algorithm for block motion estimation. IEEE Trans. on Circuits and Systems for Video Technology 4(4), 438–442 (1994)
5. Chen, M.J., Chen, L.G., Chiueh, T.D.: One-dimensional full search motion estimation algorithm for video coding. IEEE Trans. on Circuits and Systems for Video Technology 4(5), 504–509 (1994)
6. Po, L.M., Ma, W.C.: Novel four-step search algorithm for fast block motion estimation. IEEE Trans. on Circuits and Systems for Video Technology 6(3), 313–317 (1996)
7. Zhu, S., Ma, K.K.: A new diamond search algorithm for fast block-matching motion estimation. IEEE Trans. on Image Processing 9(2), 287–290 (2000)
8. Ce, Z., Lin, X., Chau, L.P.: Hexagon-based search pattern for fast block motion estimation. IEEE Trans. on Circuits and Systems for Video Technology 12(5), 349–355 (2002)
9. Tham, J.Y., Ranganath, S., Ranganath, M., Kassim, A.A.: A novel unrestricted center-biased diamond search algorithm for block motion estimation. IEEE Trans. on Circuitsand Systems for Video Technology 8(4), 369–377 (1998)
10. Liu, L.K., Feig, B.: A block-based gradient descent search algorithm for block motion estimation in video coding. IEEE Trans. on Circuits and Systems for Video Technology 6(4), 419–422 (1996)
11. Cheung, C.H., Po, L.M.: A novel cross-diamond search algorithm for fast block motion estimation. IEEE Trans. on Circuits and Systems for Video Technology 12(12), 1168–1177 (2002)
12. Jing, X., Chau, L.P.: An efficient three-step search algorithm for block motion estimation. IEEE Trans. Multimedia 6(3), 435–438 (2004)
13. Lai, Y.J., Li, C.H., Leu, S.W.: High efficient new intersected diamond search algorithm for fast motion estimation video coding. In: International Conference of Digital Technology and Innovation Management, pp. 1033–1041 (2006)
14. Goel, S., Bayoumi, M.A.: Multi-path search algorithm for block-based motion estimation. In: IEEE International Conference on Image Processing, pp. 2373–2376 (2006)
15. Christopoulos, V., Cornelis, J.: A center-biased adaptive search algorithm for block motion estimation. IEEE Trans. on Circuits and Systems for Video Technology 10(3), 423–426 (2000)
16. Nisar, H., Choi, T.S.: An advanced center biased three step search algorithm for motion estimation. In: IEEE International Conference on Multimedia and Expo, vol. 1, pp. 95–98 (2000)
17. Chien, F.-Y.: An improved fast motion estimation algorithm based on adaptively thresholding and architecture design. Master Thesis, National Chiao-Tung University, Taiwan (2005)

Moving Objects Detection Based on Hysteresis Thresholding*

Hsiang-Erh Lai[1], Chih-Yang Lin[2,**], Ming-Kai Chen[1], Li-Wei Kang[3], and Chia-Hung Yeh[1]

[1] Department of Electrical Engineering, National Sun Yat-Sen University, Kaohsiung, Taiwan
B983012011@student.nsysu.edu.tw, e123642261@gmail.com
yeh@mail.ee.nsysu.edu.tw
[2] Dept. of Computer Science & Information Engineering, Asia University, Taichung, Taiwan
andrewlin@asia.edu.tw
[3] Institute of Information Science, Academia Sinica, Taipei, Taiwan
lwkang@iis.sinica.edu.tw

Abstract. Background modeling is the core of event detection in surveillance systems. The traditional Gaussian mixture model has some defects when encountering some situations like shadow interferences, lighting changes, and other problems causing foreground image broken. All of these cases will result in deficiencies of event detection. In this paper, we propose a new background modeling method to solve these problems. The model features of our method are the combination of texture and color characteristics, hysteresis thresholding, and the motion estimation to recover broken foreground objects.

Keywords: background modeling, moving objects detection, hysteresis thresholding.

1 Introduction

In these days, people are more and more concerned about the importance of environment security. Not only working space but also residences are commonly equipped with surveillance systems. In addition, foreground detection is also a milestone of the surveillance system. By this way, we can save lots of time of focusing on the monitor. In fact, background modeling is used to distinguish foreground and background. Thus, a robust background modeling method is needed when detecting moving objects.

There are a number of methods for moving objects detection but most of them are based on color information. For instance, a statistical approach based on color information [1] built a background model and reduce the shadow interference. In addition, Wren et al. [2] proposed a one-Gaussian method to strengthen the background flexibility. However, one-Gaussian method has a defect for a dynamic background, such

---

* This work was supported in part by the National Science Council, Taiwan, under Grants NSC101-2221-E-468-021.
** Corresponding author.

J.-S. Pan et al. (Eds.): *Advances in Intelligent Systems & Applications*, SIST 21, pp. 289–298.
DOI: 10.1007/978-3-642-35473-1_29 © Springer-Verlag Berlin Heidelberg 2013

as swaying trees, ripples, and the blink of a screen. Thus, Stauffer and Grimson proposed the Gaussian mixture model (GMM) method [3]. For each pixel, GMM used more than one Gaussian to model the background [3, 4, 5]. If the pixel does not match the GMM model, it will be regarded as a foreground pixel. There is an example of GMM method for traffic monitoring [6] and others can be found in [7] and [8].

Though GMM improves Wren et al.'s method a lot, it still suffers from shadow interference and illumination changes. Therefore, Heikkilä and Pietikäinen proposed the texture-based background model with local binary patterns (LBPs) [9, 10]. This method has the tolerance to illumination changes. However, LBPs are not robust. When noises or swaying trees strike the central pixel value, the corresponding LBPs histogram would be interfered and increase the possibilities of false positive and false negative cases.

In this paper, we proposed a new background modeling method. Our main contributions are as follows: (1) Our proposed method is based on hysteresis thresholding. Hysteresis thresholding has never been used for background modeling and it can greatly alleviate the cavity problem in foreground objects. (2) We proposed a new texture descriptor derived from our previous work [11]. With this descriptor, we can enhance the tolerance to illumination changes and shadow interference. (3) Our method combines the information of both texture and color to reduce shadow problems while improving the shapes of foreground objects. (4) The motion estimation technique is also applied in our method to recover broken foreground objects caused by motion problems, such as moving too slow or walking toward the camera.

The following parts of this paper are organized as follows: in Section 2, we take a brief review on the Gaussian mixture model. Our proposed method is presented in Section 3. The results of our experiments are discussed in Section 4. Finally, conclusions are provided in Section 5.

2   Preparation Work

In this section, we will explain how a mixture of Gaussians model works. This method was first proposed by Grimson and Stauffer [1, 2]. They model each background pixel into a K-Gaussians mixture model (GMM), where K is between 3 and 5. The weight of each Gaussian distribution represents the portion of the data accounted for that Gaussian.

First, each pixel is modeled by a mixture of K Gaussian distributions. The probability of observing the current pixel value is:

$$P(X_t) = \sum_{j=1}^{K} \omega_{j,t} * \eta(X_t, \mu_{j,t}, \sum_{j,t}),$$ (1)

where X_t is the current pixel value at time t, K is the number of Gaussian distributions, $\omega_{j,t}$ is the weight estimation of the jth Gaussian in the mixture at time t, $\mu_{j,t}$ and $\sum_{j,t}$ are the mean value and covariance matrix respectively, of the jth Gaussian in the mixture at time t, and η is a Gaussian probability density function (pdf).

After the model is built, each incoming pixel of following frames is compared with the existing model components. In the case that the input pixel fits one of the weighted Gaussian distributions, it means that its pixel value is within 2.5 standard deviations of the matched distribution. Once the pixel is matched, the update process will be invoked to fine-tune the corresponding model; otherwise, we will replace the distribution, which has the lowest weight, with a new distribution using the current incoming pixel as its mean value, an initial high variance, and a low prior weight.

In order to select the best Gaussians for each pixel, the K distributions are sorted based upon the value ω/σ. Only the first B distributions are selected as the background model of a pixel for the scene and denoted as:

$$B = \arg\min_b \left(\sum_{k=1}^{b} \omega_k > T_B \right), \tag{2}$$

where T_B is a predefined threshold and usually set to about 90%, ω_k is the weight parameter of the kth model component and b indicates the number of background distributions.

At last, the update process will change the weights of K Gaussian distributions as follows:

$$\omega_{k,t} = (1-\alpha)\omega_{k,t-1} + \alpha(M_{k,t}), \tag{3}$$

where α is the learning rate and $M_{k,t}$ is 1 for the matched distribution and 0 for the unmatched distributions. In addition, weights of distributions should be renormalized. If the new pixel matches a Gaussian distribution, the values of mean and variance of this distribution are updated as follows:

$$\mu_t = (1-\rho)\mu_{t-1} + \rho X_t, \tag{4}$$

$$\sigma_t^2 = (1-\rho)\sigma_{t-1}^2 + \rho(X_t - \mu_t)^T (X_t - \mu_t), \tag{5}$$

where

$$\rho = \alpha\eta(X_t \mid \mu_k, \sigma_k). \tag{6}$$

3   Proposed Method

In this section, we describe the proposed method with texture descriptor, texture-based background modeling, hysteresis thresholding, and motion estimation for foreground recovery.

3.1   Texture Descriptor

In the beginning, we divide the input frame into several non-overlapping blocks with a size of $n \times n$ pixels. For each block, mean value m of the block is calculated by:

$$m = \frac{1}{n \times n} \sum_{i=1}^{n} \sum_{j=1}^{n} x_{ij}, \qquad (7)$$

where x_{ij} indicates the pixel value in the position (i, j) of the block.

With the mean value, we can build a binary map (BM) for the block by the following equation.

$$b_{ij} = \begin{cases} 1, & \text{if } x_{ij} \geq m, \\ 0, & \text{otherwise,} \end{cases} \qquad (8)$$

where bit "1" denotes that the pixel value is greater than the mean value m of that block; Otherwise, the bit is "0".

Here is an example for the texture descriptor. In this case, the block is set to 3×3 as shown in Fig. 1. In Fig.1, the mean of this block is 51.56 and the bitmap is built accordingly.

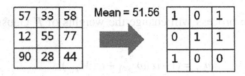

Fig. 1. The process of building a binary bitmap

3.2   Texture-Based Background Modeling

In the previous section, each block has been transformed into a binary bitmap. However, if there is a smooth block, where the pixels are either barely larger or barely smaller than the block mean, the corresponding bitmap would result in an interlaced 0/1 pattern. This situation gives rise to an unstable background model since smooth blocks and non-smooth blocks cannot be distinguished. To solve this problem, Eq.(8) is changed to Eq.(9) with a threshold TH_{smooth} to solve this problem, and TH_{smooth} is set as 8 according to our experiments.

$$b_{ij} = \begin{cases} 0, & \text{if } x_{ij} < m + TH_{smooth}, \\ 1, & \text{otherwise.} \end{cases} \qquad (9)$$

Note that in the process of the bitmap generation, the input frame captured by a camera is transformed into a grayscale image by Eq.(10)in order to improve the efficiency. Fig. 2 illustrates the result of the bitmap generation, which shows the validity of the proposed texture descriptor.

$$\text{Gray} = 0.299R + 0.587G + 0.114B \qquad (10)$$

(a) Original image   (b) Texture description

Fig. 2. The proposed texture descriptor

The proposed background modeling is a pixel-based model. Therefore, each pixel has its own texture description, *i.e.*, the corresponding BM. The background model based on the proposed texture descriptor consists of K weighted bitmaps, $\{BM_1, BM_2, ..., BM_K\}$, where each weight is between 0 and 1, and the summation of the weights is 1. The weight of the kth bitmap is denoted as w_k. When a new BM_{new} is captured, it is compared with the K bitmaps by the following Hamming distance equation, where m is in the range of $[1, K]$:

$$Dist(BM_{new}, BM_m) = \sum_{i=1}^{n} \sum_{j=1}^{n} (b_{ij}^{new} \oplus b_{ij}^{m})$$ (11)

If $\min_{m} Dist(BM_{new}, BM_m)$ is smaller than a threshold predefined, the BM_{new} matches the background model, and then the update process will be invoked; otherwise, BM_{new} is regarded as a foreground pattern, and the unmatched process will be applied. The process of the model maintenance can be referred to our previous work [11].

3.3   Hysteresis Thresholding

In the traditional GMM, whether a pixel is a background pixel or not is determined by a single threshold TH(*i.e.*, 2.5 standard deviations of a Gaussian) as follows:

$$\text{input} = \begin{cases} \text{foreground,} & \text{if } \min Dist \geq TH, \\ \text{background,} & \text{otherwise.} \end{cases}$$ (12)

This approach brings about a serious problem. If TH is too small, the output image will contain lots of noises; on the contrary, when TH is too large, the foreground objects may contain a lot of cavities.

In this paper, hysteresis thresholding is proposed to solve this problem, where double thresholds TH_{high} and TH_{low} are used to enhance the foreground estimation. The threshold TH_{high} is responsible for generating "strong" information and TH_{low} is responsible for gathering "weak" information. The strong information means these generated foreground pixels are very robust but may result in breaks or cavities in the foreground objects. On the other hand, the weak information will generate more complete shapes of foreground objects but involve more noise as well.

We apply hysteresis thresholding on both of the original color GMM and the proposed texture-based background modeling to generate four binary maps, called strong color, weak color, strong texture, and weak texture maps. The pixels in the strong texture map are called real foreground pixels and the pixels in the remaining maps are called pseudo foreground pixels. The foreground objects generation starts from the strong texture map, and traces weak texture, strong color, and weak color maps to gradually compensate or mend the shapes of the foreground objects. More specifically, if a pixel belongs to one of the strong color, weak color, and weak texture maps, and the pixel is connected with a strong texture pixel, then the pixel is identified as a real foreground pixel and will be treated as a new strong texture pixel in the next iteration. The process will continue until all the pseudo foreground pixels have been tested.

The combination of the color-based GMM and the proposed texture-based model has higher tolerance to shadow interference and illumination changes. In addition, using hysteresis thresholding can fix the cavities problem caused in the strong texture map and get more complete shapes of foreground objects. The noise in the proposed scheme can be nearly removed because noise is usually not connected with strong maps. Fig. 3 shows an example of the four maps.

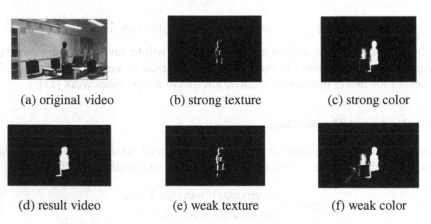

| (a) original video | (b) strong texture | (c) strong color |
| (d) result video | (e) weak texture | (f) weak color |

Fig. 3. The results of four maps based on hysteresis thresholding

3.4   Motion Estimation

In addition to the above four maps discussed in the previous section, in this section, another binary map, called motion map, is generated based on motion estimation. When there is an object that moves too slowly or moves toward the camera, some pixels of this object will be gradually becoming background due to the effects of Eqs. (3) to (6), resulting in the foreground object fractured. With the help of the motion map, this problem can be greatly alleviated.

The motion map is generated as follows. Set the binary image of the ith frame using the four maps mentioned in Section 3.3 be denoted as R_i, and set P_{i-1} be the result after applying the four maps and the motion map of the $(i-1)^{th}$ frame. The motion map for the ith frame is the difference between R_i and P_{i-1}, denoted as D_i. To mend cavities in R_i, each pixel in R_i will find its 5×5 neighbors. If some of these 5×5 neighbors belong to D_i, these pixels will be included in the foreground objects. The flow chart of generating the motion map is shown in Fig. 4 and an example of applying the motion map is presented in Fig. 5.

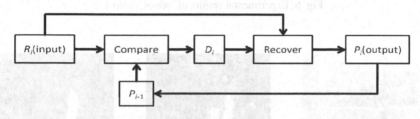

Fig. 4. Flow chart of motion map

(a) Original video   (b) Result video (R_i)   (c) Final result video(P_i)

Fig. 5. Final result of using the motion map

4   Experimental Results

4.1   Detecting Results

The following are our experimental results. Figs. 6 and 7 show the indoor and outdoor scenes with shadow interference. The results show that the proposed method has higher tolerance to the shadow interference than the original GMM.

Fig. 8 shows an outdoor scene that a person walks toward the camera. In this video, our proposed method shows the repair ability of broken image, which cannot be achieved in the traditional GMM.

Fig. 9 shows a person walking toward an indoor camera. This figure clearly reveals that the proposed method not only removes the shadow but also successfully mends most of the cavities in the body.

(a) Original image   (b) GMM method   (c) Proposed method

Fig. 6. Experimental results of indoor video 1

 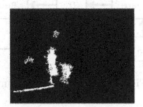 

(a) Original image   (b) GMM method   (c) Proposed method

Fig. 7. Experimental results of outdoor video 1

(a) Original image   (b) GMM method   (c) Proposed method

Fig. 8. Experimental results of outdoor video 2

(a) Original image   (b) GMM method   (c) Proposed method

Fig. 9. Experimental results of indoor video 2

4.2   Quantitative Results

We compare our method with the traditional GMM in a quantitative way. The simulation environment is equipped with a 2.93 GHz Core 2 Duo Intel processor and 4 GB of memory. All algorithms were implemented in C++. The parameters, α and K, used in the experiments are set to 0.005 and 3, respectively.

Fig. 10 shows the accuracy comparison of the proposed method, GMM method and the ground truth. There are three items in this evaluation: False positive (FP) is the number of background pixels which are mistaken for foreground; False negative is the number of foreground pixels which are mistaken for background; Total Error (TE) is the sum of FP and FN.

(a) Ground truth   (b) GMM method   (c) Proposed method

Fig. 10. Comparison on indoor video

From Table 1, it clearly reveals that the proposed method has much lower FP than that in the GMM due to the shadow removing, and FN in the proposed scheme is greatly reduced for the reason of our mending technique.

Table 1. Accuracy comparison

Items	GMM	Proposed method
FP	2376	556
FN	749	238
TE	3125	794

5   Conclusions

In this paper, we proposed a new background modeling method based on hysteresis thresholding. The proposed method has the following advantages: (1) tolerance to shadow interference and illumination change due to the texture characteristic; (2) resistance to noise and shape fracturing because of hysteresis thresolding; (3) repairing the foreground objects with the help of the mostion estimation technique. The expirement results show that FP and FN of the proposed method are much better than those in the original GMM method. Our future work will improve our efficency for real time surveillance applications.

References

1. Wixson, L.: Detecting Salient Motion by Accumulating Directionally-Consistent Flow. IEEE Transactions on Pattern Analysis and Machine Intelligence 22(8), 774–780 (2000)
2. Wren, C.R., Azarbayejani, A., Darrell, T., Pentland, A.P.: Pfinder: Real-Time Tracking of the Human Body. IEEE Transactions on Pattern Analysis and Machine Intelligence 19(7), 780–785 (1997)
3. Stauffer, C., Grimson, W.E.L.: Adaptive Background Mixture Models for Real-Time Tracking. In: Proceedings of the IEEE Computer Society Conference on Computer Vision and Pattern Recognition, Fort Collins, Colorado, USA, pp. 246–252 (1999)
4. Stauffer, C., Grimson, W.E.L.: Learning Patterns of Activity Using Real-Time Tracking. IEEE Transactions on Pattern Analysis and Machine Intelligence 22(8), 747–757 (2000)
5. Zivkovic, Z.: Improved Adaptive Gaussian Mixture Model for Background Subtraction. In: Proceedings of the 17th International Conference on Pattern Recognition, Cambridge, UK, pp. 28–31 (2004)
6. Friedman, N., Russell, S.: Image Segmentation in Video Sequences: A Probabilistic Approach. In: Proceedings of the 13th Conference on Uncertainty in Artificial Intelligence, San Francisco, pp. 175–181 (1997)
7. Dedeoğlu, Y., Töreyin, B.U., Güdükbay, U., Çetin, A.E.: Silhouette-Based Method for Object Classification and Human Action Recognition in Video. In: Huang, T.S., Sebe, N., Lew, M., Pavlović, V., Kölsch, M., Galata, A., Kisačanin, B. (eds.) ECCV 2006 Workshop on HCI. LNCS, vol. 3979, pp. 64–77. Springer, Heidelberg (2006)
8. Power, O.W.: Understanding Background Mixture Models for Foreground Segmentation. In: Proceedings of Image and Vision Computing, Auckland, New Zealand, pp. 267–271 (2002)
9. Heikkilä, M., Pietikäinen, M.: A Texture-Based Method for Modeling the Background and Detecting Moving Objects. IEEE Transactions on Pattern Analysis and Machine Intelligence 28(4), 657–662 (2006)
10. Heikkilä, M., Pietikäinen, M., Heikkilä, J.: A Texture-Based Method for Detecting Moving Objects. In: Proceedings of British Machine Vision Conference, British, pp. 187–196 (2004)
11. Lin, C.Y., Chang, C.C., Chang, W.W., Chen, M.H., Kang, L.W.: Real-Time Robust Background Modeling Based on Joint Color and Texture Descriptions. In: Proceedings of The Fourth International Conference on Genetic and Evolutionary Computing (ICGEC 2010), Shenzhen, China, December 13-15, pp. 622–625 (2010)
12. Fisher, R., Santos-Victor, J., Crowley, J.: EC Funded CAVIAR project/IST 2001 37540, http://homepages.inf.ed.ac.uk/rbf/CAVIAR/
13. Lumia, R., Shapiro, L., Zuniga, O.: A New Connected Components Algorithm for Virtual Memory Computers. Computer Vision, Graphics, and Image Processing 22(2), 287–300 (1983)
14. Chang, F., Chen, C.J., Lu, C.J.: A Linear-Time Component-Labeling Algorithm Using Contour Tracing Technique. Computer Vision and Image Understanding 93(2), 206–220 (2004)

Smart Video Camera Design – Real-Time Automatic Person Identification

Chen-Ting Ye, Tzung-Dian Wu, You-Ren Chen, Pei-An He, Pei-Qi Xie,
Yuan-Yi Zhang, Shih-Meng Teng, Yen-Ting Chen, and Pao-Ann Hsiung

Department of Computer Science and Information Engineering
National Chung Cheng University, Taiwan, ROC.
{yct98u,wtt98u,cyje98u,hpa98u,hpc98u,cyy98u,tsm100m,
cyt100m,pahsiung}@cs.ccu.edu.tw

Abstract. Automatic person identification has wide applications in various situations. For example, a wanted criminal at an international airport, a lost child in a large park, and an elderly person lost in a metropolitan city are situations in which automatic person identification must be performed in real-time. Parameters normally employedinclude a person's face, height, stride, and full-body color histogram.In this work, the problem we are solving is how to identifya lost child in a video, given the characteristics of a single still image captured when the child entered the location (park) where it was lost. We built a smart camera system that can automatically identify the lost child using a novel method called *Real-time Automatic Person Identification* (RAPId) that employs a target object segmentation with overlap method to increase accuracy in identification and reference lines for height estimation.

Keywords: Foreground-Detection, Recognition, Object-Tracking.

1   Introduction

Smart camera design incorporates intelligence into the traditional passive image capturing such that only the regions of interest (ROI) can be searched for target objects. Compared to the popular face recognition now used widely, automatic person identification has received little attention. However, automatic person identification has wide usage in different domains, especially in safety related situations such as quickly searching for a lost child in a park, a lost aged person in a city, or a runaway criminal. Though the problem seems to be a simple one, it is more difficult than simple face recognition because the characteristics of a target person might be indistinguishable when he/she is moving, in different postures, and under different illuminations. In this work, we address the problem of automatically identifying a lost child in a large amusement park from a real-time video recording, given a single still image that was captured when the child entered the park. The problem is solved by proposing a method to effectively detect, recognize, and track the target in real-time.

The current state-of-the-art solution used in such situations is mostly manual and raw-eye based which is very cumbersome, erratic, and time consuming. For example,

J.-S. Pan et al. (Eds.): *Advances in Intelligent Systems & Applications*, SIST 21, pp. 299–309.
DOI: 10.1007/978-3-642-35473-1_30   © Springer-Verlag Berlin Heidelberg 2013

a lost child's features collected from the parents' verbal descriptions or memory are used to find the lost child manually in a video. However, the features may be ambiguous leading to inaccurate time-consuming manual detection. Hence, we propose a smart camera system to find the correct target automatically in real-time.

Existing methods cannot be directly applied to the target problem in this work because of various reasons as described in the following. Everingham and Zisserman [1] proposed to use 3D model rendering of faces at different poses to identify persons via face recognition. In our work, faces can be captured when the lost child enters the park, but when he/she is roaming in the large park, faces are very difficult to detect due to very small target objects and thus even smaller faces that are not distinguishable. BenAbdelkarder et al. [2] used a person's height and stride to identify a person. This method cannot be applied because it is usually quite difficult to estimate a child's stride appropriately when he/she enters the park. Balcan et al. [3] used a graph-based *semi-supervised learning* algorithm to identify persons. However, since our input is a single still image, we do have the graph required for semi-supervised learning. Orten et al. [4] used different combinations of experts (classifiers) to identify persons, which need to be trained. The single image in our target problem is insufficient data for training the experts. Thus, a new method is proposed in this work that can extract moving persons from the foreground, recognize, and track them.

To detect a moving object in an image, there are typically two methods, namely *temporal difference* and *background subtraction*. The former is simple but not very accurate, while the latter is more complex but also more accurate. In this work, besides adopting the background subtraction method, we also increase accuracy via noise processing, shadow processing, and background reconstruction.

After detecting the foreground objects, we try to recognize them using two features of a target person, including his/her full-body color and height. The conventional method is to compare the color histogram of a given labeled image and that of a real-time detected object. However, accuracy becomes quite low due to low quality images or when the detected object is very small in size. We propose a new segmentation with overlap method to increase the accuracy of recognition. Then, back-projection is used for tracking the object.

To estimate target person's height from the video, several intrinsic parameters related to camera and extrinsic parameters related to 3D real world to 2D image mapping need to be considered [5]. However, it is difficult to determine the parameters in our setting environment. Lee et al. [6] provide an efficient way to measure theheight without knowing exactly the parameters. It uses the coordinate of six non-coplanar points in real world to determine a linear transform. But it is hard to accurately determine the coordinates without precision instruments. In this work, we improved the methods from [7], [8] by finding the vanishing points and line in a real-time system.

The article is organized as follows. Section 2 describes the proposed framework. Section 3 presents experimental results. Conclusions and future work are in Section 4.

2   Real-Time Automatic Person Identification Framework

We propose a new framework called *Real-time Automatic Person Identification* (RAPId), which consists of two parts, namely the *entrance* camera video processing and the *monitor* cameras video processing. The entrance camera is placed near the entrance of an amusement park. Whenever a person goes through the entrance, its close range image is captured. The monitor cameras are installed within the amusement park at various critical positions and captures video images.

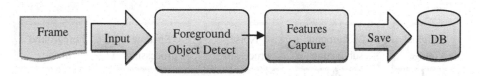

Fig. 1. Entrance camera: image capture and feature extraction

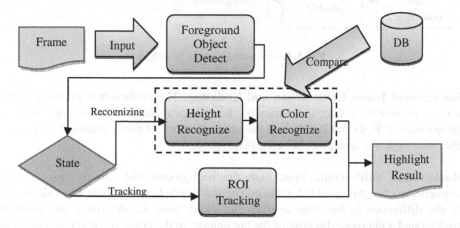

Fig. 2. Monitor camera: person detection, identification, and tracking

Fig. 1 shows how the images captured by the entrance camera are processed in RAPId. Each input frame is processed for foreground object detection and features capturing that includes color and height. Finally, the two features along with the image are stored in a database. Sections 2.1, 2.2 and 2.3 describe the foreground object detection, color recognition, and height measurement, respectively.

Fig. 2 illustrates how the images captured by the monitor cameras are processed in RAPId. Each input frame first undergoes foreground object detection and then based on whether the state is currently recognizing or tracking the image is processed differently. Initially, the state is recognizing. In the recognizing state, the two features are extracted again, that is, height and color, which are then compared with the records in the database. When the similarity between the person detected in a monitor image and an existing record of features reaches a given threshold, the results are highlighted and the state is switched to tracking. In the tracking state, a region of interest (ROI) is created and the recognized target person is tracked.

2.1   Foreground Object Detection

As shown in Fig. 3, foreground object detection is achieved through four steps, including background construction and updating, background subtraction, noise processing, and shadow processing, each of which are described as follows.

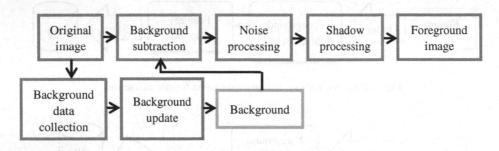

Fig. 3. Foreground detection flowchart

Background Image Construction and Updating. The construction of background images is done by collecting a series of chronological images, counting the highest proportions of R, G and B values in the same positions of these images to determine the color value for each pixel in background images.

Background Subtraction. First, both the background and current images were converted to the gray-level color space. The difference between them was calculated. If the difference is less than a given threshold, then we determine the pixel is background. Otherwise, because of the big change in the color, the pixel should be in a moving foreground object.

Noise and Shadow Processing. Variations in intensity and jitter by the camera may result in a small amount of noise in the subtraction. To solve this problem, we analyze the region near a pixel. Since the foreground objects are connected components, a pixel belongs to foreground if most of its neighbor pixels belong to the foreground. Therefore, we counted the occurrences of the state of neighbor pixels to decide the state of the pixel, based on a threshold.

And another part of noise is shadow. The shadow which follows objects is always counted into the foreground, which causes detection error. To solve this problem, we used the characteristics that a shadow has lower intensity and fewer changes in hue, and then design a shadow mask to remove shadow.

2.2 Color Recognition

Vertically Divide the Color Features of a Person.
To get a more accurate color features from each part of
one person, we vertically divided a person into six
parts.

Fig. 4 shows the six parts roughly divided people in
these sections which shows different colors. Moreover
this method combined with the position information,
so we won't compare some person's color of shirt to
another one's color of pants. However, the more
divisions we divide, the more accuracy we will
require from the foreground object detection step.

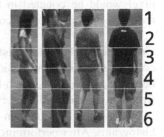

Fig. 4. Vertically divide the color features

Convert RGB to HSV. The RGB color space is a common color space to describe an
image, however, it is influenced by intensity too much, and we can't distinguish
similar color into different color group easily.

The HSV color space expresses each pixel in 3 channels: Hue, Saturation and
Value (Brightness), in the following article, we use H, S and V to represent them for
short. The hue tells us what color it belongs to; the saturation means the amount of
this color (pale or pure); the Value gives us the luminance information (dark or light).

Calculation of the Histogram. Although we can count the most frequent color in
each part, that won't apply to the case that clothes not in only one pure color. Thus,
we counted every pixel's H and S in a 2D-histogram to make it more general.

In order to get reliable similarity measures, we reduced the color, the hues are
grouped in eighteen bins, while the saturation grouped in six bins. The Fig. 5
demonstrated what color
feature this histogram stands
for.

We count a pixel as a
gray-scale color if it has too
low saturation to distinguish
its color. We choose pixels
with S under 32 (overall
range from 0 to 255), namely
the first row in Fig. 5, are

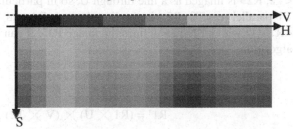

Fig. 5. The feature colors used in histogram

belong to gray-scale. Instead of counting their hues, we counted their luminance,
namely a gray-scale histogram. Since intensity may vary when lightness changes and
the shadow would make an influence, we used six bins here.

Pixels with extreme luminance, too dark/light to distinguish its color, still have
saturation above 32 in HSV sometimes. Thus, we alter the saturation of these black
and white pixels to zero.

We also used foreground map from previous step as a mask, only foreground pixel would be counted in histogram, lest the colors of background pixels would reduce the precision of the color feature histogram.

Compare Correspond Histograms. Now that we had histograms that stands for color features of each division of a person, we could compare correspond histogram between the two objects bin-by-bin to get a similarity value. The color feature loaded from our database will be served as a reference and compare to the person in screen.

We used built-in function *cv::compareHist* with the correlation method to compare histograms. After performing one-to-one comparison for each part, we used weighted sum to generate a similarity value. For the parts are usually similar among people, like first (head) part and sixth (shoes) part in Fig. 4,will get lower weights.

2.3   Height Measurement

Horizontal Vanishing Points. It can be determined by two sets of parallel lines on the reference plane, with different direction.

Vertical Vanishing Point. It can be determined by all vertical world lines intersect in this vanishing point.

Vanishing Line of the Reference Plane. By the horizontal points, we can determine the vanishing line of the reference plane.

Compute the Projective Transformations. According to the Fig. 6, we can determine point U by the intersection of <T2, R2> with L. Any scene line parallel to <T2, R2> is imaged as a line through U, so in particular the image of the line through T1 parallel to <T2, R2> is the line through R1 and U. The intersection of the line <R1, U> with <T1, T2> defines the image R1'. It can be determined by the following algorithm:

$$U = (T2 \times R2) \times L . \tag{1}$$

$$R1' = (R1 \times U) \times (V \times T2) . \tag{2}$$

Then represent the R1', T1, T2 and V by their distance from T2, as r1', t1, 0 and v respectively. The distance ratio is given by:

$$D1 / D2 = r1'(v-t1) / t1(v-r1') . \tag{3}$$

2.4   Tracking by Back Projection

Initial Settings. Once the similarity of the recognitions reached a threshold, we set the detected rectangle as ROI. And then defined the search window to be a larger surrounding rectangle that away from the ROI in the distance: one sixth of the height of the ROI. We then try to trace the new ROI in the search window in following frames.

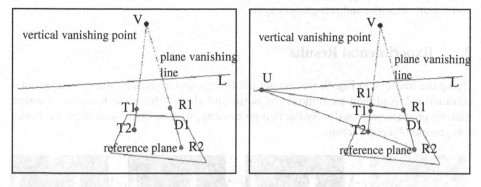

Fig. 6. The reference among the reference plane, vanishing points and vanishing line

Back-Projection. Back-projection is a method to find an object that we are looking for in an image by color histogram. Since we had color feature histogram from color recognition step, we used it to create the back-projection image.

The back-projection image will be as the same size as the current search window. For each pixel, we will give it the occurrence count from histogram that we can find by the color (H and S) of the correspond pixel in search window. Note that for low saturation less than 32, we will use the luminance rather than the hue.

Finally, we used foreground mask to eliminate the value of the pixel in the back-projection that belongs to background.

Position Masks. It is difficult to distinguish between those wearing the same color but in different position if we use the histogram of the full rectangle of a person.

To solve this problem, since we have 6 histograms for every person, we can get the 6 back-projection of our search window separately, and then we use the AND maskinFig. 7 to get rid of the pixels on the positions which are too far for each division to move to at the next frame. At last, we used OR mask to combined the 6 sub-back-projection to form our back-projection of current search window. Therefore, we can find the new ROI by the back-projection image as Fig. 8 shows.

Fig. 7. Six AND masks to use position information

Fig. 8. Find new ROI by back-projection

Updates of State. After we found the new ROI, we would use it to open up a new search window and start another tracking at next frame. However, if we could not find this new ROI, we would count it as a miss. Once miss times reaches 5, our target might have left the region of the search window. In such a case, we will stop tracking and restart detection and recognition again.

3   Experimental Results

Using the frame A in Fig. 9, we can get the foreground objects by using "background subtraction". And then we remove the noise and shadow from the foreground image that we produce previously. By the two processing, we can finally complete the better foreground object detection.

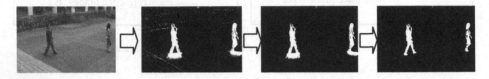

Fig. 9. Procedure of generating the foreground map

Fig. 10 shows the results histograms we got. We used intensity to represent the occurrence for any color tile. We can find that the histograms demonstrate the color feature for every vertical division of a person. For example, the histogram 1A has the brightest color in black section (see Fig. 5), which is the color of her hair color. In the same way, the histogram 1D shows the color of person her pants.

We captured the color features histograms as references, and then we compare the foreground object we detected in previous step. For instance, we used histograms in

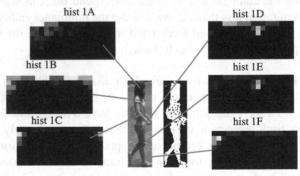

Fig. 10. The color feature histograms of a person

Fig. 11 as our references, and tried to use it to do recognitions in our test videos. We defined the states in Table 1 to show our results.

Table 1. The Highlight Result of the RAPId

States	Descriptions	Highlight Color
Detected	foreground, moving object	gray scale
Recognized	color similarity ≥0.66	green
Traced	color similarity ≥0.83	red

In video 1, we successfully distinguish the two people, because of their differences in their color features; it makes a large difference between them. In video 2, we distinguished the 2 people, too. Nevertheless their differences are not as large as last video due to their similarity at the color of their clothes. In video 3, we can find that if one dressed in gray (black/white) or pale colors, the correctness of the color recognition will decrease because these colors are more depend on the intensity.

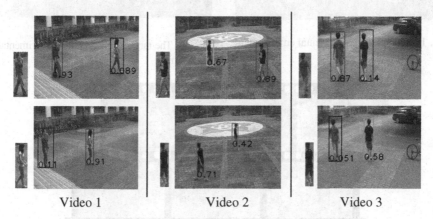

| Video 1 | Video 2 | Video 3 |

Fig. 11. Color recognition results of our test videos

The experimental environment is shown in Fig. 12. The white cross is a mark helping us to correct the vanishing line. We can choose the ROI to decrease the time of finding vanishing line. As shown in Fig. 13. By the algorithm we discuss before, we can determine the vanishing line, and the final experimental environment as shown in Fig. 14. And by indicate the target image coordinate, we can measure the height of target.

Fig. 12. Experimental environment   **Fig. 13.** Mark(Region of interest)

The Fig.16 shows the screenshots for a test video. In this figure, we used blue rectangle to represent search window and the red for ROI. We can find out that the tracking is more powerful than only detection and recognition.

Fig. 14. Complete experimental environment   **Fig. 15.** The target walks into the experimental environment

Fig. 16. Tracking results of our test videos

Only the region near ROI will be searched, and that reduced the search time and increased the precision. If we missed our ROI, we would go back to detection step, and try to find another recognized ROI to start another tracking.

In our test videos, there is no error tracking that traces wrong target, but sometimes we miss our ROI because of objects overlapping, unusual movement such as squatting down.

4   Conclusion and Future Work

Currently we have implemented a system that is a close approach to our initial idea. Our detection is able to detect the foreground moving objects and provides a detail contours. We can distinguish objects have difference in color by providing a reliable similarity value from our color recognition.In height recognition, we implement the algorithm in [8], and in the environment without reflections underground, it measures well. We begin tracing if an object on the screen reaches the threshold. We lose our target in few cases, unusual movements like sitting or jumping and error in foreground object detection, but we can still start a new trace at next frames. Userscan find the target efficiently without losing correctness with the assistance of the RAPId.

In the future, we will combine the MySQL database system to store the information we analyzed. The Pandaboard ES will be used to capture the video and encode it by using H.264 encoder, and upload to server for processing. Also, we will increase the precision of the RAPId by using more other features in our recognition, like gender, stride, body-shape, and body-proportions.

Another issue is that we will refine our results to detect and track the object we want more efficiently. Due to the cost of tracking and detecting time we spent, we are looking forward to some parallelism solutions which can help us reduce the cost of time. The OpenCV GPU module is a set of classes and functions to utilize GPU computational capabilities. It's implemented using Compute Unified Device Architecture (CUDA), so we will try to perform our algorithm in CUDA architecture to pursue better performance. We will also use the Intel· Threading Building Blocks (TBB) to achieve parallelism in our program.

References

1. Everingham, M., Zisserman, A.: Automated Person Identification in Video. In: Enser, P.G.B., Kompatsiaris, Y., O'Connor, N.E., Smeaton, A., Smeulders, A.W.M. (eds.) CIVR 2004. LNCS, vol. 3115, pp. 289–298. Springer, Heidelberg (2004)
2. BenAbdelkader, C., Cutler, R., Davis, L.: Person identification using automatic height and stride estimation. In: Proc. of the 16th International Conference on Pattern Recognition, ICPR (2002)
3. Balcan, M.-F., Blum, A., Choi, P.P., Lafferty, J., Pantano, B., Rwebangira, M.R., Zhu, X.: Person identification in webcam images: an application of semi-supervised learning. In: Proc. of the 22nd ICML Workhop on Learning with Partially Classified Training Data (2005)
4. Orten, B.B., Soysal, M., Alatan, A.A.: Person identification in surveillance video by combining MPEG-7 experts. In: Proc. of the IEEE 13th Signal Processing and Communications Application Conference, pp. 352–355 (May 2005)
5. De Angelis, D., Sala, R., Cantatore, A., Poppa, P., Dufour, M., Grandi, M., Cattaneo, C.: New method for height estimation of subjects represented in photograms taken from video surveillance systems. International Journal of Legal Medicine 121, 489–492 (2007)
6. Lee, J., Lee, E.-D., Tark, H.-O., Yoon, D.Y.: Efficient height measurement method of surveillance camera image. Forensic Science International 177(1), 17–23 (2008)
7. Criminisi, A., Zisserman, A., Van Gool, L., Bramble, S., Compton, D.: A new approach to obtain height measurements from video. In: Proc. of SPIE, vol. 3576, pp. 1–6 (November 1998)
8. Hartley, R.I., Zisserman, A.: Multiple View Geometry in Computer Vision. Cambridge University Press (2000)
9. Chiou, J.-J.: Using Temporal-spatial Analysis and Background Subtraction Method to Detect Moving Objects in the Video Sequence, Master Thesis, Department of Mechatronic Technology, National Taiwan Normal University, Taipei, Taiwan (August 2009)
10. OpenCV Tutorials,
    http://docs.opencv.org/doc/tutorials/tutorials.html
11. Laganière, R.: OpenCV 2 Computer Vision Application Programming Cookbook. Packt Publishing (2011)

A High Performance Parallel Graph Cut Optimization for Depth Estimation

Bo-Yen Chen and Bo-Cheng Charles Lai

Department of Electronics Engineering, National Chiao Tung University
Hsinchu 30010, Taiwan
thegodofdante@hotmail.com, bclai@mail.nctu.edu.tw

Abstract. Graph-cut has been proved to return good quality on the optimization of depth estimation. Leveraging the parallel computation has been proposed as a solution to handle the intensive computation of graph-cut algorithm. This paper proposes two parallelization techniques to enhance the execution time of graph-cut optimization. By executing on an Intel 8-core CPU, the proposed scheme can achieve an average of 4.7 times speedup with only 0.01% energy increase.

Keywords: graph cut, parallelization, stereo correspondence, depth estimation.

1 Introduction

Images are one of the most intuitive forms to deliver and obtain information in mass media. 3D (Three Dimensional) images augment the depth information to the conventional images, and make the viewers feel like watching real objects in the real life. Due to the rapid advances of the stereo display technology [19], 3D contents are becoming an indispensable part of people's daily life.

Generating images with depth information is an enabling technology to create 3D images. One of the most widely used methods today is based on a multi-view scheme. The target scene is captured by several cameras where each camera is located at a different view point. The depth information of the target scene is then calculated by combining the image information from multiple view points and calculating the geometry [1] with a depth estimation algorithm [2]. The generated depth information, which is usually referred as a depth map, is augmented to the conventional image. The 3D display would then use this information to synthesize the 3D contents and present the contents in stereoscopic way.

Obtaining the depth map by the simple basic matching [5] is often inappropriate and could result in broken edges in an image. An optimization procedure, such as graph cut, is used to enhance the visual quality of a 3D image. However, the optimization of the depth map is computation intensive and requires a long latency. It is estimated to take 50 seconds to finish the depth map processing for a 434X383 image with 20 layers of depth. The time consuming computation makes 3D technology too slow to be applied onto interactive applications, such as robot vision or virtual reality.

J.-S. Pan et al. (Eds.): *Advances in Intelligent Systems & Applications*, SIST 21, pp. 311–320.
DOI: 10.1007/ 978-3-642-35473-1_31   © Springer-Verlag Berlin Heidelberg 2013

Thus how to achieve the highest quality of 3D images within the stringent time requirement is a great challenge in this application.

This paper focuses on the design of a highly parallel graph-cut algorithm for depth estimation optimization. Graph-cut (GC) is a global optimization algorithm [11] which solves the problem in Markov random field. With its superior quality, GC is very popular in various image processing applications, such as image segmentation, stereo matching, image restoration, texture synthesis, and etc. Kolmogorov and Zabih [3] proposed an approach to minimize the disparity energy of 3D contents by adopting the graph-cut optimization. It has been shown that graph-cut can return the depth information with better quality than other optimization algorithms [5-9]. There are two commonly used implementations of GC. The first one uses augmenting paths proposed by Ford-Fulkurson [4], and the other one adopts a push-relabel scheme [10]. After finding the maximum flow of the graph, one can find the location of the minimum cuts with minimum energy through a breadth-first search. However, these two approaches still have significant computation complexity and require long execution time.

Leveraging the computation capability of the current parallel computation platforms has been proved as a effective solution to boost the system performance. A. Delong and Y. Boykov [12] proposed a parallel approach to enhance the performance of graph cut. In [12], each thread can use only the information of the selected range with other parts in the process without the need for data exchange. This scheme can effectively reduce the use of memory space and increase the speed with nearly linear acceleration when increasing the number of processing cores. Jiangyu Liu and Jian Sun [13] published a parallel approach based on augmenting-path algorithm where every thread calculates each block independently and then selects another completed nearby block to merge. When the memory usage does not exceed the calculated machine, the approach in [13] can achieve better performance than the former schemes. However, due to the processing overhead, these approaches can show performance enhancement only when the processing images are huge enough. We will discuss the reason on later section. When dealing with images with sizes commonly used today, the previously proposed parallel approaches cannot provide performance advantages when compared with the general depth map optimization on a sequential computation platform.

This paper proposes a highly parallel design of graph-cut optimization for the depth estimation of 3D images. The proposed approach identifies the independent processing loops in graph-cut and separate the computation of each loop. An image is then chopped into smaller image sections which will be calculated in parallel. This paper uses the total energy as the way to evaluate the quality of a 3D image. With the slightly compromised quality, the proposed approach enables great execution time speedup. The experiments have shown that the proposed approach achieves an average of 4.7 times performance speedup while increasing the overall energy by only 0.01%.

The rest of this paper is organized as follows. Section II will overview the entire stereo matching process and Graph-Cuts algorithm steps. Section III will describe the proposed parallel approach and Section IV will give a comparative analysis of the experimental data. Conclusions will be given in Section V.

2   Stereo Matching Flow and Graph-Cut Steps

2.1   Stereo Matching Flow

The disparity estimation processes generally have the following four steps. 1) **Initial matching cost calculation**. This step is to obtain a list of the cost of a three-dimensional x-y axis corresponding to the target image pixel, and z-axis corresponding to the difference of pixels between the target image and the comparison image on horizontal direction. This three-dimensional matrix (x, y, d) stores the color information difference between pixels of comparison image (x + d, y) and pixels of target image (x, y). In other words, the smaller the difference value, the closer the colors of two pixels. This difference value is used as the basis to choose the disparity. 2) **Cost aggregation**. The raw cost of the current graph needs to be processed by a simple filter to reduce noise interference. 3) **Disparity computation and optimization**. This step would determine the initial disparity map first. The raw disparity map is usually scattered because of the correspondence mismatch. An optimization algorithm would be introduced to integrate the scattered pars. 4) **Disparity refinement.** This step is to adjust and increase the resolution of the disparity map. This step is usually an option based on the requirement of an application.

The disparity map optimization in step (3) is by far the most time consuming part of the whole process. It usually takes more than 90% of the overall execution time. Therefore this paper focuses on shortening the execution time of the optimization of disparity maps. Although there exist various schemes that could be applied by each step, this paper chooses relatively simple schemes with low complexity. In step (1), the raw matching cost is obtained by absolute intensity difference [2]. The cost aggregation of step (2) uses a box filter to smooth the raw matching costs from the previous step. In the disparity computation and optimization step, we use "Winner Takes All" approach to obtain raw depth map [ref]. The graph-cut algorithm is used to optimize the results. This paper did not applied any optimization for the disparity refinement in order to maintain the low computation requirement.

2.2   Graph-Cut Algorithm

Graph-Cuts algorithm is an energy minimization algorithm. The worst case running time for GC is O $(v^2 \varepsilon)$, where v is the number of nodes and ε is the number of edges. Fig.1 shows the equations of the objectives of energy minimization [15]. Every node $d(x, y)$ in a graph can choose one of the two labels, alpha or beta. E_{data} is the appropriate degree of node d with respect to the two labels without considering the correlation with the other nodes. In equation (2), C_{AM} is the aggregated matching cost obtained in step two; E_{smooth} is the disparity discontinuity of two connected nodes. In equation (3), ρ is 1 when selected labels are different and 0 otherwise; λ is the weight proportion between E_{data} and E_{smooth}.

$$E_{total}(d) = E_{data}(d) + \lambda E_{smooth}(d) \tag{1}$$

$$E_{data}(d) = M(x, y, d(x, y)) \tag{2}$$

$$E_{smooth}(d) = \sum_{(x,y)} \rho((d(x, y) - d(x + 1, y)) + \rho((d(x, y) - d(x, y + 1)) \tag{3}$$

Fig. 1. Function of graph-cuts algorithm

Fig.2 is an example of graph-cut energy minimization on an one-dimensional image. Label α and β are two disparities of depth that we choose to be allocated. Node set {p, q ..., r, s, t} contains the points that need to be assigned to the depth α or β. c^{α} and c^{β} are the costs of the pixel (E_{data}) in disparity α and β respectively. ρ is the disparity discontinuity between these two nearby pixels (E_{smooth}). The energy represented in this graph is a line between α and β. This line seperates the graph into two parts, α or β, and assigns every node to one of the parts. The aggregated cost of the edges cut through by this line is called the energy of this graph. The graph-cut problem is to find a cut line with minimum sum of energy.

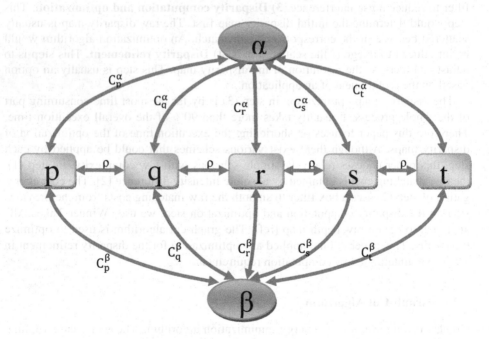

Fig. 2. A sample graph for 1-D image

To find the cut line with minimum energy, it is impractical by exhaustively searching all the possible cut lines due to the huge time complexity. The answer could be found by finding the maximum flow in this graph. Because of the maximum flow must be limited by the capacity of edges, the energy value of the minimum cut is equal to result of a maximum flow. The approach of finding the maximum flow generally uses two algorithms, augmenting path and push-relabel. The augmenting path repeatedly performs DFS (Depth-First-Search) to find all the flow paths, while the push-relabel adopts the concept of water flowing from high to low to identify all the flow paths. After finding the maximum flow, a BFS (Breadth-First-Search) is performed starting from the sink node, and assigns nodes to label α or β.

3   Proposed Parallel Graph-Cuts Algorithm

3.1   Applying Graph-Cut on Depth Map Optimization

A general graph-cut algorithm can be used to create two disparities. However, in a 3D image, multiple depths need to be compared and decided. Therefore the algorithm needs to optimize more than two disparities. In order to perform optimization with more than two disparities, a α-β swap approach was proposed by [14][15][16]. The α-β swap would randomly pick two disparities and perform the graph-cut on these two disparities. The procedure would perform on all the combinations of disparities. For each disparity-pair, the graph-cut algorithm would try to minimize the energy at every processing loop. The processing loop would be repeated until no more energy reduction can be found. Since there exists a huge amount of processing loops that can be independently performed, the parallelization of these loops becomes a good solution to achieve high performance without affecting the quality of the optimization.

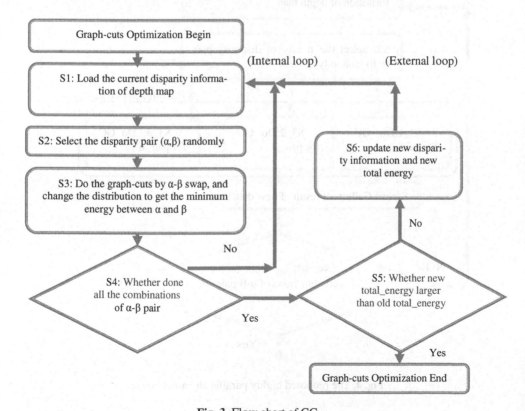

Fig. 3. Flow chart of GC

3.2   Alpha-Beta Swap Parallelization

With the observation of loop parallelism from the previous section, this section proposes a parallel design for depth map optimization. Fig.3 shows the flow chart of applying graph-cut on depth map optimization. One can see that the optimization flow is mainly composed of two kinds of loops, the internal loop and external loop. Here we choose to parallelize the internal loop. The reason is that the external loop uses the previously completed result as the initial solution. Therefore there exists a dependency between two external loops, and the two consecutive external loops cannot be separated easily without changing the results of the algorithm. On the other hand, for the internal loop, the computation of α-β swap on different pairs of disparities can be performed concurrently. This approach has expanded the computation parallelism to d/2, where d is the number of disparities. It means that in an ideal condition, the optimization of depth map can be accelerated by a factor of d/2 without loss of quality.

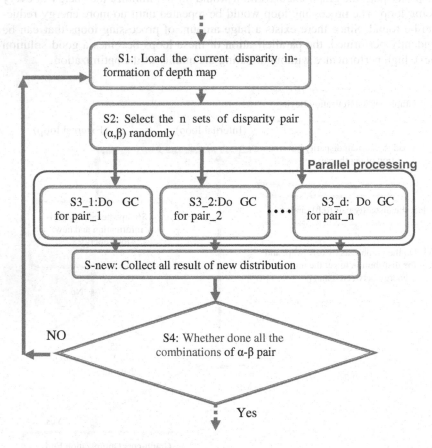

Fig. 4. The proposed highly parallel alpha-beta swap

Here we use 8 threads to experiment this approach for our 8-core CPU. The results of improvement are in Table 1.

Table 1. Computation time of α-β swap parallelization (8-thread)

Time(sec)	Original	α-β swap	Improvement(%)
Map	20.12	9.23	117.98%
Sawtooth	35.32	22.72	55.45%
Tsukuba	16.95	12.37	37.02%
Venus	50.84	31.39	59.22%
Average	**30.81**	**18.93**	**62.75%**

3.3   Image Segment Parallelization

This section further extends the parallelization on the image domain. An image is segmented into N parts, and the depth map computation is performed on each part at the same time. After the depth map segmentation, there requires no need to exchange information when performing the graph-cut optimization. In this way, we can complete the optimization without synchronizing information during the process. The computation parallelism can be significantly enhanced with this approach. Because the computation complexity of GC O ($v^2 \varepsilon$), after segmenting into N blocks, the single computing time can be greatly decreased.

This segmentation approach is commonly applied by many image processing applications. However, due to the segmented image, the depth map computation and graph-cut optimization are confined into each segment. This would change the flow of the original algorithm which work on the image as a whole, and might return different quality results.

4   Analysis of Experimental Results

Table 2 compares the computation time of the original sequential scheme and the proposed scheme. Table 3 shows the resulting energy after the graph-cut optimization. By running on the Intel (R)core$^{(TM)}$2 Quad CPU workstation, an average of 4.7X computation time enhancement can be reached with less than 0.01% of energy increase.

Table 2. Computation time of Image segment parallelization (8-thread)

Time(sec)	Original	segment	Improve (%)
Map	20.12	2.55	689.01%
Sawtooth	35.32	6.89	412.62%
Tsukuba	16.95	2.99	466.89%
Venus	50.84	9.12	457.45%
Average	**30.81**	**5.39**	**471.61%**

318   B.-Y. Chen and B.-C.C. Lai

Table 3. Energy of Image segment parallelization

Energy	Original	segment	Diff. (%)
Map	595155	595210	0.0092%
Sawtooth	2489680	2489802	0.0047%
Tsukuba	1009065	1009104	0.0039%
Venus	2077381	2077687	0.0147%
Average	**1542820**	**1542950**	**0.0084%**

Table 4. Computation time of the combination of n-d grid approaches

Time(sec)	Original	n-d [17]	Improve
Map(10 layers)	4.19	589.92	-139.79X
Sawtooth (5 layers)	7.94	319.27	-39.21X
Tsukuba(5 layers)	3.29	52.97	-15.1X
Cones(7 layers)	13.65	1103.03	-79.8X
Average	**7.27**	**516.3**	**-71.02X**

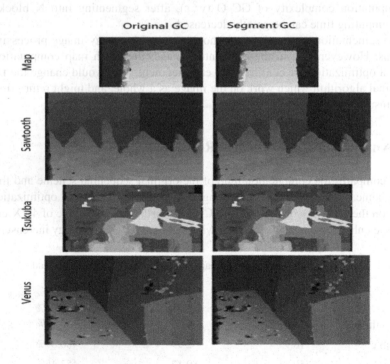

Fig. 5. The resultant disparity maps

Fig.5 shows the results after applying the image segmentation parallelization. Since this approach focuses on the optimization of a local segment, the outlines of local blocks become more clear, and the tattered conditions of the raw depth map is almost eliminated.

The two parallelism schemes proposed in this paper were adopted at the same time. As shown in Table 2 and Table 3, an average of 4.7 times of execution time acceleration has been achieved with only 0.01% energy increase.

The n-d grid parallel implementation of graph-cut [17] is also implemented. It adopted the similar concept of parallelization by segmenting an image to small regions. The results of n-d grid approaches with 75*75 regions are shown in Table 4. We can observe that execution time of the n-d grid is significantly worse than the original sequential scheme. The main reason is because this parallel approach has paid a significant cost on synchronization between parallel tasks. If the image size is not huge enough, these approaches cannot benefit from the parallel computation. The largest image size used in this paper is 450*375 pixels.

5   Conclusion

Graph-cut has been proved to return good quality on the optimization of depth estimation. However, the intensive computation requirement makes it a hurdle to apply graph-cut on applications with stringent timing requirements. Leveraging the parallel computation has been proposed as a solution to achieve high performance. This paper proposes two parallelization techniques to enhance the execution time of   graph-cut optimization. By executing on an Intel 8-core CPU, the proposed scheme can achieve an average of 4.7 times speedup with only 0.001% energy increase.

References

1. Hartley, R., Zisserman, A.: Multiple View Geometry in Computer Vision, 2nd edn. Cambridge University Press (March 2004)
2. Scharstein, D., Szeliski, R.: A taxonomy and evaluation of dense two-frame stereo correspondence algorithms. International Journal of Computer Vision (IJCV)
3. Kolmogorov, V., Zabih, R.: What energy Functions Can Be Minimized via Graph cuts? IEEE Trans. on Pattern Analysis and Machine Intelligence 26(2) (February 2004)
4. Ford, L., Fulkerson, D.: Flow in Networks. Princeton University Press (1962)
5. Boykov, Y., Veksler, O., Zabih, R.: Fast Approximate Energy Minimization via Graph Cuts. IEEE Transaction on Pattern Analysis and Machine Intelligence (T.PAMI) 23(11), 1222–1239 (2001)
6. Birchfield, S., Tomasi, C.: Multiway Cut for Stereo and Motion with Slanted Surfaces. In: Proc. Int'l Conf. Computer Vision (ICCV), pp. 489–495 (1999)
7. Kolmogorov, V., Zabih, R.: What Energy Functions Can Be Minimized via Graph Cuts? IEEE Transaction on Pattern Analysis and Machine Intelligence (T.PAMI) 26(2), 147–159 (2004)
8. Boykov, Y., Kolmogorov, V.: An experimental comparison of min-cut/max-flow algorithms for energy minimization in vision. IEEE Transactions on Pattern Analysis and Machine Intelligence (T.PAMI) 26(9), 1124–1137 (2004)

9. Worby, J., MacLean, W.J.: Establishing Visual Correspondence from Multi-Resolution Graph Cuts for Stereo-Motion. In: Proc. 4th Canadian Conf. Computer and Robot Vision (CRV), May 28-30 (2007)
10. Glodberg, A., Tarjan, R.: A New Approach to the Maximum Flow Problem. Journal of the Association for Computing Machinery 35(4), 921–940 (1988)
11. Geman, D.: Random fields and inverse problems in imaging. Lecture Notes in Mathematics, vol. 1427, pp. 113–193. Springer (1990)
12. Delong, A., Boykov, Y.: A scalable graph-cut algorithm for n-d grids. In: Proceedings of CVPR (2008)
13. Liu, J., Sun, J.: Parallel Graph-cuts by Adaptive Bottom-up Merging. In: Proceedings of CVPR (2010)
14. Boykov, Y., Veksler, O., Zabih, R.: Fast Approximate Energy Minimization via Graph Cuts. IEEE Transaction on Pattern Analysis and Machine Intelligence (T.PAMI) 23(11), 1222–1239 (2001)
15. Scharstein, D., Szeliski, R.: Middlebury college – stereo vision page research, http://vision.middlebury.edu/stereo/
16. Scharstein, D., Szeliski, R.: A taxonomy and evaluation of dense two-frame stereo correspondence algorithms. International Journal of Computer Vision (IJCV) 47, 7–42 (2002)
17. Delong, A., Boykov, Y.: A scalable graph-cut algorithm for n-d grids. In: Proceedings of CVPR (2008)
18. Liu, J., Sun, J.: Parallel Graph-cuts by Adaptive Bottom-up Merging. In: Proceedings of CVPR (2010)
19. Fehn, C.: A 3D-TV system based on video plus depth information. In: Proc. of Asilomar Conference on Signals, Systems and Computers, vol. 2, pp. 1529–1533 (2003)

An Interactive 3D Modeling System Based on Fingertip Tracking

Jia-Wei Hung, I-Cheng Chang[*], and Jiun-Wei Yu

Department of Computer Science and Information Engineering,
National Dong Hwa University, Hualien, Taiwan
icchang@mail.ndhu.edu.tw

Abstract. This work presents a real time 3D modeling system which integrates a fingertip tracking technique and a 3D editing module. Therefore, the user can build up a 3D model by moving his fingertips instead of using mouse on the computer. We try to break out the gap between users and 3D editing software and expect that artists or designers who are not familiar with computers can create a digital 3D prototype model easily and conveniently. Some improvement algorithms like tracking noise removal and response time reduction are also proposed to increase the accuracy. The experiment results show that the proposed system is capable of producing the 3D objects of various shapes and users can quickly create a 3D prototype model with our system.

Keywords: Fingertip tracking, Human-Computer Interaction (HCI), deformation, model editing.

1 Introduction

With the development of Human Computer Interaction (HCI) technologies, people pay more attention to the interactive system controlled by human body motion. For instance, some multimedia extending devices such as Kinect, Wii remote controller and PlayStation Move, provide the intuitive controlling ability. With these devices, users can play video games by waving their hands and moving bodies. Some mobile devices like the iPhone and iPad provide a touching screen interface, and users can control these devices with their fingers by touching or sliding on the screen. Instinctive and natural operating way becomes a tendency in HCI related area. In this work we use the marker based gestural HCI as the input device of a 3D modeling system. Compared to bared hand tracking, marker based hand tracking can get a more accurate result. Park 1 used LED gloves as their tracking device to obtain the hand position information. But the moving space of the hand is under the limitation of the wired device. To overcome this drawback, Bainbridge 2 extracted tracking data from Radio Frequency Identification (RFID). However, it is uncomfortable for users, and the sensor tag power is another problem for the application. Grossman 3 and Sheng 4

---

[*] Corresponding author.

J.-S. Pan et al. (Eds.): *Advances in Intelligent Systems & Applications*, SIST 21, pp. 321–329.
DOI: 10.1007/978-3-642-35473-1_32   © Springer-Verlag Berlin Heidelberg 2013

located a user's palm and fingers with Vicon motion capture system. With these tracking position data, Sheng 4 proposed an interactive 3D model modeling system in virtual reality (VR). The tracking result of the motion capture system is accurate and robust, but the system is expensive and needs a specific environment. In the field of color marker based hand tracking, Wang 5, Fredriksson 6, Mistry 7 and Schwaller 8 used wearable stuffs like color gloves and finger stall to help them locate the user's hand position, and apply the tracking result to other applications. For example, Wang 5 employed nearest-neighbor technique to track and recognize a hand gesture, and apply the observation of hand to three applications such as character animation, physical simulation and sign language. Mistry 7 proposed a wearable digital information system with a simple camera and tiny projector. User can interact with the system in augmented reality (AR) by some specific gestures, and the gesture is tracked with color rings on user's fingertip. Schwaller 8 proposed a wearable system similar to 7 but with different sensor locations. Our tracking system is inspired by 7 and 8.

In art industry, artists and designers can quickly knead a prototype model with clay or gypsum. However, to create a digital model with editing software like Maya and 3dStudio is much more complex than usual case. Besides, it is also unnatural and hard for designers to use a 2D device like a mouse and a graphics tablet to edit a digital 3D model in a computer. Cherlin 9 referred several concepts of the art design to propose a 3D modeling system. We modify the rotational blending surface idea from 9 to build our modeling function. Chen 10 deformed a target model in real time with a Wii remote controller. Andre 11 designed an editing technique that can support simple deformation of a planar surface. But it is not able to process a curved surface. Igarashi 12 proposed a sketch based 3D editing system that user can sketch a shape contour to create a rough 3D model quickly and simply. Nealen 13 extended and modifies the editing function of 12. In 2008, Gingold 14 designed a system for 3D editing that modifies a 3D model by altering the shade image of target model. In 2009, Gingold and Igarashi 15 presented a 3D modeling system that creates a model with 2D sketching method. Although the operating concepts of the approaches described previously are intuitive and natural in modeling, their control interfaces are still the 2D device like a mouse or a digital pen.

Our system has two features: an intuitive user interface with less constraint and an effective modeling module. We try to break out the gap between user and system, and hope artists and designers who are not familiar with computer art designing can create a digital 3D prototype model easily and conveniently with our system. We integrate the color based fingertip tracking as the interface, which uses the image and 3D information from Kinect. The proposed system is cheaper than other related works. Besides, we also provide some effective editing functions in the modeling module. Fig.1 describes the concept of our system. The remainder of the paper is organized as follows. Section 2 presents the color based fingertip tracking and Section 3 describes the 3D modeling functions. Experimental results are shown in Section 4. The conclusions are drawn in Section 5.

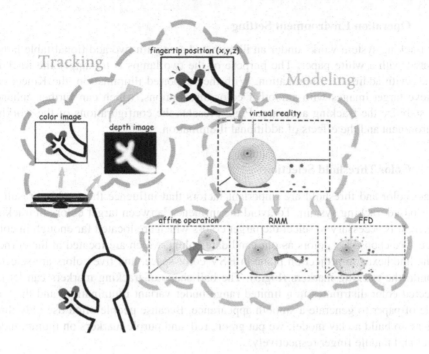

Fig. 1. The flowchart of our interactive 3D modeling system

2   Color-Based Fingertip Tracking

This chapter introduces the color based fingertip tracking system that is used as the user interface in our system. With the tracking system, our work can extract the spatial information of a user's fingertips and use the captured information to interact with the 3D environment.

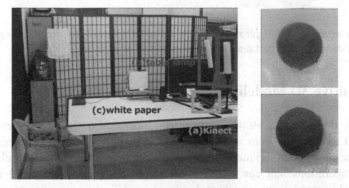

Fig. 2. Tracking environment. The right-top figure is the tracking target with additional illumination and the right-down one is what without illumination.

2.1   Operation Environment Setting

Our tracking system works under an illuminated area with two additional table lamps covered with a white paper. The purpose of the two lamps is to supply the tracking targets with additional illumination. With the supported illumination, the Kinect can retrieve target images with smoother color distributions, which can further enhance and stabilize the tracking accuracy. Fig.2 presents the configurations of the working environment and the effects of additional illumination.

2.2   Color Threshold Selection

Target color and threshold are important factors that influence the tracking result in color based tracking system. To avoid the influence between target colors in tracking system, it is necessary to select the target colors which are located far enough in color space. We choose six colors as our candidate colors, which are located at the corners of the hue hexagon projection plane in HSV color space, and five colors are selected to build the tracking markers (Fig.3). The ball shape of tracking markers can let the reflected color distributes in a limited range under variant illumination, and they are made of paper to generate a smooth appearance. Because people often use only three fingers, to build a clay model, we put green, red, and purple markers on thumb, index finger and middle finger respectively.

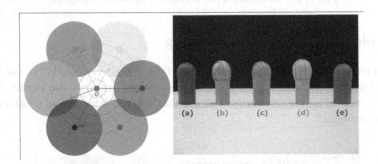

Fig. 3. Color selection and finger markers. The (a) red, (b) brown, (c) green, (d) cyan and (e) purple finger stalls. The left part diagram is referred from Wikipedia.

3   Intuitive 3D Modeling

Along with the tracking strategy mentioned previously, this work proposes an intuitive 3D modeling system which is aimed to give an example of integrating the fingertip tracking system. Hence, in this work, we mainly focus on developing the interactive operations that can intuitively manipulate or construct a 3D model. In the proposed system, there are three kinds of modeling operations, namely affine transform operation, ring-shape based mesh modeling, and free form deformation.

3.1   Affine Transform Operation

In the proposed modeling system, two basic operations, translation and rotation, are used to adjust the 4x4 transformation matrix that manipulates the orientation and position of a 3D object (Fig.4). While interacting with the user, the system keeps detecting the user's fingertips using the proposed tracking strategy and analyzing the relations among the detected results to adjust the 4x4 transformation matrix.

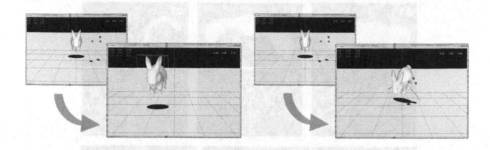

Fig. 4. State modification with affine transform operation

3.2   Ring-Shape Based Mesh Modeling (RMM)

In this section, we introduce the ring-shape based mesh modeling (RMM) function which is inspired by the model construction method in 16. This operation provides a process to dynamically construct a continuous mesh. With this operation, a user has the ability to construct a mesh with free-styled shapes.

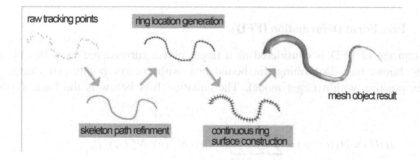

Fig. 5. Flowchart of RMM

The whole RMM is constructed by three sequential modules, namely skeleton path refinement, ring location generation, and continuous ring surface construction (Fig.5). First of all, in the module of skeleton path refinement, the system generates a smoothed path by applying the B-spline function to the tracking results. And in ring location generation, the system samples ring locations on the skeleton path. Finally, the system generates the ring meshes based on the sampled locations. During the ring mesh construction process, three parameters are defined to control and generate meshes for different shapes. Fig.6 gives some cases generated by RMM under different parametrical configurations.

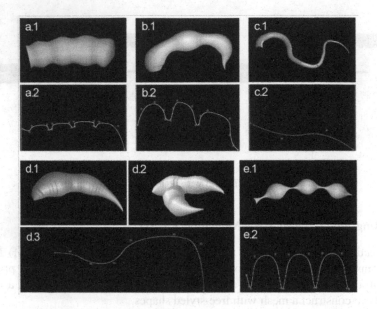

Fig. 6. Several RMM demonstration models with a ring radius set

3.3 Free Form Deformation (FFD)

The concept of FFD is considered as a target object surrendered by a flexible and elastic bound box. Deforming the bound box with control points can change the vertex position within target model. The equation lists below is the basic form of FFD.

$$H(U) = H(u,v,w) = \sum_{i=0}^{l}\sum_{j=0}^{m}\sum_{k=0}^{n} N_i^l(u) \cdot N_j^m(v) \cdot N_k^n(w) \cdot P_{i,j,k} \tag{1}$$

Considering the computation speed, we use linear polynomial as the blending function instead of the Bernstein polynomial in FFD. With the editing ability to 3D model of FFD, our modeling system can provide users a user-friendly and intuitive modeling function. Fig.7 shows a FFD function.

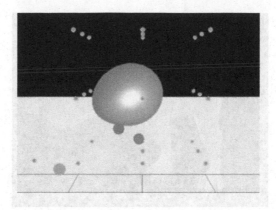

Fig. 7. FFD function. The small green nodes are the controlling points.

4 Experimental Results and Discussion

Our system works on a computer with1.9Hz CPU and 4 GB memory. In the procedure relating to digital image processing and 3D representation, we use OpenCV, OpenGL, and OpenNI to support our work. All experiments are operated interactively in real time without the calculation support of GPU. Fig.8 shows a creature, squid, which is modeled by our interactive system. This model is composed of several objects like head, eye and tentacles which are created by RMM or FFD, and all of them have the deformation of high degree and smooth surface. Furthermore, the proposed system can load other 3D models and add external parts to modify it (Fig.9). To evaluate the speed performance, we invite three people to model three creatures, a squid, a snail as well as an ant, by different editing tools and compare the consuming time (Table.1).

Fig. 8. A squid model with 16880 vertices and 33144 facets

Fig. 9. Modified Tyrannosaurus model. The red part is original and the green part is external.

Table 1. Comparison of Consuming time

		Ours	SketchUp	3dStudio
	head	3m23s	11m31s	9m15s
	tentacle	5m11s	17m52s	24m27s
	total	10m44s	29m46s	36m41s
	shell	2m53s	5m3s	3m17s
	body	11s	2m19s	9m34s
	total	4m20s	10m52s	13m11s
	head+body	3m25s	9m41s	12m31s
	teeth+feet	11m9s	21m23s	18m40s
	total	17m27s	37m50s	42m16s

5   Conclusions

In the paper, we propose an easy-to-use and interactive 3D modeling system using kinect as the input device. A color based fingertip tracking system is integrated to construct a user-friendly interface. Furthermore, we apply two modeling techniques, ring-shape mesh modeling (RMM) and free form deformation (FFD), to provide a 3D editing tool. With the interactive 3D modeling system, users like artists and designers can construct or edit a digital prototype model easily without considering the complicated computer instructions. In the future, we plan to use more features to improve the tracking accuracy, and try to apply multiple cameras to reduce the effect of occlusion. Besides, more editing functions are planned to be included in the modeling system.

References

1. Park, J., Yoon, Y.L.: LED-Glove Based Interaction in Multi-Modal Displays for Teleconferencing. In: Proceedings of the 16th International Conference on Artificial Reality and Telexistence, pp. 395–399 (2006)
2. Bainbridge, R., Paradiso, J.A.: Wireless Hand Gesture Capture Through Wearable Passive Tag Sensing. In: International Conference on Body Sensor Networks, pp. 200–204 (2011)
3. Grossman, T., Wigdor, D., Balakrishnan, R.: Multi-Finger Gestural Interaction with 3D Volumetric Displays. ACM SIGGRAPH 24, 931–931 (2005)
4. Sheng, J., Balakrishnan, R., Singh, K.: An Interface for Virtual 3D Sculpting via Physical Proxy. In: Proceedings of the 4th International Conference on Computer Graphics and Interactive Techniques in Australasia and Southeast Asia, pp. 213–220 (2006)
5. Wang, R.Y., Popovic, J.: Real-Time Hand-Tracking with a Color Glove. ACM SIGGRAPH 28(3), 631–638 (2009)
6. Fredriksson, J., Ryen, S.B., Fjeld, M.: Real-Time 3D Hand-Computer Interaction: Optimization and Complexity Reduction. In: Proceedings of the 5th Nordic Conference on Human-Computer Interaction: Building Bridges, pp. 133–141 (2008)
7. Mistry, P., Maes, P., Chang, L.: WUW – Wear Ur World: A Wearable Gestural Interface. In: Proceedings of the 27th International Conference Extended Abstracts on Human Factors in Computing Systems, pp. 4111–4116 (2009)
8. Schwaller, M., Lalanne, D., Khaled, O.A.: PyGml: Creation and Evaluation of a Portable Gestural Interface. In: Proceedings of the 6th Nordic Conference on Human-Computer Interaction: Extending Boundaries, pp. 773–776 (2010)
9. Cherlin, J.J., Samavati, F., Sousa, M.C., Jorge, J.A.: Sketch-Based Modeling with Few Strokes. In: Proceedings of the 21st Spring Conference on Computer Graphics, pp. 137–145 (2005)
10. Chen, K.H., Ou, Y.M.: A Three Dimensional Mesh Deformation System Using Wii Remote and MotionPlus. NTU Master Thesis, Taiwan (2010)
11. Andre, A., Saito, S., Nakajima, M.: CrossSketch: Freeform Surface Modeling with Detail. In: Proceedings of the 4th Eurographics Workshop on Sketch-Based Interfaces and Modeling, pp. 45–52 (2007)
12. Igarashi, T., Matsuoka, S., Tanaka, H.: Teddy: a Sketching Interface for 3D Freeform Design. ACM SIGGRAPH (21), 409–416 (2007)
13. Nealen, A., Igarashi, T., Sorkine, O., Alexa, M.: FiberMesh: Designing Freeform Surface with 3D Curves. ACM SIGGRAPH 26(3-41) (July 2007)
14. Gingold, Y., Zorin, D.: Shading-Based Surface Editing. ACM SIGGRAPH 27(95) (2008)
15. Gingold, Y., Igarashi, T., Zorin, D.: Structured Annotations for 2D-to-3D Modeling. ACM SIGGRAPH Asia 28(5), 148 (2009)
16. Guptill, A.L.: Rendering in Pencil. Waston-Guptill Publications (1977)

References

[content faded and illegible]

Comprehensive Evaluation for HE Based Contrast Enhancement Techniques

Ming-Zhi Gao[1], Zhi-Gang Wu[1], and Lei Wang[2]

[1] Department of Information Engineering and Computer Science, Feng Chia University, Taiwan
{pig07153,jackal54039}@hotmail.com
[2] Department of Electrical Engineering, Feng Chia University, Taiwan
leiwang@fcu.edu.tw

Abstract. The principle of image enhancement is based on increasing the contrast between adjacent pixels, enabling viewers to visually perceive images with greater detail in the textures and edges. Many contrast enhancement methods have been proposed to improve the quality of images and most of these methods are based histogram equalization (HE); however, the actual results remain uncertain due to the lack of an objective evaluation procedure with which to measure them. This paper proposes a quantitative analysis for the assessment of image quality based on several subjective and objective evaluation metrics. Furthermore, there are 11 different HE based contrast enhancement techniques are evaluated in this paper.

Keywords: image quality assessment, histogram equalization, image contrast enhancement, structural similarity.

1 Introduction

Contrast plays an essential role in determining image quality. However, many factors associated with the capture of images, such as the environment in which photographs are taken, may cause degradation in contrast, thereby compromising the overall quality of the image. Many algorithms have been developed to enhance contrast. Of these, histogram equalization (HE) [1] is widely used in display applications, such as medical diagnosis, character recognition and multimedia image displays. However, it is prone to various shortcomings. HE produces an increase in contrast throughout the entire image and minor details are easily lost after post-equalization. Furthermore, HE often causes excessive sharpening in gray areas of the updated histogram, resulting in unnatural images. The average brightness of images processed with HE also often differs from the original, which detracts from the fidelity of the image.

Many improved techniques been proposed in the past years. These techniques can be divided into three major classes: they are Improved He class, Spatial Processing class, and Reshaping-PDF class.

The Improved HE class uses the gray area of the original image to conduct further equalization of the image histogram. Depending on the method used to determine gray area, this type of technique can be further divided into two sub-categories: static

J.-S. Pan et al. (Eds.): *Advances in Intelligent Systems & Applications*, SIST 21, pp. 331–338.
DOI: 10.1007/978-3-642-35473-1_33   © Springer-Verlag Berlin Heidelberg 2013

range and dynamic range. The static range category uses the mean brightness or medium brightness of the original image to divide the histogram into segments. Equalization is performed on each segment, separately. Some well-known methods that fall into this category include Bi-Brightness Histogram Equalization (BBHE)[2], Dualistic Sub-Image Histogram Equalization (DSIHE)[3], Statistic Separate Tri-histogram Equalization (SSTHE)[4], Recursive Mean-Separate Histogram Equalization (RMSHE)[5], Recursive Sub-Image Histogram Equalization (RSIHE)[6], Contrast Stretching Recursively Separated Histogram Equalization (CSRSHE)[7], and Bi-Histogram Equalization with a Plateau Limit (BHEPL) [8].

Depending on the appearance of the original image histogram, dynamic range methods use the grayscale pixel distribution to determine the maximum and minimum values of the histogram in different areas. These values are fixed as segmentation points, from which sub-histograms of various sizes are derived. Finally, equalization is applied to each sub-histogram, individually. Well-known methods in this category include the Multi-peak Histogram Equalization with Brightness Preserving (MPHEBP) [9], Dynamic Histogram Equalization (DHE)[10], Brightness Preserving Dynamic Histogram Equalization (BPDHE)[11] and Adaptive Histogram Separation and Mapping (AHSM)[12]. In theory, the methods in this category are better able to appropriately segment and process different images, albeit at a higher computational cost.

The methods in the Spatial Processing class consider both the location of pixel points and pixel grayscale values. These processing methods can partially alleviate the problem of the over-sharpening that is often caused by conventional HE methods. Well-known methods in this category include Contrast Limited Adaptive Histogram Equalization (CLAHE)[13], Partially Overlapped Sub-Block Histogram Equalization (POSHE)[14] and Non-Overlapped Sub-blocks and local Histogram Projection (NOSHP)[15]. These methods involve the spatial segmentation of the original image into numerous, equally sized sections and the application of limited pixel interpolation to each section, in order to obtain new pixel values. Lastly, HE processing is applied to each section.

The methods used for the Reshaping-PDF class restrict or modify the PDF of the original image to derive a new image PDF and then use a transition function to obtain a new contrast-enhanced image. Well-known methods in this category include Histogram Equalization with Bin Underflow and Bin Overflow (BUBO)[16], Adaptively Modified Histogram Equalization (AMHE)[17], Weighted Thresholded Histogram Equalization (WTHE)[18], Adaptively Increasing the Value of the Histogram (AIVHE) [19] and the Weighted Histogram Equalization (WHE)[20]. A common feature of these methods is that they all provide the user with one or more modification parameters. Although parameter modification can be used to effectively adjust the degree of contrast enhancement, depending on user requirements, this also increases the operational complexity for the user.

The results of image processing using the contrast enhancement techniques described above often influence post-enhancement fidelity and quality. Therefore, the approach to assess the results of image quality and analyze the effectiveness of various contrast enhancement techniques presents an interesting research problem. This study collates a number of well-known techniques for the enhancement of image contrast and compares various types of subjective/objective quality assessment indicators. An integrated assessment index is then proposed, which provides a more objective and comprehensive method of evaluating the effectiveness of image enhancement techniques.

In Section 2, a quantitative evaluation method for the assessment of image quality based on several subjective and objective evaluation metrics is present. Section 3 describes the results of incorporating the method. Lastly, in Section 4, the works present in previous sections are briefly reviewed and summarized.

2   A Quantitative Evaluation Method

To combine the principles of the Subjective Fidelity Criteria and the Objective Fidelity Criteria to assess the effectiveness of various contrast enhancement techniques. The Subjective Fidelity Criteria method uses the mean opinion score (MOS) as the standard to produce assessment results, depending on the visual acuity of the observer. The Objective Fidelity Criteria method includes the features of absolute indicators and relative indicators, and conducts an integrated assessment of image quality with reference to four indicators: Relative Entropy Error (REE), Relative Contrast Error (RCE), Relative Mean Brightness Error (RMBE) and Relative Structural Similarity (RSS).

The assessment range of each indicator was set between 0 and 1. The results of numerous experiments, using various sample images, and cross-comparison analysis of the five main image assessment indicators were used to analyze the effectiveness and weaknesses of each technique used for the enhancement of image contrast.

The mean opinion score (MOS) is a measure of image assessment dependent on human visual acuity; a higher score indicates higher acuity. A mean opinion score (MOS) of 1.0 indicates that the image quality is flawless, fully achieving the aim of strengthening resolution while still retaining the content of the original image. A MOS of 0.8 indicates that the image quality is improved, but that the improvement is not obvious. A MOS of 0.6 indicates that the image quality is improved, but there are other consequent defects, such as over-sharpening, too much adjustment of the brightness, etc. A MOS of 0.4 indicates that the image quality is not significantly improved; A score of 0.2 implies that processing has lowered the image quality.

The calculation of REE involves the use of quantified entropy values to assess the degree of data enhancement between the contrast-enhanced image and the original image. The equation is shown as Eq. (1), where $E_{original}$ indicates the entropy of the original image and E_{new} indicates the entropy of the contrast-enhanced image. Because this study fixed the quantitative value of each assessment indicator between 0 and 1, calculating REE involves adding 0.5, to reach the required range. A value of REE greater than 0.5 indicates that the image data has been increased; On the contrary, an REE value of less than 0.5 indicates that image has deteriorated.

$$REE = \frac{E_{new} - E_{original}}{2 \times \log_2 L} + 0.5 \tag{1}$$

As shown in Eq. (2), the calculation for RCE involves the use of image standard deviation (Std.) to determine the degree of contrast enhancement between the original image and the enhanced image. The symbol, $Std_{original}$, indicates the contrast of the original image and the symbol, Std_{new}, indicates the contrast of the enhanced image. Since this study fixed the quantitative value of each assessment indicator between 0 and 1, the addition by constant 5 is added in the equation to reach the required range. When the image contrast has reached an optimal level, RCE = 1; If RCE equals 0.5, the contrast of the image has not been enhanced. When image contrast deteriorates, RCE is less than 0.5.

$$RCE = \frac{Std_{new} - Std_{original}}{L-1} + 0.5 \tag{2}$$

An important feature shown by RMBE is the effectiveness for maintaining the mean brightness of the original image when broadcasting continuous dynamic images. The equation of RMBE is defined in Eq. (3), where $M_{original}$ indicates the mean brightness of the original image and M_{new} indicates the mean brightness of the image to be measured. When the mean brightness of the contrast-enhanced image remains unchanged, RMBE=1, but the greater the difference in mean brightness between the original and the enhanced image, the closer RMBE is to zero.

$$RMBE = 1 - \frac{|M_{original} - M_{new}|}{L-1} \tag{3}$$

RSS uses the quantitative values of RMSE to determine the degree of structural similarity between the original image and the contrast-enhanced image. As shown in Eq. (4), the range of RSS also falls between 0 and 1. The greater the structural similarity between the images, the closer RSS is to 1 and greater structural difference between the images causes RSS to approach 0.

$$RSS = 1 - \frac{RMSE}{L-1} \tag{4}$$

Although various contrast enhancement techniques can enhance the contrast of most images, these methods also cause some images to be distorted. PIQI classifies images according to image characteristics and combines all possible image conditions to assess the effectiveness of various contrast enhancement methods. Fig. 1(a)-(d) show the original image samples used for this study. Fig. 1(a) shows an image type with lower mean brightness. The PIQI assesses the ability of an HE-based enhancement technique to maintain the mean brightness of the original image. Another concern related to this type of image is that after the image has undergone contrast enhancement, and some of the grayscale pixels in a darker state are redistributed to pixels in a lighter state, image detail can easily be lost in certain regions.

Fig. 1(b) shows an image with higher mean brightness. The observational points for this image are similar to those of the previous image. The difference is that when some of the brighter grayscale pixels are redistributed to pixels in a darker state, over-sharpening of image regions can easily occur. Fig. 1(c) shows an image with better contrast. In such case, the mean brightness is usually close to the medium brightness. This type of image (in which the original degree of contrast is good) is used to assess the contrast expansion and the ability to maintain image detail. Fig. 1(d) shows an image of a single subject with dark background. This type of image is chosen because most contrast enhancement methods use histograms to conduct equalized expansion and distribution for enhancing image contrast. However, such methods do not consider the relevance of the pixels to the image, which often results in noise being produced at the edges of the subject in the enhanced image. Insufficient adaptation capacity can then lead to unnatural results in the overall image. Therefore, when dealing with the type of image shown in Fig. 1(d), the adaptive modification capacity of each contrast enhancement technique will be verified.

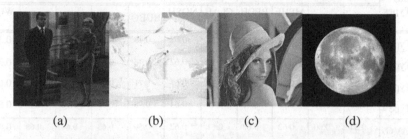

(a)   (b)   (c)   (d)

Fig. 1. Types of original images: (a) average brightness is low; (b) average brightness is high; (c) contrast is good; (d) most of the image comprises only the subject.

3   Evaluation Results

Table 1 shows the evaluation results by using the five indicators to assess the four types of images previously described. As shown in the MOS column of Table 1, by observing the Figures enhanced by various contrast enhancement methods of the 4 image samples, subjective assessment results were obtained by 119 students of Feng Chia University with a fundamental knowledge of image processing technology. On average, most of the methods introduced by Reshaping-PDF Class perform better than other methods. This is consistent with the evaluation results calculated by the average values of the five metrics. By observing the evaluating values produced by the Objective Fidelity Criteria methods, we can find that they all perform well but not best for most of issues except for RSS. This observation indicates that a good enhancement should not be designed by focusing on special topic of image quality. As a negative example, CLAHE is a famous and special technique that is designed to support the use on medical images. The excellent performance achieved from REE of view shows the enhanced figure will exhibit more detail of an image. However, it lost the quality on other aspects such as brightness and contrast effect.

The average REE results in Table 1 show that most of the techniques are capable of retaining the original level of data richness. However, the scores for BPDHE and WTHE were lower than those for other techniques. The main reason for this is that these two methods use brightness normalization to retain the original image brightness, which sacrifices image content to some degree. It is worthy to note that the REE result achieved by the Spatial Processing Class is much better than other methods.

The average RCE results in Table 1 show that the techniques of the Static Range provide better contrast enhancement. However, comparison with the results in the MOS column shows that the images enhanced with this type of technique were less acceptable to viewers conducting subjective visual assessment. On the contrary, the results of techniques in the Reshaping-PDF Class show that the contrast enhancement effects produced by the techniques that use adaptive modification were more visually acceptable.

Table 1. Results of indicators for the different types of sample images

Methods Image Types		Improved HE				Spatial Processing		Reshaping -PDF				
		BBHE	BHEPL	MPHE BP	BPDH E	CLAH E	NOSH P	BUBO	AMHE	WTHE	AIVHE	WHE
MOS	Type 1	0.50	0.50	0.84	0.83	0.28	0.62	0.46	0.60	0.68	0.71	0.33
	Type 2	0.50	0.44	0.74	0.51	0.47	0.62	0.44	0.45	0.46	0.53	0.38
	Type 3	0.59	0.69	0.36	0.74	0.79	0.61	0.71	0.70	0.69	0.62	0.67
	Type 4	0.58	0.77	0.53	0.40	0.49	0.64	0.78	0.75	0.72	0.74	0.59
	AVG.	0.54	0.60	0.62	0.62	0.51	0.62	0.59	**0.63**	**0.64**	**0.65**	0.49
REE	Type 1	0.50	0.50	0.49	0.48	0.58	0.60	0.50	0.50	0.45	0.50	0.50
	Type 2	0.49	0.50	0.49	0.49	0.63	0.58	0.50	0.50	0.46	0.50	0.50
	Type 3	0.49	0.50	0.49	0.49	0.51	0.52	0.50	0.50	0.50	0.50	0.50
	Type 4	0.49	0.50	0.49	0.48	0.60	0.59	0.50	0.50	0.50	0.50	0.50
	AVG.	0.49	0.50	0.49	0.48	**0.58**	**0.57**	0.50	0.50	0.48	0.50	0.50
RCE	Type 1	0.65	0.59	0.56	0.46	0.54	0.62	0.52	0.54	0.58	0.57	0.50
	Type 2	0.69	0.63	0.53	0.53	0.59	0.66	0.52	0.52	0.56	0.58	0.50
	Type 3	0.60	0.58	0.52	0.52	0.51	0.56	0.53	0.55	0.56	0.56	0.58
	Type 4	0.40	0.46	0.33	0.34	0.40	0.49	0.50	0.49	0.47	0.51	0.50
	AVG.	**0.58**	0.56	0.48	0.46	0.51	**0.58**	0.52	0.53	0.54	0.56	0.52
RMBE	Type 1	0.91	0.97	0.93	1.00	0.82	0.87	0.96	0.92	0.98	0.91	0.98
	Type 2	0.89	0.96	0.99	1.00	0.83	0.88	0.98	0.97	0.98	0.93	1.00
	Type 3	0.97	0.99	0.99	1.00	1.00	0.99	0.98	1.00	1.00	0.99	0.99
	Type 4	0.85	0.94	0.74	1.00	0.99	0.98	1.00	0.99	1.00	0.98	1.00
	AVG.	0.91	0.97	0.91	**1.00**	0.91	0.93	0.98	0.97	0.99	0.95	0.99
RSS	Type 1	0.80	0.89	0.90	0.92	0.81	0.80	0.95	0.90	0.90	0.87	0.97
	Type 2	0.77	0.87	0.95	0.90	0.79	0.79	0.96	0.96	0.93	0.89	1.00
	Type 3	0.89	0.92	0.97	0.91	0.89	0.93	0.96	0.94	0.94	0.93	0.91
	Type 4	0.79	0.91	0.67	0.83	0.86	0.89	0.98	0.99	0.94	0.97	1.00
	AVG.	0.82	0.90	0.87	0.89	0.84	0.85	**0.96**	**0.95**	**0.93**	**0.92**	**0.97**
Total of AVG		3.34	3.53	3.37	3.45	3.35	3.55	3.55	**3.58**	**3.58**	**3.58**	3.47

In general, techniques of the Reshaping-PDF Class have better RMBE capacity. When applied to different image types, conventional HE methods and static range techniques show unstable performance in RMBE. The BHEPL method, which uses a threshold level to prevent over-sharpening of the image, achieves more stable RMBE results. The RMBE values for WTHE and BPDHE utilizing brightness normalization are almost perfect.

The RSS results in Table 1 show that conventional HE techniques and static range methods always cause more damage to image structure. Because BHEPL compensates for the weaknesses of techniques in this category, it also improves the structural completeness of images. When applied to image type 4, dynamic range techniques induce more damage to the image structure; however, when applied to other image samples, these techniques achieve good RSS values. In conclusion, techniques in the Reshaping-PDF Class were more capable of retaining RSS than other techniques.

4   Conclusion

In recent years, many contrast enhancement techniques have been developed for various multimedia applications. However, the characteristics and performance of these techniques have not been accurately defined, due to the lack of a comprehensive evaluation standard. This research principally sought to determine the means by which quantitative quality assessment methods could be used to evaluate the effectiveness of contrast enhancement techniques.

To integrate conventional subjective/objective assessment methods, five subjective/objective indicators were defined for different aspects of image quality assessment for enhanced images. By classifying different type of enhancement methods and observing the enhanced effects of sample images created by various techniques. Some interesting characters introduced by the methods are observed.

The testing results are only the beginning for the study. By evaluating more contrast enhancement techniques and comparing the results among the various methods, some characters of methods will be figured out for the improvement of more mature techniques.

References

[1] Crane, R.: Simplified Approach to Image Processing. Prentice Hall, New Jersey (1994)
[2] Kim, Y.T.: Contrast enhancement using brightness preserving bi-histogram equalization. IEEE Trans. Consumer Electronics 43(1), 1–8 (1997)
[3] Wang, Y., Chen, Q., Zhang, B.: Image enhancement based on equal area dualistic sub-image histogram equalization method. IEEE Trans. Consumer Electronics 45(1), 68–75 (1999)
[4] Lin, P.H., Lin, C.C., Yen, H.C.: Tri-Histogram Equalization Based on First Order Statistics. In: IEEE 13th International Symposium on Consumer Electronics, Kyoto, pp. 387–391 (May 2009)

[5] Chen, S.D., Ramli, A.R.: Contrast enhancement using recursive mean-separate histogram equalization for scalable brightness preservation. IEEE Trans. Consumer Electronics 49(4), 1301–1309 (2003)

[6] Sim, K.S., Tso, C.P., Tan, Y.Y.: Recursive sub-image histogram equalization applied to gray scale images. Pattern Recognition Letters 28(10), 1209–1221 (2007)

[7] Jagatheeswari, P., Rajaram, M.: Enhanced image transmission with error control. In: Proc. of 2011 International Conference on Emerging Trends in Electrical and Computer Technology (ICETECT), Nagercoil, India, pp. 489–495 (March 2011)

[8] Chen, H.O., Kong, N.S.P., Ibrahim, H.: Bi-histogram equalization with a plateau limit for digital image enhancement. IEEE Trans. Consumer Electronics 55(4), 2072–2080 (2009)

[9] Wongsritong, K., Kittayaruasiriwat, K., Cheevasuvit, F., Dejhan, K., Somboonkaew, A.: Contrast Enhancement Using Multipeak Histogram Equalization With Brightness Preserving. In: The 1998 IEEE Asia-Pacific Conference on Circuits and Systems, Chiangmai, Thailand, pp. 455–458 (November 1998)

[10] Abdullah-Al-Wadud, M., Kabir, M.H., Dewan, M.A.A., Oksam, C.: A Dynamic Histogram Equalization for Image Contrast Enhancement. IEEE Trans. Consumer Electronics 53(2), 593–600 (2007)

[11] Ibrahim, H., Kong, N.S.P.: Brightness preserving dynamic histogram equalization for image contrast enhancement. IEEE Trans. Consumer Electronics 53(4), 1752–1758 (2007)

[12] Zhang, Q., Inaba, H., Kamata, S.: Adaptive Histogram Analysis for Image Enhancement. In: Proc of 2010 Fourth Pacific-Rim Symposium on Image and Video Technology (PSIVT), Singapore, pp. 408–413 (November 2010)

[13] Pizer, S.M., Amburn, E.P., Austin, J.D., et al.: Adaptive histogram equalization and its variations. Computer Vision, Graphics and Image Processing 39(3), 355–368 (1987)

[14] Kim, J.Y., Kim, L.S., Hwang, S.H.: An Advanced Contrast Enhancement Using Partially Overlapped Sub-block Histogram Equalization. IEEE Trans. Circuits and Systems for Video Technology 11(4), 475–484 (2001)

[15] Liu, B., Jin, W., Chen, Y., Liu, C.L., Li, L.: Contrast enhancement using non-overlapped sub-blocks and local histogram projection. IEEE Trans. Consumer Electronics 57(2), 583–588 (2011)

[16] Yang, S., Oh, J., Park, Y.: Contrast Enhancement Using Histogram Equalization with Bin Underflow and Bin Overflow. In: IEEE Proc. of 2003 International Conference on Image Processing, IEEE ICIP 2003, pp. 881–884 (September 2003)

[17] Kim, H.-J., Lee, J.-M., Lee, J.-A., Oh, S.-G., Kim, W.-Y.: Contrast Enhancement Using Adaptively Modified Histogram Equalization. In: Chang, L.-W., Lie, W.-N. (eds.) PSIVT 2006. LNCS, vol. 4319, pp. 1150–1158. Springer, Heidelberg (2006)

[18] Wang, Q., Ward, R.K.: Fast Image/Video Contrast Enhancement Based on Weighted Thresholded Histogram Equalization. IEEE Trans. Consumer Electronics 53(2), 757–764 (2007)

[19] He, K.J., Chen, K.Y., Wang, L.: A Contrast Enhancement Method for video images. In: Proc. 2010 Cross-Strait Conference on Information Science and Technology, Qinhuangdao, China, pp. 21–24 (July 2010)

[20] Yun, S.H., Kim, S., Kim, J.H., Rajeesh, J.: Contrast Enhancement using a Weighted Histogram Equalization. In: 2011 IEEE International Conference on Consumer Electronics (ICCE), Las Vegas, USA, pp. 203–204 (January 2011)

Significance-Preserving-Guided Content-Aware Image Retargeting[*]

Yu-Hsien Sung[1], Wen-Yu Tseng[1], Pao-Hung Lin[1], Li-Wei Kang[2],
Chih-Yang Lin[3,**], and Chia-Hung Yeh[1]

[1] Department of Electrical Engineering, National Sun Yat-sen University
Kaohsiung, Taiwan
facetoface9999@gmail.com
{d983010033,m983010117}@student.nsysu.edu.tw
yeh@mail.ee.nsysu.edu.tw
[2] Institute of Information Science, Academia Sinica
Taipei, Taiwan
lwkang@iis.sinica.edu.tw
[3] Department of Computer Science and Information Engineering, Asia University
Taichung, Taiwan
andrewlin@asia.edu.tw

Abstract. With the rapid development of multimedia and network technologies, sharing image contents through heterogeneous devices of different capabilities has been popular. A variety of displays provide different display capabilities ranging from high-resolution computer/TV monitors to low-resolution mobile devices, where images are usually required to be changed in size or aspect ratio to adapt to different screens. Based on the fact that straightforward image resizing operators (e.g., uniform scaling) cannot usually produce satisfactory results, content-aware image retargeting, which aims to arbitrarily change image size while preserving visually prominent features, has been a popular research topic. In this paper, we present a robust and computationally-efficient content-aware image retargeting framework based on seam carving subject to gradient energy and saliency-preserving constraint. In the proposed method, the significance map derived from adaptively integrating both the gradient and saliency maps of an image is used to accurately identify the most important area(s) to be preserved while retargeting this image. The proposed significance map can well compensate the drawbacks induced by only either gradient-based or saliency-based map is used. As a result, an image can be flexibly adapted to arbitrary sizes. Experimental results demonstrate the efficacy of the proposed algorithm.

1   Introduction

With the rapid development of multimedia and network technologies, sharing multimedia contents through heterogeneous devices of different capabilities has become

---

[*] This work was supported in part by the National Science Council, Taiwan, under Grants NSC101-2221-E-110-093-MY2 and NSC100-2218-E-001-007-MY3.
[**] Corresponding author.

more and more popular. An emerging field, multimedia content adaptation, offers a rich body of knowledge and techniques handling various resource constrains (e.g., display capability, bandwidth, processing speed, and power consumption) of different devices. In this paper, we consider sharing or exchanging image contents among devices (e.g., handheld devices, TV, or computer monitor) with different display capabilities, as examples shown in Fig. 1. The display size of a handheld device is typically much smaller than that of TV and the aspect ratio of a film is usually different for different types of device. Adaptive image scaling is therefore required to adapt visual content to fit different display formats. In addition, dynamically changing the layout of web pages in browsers should also take into account the distribution of the images in them, which should be adaptively resized for better visual quality [1]. Traditional image down-sampling approach (e.g., uniform down-scaling) usually makes the object of focus in an image too small or out of proportion, while traditional up-sampling approach (e.g., uniform up-scaling or interpolation) often makes images overly blurry, lacking important details. In this paper, we focus on that the target size is smaller than the original size of an image.

Fig. 1. Examples of image retargeting among devices with different display capabilities

As examples shown in Fig. 2, straightforward image resizing operators (e.g., uniform scaling or fixed-window cropping) cannot usually produce satisfactory results for such diversity of image devices. Hence, a new challenge has been introduced that image should be adaptively and optimally resized to fit different display conditions or applications. Content-aware image retargeting [1]-[18], which aims to arbitrarily change image size while preserving visually prominent features, has been therefore a popular research topic recently.

In most recent image retargeting or resizing works, the three main objectives for retargeting are defined as [1]: (i) the important content of an image should be preserved in its retargeted version; (ii) the important structure of an image should be preserved in its retargeted version; and (iii) the retargeted version should be free of visual artifacts. To meet the above-mentioned three requirements, the majority of image retargeting techniques follows a similar process. The first step is usually the computation of an importance map of an image, which quantifies the importance of each pixel in the image. The importance map of an image can be usually obtained by calculating the gradient energy of the image or performing saliency detection [19]-[20] for the image. Finally, given the importance map of an image and the constraints induced by preserving the structure or visual artifacts-free, a retargeting operator is performed to the image, changing its size while taking into account the importance map and the constraints. In the following two subsections, we briefly review the most popular content-aware image resizing approach, called the seam carving technique [4] and other recent image resizing approaches, and give an overview of the proposed image retargeting framework, respectively.

(a)

(b)   (c)

Fig. 2. Examples of the straightforward image resizing operators for resizing an image: (a) the original image; and the resized versions of (a) via (b) uniform scaling; and (c) fixed-window cropping operators, respectively.

1.1   Overview of Recent Approaches in Content-Aware Image Retargeting

Seam carving technique proposed in [4] is a popular approach for content-aware image retargeting, where the key idea is to decrease the image width (or height) one pixel at a time, by removing a seam of minimal importance. As an example illustrated in Fig. 3 [4], a seam is defined as an optimal eight-connected path of pixels in an image from top to bottom (or left to right) that contains only one pixel per row (or column), where the optimality is defined by an image energy function. The optimal seams are then computed using dynamic programming. By repeatedly carving out or inserting seams in one direction, one can change the aspect ratio of an image, while by applying these operators in both directions, one can retarget the image to a new size. After removing a seam from an image, the image is then readjusted by shifting pixels left or up to compensate for the removed seam. The image changes only at the seam region, while the other regions remain the same. One of the main issues here is the selection and order of seams to protect the important image content, which is defined by the employed energy function. In [4], it is observed that using the gradient energy function to derive the importance map (gradient map) of an image can usually give satisfactory results, but other importance measures could be used, such as saliency map [19]-[20].

Fig. 3. An example [4] illustrating the seams to be removed in an image (to be resized), decided by the seam carving algorithm [4]

Based on the seam carving technique proposed in [4], several improvements have been also proposed recently [6]-[7], [11], [13], [14], [15], [20]. More specifically, some works [7], [13], [14], [20] show that using grayscale intensity gradient map for identifying the important content for an image in the seam carving process [4] may suffer from some drawbacks. That is, the gradient map usually shows higher energy only at edges of objects, is sensitive to noise, and may result in deforming salient objects. Hence, the saliency map derived from saliency detection techniques [14], [19], [20] has been shown to outperform the gradient map in seam carving application [7], [13], [20].

Other recent image retargeting techniques include content-aware cropping [17]-[18], important details-preserving scaling [16], [21], rapid serial visual presentation (RSVP) [3], segmentation-based approach [2], warping-based approach [5], [8], [9]-[10], [12], patch-based approach [22], and multi-operator-based approach [6], [11], [13], [23]. More comprehensive survey of content-aware image retargeting can be found in [1].

1.2   Overview of Proposed Content-Aware Image Retargeting Framework

In this paper, we present a significance-preserving-guided content-aware image retargeting framework based on seam carving. In our framework, we propose a significance map, characterizing the visual attractiveness of each pixel, obtained by integrating gradient map and saliency map to improve the original seam carving algorithm using only gradient map and other extensions using only saliency map. Based on our significance map derived for an image to be retargeted, we can more accurately detect the important area(s) in the image to prevent the area(s) from being carved out in performing seam carving. The major contribution of this paper is to propose a principle to properly integrate the gradient and saliency maps based on energy projection and image complexity analysis to accurately detect the importance area(s) for an image. The rest of this paper is organized as follows. In Sec. 2, we present the proposed content-aware image retargeting framework. In Sec. 3, experimental results are demonstrated. Finally, Sec. 4 concludes this paper.

2 Proposed Content-Aware Image Retargeting Framework

In the proposed image retargeting framework, for an input image to be retargeted, we first compute both the gradient map and the saliency map for the image, followed by integrating them and identifying the important area(s) of the image. We then perform the seam carving algorithm [4] to resize the image to the desired image size while preserving the identified important area(s).

2.1 Problem Formulation and Significance Detection

In this paper, we formulate the problem of content-aware image retargeting as a significance-preserving problem described as follows. Given an image I to be retargeted of size $m \times n$ and a target image size of $m' \times n'$, where $m' \times n' \leq m \times n$, the goal is to produce a novel image I' of size $m' \times n'$, where I' is a good representative of I, subject to that the significance of I should be well preserved in I'.

To perform significance map detection for an image $I(x, y)$, $1 \leq x \leq m$, $1 \leq y \leq n$, we first calculate the gradient map $Grad_I(x, y)$ for I as:

$$Grad_I(x,y) = \left| \frac{\partial}{\partial x} I(x,y) \right| + \left| \frac{\partial}{\partial y} I(x,y) \right|, \tag{1}$$

where $Grad_I(x, y)$ denotes the gradient energy value of the (x, y)-th pixel, as an example shown in Fig. 4(b).

Moreover, we also calculate the saliency map $Sali_I(x, y)$ for I based on the saliency-based visual attention model proposed in [19], as an example shown in Fig. 4(c). The visual attention model [19] computes a saliency map topographically encoding for saliency at each location in I that simulates which elements of a visual scene are likely to attract the attention of human observers.

Finally, to overcome the above-mentioned drawbacks of the gradient map and improve the saliency map, we define the significance map $S_I(x, y)$ of the image $I(x, y)$ as the weighted combination of the gradient map $Grad_I(x, y)$ and the saliency map $Sali_I(x, y)$ of I as:

$$S_I(x,y) = W_G \times Grad_I(x,y) + W_S \times Sali_I(x,y), \tag{2}$$

where W_G and W_S are the weights of the gradient map and the saliency map, respectively, as an example shown in Fig. 4(d). The proposed significance map characterizes the visual attractiveness of each pixel and is less sensitive to noise. After determining the significance map for an image, we will identify the important area(s) in the image, followed by performing seaming carving to achieve significance-preserving retargeting, described in Sec. 2.2.

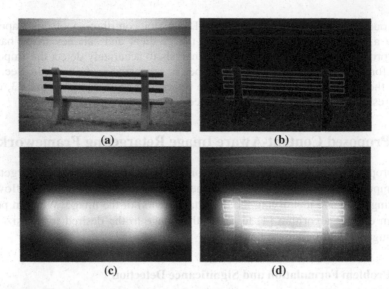

(a)
(b)
(c)
(d)

Fig. 4. An example of the significance map for an image: (**a**) the original image; (**b**) the gradient map of (a); (**c**) the saliency map of (a); and (**d**) the significance map of (a).

2.2 Important Area(s) Identification and Image Retargeting via Seam Carving

Based on the fact that the seam carving algorithm [4] decreases the image width (or height) one pixel at a time by removing a seam with minimal energy, to protect the important area(s) in the image I, we identify the important area(s) of I as follows. We first calculate the projects in both vertical and horizontal directions of the significance map of I as:

$$P_{ver}(y) = \sum_{x=1}^{m} S_I(x,y), 1 \leq y \leq n,\tag{3}$$

$$P_{Hor}(x) = \sum_{y=1}^{n} S_I(x,y), 1 \leq x \leq m.\tag{4}$$

Then, we calculate the mean value for each projection direction as:

$$\lambda_{ver} = \frac{1}{n}\sum_{y=1}^{n} P_{ver}(y),\tag{5}$$

$$\lambda_{Hor} = \frac{1}{m}\sum_{x=1}^{m} P_{Hor}(x).\tag{6}$$

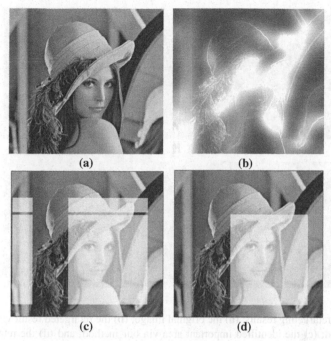

(a)   (b)

(c)   (d)

Fig. 5. An example for identifying the important area(s) for an image: **(a)** the original image; **(b)** the significance map of (a); **(c)** the initially detected important areas (green regions) of (a); and **(d)** the finally detected important area (green region) of (a).

We then identify each (x, y)-th pixel in I with significance map value greater than λ_{ver} or λ_{Hor}, i.e., $S_I(x, y) \geq \lambda_{ver}$ or λ_{Hor}, to be important pixel to find the important area(s) in I. If several important areas are identified in I, we analyze the complexity for each area by detecting its edge information, and determine the most important area(s), as an example shown in Fig. 5. Finally, we apply the seam carving algorithm to I to obtain I' of size $m' \times n'$, while preserving the identified important area(s) to achieve significance-preserving content-aware image retargeting for I.

3   Experimental Results

To evaluate the performance of the proposed content-aware image retargeting method, we conducted the experiments on several natural images by retargeting each image into arbitrary sizes. We also compared our method with the baseline approach, i.e., the uniform scaling method. Figs. 6-9 show the image retargeting results obtained by the uniform scaling and our methods, as well as the important area(s) identified by our method. It can be found from Figs. 6-8 that the important area for each image can be usually accurately identified by our method, which can be beneficial to produce visually pleasing retargeting results via seam carving. However, the uniform scaling method cannot provide acceptable results in most cases, where the main problem is serious object deformation, especially in the situation that the aspect ratio is changed.

Fig. 6. Image retargeting results: **(a)** the original image; **(b)** the retargeted result via the uniform scaling method; **(c)** the identified important area via our method; and **(d)** the retargeted result via our method.

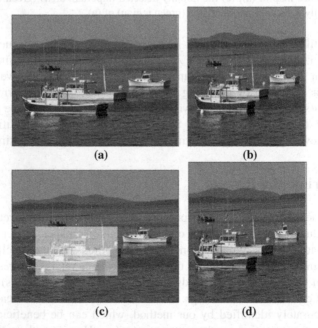

Fig. 7. Image retargeting results: **(a)** the original image; **(b)** the retargeted result via the uniform scaling method; **(c)** the identified important area via our method; and **(d)** the retargeted result via our method.

Fig. 8. Image retargeting results: (a) the original image; (b) the retargeted result via the uniform scaling method; (c) the identified important area via our method; and (d) the retargeted result via our method.

Fig. 9. Image retargeting results: (a) the original image; and (b) the result via our method

4   Conclusions

In this paper, we have proposed a novel content-aware image retargeting framework, where we propose to derive the significance map for an image by integrating its gradient and saliency maps. Then, we can accurately identify the important area(s) of the image. As a result, when performing the seam carving algorithm for image retargeting, the important area(s) can be well preserved in the retargeted result. Our experimental results have shown that the proposed method achieves better performance than the baseline algorithm. For future works, we will extend our image-based retargeting method to video retargeting and investigate possible applications of retargeting techniques for video coding, especially for content-adaptive spatial scalability in scalable video coding (SVC).

References

1. Vaquero, D., Turk, M., Pulli, K., Tico, M., Gelfand, N.: A Survey of Image Retargeting Techniques. In: Proc. SPIE, vol. 7798, pp. 779–814 (2010)
2. Setlur, V., Takagi, S., Raskar, R., Gleicher, M., Gooch, B.: Automatic Image Retargeting. In: Proc. ACM Int. Conf. on Mobile and Ubiquitous Multimedia, pp. 59–68 (2005)
3. Liu, H., Jiang, S., Huang, Q., Xu, C., Gao, W.: Region-based Visual Attention Analysis with Its Application in Image Browsing on Small Displays. In: Proc. ACM Int. Conf. on Multimedia, pp. 305–308 (2007)
4. Avidan, S., Shamir, A.: Seam Carving for Content-aware Image Resizing. ACM Trans. on Graphics 26 (2007)
5. Wang, Y.S., Tai, C.L., Sorkine, O., Lee, T.Y.: Optimized Scale-and-stretch for Image Resizing. ACM Trans. on Graphics 27 (2008)
6. Hwang, D.S., Chien, S.Y.: Content-aware Image Resizing Using Perceptual Seam Carving with Human Attention Model. In: IEEE Int. Conf. on Multimedia and Expo., pp. 1029–1032 (2008)
7. Achanta, R., Susstrunk, S.: Saliency Detection for Content-aware Image Resizing. In: Proc. IEEE Intl. Conf. on Image Processing (2009)
8. Wang, S.F., Lai, S.H.: Fast Structure-preserving Image Retargeting. In: Proc. IEEE Int. Conf. on Acoustics, Speech and Signal Processing, pp. 1049–1052 (2009)
9. Ren, T., Liu, Y., Wu, G.: Image Retargeting based on Global Energy Optimization. In: Proc. IEEE Int. Conf. on Multimedia and Expo., pp. 406–409 (2009)
10. Ren, T., Liu, Y., Wu, G.: Image Retargeting Using Multi-map Constrained Region Warping. In: Proc. ACM Int. Conf. on Multimedia, pp. 853–856 (2009)
11. Dong, W., Zhou, N., Paul, J.C., Zhang, X.: Optimized Image Resizing Using Seam Carving and Scaling. ACM Trans. on Graphics 28, 1–10 (2009)
12. Guo, Y., Liu, F., Shi, J., Zhou, Z.-H., Gleicher, M.: Image Retargeting Using Mesh Parametrization. IEEE Trans. on Multimedia 11, 856–867 (2009)
13. Liu, Z., Yan, H., Shen, L., Ngan, K.N., Zhang, Z.: Adaptive Image Retargeting Using Saliency-based Continuous Seam Carving. Optical Engineering 49 (2010)
14. Fang, Y., Lin, W., Chen, Z., Lin, C.W.: Saliency Detection in the Compressed Domain for Adaptive Image Retargeting. IEEE Trans. Image Processing 21, 3888– –3901 (2012)
15. Rubinstein, M., Shamir, A., Avidan, S.: Improved Seam Carving for Video Retargeting. ACM Trans. on Graphics 27, 1–9 (2008)
16. Yen, T.C., Tsai, C.M., Lin, C.W.: Maintaining Temporal Coherence in Video Retargeting Using Mosaic-guided Scaling. IEEE Trans. Image Processing 20, 2339– –2351 (2011)
17. Ciocca, G., Cusano, C., Gasparini, F., Schettini, R.: Self-adaptive Image cropping for Small Displays. IEEE Trans. on Consumer Electronics 53, 1622–1627 (2007)
18. Nishiyama, M., Okabe, T., Sato, Y., Sato, I.: Sensation-based Photo Cropping. In: Proc. ACM Int. Conf. on Multimedia, pp. 669–672 (2009)
19. Itti, L., Koch, C., Niebur, E.: A Model of Saliency-based Visual Attention for Rapid Scene Analysis. IEEE Trans. on Pattern Analysis and Machine Intelligence 20, 1254–1259 (1998)
20. Goferman, S., Zelnik-Manor, L., Tal, A.: Context-aware Saliency Detection. In: IEEE Conf. on Computer Vision and Pattern Recognition (2010)
21. Munoz, A., Blu, T., Unser, M.: Least-squares Image Resizing Using finite differences. IEEE Trans. on Image Processing 10, 1365–1378 (2001)
22. Barnes, C., Shechtman, E., Finkelstein, A., Goldman, D.: Patch Match: a Randomized Correspondence Algorithm for Structural Image Editing. ACM Trans. on Graphics 28 (2009)
23. Rubinstein, M., Shamir, A., Avidan, S.: Multi-operator Media Retargeting. ACM Trans. on Graphics 28, 1–11 (2009)

Identifying Device Brand by Using Characteristics of Color Filter Array

Tang-You Chang[1], Guo-Shiang Lin[2], and Shen-Chuan Tai[3]

[1,3] Institute of Computer and Communication Engineering,
National Cheng Kung University
e2490668@gmail.com

[2] Department of Computer Science and Information Engineering, Da-Yeh University
khlin@mail.dyu.edu.tw

Abstract. In this paper, we propose a passive scheme for photorealistic computer generated image (PRCG) identification and device brand classification. To this end, a periodic phenomenon of variance of pixel values resulting from demosaicing process is analyzed in RGB color space. Based on the phenomenon, the proposed scheme is composed of color space transformation, high-pass filtering, feature extraction, and classification. To make the phenomenon obvious, a high-pass filter is performed. The periodic phenomenons in different directions are measured as features in the Fourier domain and combined with a hierarchical classifier for PRCG detection and brand identification simultaneously. Experimental results show that our proposed scheme can not only detect photographic images (PIM) but also determine what brand of a camera is used.

Keywords: Color filter array; De-mosaicing algorithm; Brand identification.

1 Introduction

In past few years, a large number of digital image content has been generated by using low-cost multimedia devices, e.g., digital camera, cellphone camera, and camcorder. As we know, digital content has become easier to copy and manipulate such content without degrading the quality by using digital processing tools. Moreover, some tools are able to create photorealistic computer generated images (PRCG). So it is expected that the information such as image origin and acquisition source is useful for image's legitimacy and copyright. For example, the authenticity of an image can be determined by identifying whether it is a photographic image (PIM) or not. In addition, the source type of an image is also useful in determining its authenticity. Therefore, we aim to devise a method which is not only determine whether an image is PIM but also identify which device is used to image acquisition.

There are some methods [4],[6] proposed for distinguishing PRCG from PIM images. For example, the properties of the demosaicing process was analyzed and extracted as features. The extracted features were combined with a simple classifier to perform PRCG detection.

J.-S. Pan et al. (Eds.): *Advances in Intelligent Systems & Applications*, SIST 21, pp. 349–358.
DOI: 10.1007/978-3-642-35473-1_35   © Springer-Verlag Berlin Heidelberg 2013

As for device brand classification, some methods [1],[2] was proposed. Swamina-than et al. [1] analyze color interpolation coefficients and image noise for feature extraction and then determine the source of each image, phone camera, scanner or PRCG. Kang et al. [2] proposed a method to achieve source camera identification. In the proposed method [2], sensor pattern noise was extracted from images as unique information for source camera identification.

As mentioned in [1], Color filter array (CFA) is an important component in digital cameras. Due to nature of CFA, a demosaicing process is necessary to produce a color image. However, the demosaicing process is not necessary to generate a PRCG im-age. Moreover, different CFAs are used in different brand cameras. Therefore, it mo-tivates us to analyze characteristics of CFA to devise a method which is able to achieve PIM detection and brand identification simultaneously. Figure 1 illustrates the proposed passive method. Based on the previous work [6], we focus on device brand identification here.

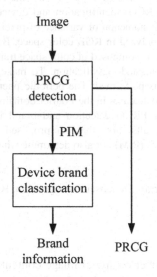

Fig. 1. Block diagram of the proposed method

2   Proposed Method

In [6], a phenomenon resulting from the demosaicing process in cameras was ana-lyzed. Since interpolation is used to generate all of pixel values in RGB color space in a demosaicing algorithm, the phenomenon that periodic signals exist in the output image after demosaicing was observed. Figure 2 shows Fourier spectrums of variance of pixel values in diagonal, vertical, and horizontal directions for PIM and PRCG images. These PIM images are captured by using three popular brand cameras, Ca-non, Nikon, and Sony. As we can see in Fig. 2(a), 2(b), and 2(c), local peaks may exist in some frequencies for PIM images and no local peaks occur for PRCG images.

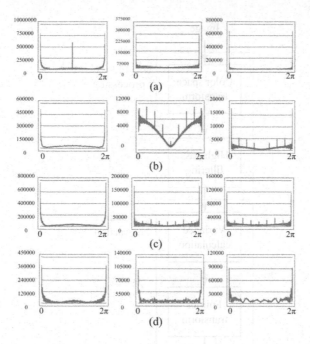

Fig. 2. Fourier spectrums of variance of pixel values in diagonal, vertical, and horizontal direction: (a) CANON, (b) NIKON, (c) SONY (d) PRCG

In addition, the positions of peaks for Canon are obviously different from the others. These phenomenons can be measured as useful information to achieve our goal.

In [6], we proposed a feature-based scheme to distinguish PIM images from PRCG images based on properties of CFA. Referring to [6], we also develop a feature-based method to achieve device brand identification. Figure 3 illustrates the block diagram of preprocessing and feature extraction in the proposed scheme. We introduce each part of pre-processing and feature extraction in the following.

1. Color Space Transformation

Since the periodic phenomenon exists in RGB color space, we need to perform color space transform before further processes if necessary.

2. High-Pass Filtering

According to the analysis in [4], the variance of pixel values for interpolated pixels is less than that of non-interpolated pixels. To enhance the phenomenon, a high-pass filter is adopted. Here we use the green channel as an example. Assuming that the pixel value is $I^G(x, y)$ at the (x, y) coordinate in green channel. After high pass filtering, the pixel value $I^{Gh}(x, y)$ is then expressed as follows:

$$I^{Gh}(x,y) = I^G(x-1,y) + I^G(x+1,y) + I^G(x,y-1) + I^G(x,y+1) - 4I^G(x,y), \qquad (1)$$

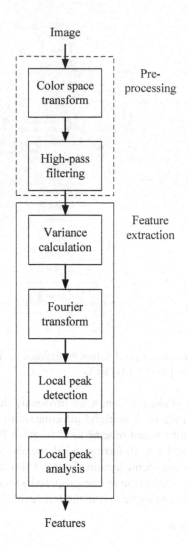

Fig. 3. Block diagram of pre-processing and feature extraction

3. Feature Extraction

As shown in Fig.3, feature extraction is composed of four parts: variance calculation, Fourier transform, local peak detection, local peak analysis. We explain each part in the following.

- **Variance Calculation**

  To deal with different kinds of CFAs, we calculate the variance of pixel values in three directions, vertical, horizontal, and diagonal. In addition, the periodic phenomenon may exist in each channel of RGB color space. We perform variance calculation in three directions of each channel in RGB color space. Figure 4 is an example of variance calculation along the vertical direction in the green color channel. Then $\sigma_R^V(i)$ can be obtained, where $\sigma_R^V(i)$ denote the variance of pixel values in the i-th column of green channel. As we can see in Fig. 4, a vector σ_R^H of variance of pixel values along the vertical direction can be obtained after variance calculation, i.e., $\sigma_R^H = [\sigma_R^V(1)\ \sigma_R^V(2)\ \sigma_R^V(3)...]$. In the same manner, we can obtain a total of 9 vectors, σ_j^H, σ_j^V, σ_j^D, $j \in \{R,G,B\}$, where σ_j^H, σ_j^V, and σ_j^D denotes the vectors of pixel values in the horizontal, vertical, and diagonal directions, respectively.

- Fourier Transform

  As we can see in Fig. 4, there are periodic patterns in these CFAs. It is expected that periodic components may exist in the vectors of variance of pixel values. To analyze and extract the periodic components in the vectors of variance of pixel values, Fourier transform (FT) is adopted here.

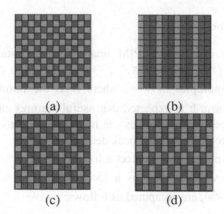

Fig. 4. Examples of CFA

Assume Ψ_j^H, $j \in \{R,G,B\}$, denotes the energy density spectrum (EDS) of σ_j^H and can be obtained by

$$\sigma_j^H \xrightarrow{\ F\ } \Psi_j^H \tag{2}$$

$$\left|\Psi_j^H\right| = \mathrm{Re}^2\left(\Psi_j^H\right) + \mathrm{Im}^2\left(\Psi_j^H\right) \tag{3}$$

where $\mathrm{Re}(\cdot)$ and $\mathrm{Im}(\cdot)$ denote the real and imaginary parts of a complex number respectively; F denotes Fourier transform. After Fourier transform, we can then obtain $\left|\Psi_j^H\right|$, $\left|\Psi_j^V\right|$, and $\left|\Psi_j^D\right|$, $j \in \{R,G,B\}$.

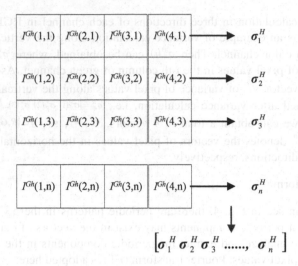

Fig. 5. Illustration of variance calculation in the vertical direction

- **Local Peak Detection**

    Based on our observation, PIM images often contain the local peaks at the multiples ($(1/4)\pi$, $(2/4)\pi$, $(3/4)\pi$, π, $(5/4)\pi$, $(6/4)\pi$, $(7/4)\pi$) of $(1/4)\pi$ in the Fourier domain, i.e., multiples of $(N/8)$, where N is the number of points for Fast Fourier transform. It is expected that useful features can be extracted if local peaks at the specific frequencies can be detected. Thus, before feature extraction, we need to perform local peak detection.

    Here we explain how to detect a local peak at $(N/8)$ in $\left|\Psi_j^D\right|$ as an example. To determine whether there is a local peak at $(N/8)$, three values, $D_1(N/8)$, $D_2(N/8)$, and $D_3(N/8)$, are computed as follows:

$$D_1(N/8) = \delta\left(\left(\max_{i \in \left[\frac{N}{8}-5, \frac{N}{8}+5\right]}\left|\Psi_j^H\right|(i)\right) - \left|\Psi_j^H\right|(i)\right) \tag{4}$$

$$D_2(N/8) = U\left(\frac{\left|\Psi_j^H\right|\left(\frac{N}{8}\right) - \left|\Psi_j^H\right|\left(\frac{N}{8}+i\right)}{\left|\Psi_j^H\right|\left(\frac{N}{8}\right)} - T_2\right), \quad i = \pm 1 \tag{5}$$

$$D_3(N/8) = U\left(\frac{\left|\Psi_j^H\right|\left(\frac{N}{8}\right) - \left|\Psi_j^H\right|\left(\frac{N}{8}+i\right)}{\left|\Psi_j^H\right|\left(\frac{N}{8}\right)} - T_3\right), \quad i = \pm 2 \tag{6}$$

where $\delta(\cdot)$ denotes the impulse function ($\delta(x)=1$ for $x=0$ and $\delta(x)=0$ for $x \neq 0$); $U(\cdot)$ denotes the unit step function ($U(x)=1$ for $x>0$ and $U(x)=0$ for $x \leq 0$). T_2 and T_3 are pre-defined thresholds and set to 0.5 and 0.4, respectively.

After obtaining $D_1(N/8)$, $D_2(N/8)$, and $D_3(N/8)$, a decision rule for determining whether there is a local peak at $(N/8)$ can be expressed as follows:

$$\text{local peak exists if } D_1(N/8) \cdot D_2(N/8) \cdot D_3(N/8) = 1.$$

In the same manner, local peaks at other frequencies can be detected and local peak detection is performed in each direction of each channel. Thus, after local peak detection, each channel of the input image has 3 binary sequences, s_j^H, s_j^V, and s_j^p, $j \in \{R,G,B\}$. Each binary sequence contains seven components and each component in the binary sequence indicates whether a local peak at a specific frequency exists or not.

- Local Peak Analysis

    After obtaining s_j^H, s_j^V, and s_j^p, $j \in \{R,G,B\}$, a local peak analysis is performed to extract useful features for distinguishing PIM and CG images. For a given image I^t and the corresponding sequences, s_j^H, s_j^V, and s_j^p, $j \in \{R,G,B\}$, the features $c_i(\cdot)$ (i=1,2,3) is measured according to the following rules:

    (i) For s_j^p, if there is a peak at π, $c_1(I^t)$ adds 2. For s_j^H and s_j^V, if there is a peak at π, increase $c_1(I^t)$ by 1.

    (ii) For s_j^H, s_j^V, and s_j^p, there are seven peaks at the specific frequencies, increase $c_2(I^t)$ by 1.

    (iii) For s_j^H, s_j^V, and s_j^p, there are seven peaks at the specific frequencies except π, increase $c_3(I^t)$ by 1.

    After obtaining features, $c_1(I^t)$, $c_2(I^t)$ and $c_3(I^t)$, a classifier is used to achieve not only determine whether an image is PIM but also identify which device is used to image acquisition.

- Device Brand Classification

    After measuring the features, we extract the properties of features to determine the information of camera brand. Currently we develop a hierarchical classification algorithm to distinguish PIM from PRCG and determine which brand camera is used. In the hierarchical algorithm, we first detect PRCG images and then perform brand identification. According to our experiences, images captured by Canon cameras have different properties compared with those captured by Nikon and Sony cameras. For brand identification, we first decide whether the brand is Canon and then a further analysis is used to distinguish Nikon from Sony.

    The steps of the hierarchical classification algorithm for PRCG detection and brand identification can be expressed as follows:

(i) If $\displaystyle\sum_{i=1,2,3} C_i(I^t) < T_4$, I^t is PRCG.

(ii) If $\displaystyle\sum_{i=1,2,3} C_i(I^t) > T_4$ and $C_1(I^t) > T_5$, I^t comes from a Canon camera.

(iii) If $\displaystyle\sum_{i=1,2,3} C_i(I^t) > T_4$ and $C_1(I^t) < T_5$, I^t is captured by using a Nikon or
     Sony camera.

where the pre-defined thresholds, T_4 and T_5, are 5 and 4, respectively. For determining an image from Nikon or Sony cameras, a further decision rule is described below:

If $C_1(I^t) > C_2(I^t)$, I^t the device brand is Nikon; otherwise, Sony.

3   Experimental Results

3.1   Preparation of Test Images

To evaluate the performance of the proposed algorithm, we collect many PIM and PRCG images for testing. These test PIM images are captured from three popular brands of digital cameras, Sony, Canon, and Nikon. The PRCG images come from Columbia's ADVENT data set [7], and two well-known computer graphic web sites: [8] and [9]. In our experiment, there are a total of 400 JPEG images for testing: 100 for each camera brands and 100 for PRCG. The size of each PIM image is 3000×2000 pixels. The average resolution of each PRCG image is 800×600 pixels, which is the common size on web sites.

3.2   Performance Analysis

First, we evaluate the performance of the proposed method for PRCG detection. The detection and false alarm rates of our proposed method for PRCG detection are 91% and 8%, respectively. The results show that the proposed scheme can achieve PRCG detection.

After PRCG detection, we evaluate the performance of the proposed method for brand identification. The result of the first-level brand identification is shown in Table 1. In the first-level classification, the classification rates for "Canon" and "Sony & Nikon" are 95% and 91.5%, respectively. The result shows that the proposed scheme can function well in the first-level classification, i.e., distinguish Canon from Sony and Nikon.

Table 1. The first-level classification rates of the proposed method for brand identification

	Canon	Sony & Nikon
Canon	95%	2%
Sony & Nikon	6%	91.5%

The second-level classification result is shown in Table 2. As we can see in Table 2, only 55% of the test images from Sony cameras can be correctly classified and 39% are mis-classified as Nikon. The result shows that the performance of our method for distinguish Nikon from Sony is not good enough. The main reason is that the feature for distinguishing Nikon from Sony is not effective enough.

Table 2. The second-level classification rates of the proposed method

	Canon	Sony	Nikon	PRCG
Canon	95%	0%	2%	3%
Sony	1%	55%	39|%	5%
Nikon	12%	10%	78%	0%
PRCG	2%	7%	0%	91%

4   Conclusion

In this paper, we propose a passive scheme for PRCG detection and device brand classification. A periodic phenomenon of variance of pixel values resulting from demosaicing process exists only in PIM images. To analyze the phenomenon, the proposed scheme is composed of color space transformation, high-pass filtering, feature extraction, and classification. To make the phenomenon obvious, a high-pass filter is performed. The periodic phenomenons in different directions are measured as features in the Fourier domain feature. Based on the features, a hierarchical classifier is used for PRCG detection and device brand classification.

To evaluate the performance, a lot of PIM and PRCG images are collected for testing. The PRCG detection of the proposed method is 91%. For brand identification, the classification rate of the proposed method is 95% to distinguishing Canon from Sony and Nikon. Experimental results show that our proposed scheme can not only detect PRCG images but also determine what brand of a camera is used.

In future work, we could a non-linear classifier such as neural network to improve the performance on differentiating Sony from Nikon.

References

1. Swaminathan, A., Gou, H., Wu, M.: Image acquisition forensics: Forensic analysis to identify imaging source. In: Proc. IEEE International Conference on Acoustics, Speech and Signal Processing, pp. 1657–1660 (2008)
2. Kang, X., Li, Y., Qu, Z., Huang, J.: Enhancing Source Camera Identification Performance with a Camera Reference Phase Sensor Pattern Noise. IEEE Transactions on Information Forensics and Security 7(2), 393–402 (2012)
3. Popescuand, A.C., Farid, H.: Exposing Digital Forgeries in Color Filter Array Interpolated Images. IEEE Transactions on Signal Processing 53(10), 3848–3959 (2005)
4. Gallagher, A.C., Chen, T.: Image authentication by detecting traces of demosaicing. In: Proc. IEEE Computer Society Conf. Computer Vision and Pattern Recognition, pp. 1–8 (2008)

5. Gallagher, A.C.: Eastman Kodak Company Detection of Linear and Cubic Interpolation in JPEG Compressed Images. In: Proc. The 2nd Canadian Conference on Computer and Robot Vision, pp. 65–72 (May 2005)

6. Chang, T.-Y., Lin, G.-S., Tai, S.-C.: Distinguishing Photographic Images and Computer Graphics by Using Characteristics of Color Filter Array. In: Proc. 2012 International Conference on Business and Information, pp. H1044–H1056 (July 2012)

7. Wang, Y., Moulin, P.: On discrimination between photorealistic and photographic images. In: Proc. IEEE Int. Conf. Acoustics, Speech, and Signal Processing (ICASSP), pp. II-161–II-164 (2006)

8. http://www.irtc.org/

9. http://www.raph.com/3dartists/

10. Lin, G.S., Chang, M.K., Chiu, S.T.: A feature-based Scheme for detecting and classifying video-shot transitions based on spatio-temporal analysis and fuzzy classification. International Journal of Pattern Recognition and Artificial Intelligence 23(6), 1179–1200 (2009)

Robust Video Copy Detection Based on Constrained Feature Points Matching

Duan-Yu Chen and Yu-Ming Chiu

Department of Electrical Engineering, Yuan Ze University, Chung-Li, Taiwan
dychen@saturn.yzu.edu.tw, s994622@mail.yzu.edu.tw

Abstract. In this paper, to efficiently detect video copies, focus of interests in videos is first localized based on 3D spatiotemporal visual attention modeling. Salient feature points are then detected in visual attention regions. Prior to evaluate similarity between source and target video sequences using feature points, geometric constraint measurement is employed for conducting bi-directional point matching in order to remove noisy feature points and simultaneously maintain robust feature point pairs. Consequently, video matching is transformed to frame-based time-series linear search problem. Our proposed approach achieves promising high detection rate under distinct video copy attacks and thus shows its feasibility in real-world applications.

Keywords: Visual attention, video copy detection, feature point.

1 Introduction

While innovative and light-weight applications have been developed for mobile platforms, huge volume of data, particularly video clips, is easily shared between users and even edited and reused with or without copyright authentication. Therefore, the issue about how to effectively detect video copies is raised again recently. In this paper, we focus on content-based video copy detection. The research field has been investigated in several years. However, how to extract representative feature(s) for video matching is still a challenging problem. In the literature, feature descriptors used such as Harris corner detector [19], SIFT [23], SURF [17], bag-of-words [13][26] etc. are used for image or video matching. However, feature point descriptors such as SIFT could work well in object recognition. For content-based video matching, the property of inexact matching of video content makes the similarity computing based on feature point matching even more challenging. In order to overcome this problem, a geometric-constraint is applied for feature point matching. In addition, based on our observations, intended editing of video copies would preserve focus of interests in common. That means computing similarity of video clips in visual attention regions would benefit video copy detection. Detecting visual saliency successfully can substantially reduce the computational complexity of the process of video copy detection. According to a study conducted by cognitive psychologists [10], the human visual system picks salient features from a scene. Psychologists believe this process emphasizes the salient parts of a scene and, at the same time, disregards irrelevant

J.-S. Pan et al. (Eds.): *Advances in Intelligent Systems & Applications*, SIST 21, pp. 359–368.
DOI: 10.1007/978-3-642-35473-1_36　　© Springer-Verlag Berlin Heidelberg 2013

information. However, this raises the question: What parts of a scene should be considered "salient"? To address this question, several visual saliency (or attention) models have been proposed in the last decade [1-5, 7-8]. Based on the type of attention pattern adopted, the models can be roughly categorized into two classes: bottom-up approaches, which extract image-based saliency cues; and top-down approaches, which extract task-dependent cues. Usually, extracting task-dependent cues requires a priori knowledge of the target(s). However, a priori knowledge of attended objects is usually difficult to obtain. Therefore, we focus on a bottom-up approach in this work.

In [8], Itti and Koch reviewed different computational models of visual attention, and presented a bottom-up, image-based visual attention system. Previously, Itti et al. [5] had proposed one of the earliest saliency-based computational models for determining the most attractive regions in a scene. The contrasts in color, intensity and orientation of images are used as clues to represent local conspicuity in images. Ma et al. [2] used the motion vector fields in MPEG bitstreams to build a motion attention map directly. In addition, some approaches have extended the spatial attention model from images to videos in which motion plays an important role. To find motion activity in videos, such models compute structural tensors [1], or estimate optical flows in successive frames directly [3].

Zhai and Shah [3] construct both spatial and temporal saliency maps and fuse them in a dynamic fashion to produce an overall spatiotemporal attention model. The model first computes the correspondence between points of interest, and then detects the temporal saliency using homography. Li and Lee [1] proposed a model of spatiotemporal attention for shot matching. Under this approach, a motion attention map generated from temporal structural tensors and a static attention map are used simultaneously to determine the degree of visual saliency. The weights of the feature maps are varied according to their contrast in each video frame. However, with this mechanism, the degree of visual saliency could be biased by the static attention map when the static features have higher contrast than the motion features. Therefore, in this work our proposed approach for effectively modeling video dynamics is used.

The remainder of this paper is structured as follows. In the next section, we introduce the proposed visual saliency model. Section 3 describes the geometric-constraint measurement for feature point matching. Section 4 presents similarity measurement. Section 5 details the experiment results. Finally, we present our conclusion in Section 6.

2   Visual Attention Modeling

To locate the salient regions in a video efficiently, we first detect the salient points in the video's corresponding spatiotemporal 3D volume. The points are then used as seeds to search the extent of the salient regions in the constructed motion attention map, in which the extent of the salient regions is determined by finding a motion map that corresponds to the maximum entropy. In the following, we describe the process for detecting spatiotemporal salient points, and explain how we generate a motion attention map. We then introduce the proposed selective visual attention model, which is based on finding the maximum entropy.

2.1   Detection of Spatiotemporal Salient Points

The spatiotemporal Harris detector, proposed by Laptev and Lindeberg [4], extends Harris and Stephens' corner detector [6] to consider the time axis. Similar to the operation performed in the spatial domain for a given spatial scale σ_l and temporal scale τ_l, the spatiotemporal Harris detector is based on a 3×3 second moment matrix μ. The matrix is composed of first order spatial and temporal derivatives averaged with a Gaussian weighting function $g\left(x, y, t : \sigma_i^2, \tau_i^2\right)$, i.e.,

$$\mu = g(x,t,t : \sigma_i^2, \tau_i^2) * \begin{pmatrix} L_x^2 & L_x L_y & L_x L_t \\ L_x L_y & L_y^2 & L_y L_t \\ L_x L_t & L_y L_t & L_t^2 \end{pmatrix}, \tag{1}$$

where the integration scales are $\sigma_i^2 = s\sigma_l^2$ and $\tau_i^2 = s\tau_l^2$; L_κ is a first-order Gaussian derivative through the κ axis; and the spatiotemporal separable Gaussian kernel is defined as

$$g(x, y, t : \sigma_l^2, \tau_l^2) = \frac{\exp(-(x^2 + y^2)/2\sigma_l^2 - t^2/2\tau_l^2)}{\sqrt{(2\pi)^3 \sigma_l^4 \tau_l^2}}. \tag{2}$$

To detect points of interest, regions that have significant corresponding eigenvalues λ_1, λ_2, λ_3 of μ are considered salient. The saliency function is defined as

$$H = \det(\mu) - k \, trace^3(\mu) = \lambda_1 \lambda_2 \lambda_3 - k(\lambda_1 + \lambda_2 + \lambda_3)^3, \tag{3}$$

where k is a tunable sensitivity parameter. The salient points detected by Eq. 3 are illustrated in Fig. 1b. We observe that the majority of salient points are located near the boundary of an attended object due to the intrinsic nature of corners. In contrast, points with relatively low saliency are located inside the moving objects, and usually correspond to consistent motion. To generate effective seeds for searching the appropriate extent of moving regions, we employ the centroid of every salient region instead of the commonly used local maxima points. The map shown in Fig. 1b is first binarized that is thresholded by the mean of the map and then be processed by performing morphological operations; and the regions that correspond to those salient points are shown in Fig. 1c. The centroid of each region shown in Fig. 1c is then taken as a seed for the subsequent search task.

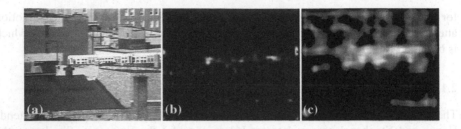

Fig. 1. (a) the original video frame; (b) points of interest detected by using 3D Harris corner detection; (c) the binarized salient map of (b)

2.2   Motion Attention Map

After detecting the seeds, we compute a motion attention map to determine the extent of searching. Based on the observation in Section 2.2 that the regions located inside moving objects usually correspond to consistent motion, our goal is to find regions that contain consistent motion near the search seeds. The optical flow (u,v,w) neighboring a search seed can be estimated by solving the following structural tensor [9]:

$$\mu \cdot \begin{bmatrix} u & v & w \end{bmatrix}^{T} = 0_{3\times 1} \tag{4}$$

where μ is the matrix defined in Eq.(1). Ideally, there would be multiple motions within the Gaussian smoothed neighborhood with spatial scale σ_i and temporal scale τ_i when $rank(\mu)$ is three. If there is consistent motion in a windowed area, $rank(\mu)$ would be equal to two. In the other two cases, i.e., $rank(\mu)$ equals 0 or 1, the motion of an image structure cannot be derived by Eq.(4) directly. However, in real videos, μ always has a full rank. Therefore, the normalized and continuous measure defined in [1] is used to quantify the degree of deficiency of a matrix. Let the eigenvalues of μ be $\lambda_1 \geq \lambda_2 \geq \lambda_3$. The continuous rank-deficiency measure d_μ is defined as

$$d_\mu = \begin{cases} 0, & trace(\mu) < \gamma \\ \lambda_3^2 / \left(\frac{1}{2}\lambda_1^2 + \frac{1}{2}\lambda_2^2 + \varepsilon \right), & otherwise \end{cases} \tag{5}$$

where ε is a constant used to avoid division by zero. The threshold γ is used to handle cases when $rank(\mu) = 0$. Since we want to find the regions with consistent motions, regions with high and low values of d_μ are not considered attended regions. In other words, regions with median values are the targets of interest. Therefore, we use a median filter f_μ to filter out regions with multiple motions and keep regions with consistent motions. As a result, we obtain the motion attention map $d_\mu^{f_\mu}$ for further processing. Clearly, the motion attention map $d_\mu^{f_\mu}$ complements the saliency map generated by H defined in Eq. (3). Hence, a proper combination of these two maps can determine the real extent of a salient area. To combine two salient maps that complement each other, the seeds produced in the first step, described in Section 2.2, are used as the starting points to search the appropriate extent in the motion attention map.

2.3   Visual Attention Modeling

To find the extent of attented regions, we take the configuration of a motion attention map that corresponds to the maximum entropy as our target. For each motion attention map, the most appropriate scale, S_e, for each region centered at the seed (x_s, y_s) is obtained by

$$S_e = arg\ max_e \left\{ H_{exp}\left(e, (x_s, y_s), t\right) \times W_{exp}\left(e, (x_s, y_s), t\right) \right\}, \tag{6}$$

where

$$H_{exp}\left(e, (x_s, y_s), t\right) = \sum_{v \in D} p_{v,e,(x_s,y_s),t} \times exp\left(1 - p_{v,e,(x_s,y_s),t}\right), \tag{7}$$

and

$$W_{exp}\left(e, (x_s, y_s), t\right) = \left| H_{exp}\left(e, (x_s, y_s), t\right) - H_{exp}\left(e-1, (x_s, y_s), t\right) \right|, \tag{8}$$

D is the set of all values that contains the values of $d_\mu^{f_\mu}$, which correspond to the histogram distribution in a local region e around a seed (x_s, y_s) in a motion attention map at time t. The probability mass function, $p_{v,e,(x_s,y_s),t}$, is obtained from the histogram of pixel values at time t for scale e in position (x_s, y_s) and the value v, which belongs to D. The efficacy of the proposed approach can be observed. The green bounding boxes are derived from the map measured by the salient function H, while the red boxes are derived by using the entropy maximization process to search for the proper scale in the motion attention map. Fig. 2(b) shows a saliency map generated by exponential entropy maximization. The attended regions are determined by finding the extent that covers the moving target well.

Fig. 2. Attended regions detected by the proposed approach. (a) the original video frame; (b) motion attention map.

3   Geometric-constraint Feature Points Measurement

As feature points are detected in Eq.(1), to achieve robust point matching we adopt the Bi-directional matching for selecting seed point pairs (stable matches) for the rest of tasks [16], which is based on the observation that true positive matches of distinctive features between the source image and the target image are more likely to be bi-directional, i.e., from source to target, and vice versa. In other words, if the relationship between matches is unidirectional, then the match could be unstable or incorrect that needs to be removed. Therefore, we assume that if there are few stable matches, the corresponding frames are very likely to be irrelative. The features finally matched by the bi-matching method are considered as seed points, *e.g.* the blue points demonstrated in Fig. 3. The geometric-constraint measurement is invariant to translation, rotation and scale transformations due to the robust feature points obtained from the Bi-directional matching. It is also invariant to affine transformation and robust to partially perspective transformation. The detail can be found in [16].

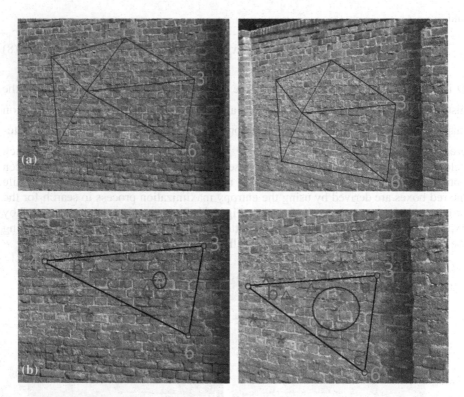

Fig. 3. Example of the Triangle-Constraint Measurement. (a) Illustration of the Delaunay algorithm; (b) Illustration of the geometric-constraint.

4   Video Similarity Measurement

Let $Q = \{q_i \mid i = 1, 2, \ldots, n_Q\}$ be a query sequence with n_Q frames, where q_i is the i-th query frame; and let $T = \{t_j \mid j = 1, 2, \ldots, n_T\}$ be a target sequence with n_T frames, where t_j is the j-th target frame, and $n_Q \ll n_T$. A sliding window is used to scan over T to search for a subsequence whose content is identical or similar to Q. Let $W = \{t_j, t_{j+1}, \ldots, t_{j+n_w-1}\}$ be a subsequence of T extracted by a sliding window with n_w frames, denoted as a window sequence.

Let $q_i p = \{q_i p_1, q_i p_2, \ldots, q_i p_n \mid q_i p_n = (X_n, Y_n)\}$ denotes n matched point-pairs of ith frame in query sequence. Meanwhile, in sliding window we have $w_i p = \{w_i p_1, w_i p_2, \ldots, w_i p_n \mid w_i p_n = (X'_n, Y'_n)\}$. The centroids $q_i p_c$ and $w_i p_c$ are then computed respectively using $q_i p$ and $w_i p$. Finally, the histogram $q_i h = \{q_i h_1, q_i h_2, \ldots, q_i h_n\}$ for each frame is obtained based on the distance between $q_i p_c$ and each point in $q_i p$.

Similarly, the histogram $w_i h = \{w_i h_1, w_i h_2, \ldots, w_i h_n\}$ can be obtained based on the distance between $w_i p_c$ and each point in $w_i p$. Finally, the similarity measurement between $q_i h$ and $w_i h$ is defined by the Jaccard coefficient as

$$J(q_i h, w_i h) = \frac{|q_i h \cap w_i h|}{|q_i h \cup w_i h|} = \frac{\sum_{n=1}^{N} \min(q_i h_n, w_i h_n)}{\sum_{n=1}^{N} \max(q_i h_n, w_i h_n)}. \tag{9}$$

5   Experimental Results

In the experiment, we select MUSCLE-VCD-2007 dataset, which is the benchmark dataset of the TRECVID video copy detection task[15], as demonstrated in Fig.4, which include distinct video types, such as human subject, animals, plants, sports, buildings, outdoor scene, etc. Totally 100 video sequences are selected and eight types of video copy attacks are edited, as presented in Table I. All the video sequences were converted into 320×240 pixels, and are re-sampled to 2 fps since usually frames in a scene would be near-duplicate. The threshold for Eq.(9) to determine if a video copy is detected is set as 0.9 through all types of video copy attacks.

For performance comparison, a state-of-the-art work proposed by Chiu and Wang [13] is selected. In [13], a SIFT codebook with 1024 codewords is first constructed and then compute a histogram for video frames. Finally, the min-hash approach is used to detect video copies. The quantitative evaluation is presented in Table II. It is worth noting that among all video attacks our approach outperform Chiu and Wang's method particularly in precision measurement. The good performance is benefited from the robust feature point matching using geometric-constraint in visual attention regions because we can focus on evaluate similarity between source and target video sequences in focus of interests for general users. Frame matching out of these regions could result in false alarms.

Fig. 4. Demonstration of the selected video dataset [15]

Table 1. Eight video copy attacks applied in the experiment

Type	Description
Brightness	Enhance the brightness by 20%.
Compression	Set the compression quality at 50%.
Noise	Add random noise (10%).
Resolution	Change the frame resolution to 120×90 pixels.
Cropping	Crop the top and bottom frame regions by 10% each.
Zoom-in	Zoom in the frame 10%.
Slow motion	Halve the video speed.
Fast forward.	Double the video speed

Table 2. Performance comparison using precision(P) and recall(R)

Type		[13]	Ours
Brightness	R	1.0000	0.9895
	P	0.9394	1.000
Compression	R	0.9677	0.9903
	P	0.9091	1.000
Noise	R	1.0000	0.9666
	P	1.0000	1.000
Resolution	R	0.8065	0.9702
	P	0.9259	1.000
Cropping	R	0.9677	0.9967
	P	0.9375	1.000
Zoom-in	R	0.9355	0.9926
	P	0.9355	1.000
Slow motion	R	0.9355	1.000
	P	1.0000	1.000
Fast forward.	R	1.0000	1.000
	P	0.8378	1.000

6   Conclusion

In this paper, a robust video copy detector has been proposed. To efficiently detect video copies, focus of interests in videos is localized based on 3D spatiotemporal visual attention modeling. Salient feature points are then detected in visual attention regions. Prior to evaluate similarity between source and target video sequences using feature points, geometric constraint measurement is employed for conducting bi-directional point matching in order to remove noisy feature points. Consequently, video matching is transformed to frame-based time-series linear search problem. Comparing to the state-of-the-art work, our proposed approach has achieved promising high detection rate under distinct video copy attacks and thus shows its feasibility in real-world applications.

References

1. Li, S., Lee, M.C.: An Efficient Spatiotemporal Attention Model and Its Application to Shot Matching. IEEE Trans. on Circuits and Systems for Video Technology 17(10), 1383–1387 (2007)
2. Ma, Y.F., Lu, L., Zhang, H.J., Li, M.: A User Attention Model for Video Summarization. In: Proc. ACM Multimedia, pp. 533–541 (December 2002)
3. Zhai, Y., Shah, M.: Visual Attention Detection in Video Sequences Using Spatiotemporal Cues. In: Proc. ACM Multimedia, pp. 815–824 (2006)
4. Laptev, L., Lindeberg, T.: Space-Time Interest Points. In: Proc. IEEE International Conference on Computer Vision, pp. 432–439 (October 2003)
5. Itti, L., Koch, C., Niebur, E.: Model of Saliency-Based Visual Attention for Rapid Scene Analysis. IEEE Trans. on Pattern Analysis and Machine Intelligence 20(11), 1254–1259 (1998)
6. Harris, C., Stephens, M.: A Combined Corner and Edge Detector. In: Alvey Vision Conference, pp. 147–151 (1988)
7. Navalpakkam, V., Itti, L.: An Integrated Model of Top-Down and Bottom-Up Attention for Optimizing Detection Speed. In: Proc. IEEE CVPR, vol. 2, pp. 2049–2056 (2006)
8. Itti, L., Koch, C.: Computational Modeling of Visual Attention. Neuroscience 2, 1–11 (2001)
9. Lucas, B., Kanade, T.: An Iterative Image Registration Technique with an Application to Stereo Vision. In: Proc. International Joint Conference on Artificial Intelligence, pp. 674–679 (1981)
10. James, W.: The Principles of Psychology. Harvard Univ. Press, Cambridge (1981)
11. Shih, C.C., Tyan, H.R., Liao, H.Y.M.: Shot Change Detection Based on the Reynolds Transport Theorem. In: Shum, H.-Y., Liao, M., Chang, S.-F. (eds.) PCM 2001. LNCS, vol. 2195, pp. 819–824. Springer, Heidelberg (2001)
12. Su, C.W., Mark Liao, H.Y., Tyan, H.R., Fan, K.C., Chen, L.-H.: A Motion-Tolerant Dissolve Detection Algorithm. IEEE Trans. on Multimedia 7(6) (December 2005)
13. Chiu, C.Y., Wang, H.M.: Time-Series Linear Search for Video Copies Based on Compact Signature Manipulation and Containment Relation Modeling. IEEE Transactions on Circuits and Systems for Video Technology 5604280, 1603–1613 (2010)
14. Chen, D.Y.: Modelling salient visual dynamics in videos. In: Multimedia Tools and Application, pp. 271–284 (2011)

15. TRECVID, Guidelines (2010)
16. http://www-nlpir.nist.gov/projects/tv2010/tv2010.html#ccd
17. Guo, X., Cao, X.: Triangle-Constraint for Finding More Good Features. In: International Conference on Pattern Recognition, No. 5597550, pp. 1393–1396 (2010)
18. Bay, H., Tuytelaars, T., Van Gool, L.: Surf:speeded up robust features. Lecture Notes in Computer Science, pp. 404–417 (2006)
19. Brown, M., Lowe, D.: Recognising panoramas. In: IEEE International Conference on Computer Vision, pp. 1218–1227 (2003)
20. Harris, C., Stephens, M.J.: A combined corner and edge detector. In: Alvey Vision Conference, vol. 20, pp. 147–152 (1988)
21. Jiang, H., Yu, S.: Linear solution to scale and rotation invariant object matching. In: IEEE Computer Society Conference on Computer Vision and Pattern Recognition Workshops, pp. 2474–2481 (2009)
22. Lee, S., Liu, Y.: Curved glide-reflection symmetry detection. In: IEEE Computer Society Conference on Computer Vision and Pattern Recognition Workshops, pp. 1046–1053 (2009)
23. Leordeanu, M., Hebert, M.: A spectral technique for correspondence problems using pairwise constraints. In: IEEE International Conference on Computer Vision, pp. 1482–1489 (2005)
24. Lowe, D.: Distinctive image features from scale-invariant Keypoints. In: International Journal of Computer Vision, pp. 91–110 (2004)
25. Rabin, J., Delon, J., Gousseau, Y.: Circular earth mover's distance for the comparison of local features. In: International Conference on Pattern Recognition, pp. 1–4 (2008)
26. Tuytelaars, T., Van Gool, L.: Matching widely separated views based on affine invariant regions. International Journal of Computer Vision, 61–85 (2004)
27. Zhang, S., Tian, Q., Hua, G., Huang, Q., Li, S.: Descriptive visual words and visual phrases for image applications. In: Proc. ACM Int'l Conf. Multimedia, Beijing, China, October 19-24, pp. 75–84 (2009)

A Mass Detection System in Mammograms Using Grey Level Co-occurrence Matrix and Optical Density Features

Shen-Chuan Tai[1], Zih-Siou Chen[1,*], Wei-Ting Tsai[1],
Chin-Peng Lin[1], and Li-li Cheng[2]

[1] Institute of Computer and Communication Engineering, Department of Electrical Engineering, National Cheng-Kung University, Tainan 701, Taiwan, R.O.C.
{sctai,czs99d,twt99m,lcp100m}@dcmc.ee.ncku.edu.tw
[2] National Cheng-Kung University Hospital, 138, Sheng Li Road, Tainan 704, Taiwan, R.O.C.
lili_112005@yahoo.com.tw

Abstract. For the radiologists, it is difficult to identify the mass on a mammogram since the masses are surrounded by the mammary gland and blood vessel. In current breast cancer screening, about 10% - 30% of tumors are often missed by radiologists owing to the ambiguous margins of lesions and the visual fatigue of radiologists resulting from the long-time diagnosis. For these reasons, many computer-aided detection (CADe) systems have been developed to aid radiologists in detecting mammographic lesions that may indicate the presence of breast cancer. The purpose of this study is to construct an automated CADe system using a new feature extraction method for mammographic mass detection. In this system, some adaptive square regions of interest (ROIs) are segmented according to the size of suspicious areas. Then a new feature extraction method adopting grey level co-occurrence matrix and optical density features called GLCM-OD features is applied to each ROI. The GLCM-OD features describe local grey level texture characteristics and the whole photometric distribution of the ROI. Finally, the stepwise linear discriminant analysis is applied to classify abnormal regions by selecting and rating individual performance of each feature. This system is trained and tested by 358 mammographic cases from the digital database of screening mammography (DDSM). The proposed system averagely provides sensitivity of 97.3% with 4.9 false positives per image and the Az is 0.981. The results prove that the proposed system achieves satisfactory detection performance.

Keywords: computer-aided detection system, grey level co-occurrence matrix, optical density, GLCM-OD features.

1 Introduction

Breast cancer is one of the most devastating and deadly diseases for women. It is the second major cause of death after lung cancer among women. According to statistics,

* The work described in this paper was conducted under the research project "A CAD System for Automated Detection and Classification of Mass Lesions in Digital Mammography" (D101-15105 and B101-MD08), supported by National Cheng Kung University and Delta Electronics.

J.-S. Pan et al. (Eds.): *Advances in Intelligent Systems & Applications*, SIST 21, pp. 369–376.
DOI: 10.1007/978-3-642-35473-1_37 © Springer-Verlag Berlin Heidelberg 2013

breast cancer has become a major health problem in both developed and developing countries over the past 50 years, and its incidence has increased in recent years. In 2012, there were estimated 229,060 new cases and 39,920 deaths from breast cancer in the United States [1]. Currently, there are no effective ways to prevent breast cancer, because its cause remains unknown.

Mammograms are X-ray images of breast region. Screening mammography is currently reliable, nonpalpable and potentially curable breast cancer technique for early detection. There are two screening views of each breast; the craniocaudal (CC) view, which is a top-to-bottom view, and a mediolateral oblique (MLO) view, which is a side view taken at 45° angle. Examples of the MLO and CC views are shown in Fig. 1. The images were obtained from the digital database for screening mammography (DDSM) [2]. Radiologists visually search mammograms for specific abnormalities. The Radiologists look for the important signs of breast cancer, such as clusters of microcalcifications, and masses. A mass is defined as a space-occupying lesion seen in at least two different projections [3]. Masses are described by their shape and margin characteristics. Calcifications are tiny deposits of calcium, which appear as small bright spots on the mammogram. They are characterized by their type and distribution properties.The Computer-Aided Detection (CADe) system integrates diagnostic imaging with computer science, image processing, pattern recognition, and artificial intelligence technologies [4]. A radiologist uses the output of CADe system from computerized analysis of medical images as a " second opinion " in detecting lesions and making diagnostic decisions. The goal of diagnosis is to distinguish between normal and cancer image. A good CAD system can be identified by low false negative and false positive rate.

Researches on CADe systems and related techniques have attracted great attention. There are several papers published [5–7] and at least three commercial CADe systems available on the market in the United States, including the R2 system, the iCAD system, and Kodaks system. However, there is still a long way to go before CADe systems become widely used in clinics and screening centers. Researches [8], [9], and [10] have

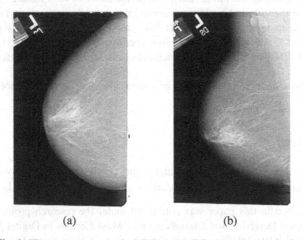

(a)   (b)

Fig. 1. The mammograms in DDSM. (a) CC view. (b) MLO view.

shown that CADe represents a useful tool for the detection of breast cancer. Furthermore, the results from a few recent studies, [11] and [12] show that the performance of current commercial CADe systems still need to be improved to meet the requirements of clinics and screening centers. Thus, improving the performance of CADe systems remains to be an important issue for future research and development. In fact, breast masses are more difficult to identify than microcalcifications due to the abundant appearances and ambiguous margins [13]. In this paper, a new feature extraction method is proposed to improve the sensitivity for mammographic mass detection.

2   Materials and Methods

This paper proposes a mass detection method composed of three stages. The entire block diagram of the proposed algorithm is shown in Fig. 2. In the preprocessing stage, the breast region is kept by some simple methods, such as contrast stretching method, to remove tags, background, and the pectoral muscle. Removing such useless area can improve the processing efficiency of the system. Then, the morphological filters are used to suppress blood vessels and mammary gland on the breast region. Since masses are often hidden by some structural noises, suppressing the noises can make a reliable input for the following step. Fig.3 revels the final segmented breast image after preprocessing step. Xu *et al.* [14] proposed a hierarchical matching method to detect the suspicious breast masses on mammograms. According to [14], three types of templates were compared. By taking into account the performances of each template, the Sech template was selected to match the suspicious area. In this system, the Sech template is also used and defined as

$$S(x, y) = \frac{2}{\exp\left(\beta * \sqrt{x^2 + y^2}\right) + \exp\left(-\beta * \sqrt{x^2 + y^2}\right)} \tag{1}$$

where x, y represents the coordinate of the template and various templates could be obtained by different β. Two different size templates, the part-based (33×33) and the

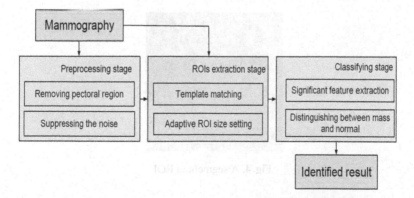

Fig. 2. Block diagram of the proposed mass detection algorithm

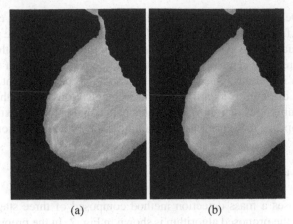

(a)   (b)

Fig. 3. (a) The segmented breast region. (b) The filtered breast region.

complete (65×65) template, are used to calculate the similarity between the breast region and the template. The correlation measurement which is applied to describe the similarity is defined as

$$cor(T, I) = \frac{E[(I - \mu_I) - (T - \mu_T)]}{\sigma_I \sigma_T} \tag{2}$$

where T denotes the template, I represents the region of the breast, μ_I is the average of I, μ_T is the average of T, σ_I is the standard deviation of I, and σ_T is the standard deviation of T. Two correlation maps would be obtained by the correlation measurement. Afterwards two correlation maps are converted into an averaging correlation map to segment the suspicious regions. The roughly suspicious regions are found by providing appropriate threshold in the averaging correlation map. Since the masses vary much in size, an adaptive square region of interest (ROI) is determined according to the size of suspicious area. Fig.4 shows a mass segmented result of the ROI extraction stage.

Fig. 4. A segmented ROI

After the suspicious regions are segmented, some features are extracted from the characteristics of the ROI. We proposed a new feature extraction method adopting grey level co-occurrence matrix and optical density features called "GLCM-OD features". The intensity distribution of masses is an important characteristic for mass detection. For this reason, some pattern recognition method use GLCM to extract the features. The idea behind GLCM is to describe the textures by a matrix of pair grey level probabilities. In our scheme, four co-occurrence matrices are computed by four directions: left diagonal, right diagonal, vertical, and horizontal. The distance between each pair of pixels is one. Further, Haralick *et al.* [15] described fourteen statistics that can be calculated from a co-occurrence matrix with the intent of describing the texture of the image. The fourteen texture features are defined as follows: *Entropy, Energy, Local homogeneous, Contrast, Intensity, Correlation, Inverse difference moment, Sum average, Sum of squares variance, Sum entropy, Difference entropy, Inertia, Cluster Shade* and *Cluster Prominence*. Moreover, Sameti *et al.* [16] proposed a feature extraction method used in early detection of malignant masses in digital mammogram studies. There were sixty-two statistics of five categories used in the study. In this paper, one of the categories, the discrete texture feature with twenty characteristics, is added to complete the characteristic description. In discrete texture feature category, the background information is considered by transforming the grey level into optical density value. Finally, the GLCM-OD features containing seventy-six statistics are applied to the stepwise linear discriminant analysis to train the discriminant functions.

3   Experimental Results and Analysis

In the proposed system, the data for experiment are taken from the Digital Database for Screening Mammography (DDSM) [2], provided by South Florida University. The DDSM database includes about 2500 cases. Our study focused on the set of mammograms digitized by the Lumisys scanner at 0.5 mm, and Howtek scanner at 0.435 mm pixel size, 12 bits per pixel. And only MLO view is analyzed for this thesis. In addition, thanks to [17] and [18] the ljpeg format of the mammograms can be translated into png format for the sake of being processed in the Microsoft Windows System. The system is evaluated by 358 mammograms which are randomly selected in each density rating from the DDSM. Table 1 contains the relevant statistics for the test sets. Among the 358 testing cases, 50 cases including 30 malignant masses and 20 benign masses are selected for the training set.

After the training stage, the discriminant functions contain four features: $Sum\ Entropy, Local\ Homogeneous, Med - Avg - Dst$, and $Highmed - Avg - Dst$. The former two are selected from the GLCM features, and the others are picked from the discrete texture feature in optical density value. Table 2 list the calculated statistics to compare mass ROIs with normal ones. *Local Homogeneous* measures the textural uniformity and detects disorders in textures. This parameter indicates how much homogeneous the texture is, i.e. the GLCM contains values distributing fairly uniform over all grid. It is high when the GLCM has few entries of large magnitude, low when all entries are almost equal. It is well known that entropy measures information of

Table 1. Statistics of mammograms for performance test

	Benign	Cancer	Normal	Total
Density 1	17	39	12	68
Density 2	66	53	20	139
Density 3	36	34	10	80
Density 4	9	4	8	21
Total	128	130	50	308

an image. Moreover, $Sum\ Entropy$ is computed by the probability of the sum of pixel pair, and indicates the complexity of the sum of pixel pair. In an image, it also relates to the amount of edged information. $Med - Avg - Dst$ and $Highmed - Avg - Dst$ are discrete texture features. $Med - Avg - Dst$ represents the average distance between medium density pixels and the center of the object circle. In the same manner, $Highmed - Avg - Dst$ is calculated from the high and medium density pixels. If a ROI contains a mass, the general medium density pixels will locate around the rim of the mass and some medium density pixels randomly emerge in the form of structural noise. In Table 2, if the difference in value between two features is larger, it represents that the high density pixels concentrates on the center of ROI. In another word, the ROI may contain a mass. On the other hand, small difference represents more probable to be a normal ROI.

Table 2. The calculated value of features for mass and normal ROIs

Type	Local Homogeneous	Sum Entropy	Med − Avg − Dst	Medhi − Avg − Dst
Mass-1	0.41922	5.2849	0.84557	0.69387
Mass-2	0.57398	4.7448	0.99525	0.74728
Mass-3	0.49267	5.4491	0.90937	0.70384
Mass-4	0.4864	4.8182	0.95711	0.73557
Normal-1	0.48198	3.7794	0.63222	0.61055
Normal-2	0.48683	3.1945	0.73541	0.74692
Normal-3	0.40886	3.7766	0.55165	0.57809
Normal-4	0.39561	3.9768	0.64144	0.66190

In order to compare our results with previously published study, the performance of our system is evaluated by the free response operating characteristic (ROC) analysis. Table 3 presents the performance of the proposed system in overall cases, and the ROC curve is shown in Fig.5. The area under the ROC curve (Az) is 0.981. Fig.5 shows that the proposed system averagely provides sensitivity of 97.3% with 4.9 false positives per image.

Table 3. The average performance of the proposed system at various operating points

$Sensitivity(\%)$	FPs/I
100(258/258)	11.4(3512/308)
99(255/258)	8.1(2501/308)
98.3(254/258)	5.8(1799/308)
97.3(251/258)	4.9(1508/308)
96.3(248/258)	3.7(1148/308)
95.7(247/258)	3.7(1131/308)
92.6(239/258)	3(925/308)
91.6(236/258)	2.8(874/308)
91(235/258)	2.8(857/308)
90.3(233/258)	2.7(839/308)
89.3(230/258)	2.6(805/308)
88.3(228/258)	2.4(754/308)
87.3(225/258)	2.3(702/308)
85.3(220/258)	2.1(634/308)
83.9(216/258)	1.9(600/308)
82.3(212/258)	1.9(582/308)
81.3(210/258)	1.8(565/308)

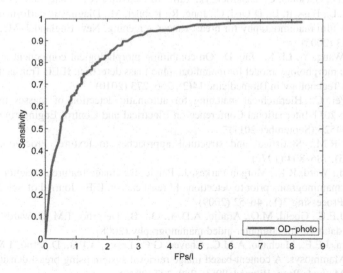

Fig. 5. The ROC curve of whole average performance

References

1. NCI Cancer Fact Sheets (2012),
   http://www.cancer.gov/cancertopics/types/breast
2. Heath, M., Bowyer, K., Kopans, D., Moore, R., Kegelmeyer Jr., P.: The digital databse for screening mammograpy (2000)
3. A. C. of Radiology, ACR BI-RADS–Mammography, Ultrasound & Magnetic Resonance Imaging, 4th edn. American College of Radiology, Reston, VA (2003)
4. Tang, J., Rangayyan, R., Xu, J., El Naqa, I., Yang, Y.: Computer-aided detection and diagnosis of breast cancer with mammography: Recent advances. IEEE Transactions on Information Technology in Biomedicine 13(2), 236–251 (2009)
5. Sampat, M.P., Markey, M.K., Bovik, A.C.: Computer-aided detection and diagnosis in mammography. In: Bovik, A.C. (ed.) Handbook of Image and Video Processing, 2nd edn., Academic, New York (2005)
6. Rangayyana, R.M., Ayresa, F.b.J., Desautels, J.L.: A review of computer-aideddiagnosis of breast cancer: Toward the detection of subtle signs. Journal of the Franklin Institute 344(3-4), 312–348 (2007)
7. Yoon, H.J., Zheng, B., Chakraborty, D.P.: Evaluating computer-aided detection algorithms. Med. Phys. 34(6), 2024–2048 (2007)
8. Brem, R., Rapelyea, J., Zisman, G., Hoffmeister, J., Desimio, M.: Evaluation of breast cancer with a computer-aided detection system by mammographic appearance and histopathology. Cancer 104(5), 931–935 (2005)
9. Berman, C.G.: Recent advances in breast-specific imaging. Cancer Control 14(4), 338–349 (2007)
10. Brem, R.F., Hoffmeister, J.W., Zisman, G., DeSimio, M.P., RogersSeptember, S.K.: A computer-aided detection system for the evaluation of breast cancer by mammographic appearance and lesion size. Amer. J. Roentgenol 184(3), 893–896 (2005)
11. Ciatto, S., Houssami, N., Gur, D., Nishikawa, R., Schmidt, R., Metz, C., Ruiz, J., Feig, S., Birdwell, R., Linver, M., Fenton, J., Barlow, W., Elmore, J.: Computer-aided screening mammography. New England J. Med. 357(1), 83–85 (2007)
12. Pisano, E.D., Gatsonis, C., Hendrick, E., Yaffe, M., Baum, J.K., Acharyya, S., Conant, E.F., Fajardo, L.L., Bassett, L., D'Orsi, C., Jong, R., Rebner, M.: Diagnostic performance of digital versus film mammography for breast-cancer screening. New England J. Med. 353(17), 1773–1783 (2005)
13. Gao, X., Wang, Y., Li, X., Tao, D.: On combining morphological component analysis and concentric morphology model for mammographic mass detection. IEEE Transactions on Information Technology in Biomedicine 14(2), 266–273 (2010)
14. Xu, S., Pei, C.: Hierarchical matching for automatic detection of masses in mammograms. In: 2011 International Conference on Electrical and Control Engineering (ICECE), pp. 4523–4526 (September 2011)
15. Haralick, R.M.: Statistical and structural approaches to texture. Proceedings of the IEEE 67(5), 786–804 (1979)
16. Sameti, M., Ward, R.K., Morgan-Parkes, J., Palcic, B.: Image feature extraction in the last screening mammograms prior to detection of breast cancer. IEEE Journal of Selected Topics in Signal Processing 3(1), 46–52 (2009)
17. Oliveira, J.E.E., Gueld, M.O., Araújo, A.D.A., Ott, B., Deserno, T.M.: Towards a standard reference database for computer-aided mammography (2008)
18. de Oliveira, J.E.E., Machado, A.M.C., Chavez, G.C., Lopes, A.P.B., Deserno, T.M., Araújo, A.de.A.: Mammosys: A content-based image retrieval system using breast density patterns. Comput. Methods Prog. Biomed. 99(3), 289–297 (2010),
    http://dx.doi.org/10.1016/j.cmpb.2010.01.005

Automatic Evaluation of Choroidal Neovascularization in Fluorescein Angiography

Kai-Shun Lin[1], Chia-Ling Tsai[2], Shih-Jen Chen[3], and Wei-Yang Lin[1]

[1] National Chung Cheng University, Chiayi 621, Taiwan
links@cs.ccu.edu.tw
[2] Iona College New Rochelle, NY 10801, U.S.A.
[3] Taipei Veterans General Hospital, Taipei 11217, Taiwan

Abstract. In this paper, we propose a system for generating Choroidal NeoVascularization (CNV) lesion severity map and evaluating the extent of classic CNV, occult CNV or Pigment Epithelial Detachment (PED) in a Fluorescein Angiography (FA) sequence. Given an FA sequence, our system first warps each frame to a common coordinate system. For each pixel location in the common coordinate system, we use intensity values, intensity changes and area of fluorescence leakage as feature vector for SVM classification. Then, our system generates a severity map associated with the input FA sequence. The severity map will serve to help ophthalmologist perform evaluation of CNV lesion in FA sequence. We have validated our system using patient data from Taipei Veterans General Hospital, Taiwan. The testing data contain 3 classic CNVs, 2 occult CNVs and 2 PEDs. The preliminary results show that our method achieves a promising recognition rate of 85.7%.

1 Introduction

Age-related macular degeneration (AMD) with the devastating complication of choroidal neovascularization (CNV) is the leading cause of legal blindness in the Western world [1]. The vascular nature of the CNV with hemorrhage, subretinal fluid, and retinal edema can be identified with the clinical diagnostic tools of fluorescein angiography (FA) [2]. Fluorescein angiography is a powerful imaging modality to identify the presence, location and size of the choroidal neovascular lesion. Many clinical trials of AMD utilize FA characterization for treatment endpoints. However, analysis and interpretation of FA sequences are largely performed by skilled observers , and analysis is usually performed on single angiographic frames. The process is also subjective with considerable observer variability [3,4,5,6] for image comparison. For facilitate accurate diagnosis of CNV, digital image processing methodologies are used for more efficient and accurate processing of fundus images [7,8].

According to the Macular Photocoagulation Study Group [9], components of CNV exhibit distinct temporal leakage patterns. For classic CNV, distinct areas of hyperfluorescence (leakage) become noticeable within 30 seconds and expand and intensify toward the end of the angiogram. For occult CNV, the leakage is detected as indistinct areas of hyperfluorescence which can appear either early or late and intensify over time. The leakage characteristics of both types of CNV are confirmed again by Berger [10]

J.-S. Pan et al. (Eds.): *Advances in Intelligent Systems & Applications*, SIST 21, pp. 377–382.
DOI: 10.1007/978-3-642-35473-1_38   © Springer-Verlag Berlin Heidelberg 2013

using computerized, spatiotemporal image analysis. By learning the typical leakage pattern of a certain CNV lesions type, our system can map the intensity variation of a set of spatially corresponding pixels of the same physical point across the sequence to the targeted leakage pattern for computation of severity.

In this paper, we propose an algorithmic approach to provide retina specialists a computer-aided tool for accurate diagnosis of choroidal neovascularization (CNV). For the vascular nature of CNV, fluorescein angiography (FA) is a special kind of photograph often used by physicians, and the interpretation is largely performed manually by skilled observers on single angiographic frames. To reduce observer variability and to exploit the characteristics of the temporal profile of fluorescence leakage, we propose a system which generates the severity map and determine the extent of classic CNV, occult CNV or pigment epithelial detachment (PED) by delineating areas of high severity for a given FA sequence, based on the leakage characteristics of CNV lesion.

2   Materials and Methods

The image sequences are provided by the Taipei Veterans General Hospital, Taipei, Taiwan. All sequences are categorized as CNV lesion and contain multiple images in each of the early, mid, and late phases. Image frames of a FA sequence are first mapped to a global space to compensate for the saccadic eye motions for a wide field of view. The alignment technique we adopt is Edge-Driven Dual-Bootstrap Iterative Closest Point [11]. To automatically determine the severity of each pixel in the global space and the extent of CNV, our system matches the intensity variation of its set of spatially-corresponding pixels across the sequence with the targeted leakage pattern, learned from a sampled population graded by a retina specialist.

The learning strategy is Support vector machine (SVM) [12] and the fluorescence leakage are described using 18 features. The intensity variation of the fluorescence leakage is 13 features: the intensity value and the intensity change in 6 time labels, and the slope and intercept of the straight line from linear least-squares regression of the time-intensity data points from the entire sequence, as shown in Figure 1(d). The temporal fluorescence profile of each sample point is divided into 6 labels: 30, 60, 180, and 300 to 600, and after 1200 seconds. Each interval provides two features: the intensity value and the intensity change. The fluorescence leakage area is 5 features: the size (pixels) of the leak area from the binary images of fluorescence leakage area, as shown in Figure 2(d). In summary, a 18-tuple feature vector is associated with a set of spatially corresponding pixels:

$$x = \{s_1, s_2, s_3, s_4, s_5, i_1, i_2, i_3, i_4, i_5, i_6, a, b, r_1, r_2, r_3, r_4, r_5\}, \tag{1}$$

where s_w and i_w are the intensity change(change in the 5 interval) and intensity value, respectively, in interval w, a and b are the slope and intercept, r_w is the size of the leak area. All features are associated with a binary classifier, determining if a feature belongs to classic CNV, occult CNV or pigment epithelial detachment (PED). SVM builds the classifiers. Consider the set l vectors $\{x_i\}, x_i \in R^n, 1 \leq i \leq l$, representing input samples and set of labels $\{y_i\}, y_i \in \pm1$, that divide input samples into two classes.

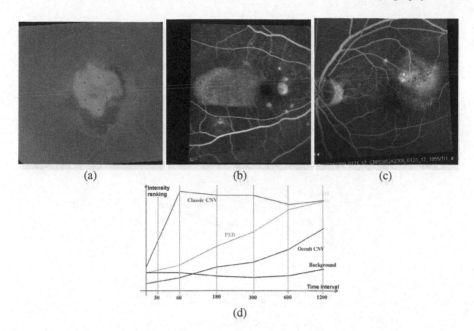

(a)   (b)   (c)

(d)

Fig. 1. The intensity variation of the fluorescence leakage features.(a) Samples are chosen from the classic CNV lesion (red dots). (b) The pigment epithelial detachment samples (green dots). (c) The occult CNV samples (blue dots). (d) The temporal fluorescence profile of each sample point is divided into 6 labels: 30, 60, 180, and 300 to 600, and after 1200 seconds. Each interval provides two features: the intensity value and the intensity change. The red plot is average drawn from the classic CNV pixels, the green plot from the pigment epithelial detachment pixels, and the blue plot from the occult CNV pixels.

There exists a separating hyperplane (w, b) defining the function

$$f(x) = < w \cdot x > +b, \qquad (2)$$

and $sgn(f(x))$ shows on which side of the hyperplane x rests. Vector w of the separating hyperplane can be expressed as a linear combination of x_i with weights α_i:

$$w = \sum_{1 \leq i \leq l} \alpha_i y_i x_i. \qquad (3)$$

The dual representation of the decision function f(x) is then:

$$f(x) = \sum_{1 \leq i \leq l} \alpha_i y_i < x_i \cdot x > +b. \qquad (4)$$

Training a linear SVM means finding the strengths $\{\alpha_i\}$ and offset b such that hyperplane (w, b) separates positive samples from negatives ones with a maximal margin. A kernel function $K(x, y)$ computes the dot product, $K(x, y) = < x \cdot y >$. In our implementation we use the Gaussian radial basis function $K(x, y) = exp(\frac{-\|x-y\|^2}{2\sigma^2})$.

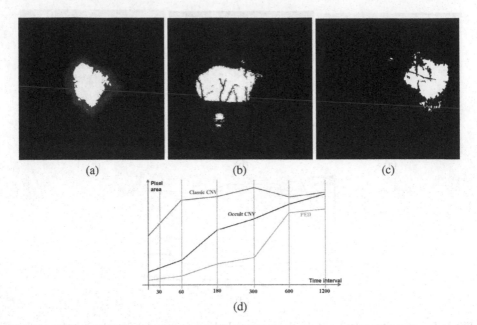

Fig. 2. The binary images of fluorescence leakage area.(a)The classic CNV leakage area. (b)The pigment epithelial detachment leakage area. (c)The occult CNV leakage area. (d) Each interval is the size value (pixels) of the leak area from temporal fluorescence leakage area.

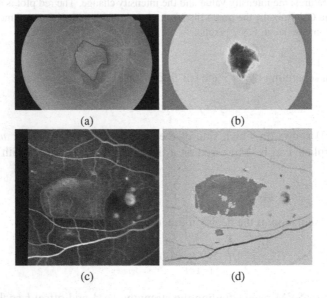

Fig. 3. Information generated by the proposed system for CNV lesion. (a) The region of classic CNV lesion is indicated in red line by the retina specialist. (b) The classic CNV severity map using SVM classification displayed with a color map. (c) The region of PED lesion. (d) The PED severity color map.

At completion of SVM training, each classifier is assigned an optimal threshold value and the parameters for computation of the contribution. To compute the severity value for a point in the global space, the contribution from SVM classifiers are added up and normalized to the range of [0,1]. Figure 3(a) is the image with classic CNV delineated by a retina specialist, and Figure 3(b) is the corresponding severity map generated using SVM classification. Figure 3(c) is the image with PED and Figure 3(d) is severity color map for PED.

3   Results

This study includes an experiment using 7 FA sequences. The dataset was obtained from the Taipei Veterans General Hospital and all sequences contain CNV lesion. The performance of the system is analyzed using leave-one-out cross validation strategy - each sequence is tested with the SVM classifier trained using the other 6 sequences. The test data contains 3 classic CNV, 2 occult CNV and 2 pigment epithelial detachment (PED). The performance is measured the accuracy rate as the percentage of correctly identified class over the total number of case. The accuracy is 85.7%. The classic CNV and the pigment epithelial detachment are correctly identified. In Figure 4, this is a case of occult CNV with classic CNV and pigment epithelial detachment. The occult CNV is incorrectly identified as the leakage pattern of classic CNV.

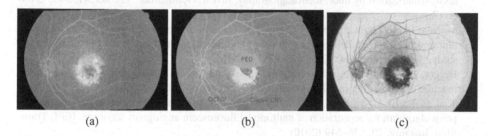

| (a) | (b) | (c) |

Fig. 4. The occult CNV is incorrectly identified as the leakage pattern of classic CNV. (a) The original image. (b) The groundtruth. (c) The classic CNV severity map.

4   Discussion and Conclusions

In this paper, we propose a system to generate a CNV lesion severity map in fluorescein angiography sequences. The severity map can serve as references to reduce grader variability and to improve the accuracy of the diagnosis. This system can be particularly useful for less experienced clinicians since the system can learn the temporal profile of the chosen type of CNV using sequences examined by more experienced clinicians. The algorithm achieves 85.7% accuracy in our preliminary study. We are also interested in studying the correlation between the visual functions and the measurements derived from a system for evaluation of a treatment. Measurements can be easily derived from the severity map areas to correlate with the visual functions for assessment of treatment and to facilitate clinical studies. Such findings are important in the development of surgical methods and pharmaceutical products.

References

1. Bressler, N.M., Bressler, S.B., Fine, S.L.: Age-related macular degeneration. Surv. Ophthalmol. 32, 375–413 (1988)
2. Amitha, D., Danis Ronald, P.: Fluorescein angiography in neovascular amd. Review of Ophthalmology 15, 56 (2008)
3. Hogg, R., Curry, E., Muldrew, A., Winder, J., Stevenson, M., McClure, M., Chakravarthy, U.: Identification of lesion components that influence visual function in age related macular degeneration. Br. J. Ophthalmol. 85, 609–614 (2003)
4. Jaakkola, A., Tommila, P., Laatikainen, L., Immonen, I.: Grading choroidal neovascular membrane regression after strontium plaque radiotherapy: masked subjective evaluation vs. planimetry. Eur. J. Ophthalmol. 11, 269–276 (2001)
5. Tsai, C.-L., Yang, Y.-L., Chen, S.-J., Lin, K.-S., Chan, C.-H., Lin, W.-Y.: Automatic characterization and segmentation of classic choroidal neovascularization using adaboost for supervised learning. Invest. Ophthalmol. Vis. Sci. 52, 2767–2774 (2011)
6. Fahmy, A.S., Abdelmoula, W.M., Mahfouz, A.E., Shah, S.M.: Segmentation of choroidal neovascularization lesions in fluorescein angiograms using parametric modeling of the intensity variation. In: ISBI, pp. 665–668 (2011)
7. Chakravarthy, U., Walsh, A.C., Muldrew, A., Updike, P.G., Barbour, T., Sadda, S.R.: Quantitative fluorescein angiographic analysis of choroidal neovascular membranes: validation and correlation with visual function. Invest. Ophthalmol. Vis. Sci. 48, 349–354 (2007)
8. Shah, S.M., Tatlipinar, S., Quinlan, E., et al.: Dynamic and quantitative analysis of choroidal neovascularization by fluorescein angiography. Invest. Ophthalmol. Vis. Sci. 47, 5460–5468 (2006)
9. Macular Photocoagulation Study Group. Subfoveal neovascular lesions in age-related macular degeneration: guidelines for evaluation and treatment in the macular photocoagulation study. Arch. Ophthalmol. 109, 1242–1257 (1991)
10. Berger, J.: Quantitative, spatio-temporal image analysis of fundus features in age-related macular degeneration. In: SPIE Ophthalmic Technol., pp. 48–53 (1998)
11. Tsai, C.-L., Li, C.-Y., Yang, G., Lin, K.-S.: The edge-driven dual-bootstrap iterative closest point algorithm for registration of multimodal fluorescein angiogram sequence. IEEE Trans. Med. Imaging. 29, 636–649 (2010)
12. Vapnik, V.: Statistical Learning Theory. Wiley, NY (1999)

3D Spinal Cord and Nerves Segmentation from STIR-MRI[*]

Chih Yen[1], Hong-Ren Su[1], Shang-Hong Lai[1],
Kai-Che Liu[2], and Ruen-Rone Lee[1]

[1] Department of Computer Science, National Tsing Hua University, Hsinchu, Taiwan
[2] Chang Bing Show Chwan Memorial Hospital, Taiwan
{charlesx,suhongren}@gmail.com, lai@cs.nthu.edu.tw

Abstract. In this paper, we present a system for spinal cord and nerves segmentation from STIR-MRI. We propose an user interactive segmentation method for 3D images, which is extended from the 2D random walker algorithm and implemented with a slice-section strategy. After obtaining the 3D segmentation result, we build the 3D spinal cord and nerves model for each view using VTK, which is an open-source, freely available software. Then we obtain the point cloud of the spinal cord and nerves surface by registering the three surface models constructed from three STIR-MRI images of different directions. In the experimental results, we show the 3D segmentation results of spinal cord and nerves from the STIR-MRI (Short Tau Inversion Recovery - Magnetic Resonance Imaging)images in three different views, and also display the reconstructed 3D surface model.

Keywords: STIR-MRI, spinal cord segmentation, random walker algorithm, 3D point set registration, 3D affine Fourier transform, surface reconstruction.

1 Introduction

Nowadays, due to the higher medical treatment level, in the human life expectancy has increased steadily in the past few decades. The rapid rise in the elderly population leads to the increase of the cases of spinal cord diseases , such as the degenerative and osteoporosis diseases. To treat these diseases, surgery is usually needed. Recently, Minimally Invasive Surgery (MIS) becomes more popular, because it does only make small injuries and cause less pain to the patient, thus taking shorter time for recovery. When doing MIS of spinal cord, an accurate Computer Aided Diagnosis (CAD) is very important, such as the patients 3D organ model, which can help the physician avoid hurting the spinal cord and nerves. Computer Tomography (CT), Magnetic resonance imaging (MRI) and STIR-MRI provide vital information separately, and these images can be used for 3D model reconstruction of organs or regions of interest. In this paper, we

---

[*] This work is supported by the National Science Council in Taiwan under the Grant NSC100-2221-E-007-078.

J.-S. Pan et al. (Eds.): *Advances in Intelligent Systems & Applications*, SIST 21, pp. 383–392.
DOI: 10.1007/978-3-642-35473-1_39   © Springer-Verlag Berlin Heidelberg 2013

especially focus on the STIR-MR images, which contains more spinal cord and nerves information, and try to build a 3D surface model from the segmented region.

This paper is organized as follows. We review some related works in Section 2. The 3D segmentation method based on Random Walker Algorithm [1] and 3D registration method are given in Section 3. We present the segmentation results and display the 3D surface model reconstruction in Section 4. In the end, we conclude this paper in Section 5.

2   Related Works

Since medical image segmentation has been developed for a long time, there are three main approaches; i.e. threshold-based, recognition-based , and contour-based approaches. But most of these method are concentrated on some specialized regions; for example, brain, heart, and other internal organs. The research on spinal cord and nerves segmentation is not common. Spinal cord and nerves has complicated shape and structure. It divides into branches outside the vertebral column, and the branches are really tiny. Besides, the intensity, shape, and position of spinal cord and nerves may be so different from one slice to the next. Because of the above reasons, accurate spinal cord and nerves segmentation is still a challenging problem.

Most of the segmentation techniques for medical image segmentation may include an interactive mechanism. Statistical shape models (SSMs) based methods are popular for medical image segmentation [3]. Smyth et al. [4] proposed a active shape models (ASMs) based method in lateral Dual Energy X-ray Absorptiometry (DXA) images of the spine, which can accurately and robustly locate vertebrae. Brejl and Sonka [5] provided the methodology for fully automated model-based image segmentation, using the mean shape model and shape-variant Hough transform to determine an approximate location of target objects. Carballido-Gamio et al. [6] applied Normalized Cuts [7] with the Nyström approximation method, which admits combinations of different features. Huang et al. [8] presented an automatic detection method for vertebra regions, followed by an iterative normalized-cut segmentation algorithm for vertebra segmentation.

3   3D Segmentation and Surface Model Reconstruction

3.1   3D Random Walker Segmentation Algorithm

For medical image segmentation, it is hard to use an automatic method to obtain precise results. Even using learning-based methods, due to the variations between patients, it is still a difficult job without human labeling. Therefore, we use a semi-automatic segmentation method, Random Walker Segmentation proposed by Grady [1], in this work. In this method, an user has to initially give some labels as input seed points for background and foreground. The probabilities of all unlabeled points reaching the labeled points, are computed by this algorithm.

Then the segmentation results can be obtained by assigning each pixel to the label with the highest probability. The algorithm overview is described below, and more details can be found in [1]. Based on graph theory, a graph consists of a pair $G = (V,E)$, with vertices set $v \in V$ and edge set $e \in E$. Between neighboring vertices v_i and v_j, there is an edge, e_{ij}. The weight of it is denoted as w_{ij}, depending on the difference between intensity g_i and g_j. The weighting function is shown below

$$w_{ij} = exp(-\beta(g_i - g_j)^2) \tag{1}$$

The Dirichlet integral for a field u in region Ω is defined as

$$D[u] = \frac{1}{2}\int_\Omega |\nabla u|^2 d\Omega \tag{2}$$

$D[u] = 0$ if u is in smooth area, therefore, e.q. (2) can be rewritten as

$$D[x] = \frac{1}{2}x^T LX \tag{3}$$

where the Laplacian matrix L is defined as

$$L_{ij} = \begin{cases} d_i, & \text{if } i = j \\ -w_{ij}, & \text{if } v_i \text{ and } v_j \text{ are adjacent nodes} \\ 0, & \text{otherwise} \end{cases} \tag{4}$$

where d_i is the sum of weight that related to v_i, L_{ij} is indexed by v_i and v_j .

User labels some points as initial seed points, x_M, and the unlabeled points x_U can be solved by minimizing the following function

$$
\begin{aligned}
D[x_U] &= \frac{1}{2} \begin{bmatrix} x_M^T & x_U^T \end{bmatrix} \begin{bmatrix} L_M^T & B \\ B^T & L_U^T \end{bmatrix} \begin{bmatrix} x_M \\ x_U \end{bmatrix} \\
&= \frac{1}{2}\left(x_M^T L_M x_M + 2x_U^T B^T x_M + x_U^T L_U x_U \right)
\end{aligned} \tag{5}
$$

L_M and L_U is the Laplacian matrix for labeled and unlabled points, and B is the matrix of the left elements of the Laplacian matrix. By minimizing the above equation, the solution can be found by solving the following linear system.

$$L_U x_U = -B^T x_M \tag{6}$$

By solving a sparse linear system, the probabilities of all points for all labels are known, and the segments for all labels can be obtained.

The real STIR-MRI spinal cord and nerves images are in 3D space. When the 2D random walker algorithm is employed, user can only give labels and apply the segmentation algorithm on one image once a time. Thus, it is more efficient to extend the original 2D random walker algorithm to 3D image segmentation [9]. In 2D random walker algorithm, the neighboring vertices in Laplacian matirx are considered in 4-connected neighborhood. To implement 3D random walker segmentation, the definition of Laplacian matrix is changed in the way that the number of neighboring vertices are considered in 3D space. It is increased from four to six by adding the vertices of upper and lower position.

When solving the probability map, it is difficult to process all the 3D data at once, due to the huge number of unlabeled points which requires a considerable amount of memory. So we apply a slice-section strategy, dividing whole slices into many sections, each contains five images. We apply the 3D random walker segmentation once for 5 slices. To reduce the times of user labeling of seed points, we only give initial seed points on the middle slice, which contains more average information of neighboring slices.

There are three views of STIR-MRI spinal cord and nerves images, and the shape and position are different in each view, as depicted in Fig. 1. In Axial view, spinal cord is usually in the center of image, and the change of intensity is not significant. In Coronal view, spinal cord is in the middle-upper position, and the shape varies a lot. Because of the middle slices show more complete spinal cord region than the other slices, we start doing segmentation from these silces to the beginning and the end of slices. In Sagittal view, spinal cord is S-shaped and also in the center of slices. Spinal cord in Sagittal view is similar to Coronal view, the middle slices contain more information, and the region of other slices diminished from it. Therefore, we also start the segmentation process from the middle slices.

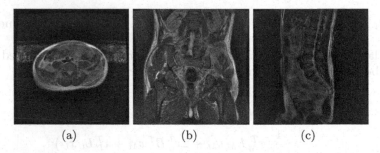

(a)   (b)   (c)

Fig. 1. Three different views of spinal cord and nerves. (a) Axial view. (b) Coronal view. (c) Sagittal view.

In our method, we just focus on the spinal cord and nerves segmentation. Therefore, user only has to give two kinds of seed points, foreground seeds for spinal cord and nerves region, setting the probabilities to 1, and background seeds for the other region we do not care about, thus setting the probabilities to 0. Some examples of input scribbled images are shown in Fig. 2. Moreover, because of the primary position of spinal cord is regular, we use a binary mask to set the probabilities of the region, which surely does not belong to the spinal cord, to 0. It not only decreases the computation time, but also reduces the time for scribbling background seeds.

The intensity, position and shape of spinal cord and nerves will change significantly from slice to slice in 3D STIR-MRI dataset, making the segmentation work more difficult. So after obtaining the results of 3D random walker, we use them as new initial seed points, and do 2D random walker segmentation on each slice, getting a better solution. We also provide an interface for modifying the outputs. Based on the previous results, user can give additional labels on the image, and apply the 2D random walker segmentation to generate more accurate segments.

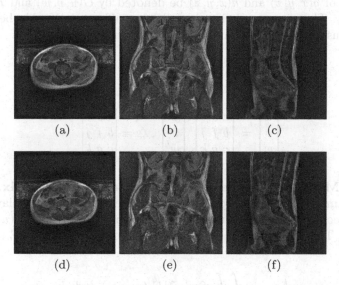

(a)   (b)   (c)

(d)   (e)   (f)

Fig. 2. Three different views of scribbled spinal cord and nerves images. Foreground seeds are scribbled by red line; background seeds are scribbled by blue line. (a) Axial view. (b) Coronal view. (c)Sagittal view. (d)-(f) Segmentation result of (a)-(c).

3.2   3D Point Set Registration

There are three views of 3D STIR-MRI spinal cord and nerves images - axial, coronal and sagittal. Each view has higher resolution in the intra-slice and poorer resolution in the inter-slice so that they are not the same segmentation results from different views. In order to improve the accuracy of the segmentation results, we integrate the segmentation results from the STIR-MRI of three different views into one to obtain more complete 3D reconstruction. We adopt the Fourier moment matching method [10] for the 3D registration of the three surface point data sets. The method is based on the fact that the affine transform between two images corresponds to a related affine transform between their Fourier spectrums, and the moments for the corresponding Fourier spectrum distributions can be calculated as probability density function. Then the affine registration parameters can be estimated by minimizing the affine relationship between the moments for the Fourier spectrums of the two images. The translation vector can be also estimated by the cross power spectrum. In order to further extend the algorithm to solve point set registration problem, the algorithm represents the 3D segmentation surface data as point sets in the 3D space as a binary image. In addition, the algorithm employs a distance weighting scheme to reduce the influence of outlier data points to achieve better robustness.

3D Affine Fourier Transform. Consider two 3D image functions $g(x, y, z)$ and $h(x, y, z)$, which are related by an affine transformation, i.e. $h(x, y, z) = g(ax + by + cz + d, ex + fy + gz + h, ix + jy + kz + l)$. Assume the Fourier transforms of $g(x, y, z)$ and $h(x, y, z)$ be denoted by $G(u, v, w)$ and $H(u, v, w)$, respectively. Then, we can derive the following affine relationship between the Fourier transforms $G(u, v, w)$ and $H(u, v, w)$ given as follows:

$$|G(u, v, w)| = \frac{1}{|\triangle|}|H(u', v', w')| \tag{7}$$

where

$$\begin{bmatrix} u \\ v \\ w \end{bmatrix} = \begin{bmatrix} a & e & i \\ b & f & j \\ c & g & k \end{bmatrix} \begin{bmatrix} u' \\ v' \\ w' \end{bmatrix}, \triangle = \begin{vmatrix} a & e & i \\ b & f & j \\ c & g & k \end{vmatrix} \tag{8}$$

Moment Matching Approach to Estimating Affine Matrix. The 3D affine parameters $(a, b, c, e, f, g, i, j, k)$ can be estimated by similar moment matching technique from the moments of the Fourier spectrums $F_1(u, v, w)$ and $F_2(u, v, w)$. The $(\alpha + \beta + \gamma)$-th moment for the Fourier spectrum $|F_n(u, v, w)|$ is defined as

$$m_{\alpha,\beta,\gamma}^k = \int \int u^\alpha v^\beta w^\gamma |F_k(u, v, w)| du dv dw \tag{9}$$

Thus, we have the following relationship for the first-order moments in the Fourier spectrums.

$$\begin{bmatrix} m_{1,0,0}^1 \\ m_{0,1,0}^1 \\ m_{0,0,1}^1 \end{bmatrix} = \begin{bmatrix} a & e & i \\ b & f & j \\ c & g & k \end{bmatrix} \begin{bmatrix} m_{1,0,0}^2 \\ m_{0,1,0}^2 \\ m_{0,0,1}^2 \end{bmatrix} \tag{10}$$

For the second-order Fourier moments, we can derive the following relationship.

$$\begin{bmatrix} m_{2,0,0}^1 \\ m_{0,2,0}^1 \\ m_{0,0,2}^1 \\ m_{1,1,0}^1 \\ m_{1,0,1}^1 \\ m_{0,1,1}^1 \end{bmatrix} = \begin{bmatrix} a^1 & e^2 & i^2 & 2ae & 2ai & 2ei \\ b^1 & f^2 & j^2 & 2bf & 2bj & 2fj \\ c^1 & g^2 & k^2 & 2cg & 2ck & 2gk \\ ab & ef & ij & af+be & aj+bi & ej+fi \\ ac & eg & ik & ag+ce & ak+ci & ek+gi \\ bc & fg & jk & bg+cf & bk+cj & fk+gj \end{bmatrix} \begin{bmatrix} m_{2,0,0}^2 \\ m_{0,2,0}^2 \\ m_{0,0,2}^2 \\ m_{1,1,0}^2 \\ m_{1,0,1}^2 \\ m_{0,1,1}^2 \end{bmatrix} \tag{11}$$

The relationship of the first-order and second-order Fourier moments, given in e.q. (10) and (11), can be used for the least-squares estimation of the above nine 3D affine parameters.

4   Experimental Result

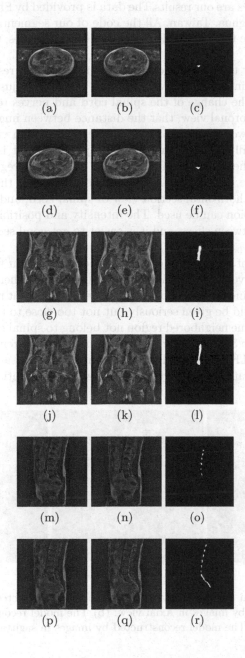

Fig. 3. The segmentation result of spinal cord and nerves . The first column is the original image. The second column is our result, outlined with green line. The third column is the binary result, white for foreground and black for background. (a)-(f) is in axial view; (g)-(l) is in coronal view; (m)-(r) is in sagittal view.

We test our 3D segmentation method on three different STIR-MRI scan datasets. The following images are our results.The data is provided by Show Chwan Health Care System, Changhua, Taiwan. All the code of our segmentation work is written in Matlab. Since the STIR-MRI datasets are real images, there is no ground truth of them.

We find out that the shape and intensity of segmented results really be significantly different in sagittal view. Although Fig.3(m) is just 5.5mm in z-axis between Fig.3(p), the change of the spinal cord and nerves region is obviously; it also happens in coronal view, that the distance between images is only 4.5mm in z-axis. In these two views, there both are vertical section view of spinal cord. Because of the small cross sectional area of spinal cord, it is just a few information of it, and the variation of spinal cord and nerves region in serial slices makes the job more difficult. In Axial view, even though the gap is 10mm in axial view, it is the horizontal section view of spinal cord, and due to the height of it, more information can be used. The intensity and position of spinal cord do not change a lot between slices, and it is easier to get good segmentation results in this view.

In our experiment, we start the segmentation work from the middle part of spinal cord and nerves, which consists most information, then to the beginning and ending parts. Since the label positions play an important role in the segmentation work, it should be given seriously but not too close to the edge, or the result may include some neighbored region not belong to spinal cord and nerves. In Fig.4, it shows three 3D spinal nerves surface model of different datasets of view, reconstructed by VTK [2]. After 3D model is built, it can be saved as STL (STereoLithography) format, and displayed by visualization application, like 'Paraview'.

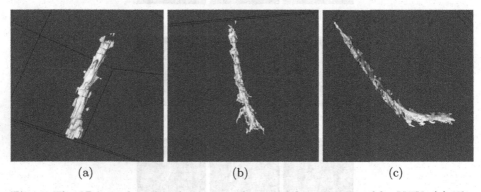

| (a) | (b) | (c) |

Fig. 4. The 3D spinal cord and nerves surface model reconstructed by VTK. (a) The model reconstructed by images in Axial view. (b) The model reconstructed by images in Coronal view. (c) The model reconstructed by images in sagittal view.

After these 3D surface model reconstructed, we can easily extract the point cloud of each model, and use them for 3D point set registration. We use the point cloud of sagittal view as reference data, and the others two as target data. The point cloud registration results are shown in Fig.5.

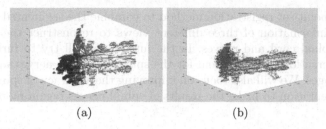

(a)   (b)

Fig. 5. The result of 3D point set registration. (a) The data distribution before registration. (b) The data distribution after registration. In (a)-(b), the blue points are of reference data, and red points are of target data, which is of axial view.

After the transformation function obtained, we resample the target data to the geometry of the reference data, and get a new point cloud which contains the spinal cord and nerves information of three views. We use the new point cloud to reconstruct the 3D surface model by bipartite polar classification [11], as shown below.

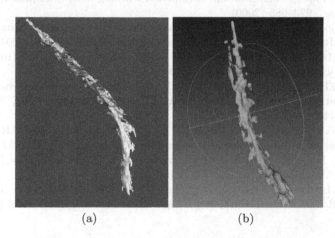

(a)   (b)

Fig. 6. The comparison of surface model that doing 3D registration before and after. (a) The 3D surface model of sagittal view before 3D registration. (b) The 3D surface model contains information of three views after 3D registration.

Before 3D point set registration, the surface model of spinal cord and nerves is flat, which is not the actual shape. After registration, the surface model becomes more three-dimensional, and the tube-shaped structure of spinal cord is more obvious.

5   Conclusion

In this paper, we present an interactive 3D random walker segmentation system for segmenting spinal cord and nerves from STIR-MRI images. In addition, we apply a 3D point-set registration method to combine the segmented spinal cord and nerves information of three different views to reconstruct the 3D surface model for spinal cord and nerves. In the future, we will try to further reduce the user interaction in our system for the spinal cord and nerves segmentation from 3D images. We will also focus on improving the accuracy of spinal cord and nerves segmentation to make the results of spinal nerves region more accurate.

References

1. Grady, L.: Random walks for image segmentation. IEEE Trans. on Pattern Analysis and Machine Intelligence 28(11), 1768–1783 (2006)
2. http://www.vtk.org/
3. Heimann, T., Meinzer, H.P.: Statistical shape models for 3D medical image segmentation: A review. Medical Image Analysis 13(4), 543–563 (2009)
4. Smyth, P.P., Taylor, C.J., Adams, J.E.: Automatic measurement of vertebral shape using active shape models. Image and Vision Computing, 705–714 (1996)
5. Brejl, M., Sonka, M.: Object localization and border detection criteria design in edge-based image segmentation: Automated learning from examples. IEEE Trans. Med. Imag., 973–985 (2000)
6. Carballido-Gamio, J., Belongie, S.J., Majumdar, S.: Normalized cuts in 3-D for spinal MRI segmentation. IEEE Trans. Med. Imag. 23(1), 36–44 (2004)
7. Shi, J., Malik, J.: Normalized cuts and image segmentation. IEEE Trans. Pattern Anal. Mach. Intell. 22(8), 888–905 (2000)
8. Huang, S.H., Chu, Y.H., Lai, S.H., Novak, C.L.: Learning-Based Vertebra Detection and Iterative Normalized-Cut Segmentation for Spinal MRI. IEEE Trans. Med. Imag. 28(10), 1595–1605 (2009)
9. Cheng, Y.Y., Chang, H.M., Su, H.R., Lai, S.H., Liu, K.C., Lin, C.H.: 3D Liver Segmentation and Model Reconstruction from CT Images. iCBEB, 654–657 (2012)
10. Su, H.R., Lai, S.H.: CT-MR Image Registration in 3D K-Space Based on Fourier Moment Matching. PSIVT (2), 299–310 (2011)
11. Chen, Y.L., Lee, T.Y., Chen, B.Y., Lai, S.H.: Bipartite Polar Classification for Surface Reconstruction. Computer Graphics Forum 30(7), 2003–2010 (2011)

Speeding Up the Decisions of Quad-Tree Structures and Coding Modes for HEVC Coding Units

Shen-Chuan Tai, Chia-Ying Chang, Bo-Jhih Chen, and Jui-Feng Hu

Institute of Computer and Communication Engineering, Department of Electrical Engineering
National Cheng Kung University
sctai@mail.ncku.edu.tw

Abstract. High Efficiency Video Coding (HEVC) is being developed by the joint development of ISO/IEC MPEG and ITU-T Video Coding Experts Group (VCEG) and is expected to be a popular next-generation video codec in the future. HEVC can provide higher compression ratio compared to H.264/AVC standard; however, the coding complexity is dramatically increased as well. In this thesis, a fast algorithm for coding unit decision is proposed to reduce the burden of the encoding time in HEVC. The proposed algorithm exploits the temporal correlation in the neigh-boring frames of a video sequence to avoid the unnecessary examinations on CU quad-trees. In addition, based on an adaptive threshold, the best prediction mode is early determined to SKIP mode for reducing the exhaustive evaluations at prediction stage. The performance of the proposed algorithm is verified through the test model for HEVC, HM 5.0. The experimental results show that the proposed algorithm can averagely achieve about 27%, 33%, 20%, and 21% total time encoding time reduction under Low-Delay High Efficiency, Low-Delay Low Complexity, Random-Access High Efficiency, and Random-Access Low Complexity configurations respectively with a negligible degradation of coding performance.

The rest of this thesis is organized as follows. Section 1 gives a brief introduction to the HEVC encoder, includes overview of HEVC coding standard. Simultaneously, some previously proposed methods for fast CU decision are also investigated in this chapter. Section 2 proposes a new early termination algorithm for CU decision. Section 3 demonstrates the experimental results verified through the test model for HEVC, HM 5.0 [4]. Section 4 concludes the studies presented in this thesis.

Keywords: High Efficiency Video Coding, Coding Unit, Quad-Tree Structure, Coding Mode, Fast Algorithm.

1 Introduction

1.1 Overview of Video Compression

In recent years, many video compression standards have been proposed, such as MPEG-2/-4 [1], H.261/H.263 and H.264/AVC [2]. Some of these standards produce huge commercial interest and gain popular acceptance in the marketplace. For example, the success of digital TV and Digital Versatile Disk (DVD) is exactly based upon

J.-S. Pan et al. (Eds.): *Advances in Intelligent Systems & Applications*, SIST 21, pp. 393–401.
DOI: 10.1007/978-3-642-35473-1_40   © Springer-Verlag Berlin Heidelberg 2013

MPEG-2. After the successful experience of MPEG-2, Joint Video Team (JVT), a collaborative group of ITU-T Video Coding Experts Group (VCEG) and Moving Picture Experts Group (MPEG), jointly proceeded to develop a new compression standard known as H.264/AVC. It is reported that H.264/AVC provides significantly higher performance in both visual quality and data compression than MPEG-2. Nowadays, a large number of video applications have been dominated by H.264/AVC. These applications utilize highly integrated semiconductor solutions to reduce costs and exhibit the benefit of high-efficiency compression and high-speed decompression in H.264/AVC.

1.2   Overview of HEVC

The encoding layer of HEVC is still based on traditional approach as founded in previous standard designs, including block-based motion-compensated prediction, spatial redundancy prediction, 2D transformation of residual difference signals, and adaptive entropy coding. HEVC integrates many efficient coding tools [9] and provides higher coding performance than earlier video coding standards such as H.264/AVC. The general structure of the HEVC encoder is similar to H.264/AVC. However, there are a number of distinguishing features in HEVC.

Different from previous standards, large sized blocks with flexible structure of quad-tree are applied in HEVC. For this, the HEVC draft specification [6] adopts variable sized coding units (CUs), which define a sub-square region in a frame. HEVC replaces the macroblock scheme as known in previous video coding standards by CU-based quad-tree structure. The encoding process recursively investigates the CUs on the quad-tree. Each CU contains one or several variable sized prediction units (PUs) and transform units (TUs).

The following sub-sections present some specific features in HEVC mentioned above.

Coding Unit (CU)
Coding unit (CU) is considered to be the fundamental processing unit just as macroblock in H.264/AVC. CUs are restricted to be square in shape but conserve the characteristic of variable sizes. CUs vary in size with a wide range corresponding to the depth of the CU quad-tree. A CU with the largest size is called Largest Coding Unit (LCU) and the Smallest Coding Unit (SCU) is defined from the size of the LCU and the maximum depth of the CU quad-tree.

Prediction Unit (PU)
Prediction Unit (PU) is the basic unit used for carrying information related to prediction stage. Each CU on a quad-tree enters its own prediction stage during the encoding process. PUs are not restricted to be square in shape in order to facilitate the partitions which match the boundaries of real objects in video frames.

Transform Unit (TU)
Transform Unit (TU) is used for the transform and quantization of residual signals resulting from prediction stage. TUs must be smaller than or equal to the corresponding CU. Each CU may contain one or more TUs which vary in size from 4×4 up to 32×32.

The TUs in a CU are arranged in quad-tree structure known as residual quad-tree (RQT) just similar to the CU quad-tree. For each CU, a residual quad-tree must be defined, which means that several nested TU quad-trees are embedded in a CU quad-tree. Moreover, under HE configurations, non-square transform (NSQT) is applied for transform step to achieve further coding performance.

2   Proposed Algorithm

Consider previously proposed fast CU decision algorithms. Gweon [8] and Choi [7] early terminate the encoding process within the CU quad-trees based on predefined conditions. Leng [11] and Kim [10] have a lot in common such as referring to spatially neighboring CUs or temporally co-located CUs. The proposed algorithm combines CU quad-tree pruning method (CUQTP) at CU level and early SKIP mode decision at the prediction stage to reduce the computational complexity of the encoding process. At the CU level, some specific depths of a CU quad-tree can be eliminated by referring the coding information of the co-located CUs and CUs adjacent to the co-located CUs. At the prediction stage, the best prediction mode can be early determined to SKIP mode based on adaptive thresholds, and the exhaustive evaluations of INTER modes and INTRA modes is omitted. The overall flowchart of the proposed algorithm is shown in Figure 1. The proposed CU quad-tree pruning method is introduced in Section 2.1, and the proposed early SKIP mode decision is introduced in Section 2.2.

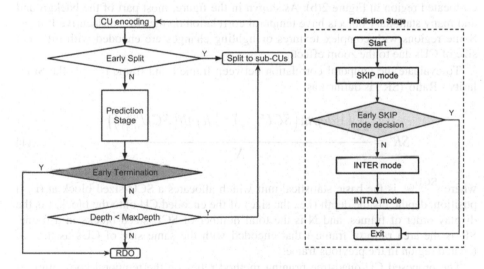

Fig. 1. The overall flowchart of the proposed algorithm

2.1   Coding Unit Quad-Tree Pruning Method

As discussed in Section 1.2, homogeneous textured or low-motion regions tend to be encoded with large CUs. On the other hand, complex textured or high-motion regions tend to be encoded with small CUs. The information of a CU consists of residual signals, motion vectors, and other side information. For several consecutive frames, some features stay the same such as the video resolution, the stationary background and the speed of moving object, and the complex regions and the homogeneous regions stay the same too. Hence, there exists the coding information correlation among consecutive frames [11]. According to these features, the examinations of CUs in specific depth can be passed over by analyzing the coding information in neighboring frames during the encoding process.

(a) The encoding result of the 6th frame   (b) The encoding result of the 7th frame

Fig. 2. An example of temporal correlation between neighboring frames in PartScene

Figure 2(a) and Figure 2(b) show two consecutive B-frames in the sequence. The yellow region in Figure 2(b) is encoded with the same size of CUs as the temporally co-located region in Figure 2(b). As shown in the figure, most part of the background and many stationary objects have temporal correlation on the two consecutive frames. Some regions with complex textures or lighting changes are encoded with different size of CUs due to the zoom effect.

To evaluate the temporal correlation between frame t and frame t − 1, the Similarity - Ratio^t (SR^t) is defined as:

$$SR^t = \frac{\sum_N \left(1 \middle| depth\left(SCU^t_{(i,j)}\right) = \middle| depth\left(SCU^{t-1}_{(i,j)}\right)\right)}{N} \tag{1}$$

where $SCU^t_{(i,j)}$ is the basic statistical unit which allocates a SCU-sized block at (i , j) position, depth(.) is the depth (i.e., the size) of the encoded CU onto the block, t is the display order of frames, and N is the total number of SCU-sized blocks in a frame. SR^t is the area ratio of frame t that encoded with the same size of CUs as the co-located region in the previous frame.

The proposed CU quad-tree pruning method relies on the temporal correlation of CU quad-tree. A cluster of CUs (CU$_R$), which consists of the temporally co-located CU and CUs adjacent to the co-located CU in the encoded neighboring frame. Figure 3 shows the relationship between the current CU (CU$_{current}$) and the CU$_R$. There are uncertain numbers of CU$_s$ in the CU$_R$. Before the CU$_{current}$ enters the prediction stage, the coding information of the CU$_R$ is used for determining whether to perform the prediction stage or not. There are two additional conditions, "Early Split" and "Early Termination", included in the general encoding process.

First, the Early-Split condition is defined as Figure 3.

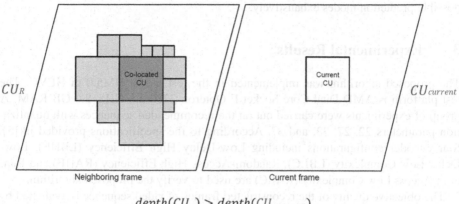

$$depth(CU_x) > depth(CU_{current})$$
$$\vee$$
$$depth(CU_x) = depth(CU_{current}) \; \wedge \; mode(CU_x)! = SKIP,$$
$$\text{for all } CU_x \in CU_R$$

Fig. 3. Condition for early split

Second, the Early-Termination condition is defined as Figure 4.

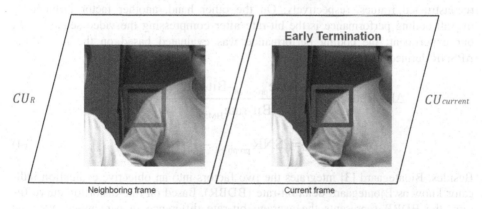

$$depth(CU_x) < depth(CU_{current}) \; \vee \; mode(CU_x) = SKIP,$$
$$\text{for all } CU_x \in CU_R$$

Fig. 4. Condition for early termination

2.2 Early Skip Mode Decision

As aforementioned, the encoder computes the R-D costs of all possible inter prediction modes and intra prediction modes to decide the best prediction mode. Since each of them entails high computational complexity, it is very desirable if the encoder can decide the best prediction mode at the earliest possible stage without evaluating all possible prediction modes exhaustively.

3   Experimental Results

The proposed algorithm was implemented in the test model HM5.0 of HEVC. The test platform is AMD Dual-Core Socket F Opteron 2220 2.8 GHz, 8.0 GB RAM. A group of experiments were carried out on the recommended sequences with quantization parameters 22, 27, 32, and 37. According to the specifications provided in [5], four encoder configurations including Low-Delay High Efficiency (LBHE), Low-Delay Low Complexity (LBLC), Random-Access High Efficiency (RAHE) and Random-Access Low Complexity (RALC) are used to verify the proposed algorithm.

The objective quality of the reconstructed frames of video sequence is evaluated by the peak signal-to-noise ration (PSNR), which is defined as:

$$PSNR = 10 \times \log_{10} \frac{255^2}{\frac{1}{M}\sum_{n=1}^{M}(o_n - r_n)^2} \tag{2}$$

where M is the number of samples, and o_n and r_n are the gray level of the original and reconstructed frames, respectively. On the other hand, another factor influencing overall coding performance is the bit-rate after compressing the video sequence. In our experiment, the coding performance was evaluated based on the ΔBit-Rate, ΔPSNR defined as follows:

$$\Delta\text{Bi-Rate}(\%) = \frac{\text{Bit-rate}_{proposed} - \text{Bit-rate}_{HM5.0}}{\text{Bit-rate}_{HM5.0}} \times 100\% \tag{3}$$

$$\Delta\text{PSNR}(dB) = \text{PSNR}_{proposed} - \text{PSNR}_{HM5.0} \tag{4}$$

Besides, Bjontegaard [3] integrates the two factors into an objective evaluation indicator know as Bjontegaard delta bit-rate (BDBR). Based on rate-distortion curve fitting, the BDBR represents the average bit-rate difference in percentage over the whole range of PSNR. This evaluation indicator is also applied in our experiment. For the complexity evaluation, the total execution time of the proposed algorithm is assessed in comparison to that of HM 5.0. The time reduction in computational was evaluated based on ΔEncT(%) as follows:

$$\Delta\text{EncT}(\%) = \frac{\text{Time}_{proposed} - \text{Time}_{HM5.0}}{\text{Time}_{HM5.0}} \times 100\% \tag{5}$$

Thirteen benchmark test sequences are selected to be encoded with four different quantization parameters (QPs: 22, 27, 32, 37) under different encoder settings.

Table 1 shows the detailed information of the test sequences. These sequences have various characteristics including motion types, texture types, and resolutions.

Table 1. Benchmark test sequences

Class	Sequence	Frames	FrameRate
B 1920×1080	Kimono	240	24 fps
	ParkScene	240	24 fps
C 832×480	BasketballDrill	500	50 fps
	BQMall	600	60 fps
D 416 ×240	BasketballPass	500	50 fps
	BQSquare	600	60 fps
	BlowingBubbles	500	50 fps
	RaceHorses	300	30 fps
E 1280× 720	Vidyo1	600	60 fps
	Vidyo3	600	60 fps
	Vidyo4	600	60 fps

Figure 5 shows the worst cases of R-D curves compared with that of HM 5.0. As shown in the figures, the RD curves almost overlap in each case. The results indicate that the proposed algorithm has almost the same coding performance as that of HM 5.0.

Table 2 summarizes results of the proposed algorithm with the BDBR and average time reduction. It should be noted that positive values represent bit-rate increase (i.e., PSNR degradation). The results of average BDBR are 0.39%, 0.47%, 0.10% and 0.10% for LBHE, LBLC, RAHE, RALC, respectively, and the average time reductions are -27%, -33%, -20% and -22% for the four configurations.

4   Conclusions

In this thesis, a fast algorithm for speeding up the decision of quad-tree structures and coding modes for HEVC coding units is proposed. The CU-based quad-tree structure is responsible for the complexity of the encoding process. The proposed algorithm accelerates the encoding process on two aspects including CU level and prediction stage.

At the CU level, the proposed algorithm prunes the CU quad-tree to a condensed shape based on the temporal correlation between neighboring frames. At prediction stage, the best prediction mode is early determined to SKIP mode based on an adaptive threshold. The main contribution of this thesis is to provide a simple and efficient fast algorithm with very negligible loss of coding performance compared to the original HEVC encoder. On the other hand, the proposed algorithm can be easily combined with the early termination method of mode decision [10] which was adopted in the HM5.0 anchor. In terms of the complexity reduction, the combined algorithm can speed up the encoding process considerably with reasonable degradation of coding performance.

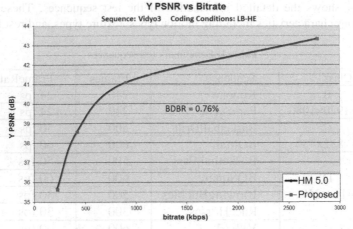

(a) RD curve of Vidyo3 under LBHE

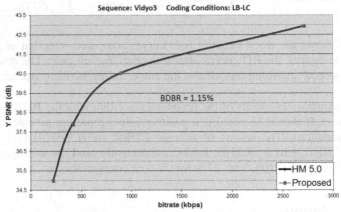

(b) RD curve of Vidyo3 under LBLC

Fig. 5. The R-D performance of the proposed algorithm with Vidyo3

Table 2. Overall experimental results with BDBR (%) and ΔEncT (%)

Class	Sequence	LBHE		LBLC		RAHE		RALC	
		BDBR	ΔEncT	BDBR	ΔEncT	BDBR	ΔEncT	BDBR	ΔEncT
B	Kimono	0.33	-24	0.35	-26	-0.10	-15	0.06	-16
	ParkScene	0.50	-25	0.79	-29	-0.05	-15	0.09	-17
C	BasketballDrill	0.36	-15	0.31	-21	0.07	-10	0.05	-11
	BQMall	0.38	-18	0.50	-25	0.14	-14	0.15	-15
	PartyScene	0.31	-14	0.45	19	0.11	-13	0.11	-13
	RaceHorse	0.31	-14	0.45	19	0.11	-13	0.11	-13
D	BasketballPass	0.21	-14	0.21	-19	0.20	-12	0.02	-12
	BQSquare	0.25	-15	0.30	-21	0.18	-14	0.17	-15

Table 2. (*continued*)

	BlowingBubbles	0.42	-15	0.34	-20	0.22	-12	0.12	-12
	RaceHorse	0.21	-11	0.16	-15	0.24	-12	0.23	-11
	Vidyo1	0.69	-36	0.90	-43	0.09	-28	0.11	-31
E	Vidyo3	0.76	-37	1.15	-43	0.07	-28	0.06	-31
	Vidyo4	0.50	-37	0.44	-44	0.04	-29	0.04	-32
	Average	0.39	-27	0.47	-33	0.10	-20	0.10	-21

References

1. Information tecchonology: Generic coding of moving pictures and associated audio information, Std. (1995)
2. Advanced video coding for generic audiovisual services, Std. (2007)
3. Bjøntegaard, G.: Calculation of average PSNR differences between RD-curves, ITU-T VCEG-M33 (2001)
4. Bossen, F.: HM reference software 5.0 (2011),
   https://hevc.hhi.fraunhofer.de/svn/svnHEVCSoftware/
5. Bossen, F.: Common HM test conditions and software reference configurations (January 2012), http://phenix.int-evry.fr/jct/docenduser/documents/7Geneva/wg11/JCTVC-G1200-v2.zip
6. Bross, B., Han, W.-J., Sullivan, G.J., Ohm, J.-R., Wiegand, T.: High Efficiency Video Coding (HEVC) text specification Working Draft 5 (February 2012),
   http://phenix.int-evry.fr/jct/docenduser/documents/7Geneva/wg11/JCTVC-G1103-v12.zip
7. Choi, K.H., Jang, E.S.: Coding tree pruning based cu early termination (July 2011), http://phenix.int-evry.fr/jct/docenduser/documents/6Torino/wg11/JCTVC-F092-v3.zip
8. Gweon, R.H., Lee, Y.L., Lim, J.Y.: Early termination of cu encoding to reduce hevc complexity (July 2011), http://phenix.int-evry.fr/jct/docenduser/documents/6Torino/wg11/JCTVC-F045-v1.zip
9. Jeong, S.Y., Lim, S.C., Lee, H.Y., Kim, J.H., Choi, J.S., Choi, H.C.: Highly efficient video codec for entertainment-quality. ETRI Journal 33, 145–154 (2011)
10. Kim, J.H., Jeong, S.Y., Cho, S.H., Choi, J.S.: Adaptive coding unit early termination algorithm for hevc. In: 2012 IEEE International Conference on Consumer Electronics (ICCE), pp. 261–262 (January 2012)
11. Leng, J., Sun, L., Ikenaga, T., Sakaida, S.: Content based hierarchical fast coding unit decision algorithm for hevc. In: 2011 International Conference on Multimedia and Signal Processing (CMSP), vol. 1, pp. 56–59 (May 2011)

Optimal GOP Size of H.264/AVC Temporal Scalable Coding

Wei-Lune Tang and Shih-Hsuan Yang

Department of Computer Science and Information Engineering
National Taipei University of Technology
1, Sec. 3, Chung-Hsiao E. Rd., Taipei, Taiwan
allen08312002@hotmail.com
shyang@ntut.edu.tw

Abstract. Scalable video coding (SVC) encodes image sequences into a single bit stream that can be adapted to various network and terminal capabilities. The H.264/AVC standard includes three kinds of video scalability, spatial scalability, quality scalability, and temporal scalability that renders different frame rates from coded bit streams. The set of pictures from one temporal base layer to the next is referred to as a group of pictures (GOP). In this paper, we investigate the GOP size that achieves the best rate-distortion performance. The factors that affect the optimal GOP size, such as video characteristics and quantization parameter (QP), are explored as well.

Keywords: scalable video coding, rate-distortion performance, coding complexity, H.264/AVC, GOP size.

1 Introduction

Video is used in diversified situations. The same video content may be delivered in different and variable transmission conditions (such as bandwidth), rendered in various terminal devices (with different resolution and computational capability), and served for different needs. Adaptation of the same video content to every specific purpose is awkward and inefficient. Scalable video coding (SVC), which allows once-encoded content to be utilized in flexible ways, is a remedy for using video in the heterogeneous environments [1].

Video scalability refers to the capability of reconstructing lower-quality video from partial bit streams. An SVC signal is encoded at the highest quality (resolution, frame rate) with appropriate packetization, and then can be decoded from partial streams for a specific rate or quality or complexity requirement. There are three common categories of scalability in video: spatial (resolution), temporal (frame rate), and quality (fidelity). The major expenses of SVC, as compared to state-of-the-art non-scalable single-layer video coding, are the gap in compression efficiency and increased encoder and decoder complexity. The H.264 standard, also known as MPEG-4 AVC (Advanced Video Coding) [2], has been dominating the emerging video applications including digital TV, mobile video, video streaming, and Blu-ray discs. The wide

J.-S. Pan et al. (Eds.): *Advances in Intelligent Systems & Applications*, SIST 21, pp. 403–412.
DOI: 10.1007/978-3-642-35473-1_41   © Springer-Verlag Berlin Heidelberg 2013

adoption and versatility of H.264/AVC leads to the inclusion of scalability tools in its latest extension [3].

This paper investigates the GOP (group of pictures) size that achieves the best rate-distortion performance for H.264/AVC temporal scalable video coding. There are several advantages using smaller GOP size, such as increasing the key picture, easy controlling drift, reducing the encoding complexity and, reducing the delay time in decoding. However, the coding performance may degrade for small GOP size due to limited temporal reference. The best GOP size under different video characteristics and quantization parameters (QP) is thus worth exploring. The rest of the paper is organized as follows. Section 2 provides some background information about SVC. Experiments on H.264 temporal scalability with various GOPSize configurations of the JSVM (official H.264/AVC reference software) are given in Section 3. Conclusion and future work are given in Section 4.

2   H.264 Scalable Video Coding

2.1   Overview of H.264 Scalable Video Coding

H.264 includes two layers in structure: video coding layer (VCL) and network abstraction layer (NAL). Based on the core coding tools of the non-scalable H.264 specification, the SVC extension adds new syntax for scalability [2]. The representation of the video source with a particular spatio-temporal resolution and fidelity is referred to as an SVC layer. Each scalable layer is identified by a layer identifier. In JSVM, three classes of identifiers, T, D, and Q, are used to indicate the layers of temporal scalability, spatial scalability, and quality scalability, respectively. A constrained decoder can retrieve the necessary NAL units from an H.264 scalable bit stream to obtain a video of reduced frame rate, resolution, or fidelity. The first coding layer with identifier equal to 0 is called the base layer, which is coded in the same way as non-scalable H.264 image sequences. To increase coding efficiency, encoding the other enhancement layers may employ data of another layer with a smaller layer identifier.

2.2   Temporal Scalability

Temporal scalability, which is the main concern of this paper, provides coded bit streams of different frame rates. The temporal scalability of H.264 SVC is typically structured in hierarchical B-pictures, as shown in Fig 1. In this case, each added temporal enhancement layer doubles the frame rate. These dyadic enhancement layer pictures are coded as B-pictures that use the nearest temporally available pictures as reference pictures. The set of pictures from one temporal base layer to the next is referred to as a group of pictures (GOP). It is found from experiments that the GOP size of 8 or 16 usually achieves the best rate-distortion performance [3]. In [7] the authors propose an algorithm for dynamic GOP structures, but the result does not have any noticeable impact in full frame rate. Note that the GOP size also determines the total number of temporal layers (no. of temporal layers = (log2 GOPSize) + 1).

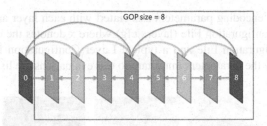

Fig. 1. Hierarchical B structure and the GOP size = 8

2.3   Spatial Scalability

Each layer of H.264 spatial scalability corresponds to a specific spatial resolution. In addition to the basic coding tools of non-scalable H.264, each spatial enhancement layer may employ the so-called interlayer prediction, which employs the correlation from the lower layer (resolution). There are three prediction modes of interlayer coding: interlayer intra prediction, interlayer motion prediction, and interlayer residual prediction. Accordingly, the up-sampled reconstructed intra signal, the macroblock partitioning and the associated motion vectors, or the up-sampled residual derived from the colocated blocks in the reference layer, are used as prediction signals. The interlayer prediction shall compete with the intra-layer temporal prediction for determining the best prediction mode.

2.4   Quality Scalability

Quality scalable layers have identical spatio-temporal resolution but different fidelity levels. H.264 offers two options for quality scalability, CGS (coarse-grain quality scalable coding) and MGS (medium-grain quality scalability). An enhancement layer of CGS is obtained by requantizing the (residual) texture signal with a smaller quantization step size (quantization parameter, QP). CGS incorporates the interlayer prediction mechanisms very similar to those used in spatial scalability, but with the same picture sizes for the base and enhancement layers. Besides, the up-sampling operations and the interlayer de-blocking for intra-coded reference layer macroblocks are omitted. Also, the interlayer intra and interlayer residual predictions are directly performed in the transform domain. SVC supports up to 8 CGS layers but the interlayer prediction is constrained to at most three CGS layers including the required base layer. Usually, a significant difference in QP, which corresponds to largely deviated bit rates, is expected in order to achieve good RD performance [3][4].

3   Investigation of Temporal Scalability

We comprehensively evaluate the rate-distortion performance for H.264 Temporal Scalability, with focuses on GOP size. Experiments were conducted with JSVM JSVM 9.19.13 [6], on nine test sequences with the frame rate of 30Hz, shown in Fig. 2. In JSVM, the primary encoding parameters are specified in the Main Configuration File (main.cfg), and the encoding parameters associated with each layer are specified in the individual Layer Configuration File (layerx.cfg) where x denotes the dependency_id. A typical Main Configuration File and a typical Layer Configuration File are shown in Table 1, where only the parameters important to our evaluations are listed.

(a)   (b)   (c)

(d)   (e)   (f)

(g)   (h)   (i)

Fig. 2. Test sequences, (a) Foreman (289), (b) Mobile (289), (c) Tempete (257), (d) City (289), (e) Bus (145), (f) Flower (241), (g) Soccer (289), (h) Football (257), (i) Harbour (257). The number in the parentheses indicates the number of frames to be encoded for simulation.

Table 1. Encoding parameters

(a) main.cfg

Parameter	Value	Remarks
FrameRate	30.0	
FramesToBeEncoded	289	No. of frames
GOPSize	1-64	
SearchMode	4	FastSearch
SearchRange	32	In full pels
NumLayers	1	CGS layers
LayerCfg	layer0.cfg	Layer configuration file

(b)layerx.cfg

Parameter	Value	Meaning
SourceWidth	352	Input frame width
SourceHeight	288	Input frame height
InterLayerPred	2	Inter-layer Prediction (0: no, 1: yes, 2:adaptive)
QP	20	Quantization parameters

The GOPSize is the main variable of experiment. In JSVM, the parameter GOPSize, whose minimum value is 1 and maximum value is 64, must be equal to a power of 2. The GOPSize depends on the value of FrameRateOut in the layer configuration file. Hence, the GOPSize must be less than FrameRateOut. For example, if frame rate is equal to 30Hz, the GOPSize must be less than 30, which may be 1, 2, 4, 8 or 16. We thus set the number of frames to be encoded as a multiple of GOPSize plus 1, to its maximum. (The extra one is added for accomplishing the reference pictures of the hierarchical B structure.) Hence, 289 frames are used for image sequences of 300 frames.

In the following, we compared the results of different GOP size with QP equal to 20. Due to the page limit, we primarily present the results for the three sequences: Foreman, Bus, and Flower. The results are shown in Fig. 3. Because the frame rate of each test sequence is equal to 30Hz, the GOP size must be less than 30. In Fig. 3, the rightmost point at each curve means the maximum frame rate. There are two temporal layers in GOPSize=2. The frame rate of the left point, the first point, is 15Hz, and the right point, the second point, is 30Hz. The GOPSize=4 has three temporal layers. The frame rate of the leftmost point is 7.5Hz, the middle point is 15Hz, and the rightmost point is 30Hz, and so on.

The Fig. 3 shows that the rate distortion of lower layer is better than higher layer because the temporal scalability of H.264 SVC is typically structured in hierarchical B-pictures. As the prediction times increase, the rate distortion of B-picture decreases. Therefore, getting more layers means the bit stream includes more B-pictures, and the average PSNR will decline. The rate distortion of GOPSize=2 and GOPSize=4 is inefficient using too much bit rate. The last two points of GOPSize=16 and GOPSize=8 are similar, which means the rate distortion at 15Hz and 30Hz is alike, but the first two points of rate distortion at GOPSize=8 are not good. In anthropology, humans can see a continuous image with frame rate above 10Hz. Hence, we are more concerned about the higher layer, the frame rate above 15Hz. In these experiments, the conclusion is that the performance of GOPSize=8 at the frame rate above 15Hz is similar to GOPSize=16, therefore, we recommend that the GOP size can be configured to 8. The next experiment, we would compare different values of QP at GOP size equal to 8 and 16.

Then, in the following experiment, we focus on GOPSize=8 and GOPSize=16, setting value of QP equal to 20, 24, 26, 28, 30, 32, 34, and 36. The results for Foreman are shown in Fig. 4. At the low QP (less than 28), the last two points for each curve are similar, but the distance between these two curves becomes far at QP equal to 30. The curve of GOPSize=8 is above GOPSize=16 at the value of QP less than 30, however, when the value of QP is larger than 30, the curve of GOPSize=8 is under GOPSize=16. We call this changing timing, QP=30, as a cross-point. The results for Mobile are shown in Fig. 5. The cross-point is at QP equal to 32 for Mobile. Due to the page limit, we present the results for the Foreman and Mobile. The other test sequences have the same curve, but the cross point is at different QP. Then, we calculated the cross-point for each test sequences, as shown in Table 2. When the value of QP is bigger, the cross-point will appear. The last two points are close at low QP, however, after the cross-point, these two curves will cross and then swap.

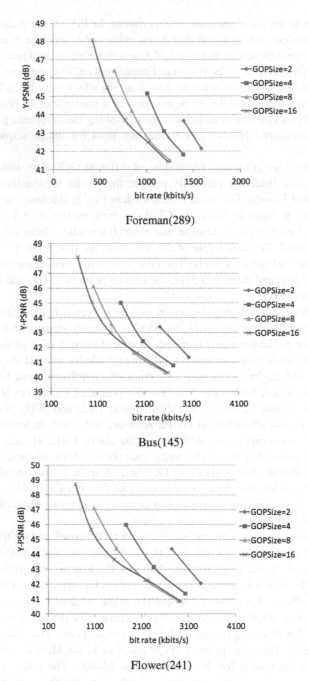

Fig. 3. Rate-distortion results, different GOP size, with JSVM 9.19.13

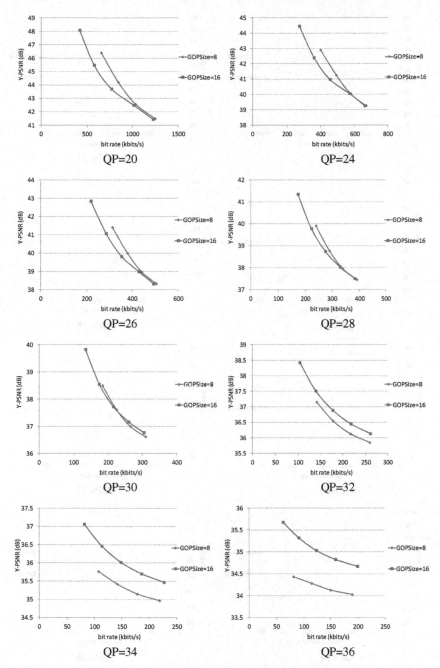

Fig. 4. Foreman(289), different value of QP at GOPSize=8 and 16

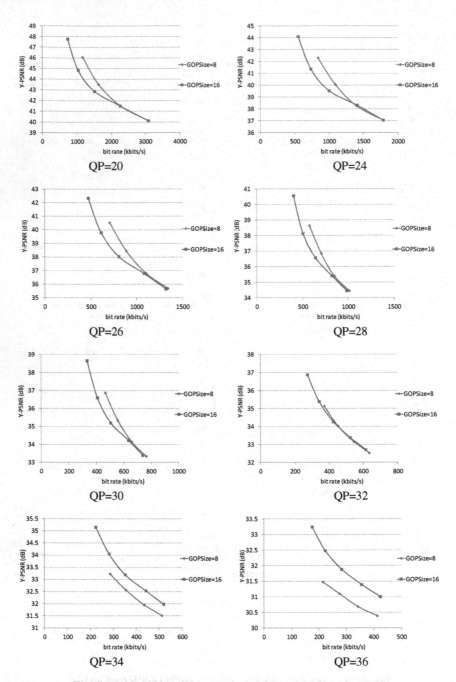

Fig. 5. Mobile(289), different value of QP at GOPSize=8 and 16

Table 2. The cross point for each test sequences

Test sequence	Value of QP
Harbour	33
Flower	33
Mobile	32
Tempete	32
Foreman	30
Football	30
Bus	30
Soccer	29
City	28

The BD-rate values with low QP presented in Table 3 are for GOPSize=8 compared to the reference GOPSize=16. Therefore, positive values of BD-rate imply bit rate savings at constant quality for GOPSize=8 with respect to GOPSize=16. There are two parts, 15Hz frame rate and 30Hz frame rate, as shown in Table 3. Most of the results are positive value or close to zero. GOPSize=8 saves about 0.59% bit rate of average with 15Hz frame rate, and saves about 0.49% bit rate of average with 30Hz frame rate. The results show that using GOPSize=8 is similar to GOPSize=16 or a little better than it in most cases with low QP.

The results of BD-rate values with high QP are shown in table 3. The BD-rate with high QP is obviously lower than the BD-rate with low QP. Negative values of BD-rate mean GOPSize=16 can save more bit rate than GOPSize=8. Most of the results are negative value, only Bus and Flower having positive value. These two video sequences are more complex than other sequences. GOPSize=16 saves about 2.96% bit rate of average with 15Hz frame rate, and saves about 3.02% bit rate of average with 30Hz frame rate; hence, we recommend using GOPSize=16 at high QP.

Table 3. The BD-rate of GOPSize=8 and GOPSize=16 with low QP and high QP

Sequence	Low QP=20,24,26,28		High QP = 30,32,34,36	
	BD-rate (15Hz)	BD-rate (30Hz)	BD-rate (15Hz)	BD-rate (30Hz)
Bus	5.4%	4.7%	5.6%	1.5%
Flower	2.8%	2.4%	0.2%	0.5%
Football	0.8%	0.8%	-1.3%	-1.2%
Harbour	0.2%	-0.1%	-2.9%	-2.4%
Soccer	0%	0.7%	-1.4%	-1.6%
Mobile	-0.3%	0%	-8.1%	-7.4%
Tempete	-0.3%	-0.2%	-5.1%	-4.6%
City	-1.4%	-2.6%	-7.5%	-6.6%
Foreman	-1.9%	-1.3%	-6.1%	-5.4%
Average	0.59%	0.49%	-2.96%	-3.02%

4   Conclusion and Future Work

The effects of GOP size on H.264/AVC temporal scalable video coding are investigated in this paper. Although a smaller GOP size reduces the drifting phenomenon, encoding complexity, and coding delay, it may not achieve good rate-distortion. Previous efforts show that the GOP size of 8 or 16 usually achieves the best rate-distortion performance; however, the determination mechanism is not clear. We found from experiments that a GOP size of 8 gives better coding performance for low QP (less than 28); and yet a GOP size of 16 should be chosen for high QP (larger than 32). For QP values in between, the optimal GOP size depends on video characteristics, and a smaller GOP size (8) is preferable for videos with complex scene and high motion. The reasons behind the cross-point are under study. In the future, we will test more situations (adding spatial and quality scalability, more rate points, etc), and continue to thoroughly investigate temporal scalability.

Acknowledgement. This work is supported in part by the National Science Council, Taiwan, under the Grant NSC 101-2219-E-027-002.

References

1. Ohm, J.-R.: Advances in Scalable Video Coding. Proceedings of the IEEE 93, 42–54 (2005)
2. ITU-T Rec. H.264 (MPEG-4 AVC), Fifth Edition (including SVC and MVC extensions) (2009)
3. Schwarz, H., Marpe, D.: Overview of the Scalable Video Coding Extension of the H.264/AVC Standard. IEEE Trans. Circuits Syst. Video Technol. 17, 1103–1102 (2007)
4. Pulipaka, A., Seeling, P., Reisslein, M., Karam, L.J.: Overview and Traffic Characterization of Coarse-Grain Quality Scalable (CGS) H.264 SVC Encoded Video. In: Consumer Communications and Networking Conference, pp. 1–5 (2010)
5. Görkemli, B., Şadi, Y., Tekalp, A.M.: Effects of MGS Fragmentation, Slice Mode and Extraction Strategies on the Performance of SVC with Medium-Grained Scalability. In: IEEE International Conference on Image Processing, pp. 4201–4204 (2010)
6. JSVM Software Manual, Version 9.19.9 and Version 9.19.13 (2010/2011)
7. Ferreira, L., Cruz, L., Assuncao, P.: Efficient scalable coding of video summaries using dynamic GOP structures. In: IEEE EUROCON - International Conference on Computer as a Tool, pp. 1–4 (2011)

Motion Estimation and DCT Coding Combined Scheme for H.264/AVC Codec

Wei-Jhe Hsu[1], Hsueh-Ming Hang[1], and Yi-Fu Chen[2]

[1] Department of Electronics Engineering, National Chiao-Tung University, Hsinchu, Taiwan
hsu761001@gmail.com, hmhang@mail.nctu.edu.tw
[2] Telecommunication Laboratories, Chunghwa Telecom Co., Ltd., Taoyuan, Taiwan

Abstract. A typical H.264 video encoder (such as JM) selects the best motion vector based on the sum of absolute difference (SAD) and the sum of absolute transformed difference (SATD) in different accuracy layers. In this paper, we propose a jointly optimal approach that selects the best motion vector that minimizes the rate-distortion cost of the quantized transform coefficients. We test the proposed scheme on a number of sequences. The results indicate that our scheme provides a bit-rate gain up to 4% for P pictures.

Keywords: Motion Estimation (ME), H.264/AVC, Rate-Distortion (R-D) Optimization, Hadamard Transform.

1   Introduction

The H.264/AVC video coding standard [1] provides a rather high coding efficiency. A typical H.264 video coder contains a motion vector (MV) selection module and a coding mode selection module. Typically, such as the standard committee reference software JM 18.0 [2], picks up the best motion vectors and the best coding modes in two separate steps. In this paper, we focus on picking the best motion vectors for the best final coding results and thus they improve the overall rate-distortion (R-D) performance. The remaining sections of this paper are organized as follows. Section 2 introduces the typical R-D optimization in the inter procedure for H.264/AVC and the related work. Our proposed algorithm is described in Section 3. Section 4 presents the experimental results and the discussion. Section 5 concludes our work.

2   Motion Estimation for H.264/AVC and Related Work

The encoder of H.264/AVC uses the transform coding technique to encode the motion-compensated prediction errors. A residual block is produced by subtracting the prediction from the current block. Then, the residual block is transformed by the 4x4 separable integer DCT (IDCT) [3] or the 4x4 Hadamard transform (H matrix) as shown in (1) which is an approximation form of IDCT for low complexity. For more details of encoding procedure, please refer to [1], [4].

J.-S. Pan et al. (Eds.): *Advances in Intelligent Systems & Applications*, SIST 21, pp. 413–421.
DOI: 10.1007/978-3-642-35473-1_42   © Springer-Verlag Berlin Heidelberg 2013

2.1   R-D Optimization in Inter Procedure

We use the reference software JM 18.0 as the platform, which is widely recognized as one of the best H.264 encoder from the coding performance viewpoint. The general R-D cost function for video coding is expressed by (2), the so-called Lagrange cost function. In (2), symbol D denotes the distortion, which is often the absolute difference between the processed image block and the original block. Symbol R denotes the rate, which is the bit rate needed to send the processed information. How to select the optimal Lagrange multiplier λ is a difficult problem in practice; often, an empirical formula is in use, as described in [5], [6].

$$H = \begin{bmatrix} 1 & 1 & 1 & 1 \\ 1 & 1 & -1 & -1 \\ 1 & -1 & -1 & 1 \\ 1 & -1 & 1 & -1 \end{bmatrix} \tag{1}$$

$$J = D + \lambda R \tag{2}$$

A traditional H.264/AVC encoder splits the cost function optimization process into two steps, and the 2-step process is illustrated by Fig. 1. We describe below the R-D optimization scheme in the original setting of JM18.0 [2]. In the first step, we find the MVs with the least residual distortion and the MV coding bits. Based on the motion R-D cost function in (3), the motion estimation step finds the vector with the least cost for various block sizes. Given the current and the reference frames and the Lagrange multiplier λ_{motion}, ME search looks for the best MV for each partition block s_i to minimize (3).

$$J_{motion} = D\left(s_i, m\right) + \lambda_{motion} R_{motion}\left(s_i, m\right) \tag{3}$$

where m is the set of all possible vectors. In (3), R_{motion} is the number of bits for transmitting MV, and $D\ (s_i,m)$ is the term of   SAD given by

$$SAD = \sum_{x,y} \left| Dblock(x, y) \right| \tag{4}$$

The symbols in (4), x and y, are the pixel location in a block, and $Dblock$ is the difference between the referenced candidate block and the original block. It should be noted that the best sub-pixel MV (half and quarter accuracy) is decided according to the Hadamard transform consideration in $D\ (s_i,m)$ in (3); that is, the term, $D\ (s_i,m)$, is SATD defined by (5), where H is the Hadamard matrix in (1).

$$SATD = \sum_{x,y} \frac{1}{2} \left| H * Dblock * H \right| \tag{5}$$

In the second step of the inter mode encoding process, the encoder also applies the Hadamard transform to the motion-compensated residual signals of each inter mode, and then we choose the best MB coding mode by minimize cost function (2).

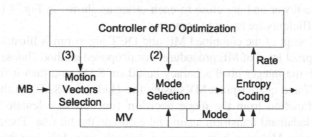

Fig. 1. R-D optimization for selecting MV and mode in JM

2.2   Related Work

In [7], the effect of SATD on ME at different layers is discussed and tested. The encoder uses SATD for searching for integer MVs, and averagely achieves 1.85% bit rate saving but with 781% encoding time increase when the sub-pixel motion search is disabled. However, the same method leads a small amount of coding loss, about 0.39% BD rate [8], when the sub-pixel MV is enabled. The reason is that SATD tries to match frequency-domain patches rather than the pixel-domain patches. The interpolated pixels at sub-pixel accuracy seem to have negative effect. The report in [7] is interesting but there are 2 undesirable points. First, $(2 \times \text{search range} + 1)^2$ searching points with SATD require high complexity. Second, the experiment indicates that SAD is a better criterion in finding MV at integer pixel level. Therefore, our algorithm is designed to improve the above problems.

3   Combined Motion Estimation and DCT Algorithm

In this section, we describe the principle behind the proposed combined ME and DCT algorithm and its implementation step by step. In the traditional H.264/AVC encoder, the ME procedure chooses the integer pixel vector that minimizes (3) with SAD criterion. However, (3) does not truly reflect the final distortion and the bit rate of the encoded block. Therefore, we include (2) into the ME procedure in selecting MVs to improve overall coding performance. That is, we combine (2) and (3) in the integer ME procedure.

The motivation is as follows. Although a selected MV is not the best candidate in the MV decision at the integer pixel level, its residual DCT may have fewer large transform coefficients and thus produces fewer bits in entropy coding in the final stage. Figs. 2–4 show an image example. Fig. 2 shows the difference between the JM-coded frame and the coded frame using the proposed method. In Figs. 3–4, we compare the residual MBs produced by two MVs on the second frame of the FOREMAN sequence. The comparison is done in both the spatial domain and the frequency domain. Our proposed algorithm chooses a different MV in the final stage (called Motion RDcost#2 means the 2nd best MV in the integer ME step). The resultant residual block has a more clustered frequency domain distribution; that is, the large magnitude coefficients are fewer and are close to each other as shown in Fig. 4 (right). Therefore, these coefficients are easier to compress.

The core concept of the combined ME and DCT algorithm is illustrated by Fig. 5. In the integer-pixel level of ME procedure, our proposed method chooses 5 top candidate MVs with the integer-pixel accuracy based on SAD, and then it finds their corresponding half and quarter-pixel MVs using the Hadamard SAD. At the end, we use the modified function from the mode decision function to calculate the distortion based on the Hadamard transform again and estimate the bit rate. Therefore, in addition to the integer MV search, we increase the sub-pixel MV searches (SATD) for about $5 \times [2 \times$ (sub search points) $+1]$ times. After our proposed scheme, we get the best integer vector of each partitioned block, and then use it to the following steps as the original JM, such as sub-pixel ME and mode decision.

Fig. 2. The left decoded picture is coded by our proposed algorithm, and the right residual picture is the difference between the left picture and the coded picture by JM at QP = 22. The differences are 10 times magnified.

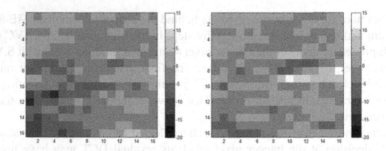

Fig. 3. Spatial domain: The residual MBs of Inter-16x16 mode on the second frame. The MB location (upper-left corner) is (80,160). Gray values are adjusted to show a range from 15 to -20 (the maximum and minimum pixel values). (left) The residual block produced by the MV with Motion RDcost#1. (right) The residual block produced by the MV with Motion RDcost#2.

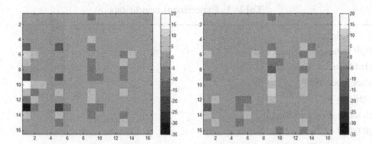

Fig. 4. Frequency domain: The transformed and quantized residual MBs of Fig. 3. Coefficients are produced by 4x4 integer DCT with QP 22. Gray values are adjusted to show a range from 20 to -35. (left) A residual transform block produced by the MV with Motion RDcost#1. (right) A residual transform block produced by the MV with Motion RDcost#2.

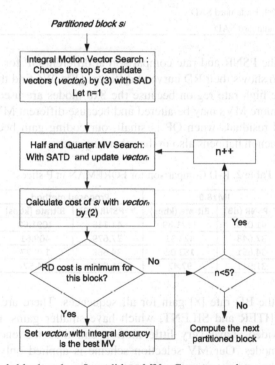

Fig. 5. For each sub-block, select 5 candidate MVs. Compute and compare their R-D costs to decide the best integer MV.

4   Simulation Results

To verify the effectiveness of our proposed motion estimation and DCT combined algorithm, we implement it on the software JM 18.0 [2], which is the reference software of the H.264/AVC encoder. We compare its performance with that of the original JM encoder. The experimental conditions are listed in Table 1, and the test sequences are from [9] (MPEG test video).

Table 1. Experiment conditions

Profile: Baseline
Used QP values : 22, 27, 32, and 37
Encoded Frames : 32
Sequence type : IPPP
Intra Period : 16
Search mode: Fast Full Search
Search range: ±32
Reference frame: The previous frame
Entropy Coding: CAVLC
RD-Optimization: High complexity
ME-Distortion-FPel: SAD
ME-Distortion-HPel: Hadamard SAD
ME-Distortion-QPel: Hadamard SAD
MD-Distortion: Hadamard SAD

Table 2 shows the PSNR and rate comparison at different QP for the FOREMAN sequence and Fig. 6 shows their RD curve at different QPs. We find that the curve has a larger gain in the high rate region because the 8x8 modes are used more often. In this case, because more MVs may be altered and because different MVs may result in different quantized residuals when QP is small, our coding gain becomes more obvious. This phenomenon happens also in the other sequences.

Table 2. R-D Comparison for FOREMAN in P slices

FOREMAN	JM18.0		Proposed method		Y BD rate
	Y_PSNR (dB)	Bitrate (kbps)	Y_PSNR (dB)	Bitrate (kbps)	
QP=22	41.078	1121.89	41.115	1091.63	
QP=27	37.648	423.31	37.679	409.61	-3.4%
QP=32	34.651	183.02	34.668	179.77	
QP=37	31.911	97.47	31.924	94.57	

Table 3 shows the BD rate [8] gain for all sequences. There are two sequences, MOTHER_DAUGHTER and SILENT, which have smaller gains at about 1% because these two videos have very little motion and thus the encoder frequently chooses the skip modes. Our MV selection scheme is applied only to the motion-compensated blocks, whose number is now small. Another factor affects the performance is image contents (patterns). In some sequences, such as CITY and MOBILE, our method provides more gain because they contain a number of fine edges, and thus our method has more chances to manipulate the residual distribution patterns. In summary, two factors seem to have major impact on our algorithm performance. One is the percentage of motion-compensated modes in P-slices, and the other one is the texture pattern of the residual blocks.

We collect the final MV choices in our method in Table 4. It shows that the best motion R-D cost vector is chosen with higher probability when QP is large. In this case, because the number of transform coefficients is small, it thus makes little difference on the residual blocks produced by different MVs. On the average, the probability of choosing the fifth candidate MV is less than 5%. Thus, retaining more than 5 candidate MVs does not seem to offer much improvement.

Finally, we like to know how many "different" MVs at the integer-pixel level are chosen at the end using this approach (versus JM 18.0). We examine both the numbers of sub-blocks and their area. Table 5 shows the sub-block numbers and the area ratio of the changed MVs due to the adaptation of our algorithm.

Table 3. BD rate improvement in P slice of all sequences

	Test sequences	Y BD rate	Encoding Time
CIF	FOREMAN	-3.4%	+43.2%
	BUS	-2.6%	+46.6%
	FOOTBALL	-1.9%	+49.6%
	MOBILE	-2.4%	+48.9%
	NEWS	-2.7%	+43.0%
	ICE	-4.2%	+39.8%
	PARIS	-1.6%	+45.3%
	MOTHER_DAUGHTER	-1.3%	+41.2%
	SILENT	-1.1%	+43.2%
4CIF	HARBOUR	-2.2%	+47.0%
	CITY	-2.9%	+45.9%
	SOCCER	-1.8%	+46.1%
	CREW	-1.7%	+45.2%
Average		-2.3%	+45.0%

Table 4. Final MV selected from candidate MVs and percentages

FOREMAN	QP=22	QP=27	QP=32	QP=37
Motion RDcost1	53.4%	56.7%	61.4%	67.3%
Motion RDcost2	21.8%	21.3%	19.9%	17.4%
Motion RDcost3	11.8%	10.5%	9.1%	7.5%
Motion RDcost4	7.6%	6.7%	5.7%	4.5%
Motion RDcost5	5.5%	4.7%	3.9%	3.3%
ICE	QP=22	QP=27	QP=32	QP=37
Motion RDcost1	70.3%	73.9%	75.7%	77.2%
Motion RDcost2	14.2%	13.1%	12.4%	11.8%
Motion RDcost3	7.2%	6.1%	5.6%	5.3%
Motion RDcost4	4.7%	4.0%	3.6%	3.3%
Motion RDcost5	3.6%	3.0%	2.8%	2.5%
SILENT	QP=22	QP=27	QP=32	QP=37
Motion RDcost1	83.4%	84.0%	85.8%	89.1%
Motion RDcost2	7.8%	7.8%	7.0%	5.6%
Motion RDcost3	4.2%	4.0%	3.4%	2.5%
Motion RDcost4	2.7%	2.5%	2.2%	1.6%
Motion RDcost5	2.0%	1.7%	1.5%	1.2%

Table 5. Changed MV partitioned sub-blocks and the area ratio used changed vector

FOREMAN	Changed MV Blocks	Partitioned Blocks	Changed Area Ratio
QP=22	16223	36196	35.38%
QP=27	9739	23969	31.60%
QP=32	5780	17047	26.52%
QP=37	3559	14158	20.05%
ICE	**Changed MV Blocks**	**Partitioned Blocks**	**Changed Area Ratio**
QP=22	11279	28617	23.91%
QP=27	9182	24598	21.35%
QP=32	6936	20481	20.23%
QP=37	4787	16719	18.53%
SILENT	**Changed MV Blocks**	**Partitioned Blocks**	**Changed Area Ratio**
QP=22	4534	21115	9.27%
QP=27	3016	16681	8.93%
QP=32	1932	14066	7.92%
QP=37	1214	12743	6.62%

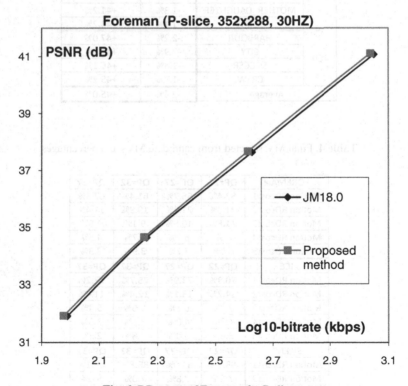

Fig. 6. RD curve of Foreman for P slice

5   Conclusion

In this paper, we propose a 2-pass motion estimation method to enhance the coding R-D performance by combining motion estimation and DCT for the H.264/AVC encoders. The algorithm considers the transform coding effect in choosing the best motion vectors from integer to quarter pixel accuracy. Based on the multiple sequences tests, we demonstrate that the proposed algorithm can achieve 2.3% average bit-saving without changing the syntax of the AVC/H.264 standard. There is a trade-off between coding efficiency and computational complexity. Although we reduce the SATD operations significantly comparing to [7], the encoding time is still increased by about 45%. Acceleration of our scheme is one of possible future work items.

Acknowledgments. This work was supported in part by the Chung-Hwa Telecomm, Taiwan under Grants TL-100-G113 and by the NSC, Taiwan under Grants 98-2221-E-009 -076.

References

1. Wiegand, T., et al.: Draft ITU-T Recommendation and Final Draft International Standard of Joint Video Specification (ITU-T Rec. H.264 | ISO/IEC 14496-10 AVC). ISO/IEC JTC/SC29/WG11 and ITU-T SG16 Q.6, JVT-Go50r1 (March 2003)
2. H.264/AVC Codec, http://iphome.hhi.de/suehring/tml/download/
3. Malvar, H.S., et al.: Low complexity transform and quantization in H.264/AVC. IEEE Trans. Circuit Syst. Video Technol. 13(7), 598–603 (2003)
4. Richardson, I.E.G.: H.264 and MPEG-4 Video Compression: Video Coding for Next-Generation Multimedia. Wiley
5. Wiegand, T., et al.: Rate-constrained coder control and comparison of video coding standards. IEEE Trans. Circuit Syst. Video Technol. 13(7), 688–703 (2003)
6. Wiegand, T., Girod, B.: Lagrange multiplier selection in hybrid video coder control. In: Proc. Int. Conf. Image Proc., pp. 542–545 (October 2001)
7. Abdelazim, A., et al.: Effect of the Hadamard transform on motion estimation of different layers in video coding. In: International Archives of Photogrammetry, Remote Sensing and Spatial Information Sciences, Part5 Commission V Symposium, Newcastle upon Tyne, UK, vol. XXXVIII (2010)
8. Bjontegaard, G.: Calculation of Average PSNR Differences between RD-curves. Document VCEG-M33 (April 2001)
9. Test sequences, http://media.xiph.org/video/derf/

Free View Point Real-Time Monitor System Based on Harris-SURF

Tzu-Ti Chang[1], Fang-Yi Yu[1], Wei-Tsong Lee[1], Feng-Yu Chang[2], and Jason Wu[3]

[1] Dept. of Electrical Engineering, Tamkang University, Tamsui, Taiwan (R.O.C)
[2] Chung-Shan Institute of Science and Technology, Taiwan, R.O.C
[3] LEED Technology Ltd., Taiwan (R.O.C)
julia30713@hotmail.com, 600440258@s00.tku.edu.tw,
wtlee@mail.tku.edu.tw, incub@csistdup.org.tw,
jason@surewin.com.tw

Abstract. This paper describes a novel feature detector on free view point real-time monitor system, we named it Harris-SURF detector. Due to Harris[1][2] and SURF[3][4] are both unreliable on free view point real-time monitor system if we use one of them along, we combined their advantages as Harris-SURF detector . The first step is to extract corners from original images by using Harris corner detector. The second step is using SURF algorithm to give corners main factor and descriptors to match them from different images.

1 Introduction

Due to Internet of things increasingly mature, future monitoring has become not only simply display 3D images, but also be able to interact with users. In this paper, we present a fast free view point monitor system without rebuild 3D module tardily.

Images lost their depth information after they are captured by cameras, recalculating their coordinates in real world are inefficient and usually easily been distorted. In order to achieve the goal of free view point real-time processing, parallax of images become an very important information to us.

Computing time plays an important role in free view point real-time monitor system. Tradition 3D modeling algorithms usually have high accuracy but low performance, speeding up the system is the first problem we face. Instead of reconstructing 3D models, we put our focus on simulating users' point of view in our new algorithm. Our experimental environment requires multiple cameras focus on one object in different angle. After images are captured by cameras, we'll find feature points on each image with Harris corner detector. The second step is matching these corner points, finding relations between different images. After matching feature points, the third step is triangle meshing. By using feature points as vertex, the images are segmented into several triangles. Meshed triangle images transformed into the user's view point and recover texture[6] on simulation image in the last step. System repeats step one to three until user has new view point commands.

SURF is very good at handling scale changing and image twisting, but feature points found by SURF are no corners, without corner information, it is hard to

J.-S. Pan et al. (Eds.): *Advances in Intelligent Systems & Applications*, SIST 21, pp. 423–430.
DOI: 10.1007/978-3-642-35473-1_43   © Springer-Verlag Berlin Heidelberg 2013

simulate users' view point. Harris corner detector is well known of its good performance and stabilization, that is why we combined these two algorithms in our research. In this paper, we'll take a short view on SURF and Harris corner detector before introducing our algorithm.

2   Speed Up Robust Feature

There are four steps in SURF[3][4]. First, extract edge information on gray-scaled image. Second step, calculate different scale space information by Hessian matrix[3]. Third step, locate feature points, collect feature points in to matrix. Establish main vector and descriptors on feature points.

2.1   Integral Images

In order to speed up the system, integral image is used in SURF algorithm:

$$I_\Sigma(x) = \sum_{i=0}^{i \leq x} \sum_{j=0}^{i \leq y} I(i, j) \tag{1}$$

Where $I_\Sigma(x)$ is the sum of all pixels in the input image I within a rectangular region formed by the origin and x, where $x = (x, y)^T$. Once the integral image has been computed, it takes only three addition steps to calculate sum of intensities (see Fig. 1), the performance with larger filter has obviously improved.

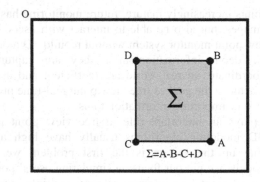

Fig. 1. Only three addition steps are needed to calculate the sum of intensities

2.2   Feature Point Detector

Feature point detector of SURF is based on Hessian matrix:

$$H(x, \sigma) = \begin{pmatrix} L_{xx}(x, \sigma) & L_{xy}(x, \sigma) \\ L_{xy}(x, \sigma) & L_{yy}(x, \sigma) \end{pmatrix} \tag{2}$$

Where x is a point (x, y) in an image I. $H(x, \sigma)$ is the Hessian matrix in x at scale σ. $L_{xx}(x, \sigma)$, $L_{xy}(x, \sigma)$ and $L_{yy}(x, \sigma)$ are the convolution of Gaussian derivative $\dfrac{\partial^2 g(\sigma)}{\partial x^2}$, $\dfrac{\partial^2 g(\sigma)}{\partial x \partial y}$ and $\dfrac{\partial^2 g(\sigma)}{\partial y^2}$.

In SURF, they approximated the Guassian matrix with box filters(see Fig. 2). These approximation increase the convolution speed with performance slightly lose.

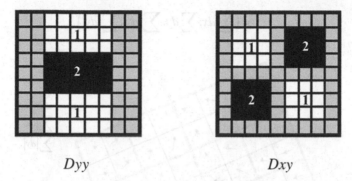

$$Dyy \qquad\qquad Dxy$$

Fig. 2. The approximation for Guassian matrix in SURF algorithm

The box filter in Fig. 2 are approximations of Gaussian with $\sigma = 1.2$. D_{xx}, D_{yy} and D_{xy} are denoted. The Hessian matrix approximation are defined as:

$$\det(H_{approx}) = D_{xx}D_{yy} - (\omega D_{xy})^2 \tag{3}$$

Weight ω is related with dimension of image scale-space. But in practice, the value of ω is usually static, so we redefine it as constant 0.9.

2.3   Extract Main Vector and Descriptor

Main vector and descriptor give feature points the ability to resist image rotation. Fist, calculate the Harr wavelet responses in x and y direction within radius $6s$, where s is the scale which the feature point was detected. The size of the wavelets are set to a side length of $4s$.

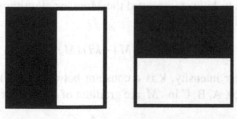

Fig. 3. Harr wavelet filters in x and y direction. The dark parts have the weight -1 and the light part +1.

The sum of all responses within a 60-degree sector are calculated to be the direction of this feature point. The maximum direction is defined as the main vector.

To build the descriptor, make the feature point as the center of a $20\,s$ length rectangle region. Separate the rectangle into multiple 4*4 square sub-region. For each square, the Harr wavelet responses are computed from 5*5 samples. Compute relatively to the orientation of the grid by collect the sums dx, $|dx|$, dy and $|dy|$ (see equation 4, Fig. 4).

$$v = \left(\sum dx, \sum dy, \sum |dx|, \sum |dy| \right) \tag{4}$$

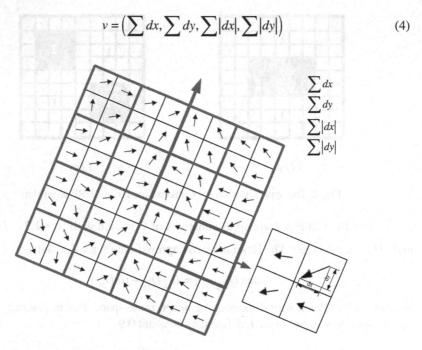

$$\sum dx$$
$$\sum dy$$
$$\sum |dx|$$
$$\sum |dy|$$

Fig. 4. Build descriptor by establishing a rectangle with multiple 4*4 square sub-region. Wavelet responses are computed from 5*5 samples. For illustrative purposes, only 2*2 are shown in this fig.

3   Harris Corner Detector

Harris implemented a technique improved the Moravec algorithm[10]. An equation of corner detect was defined:

$$R = Det(M) - kTr(M)^2 \tag{5}$$

Where R is the corner intensity, k is a constant between 0.04 to 0.06, M is a 2*2 matrix (see equation 6). A, B, C in M are gradient of direction x, y and xy.

$$M = \begin{pmatrix} A & C \\ C & B \end{pmatrix} \tag{6}$$

$$Tr(M) = \alpha + \beta = A + B \qquad (7)$$

$$Det(M) = \alpha\beta = AB - C^2 \qquad (8)$$

α and β are the eigenvalues of R. The points are defined as corner if R is positive and above the threshold.

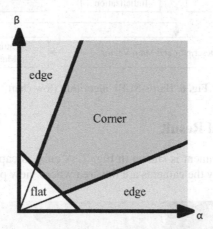

Fig. 5. Corners are defined when R is positive and above the threshold

4   Harris-SURF Detector

SURF is a strong algorithm on image recognization, but the feature points found by SURF are difficult to identify the object structural. To overcome this problem, we replace feature point detector in SURF with Harris corner detector.

There are several steps in our algorithm(see Fig. 6):

4.1   Image Initialization

Transform input image into gray scale and calculate image integral.

4.2   Extract Feature Points with Harris Detector

SURF feature point detector is replaced by Harris corner detector in our algorithm, Hessian matrix approximation are abandoned.

4.3   Find Main Vactor and Descriptors of Feature Points

By adding 64-dimension vectors, feature points found by Harris detector now have the ability to resist image rotation.

4.4   Matching Feature Points

Build a k-d tree with feature points' descriptors, continuously searching the nearest point to complete matching[8][9].

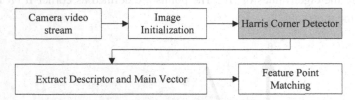

Fig. 6. Harris-SURF algorithm flow chart

5   Experimental Result

Our experiment environment is shown in Fig. 7. A cube is captured by four cameras. The square been built by the cameras are the area where view points can be simulated.

Fig. 7. Real-time free view point monitor experiment environment

Fig. 8. Harris-SURF performance. Numbers on cube are the corners which matched successfully.

For real-time purpose, we do not reconstruct a 3D module. After find feature points and complete matching by Harris-SURF. In Fig. 8, we can see corners been found on the cube are matched successfully. After matching, we simulate the location of corner points by calculating their interpolation.

6   Harris Corner Detector

This paper presents a fast free view point monitoring system which is useable on real-time purpose. We find feature points with Harris corner detector and match these corners with SURF-descriptors. Simulate the location of corner by calculating their interpolation. At last, rebuild the image with these simulated corner points. Unlike traditional 3D reconstruction, rebuild the module usually take seconds, or even minutes. Harris-SURF provide a more simply and facility way to do so.

Fig. 9. Several view points simulated by our system

References

1. Harris, C., Stephons, M.: A Combined Corner and Edge Detector. In: Proceeding of the Fourth Alvey Vision Conference, pp. 147–151 (1988)
2. Yun, G.-C.: Corner Detection Approach to the Building Footprint Extraction from Lidar Data (July 2008)
3. Bay, H., Ess, A., Tuytelaars, T., Van Gool, L.: Speeded-Up Robust Features (SURF). Computer Vision and Image Understanding 110, 346–359 (2008)
4. Luo, I.-C.: Image Feature-Based Omni-directional Visual Odometry Design, 淡江大學電機工程學系研究所碩士論文 (2011)
5. He, Z.-G.: 3D Facial Animation, 義守大學工程學系研究所碩士論文 (June 2005)
6. Liu, C.W.: Active Object Movie, 國立東華大學資訊工程學系碩士論文 (July 2003)
7. Haar, A.: Zur Theorie der orthogonalen Funktionensysteme. Mathematische Annalen 69, 331–371 (1910)

8. Beis, J.S., Lowe, D.G.: Shape indexing using approximate nearest-neighbor search in high-dimensional spaces. In: Computer Vision and Pattern Recognition, pp. 1000–1006 (June 1997)
9. Hung, C.-H.: Design and Implementation of Vision System with Feature Recognition for Humanoid Robot, 淡江大學電機工程學系研究所碩士論文,　(June 2010)
10. Moravec, H.P.: Towaeds Automatic Visual Obstacle Avoidance. In: Proc. Int'l Joint Conf. Artificial Intelligence, p. 584 (1977)

Virtual Multiple-Perspective Display Using Pyramidal or Conical Showcase

Yu-Tsung Chiu and Mau-Tsuen Yang

National Dong Hwa University, Taiwan
610021009@ems.ndhu.edu.tw, mtyang@mail.ndhu.edu.tw

Abstract. We implement a virtual display system using pyramidal or conical showcase. For image acquisition, a real object is placed inside the showcase made of mirror coated surfaces and a multiple-perspective image (or video) is captured by a bird-view camera. The captured image (or video) is processed to remove outliers in the background and compensate distortion. For virtual display, the processed image (or video) is shown on a flat screen, and the transparent surfaces of the showcase reflect the contents on the screen so that a hologram seems to appear inside the virtual showcase. The proposed system can be used to display any expensive, fragile, sensitive, or unique objects while the genuine objects are stored in a secured or remote location. Some possible applications of the proposed virtual showcase include digital museum, online exhibition, jewelry show and commercial display.

Keywords: virtual showcase, multiple-perspective display.

1   Introduction

Display technology in the field of computer vision evolves from the traditional 2D to modern 3D stereoscopic display. Users can not only watch the flat images, but also feel the realistic three-dimensional effects in a space. The 3D displays can be divided into two types: glasses and glasses-free (autostereoscopy). In the glasses type, users require to wear extra glasses to watch three-dimensional effects. Long-term usage of the glasses is cumbersome and not intuitive. Therefore, autostereoscopic type gradually becomes popular.

Compared to the stereoscopic display technology, volumetric display technology tries to construct a volume of space to create stereoscopic effects. This display style is ideal to make a virtual showcase that looks similar to a real transparent showcase. Particularly, in a pyramid-shaped showcase, each different plane of the side of the pyramid can display a corresponding view of a virtual object. Thus the pyramidal showcase is capable to achieve a simultaneous virtual display of four perspectives. However, when an observer moves from one perspective to another, two eyes see two different images (one reflected by the original plane and the other reflected by a neighboring plane), resulting in the rendered object to appear inconsistent and disconnected. On the other hand, a cone-shaped showcase can solve this problem by achieving a virtual display of multiple perspectives [5]. An observer can move from one

J.-S. Pan et al. (Eds.): *Advances in Intelligent Systems & Applications*, SIST 21, pp. 431–438.
DOI: 10.1007/978-3-642-35473-1_44   © Springer-Verlag Berlin Heidelberg 2013

perspective to another seamlessly. However, since the virtual images are reflected by a curved surface instead of a flat plane, more complicate conversion and warping are required to handle the distortion images properly.

The Augmented Reality (AR) technology can combine real and virtual objects to achieve more vivid and realistic effects. The Computer Vision (CV) technology can provide interaction to add more diversification and functionality. We propose to integrate both the AR and CV technologies to develop a realistic and interactive showcase. For image acquisition, a real object is placed inside the showcase made of mirror coated surfaces and a multiple-perspective image (or video) is captured by a bird-view camera. The captured image (or video) is processed to remove outliers in the background and compensate distortion. For virtual display, the processed image (or video) is shown on a flat screen, and the transparent surfaces of the showcase reflect the contents on the screen so that a hologram seems to appear inside the virtual showcase. The proposed showcase can be used to display any expensive, fragile, sensitive, or unique objects while the genuine objects are stored in a secured or remote location.

2   Related Work

Two main areas related to our approach are stereoscopic display technology and volumetric display technology.

2.1   Stereoscopic Display Technology

The goal of stereoscopic displays is to fulfill of the stereo vision of humans. Human has two symmetrical eyes with a horizontal gap between eyes. When we watch at an object, two eyes see the object from two different angles, and two different images are sensed. The image pair is processed and combined by our brain to produce the stereoscopic and depth sense.

A pioneer of the stereoscopic display is Charles Wheatstone who invented the stereoscope [1] in 1833, and then became the major promoter of the development of three-dimensional displays. Three-dimensional stereoscopic displays can be divided into glasses-required and glasses-free (autostereoscopic). Users of the first type need to wear additional equipment to achieve stereoscopic effects. This requirement is inconvenient and not intuitive. On the other hand, autostereoscopic technology conveys a pair of stereoscopic images to the eyes of views through lenticular lens or parallax barriers that are placed in front of the screen instead of the user. Autostereoscopic displays generate stereoscopic effects without the need for users to wear any special equipment. Therefore, autostereoscopic technology becomes one of the most important directions of the three-dimensional display development.

2.2   Volumetric Display Technology

Volumetric displays can be divided into a number of different types such as tomography display, sweep volume display, and static volume display. The tomography display [2], use of multiple projectors in diffuse water vapor or smoke space, projected three-dimensional object, so that the observer sees the stereoscopic image floating in space, but need multiple projectors and large display space, so rendered not easy. Sweep volume of display [3], using the fast way, let to user produced persistence of vision, which produce three-dimensional feeling to the objects, but need more secret equipment or drive motor, implement a process to render stereoscopic effect. Compared to above two types of stereoscopic presentation modalities, static volume display [4], without too much display space, but also the support is not limited to the required precision equipment, we can achieve rendering the sense of depth stereoscopic vision. Even though the ways of the presentation and implementation of these volumetric display types and are different, but all of them can create senses of stereoscopic objects or scenes. However, it is usually sophisticated and expensive to make a volumetric display. Hence practical applications are limited.

3   System Design

This section describes the implementation of the proposed pyramidal and conical showcases in details. For image acquisition, a real object is placed inside the showcase made of mirror coated surfaces and a multiple-perspective image (or video) is captured by a bird-view camera. A shown in Fig. 1, the setup of the camera affects the quality of the recorded images. Camera calibration is performed to alleviate skew, tilt and deformation problems. Then image segmentation is applied to separate the foreground object from the background image. For virtual display, the processed image (or video) is shown on a flat screen, and the transparent surfaces of the showcase reflect the contents on the screen so that a hologram seems to appear inside the virtual showcase. The complete flowchart is shown in Fig. 2.

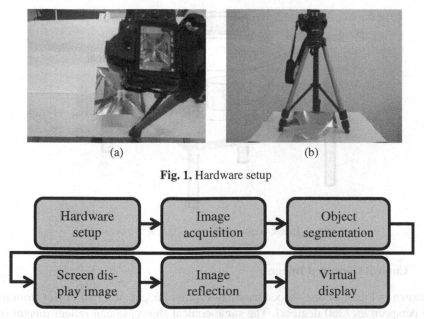

(a)   (b)

Fig. 1. Hardware setup

Hardware setup	→	Image acquisition	→	Object segmentation
Screen display image	→	Image reflection	→	Virtual display

Fig. 2. System flowchart

3.1   Pyramidal Captured Images

As shown in Fig. 3, a pyramid-shaped case with mirror surfaces coated inside can collect images for the four sides (front、back、left and right) of a real object. The captured image is processed and displayed on a flat screen. The same pyramidal showcase with transparent and reflective surfaces can be placed on top of the screen to reflect screen's contents to generate an illusion of the virtual object. Also, other real objects can be added to the showcase so observers can see both virtual and real objects in the showcase concurrently.

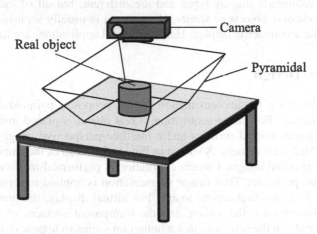

Fig. 3. Image acquisition using pyramidal showcase

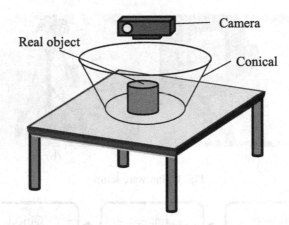

Fig. 4. Image acquisition using conical showcase

3.2   Conical Captured Images

As shown in Fig.4, a cone-shaped case can capture images of a real object from multiple perspectives (360 degrees). The same conical showcase can reflect images on a flat screen to generate an illusion of the virtual object inside the showcase. However, due to the images reflected by the curved surface of the conical showcase are distorted, the captured images should be processed to correct the distortion so the virtual objects can be rendered realistically.

4   Result

A virtual display system using pyramidal or conical showcase has been implemented. Experimental environment and hardware specification are described in Table 1.

Table 1. Experimental environment

Name	Specification
Camera	Nikon D5100
Image Resolution	4928*3264 pixels
Screen	Acer AL503
Screen Resolution	1024*768 pixels
Pyramidal Showcase	Material: mirror Size: 3.5*21*9.0(cm)
Conical Showcase	Material: mirror Size: 2.0*16*6.0(cm)

4.1   Pyramidal Display

By turning a pyramid-shaped case upside down and placing a real object in the middle of the case, a camera installed directly above the case shoots an overhead image. As shown in Fig. 5(a), the captured image contains four different sub-images, each sub-image represents a typical view of the object (front, back, left, and right). After the foreground-background segmentation, only foreground pixels are kept and all background pixels are cleared (as shown in Fig. 5(b)).

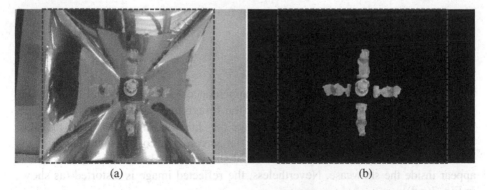

(a)   (b)

Fig. 5. Image acquisition for pyramidal display. (a) original image (b) segment image.

By showing the processed image on a flat screen and placing the same pyramidal showcase on top of the screen, the screen contents are reflected by the side planes of the showcase and a virtual object seems to appear inside the showcase (as shown in Fig. 6). Four perspectives (front, back, left, and right) of the object are displayed correctly. However, when an observer moves from one perspective to another, two eyes see two different images (one reflected by the original plane and the other reflected by a neighboring plane), resulting in the rendered object to appear inconsistent and disconnected.

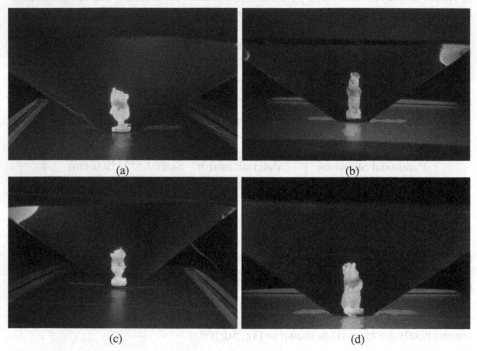

Fig. 6. Pyramidal display. (a) left viewpoint (b) front viewpoint (c) right viewpoint (d) back viewpoint.

4.2   Conical Display

The conical case can seamlessly capture appearances of a 3D object from multiple perspectives (as shown in Fig. 7(a)). After the foreground-background segmentation, the captured image becomes a circular image (as shown in Fig. 7(b)). Then the same conical showcase is placed on the top of a flat screen. The contents on the screen are reflected by the curved surface of the conical showcase, and a virtual object seems to appear inside the showcase. Nevertheless, the reflected image is distorted (as shown in Fig. 7(c,d)).

(a)

(b)

(c)

(d)

Fig. 7. Conical display. (a) original image (b) segment image (c) front viewpoint (d) back viewpoint.

Due to the distortion caused by the curved surface, the reflected stereoscopic effects are not ideal. We made another experiment to address this problem. As shown in Fig. 8, two conical showcases are installed and aligned vertically. The lower showcase captures an image of a real object, while the upper showcase reflects the captured image to generate an illusion of the virtual object. In this case, due to the parallelism of the two curved surfaces, the distortion is corrected and the virtual object is displayed properly as shown in the red rectangles in Fig. 8.

(a)

(b)

Fig. 8. Expected results. (a) battery (b) toy.

5   Conclusion and Future Work

The proposed system shoots a real object to capture a multi-perspective view of 360 degrees, and use of this image to implement a virtual showcase for remote demonstrations. Using the proposed method, the genuine object can be stored securely to avoid damages, while several showcases can be displayed simultaneously at remote positions. Especially, since the showcase is displayed by a computer, it is possible to integrate computer vision technology to develop human-computer interactions to make an interactive showcase.

In the future, we plan to find more robust, transparent, and reflective materials to make the showcase, and to develop more accurate and efficient ways to capture and display the virtual image. Finally, the conical displays need more correction and conversion to remove distortion to achieve a realistic and practical showcase.

Acknowledgement. This research is supported by the National Science Council, Taiwan.

References

1. IJsselsteijn, W.A., Seuntiëns, P.J.H., Meesters, L.M.J.: State-of-the-art in human factors and quality issues of stereoscopic broadcast television. In: Advanced Three-dimensional Television System Technologies (ATTEST), pp. 1–10 (2002)
2. Lee, C., Di Verdi, S., Hollerer, T.: An Immaterial Depth-Fused 3D Display. In: ACM Symposium on Virtual Reality Software and Technology, pp. 191–198 (2007)
3. Jones, A., McDowall, I., Yamada, H., Bolas, M., Debevec, P.: Rendering for an Interactive 360 Light Field Display. In: ACM SIGGRAPH (2007)
4. Bimber, O., Miguel Encarnação, L., Schmalstieg, D.: The Virtual Showcase as a new Platform for Augmented Reality Digital Storytelling. In: EGVE, pp. 87–95 (2003)
5. Bimber, O., Fröhlich, B., Schmalstieg, D., Miguel Encarnação, L.: The Virtual Showcase. In: SIGGRAPH (2001)

Stroke Rehabilitation via a Haptics-Enhanced Virtual Reality System

Shih-Ching Yeh[1], Si-Huei Lee[2], Jia-Chi Wang[2], Shuya Chen[3], Yu-Tsung Chen[1], Yi-Yung Yang[2], Huang-Ren Chen[1], Yen-Po Hung[1], Albert Rizzo[4], and Te-Lu Tsai[5]

[1] Department of Computer Science and Information Engineering, National Central University
shihchiy@csie.ncu.edu.tw
[2] Taipei Veterans General Hospital
[3] Department of Physical Therapy, China Medical University
[4] Institute for Creative Technologies, University of Southern California
[5] Institute for Information Industry

Abstract. Stroke is one of the major diseases causing brain injury, its sequela will, depending on persistent nervous injury, derive different types of limb and body exercise barriers, which will cause large challenge to the daily life of the patient and will seriously affect the quality of life of the patient. Along with the development and popularity of technology, scholars in the medical care and rehabilitation fields are trying to integrate all kinds of new technologies to perform the development of new rehabilitation training system.

In this study, for the rehabilitation of upper extremity, trainings are provided respectively for fore arm, for the endurance, stretching and flexibility of the wrist. Here game technology, force feedback technology and stereo image technology are associated to develop virtual reality body perceptive training task. In the rehabilitation process, multi-dimensional experimental results are acquired, for example, clinical test assessment, task performance, exercise track historical data and psychological emotional data. The research objectives are to verify the functionality of the system, to verify the effectiveness of the system on rehabilitation, to develop new assessment method and to investigate topics related to human machine interaction.

After initial pilot test is done on stroke patient, the experimental result has verified the functionalities of this rehabilitation training system in several aspects. Meanwhile, it can acquire reliable and valuable information successfully, for example, through the exercise analysis using exercise track historical data and using the statistical analysis of the task performance in the past therapeutic sessions, the medical therapeutic effect can be verified in the future, and new clinical assessment method can be developed. Not only so, according to the measured psychological emotional data as perceived subjectively, this system indeed can urge the patient to engage continuously rehabilitation therapeutic session that is based on this training system and enjoy it, besides, the authors are very confident on the possibly generated rehabilitation effect of these two training tasks.

J.-S. Pan et al. (Eds.): *Advances in Intelligent Systems & Applications*, SIST 21, pp. 439–453.
DOI: 10.1007/978-3-642-35473-1_45   © Springer-Verlag Berlin Heidelberg 2013

1   Introduction

According to clinical data, it can be seen that within six months after the stroke, 88% of the acute stroke patient will have upper extremity hemiplegia, which includes the lack of strength in the arm muscle, incapability of stretching, heteronomous spasm, the loss of original acting scope of the wrist and palm, the loss of the capability to catch and take, sometimes, the loss of muscle strength or incapability of normal movement of the finger due to abnormal spasm might happen [7]. When the patient performs Activities of Daily Living, for example, the button-up of the clothes, the hanging of the clothes and dining, etc., all these actions will cause serious challenge to the patient, not only the living quality of the patient will be seriously impacted, the social cost accompanied due to the medical care need, for example, the human resource, material and medical resources needed for the medical care system, will also be pretty large.

Neurological exercise disorder originates the injury on the cortex exercise area of the brain, however, the cortex of the brain of human beings and the related nervous system are always in the plasticity state [6] and nervous re-organization processes [10], which in turn will affect and accelerate the learning (recovery) processes of the exercise function, and related researches also prove that systematic and group exercise rehabilitation model can indeed assist the enhancement of rehabilitation therapy, in addition, some researches also pointed out that for the learning of a new exercise technical model, the providing of the behavioral performance of the user to be used as expanded feedback mechanism is one of the important rings in the learning principles of enhancing the learning effectiveness. Similarly, exercise rehabilitation of brain injury can be seen as one type of exercise learning process, and the above principle can also be used to provide continuously the patient with rehabilitation performance as feedback [6][10], in another view, the feedback of vision and hearing is also one of the important factors to keep the exercise function [11].

However, for the action training method and therapeutic session design used in traditional physical therapy or occupational therapy, no matter in practice or economically, the above goals will all be difficult to be reached, and some inherent limitations do exist. In the mean time, the effect of rehabilitation therapeutic session will to certain extent be dependent on the level of engagement of the patient. Since the rehabilitation session is tedious, lots of external factors might reduce the participation or the motivation to complete the session from the patient, for example, when the therapeutic session content is too repeated or boring, or the traffic inconvenience factor, etc.

The constant advancement of 3D animation technology and internet technology not only provides technological enhancement, but also provides economic popularization, hence, lots of scholars performing medical rehabilitation related researches and the front line doctor or therapist in the world have tried to integrate the above technologies, and they try to use virtual reality, Augmented Reality and mixed virtual reality e as the theoretical basis, meanwhile, User Centered design concept is also put in, furthermore, user's perception on the system and usability and immersion of the system are also considered, and finally, the interactive model and strategy provided by the human machine interface are also used to perform the development of all kinds of new rehabilitation therapeutic sessions and rehabilitation technologies. Through the use of new technologies and new rehabilitation method as well as systematic digital management method, optimal rehabilitation efficiency and good rehabilitation quality can both be obtained, and the second injury in the rehabilitation process can then be reduced. Eventually, the effect of rehabilitation training can be enhanced, and the burden on the medical care personnel and the patient's family can be reduced, finally, a tool that allows the patient to do independent and autonomous rehabilitation can be developed. In the tool, game model is used to increase the user's aggressiveness and participation on rehabilitation training, moreover, a game of more fun is used to urge the patient to do rehabilitation, and the patient can, through this, obtain the feeling of entertainment and achievement from rehabilitation training. Not just this, the gaming process can also be recorded in digital data way to facilitate the tracking of the rehabilitation effect by the medical care personnel.

In addition, in order to assess the current status of the patient effectively so that medical personnel and the physical therapist can seize the current status of the patient more accurately and set up the rehabilitation goal. Moreover, in order to let the medical personnel be able to perform personalized therapeutic session design and provide the fittest rehabilitation therapeutic schedule and related rehabilitation strength design and to achieve the rehabilitation goal, effective physiological assessment tool is very important. Currently, there are lots of assessment tools designed for different parts of the bodies, for different functions and for different objectives, which include physical assessment and self assessment. Clinically, the frequently used physical assessment tools include Fugl-Meyer Assessment table, Wolf Motor Function. Test (WMFT) and TEMPA(Upper Extremity Performance Evaluation Test for The Elderly), and they will perform respective assessment on the exercise function of the upper extremity, the usage capability of the wrist and palm and whether daily actions can be finished smoothly or not. In self assessment aspect, EXCITE MAL Score Sheet, CONFIDENCE IN ARM AND HAND MOVEMENTS (CAHM) and Stroke Impact Scale (SIS)[12][13][14][15][16] are frequently adopted to make self assessment on the usage injured part of the patient for confidence index and the actions needed for daily life after the carry-out of rehabilitation training, and the assessment includes whether the action needed for the life can be finished or not, the confidence index and the cause the action cannot be finished.

In the mean time, the behavioral performance values measured through digital system are very diversified in its type, and the data are also very rich in its quantity, hence, how to, from massive and diversified information, explore or delve out valuable clue and set up new medical assessment method, and how to act in accordance with the existed assessment method and to provide more accurate and faster assessment method for clinical medical diagnosis is going to be a direction with great potential.

During the rehabilitation process, in addition to the physiological state of the patient, the emotional factor of the patient is also very important. For the psychological emotional behavioral model of stroke patient, researchers including Maclean[1] [2], Colombo[8] and Paolucci[9], have made studies regarding the correlation between the motive factor of stroke patient and rehabilitation effectiveness, hence, and the patient's acceptance on new technology includes:

Factors such as whether the system is easy to be operated, whether the patient thinks this system is helpful or not, the level of pleasure provided by the system, whether strong focus is put on the system during the experimental process and whether it is easy to adapt to the environment set up by the system might all affect the patient's participation and the willingness and motive in the rehabilitation therapeutic session.

In this study, trainings are to be provided for upper extremity rehabilitation items, for example, the endurance, stretching and flexibility of the fore arm and wrist; meanwhile, gaming technology, force feedback technology and stereo imaging technology are associated to develop virtual reality body perceptive training task. In the mean time, this research is going to perform clinical experiment, to recruit right handed patient with Fugl-Meyer Assessment score reaching the range of 40 to 50 to carry out a series of gaming type rehabilitation training therapeutic sessions that are based on virtual reality in association with physical-based equipment. There are a total of 12 therapeutic sessions, and each session lasts for two hours, and before and after the training, functional assessments will be performed, which include physical assessment and self assessment. In the training process, the exercise track and task performance of the patient is going to be completely measured and recorded, then after the finishing of the training, patient's technological acceptance on the new rehabilitation system is going to be measured. Based on the above measurement result, exercise analysis and statistical analysis will be performed so as to verify the functionality of the system, to verify the system's effectiveness on rehabilitation, to develop new assessment methods, and to investigate the topics related to human machine interactions. The entire research architecture is as shown in Fig. 1.

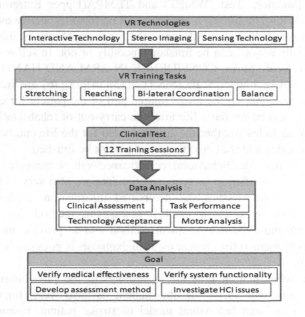

Fig. 1. Research architecture

2   Literature Review

In recent years, lots of research teams applied virtual reality in the rehabilitation of medical care [14-16], and after long term and repeated experiments, this technology has been proved to be effective in the rehabilitation training of the stroke patient. Virtual reality can display the realized training action in gaming way in the virtual environment, and in the gaming process, all kinds of rehabilitation actions can be implemented. Since all the objects interacted with the user are all virtual objects, not only the usage timing with the object can be controlled, the mutual accuracy with the user can be controlled, the feedback for achieving the task can be displayed, but also the safety of the rehabilitation training environment can be ensured. Moreover, real time gaming feedback let the user confirm his own rehabilitation progress, and it can also display the successful reaching of the goal and recognize the user's self capability and enhance the user's confidence. The application of virtual reality not only can let the user learn all kinds of learning skills, but also can train the recognition functions of the patients.

In this research, virtual reality with gaming characteristic is used as rehabilitation tool to give force feedback such as pushing force or dragging force to the hand. In the past, a team formed by scholar Lauri [17] has designed a set of system to give force to the hand through the wearing of pneumatic gloves. Here the glove is filled with air, and air pressure is used to give pushing force or dragging force so as to simulate the haptics or the weight of an object. In addition, through the optical mark installed on the body, the arm position is traced, and head-wearing type display is used to create the environment of virtual reality. The task of the game of the system is to operate in the virtual reality environment to let the palm move to the designated position, at different check points, tasks such as grasping, pinching or holding the object for a distance do not need to be achieved additionally, besides, the moving scope of the hand, the size of the force for seizing the object and whether supplemental force is added can be designed.

For this experiment, although certain positive result has been acquired in rehabilitation training, yet the equipment needed and the setting of the environment used is not so easy to general medical care personnel, moreover, it might brings up certain difficulty for the wearing of the gloves when considering the level of spasm of the hand of the patient, besides, the wearing of optical marks on the body of the patient for positioning might consume lots of time for pasting the optical marks onto correct positions. Besides, helmet type display might bring dizziness or psychological discomfort such as oppression to the patient due to long time of wearing. Therefore, we hope that simpler equipment and more comfortable operation environment can be used to achieve better training result. Not only it will become easier to use, but also it will make the rehabilitation process smoother and more efficient. In the mean time, the cost needed to purchase massive equipment and for maintenance can then be saved.

3   Research Method: System Design

In this experiment, traditional upper extremity rehabilitation items have been referred to, and trainings have been provided for the endurance, stretching and flexibility of fore arm and wrist, and two training tasks are designed respectively with each task contains three types of force feedback models for selection. Moreover, the training content is performed with task simulation and design using the game engine Unity so as to urge the patient to implement the setup target action to complete the task; furthermore, virtual reality is constructed using the product 3D-Vision of Nvidia corporation, and the advantages are that this product is supported by the current mainstream stereo display and stereo projector display equipment, besides, 3D-Vision has price lower than other product and is easier to get, meanwhile, the equipment only needs USB to be connected to PC, then through simple setup, it can be applicable to any software supported by Nvidia corporation. Furthermore, the force feedback device Falcon as launched by Novint is selected. This device is single point haptics virtual equipment. Its mechanical arm is movable along six axes, and its updating frequency is 1000 Hz. In each update, Falcon will provide the coordinate of the mechanical arm currently in the space. In the mean time, during the exaggeration process of the game, the feedback vector and size can be provided simultaneously to the Falcon, and the haptics simulation or dynamic physical force feedback can then be achieved smoothly and stably, hence, through this machine, the interaction to the object in the virtual space can then be activated. In its utilization, USB is ready for plug and play. In the followings, two training tasks and force feedback models will be introduced respectively:

3.1   Wrist Exercise

The task is designed as simulating a task of "Flying and moving across the barrier". The rehabilitation patient uses wrist exercise (inner rotation or outer rotation) to control the rolling of the airplane in virtual reality so as to move across a series of square and hollow frames, and these square and hollow frames possess respectively different rolling.

The system is designed in a way to use 3D game engine to construct an interactive virtual flying scene. Meanwhile, 3D stereo screen accompanied with polarized glasses are used to provide stereo depth of field; in the mean time, dual-haptics devices are used to construct flying controller, and through the characteristic of haptics simulator, the impact vibration can be simulated to be provided to the rehabilitation patient for force perception feedback. This system is also assisted with the measurement of brain wave to understand the correlation between the inner rotation or outer rotation exercise of the elbow and the brain wave under different torsion models.

Adjustable difficulty design uses two parameters, that is, "torsion amplitude" and "torsion model" to guide the rehabilitation patient to perform elbow exercise training of different difficulties: (1) Torsion amplitude: It set up the rolling of the square and hollow obstacle, and it will decide the torsion amplitude of the elbow of rehabilitation patient for the completion of the task. (2) Torsion model: Based on the injury condition of the rehabilitation patient, this system has designed three different torsion models as in the Fig. 3 to be used for the rehabilitation strategy planning: a. Guiding model: The flying controller will provide guiding and dragging force according to the rotational angle of the airplane and the angle of the target frame. The dragging force will become smaller when the difference of the angle between the airplane and the frame. b. Natural model: The flying controller does not provide any force. c. Resistance force model: Flying controller will provide resistant dragging force according to the rotational angle of the airplane and the angle of the target frame. The dragging force will become larger as the difference of the angle between the airplane and the frame becomes closer.

3.2   Fore Arm Exercise

The task is designed to use the task activity of the above mentioned "Flying across the barrier". Rehabilitation patient uses forearm exercise (Curving, stretching and left and right deviation) to control the left and right position as well as the height of the airplane in the virtual reality so as to move across a series of square hollow obstacles. These square and hollow obstacles possess respectively different left and right position and different height.

In the system design, 3D game engine is used to construct an interactive virtual flying scene. Meanwhile, 3D stereo screen accompanied with polarized stereo glasses are used to provide stereo depth of field. Moreover, double haptics simulation equipment is used to construct flying controller, through the characteristic of haptics simulator, collision and vibration can be simulated so as to provide force perception feedback to rehabilitation patient. Moreover, this system is accompanied with the measurement of brain wave to understand the correlation between wrist exercise (curving, stretching and left and right deviation) and brain wave under different force models.

In the adjustable difficulty design, two parameters are applied: "Deviation amplitude" and "Deviation force model" are used to guide the rehabilitation patients to perform wrist exercise training of different difficulties: (1) Deviation amplitude: It is to set up the deviation amount of the square and hollow obstacle, and it will decide the deviation amplitude of the wrist of the rehabilitation patient so as to complete the task. (2) Deviation force model: This system, based on the injury condition of the rehabilitation patients, has designed three different deviation force model as shown in the Fig. 5. It is to be used for the rehabilitation strategy planning: a. Guiding model: Flying controller will, according to the distance between the airplane and the target frame, give guiding and dragging force, and the guiding and dragging force will become smaller as the airplane gets closer to the frame. b. Natural model: Flying controller does not provide any force. c. Resistant force model: Flying controller will, based on the distance of the airplane to the target frame, give resistant force, and the resistant force will get larger as the distance between the airplane and the frame becomes closer.

In the above two exercise models, the force feedback output size of the guiding and dragging force are all calculated using linear algorithm. In the guiding model of the forearm exercise, to avoid sudden large force output in the guiding model, the maximal value of the guiding force will get increased gradually from small to large along with the closer straight line distance to the target frame. Then mixed with the frame planar distance, maximal force output proportion calculation is made. When the planar distance gets closer to the target, the force exertion will become smaller so as to avoid the generation of unnecessary bouncing and the subsequent rehabilitation injury due to over-sensitivity of the equipment. In addition, under the resistant force model, the force feedback maximal value and frame planar distance will be taken directly for force output proportion calculation. Since the force size is controlled by the user under the user's controllable range, there is thus no rehabilitation injury issue. Similarly, the force feedback algorithm design of the wrist exercise is the same as that of the forearm exercise, which is as shown in Fig. 6.

Fig. 2. Game design and hardware equipment for wrist exercise

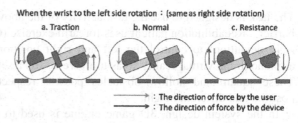

Fig. 3. It illustrates the force applied directions of three different torsion models

Fig. 4. Game design and hardware equipment for forearm exercise

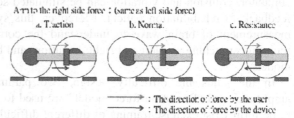

Fig. 5. It illustrates the force applied direction of three different types of forced deviation models

4 Research Method: Experimental Design

4.1 The Received Case and Case Receiving Standard

The received case contains nine members, namely, three stroke patients in acute, sub-acute and chronic case respectively. The functions of the upper extremity of the patients all reach Fugl-Meyer Assessment with score in the range from 40 to 50, and all the patients are right handed.

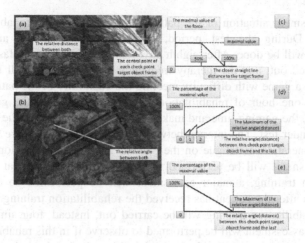

Fig. 6. (a) The relative distance of forearm exercise, (b) The relative angle of wrist exercise, (c) The relationship between the force and the closer straight line distance, (d) The relationship between the relative angle(distance) and the force in guiding model, (e) The relationship between the relative angle(distance) and the force in resistant force model

The selection standard is the unilateral brain stroke patient of the first onset of the disease, and the period between stroke onset time and the participation time should not be more than one year, and the age is in the range from 20 to 85 years old. Meanwhile, the near end action functions of the upper extremity of the disease side should all be above the fourth term (included) of Brunnstrom's stage) and the patient should have exercise disorder of the upper extremity. Moreover, the patient should have no significant recognition function loss to follow simple command, to understand experimental objective and to act in accordance with research procedure and is willing to sign the agreement for person under test.

Each patient will be recorded with his age, gender, left handed or right handed, existed diseases (including hypertension, diabetes, hyperlipemia, arrhythmi, epilepsy), the location of the brain injury, stroke type (infarct type or hemorrhagic type), date of stroke onset, number of months of rehabilitation, and the score of Barthel's Index. A professional assessment person will be responsible for the assessment of the above data and scale.

4.2   Experimental Process

This experiment is going to be designed in reversal replication design. In the beginning of the experiment, the person under test will be explained with experimental process and its objective, then the person under test will be asked to sign the agreement. In addition, we have to remind the person under test not to attend any other rehabilitation training activities during the test period.

The entire experiment lasts for 31 days. The first eight days will be the situation assessment of the patient, and the physiological and psychological state of the person under test before the experiment will be recorded. The medical personnel can, based on the assessment situation of the patient, make all kinds of rehabilitation training difficulties. During the test period, two major assessments and four simple assessments will be done. After eighth day, rehabilitation training lasting for 15 days will be carried out, and the therapy frequency for the patient will be three times a week. Select a game with difficulty meeting the capability of the patient, then collect the data for one hour of rehabilitation training, then record the game played, the performance, the gaming time and make sure there is no error in the game output for the person under test. The interval between therapy should be more than 24 hours to avoid the generation of fatigue on the patient. After the completion of the therapy, simple assessment will be performed for a total of eight times, at the last time of rehabilitation training, a major assessment will be performed so as to record the effectiveness after the patient has received the rehabilitation training. In the last eight days, no rehabilitation training will be carried out, instead, four simple assessments and one last assessment will be performed to observe if in this rehabilitation way, the patient can, without carry-out of any training, keep the level previously achieved just right before the ending of a series of rehabilitation training.

4.3   Data Analysis and Performance Assessment Method

The data type and format acquired in this research includes clinical assessment tool data, task performance data, exercise raw data and the psychological emotional data of subjective perception.

Clinical assessment tools include physical assessment and self assessment. We have used the following physical assessments: FM(first assessment), WMFT, TEMPA (four major assessment), and self assessment: MAL, CAHM, SIS(second and third major assessment). Moreover, the difference between pre-test and post-test will be used to assess rehabilitation effectiveness.

Game task performance includes task success rate, task completion time, response time and task score. Furthermore, statistical analysis will be performed to assess the improvement trend of the exercise performance.

The exercise raw data is the exercise locus information, and further exercise analysis will be performed to calculate different exercise indexes, which include stability, discontinuity, efficiency, oscillation and level of loading. Furthermore, the difference between the pre-test and post-test is going to be used to assess the rehabilitation effectiveness, in the mean time, statistical analysis will be performed to assess the trend of exercise function improvement.

The psychological emotional data of subjective perception is technological acceptance survey questionnaire, which will be used to do survey on user for the followings: awareness on the game content, presentences of virtual reality, usefulness of the training task, playfulness of the training task, intension to use of the training task and ease of use of the training task, the main objective is to obtain the assessment on this system from the person under test, and it is hoped that the result can be used as better system reference in the future design. In the mean time, related analysis will be done with the above mentioned data to investigate the correlation between psychological emotional factor and rehabilitation effectiveness.

5   Case Study

In the experiment, nine patients are expected to be recruited to participate in the experiment. Currently, one person under test has been recruited, with data as: age 82 years old, female who is a patient of Ischemic Stroke, and the score obtained from FM assessment is 27, TEMPA and WMFT performances are all middle and low. According to the above mentioned experimental process flow, we have finished Pilot Test, in the followings, we are going to show all the experimental results of the person under test.

5.1   Exercise Analysis

According to the spatial status data of the operation of virtual reality object through force feedback device from the patient under measurement, exercise analysis can be done to understand further and investigate the exercise model and behavioral strategy for the patient to complete the task.

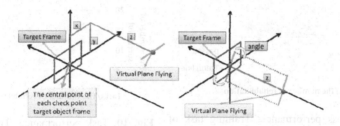

Fig. 7. The coordinates schematic

For the training task regarding the use of forearm to operate the airplane to make up and down or left and right movement, the task goal is to use force feedback device to operate the airplane and to move it to designated direction so as to pass smoothly the target object frame. The location of the airplane in the virtual space in each time point is subtracted with the spatial location of the central point of the target object frame to get the relative distance between both of them. Meanwhile, Cartesian coordinate is used as basis to cut it into three partials (x, y, z) for the representation. The units are all the coordinates defined by game engine. The central point of each check point target object frame is used as coordinate origin. Wherein x represents the left and right direction distance of the airplane relative to the central point of the target object frame, y represents the up and down direction distance of the airplane relative to the central point of the target object frame, z represents the front and rear direction distance of the airplane relative to the central point of the target object frame. When the airplane passes through the target object successfully, x, y, z values will all approach zero (The airplane size will approach but is a little bit smaller than the target object frame), which is as shown in Fig. 7(left). Since z values are constantly accumulated using fixed value, then the x, y value at each time point of the task process is plotted into time-history diagram, which represents the approaching history diagram between the airplane and the target object, which is as shown in Fig. 8. It can clearly display the behavioral model of the forearm exercise in two directions of the Cartesian coordinate for the patient in the task completion process, then a comparison can be made for the investigation of the behavioral strategy.

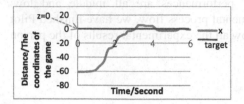

Fig. 8. The distance approaching history diagram between airplane and target object using the check point of the left and right movement training as example

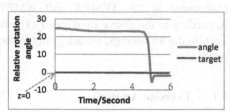

Fig. 8. The distance approaching time history diagram between airplane and target object

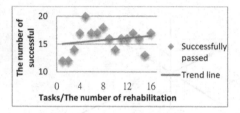

Fig. 9. Task performance: Training task of wrist exercise

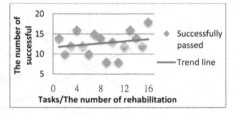

Fig. 10. Task performance: Training task of forearm exercise

For the training task regarding the use of wrist for the operation of the left and right rotation of the airplane, the task target is to use force feedback device to operate the airplane to rotate to the assigned angle so as to pass the target smoothly (frame). When we subtract the rotational angle of the airplane in the virtual space for every time point by the rotational angle of the target space, the relative angle between both of them can then be obtained; meanwhile, based on the Cartesian coordinate, Euler Angle and one partial z can be used for the representation, and the units are the coordinates defined by the game engine. Take the Euler Angle of the target object of each check point as basis, wherein the angle is the angle difference between the airplane and target object, z is the front and rear distance between the target object and the airplane, when the airplane successfully passes the target object, the relative angle and the z value will all approach zero, which is as shown in Fig. 7(right). Since z value is accumulated constantly from a fixed value, then the relative angle at each time point in the task process is plotted as time-history diagram, which then represents the approaching history diagram between the airplane and the target object as in Fig. 9. It can clearly display the behavioral model of Euler Angle of Cartesian coordinate of the wrist exercise of the patient in the task completion process. Then a comparison can be made and the behavioral strategy can be investigated.

5.2   Task Performance

According to the measured patient's task performance, accompanied with the frequency of the training sessions, statistical analysis can be made on the time axis so as to evaluate the level of improvement and the development trend of the task performance. The task performances of two training tasks at different therapeutic processes are as shown in Fig. 10 ~ Fig. 11.

5.3   Psychological Emotional Data of Subjective Feeling

The measurement result of the psychological emotional data of subjective feeling is as shown in table 1. The result shows that in two items of playfulness and intention to use, scores of more than 4 can be obtained, which shows that this system can indeed make the patient continuously involved in rehabilitation therapeutic process that is based on this training system and have fun from it. However, in the usefulness part, lower score is obtained, and the main reason is because in this part, the issues are mostly on whether the force or endurance training has brought some significant effects. However, the hemiparalysis level of this patient still cannot afford, in the rehabilitation training process, training with resistant force model, hence, lower score is obtained, and in this case, it shows that this part of survey questionnaire design has room to be improved. In the playfulness and Ease of use parts, pretty good scores are obtained, which show that such training game can increase patient's fun in the rehabilitation process, and it worth of being recommended to other patients. However, Ease of use item does not reach the score level of 4, the reason might be due to the defect in the operation hardware and equipment. Moreover, due to lower adaptation of the higher age patients on the stereo display equipment, the patients can easily feel less direct and obvious operation easiness of the system, and this is also the part of the system that needs to be improved continuously.

Table 1. Psychological Evaluation

Grading items	Awareness	Presence	Usefulness	Playfulness	Intension to use	Ease of use
Average score	4	4.72	2.53	5	4	3.56

6   Conclusions

In this research, trainings are going to be provided for the upper extremity rehabilitation items, which include the endurance, stretching and flexibility of the forearm and the wrist. Moreover, game technology, force feedback technology and stereo image technology are associated to develop body perceptive training task in virtual reality. In the mean time, this research has designed rehabilitation therapeutic session for right handed patient with Fugl-Meyer Assessment score in the range 40 to 50, and pilot test has been successfully completed. The experimental result has proved the functionality of this set of rehabilitation training task in all aspects. Through the exercise analysis of the historical data of exercise locus and the statistical analysis of the task performance of the past therapeutic sessions, this system can successfully acquire reliable and valuable information to be used in the future for verifying medical care effectiveness and for developing new clinical assessment method. In the mean time, according to the measured psychological emotional data of subjective perception, it can be seen that this system can indeed urge the patient to continue getting involved in rehabilitation therapeutic session based on this training system and enjoy it. Meanwhile, it makes the patient more confident on the rehabilitation effect possibly generated by these two training tasks.

In the future, the pilot test of this research is going to be used as the basis to perform the system improvement, and large scale clinical test is going to be conducted continuously so as to verify the medical care effectiveness of this system, in the mean time, clinical assessment method will be developed too.

Acknowledgments. We would like to thank the researchers, teachers, and students who participated in the system design, implementation, and experiment. We are also grateful for the support of the National Science Council, Taiwan, under NSC 100-2221-E-008-043- & NSC 100-2631-S-008-001. Also, this study is conducted under the "Smart terminal software and networked television platform development "of the Institute for Information Industry which is subsidized by the Ministry of Economy Affairs of the Republic of China.

References

1. Pound, P., Wolfe, C., Rudd, A.: The Concept of Patient Motivation: A Qualitative Analysis of Stroke Professionals Attitudes Niall Maclean. Stroke 33, 444–448 (2002)
2. Maclean, N., Pound, P., Wolfe, C., Rudd, A.: Qualitative analysis of stroke patients' motivation for rehabilitation. BMJ 321 (October 28, 2000)
3. Rizzo, A.A., Bowerly, T., Galen Buckwalter, J., Klimchuk, D., Roman Mitura, B.A., Parsons, T.D.: A Virtual Reality Scenario for All Seasons: The Virtual Classroom (2006)
4. Jung, Y., Yeh, S., Stewart, J.: Tailoring virtual reality technology for stroke rehabilitation: a human factors design. In: Proceedings of ACM CHI 2006 Conference on Human Factors in Computing Systems: Monteal Edited by: Mads Soegaard, April 22-27, pp. 929–934 (2006)
5. Edmans, J., Gladman, J.: Clinical evaluation of a non-immersive virtual environment in stroke rehabilitation. Clinical Rehabilitation 23, 106–116 (2009)
6. Krakauer, J.W., Mazzoni, P., Ghazizadeh, A., Ravindran, R., Shadmehr, R.: Generalization of Motor Learning Depends on the History of Prior Action. PLoS Biology 4(10), e316 (2006)
7. Fredericks, C., Saladin, L.: Pathophysiology of the Motor Systems. FA Davis, Philadelphia (1996)

8. Colombo, R., Pisano, F., Mazzone, A., Delconte, C., Micera, S., Chiara Carrozza, M., Dario, P., Minuco, G.: Design strategies to improve patient motivation during robot-aided rehabilitation. Journal of NeuroEngineering and Rehabilitation 4, 3 (2007)
9. Paolucci, S., Antonucci, G., Grasso, M.G., Morelli, D.: Post-Stroke Depression, Antidepressant Treatment and Rehabilitation Results. Cerebrovascular Diseases 12, 3 (2001)
10. Winstein, C.J., Merians, A., Sullivan, K.: Motor learning after unilateral brain damage. Neuropsychologia 37, 975–987 (1999)
11. Cheng, P.T., Wang, C.M., Chung, C.Y., Chen, C.L.: Effects of visual feedback rhythmic weight-shift training on hemiplegic stroke patients. Clin. Rehabil. 18, 747–753 (2004)
12. Ashford, S., Slade, M., Malaparade, F., Turner-Stokes, L.: Evaluation of functional outcome measures for the hemiparetic upper limb – A systematic review. Journal of Rehabilitation Medicine 40(10), 787–795 (2008)
13. Lin, J.-H., Hsu, M.-J., et al.: Psychometric comparisons of 4 measures for assessing upper-extremity function in people with stroke. Phys. Ther. 89, 840–850 (2009)
14. Farrell, T.R., Weir, R.F., Heckathorne, C.W.: The Effect of Controller Delay on Box And Block Test Performance. In: Proceedings of the Myoelectric Controls Conference (MEC 2005), Fredericton, New Brunswick, Canada, August 15-19, University of New Brunswick, New Brunswick (2005)
15. Moriello, C., Byrne, K., Cieza, A., Nash, C., Stolee, P., Mayo, N.: Mapping the Stroke Impact Scale (SIS-16) to the International Classification of Functioning, Disability and Health. Journal of Rehabilitation Medicine 40(2), 102–106 (2008)
16. Desrosiers, J., Hebert, R., Dutil, E., Bravo, G.: Development and reliability of an upper extremity function test for the elderly: the TEMPA. Can. J. Occup. Ther. 60, 9–16 (1993)
17. Connelly, L., Yicheng, J., et al.: A Pneumatic Glove and Immersive Virtual Reality Environment for Hand Rehabilitative Training After Stroke. IEEE Transactions on Neural Systems and Rehabilitation Engineering 18(5), 551–559 (2010)

Image-Based Wearable Tangible Interface

Jiung-Yao Huang[1], Yong-Zeng Yeo[1], Lin Huei[2], and Chung-Hsien Tsai[3]

[1] Department of Computer Science and Information Engineering, NTPU, Taiwan
{jyhuang,s79983201}@mail.ntpu.edu.tw
[2] Department of Computer Science and Information Engineering, TKU, Taiwan
amar0627@gmail.com
[3] Department of Computer Science & Information Engineering, NDU, Taiwan
keepbusytsai@gmail.com

Abstract. This paper presents a novel technique for the mobile TUI system which consists of a pico projector and a camera only. The proposed system can transform an arbitrary flat surface into a touch panel. It allows user to interact with a computer by hand gestures on any flat surface in anytime and anyplace. The contributions of the proposed system include the extraction of display screen from the live captured image by a RGB-based camera, shadow-based fingertips detection approach and a fast yet reliable FSM grammar to determine user's finger gesture. At the end, a prototype system is built to validate the proposed techniques and an experiment is shown in the last.

Keywords: Tangible User Interface, Computer Vision, Finger Tracking, Camera Projector.

1   Introduction

The intuitiveness is an important feature in the User Interface (UI) research [1, 2]. For a long decade, the research on UI or Human-Computer Interaction (HCI) usually is limited by the graphic display and the standard I/O interface. Recently this paradigm has changed due to the wireless, handheld and mobile devices are emerging as promising techniques. The Tangible User Interface (TUI) is the result of such trend and it pursues seamless coupling between digital information and the physical environment [3, 4]. This allows us for a much richer modal interaction between human and computer.

It would be more comfortable and effective if the user can directly control the computer in anytime and anyplace without any other hardware equipments [5]. This paper proposes an effective TUI system which is composed of a mobile computer integrated with a pico projector and a camera as illustrated in Fig. 1. As shown in Fig. 1, the proposed approach assumes the user will wear the system in the front which projects the computer display on any flat surface and allows him to manipulate the computer by hand. In other words, the proposed approach allows the user to manipulate the computer on any flat surface just like interacting with a tablet computer. The proposed system projects the display of a computer onto a flat surface similar to a virtual desktop and recognizes the user's finger gesture through the camera captured image.

J.-S. Pan et al. (Eds.): *Advances in Intelligent Systems & Applications*, SIST 21, pp. 455–464.
DOI: 10.1007/978-3-642-35473-1_46   © Springer-Verlag Berlin Heidelberg 2013

Fig. 1. Scenario for the proposed system

For the rest of the paper, the related works are presented first. The overview of the system will be presented next. The techniques of the proposed system will then be described in the following four sections. Finally, this paper is concluded with the implementation of a prototype system and a performance validation experiment.

2   Related Works

There has been a great variety of interactive tables and interactive wall researches proposed in the recent years, such as WUW – Wear Ur World[6] by Pranav Mistry (MIT), PlayAnywhere [7] by Andrew D. Wilson (Microsoft Research), and Twinkle [8] by Takumi Yoshida (The University of Tokyo).

Pranav et al. from MIT demonstrated WUW – Wear Ur World [6] at ACM CHI conference in 2009. This system was also well-known as the 6th Sense system. It allows user to access information as though user always have a computer next to him, yet, the computer is completely controlled by hand gestures. They combine a number of standard gadgets including a webcam, a projector, a mobile phone and a notebook. In its current form, the battery-powered projector is attached to a hat, the webcam is hung around the neck (or also is positioned on the hat) and the mobile phone provides the connection to the Internet.

In Play Anywhere [7] system, Andrew coupled a camera with a projector and placed it on a fixed-end of flat surface. The system was capable of detecting and re-cognizing the objects placed on the surface. The most important contribution of this system was that it did not rely on fragile tracking algorithms. Instead, it used a fast and simple shadow-based touch detection algorithm for gesture recognition.

In Twinkle [8], Takumi et al. proposed an interface for interacting with an arbitrary physical surface by using a handheld projector and camera. The handheld device re-cognizes features of the physical environment and tags images and sounds onto the environment in real time according to the user's motion and collisions of the projected images with objects. Similar to Play Anywhere, they are also employed a fast and simple motion estimation algorithm.

3   System Overview

The proposed method extracts the computer display from the projected images that are live captured by the camera. User's fingertips are identified from the extracted display and his touch gestures are then recognized accordingly. The proposed method adopts fingers touch gestures from Apple Inc [15]. Further, a FSM is proposed to infer the touch gestures.

Fig. 2 shows the pipeline of the proposed method. The first step is to compute four corner points of the computer display from the projected images that are captured by the camera. The correlation between the projected computer display and the original computer display is derived in terms of the transformation matrix in the second step. User's fingertips's position along with the amount of fingertips are then extracted from the projected image. Finally, FSM is used to determine the touch gestures of the user.

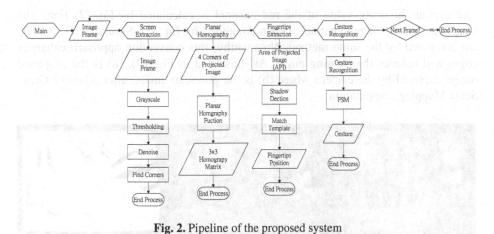

Fig. 2. Pipeline of the proposed system

4   Screen Extraction

The proposed approach wants the user can interact with the computer at any flat surface in anytime and anyplace. However, since the proposed system is aimed to be worn by the user, the field of view of the camera is constantly changed according to the user positions. In this case, most of the similar systems had provided other sensors or devices such as Kinect or FR to handle the changes of the background.

The proposed system completely relies on the image processing technique. The goal of the screen extraction stage is to extract the projected display from the camera captured image. Hence, this stage is further decomposed into four steps: gray-scaling, thresholding, morphology denoise, and four corners computation.

4.1   Convert Color Image to Grayscale

In the proposed system, the camera captured image is in RGB color space. Traditionally, there are two approaches to convert RGB image into the grayscale image which are static weighting and adaptive conversion. Since the pixels' value of the projected image has varied according to the ambient light in our operating environment, an adaptive Gray-Scale Mapping approach is adopted to convert the RGB image into the grayscale image. The conversion formula is shown as in Eq. (1).

$$I_G(x,y) = R_w R(x,y) + G_w G(x,y) + B_w B(x,y) \tag{1}$$

Where I_G denotes the grayscale image and $R(x, y)$, $G(x, y)$, $B(x, y)$ represents RGB channels of a color image respectively. Furthermore, R_w is the weight of R channel which is computed by Eq. (2) and Eq. (3) with N as the total number of pixels.

$$R_{av} = \sum \frac{R(x,y)}{N} \tag{2}$$

$$R_w = \frac{R_{av}}{R_{av} + G_{av} + B_{av}} \tag{3}$$

The average intensity value of the red channel is calculated by Eq. (2) first. The weight parameter of red channel is then computed by Eq. (3). Similarly, G_w and B_w are computed by the same method. Noteworthy, this conversion approach enhances edges and reduces the lighting noise. As illustrated in Fig. (3), (a) is the projected image captured by the camera where (b) is the grayscale image after adaptive Gray-Scale Mapping computation.

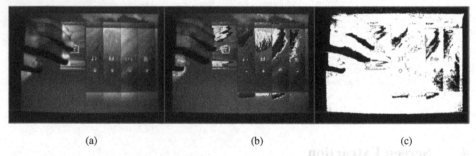

(a)   (b)   (c)

Fig. 3. (a) Camera captured image; (b) Gray-level image; (c) Thresholding and de-noises result

4.2 Adaptive Thresholding

From a grayscale image, thresholding is a pixel-by-pixel operation, as shown in Eq. (4), used to create a binary image as illustrated in Fig. 3(c). The key parameter in the thresholding process is the choice of the threshold value T.

$$I_b(x,y) = \begin{cases} 1 & if\ I_g(x,y) < T \\ 0 & otherwise \end{cases} \tag{4}$$

where I_b is resulted binary image and $I_b(x,y)$ is the value of pixel at location (x, y).

Since the proposed system has the camera and the projector bound together, the ratio of the projected computer display size to the projected images is roughly the same all the time even when the occlusion happens. On the other hand, the projected computer display area is always brighter than the surrounding area. These two conditions enable us to propose a purely intensity-base approach to obtain a rather good threshold value T.

We define Area of Projected Display (APD) as the total number of pixels of the projected display in the camera captured projection image. According to experiement, this value can be treated as a constant in the whole process. Hence, the threshold value T can be given by Eq. (5)

$$f(T) = \text{argmin}\{f(g) | f(g) = |\Sigma_{x=0}^{g} p(x) - A|, \ \forall g \in [0,255]\} \tag{5}$$

Where A is Area of Projected Display (APD) and $p(x)$ is the number of pixels in I_g at the grey-level x.

4.3 Morphology De-noise

There might be some bright pixels left in the surrounding area during the thresholding process. The proposed system uses the erosion method [9] to erode away the noise of background pixels.

4.4 Find Corners

After the processes of gray-scale, thresholding, morphology denoise are executed, the resulted image is as shown in Fig. 3(c). Furthermore, Fig. 3(c) shows that the projected computer display is close to a quadrangle within the projected image. Hence, the next step is to compute the coordinate of four corners of this quadrangle. Each corner is the pixel in the quadrangle that has longest distance with respect to the center of quadrangle. Given these four corners as N, S, E, W, and define a, b as the height and width of the camera captured image, the corner computing formula is given in Eq. (6).

$$N(i,j) = \text{argmax}\left\{\left\|\langle i,j\rangle - \langle\frac{a}{2},\frac{b}{2}\rangle\right\|^2 \mid \forall i \leq \frac{a}{2} \ j > \frac{b}{2}\right\}$$

$$W(i,j) = \text{argmax}\left\{\left\|\langle i,j\rangle - \langle\frac{a}{2},\frac{b}{2}\rangle\right\|^2 \mid \forall i < \frac{a}{2} \ j \leq \frac{b}{2}\right\}$$

$$E(i,j) = \text{argmax}\left\{\left\|\langle i,j\rangle - \langle\frac{a}{2},\frac{b}{2}\rangle\right\|^2 \mid \forall i > \frac{a}{2} \ j \geq \frac{b}{2}\right\} \tag{6}$$

$$S(i,j) = \text{argmax}\left\{\left\|\langle i,j\rangle - \langle\frac{a}{2},\frac{b}{2}\rangle\right\|^2 \mid \forall i \geq \frac{a}{2} \ j < \frac{b}{2}\right\}$$

5   Planar Homography

In this step, the relation between the original computer display and the projected display is computed.Here, we only consider the camera and projector are bound together in the proposed system. We also assume that the projected surface is a perfectly planar surface. In other word, we assume that the projected display is intacting inside the camera captured image.

5.1   Projector/Camera Calibration

To enable users to author and interact with the projected display, it is necessary to calibrate the projector and camera in an unified 3D space [10]. This research adopts Zhang's Method [11] and Falcao's Method [12] to perform camera and projector calibration.

5.2   Original Reference Image and Projected Image Homography Estimation

Since the coordinate of four corners of the projected display is already computed by Eq.(6), the homography matrix H, i.e. Eq. (8), of the original computer display and the projected display can be easily done by Eq. (7), where x' , y' are coordinates of original computer display and x, y are coordinates of the projected display.

$$\begin{bmatrix} x_1 & y_1 & 1 & 0 & 0 & 0 & -x_1x'_1 & -y_1x'_1 \\ 0 & 0 & 0 & x_1 & y_1 & 1 & -x_1y'_1 & -y_1y'_1 \\ x_2 & y_2 & 1 & 0 & 0 & 0 & -x_2x'_2 & -y_2x'_2 \\ 0 & 0 & 0 & x_2 & y_2 & 1 & -x_2y'_2 & -y_2y'_2 \\ x_3 & y_3 & 1 & 0 & 0 & 0 & -x_3x'_3 & -x_3x'_3 \\ 0 & 0 & 0 & x_3 & y_3 & 1 & -x_3y'_3 & -x_3y'_3 \\ x_4 & y_4 & 1 & 0 & 0 & 0 & -x_4x'_4 & -x_4x'_4 \\ 0 & 0 & 0 & x_4 & y_4 & 1 & -x_4y'_4 & -x_4y'_4 \end{bmatrix} \begin{bmatrix} h_{11} \\ h_{12} \\ h_{13} \\ h_{21} \\ h_{22} \\ h_{23} \\ h_{31} \\ h_{32} \end{bmatrix} = \begin{bmatrix} x'_1 \\ y'_1 \\ x'_2 \\ y'_2 \\ x'_3 \\ y'_3 \\ x'_4 \\ y'_4 \end{bmatrix} \tag{7}$$

$$\begin{bmatrix} x' \\ y' \\ 1 \end{bmatrix} = \begin{bmatrix} h_{11} & h_{12} & h_{13} \\ h_{21} & h_{22} & h_{23} \\ h_{31} & h_{32} & h_{33} \end{bmatrix} \begin{bmatrix} x \\ y \\ 1 \end{bmatrix} \tag{8}$$

6   Fingertips Extraction

The transformation matrix of Eq.(8) is then useful to compute user's finger position on the original computer display from the shadow of the user's fingers. From the computed fingers's position, the user's touch gesture can then be derived. To further speed up the derivation of user's touch gesture, the detection of user's fingers are limited to identify fingertip from the fingure's shadow. Hence, the task of this step is further divied into two sub-tasks: shadow detection and template matching.

Since the gray-scale image and binary image of the camera captured image were computed in the first stage of the proposed pipeline, Fig.(2), we can simply perform the AND operation on these two images to derive the shadow of user's fingers. A pixel of resulted image is denoted as part of shadow if its value is lower than a predefined threshold. The resulting shape will then be the finger shadow. A thining process is apply next to the derived finger shadow to locate fingertips position. The amount of thining lines is depend on the detected finger shadows. All generated lines will be processed to determine user gesture. The result of the AND operation and line thining effect are shown in Fig. 4(a). Line that exceed certain length will be identified as finger and be tracked. From the Fig. 4(a), we can clearly see that a straight finger will generate a line which is the combination of a beeline and small curve at the end of the line. In other word, we use this feature to determine fingertips location as the resulting curve is due to the shape of fingertip. Besides that, the last pixel from the line will be set as the fingertips location. Apart of that, line that only possess a straight line will be determined as "select" as a press motion will not generate a curve shape. Fig. 4(b) denote the detected fingertips by red circles. Notice that, in oder to clarly display computed result, we marked red circles on the original captured image.

(a)   (b)

Fig. 4. (a) Shadow detection (b) Fingertips Extraction

7   FSM-Based Recognition of Dynamic Fingers Gestures

Finally, the proposed system employs FSM to infer user's fingers gestures from the computed position of fingertip and the amount of fingertips. The survey by [14] shows that previous effort on gesture recognition can be categories into hidden Markov models (HMMs), particle filtering and condensation, finite-state machine (FSM), and neural network. In addition, most of FSM approaches model gesture as an ordered sequence of states in a spatial–temporal configuration space. The proposed system recognize nine gestures at this moment and, different from [13], they are modeled into one ordered sequence of states. This design offers some advantages: (1) no any specific initial gesture is required; (2) the computational complexity is reduced.

This research adopts touch gesture from Apple Inc. [14] as shown in Table 1. We define the set of states as $\{S_0, S_1, S_2, S_3, S_4, S_5, S_6, S_=, S_u, S_d, S_e, Z_m, S_h, S_l, S_r\}$ with S_0 be the initial state. Furthermore, the input alphabet, i.e. $\{f_1, f_2, f_3, f_4, d_>, d_<, d_=\}$, and gesture states, i.e. {Scroll Up, Scroll Down, Select, Zoom, Shrink, Slide Left, Slide Right}, are also defined. The context free grammar rules are shown as follows.

$$
\begin{aligned}
&S_0 \rightarrow f_1 S_1 |\ f_2 S_2\ |\ f_3 S_3\ |\ f_4 S_0 & S_u &\rightarrow \text{Scroll Up} \\
&S_1 \rightarrow f_1 S_4 |\ f_2 S_0\ |\ f_3 S_0\ |\ f_4 S_0 & S_d &\rightarrow \text{Scroll Down} \\
&S_2 \rightarrow f_1 S_0 |\ f_2 S_5\ |\ f_3 S_0\ |\ f_4 S_0 & S_e &\rightarrow \text{Select} \\
&S_3 \rightarrow f_1 S_0 |\ f_2 S_0\ |\ f_3 S_6\ |\ f_4 S_0 & Z_m &\rightarrow \text{Zoom} \\
&S_4 \rightarrow d_> S_u |\ d_< S_d |\ d_= S_= & S_h &\rightarrow \text{Shrink} \\
&S_5 \rightarrow d_> Z_m |\ d_< S_h |\ d_= S_= & S_l &\rightarrow \text{Slide Left} \\
&S_6 \rightarrow d_> S_l |\ d_< S_r |\ d_= S_6 & S_r &\rightarrow \text{Slide Right} \\
&S_= \rightarrow f_1 f_1 f_1 S_e |\ f_2 f_2 f_2 S_e
\end{aligned}
\tag{9}
$$

Table 1. Dynamic Fingers Gestures

Command	Filter		Icon
Select	Fingertips Amount	2	
	Duration	long	
	distance	=	
Select	Fingertips Amount	2	
	Duration	short	
	distance	=	
Select	Fingertips Amount	1	
	Duration	long	
	distance	=	
Zoom	Fingertips Amount	2	
	Duration	short	
	Distance	>	
Shrink	Fingertips Amount	2	
	Duration	short	
	Distance	<	
Scroll Up	Fingertips Amount	1	
	Duration	short	
	Distance	>	
Scroll Down	Fingertips Amount	1	
	Duration	short	
	Distance	<	
Slide Left	Fingertips Amount	3	
	Duration	short	
	Distance	<	
Slide Right	Fingertips Amount	3	
	Duration	short	
	Distance	>	

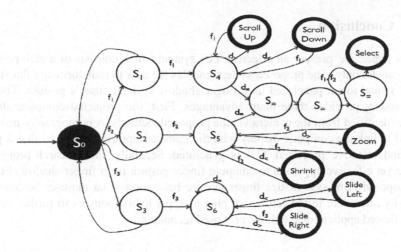

Fig. 5. The FSM of the proposed System

According to above context free grammar rules, we can use the following FSM, as shown in Fig. 5, to model gestures supported by the proposed system.

8   Implementation and Experiment

To validate the proposed approach, a prototype system is developed by binding a pico projector (CROCUS CSPP-160) and a CCD camera (LD-650DH) together. The resolution of the camera is 640x480 (VGA). The resolution of the projector is 1024x768 with the frame rate is 30 fps. All the processes are executed on a PC with Intel Core 2 CPU (2.13GHz, 2.5GB RAM) that runs 32-bit Microsoft Windows 7.

An experiment is then conducted to further confirm the performance of the protype system. In our experiment, we verify the proposed system with dynamically changed circumstances, such as turn on and off room light rapidly, move down and up the proposed device constantly, change different display surfaces continuously etc. The experimental result is depicted in Fig 6. The vertical y-axis is the execution time of each frame. Overall, our prototype obtains a mean execution time with 0.031ms each frame.

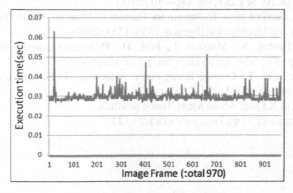

Fig. 6. Result of execution time

9   Conclusion

In this paper, we present an effective TUI system which consists of a pico projector and a camera only. The proposed system enables the user to transform any flat surface into a virtual touch panel and uses finger shadow to detect user's gesture. The propoed system provides three main advantages. First, the projected computer display can be identified from the camera via the proposed methods. A homography matrix H is used to link the computer display with the camera captured image so that a precision and effective projected TUI is generated. Secondly, this research propose an simple yet effective approach to compute finger gesture from finger shadow. Finally, the proposed FSM systematize finger gesture behavior so that the user behavior can be easily inference. In the future, we plan to apply hybrid sensors to further develop sophisticated applications with the proposed techniques.

References

1. Jeffries, R., Miller, J.R., Wharton, C., Uyeda, K.: User interface evaluation in the real world: a comparison of four techniques. In: SIGCHI 1991, pp. 4–11 (1991)
2. Kay, A.: User Interface: a personal view, pp. 121–131 (1990)
3. TUI –Wikipedia,
   http://en.wikipedia.org/wiki/Tangible_user_interface
4. Ishii, H.: The tangible user interface and its evolution. Communications of the ACM - Organic User Interfaces 51(6), 32–36 (2008)
5. Licsar, A., Sziranyi, T.: Hand Gesture Recognition in Camera-Projector System. In: International Workshop on Human-Computer Interaction, pp. 83–93 (2004)
6. Mistry, P., Maes, P., Chang, L.: WUW - Wear Ur World - A Wearable Gestural Interface. In: ACM CHI, pp. 4111–4116 (2009)
7. Wilson, A.D.: PlayAnywhere: a compact interactive tabletop projection-vision system. In: ACM UIST, pp. 83–92 (2005)
8. Yoshida, T., Hirobe, Y., Nii, H., Kawakami, N., Tachi, S.: Twinkle: Interacting with Physical Surfaces Using Handheld Projector. In: IEEE VR, pp. 87–90 (2010)
9. Gonzalez, R.C., Woods, R.E.: Digital Image Processing, 2nd edn. (2002)
10. Harrison, C., Benko, H., Wilson, A.D.: OmniTouch: Wearable Multitouch Interaction Everywhere. In: ACM UIST, pp. 441–450 (2011)
11. Zhang, Z.: A Flexible New Technique for Camera Calibration. IEEE Transactions on Pattern Analysis and Machine Intelligence, 1330–1334 (2000)
12. Falcao, G., Hurtos, N., Massich, J., Fofi, D.: Projector-Camera Calibration Toolbox (2009), http://code.google.com/p/procamcalib
13. Mitra, S., Acharya, T.: Gesture Recognition: A Survey. IEEE Transactions on Systems, Man, and Cybernetics, 311–324 (2007)
14. OS X Lion, OS X Lion: About Multi-Touch gestures,
    http://support.apple.com/kb/HT4721

The Creation of V-fold Animal Pop-Up Cards from 3D Models Using a Directed Acyclic Graph

Der-Lor Way[1], Yong-Ning Hu[2], and Zen-Chung Shih[2]

[1] Department of New media Art, Taipei National University of Arts, Taipei, Taiwan, R.O.C.
adlerway@gmail.com
[2] Department of Computer Science, National Chiao Tung University, Hsinchu, Taiwan, R.O.C.

Abstract. Pop-up cards are an interesting form of paper art with intriguing geometrical properties. It is labor-intensive and requires a high level of skill to generate two-dimensional objects that pop-up into realistic 3D objects. However, this special feature makes the design procedure of a pop-up card challenging. This paper proposes a novel algorithm to create a v-fold pop-up card from a 3D model using a directed acyclic graph. The algorithm computes a class of elements containing planar pieces and connections that approximate an input 3D geometry. Moreover, the pop up card is foldable, stable, and intersection-free when open and closed. The proposed method is demonstrated with various paper pop-ups, and experimental examples are presented.

Keywords: pop-up, digital art, geometric modeling, directed acyclic graph.

1 Introduction

The first pop-up books were created by Ernest Nister. These books were popular in Germany and Britain during the 19th century. Nowadays, people usually send them as greeting cards. Recently, many pop-up cards have become more delicate and creative. Since a v-fold pop-up card is organized by planar pieces, the shapes of these pieces are paramount for showing an object's features. The paper architecture will form a realistic 3D scene. Figure 1 displays some pop-ups designed by Sabuda [29].

Making a pop-up card is fun but labor-intensive. It is very easy to design a pop-up card with a convenience tool [1]. This paper provides a novel method for constructing a v-fold pop-up from a 3D model of an animal. The proposed method extracts typical contours from a model and creates pop-up pieces. These can be attached to the structure's base to construct an animal pop-up card. As a result, users can easily create a pop-up card from a 3D model.

The main contributions of this investigation are as follows: (1) a directed acyclic graphic (DAG) for the v-fold pop-up; (2) a novel algorithm to be applied for generating a pop-up card from an animal's 3D model; (3) how all pieces do not intersect during opening and closing. The v-fold DAG's properties ensure the stability of a pop-up structure. With these structures, a few planar pieces were used to represent a 3D model and preserve its shape. Various experimental pop-up cards were constructed of paper. The opening and closing of the card was simulated.

J.-S. Pan et al. (Eds.): *Advances in Intelligent Systems & Applications*, SIST 21, pp. 465–475.
DOI: 10.1007/978-3-642-35473-1_47   © Springer-Verlag Berlin Heidelberg 2013

Fig. 1. Pop-up cards designed by Robert Sabuda

2   Related Works

Origami is the Japanese art of folding paper into decorative shapes and figures. Its main object is foldability [9, 4, 23]. Tachi applied an algorithm that automatically generated arbitrary polyhedral surfaces [25]. Kilian proposed an algorithm that created curved folds automatically, based on the analysis of a developable surface [12]. *Paper modeling* produces 3D models using developable patches or strips. Mitani proposed a scheme for producing unfolded paper craft patterns of animals from triangulated meshes using strip-based approximations [20]. His work was followed by several interesting extensions, which approximated the shape of a 3D mesh using different types of surfaces [24, 16]. Takahashi offered a method of unfolding triangular meshes without any shape distortion [26]. *Paper architecture* is a type of pop-up that is made by cutting and folding a sheet of paper where the patches remain parallel to the two exterior pages during the opening and closing. This mechanism was the instigation for studies on the construction of paper architecture from a 3D model [21, 14, 19]. Way proposed a novel method of producing paper architecture using a directed acyclic graph [28].

Computational pop-ups are the focus of mechanisms, such as the v-fold, and the lattice and cube folds. Lee et al. [13] and Glassner [5] studied the v-fold structure and presented formulas for simulating a pop-up. Mitani et al. presented a method to design a lattice pop-up composed of pieces of paper [18]. Okamura et al. developed an interface for assisting in the design of pop-ups [22]. More recently, Li presented a geometric study of v-style pop-ups, which has more restrictions than the v-fold, and consisted of patches that have four orientations [15]. Because the configuration of a pop-up's foldability is a NP-hard problem [27], little research has been done on how to generate pop-ups automatically. Hara and Sugihara presented an algorithm to construct a two-dimensional v-fold pop-up automatically with any given polygon [7] and Li proposed an algorithm to generate a v-style pop-up from a voxelized model [15]. However, his algorithm approximates the input model into so many pieces it is hard to make a true pop-up card manually, shown as Fig. 2. In contrast, the goal of this paper is to show how to make animals using only a few pieces without losing their characteristics. Users can easily make the pop-up card by hand.

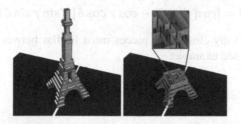

Fig. 2. The Eiffel Tower pop-up constructed by Li's V-style pop-up [15]

Planar shape abstraction addresses the problem of finding planes that are a good proxy for a 3D shape [3, 10, 11]. McCrae et al. proposed an approach for generating shape proxies based on principles inferred from user studies [17].   A paper statue can be constructed by sliding intersecting planes into slits [8]. In this paper, a principal component analysis (PCA) was adapted for obtaining the appropriate planes.

3   Algorithm Formulation

A v-fold is a rigid pop-up consisting of elements. Each element consists of two planar pieces and two supporting pieces. A general v-fold pop-up is shown in Fig. 3. To build a pop-up, a step-by-step element needs to be added which specifies two existing pieces as the supporting pieces. For example, the level 2 and 3 pieces in Fig. 3 are supported by level 1 and level 0 pieces.

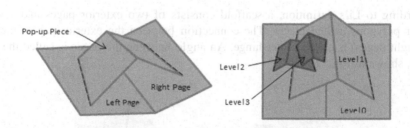

Fig. 3. A general pop-up

3.1   Pop-Up Mechanisms

Li [15] determined some geometric properties for a general pop-up. They were foldable, stable, and intersection-free. To be a foldable pop-up, the angle conditions need to be satisfied [6, 13, 15]. The fold lines must be either all parallel or all concurrent. A parallel state is approximated with a distant convergent point. The four angles between the fold lines are denoted as α, β, γ and δ, and shown as Fig. 4(a). Note that γ and δ are obtuse angles. The four angles decide the limits of angle θ when the pop-up pieces are fully stretched. The relationship between the limits of fold angle θ to the angles in the fold lines is:

$$\cos\theta = [\cos(\alpha + \beta) - \cos\gamma\cos\delta]/(\sin\gamma\sin\delta). \tag{1}$$

When the pop-up is fully closed, the pieces must lie flat between the exterior pages. Fig.4 (b) shows a closed example:

$$\gamma - \alpha = \delta - \beta > 0. \tag{2}$$

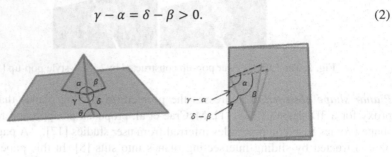

Fig. 4. (a) A open pop-up   (b) A closed pop-up

Li [14, 15] mentioned the stability of a pop-up. It is stable if pieces are placed so that each one is supported by the two preceding pieces. A pop-up can be closed flat and then opened without destroying its construction. The opening and closing of the pop-up should not need any extra force, except for holding and turning the two exterior pages. The pop-up pieces do not intersect when a card is opened and closed. When it is closed, all pieces are enclosed within the exterior pages.

3.2   Scaffold

According to Li's definition, a scaffold consists of two exterior pages and several planar polygons, called pieces. The connection between the exterior pages (the left and right pages) is called center hinge. An angle between the pages is called the fold angle, shown as Fig.5 (a).

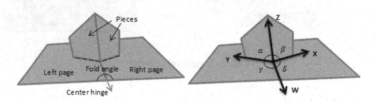

Fig. 5. (a) Scaffold   (b) The axes and angles in an element

To build a stable scaffold, two connected pieces are added. A pair of new pieces is called an element. Before adding an element, two existing connective pieces need to be selected as the support. These, in addition to the new pieces, form a simple pop-up structure. The four fold lines converge to one point. Therefore, four vectors W, X, Y and Z can represent the axes in the element's structure, as illustrated in Fig.5 (b). Every point in the element can be spanned by two axes such as $aX + bZ$ or

$aY + bZ$, and the coefficients (a, b) are the coordinates for the pop-up's simulation. The angles between the axes are denoted as α, β, γ, and δ. The angles are computed by Eqs. (1) and (2) as a foldable element. If every element is foldable, then the whole scaffold is guaranteed to be stable.

3.3   A V-fold Directed Acyclic Graphic

A v-fold digraph $G = (V, E)$ is an acyclic directed graph (DAG). There is a set of vertices $V = \{v_0, v_1, ..., v_n\}$ where v_0 represents the exterior pages, and other vertices represent the individual elements. The two pieces of element v_k are denoted as $l(v_k)$ and $r(v_k)$. A set of directed edges is $E = \{e_0, e_1, ..., e_m\}$, where $e_k = (v_i, v_j)$ denotes v_i supports v_j, $0 \le k \le m, 0 \le i, j \le n, i \ne j$. $deg^-(v_i)$ is number of directed edges entering v_i. There are some properties of a v-fold DAG:

1. $deg^-(v_0) = 0$, where v_0 is a root node at level 0.
2.   Element v_i is supported by one or two other element(s). $1 \le i \le n$.
     $\begin{cases} deg^-(v_i) = 1. \text{ if } v_i \text{ supported by one element.} \\ deg^-(v_i) = 2. \text{ if } v_i \text{ supported by two element.} \end{cases}$
3.   There is no directed cycle in a v-fold DAG.
4. v_i is at level n, where n is the number of edges on the DAG's longest path.

The first property is obvious. The second one means that every element is supported by two pieces that are located in one or two different element(s). If there is an element supported by three or more other pieces, the scaffold cannot be folded flat while maintaining its rigidity. The third property ensures that the v-fold structure is stable. Fig. 6 illustrates an example of a v-fold DAG. Vertex v_0 is at level 0, v_1 at level 1, v_2 at level 2. The longest path for vertex v_3 is 3, so v_3 belongs with level 3.

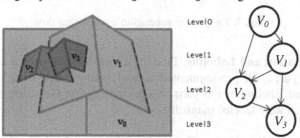

Fig. 6. A v-fold pop-up and its v-fold DAG

Li et al. also suggested the following stability proposition: A pop-up scaffold is stable if there is an order to the pieces in the scaffold $\{p_1, p_2, ...\}$ so that p_1, p_2 are the left   and right pages, and for every $k > 2$,

1. Either p_k connects with p_i and p_j where $i, j < k$, or
2. p_k connects with p_{k+1} and the two pieces connect with p_i and p_j where $i, j < k$.

Because there are no directed cycles in a v-fold digraph, the vertices can be separated into different levels. The elements are arranged from low to high according to their level. The order of the elements can transform to the order of the pieces. For example, an order of $\{v_0, v_1, v_2, v_3\}$ is calculated from the v-fold DAG in Fig. 6. Then the order of the elements $\{v_0, v_1, v_2, v_3\}$ can transform to the order of the pieces $\{l\,(v_0), r\,(v_0), l\,(v_1), r\,(v_1), l\,(v_2), r\,(v_2), l\,(v_3), r\,(v_3)\}$. This ordering satisfies Li's proposition because the lower level supports a higher level. Therefore, a v-fold DAG ensures the pop-up structure's stability.

4   Pop-Up Generation

The processing flow of a pop-up generation is shown in Fig. 7. The input mesh's pre-processing was segmented and labeled. The pop-up generation started with the left and right pages, and an element was added that conforms iteratively to a defined animal scaffold. The corresponding part of the mesh was segmented in each element's generation, and then a 2D shape was extracted from the mesh vertices. The 2D shape was transformed into the element's pieces. Finally, a pop-up card was generated using a v-fold DAG.

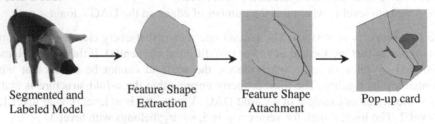

Segmented and Labeled Model → Feature Shape Extraction → Feature Shape Attachment → Pop-up card

Fig. 7. The pop-up generation's processing flow

Mesh Segmentation and Labeling. First, a 3D mesh is segmented and labeled into corresponding parts. All pre-segmented animal meshes used Chen's approach [2]. Since automated labeling of the parts was a challenge in the shape recognition, the parts of the mesh were labeled manually.

2D Feature Shape Extraction. The quality of the result planes depends on many factors such as the number of features covered, human perception, and visual aesthetics. McCrae et al. proposed a dynamic method to find a good planar shape proxy for a 3D mesh [17]. A Principal Components Analysis (PCA) plane was one of the factors under consideration. Therefore, the shape of the mesh was computed using a PCA plane during the 2D features' shape extraction.

2D Feature Shape Attachment. The 2D features' shapes were obtained by the aforementioned stages. The polygons can be rotated to the correct position to attach them to the scaffold. A rotate function $Rotate(p, \vec{v}, \theta)$ means rotating point p around an axis $\vec{v}$ by angle θ. The polygon's new point position is computed as

$$P_{goal} = Rotate(P_{ori}, \vec{r}_{axis}, \theta). \tag{3}$$

The term P_{ori} is the position of the polygon's original point, and P_{goal} is the point's desired position. The rotated axis $\vec{r}_{axis}$ can be obtained from the cross product of $\vec{n}_{goal}$ and $\vec{n}_{ori}$ i.e. the normal plane, and the original plane. Rotated angle θ is the acute angle between the polygon's plane and the desired plane. Every polygon was fitted to its corresponding element in the scaffold. After rotation, the polygons were attached, becoming pieces of the scaffold. The axes of the element span the points of the pieces, so the coordinates of the points can then be stored.

5   Pop-Up Simulation

To simulate a pop-up animation, the positions of every piece can be computed in an arbitrary fold, at angle θ, from between $0°$ to $180°$. Four unit vectors, W, X, Y, and Z, were used as the axes to represent an element's fold lines, as illustrated in Fig. 8. Since Y, Z, or X, Z can span the points of the pop-up pieces; every point position has to be calculated. However, the left page is fixed while the right page can be rotated. Therefore, Y and W are unchanged, and X and Z are rotated towards the left page. Finally, all vectors lie on the same plane at $\theta = 0$.

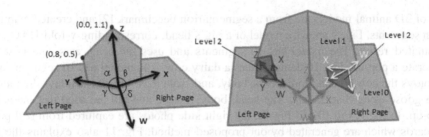

Fig. 8. The axes and angles of a pop-up element

A pop-up also consists of more than one element. According to the aforementioned concept, each element has four axes, as illustrated in Fig.8. Note now that Y and W are not always unchanged and the algorithm for computing the axes of the elements is at fold angle θ in Eq. (1).

The pop-up list is made up of elements organized into levels. Hence, the positions of pieces are updated with the traversal DAG from the root node. The position of the level 0 pages can be updated by rotating the right page. For the level 1 element, the supporting planes are the left and right pages. Axis W is unchanged since it always lies on the center hinge. For another complex element level, the axis W was obtained from normal support planes: $W = \vec{n}_{sup1} \times \vec{n}_{sup2}$. Furthermore, W can be rotated around the normal support planes to get X and Y. The negative angle means the rotation is counterclockwise: $X = Rotate(W, \vec{n}_{sup1}, -\delta)$, $Y = Rotate(W, \vec{n}_{sup2}, \gamma)$.
Moreover, Z can be calculated by X, Y, α, and β with Eq. (4). Vector P on the XY plane is the projection of Z, and P is normalized to a unit length. The angle between X

and Y is φ and the angle between X and P is ω. Vector P was obtained by rotating X with angle ω, where ω = φ · α/(α + β). Then Z could be computed with Eq. (4):

$$Z = Rotate(P, \vec{n}_{rot}, -\sigma). \tag{4}$$

where $\vec{n}_{rot}$ is the cross product of P, and the normal of the XY plane, and σ is the angle between P and Z that is given by:

$$\sigma = \cos^{-1}\left(\frac{a^2+b^2-c^2}{2ab}\right). \tag{5}$$

$$a = 1, \ b = \frac{e}{\cos \omega}, \ c = \sqrt{f^2 - d^2}, \ d = b \cdot \sin\omega, \ e = \cos\alpha, \ f = \sin\alpha.$$

where e and f can be computed, since the triangle formed by a, e, f is a right-angled triangle. Then, b and d can also be obtained. The Pythagorean theorem can be applied to get argument c, and the cosine law computed angle σ. Therefore, pieces of the element can be updated from its spanning axes. Then the updated lower level elements become the input for the more complex ones. Finally, all elements in the pop-up list are updated to their correct position at a specified fold angle.

6   Experimental Results

All of 3D animal models are from a segmentation benchmark [2] and created manually in segments. Fig. 9 shows a model of a pig's head, corresponding v-fold DAG and simulated result. The model has 10 segments and uses the head, ears, and nose to generate a pop-up. Fig. 10 demonstrates a dairy cow. The model has 10 segments and employs the cow's head, horns, ears, body, and front legs. Fig.11 displays the model of a goose using 8 segments. The head, body, wings, and tail are used to generate a pop-up. In Fig.9, Fig.10 and Fig.11, the right side photos are captured from real pop-up cards which are generated by our proposed method. Fig. 11 also explains the intermediate stages with different angles. Users can check intersection-free of all pieces during opening and closing through animations. More experimental examples are shown in Fig. 12. Most of the computation time for pop-up generation is spent trying to find the contour points in seconds. Users adjust the pieces in a few minutes and the system executes the cutting process within a second.

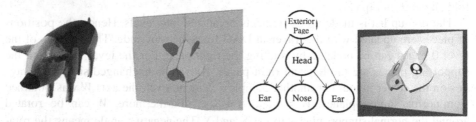

Fig. 9. A pig's head model, v-fold DAG and simulated pop-up card

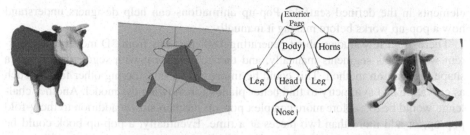

Fig. 10. A dairy cow's model, v-fold DAG and simulated pop-up card

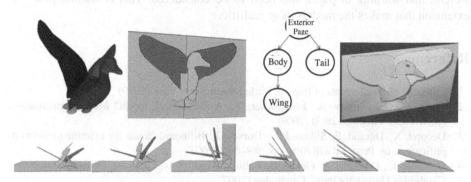

Fig. 11. A goose's model, v-fold DAG and simulated pop-up card

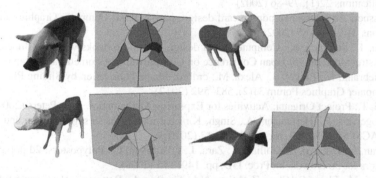

Fig. 12. Other experimental examples

7   Conclusion and Future Work

Pop-up cards are an interesting form of paper art with intriguing geometrical properties. It is labor-intensive and requires a high skill to generate two-dimensional objects that pop-up into realistic 3D objects. This paper proposes a novel algorithm to create a v-fold pop-up card from a 3D model using a directed acyclic graph. The properties of the v-fold DAG ensure that the pop-up structure is foldable, stable, and intersection-free when it is opened and closed. The input mesh is segmented and labeled allowing the animal's features to be extracted. These shapes are attached to the corresponding elements in the defined scaffold. Pop-up animations can help designers understand how a pop-up works before making it manually.

There are a few studies about generating 180° pop-ups from 3D models. The current work labels segments manually, and then combines it with segmentation and a shape recognition method. This could be improved by considering other factors such as curvature and symmetry to find better planes for showing the model. Another challenge would be to explore more complex pop-up mechanisms in addition to the v-fold pop-up, or add more than two pieces at a time. Eventually, a pop-up book could be created automatically for a story scene. Physical properties such as the thickness, weight, and warping of paper also need to be considered. This is another practical extension that makes the models more realistic.

References

1. Carter, D.: The Elements of Pop-up. Little Simon, New York (1999)
2. Chen, X., Golovinskiy, A., Funkhouser, T.: A Benchmark for 3D Mesh Segmentation. ACM Trans. Graphics 28(3) (2009)
3. Decoret, X., Durand, F., Sillion, F.X., Dorsey, J.: Billboard clouds for extreme model simplification. In: Proc. SIGGRAPH, pp. 689–696 (2003)
4. Demaine, E., O'rourke, J.: Geometric Folding Algorithms: Linkages, Origami, Polyhedra. Cambridge University Press, Cambridge (2007)
5. Glassner, A.: Interactive pop-up card design, part 1. IEEE Computer Graphics and Applications 22(1), 79–86 (2002)
6. Glassner, A.: Interactive pop-up card design, part 2. IEEE Computer Graphics and Applications 22(2), 74–85 (2002)
7. Hara, T., Sugihara, K.: Computer aided design of pop-up books with two-dimensional v-fold structures. In: 7th Japan Conference on Computational Geometry and Graphs (2009)
8. Hildebrand, K., Bickel, B., Alexa, M.: crdbrd: Shape Fabrication by Sliding Planar Slices. Computer Graphics Forum 31(2), 583–592 (2012)
9. Hull, T.: Project Origami: Activities for Exploring Mathematics. A.K. Peters (2006)
10. Kalogerakis, E., Hertamann, A., Singh, K.: Learning 3d mesh segmentation and labeling. In: ACM SIGGRAPH, pp. 102:1–102:12 (2010)
11. Kavan, L., Dobbyn, S., Collins, S., Zara, J., O'sullivan, C.: Polypostors: 2d polygonal impostors for 3d crowds. In: Proc. I3D, pp. 149–155 (2008)
12. Kilian, M., Flory, S., Chen, Z., Mitra, N.J., Sheffer, A., Pottmann, H.: Curved folding. In: ACM SIGGRAPH, vol. 27(3), pp. 75:1–75:9 (2008)
13. Lee, Y.T., Tor, S.B., Soo, E.L.: Mathematical modelling and simulation of pop-up books. Computers & Graphics 20(1), 21–31 (1996)
14. Li, X.-Y., Shen, C.-H., Huang, S.-S., Ju, T., Hu, S.-M.: Popup: automatic paper architectures from 3d models. ACM Transactions on Graphics 29(4), 111:1–111:9 (2010)
15. Li, X.-Y., Ju, T., Gu, Y., Hu, S.-M.: A Geometric Study of V-style Pop-ups: Theories and Algorithms. ACM Transactions on Graphics 30(4) (2011)
16. Massarwi, F., Gotsman, C., Elber, G.X.: Papercraft models using generalized cylinders. In: Proc. Pacific Graphics 2007, 148–157 (2010)
17. Mccrae, J., Singh, K., Mitra, N.J.: Slices: A shape-proxy based on planar sections. ACM Trans. Graph. 30(6) (2011)

18. Mitani, J., Suzuki, H.: Computer aided design for 180-degree flat fold origamic architecture with lattice-type cross sections. Journal of Graphic Science of Japan 37(3), 3–8 (2003)
19. Mitani, J., Suzuki, H.: Computer aided design for origamic architecture models with polygonal representation. In: CGI 2004: Proceedings of the Computer Graphics International, pp. 93–99. IEEE Computer Society, Washington, DC (2004)
20. Mitani, J., Suzuki, H.: Making papercraft toys from meshes using strip-based approximate unfolding. ACM Trans. Graphics 11(3), 259–263 (2004)
21. Mitani, J., Suzuki, H., Uno, H.: Computer aided design for origamic architecture models with voxel data structure. Transactions of Information Processing Society of Japan 44(5), 1372–1379 (2003)
22. Okamura, S., Igarashi, T.: An Interface for Assisting the Design and Production of Pop-Up Card. In: Butz, A., Fisher, B., Christie, M., Krüger, A., Olivier, P., Therón, R. (eds.) SG 2009. LNCS, vol. 5531, pp. 68–78. Springer, Heidelberg (2009)
23. O'rourke, J.: How to Fold It: The Mathematics of Linkages, Origami, and Polyhedra. Cambridge University Press, UK (2011)
24. Shatzi, I., Tal, A., Leifman, G.: Paper craft models from meshes. The Visual Computer 22(9), 825–834 (2006)
25. Tachi, T.: Origamizing polyhedral surfaces. IEEE Transactions on Visualization and Computer Graphics 16(2), 298–311 (2009)
26. Takahashi, S., Wu, H.Y., Saw, S.H., Lin, C.C., Yen, H.C.: Optimized Topological Surgery for Unfolding 3D Meshes. Computer Graphics Forum 30(7), 2077–2086 (2011)
27. Uehara, R., Teramoto, S.: The complexity of a pop-up book. In: 18th Canadian Conference on Computational Geometry (2006)
28. Way, D.L., Tsai, Y.S., Shih, Z.C.: Origami pop-up card generation from 3D models using a directed acyclic graph. Journal of Information Science and Engineering (accepted)
29. Sabuda, R.: http://wp.robertsabuda.com/

MagMobile: Enhancing Social Interactions with Rapid View-Stitching Games of Mobile Devices

Da-Yuan Huang[1], Tzu-Wen Chang[2], Min-Lun Tsai[1], Chien-Pang Lin[2],
Neng-Hao Yu[3], Mike Y. Chen[2], Yi-Ping Hung[1], and Chih-Hao Hsu[4]

[1] Graduate Institute of Networking and Multimedia, National Taiwan University, Taiwan, ROC
[2] Department of Computer Science and Information Engineering, National Taiwan University, Taiwan, ROC
[3] Department of Computer Science, National Chengchi University, Taiwan, ROC
jonesyu@nccu.edu.tw
[4] Institute for Information Industry
eden@iii.org.tw

Abstract. Most mobile games are designed for users to only focus on their own screens thus lack of face-to-face interaction even users are sitting together. Past work shows that the shared information space created by multiple mobile devices can encourage users communicate to each other naturally. The aim of this work is to provide a fluent view-stitching technique to help mobile phone users establish their information-shared view. We present MagMobile: a new spatial interaction technique that allows users to stitch views by simply putting multiple mobile devices close to each other. We describe the design of spatial-aware sensor module and tailor-made magnetic sensor units which are low cost, easy to be obtained into phones. We also propose two collaborative games to engage social interactions in the co-located place.

Keywords: human-computer interaction, co-located collaboration, spontaneous device sharing, view-stitching, mobile devices, cooperative games, sensors.

1   Introduction

Because mobile phones are originally designed for personal and individual use, the interaction among mobile phone users is identical regardless of whether they are remote or collocated. Andres et al. have proposed the concept of the Social and Spatial Interactions (SSI) [Lucero et al. 2010]. Mobile phones in SSI platform can detect the spatial location between devices with wireless sensors and stitch the views into an information-shared space. Some works expand this concept to several applications such as brainstorming, photo sharing, and map-based sharing interactions [Ashikaga et al. 2011; Lucero et al. 2010; Lucero et al. 2011]. As the shared view has been created, it encourages users to communicate to each other naturally therefore enhances the social interactions.

In order to create a shared information space, past works have proposed several stitching techniques such as Bumping[Hinckley 2003], onscreen gestures

J.-S. Pan et al. (Eds.): *Advances in Intelligent Systems & Applications*, SIST 21, pp. 477–486.
DOI: 10.1007/978-3-642-35473-1_48   © Springer-Verlag Berlin Heidelberg 2013

[Lucero &Holopainen, 2011] and infrared sensors [Merrill &Maes, 2007]. However, many users are afraid to damage their phones while bumping the devices. Applying gestures frequently might be cumbersome for users and decreases gaming experience. IR sensors won't work because infrared light might be occluded by hand while users grasp the phones.

In this work, we present MagMobile: that is lightweight, lowpowered and easy to be integrated in smart phones. MagMobile enables fluent dynamic view-stitching across multiple mobile devices. Users simply put their devices close to each other then the screen views are stitching together. We use this technique and develop two co-located collaborative games that can enhance the social interaction among users who are physically sitting together.

Fig. 1. MagMobile enables rapid view-stitching technique by simply putting mobile devices close to each other. We build a collaborative Snake game to demonstrate this technique.

2   Related Works

This work is related to share collocated mobile phone use and view sharing approaches.

2.1   Shared Collocated Mobile Phone Use

Andres et al proposed a SSI platform that engages phone users to collocated social interactions. They applied this interaction technique to build a MindMap prototype that supports cooperative brainstorming tool [Lucero et al. 2010]. Users can create and edit virtual notes for mind maps. Users could enlarge the scale of mind maps by stitching two devices together, the system is aware of the relative position between phones when a user applies pinch gesture on the screens. The authors further proposed Pass-them-around [Lucero et al. 2011], which is an photo-sharing prototype for mobile phone users, Pass-them- around allows user to engage collocated story-telling interactions by throwing or view-sharing digital photos between phones, this prototype also follows the SSI principles. The system supports directional communication between phones, but does not support instant short-distance view sharing mechanism. Users need apply gestures for view sharing. Ashikaga et al explored collaborative interactions for map-based applications [Ashikaga et al. 2011]. Users can share their bookmark information or synchronize the view on display. The system allows users to communicate to each other without showing their displays to other users in the process of interaction.

2.2   View Sharing Approaches

In order to achieve view-sharing mechanism, some works designed stitching gestures. Pass-them-around [Lucero et al. 2011] use pinch gesture on two device screens to stitch them together. Stitching system [Hinckley et al. 2004] asks user to draw a continuous line across different screens of devices. By analyzing the path between two devices, system could estimates the relative position between two devices. However, gestures are not suitable for collaborative tasks with fast interactive process such as cooperative games or other tasks that require continuous interactions. Applying gestures frequently might be cumbersome for users and lower down the fluency and continuity of interactive process. Gesture-free techniques can overcome this problem. In order to know the spatial information between devices, Kris et al. proposed a V-scope system [Luyten et al. 2007] to support wireless 3-dimensional location tracking by "time-of-flight" technique. Three-towers in V-scope system send infrared signals to a personal device and receive an ultrasonic back signal sent back by the device. Vscope system then calculates the 3-dimensional location of each phone. Users can treat their phones as a window above a virtual space and brows the shared digital information. However, V-scope needs an extra signal tower and force users to stay inside the sensing zone. Ken proposed bump technique that user can simply bump their devices to connect each other [Hinckley 2003] by using built-in accelerometer sensor. However, as the smart phones are expensive, many users are afraid to damage their phones while bumping the devices. Siftables[Merrill et al. 2007] is a playful distributed TUI blocks. Each Siftables equips four IrDA on each side face for sensing other neighboring blocks. Once two blocks are align together, they can communicate with each other and know the relative positions. This technique is low-cost and easy to deploy. But the character of infrared light might be occluded by hand while user grasps the phone. To avoid this situation, we use magnetic sensors for MagMobiles, which can still sense magnetic force precisely even when there is an obstacle in front of it.

3 System Design

In this section we present the design process of the MagMobile. When users want to share their views, they simply put their phones close to each other. The view will expand to both displays in less than half a second, which is sufficient for speedy view sharing applications such as playing games. Moreover, user can sense attractive force-feedback during the view-stitching process.

3.1 Basic Concept

To allow users re-arrange their shared-information as quick as possible, sensor based solutions were adopt in our prototype. We decided to use magnetic sensors and magnets to implement our sensing mechanism. The reason we adopt magnetic solution is because magnets are low-cost and magnetic force has the character of penetrability, thus the magnetic value can be sensed even the user's hand cover the magnets. In the early implementation of our prototype, we tested electronic compass embedded in a smart phone to sense the neighboring magnetic field. But build-in electronic compass can only detect one magnetic value for each axis in 3D space, which is insufficient to sense multiple mobile phones at the same place. Furthermore, we found that if there are multiple magnets nearby the built-in electronic compass, the signal output from the compass will be unstable. As a result, we decided to use Hall Effect sensors to sense the magnetic field.

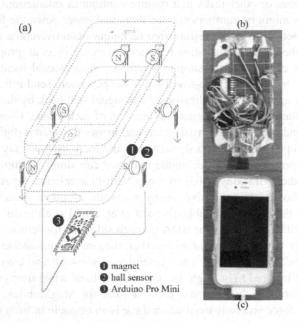

Fig. 2. (a) The design of sensor module: We attach 6 sensor units (magnet + Hall Effect sensor) on a sleeve case, (b) the backside, and (c) the front side of the MagMobile

3.2   Hardware Settings

We use UGN3503 series Hall Effect sensors which are ratiometric linear Hall Effect sensors. According to the spec of UGN3503, UGN3503 series Hall Effect sensors are sensitive and precise enough to track the small changes in magnetic flux. An Arduino pro mini is used to read the sensor values in our prototype, which transform the output voltage of sensors to a digital value, which is ranged in 1024 levels, from 0 to 1023. Fig. 2 shows our latest prototype, which is composed of six UGN3503 series Hall Effect sensors, six 10x1.5 mm cylindrical magnets, and an Arduino pro mini into one sleeve case. We stick a Hall Effect sensor and a magnet into one unit, so the unit could release magnetic force while sensing the change of neighboring magnetic field. We attach six units around the sleeve case (left, right and top). Each side is attached two units. The directions of the magnetic pole are arranged as Fig. 2a. All units are connected to Arduino pro mini. The signal processor reads the sensor values and sends them to a smart phone. Fig. 3 represents four possible spatial arrangements between two phones. The magnetic attraction force augments the feeling of stitching mechanism.

We integrate the magnetic module with an iPhone, and test the functionality of the internal digital compass, gyroscope, and accelerometer. All of them work normally. The magnets used in our prototype won't affect the performance of build-in sensors because their magnetic field strength is too weak. We have not attached the units to the bottom of sleeve case because our prototype now has to be linked to a computer as a relay station. This limitation will be eliminated in the future when the sensor module is integrated into mobile phones.

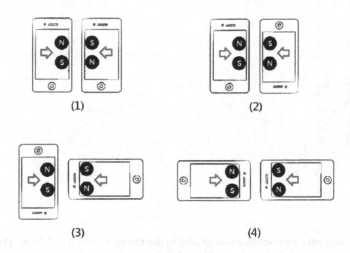

(1)   (2)

(3)   (4)

Fig. 3. Four possible spatial arrangements are all attracted to magnets. The magnetic attraction force augments the feeling of stitching between phones.

3.3   Stitching Mechanism with MagMobile Modules

The magnetic data read by Arduino are used to estimate the distance between two phones. We did an experiment to collect the data and build a lookup table for the spatial relationship between sensor module. We put two devices close in various distances and directions (see Fig. 3), and record sensor values of each unit. Fig. 4 shows the results of the experiment. At the beginning of synchronize mechanism, sensor values of a unit is calibrated by using the lookup table we built. Then the changes of sensor values are used to be the indicator of the distance between mobile phones. To judge if there is any device is getting closer in a certain direction or not, we simply accumulate the changes of sensor values on one direction. If the accumulated value is larger than a threshold, the phone assumes that there is another device nearby and stitches the screen views. Currently our prototype can sense three directional stitched position (i.e. left, right, and top), but it could sense more precise stitched position if we add more Hall Effect sensors on each side. (Fig. 7) Each module sends those stitched position to a server when the stitching edge has been detected. The server is written in Python 3. Our next plain is to run the server directly on the phone to eliminate the external server. Fig. 5 represents a view-stitching demo when two phones get close together.

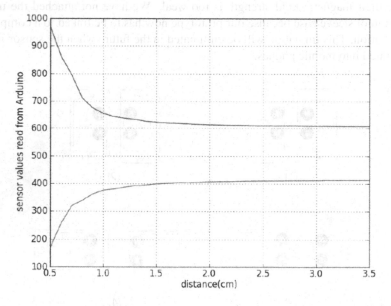

Fig. 4. The magnetic data between two phones in the distance from 0 to 3.5 cm. The blue curve represents the average sensor values of N-pole-outward unit, and another curve represents S-pole-outward unit.

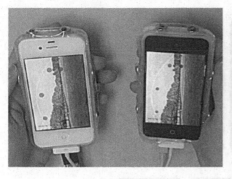

Fig. 5. Two phones with MagMobile module (left). They can detect magnetic change in short distance and stitch screens (right). Each blue dot on the screens represents magnetic force sensed by a sensor unit.

4   Applications

MagMobile stitching technique enhances the fluent dynamic view-stitching in the collaborative game and increase body interaction among collocated users. Here we propose two applications(games) to demonstrate the realistic scenarios of the use of MagMobile sensor modules.

4.1   Collaborative Snake Game

We redesign the classic Snake game to a collocated collaborative game that supports 2~6 players. The game starts while all devices connect to a local wireless network. Like a classic snake game, the game picks one device as a start point. User can use swipe gesture to direct the snake's direction. The gold are randomly showed in different players' device. So users have to move their phones and rearrange the relative position quickly to collect gold and avoid obstacles. (Fig. 1)

4.2   Collaborative Tower Defense Game

Tower defense is a very popular game genre on touch devices. Base on the rules of traditional tower defense, we added collaborative factors into games. The map is not limited by a tiny mobile display. By putting multiple mobile devices together, the piece of map can be stitched into one large map. Players need to re-arrange their phones to generate a better path on the map dynamically. (Fig. 6)

As the game starts, a map shows different path on every smartphone and the defense towers are randomly located on the map to attack the enemies. At the beginning of each round, enemies will randomly appear on one of the smartphones, and then start to move through the path on the map.

Players have to extend the path by stitching their phones so the defense towers can eliminate more enemies and get more scores. As the enemies go from one phone to another, when the enemies pass to the next phone, the previous phone is free to take

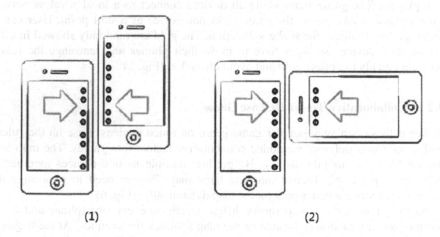

Fig. 6. Collaborative Tower Defense Game: (a) The path expands while two phones stitching together. (b) If there is no other phone be stitched in the end of the path, the enemies escape and decrease the player's life point.

Fig. 7. Increase the sensor units around the edge to expand views between mobile devices in a various way

away and change a new path to be stitched in the next round. If the enemies move to the end of the path or the edge of the screen, the enemies will escape and decrease player's life points. When the life points decrease to zero, the game is over.

In the game design, players are required to team up with each other and use their phones to create a bigger map and a better path to win the game. Thus the game encourages the social interaction and provide more fun for players who are sitting together.

5   Conclusion and Future Work

In this paper, we provide a rapid view-stitching technique to build the information-shared space and encourage users to engage co-located social interactions. We present a prototype: MagMobile, which contains magnets and magnetic sensors to sense the spatial relationship among multiple mobile devices. The main contribution of this work is to provide a fluent view sharing mechanism by just putting devices close to together. Furthermore, MagMobile is low cost, low power consumption and easy to be integrated to mobile phones. We design and implement two collaborative games that support the view-stitching interactions to demonstrate the realistic scenarios of the use of MagMobile sensor modules.

In the future, we plan to conduct a user study to observe the user's behavior with our technique in co-located collaborative interactions. We plan to discover the possibility about magnet force feedback between devices. For example, users will feel attracted force if the phones are allowed to be stitched and feel repulsive force conversely. On the other hand, we will try to increase the density of the sensor units, we plan to combine an Npole- outward unit and an S-pole-outward unit into a pair, and then the number of relative spatial arrangements will be increased. (Fig. 7).

Acknowledgements. This studywas supported by the National ScienceCouncil, Taiwan, under grant NSC101-2218-E-004-001. And the studyis conducted under the Project Digital Convergence Service Open Platform of the Institute for Information Industry which is subsidized by the Ministry of Economy Affairs of the Republic of China.

References

1. Ashikaga, E., Iwata, M., Komaki, D., Hara, T., Nishio, S.: Exploring map-based interactions for co-located collaborative work by multiple mobile users. In: GIS 2011, pp. 417–420 (2011)
2. Hinckley, K.: Synchronous gestures for multiple persons and computers. In: UIST 2003, pp. 149–158 (2003)
3. Hinckley, K., Ramos, G., Guimbretiere, F., Baudisch, P., Smith, M.: Stitching: pen gestures that span multiple displays. In: AVI 2004, pp. 1–8 (2004)
4. Lucero, A., Holopainen, J., Jokela, T.: Pass-them-around: collaborative use of mobile phones for photo sharing. In: CHI 2011, pp. 1787–1796 (2011)

5. Lucero, A., Keranen, J., Hannu, K.: Collaborative Use of Mobile Phones for Brainstorming. In: Mobile HCI 2010, pp. 337–340 (2010)
6. Lucero, A., Keranen, J., Jokela, T.: Social and spatial interactions: shared co-located mobile phone use. In: CHI 2010, pp. 3223–3228 (2010)
7. Luyten, K., Verpoorten, K., Coninx, K.: Ad-hoc co-located collaborative work with mobile devices. In: Mobile HCI 2007, pp. 507–514 (2007)
8. Merrill, D., Kalanithi, J., Maes, P.: Siftables: towards sensor network user interfaces. In: TEI 2007, pp. 75–78 (2007)
9. Rekimoto, J., Ayatsuka, Y., Kohno, M.: SyncTap: An Interaction Technique for Mobile-Networking. In: Mobile HCI 2003 (2003)

Computer-Vision Based Hand Gesture Recognition and Its Application in Iphone

Hsi-Chieh Lee[1,*], Che-Yu Shih[2], and Tzu-Miao Lin[3]

[1] National Quemoy University
Department of Computer Science and Information Engineering
No.1 University Rd, Jinning, Kinmen, 89250, Taiwan, R.O.C.
[2,3] Yuan Ze University
Department of Information Management,
135 Yuan-Tung Rd, Chung-Li, 32003, Taiwan, R.O.C.
imhlee@gmail.com

Abstract. The objective of this study is to create an effective way on communication between two deaf people and the others. The Computer-Vision based hand gesture recognition was developed. We apply Background Subtraction method to show a targeted gesture motion images. The images were transformed to YC_bC_r color space and binaries to locate the skin region. We used Morphological and Connected Component method to remove the noises produced in image process. Further, we also eliminated problems with recognizing process like a slanted hand gesture. Finally, we used Artificial Neural Network for recognizing the sign language and transmitted the out-come to handheld device such as Iphone.

Experimental results show that the accuracy of 89% in average and the processing time of 55ms in each gesture were archived in the motion hand gesture recognition. While the accuracy of 94.6% in average and the processing time of 39ms in each gesture was archived in the static hand gesture recognition.

Our results demonstrated the Computer-Vision based hand gesture recognition can be applied to next generation of Iphone.

Keywords: Gesture recognition, Pattern recognition, Human-computer interaction, Backpropagation neural network.

1 Introduction

Many new technology systems have been proposed with the advance of science and technology. The system which can easily use our voice to control the device is proposed in Smart Home Environment [1] and [2]. Automatic Recognition System is used in Car Theft Protection [3]. Dent Detection system in Car Bodies, can detected dent automatically in production process [4].A guide robot system can assisted

---

* Corresponding author.

J.-S. Pan et al. (Eds.): *Advances in Intelligent Systems & Applications*, SIST 21, pp. 487–497.
DOI: 10.1007/978-3-642-35473-1_49   © Springer-Verlag Berlin Heidelberg 2013

hearing-impaired people for going outside [5]. Moreover, some of the application using in gender detection [6], smile detection [7] etc. Today, these Human-Computer Interaction (HCI) systems are created for improving our life.

Body Language is a very important communication way, we naturally use it for more precisely representing our meaning. If we have language barrier with foreigner, we also can approximately express our meaning by using body language. Gesture Language is indispensable in Body Language. Gesture Recognition is widely used in many applications. For instance, people can definite specific gesture for controlling their action. It could help physical disabilities or someone who has been bedridden. They just assume the gesture which has been definite and they could get assistance in time. Combined Gesture Recognition and Robot Arm could avoid worker working in very dangerous environment.

Due to congenital condition, deaf people could not properly communicate with the others. They have to use Gesture Language for communicating. But it is restricted on someone, who talking with, has to have learned the Gesture Language. This situation caused those people have more difficulties than the others in society. The science and technology should be able to use the human nature, it in humanistic care. Our goal is to create an effective way on communications between deaf people and the others by using our gesture recognition system. Furthermore, we can develop recognition system to the hand-held device for disable people to communicate with others.

The following: Section 2 discusses the state of the art in hand gesture recognition. Section 3 presents the system framework. Section 4 presents results for each of the modules.

2   Review of Hand Gesture Recognition

Gesture Recognition mainly can be accomplished with three kinds of method [8]. The first method used glove-based devices. Although it can correctly recognize gesture and work timely, it is restricted that it requires the user to wear an additional device, and generally carry a load of cables to connect to a computer. The second method is called Video-based Recognition method which recognizes our gesture with camera or other devices. Because the method do not require user to wear cumbersome devices, it is widely used in recent years. The third method is called 3D Model-Based Hand method which needs to build the user's 3D Hand model. Although the method can correctly recognize gesture, it need more time consuming than other two methods. For a practical application, it is important to develop a gesture recognition system can be used in real time environment.

Jagdish Lal Raheja [9] developed a gesture recognition system and defined 10 gestures. The system use Background Subtraction method to grab gesture motion images. The images will be used Vertical Projection method for finding cut point. Finally, it will be compare with database which has been built before to find what gesture is it. The experimental results show the accuracy is 90%. But the system is restricted in the hand should enter the screen in the same direction.

Nobuharu Yasukochi [10] developed a gesture recognition system and defined 9 gestures. The system use Background Subtraction method to grab gesture motion images. The images will be transformed to RGB color space and binarized to locate the skin region. The experimental results show the accuracy is 96.7%, and the processing time is 28ms per frame. But the RGB color space is easily affected by light condition.

Liou [11] developed the gesture recognition system which used Adaptive Skin-Color Detection and Motion History Image method for distinguishing different gestures. The system will detect where the face is and adjust the user's skin extraction threshold values. He uses Background Subtraction method to grab gesture motion for realizing the gesture moving direction. He defined 6 kinds of moving gesture. The experimental results show the accuracy is 94%, and the processing time is 38.1ms per frame. However it not only must have the features to show enough face area but also there must has consistency between the hand and the face.

Tu [12] developed a gesture recognition system which use in robot behavior control He used NCC color space for extracting skin region. The system can automatically deal with some problem, such as slanted hand gesture and finding wrist cut point. He defined 11 kinds of gesture. The experimental results show the accuracy is 92.7%, and the processing time is 66.5ms per frame.

Tsai [13] developed an American Sign Language recognition system. The system grab motion images which will be detected whether the object in screen is moving or not by using Background Subtraction method. If the system has detected the moving object, the system will transfer the motion image to YC_bC_r color space for extracting skin region. And the system use Thinning Algorithm to reduce gesture feature. The system adopts 36 kinds of gesture (which is refer to A~Z, 0~9) which definite in American Sign Language. The experimental results show that the accuracy is above 90% mostly, and the processing time is 152.24ms per frame. The Thinning Algorithm has a well-known problem which is redundant branches problem. However, Thinning Algorithm can simply describe the object's outline and do not lose the object's important information as proposed in 1967. Many Thinning Algorithms have been proposed, with the main object of reducing the redundant branches problem.

3   System Architecture

Fig.1 shows the flow chart of a recognition system of gesture, in which the difference between frames is used to detect gesture. We transform the frames which contain the gesture to YC_bC_r color space for extracting skin region with specific threshold. The method might noise the gesture image and thus we use Morphological Processing and Connected Component method for reducing the noise effect. Moreover, we normalize the gesture to the same direction, since we do not know what degree the user's gesture will be taken. Wrist is usually not useful for the recognition process, and therefore the system will crop it. Finally, we use Artificial Neural Network for recognizing the gesture.

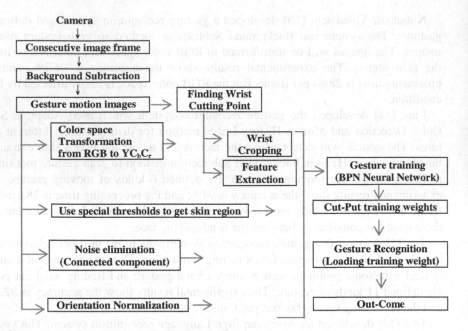

Fig. 1. System Framework

3.1   Gesture Definition

The gestures adopted in system are according to American Sign Language dictionary (ASL) show in Fig2. The American Manual Alphabet is a manual alphabet that augments the vocabulary of American Sign Language. It is usually used in spelling individual letters of a word, proper names or the titles of works.

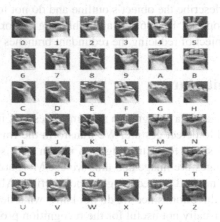

Fig. 2. Gestures in ASL

Raymond Lockton [14] adopted static gesture which can grab critical frame for re-pressing its meaning J and L for symbolizing the original J and L.

3.2   Background Subtraction

In this section, the system applies Background Subtraction method for extracting the moving gesture. The system does a subtraction which is according to the time frame. If the subtraction result is more than specific value, there is something which is mov-ing in the screen. An image which is two dimensions shows in Fig3.

Fig. 3. Background Subtraction's flow chart

The system labels four boundary points which the subtraction result's value has changed so that the system can only process the area of the rectangle. In this process, the system also applies Background Subtraction method for avoiding tiny difference.

3.3   Skin Color Extraction

The system normalizes the image which grab from camera to scale 400x300. The image is transformed to YC_bC_r color space from RGB for grabbing skin region. It applies specific threshold which is according to Chai [15]. As a result, we can get a binary image by equation (1).

$$\text{ROI}(x,y)=\begin{cases} 255, & \text{if } (77 <C_b(x,y)<127) \\ & \wedge\ (133<C_r(x,y)<173) \\ 0, & \text{otherwise} \end{cases} \tag{1}$$

Where x=0,1,2,...,400 and y=0,1,2,...,320.

Fig4 shows the extraction result. Besides the gesture, as we can see there are other noises in the processed result. These noises might affect the following recognition process, so we should properly lessen the noise.

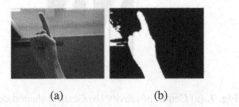

(a)   (b)

Fig. 4. (a) Original image (b) After extraction result

3.4   Noise Elimination

The Morphological Method is widely used in noises elimination in Image Processing. Fig5 (a) is a result after Erosion process. As we can see, the small noises have been removed but the gesture has also been eroded improperly. So we apply Dilation process for restoring the gesture (Fig5 (b)).

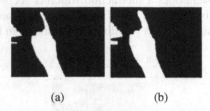

(a)   (b)

Fig. 5. (a) Erosion result (b) Dilation result

Thus, we apply Connected Component for labeling different part of region. Fig.6 is a result that we can get rid of other parts besides gesture.

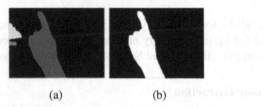

(a)   (b)

Fig. 6. (a) Applying Label (b) Labeling result

3.5   Orientation Normalization

Because we do not know what degree of gesture will be taken or if the camera position is correct, so we normalize to minimize those variations. After previous steps, we can acquire a region of interest (ROI) image. We locate a center of gravity of the gesture first by equation (2).

$$X_g = \frac{\sum_{i=0}^{N} X_i}{N}, \; Y_g = \frac{\sum_{i=0}^{N} Y_i}{N} \tag{2}$$

Where N = sum of the ROI point

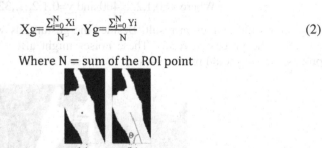

(a)   (b)

Fig. 7. (a) Center of gravity (b) Gesture slanted degree

Moreover, we locate a bottom and center of ROI and calculate these two points slope. Thus, we can get a degree of slanted gesture by equation (3).

$$\theta = \text{atan}\left(\frac{Y_m - Y_g}{X_m - X_g}\right) \times \frac{180}{PI} \tag{3}$$

The next step we use Horizontal Projection method for cropping the wrist, so we apply a rotation method in Two-Dimension Coordinate System by equation (4). As a result, we can get a gesture which has been adjusted.

$$\begin{bmatrix} x' \\ y' \end{bmatrix} = \begin{bmatrix} cos\theta & -sin\theta \\ +sin\theta & cos\theta \end{bmatrix} \begin{bmatrix} x \\ y \end{bmatrix} \tag{4}$$

Fig. 8. A adjusted gesture

3.6   Wrist Cropping

In the recognition process, the wrist is not useful for the process so it should be removed. We apply Horizontal Projection for cropping wrist. There is a characteristic that the projection number of wrist is growing until palm. We locate a point which has not been rotated regard as a Lower bound. And the most of the projection number regard as Upper bound. We find a cut point which the projection number is getting up in this two interval. A result is shows in Fig.9.

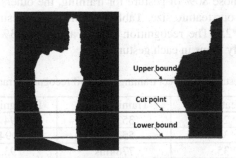

Fig. 9. Finding wrist's cut point

Finally, we extract the gesture's feature to recognition process. The process normalizes the gesture to 25x25 pixels. Every pixel is used for Back Propagation Neural Network input data.

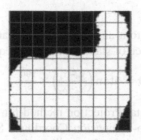

Fig. 10. Gesture feature extraction

4   Results

According to the previous steps, we have built a system which can automatically extract gesture feature, adjust slanted gesture, wrist crop, etc. for accomplishing gesture recognition task with Back Propagation.

In our system, the computer platform is Microsoft Windows 7, CPU is Intel(R) Core(TM) 2 Quad CPU Q8300, and the memory size is 4GB. Logitech Portable Webcam C170 which resolution is 640x480, and used to capture gesture. To speed up process we normalize image to 400x400 pixels. The software development environment is Visual Studio 2008 with OpenCV2.1 digital image processing library.

In the system training process, we adopt 360 images which are taken randomly in different degree and lighting environment. Although some of the gesture which is processed by our method would have cropping point is upper, lower or the binary image which performed by our skin region extraction process is some deficiency. But Neural Network has capability which can put some improper cases in training process to increase robustness. So, we put these cases in training process.

We designed different experiments to test the reliability and stability for our method. Experiment 1 is to determine and inspect how many features which we should take. We randomly chose 80% of gesture for training, the others for testing. We took 25*25 features to be our feature size. Table 1 is shown the results of gesture feature size greater than 25*25. The recognition rate is about 94.63% in average and the processing time is only 39ms in each gesture.

Table 1. Different gesture size affect training time and recognition rate (10 times average)

Gesture feature size	Training time	Recognition rate
25*25	3539ms	94.2%
30*30	5549ms	94%
35*35	7784ms	93.7%
40*40	11024ms	94.1%

Experiment 2 is to verify the result which in different lighting condition. Our result is good in different conditions.

Table 2. Different light condition affect the extraction result

Luminance Test	Image	Result	Luminance Test	Image	Result
Bright			Complex Background		
Backlight			Complex Background2		
Warm color temperature			-		-

Experiment 3 is to test recognition rate in our method. The gestures were randomly moving in front of camera and the system would recognize 100 times for calculating the recognition rate. The result is shown in Table3. The recognition rate is about 89% in average and the processing time of 55ms in each gesture.

Table 3. Recognition rate test

Gesture	0	1	2	3	4	5	6	7	8	9
Reg. rate	88%	96%	88%	93%	87%	98%	82%	86%	90%	98%
Gesture	A	B	C	D	E	F	G	H	I	J
Reg. rate	92%	94%	95%	97%	87%	91%	86%	81%	92%	84%
Gesture	K	L	M	N	O	P	Q	R	S	T
Reg. rate	81%	89%	80%	89%	81%	89%	80%	89%	80%	82%
Gesture	U	V	W	X	Y	Z	-	-	-	-
Reg. rate	98%	94%	100%	83%	100%	85%	-	-	-	-

Our results are comparable to published results (9, 10, 11, 12). The accurary was from 90% to 96.7% while the processing time was from 28ms to 66.5ms. As comparasion to the American Sign Language Recognition System developed by Tsai (13), both our accurary was in 89~90%, but our processing time was only 55ms instead of his 152.24ms.

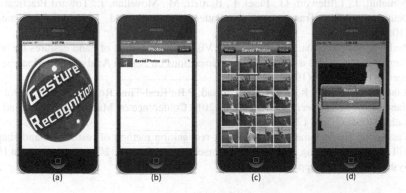

Fig. 11. Static gesture recognition system on Iphone

We also deploy the system on Handheld Device which would use IPhone for displaying. After we calculated, our system processing time is 307ms on the IPhone. Considering the time complex, static gesture recognition system which is removed Background Subtraction method would be deployed on IPhone.

5   Conclusion

According to ASL, we developed a system which can automatically recognize gestures. Our recognition result is 89% in average and the processing time of 55ms in each gesture. The system is fully automatic and it can work in real-time environment. Our result is also good in different light condition. Moreover, we have deployed the system on the hand-held device, such as Iphone. Gesture recognition system on the hand-held device would be very help for those people communicating with the others.

References

1. Jae-Han, P., Seung-Ho, B., Jaehan, K., Kyung-Wook, P., Moon-Hong, B.: A new object recognition system for service robots in the smart environment. In: ICCAS 2007 International Conference on Control, Automation and Systems, pp. 1083–1087 (2007)
2. Swaminathan, R., Nischt, M., Kuhnel, C.: Localization based object recognition for smart home environments. In: IEEE International Conference on Multimedia and Expo, pp. 921–924 (2008)
3. Santos, D., Correia, P.L.: Car recognition based on back lights and rear view features. In: 10th Workshop on Image Analysis for Multimedia Interactive Services WIAMIS 2009, pp. 137–140 (2009)
4. Lilienblum, T., Albrecht, P., Calow, R., Michaelis, B.: Dent detection in car bodies. In: 15th International Conference on Pattern Recognition Proceedings, vol. 4, pp. 775–778 (2000)
5. Moriwaki, K., Katayama, Y., Tanaka, K., Hikami, R.: Recognition of moving objects by image processing and its applications. In: ICCAS-SICE, pp. 667–670 (2009)
6. Shiqi, Y., Tieniu, T., Kaiqi, H., Kui, J., Xinyu, W.: A Study on Gait-Based Gender Classification. IEEE Transactions on Image Processing 18, 1905–1910 (2009)
7. Whitehill, J., Littlewort, G., Fasel, I., Bartlett, M., Movellan, J.: Toward Practical Smile Detection. IEEE Transactions on Pattern Analysis and Machine Intelligence 31, 2106–2111 (2009)
8. Pavlovic, V.I., Sharma, R., Huang, T.S.: Visual interpretation of hand gestures for human-computer interaction: a review. IEEE Transactions on Pattern Analysis and Machine Intelligence 19, 677–695 (1997)
9. Raheja, J.L., Shyam, R., Kumar, U., Prasad, P.B.: Real-Time Robotic Hand Control Using Hand Gestures. In: Second International 2010 Conference on Machine Learning and Computing (ICMLC), pp. 12–16 (2010)
10. Yasukochi, N., Mitome, A., Ishii, R.: A recognition method of restricted hand shapes in still image and moving image as a man-machine interface. In: 2008 Conference on Human System Interactions, pp. 306–310 (2010)

11. Liou, D.H.: A Real Time Hand Gesture Recognition System by Adaptive Skin-Color Detection and Motion History Image. A thesis of Master Degree, Department of Computer Science and Engineering, Tatung University. Taipei (2009)
12. Tu, Y.J.: Human Computer Interaction Using Face and Gesture Recognition. A thesis of Master Degree, Department of Electrical Engineering, National Chung Cheng University, ChiaYi (2007)
13. Tsai, M.K.: A Study of a Real-Time American Sign Language Recognition System Using Thinning Algorithm. A thesis of Master Degree, Department of Electrical Engineering, National Taiwan University of Science and Technology, Taipei (2008)
14. Lockton, R., Fitzgibbon, A.W.: Real-time gesture recognition using deterministic boosting. Presented at the Proc. British Machine Vision Conference, Cardiff, UK (2002)
15. Chai, D., Ngan, K.N.: Face segmentation using skin-color map in videophone applications. IEEE Transactions on Circuits and Systems for Video Technology 9, 551–564 (1999)

11. Liao, D.H.: A Real-Time Hand Gesture Recognition System by Adaptive Skin-Color Detection and Motion History Image. A thesis of Master Degree, Department of Computer Science and Engineering, Tatung University. Taipei (2009)
12. Du, Y.T.: Human Computer Interaction Using Face and Gesture Recognition. A thesis of Master Degree, Department of Electrical Engineering, National Chung Cheng University. ChiaYi (2002)
13. Tsai, M.K.: A Study of a Real-Time American Sign Language Recognition System Using Dynamic Algorithm. A thesis of Master Degree, Department of Electrical Engineering, National Taiwan University of Science and Technology. Taipei (2008)
14. Roomberi, R., Laughton, A.W.: Real-time gesture recognition using deterministic boosting. Presented at the Proc. British Machine Vision Conference Cardiff UK (2002)
15. Chai, D., Ngan, K.N.: Face segmentation using skin-color map in videophone applications. IEEE Transactions on Circuits and Systems for Video Technology, p. 551–554 (1999)

An Adaptive Video Program Recommender Based on Group User Profiles

Chun-Rong Su, Yu-Wei Li, Rui-Zhe Zhang, and Jiann-Jone Chen

EE Dept., National Taiwan Univ. Science & Tech., Taipei 10607, Taiwan
{D9607304,M9907310,M9807314,jjchen}@mail.ntust.edu.tw

Abstract. Recommender systems for personal preferences have been widely developed for applications such as Amazon online shopping website, pandora radio and netflix movie. They are developed based on recorded user preferences to estimate user ratings on new items/stuffs. To recommend TV or movies, it has to perform recommendation for group users. By simply merging preferences of group users, it can act as a single user for recommendation. However, this approach ignores individual user preferences, and user dominance in interaction, which cannot reflect practical group user interests. We proposed to estimate inter-user dominance factor through the neural network algorithm, based on practical group user rating records. In addition, both content-based and user-based collaborative filtering methods are adopted based on the inter-user dominant factors to predict group users' preference for program recommendation. The proposed adaptive program recommender based on dynamic group user profiles is evaluated from practical experiments on Movielens user rating databases. In addition, an active face recognition function has been integrated with the recommender system to provide touchless and user-friendly user interface for a home TV program recommendation prototype. Experiments showed that the proposed method can achieve higher accuracy in recommending video programs for group users, in additional to user-friendly recommendation function.

Keywords: content-based, user-based collaborative filtering, neural network.

1 Introduction

With the advance of computer and network technologies, the multimedia becomes the dominant information at current networked community. However, due to large scale data can be reached through the Internet connected computers, how to help to distribute right contents to right persons becomes important. In general, it requires media search engine or program recommender system to perform the above person-content matching function. The personalized recommender system have been developed for years and it can be approximately categorized into three approaches: (1) content-based (CB) recommendation [1],[2]; (2) collaborative-filtering (CF) [3], [4], [5]; (3) and hybrid recommendation [6]. In the CB method, it first collects and analyzes the user preferences to build the

J.-S. Pan et al. (Eds.): *Advances in Intelligent Systems & Applications*, SIST 21, pp. 499–509.
DOI: 10.1007/978-3-642-35473-1_50 © Springer-Verlag Berlin Heidelberg 2013

user profile. It then can recommend programs for users based on the correlation of content attributes recorded in the user profile. In the CF method, the user profiles are utilized to estimate correlation between user preferences, in which users will be clustered into the same group when their preferences are with high correlation. For users in the same group, each one of them can utilize other users' preference to estimate the preference ratings of a new item. For the hybrid approach, two models are established: sequential- and linear-combination [6]. For the former, it first utilizes CB to find users with similar preferences and then uses CF procedure to recommend the items for users. For the latter, it uses CB and CF to estimate for two sets of recommendation results, respectively, which will be integrated into one set through normalization and weighted processes. Although the personalized recommender system has been well-developed, it still requires to develop group-profile based recommender [7] to better perform the content-person matching function. We proposed to develop TV program recommender based on group-profile for applications that more then one users are involved for the recommendation. As shown in Fig. 1, a smart TV can provide user recognition, program recommendation and streaming control. A face recognition module was utilized to provide user identifications for the group-profile recommender engine. The group users can select programs from the recommended list, in which only interested/filtered items suitable for the group users are provided. The proposed group-profile based recommender system would help the program vendor to distribute the right content to the right user effectively.

The rest of this paper is organized as follows. In Section 2, the personalized recommendation model is introduced. In Section 3, the group-profile based recommendation model is presented. Section 4 describes the databases for personalized and group-profile based databases and the experimental setups. Section 4.2 illustrates the experimental results. Section 5 concludes this paper.

Fig. 1. A smart TV with the adaptive recommender system

2   Adaptive Recommender System

To recommend potential interesting items for a single user or a group of users, the proposed adaptive recommender system (ARS) would record and evaluate users' behaviors, and the user dominance in one group to design the recommender system, as shown in Fig. 1. In the ARS, as shown in Fig. 2, the human-machine interface module (HMIM) is designed to recognize users' identify, e.g., either a single user or a group of users. When only one single viewer was identified by the HMIM, the ARS will perform recommendation based on the user's preferences recorded by personalized recommendation model (PRM) [8]. If the ARS identified a group of members, the group recommendation module (GRM) that adopts neural network training algorithms [9], [10] will be used to recommend potential interesting items to the group users.

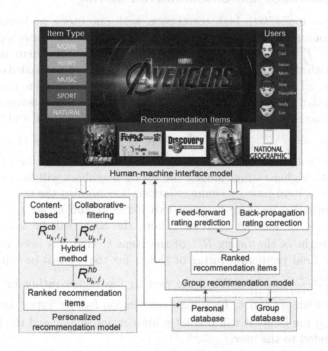

Fig. 2. The proposed adaptive recommender system configuration

2.1   Personalized Recommendation Model

In most multimedia recommender systems, the users can be, for example, TV viewers, shopping customers, or general Internet users. They are represented as $U = \{u_1, u_2, \ldots, u_{n_u}\}$, where n_u is the number of users in the system. The group G consists of n_G subgroups represented as $G = \{g_1, g_2, \ldots, g_{n_G}\}$. Each subgroup g_o consists of k users from U, denoted as $g_o = \{u_1, u_2, \ldots, u_k\}$, and $|g_o|$ denotes the number of users in g_o, i.e., $|g_o| = k$. For example, $g_o = \{u_1, u_2, u_3, u_4\}$, where g_o consists of 4 users, i.e., $|g_o| = 4$. Items can be represented as different data types, for example, television programs, commodities, or websites. The item ℓ_i denotes one movie in our experiments and is denoted as $f_j^{\ell_i} = \{f_j^{\ell_i}\}_{j=1,2,\ldots,n_t}$, where n_t is the number of feature types for the item (e.g., action, adventure, or comedy). In the proposed ARS, each user has to register their identities through the HMIM and rate the preferences for the feature types. The recommender can recommend the items to the user based on the user's preferences stored in the personal user preferences record database. The preference of the feature type for one user, u_k, can be denoted as $f_j^{u_k} = \{f_j^{u_k}\}_{j=1,2,\ldots,n_t}$, where n_t is the number of feature types. For example, $f_2^{u_k} = 3$ denotes that the 2nd feature type (e.g., adventure) of u_k's preference is 3.

2.2 Content-Based and Collaborative-Filtering

To recommend items with high correlation with one user's preference, the CB method computes the inner product between the item and the user's preferences, i.e., $R_{u_k,\ell_i}^{cb} = f_j^{u_k} \cdot f_j^{\ell_i}$. Large R_{u_k,ℓ_i}^{cb} values imply that the item is with high correlation with the user's preference. The CF method can be carried out through either user-based or item-based approach. For the former, it has to find neighbors who have similar preferences with the user. To estimate the preference similarity between two users $f_j^{u_{k_1}}$ and $f_j^{u_{k_2}}$, the Euclidean distance is used and is measured as $d(u_{k_1}, u_{k_2}) = \sqrt{\sum_{j=1}^{n_t}(f_j^{u_{k_1}} - f_j^{u_{k_2}})^2}$. K nearest neighbors of one user are selected to be with the smallest $d(\cdot, \cdot)$s. When recommending items to one user, the items with higher ratings among the user's neighbors will be recommended to the user. For item-based recommendation, it computes the correlation S_{ℓ_i,ℓ_j} by cosine similarity measure between two items ℓ_i and ℓ_j for $i \neq j$ and $j \in n_\ell$, where n_ℓ is the number of items. Then, the S_{ℓ_i,ℓ_j} will be normalized and act as the weight of the rating $\hat{R}_{\ell_j}^{u_k}$ of the item ℓ_j that has been rated by the user u_k. The final prediction rating of the ℓ_i for the u_k will be represented as $R_{u_k,\ell_i}^{cf} = \frac{\sum_{j\in n_\ell, i\neq j} S_{\ell_i,\ell_j} \cdot \hat{R}_{\ell_j}^{u_k}}{\sum_{j\in n_\ell, i\neq j} |S_{\ell_i,\ell_j}|}$. When the user u_k has higher rating for the item ℓ_j and the two items ℓ_i and ℓ_j have high correlation feature types, i.e., values of $\hat{R}_{\ell_j}^{u_k}$ and S_{ℓ_i,ℓ_j} are large, it will predict a high rating R_{u_k,ℓ_i}^{cf} and the item ℓ_i will be recommended to the user.

2.3 Hybrid and Personal Recommendation Method

To recommend potential interesting items for users in a hybrid approach, both CB and CF methods are integrated to predict the ratings R_{u_k,ℓ_i}^{cb} and R_{u_k,ℓ_i}^{cf} of the item ℓ_j for the user u_k. It can be represented as

$$R_{u_k,\ell_i}^{hb} = \frac{\alpha}{N} \times R_{u_k,\ell_i}^{cf} + \frac{N-\alpha}{N} R_{u_k,\ell_i}^{cb}, \tag{1}$$

where α is the number of days that the user used the system in the past N days. When α is larger, the CF-based rating, R_{u_k,ℓ_i}^{cf}, would become more dominant than the CB-based rating, $R^{cb}_{u_k, \ell_i}$. The parameter value of α should be determined from experiments, which would be dependent on database record.

2.4 Personal Preferences Updating

When the user is identified by the HMIM in ARS, the PRM will record the item that one user is consuming and the time duration of the consuming process. The recorded data will be represented as ratings of the items for the user and was stored in the personal user rating record database. It can be represented as $p^{u_k}_I = \{p^{u_k}_{\ell_j}\}_{j=1,2,\ldots,n_\ell}$ for $k \in [1, n_u]$, where n_ℓ is the number of items and n_u is the number of users. According to the recorded personal rating for each item $p^{u_k}_{\ell_j}$, which would help the PRM to reflect the most updated user preferences and then recommending the items that are satisfiable the user. The personal preferences updating can be represented as

$$f^{u_k}_j = \beta \times p^{u_k}_{\ell_j} \times f^{u_k}_j + (1 - \beta) \times \hat{f}^{u_k}_j, \tag{2}$$

where β is the individual's updating rate, $j \in n_t$, and $\hat{f}^{u_k}_j$ is the recorded preference of the j-th feature type for the user u_k in the past. The current personal preferences $f^{u_k}_j$ will be updated repeatedly in Eq. (2).

3   Group Recommendation Model

When the HMIM identifies a group of users in the proposed ARS, the ARS will compute the similarity for users in the group. If the computed similarity is lower than the defined threshold, it means that the users in the group have near preferences, thus, the group will receive items as single user recommended by PRM. If not, the preferences of users in the group are diverse and the group will receive items recommended by GRM. In the HMIM, it will record which items the users in the group have been used and how long the users used the items. The recorded data for the group will be represented as the truly group's ratings of the items and stored in the group database. In the GRM, the feedforward neural network (FNN) is used to predict the ratings of items for the group. If the prediction rating of items for the group is close enough to the truly group's rating, the GRM will recommend the items that are satisfiable the group. If not, the back-propagation neural network (BPNN) model will correct the group ratings predicted form FNN repeatedly converging to the truly group's ratings recorded in the database.

3.1 Group Prediction Model

In the GRM, the FNN is used to predict the ratings of items for the group and then recommending the top ranked items to the group. The FNN comprises three layers: input layer, hidden layer, and output layer. For example, as shown in Fig. 3, the inputs are three users' ratings for the item ℓ_j, i.e., $\{p^{u_k}_{\ell_j}\}_{k=1,2,3}$.

For each rating $p_{\ell_j}^{u_k}$, the user u_k rated the item ℓ_j in PRM and stored in the personal database. At hidden layer, each $p_{\ell_j}^{u_k}$ is weighted by the weight $w_{k,t}$ and then adding the bias θ_t of the neuron. It is represented as $nur_t = \theta_t + \sum_k p_{\ell_j}^{u_k} \cdot w_{k,t}$, where $t \in [4,5]$ here. For each nur_t, it will be transformed by unipolar sigmoid activation function into a new neuron impulse H_t, i.e., $H_t = \frac{1}{1+e^{-\alpha \cdot nur_t}}$, where the slope parameter is defined as $\alpha = 1$ and $H_t \in [0,1]$. At output layer, the input signals $\{H_t\}_{t=4,5}$ from the hidden layer can be weighted by $w_{t,m}$ for $m=6$ and then adding the bias θ_m of the neuron to obtain the nur_m. By transforming the nur_m through unipolar sigmoid activation function, the output prediction rating of the item ℓ_j for the group g_i, i.e., $\hat{R}_{\ell_j}^{g_i}$, will be attained and normalized to $\hat{R}_{\ell_j}^{g_i} \in [0,1]$. When the $\hat{R}_{\ell_j}^{g_i}$ is predicted with larger value, e.g., $\hat{R}_{\ell_j}^{g_i} = 0.8$, it means that the group users may be interested the item ℓ_j. While there are n_ℓ items that are required to predict the ratings for the group, i.e., $\hat{R}_I^{g_i} = \{\hat{R}_{\ell_j}^{g_i}\}_{j=1,2,\dots,n_\ell}$, the GRM will recommend the top ranked items of programs to the group g_i following the described three layers.

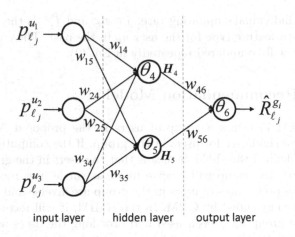

Fig. 3. A neural network example with ratings of the item ℓ_j for three users to predict the $\hat{R}_\ell^{g_i}$

3.2   Back-Propagation Neural Network Model

Although the $\hat{R}_I^{g_i}$ can be predicted in the FNN model to recommend the top ranked rated items to g_i, it may not be satisfiable for g_i due to diverse personal preference ratings. To solve this problem, the back-propagation neural network (BPNN) model is adopted and is operated based on the true group's rating recorded by HMIM, e.g., the system operation time for the group consuming the items. It will be considered as the true group preference ratings for the items, i.e., $R_I^{g_i} = \{R_{\ell_j}^{g_i}\}_{j=1,2,\dots,n_\ell}$, and $R_{\ell_j}^{g_i}$ will be normalized to the range $[0,1]$. As compared to the prediction rating $\hat{R}_I^{g_i}$, an error function is defined and is computed as $E = \frac{1}{2}\sum_{j=1}^{n_\ell}(R_{\ell_j}^{g_i} - \hat{R}_{\ell_j}^{g_i})^2$. If E is smaller than the predefined threshold Th_E, i.e., the ratings between $\hat{R}_I^{g_i}$ and $R_I^{g_i}$ are close enough to each other, the system will recommend interested items for the group. If not, the BPNN will use the steepest descent method [10] to re-adjust the weights by adding the correction, i.e., $w_{t,m} = w_{t,m} + \triangle w_{t,m}$ between two connected neurons t and m at hidden and output layers, respectively. Fig. 4 shows that each step to correct the weightings by BPNN model. For weighting correction term $\triangle w_{t,m}$, it is represented as $\triangle w_{t,m} = -\eta \frac{\partial E}{\partial w_{t,m}}$, where η is the learning rate and $\frac{\partial E}{\partial w_{t,m}}$ is the gradient. By using the chain rule [11], the $\triangle w_{t,m}$ is further represented as $\triangle w_{t,m} = \eta \cdot \delta_m^o \cdot H_t$ and can be used to correct the weighting $w_{t,m}$, where δ_m^o is the error signal of neuron m at output layer and is represented as $\delta_m^o = \hat{R}_{\ell_j}^{g_i} \cdot (1 - \hat{R}_{\ell_j}^{g_i}) \cdot (R_{\ell_j}^{g_i} - \hat{R}_{\ell_j}^{g_i})$ [12]. For the weighting $w_{k,t}$, it is connected between neurons k and t at input and hidden layers, respectively. $w_{k,t}$ also utilized the same adjusting processes to update the $w_{k,t}$ by adding the correction term $\triangle w_{k,t}$. For $\triangle w_{k,t}$, it can be represented as $\triangle w_{k,t} = \eta \cdot \delta_t^h \cdot p_{\ell_j}^{u_k}$, where δ_t^h is the error signal of neuron t at hidden layer and is represented as $\delta_t^h = [\sum_m w_{t,m} \cdot \delta_m^o] \cdot H_t \cdot (1 - H_t)$ [12]. The steepest descent approach is used to determine suitable $\triangle w_{k,t}$ and $\triangle w_{t,m}$. The weighting will be adjusted repeatedly until the $\hat{R}_I^{g_i}$ converges to $R_I^{g_i}$, i.e., $E < Th_E$.

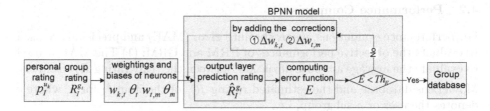

Fig. 4. Flowchart of weightings correction procedure by BPNN model

4 Experimental Study

4.1 User Rating Setup

The personal user rating records are collected from the GroupLens research lab. Massive user ratings records are collected as one database, named as MovieLens, which is open to download through their website. The MovieLens database consists of ten thousand user ratings from 943 users on 1682 movies. Each MovieLens data set comprises three files including movie item, user tag, and data rating. The movie item is represented by a 1×19 feature vector, in which each element denotes the feature type of the item, e.g, action, adventure, and animation. The group user rating can be simulated by merging personal user rating records in the MovieLens under specified strategy. First, it needs to calculate the personal opinion importance $Poi_{\ell_j}^{u_k}$ and then combines the user rating $R_{\ell_j}^{u_k}$ on the item

ℓ_j for each user u_k in the group g_i. Hence, the rating data of the item ℓ_j for the group g_i can be represented as

$$R_{\ell_j}^{g_i} = \frac{\sum_{k=1}^{|g_i|} Poi_{\ell_j}^{u_k} \times R_{\ell_j}^{u_k}}{\sum_{k=1}^{|g_i|} Poi_{\ell_j}^{u_k}}, \tag{3}$$

where $|g_i|$ is the number of members in the group and $R_{\ell_j}^{g_i} \in [0,1]$. To compute the $Poi_{\ell_j}^{u_k}$ of the item ℓ_j for the user u_k, it combines the variance of the rating for the item $var(\ell_j)$ in the group, personal authority $auth(u_k)$, and personal obedient $obe(u_k)$, i.e., $Poi_{\ell_j}^{u_k} = auth(u_k) + obe(u_k) \times var(\ell_j)$.

In the MovieLens database, we select the data sets that record the 363 users who have rated more than 100 movies. The selected data sets are divided into the training dataset T_r and the test set T_e to verify the performance of proposed personalized recommendation model (PRM). To carry out the group-based recommendation model (GRM), these 363 users are divided into three age groups: [7,25], [26,50], and [51,75]. We randomly select two or three users from the three age groups to form a subgroup. 200 subgroups are required in building our experiments. The authority and obedient values are randomly predefined to each user in the subgroup and the subgroup rating can be computed in Eq. (3).

4.2 Performance Comparisons

For performance evaluations, mean absolute error (MAE) and precision are used to evaluate the objective performances of PRM and GRM: (1) The MAE is used to evaluate the average absolute deviation between the truly rating $R_{\ell_j}^c$ from the MovieLens database and the estimated rating $\hat{R}_{\ell_j}^c$ of the item ℓ_j for c, where c denotes the user or a subgroup, i.e.,

$$MAE = \frac{\sum_{j=1}^{n_\ell} |R_{\ell_j}^c - \hat{R}_{\ell_j}^c|}{n_\ell}, \tag{4}$$

where n_ℓ is the number of items; (2) Precision is defined as the ratio of the error for the estimated ratings to the accurate ratings, i.e.,

$$P = 1 - \frac{\sum_{j=1}^{n_\ell} \frac{R_{\ell_j}^c - \hat{R}_{\ell_j}^c}{max(|R_{max}-R_{\ell_j}^c|,|R_{min}-R_{\ell_j}^c|)}}{n_\ell}, \tag{5}$$

where $R_{max} = 1$ and $R_{min} = 0$ define as the maximum and the minimum ratings in MovieLens database, respectively, and $P \in [0,1]$; (3) Normalized distance-based performance measure (NDPM) was used to compare two different weakly ordered rating rankings, i.e., the items rated by the GRM and the user to obtain the two rankings, respectively. The NDPM can be represented as

$$NDPM = \frac{C^- + 0.5C^u}{C^t} \tag{6}$$

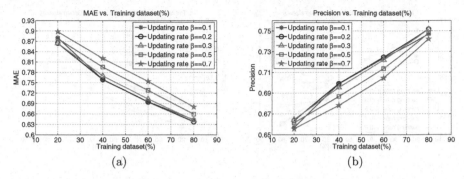

Fig. 5. Performance comparisons for personalized and group-based recommendation models under different updating rates: (a) MAE and (b) precision

where C^- is the number of contradictory preference relations between the system ranting and the group users ranting, e.g., two items ℓ_1 and ℓ_2 rated by the group users and GRM to get the ratings $R^c_{\ell_1} > R^c_{\ell_2}$ and $\hat{R}^c_{\ell_1} < \hat{R}^c_{\ell_2}$, respectively. C^u is the number of consistent preference relations for $R^c_{\ell_1} > R^c_{\ell_2}$ and $\hat{R}^c_{\ell_1} = \hat{R}^c_{\ell_2}$. C^t is total number pair of items for which one is rated higher than the other by the group users.

The recommendation performance in terms of MAE and the T_r for different update rates of β are compared and evaluated. The MAE reducing speed along the T_r for the $\beta = 0.2$ is the lowest as compared to the others, as shown in Fig. 5(a). For the updating rates $\beta = 0.5$ or $\beta = 0.7$, the system largely increased the user's preferences based on the items that the user has been used, as compared to that of $\beta = 0.2$. Although the user may not be interested in some items, the system can still update the user's preferences. Under this condition, the inaccurate preferences for the user would prevent the PRM from recommending correct items to the user based on the user's preferences. When setting a smaller updating rate for user's preferences, e.g., $\beta = 0.2$, the user's preferences can be progressively converged and the PRM can precisely recommend the items to the user, as shown in Fig. 5(b). As compared to the performance without updating user's preferences, in terms of precision and the T_r, the PRM adopts user's preferences update strategy can yield more accurate recommendation results when the ratio of the training data is more than 53%, as shown in Fig. 7(a).

To evaluate the performances of the GRM, in terms of MAE and the T_r, for different learning rates of η are compared and evaluated. The MAE reducing speed along the T_r for $\eta = 0.1$ and $\eta = 0.3$ have the lowest as compared to the other learning rates, as shown in Fig. 6(a). When using larger learning rates for $\eta = 0.7$ or $\eta = 0.9$, the weights would be iteratively over-adjusted so that the group estimating rating cannot converge to close with the target rating in the BPNN, i.e., the error function E is consistently larger than the Th_E. Comparing the performances, in terms of precision and the T_r for different learning rates, the $\eta = 0.1$ and $\eta = 0.3$ have the highest precisions, as shown in Fig. 6(b). However, adopting 80% of the T_r, the learning rate $\eta = 0.3$ could obtain the highest

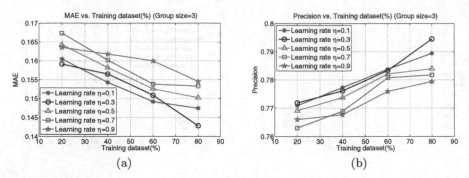

Fig. 6. Performance comparisons for personalized and group-based recommendation models under different learning rates: (a) MAE and (b) precision

precision due to enough training data and the proper learning rate under the limited adjusting times. The performances, in terms of NDPM and the T_r, for different group recommendation algorithms are compared. Previous research, an adaptive correlation-based group recommendation system (ACGRS) [13], is simulated for comparison. The NDPM along the T_r of the proposed adaptive recommendation system (ARS) was lower as compared to the ACGRS, as shown in Fig. 7(b), which demonstrated that the proposed ARS could recommend the more accurate ranking rated items to the group that close to the group users' preferences.

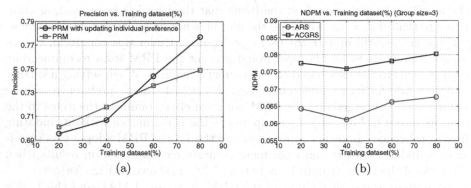

Fig. 7. (a) Personalized recommendation model with and without updating individual user's preference; (b) NDPM performance for the proposed ARS and ACGRS

5   Conclusions

An adaptive recommender system was proposed which can recommend user interested items to group users. Contributions of this paper are: 1) A hybrid personalized recommendation method that integrate both CB and CF recommendation result to yield the final recommendation result has been proposed;

2) A user profile updating mechanism was proposed, which helps the recommender to reflect the most updated user preferences for adaptive control; 3) A hybrid group-profile recommendation model was proposed, which integrate personal item rating for group item rating through the feed-forward neural network process, and then adjusted the group's prediction rating to reach the target rating iteratively in back-propagation process; 4) The recommendation precision can be improved through the preference update procedure, as compared to the one without update. For the group recommendation performances, experiments showed that it can adjust the learning rate to recommend the items to the group with high precision. This group-profile recommender system has been integrated with a media streaming server, IPMP terminal and a face recognition module to provide a video program recommender with a natural interface.

References

1. Pazzani, M.J., Billsus, D.: Content-Based Recommendation Systems. In: Brusilovsky, P., Kobsa, A., Nejdl, W. (eds.) Adaptive Web 2007. LNCS, vol. 4321, pp. 325–341. Springer, Heidelberg (2007)
2. Rokach, F.L., Shapira, B., Kantor, P.B.: Recommender system handbook, 1st edn. Springer (2010)
3. Herlocker, J.L., Konstan, J.A., Borchers, A., Riedl, J.: An algorithmic framework for performing collaborative filtering. In: Proc. of the 22nd Ann. Int. ACM SIGIR Conf. on Research and Development in Information Retrieval, pp. 230–237 (1999)
4. Konstan, J.A., Miller, B.N., Maltz, D., Herlocker, J.L., Gordon, L.R., Riedl, J.: GroupLens: applying collaborative filtering to usenet news. Communications of the ACM 40(3), 77–87 (1997)
5. Knijnenburg, B.P., Willemsen, M.C., Gantner, Z., Soncu, H., Newell, C.: Explaining the user experience of recommender systems. User Modeling and User-Adapted Interaction 22, 1–64 (2012)
6. Kim, B.M., Li, Q., Park, C.S., Kim, S.G., Kim, J.Y.: A new approach for combining content-based and collaborative filters. Journal of Intelligence Information Systems 27(1), 79–91 (2006)
7. Masthoff, J.: Group recommender systems: combining individual models, pp. 677–702. Springer (2011)
8. Chu, W., Park, S.T.: Personalized recommendation on dynamic content using predictive bilinear models. In: Proc. of the 18th Int. Conf. on World Wide Web, pp. 691–700 (2009)
9. Werbos, P.J.: Backpropagation through time: what it does and how to do it. Proceedings of the IEEE 78(10), 1550–1560 (1990)
10. Rumelhart, D.E., Hinton, G.E., Willians, R.J.: Learning representations by back-propagating errors. Nature 323, 533–536 (1986)
11. Boden, M.: A guide to recurrent neural networks and backpropagation. The DALLAS project. Report from the NUTEKsupported project AIS-8, SICS.Holst: Application of Data Analysis with Learning Systems (2001)
12. Lee, T.L.: Back-propagation neural network for long-term tidal predictions. Ocean Engineering 31(2), 225–238 (2004)
13. Lin, K.H., Chiu, Y.S., Chen, J.S.: An adaptive correlation-based group recommendation system. In: Intelligent Signal Processing and Communications Systems, pp. 1–5 (2011)

Automatic Dancing Assessment Using Kinect

Ta-Che Huang, Yu-Chuan Cheng, and Cheng-Chin Chiang

Dept. of Computer Science and Information Engineering
National Dong Hwa University
Hualien, Taiwan
{m9921103,m10021015}@ems.ndhu.edu.tw,
ccchiang@mail.ndhu.edu.tw

Abstract. This paper presents the design of an automatic assessing system for dance learners. By exploiting the Microsoft Kinect to acquire 3-D motion data of the learners, the proposed system extracts features and performs matching between the acquired data and the corresponding sample motion of a teacher. The proposed systems have several distinctive functions not appearing in other existing systems, including (1) the performance of a learner can be assessed separately in terms of accuracies in posture and rhythm and (2) the automatic identification of error motion on body articulations during the learn's exercising. To tolerate the lagged, redundant and wrong motions, which are frequently seen on a naïve learn's exercising, our assessment of posture accuracy is done by dynamic time warping (DTW) algorithm so that two motion sequences with different timing and lengths can be accurately aligned. On assessing the accuracy in rhythm, the system applies the fast Fourier transform (FFT) to extract the motion frequency which portrays the rhythm of body motions. Compared with the assessment made by human judges, the system's assessment achieves the consistency of 84% and 77% in assessing the posture accuracy and tempo accuracy, respectively.

Keywords: Dancing Assessment, Kinect, Motion Sequence Matching, Dynamic Time Warping (DTW), Fast Fourier Transform (FFT).

1   Introduction

Prior to the development of human languages, dancing has been one of the important modalities for human communications. Combining graceful motions and euphonious rhythms, dancing is the best way of presenting both the vigor and beauty of human bodies. For some ethnic groups, dancing characterizes their unique cultures. Also, dancing is conductive to our health, disposition and responsiveness. Therefore, dancing always attracts the learning interests of people at different ages. For many people, dancing had become more than an entertaining frill.

In view of the requirement that self-learning dancing learners need an assisting training system which can offer the functions of automatic assessment and error identification anytime and anywhere, our work aims to design and develop such a system and propose technical solutions to some practical issues. Thanks to the successful

J.-S. Pan et al. (Eds.): *Advances in Intelligent Systems & Applications*, SIST 21, pp. 511–520.
DOI: 10.1007/978-3-642-35473-1_51   © Springer-Verlag Berlin Heidelberg 2013

development of some low-cost 3-D motion sensing devices, such as the Kinect developed by Microsoft's, the proposed system can do the assessment using the acquired 3-D motion data. Since the posture and rhythm are the two most important ingredients of every dance, the performance of a learner's dancing should be assessed separately in these two aspects. Hence, one novel contribution of the proposed system is that we propose two novel mechanisms for independently assessing the performance of a dance learner in terms of accuracies in both posture and rhythm. To assist the learner to correct the error motions of body articulations, our system can also clearly and accurately highlight where and how incorrect the error body motions are through a rendered 3-D skeleton model of the learner's body.

2   Related Work

In the past decade, action recognition and behavior analysis were two research topics which attract the increasing interests from many researchers. Both topics involve acquiring and analyzing the motional data of human bodies. Most studies conducted on these two topics focused on inferring human motions from 2-D videos. To infer the human motions from 2-D videos, many different spatial and temporal features that could be extracted on 2-D video frames were proposed. For example, W.-Y. Chen [1] proposed the curves of the body skeleton for pose estimation and behavior matching. Y.J. Yan [2] exploited the motion-energy images (MEI) and the motion-history images (MHI) which was originally proposed by Bobick et al. [3] as the features for recognizing gymnastics motions. Shahbudin et al. proposed to integrate the principal component analysis (PCA) and the support vector machine for posture classification [4]. Hsu proposed to triangulate the shapes of human postures into triangle meshes from which two important posture features are then extracted, i.e., the body skeleton and the centroid contexts. One critical difficulty faced by these approaches is how to robustly segment the human bodies from video scenes under some tough conditions, such as partial occlusions and noisy interference of cluttered backgrounds.

Since 2010, the release of Microsoft's 3-D motion-sensing device, Kinect [7], moves the research of action recognition and behavior analyses into the new era of 3-D machine perception. Kinect now becomes a popular replacement of conventional 2-D video cameras because of its affordable price and acceptable performance. By acquiring the 3-D motional data, including both the depth data and the body skeleton, Kinect has created a wide spectrum of applications in action recognition and behavior analyses [8][9][10]. Using 3-D motional data has some advantages over the 2-D video data when we develop systems for action recognition and behavior analyses. For instance, the body of the subject is much easier to be extracted from cluttered background scenes. The extra information of pixel depths along the z-axis also enables computers to do many intelligent analyses which cannot be achieved in 2-D video approach.

Automatic assessing of dance performance has some different intrinsic characteristics compared to action recognition. Firstly, the matching algorithm must be capable of dealing with the possible lagged timing of the dance performance of most naïve learners. When imitating the dancing postures on sample videos, an inexperienced learner usually cannot keep up with the motion timing of the experienced teacher. Consequently, some matching algorithms, such as those exploited in many video games, may fail to correctly assess the learner's performance even though the learner has made the correct dance movements. Also, when the learner makes some error or redundant movements, the matching algorithm must continue the reasonable assessment of those correct movements that follow these incorrect or improper movements. Some simple matching algorithm may cease to correctly match the remaining movements of the learner with the template movements once error movements appear. The second problem we need to handle is how to define reasonable measures for evaluating the goodness of the learner's dance performance. In our design, we focus on measuring the accuracies in posture and rhythm because these two factors are key elements of dancing. The third problem to be addressed in our work is how to accurately identify the error movements in the learner's dance performance. Additionally, the identified error movements must be precise enough to tell the learner which body articulations are not in their correct positions and how wrong the positions of these articulations are.

3   Proposed Automatic Assessing System for Dance Performance

3.1   Feature Extraction

The acquired 3-D motional data could have different scales depending on the body sizes of users. The scale of each acquired 3-D skeleton of the human body needs to be normalized to reduce the size variations. The motional data is composed of the 3-D positional coordinates of each body articulation. Considering that the absolute positional 3-D coordinates are not tolerant to positional translations, we convert the 3-D coordinates of consecutive time points into relative motion vectors. Due to the low cost design of the Kinect, the accuracy of the acquired 3-D motional data cannot be as high as that of a high-end 3-D motion sensing device. Noise may appear frequently in the acquired 3-D motional data. Hence, we also need to remove the unexpected noise to reduce the negative influences on the matching results. In what follows, we present the method of size normalization, body motion vector calculation, and noise removal.

Size Normalization of Body Skeleton

The 3-D coordinates from a Kinect may vary with the body size of a dancer. Consequently, the matching between the motional data of different persons may be improperly affected by the size variations on the positional coordinates. Normalizing the size of a body skeleton is very easy. All we need to do is to rescale the distances between body articulations. To do this, we set the spine point that connects the neck and the torso as a reference point. With the reference point, the   coordinates of all body articulations are translated according to

$$W_t(x, y, z) = P_t(x, y, z) - Spine_t(x, y, z) \tag{1}$$

where $P_t(x, y, z)$ and $Spine_t(x, y, z)$ denote the coordinates of a body articulation and the spine point, respectively, and the subscript t denotes the time index. After the translation, we normalize the scale of the coordinates of $W_t(x, y, z)$ by

$$C_t(x, y, z) = \frac{W_t(x, y, z)}{\|spine_t(x, y, z) - head_t(x, y, z)\|} \tag{2}$$

where $\|spine_t(x, y, z) - head_t(x, y, z)\|$ is the distance from the spine point to the head point $head_t(x, y, z)$.

Body Motion Vector Extraction

When different persons exercise the same dance movements, the scales of the acquired movements may have different scales of motions. Thus, the motion scales of different persons need another way of normalization.

To normalize the motion scales, our first step is to convert the normalized 3-D coordinates of body articulations to motion vectors which account for the relative displacements of body articulations across two consecutive time points. Namely, the motion vector of a body articulation P at time t is

$$V_t(x, y, z) = P_t(x, y, z) - P_{t-1}(x, y, z). \tag{3}$$

At any time point t, we concatenate the motion vectors of all body articulation into a super motion vector, which is called a *body motion vector* in this paper. At any time point, evaluating the difference between the input body motion vector and a template body motion vector gives an overall goodness measurement for the body motion at this time point. It would be unnecessary to identify the error body articulations individually if the evaluated difference is small. If the evaluated difference exceeds a prescribed threshold, the error body articulations can be easily located by examining the movement differences of individual body articulations.

Converting positional coordinates to a body motion vector is just our first step to handle the variations of motion scales. Besides this conversion, we use the angle, instead of the Euclidean distance, between body motion vectors as our difference measure for assessing the accuracy of dance movements. That is,

$$Dist(V_1, V_2) = \cos^{-1}\left(\frac{V_1 \cdot V_2}{\|V_1\|\|V_2\|}\right) \tag{4}$$

The angle is less sensitive to the variations of maganitudes of the two consecutive body motion vectors, V_1 and V_2, thus reducing the bad interference of motion scales in assessing dance performance. The detailed method of our assessment will be presented later.

Noise Removal

The Kinect usually introduces two kinds of noise, jittering and beating, into the acquired data. Jittering occurs when the Kinect senses unstable positional drifts even though the user's body keeps unmoved. A simple solution to handle the jittering is to set a threshold to filter out trivial motion drifts of body articulations. The beating happens when the Kinect outputs incorrect motional data due to partial occlusions or interferrences of cluttered backgrounds. Our simple solution for removing beating noise is to filter the motional data by the following equation:

$$V_t = \begin{cases} 0, & (V_{t-1}V_t < 0) \wedge (V_t V_{t+1} < 0) \\ V_t, & \text{otherwise} \end{cases} \tag{5}$$

3.2   Posture Accuracy Assessment

The assessment of posture accuracy of a dance performance is done by matching the sequence of body motion vectors with a template one in the database. To match two sequences of unequal number of body motion vectors, we thus demand a non-linear alignment method to find the best correspondences between their elements. Finding the best alignment between two sequences of different number of elements is not trivial because the number of possible correspondences increases exponentially with the lengths of compared sequences. A feasible solution to this problem is the dynamic time warping (DTW) algorithm [16]. The DTW algorithm is a non-linear sequence alignment method based on dynamic programming. Let X $(=\{x_1, x_2, ..., x_N\})$ and Y $(=\{y_1, y_2, ..., y_M\})$ be two sequences of body motion vectors. We can compute a local distance matrix $C \in R^{N \times M}$ whose element $C(x_i, y_j)$ is the distance between x_i and y_j, i.e., the angle between x_i and y_j defined in Eq. (4). The goal of the DTW algorithm is to find a path of element-wise alignment, i.e., $P^* = (p_1, p_2, ..., p_K)$ with $p_l = (i_l, j_l) \in [1:N] \times [1:M]$ for $l \in \{1, 2, ..., K\}$, such that the accumulated global distance $DTW(X, Y) = \sum_{l=1}^{K} C(x_{i_l}, y_{j_l})$ is minimized.

To efficiently find the optimal aligning path, the DTW exploits the technique of dynamic programming (DP). Besides the local distance matrix C, we use another accumulated distance matrix $D \in R^{N \times M}$ whose element $D(i,j)$ keeps the minimal accumulated local distance corresponding to the best alignment between the two subsequences $\{x_1, x_2, ..., x_i\}$ and $\{y_1, y_2, ..., y_j\}$. The steps of the DTW algorithm are listed in the following:

1.   Initialization: $D(1, j) = \sum_{k=1}^{j} C(x_l, y_k), j \in [1, M]$
2.   Initialization: $D(i, 1) = \sum_{k=1}^{i} C(x_k, y_l), i \in [1, N]$
3.   Recursion: $D(i, j) = min\{D(i - 1, j - 1), D(i, j - 1), D(i - 1, j)\} + C(x_i, y_j)$
4.   Bookkeeping: $S(i, j) = \underset{(u,v) \in \{(i-1,j),(i,j-1),(i-1,j-1)\}}{arg\,min} \{D(u,v)\} + C(x_i, y_j)$
5.   Finalization: $TW(X, Y) = D(N, M)$.

The $DTW(X,Y)$ in Step 4 gives the minimal accumulated distances of aligning the two sequences X and Y. In Step 3, the bookkeeping of the optimal source pair (i.e., $(i-1, j-1)$, $(i, j-1)$, or $(i-1, j)$) is made on the element $S(i,j)$ of a matrix $S \in N^{N \times M}$. With the matrix S, the path of optimal alignment can be easily traced back from $S(N,M)$. With the accumulated distance computed by the DTW algorithm, we evaluate the overall score of the posture accuracy as

$$S_{posture} = \left(1 - \frac{DTW(X,Y)}{Len(P)}\right) \times 100.$$

where $Len(P)$ is the number of correspondence pairs after the DTW aligning path P. Since the local distance of each corresponding pair ranges between 0 and 1, the value of the accumulated distance $DTW(X,Y)$ is between 0 and $Len(P)$. Hence, the value of the term $DTW(X,Y)/Len(P)$ is between 0 and 1. Accordingly, the final score ranges between 0 and 100.

3.3 Rhythm Accuracy Assessment

The sequence of body motions can be regarded as a time-varying signal. As usually done for time-varying signals, the signals can be converted into signals in frequency domain. This conversion is valid for our rhythm assessment because the frequency responses of signals are tightly coupled with the rhythm of signals. Thus, we perform the fast Fourier transform (FFT) on the magnitudes of body motion vectors along the time axis. The output of the FFT is a spectrum of frequency distribution of the magnitudes of body motion vectors.

Length Normalization and Smoothing of Spectrums

Due to the efficiency consideration, applying the FFT to a sequence of signal usually requires the sequence lengths to be a power of two [15]. To meet the requirement, we perform linear interpolation [17] on the motion sequence to amend the length the motion sequence. The interpolation is done by

$$y = y_0 + (x - x_0)\frac{y_1 - y_0}{x_1 - x_0},$$ (6)

where (x, y) is the newly interpolated data point and (x_0, y_0) and (x_1, y_1) are the two reference data points in the original signal.

After applying the FFT, the obtained frequency spectrum is further smoothed by the moving average (MA) method [18]. The MA smoothing is done according to

$$y_t = \frac{y_{t-1} + y_t + y_{t+2}}{3}.$$ (7)

With the frequency spectrums of two compared motion sequence, the similarity between the two frequency spectrums can be converted into an assessed score in rhythm accuracy for the learner's performance. The formula of conversion is

$$S_{rhythm} = \frac{\sum\limits_{i=1}^{F} \min(f_{A,i}, f_{B,i})}{\sum\limits_{i=1}^{F} \max(f_{A,i}, f_{B,i})} \times 100 \tag{8}$$

where $f_{A,i}$ and $f_{B,i}$ for $i=1,2,\ldots,F$, are the ith frequency element of the two sequences and F is the length of the two spectrums. The evaluated score S_{rhythm} would be also in the range of [0,100].

4   Experimental Results

The system is implemented with the Kinect SDK and Coding4fun components using Microsoft Visual Studio 2010 an XNA Game Studio 4.0. Table 1 lists the hardware and software specifications of the system.

Table 1. Experimental environment

Hardware	CPU	Intel Core i7-2600 3.40GHz
	RAM	DDR3 1333 8GB
	Video Card	NVIDIA GeForce GTX 560 Ti
	Motion-sensing device	Kinect
Software	O.S.	Windows 7 Ultimate SP1 64bit
	Program Language	C#
	Developing Tool	Microsoft Visual Studio 2010
	Library	Microsoft Kinect 1.0 Beta2 SDK Coding4Fun.Kinect.Toolkit

For the experiments on the proposed system, we collect the motional data of five dance footworks, which have different levels of difficulties for learners, as the template motion sequence. To collect the motional data of dancing performance for assessment, five learners are asked to exercise each dance footwork, obtaining 25 sequences totally. The goal of conducted experiments is to measure the consistency between the machine-made assements and the human-made assessments. Therefore, nineteen human judges are invited to give human-made assessments on all dance performances. The consistency measurement must be evaluated with respect to the assessed posture accuracy and rhythm accuracy to verify the efficacy of the proposed assessment mechanisms.

For measuring the consistency quantitatively, a rank constancy measure (RCM) is defined. This measure requires the ranking, from both human judges and our system, on the performances of the same dance for different learners. Let $R_j(d, f)$ denote the rank given by the human judge j on the footwork f performed by the dancer d. Note that the rank sum of the five dancers, $\sum\limits_{d=1}^{5} R_j(d, f)$, is 15 (=1+2+3+4+5), for given f

and j. Let $R(d,f)$ be the rank made by the system; the sum $\sum_{d=1}^{5} R(d,f)$ is also $(1+2+3+4+5) = 15$. For the assessment of a given footwork f made by human judge j, a difference between the ranking on the five dancers made by the human judge j and the system is

$$RD(j,f) = \sum_{d=1}^{5} |R(d,f) - R_j(d,f)| . \tag{9}$$

Apparently, if we have d dancers, the maximum of the possible values, denoted as $RD_{max}(j,f)$, for $RD(j,f)$ would be $\sum_{k=1}^{d} |2k - d - 1|$. For our case, d equals 5 and thus RD_{max} is 12. Hence, the following term

$$C(j,f) = 1 - \frac{RD(j,f)}{RD_{max}(j,f)},$$

quantitifies the consistency between the ranks made by the system and the human judge j for footworks f.

When assessing the performances of a certain footwork for the five dancers, a human judge may be not able to clearly discern which dancers have better performances in some cases. Under such a situation, the rank made by this judge could become unreliable and might differ singularly from those of other judges and the system. To reach higher reasonability, the rank consistency $C(j,f)$ needs to be weighted by a confidence factor of the judge j. Let $S(j,f,d)$ be the score of assessing the accuracy in some aspect (i.e., posture or rhythm) given by the judge j for some footwork f exercised by the dancer d. The confidence factor of the judge j can be measured by the following normalized standard deviation

$$w(j,f) = \frac{std(\{S(j,f,d)\}_{d=1}^{5})}{\sum_{j=1}^{19} std(\{S(j,f,d)\}_{d=1}^{5})} . \tag{10}$$

By multiplying the rank consistency $C(j,f)$ with the confidence factor $w(j,f)$, we get a more reasonable weighted rank consistency $\bar{C}(j,f) = C(j,f)w(j,f)$. Finally, we compute the overall consistency between the assessments made by the system and the human judges by averaging all weighted rank consistency, i.e.,

$$C = \frac{1}{19 \cdot 5} \sum_{j=1}^{19} \sum_{f=1}^{5} C(j,f) \times 100.$$

According to our results, the overall consistencies evaluated for assessment the accuracies in posture and rhythm are 84% and 77%, respectively. In Fig. 1 and Fig. 2, we plot the $C(j,f)$ for each pair of judge j and footwork f in measuring the weighted rank consistency of assessments in posture and rhythm, respectively.

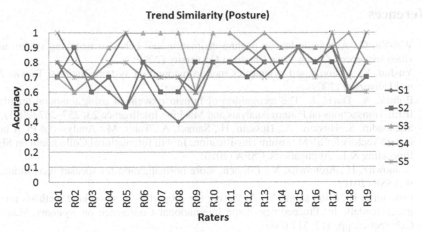

Fig. 1. Weighted rank consistency between the system and human judges for different footworks (posture)

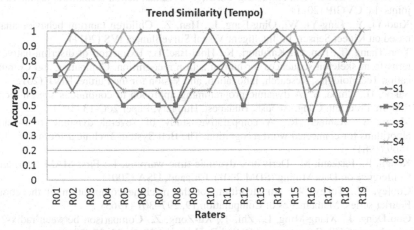

Fig. 2. Weighted rank consistency between the system and human judges for different footworks (rhythm)

5   Concluding Remarks and Future Work

The paper presents the design of an assisting system for the learners of dancing sports. The system features its capability of automatic and independent assessment of learners' dancing in terms of posture accuracy and tempo accuracy. Another useful and powerful function of the system is the accurate identification and indication of how and where the errors of learners' exercising occur. The function would be more helpful to improve the efficiency and efficacy of the learners' dance learning. In the future, we plan to extend the system as a basis for dance learning over a cloud platform. Through the cloud platform, the learners can learn dancing and participate in domestic or international dance competitions over the Internet.

References

[1] Wei-Yu, C.: Automatic human posture estimation and behavior matching system using video sequences. Chung Yuan Christian University (2009)

[2] Yu-Jun, Y.: Vision-based gymnastics motion recognition system. National Taiwan Normal University (2009)

[3] Bobick, A., Davis, J.: The recognition of human movement using temporal templates. IEEE Transactions on Pattern Analysis and Machine Intelligence 23, 257–267 (2001)

[4] Shahbudin, S., Hussain, A., Hussain, H., Samad, A., Tahir, M.: Analysis of PCA based feature vectors for SVM posture classification. In: 6th International Colloquium on Signal Processing & Its Applications, CSPA (2010)

[5] Aronowitz, H., Aronowitz, V.: Efficient score normalization for speaker recognition. In: ICASSP (2010)

[6] Feng-long, H., Ming-shing, Y.: Analyzing the properties of smoothing methods for language models. In: Proceedings of the International Conference on Systems, Man, and Cybernetics, pp. 512–517 (2001)

[7] Leyvand, T., Meekhof, C., Yi-Chen, W., Jian, S., Baining, G.: Kinect identity: technology and experience. Computer 44, 94–96 (2011)

[8] Ko-Hsin, C., Chaur-Heh, H., Chang-Chieh, W.: Human action recognition using 3d body joints. In: CVGIP (2011)

[9] Xiao-Yi, Y., Ling-Yi, W., Qing-Feng, L., Han, Z.: Children tantrum behavior analysis based on Kinect Sensor. In: Intelligent Visual Surveillance, IVS (2011)

[10] Tran-Thang, T., Fan, C., Kazunori, K., Hoai-Bac, L.: Extraction of discrimi-native patterns from skeleton sequences for human action recognition. Computing and Communication Technologies, Research, Innovation, and Vision for the Future, RIVF (2012)

[11] Sakoe, H., Chiba, S.: A dynamic programming approach to continuous speech recognition. In: Proc. Int. Cong. Acoust., Budapest, Hungary, Paper 20C-13 (1971)

[12] Myers, S., Rabiner, R.: A comparative study of several dynamic time-warping algorithms for connected word recognition. The Bell System Technical Journal 60, 1389–1409 (1981)

[13] Keogh, E., Pazzani, M.: Derivative dynamic time warping. In: First SIAM International Conference on Data Mining (SDM 2001), Chicago, USA (2001)

[14] Cooley, W., Tukey, W.: An algorithm for the machine computation of the complex Fourier series. Mathematics of Computation 19, 297–301 (1965)

[15] Guo-Dong, J., Xiang-Ming, L., Zhi, T., Xu-Zong, Z.: Comparison between radix-2 and mixed radix FFT. Computer and Digital Engineering 38(3), 25–27 (2010)

[16] Nemhauser, L.: Introduction to dynamic programming. Wiley (1966)

[17] Bresenham, J.: Algorithm for computer control of a digital plotter. IBM Systems Journal 4, 25–30 (1965)

[18] Vadakkoot, R., Shah, D., Shrivastava, S.: Enhanced moving average computation. In: World Congress on Computer Science and Information Engineering. Los Angeles, USA (2009)

A New View-Calibrated Approach for Abnormal Gait Detection

Kuo-Wei Lin, Shu-Ting Wang, Pau-Choo Chung[*], Ching-Fang Yang

Dept. of Electrical Engineering, Institute of Computer and Communication Engineering
National Cheng Kung University, Tainan 70101, Taiwan ROC
pcchung@eembox.ee.ncku.edu.tw

Abstract. Gait, or the style of walking, has been recently a popular topic in vision-based analysis. Most vision-based works about gait are devoted to the application of human recognition, but abnormal walking styles are rarely discussed. Accordingly, a vision-based method is proposed to analyze abnormal types of walking. In the proposed method, a background subtraction algorithm is applied to segment out the silhouette of the walker at each frame in a sequence. For each frame, we define a feature based on the length between two legs and the height of the individual, called aspect ratio (AR). By observing this feature value across time (or frame), a periodic wave is obtained. With this analysis, a few abnormal types of walking can be distinguished. Since an oblique camera view angle causes a distortion of the AR wave, a rectification mechanism is proposed based on a camera pinhole model to reduce the view angle effect. The experimental results show that the proposed rectification method identified abnormal walking patterns reliably irrespective of the direction in which the individual walks relative to the camera plane.

Keywords: aspect ratio, abnormal gait, AR rectification.

1 Introduction

Gait, or the style of walking, is a very common behavior we exhibit in daily life. In the normal situation, everyone walks in a similar way. Gait can be easily observed from a distance, and because of the obvious change of gait shape, it is less sensitive to segmentation noise. In the literature, there has been many works using biometric features to recognize individuals [1-6]. Lee et al. [5] fitted several ellipses to different parts of the silhouette of the walker and the parameters extracted from the ellipses such as the ellipse centroid, etc. are used as features to represent gait of a person. In [6], Cunado et al. extract gait features by fitting the movement of the thighs to an articulated pendulum-like motion model. The works mentioned previously treat gait as biometric feature, which is mainly used on application of identification instead of surveillance or healthcare.

Generally speaking, the studies above considered the use of human gait as a biometric feature in accomplishing the automatic identification and verification of human

---

[*] Corresponding author.

J.-S. Pan et al. (Eds.): *Advances in Intelligent Systems & Applications*, SIST 21, pp. 521–529.
DOI: 10.1007/978-3-642-35473-1_52 © Springer-Verlag Berlin Heidelberg 2013

individuals. However, the gait patterns also provide a useful indication of the physical condition and are often used as a good object of behavior analysis. For example, an individual who is drunk or feels unwell exhibits a notably different gait than that of an individual in good physiology condition. In addition, it has been shown that human gait dynamics provide useful insights into an individual's neural control of the locomotion function, and therefore enable a functional assessment about the impacts of aging and chronic disease on human mobility to be made [7].

Gait directly or indirectly reflects body conditions, such as someone limps or feel dizzy, there will be disturbance in his gait. Therefore, in contrast to traditional researches that the gait is applied for biometric identification, a gait analysis method is used based on stride patterns from aspect ratio (AR) for identifying abnormal walking in this paper. Also, to reinforce the performance of the recognition scheme that may be affected by variations in the camera view angle, a rectification mechanism is proposed in which the AR of the individual in the oblique angle view is converted to an equivalent AR in the canonical view. The feasibility of the proposed approach is demonstrated by performing a series of experimental trials using individuals with and without walking difficulties, respectively.

The remaining parts of the paper are organized as follows. Section 2 introduces the aspect ratio and rectification mechanism in analyzing walking patterns. Section 3 describes the features extracted from AR wave for diagnosing abnormal walking patterns. Section 4 shows the experimental results. Finally conclusions are drawn in Section 5.

2   Rectification Mechanism

To capture the dynamics of the stride pattern, the change of the foot step is analyzed in terms of the individual's aspect ratio (AR) while walking, i.e. $AR = W / H$, where W is the width between the two feet and H is the height of the individual, both are obtained from a silhouette image of the individual. In order to get the silhouette of foreground walker, a background subtraction procedure is performed to segment out the moving object [8]. The AR feature is obtained for each frame in the sequence, and the ARs are then arranged in time order to construct a so-called "AR wave".

It is generally assumed that the frames are captured under a canonical view, i.e. the walking path is parallel to the image plane of the camera. However, in practical applications, the camera may well be positioned at an oblique angle to the subject, and thus the measured AR values are distorted and some form of rectification mechanism is required.

According to perspective projection as shown in Fig. 1, the relationship between the height of an individual in a photographic image and the height of the individual in the real world is represented by

$$h = \frac{f}{z}H, \tag{1}$$

where f is the focal length of the camera, z is the depth of field, and H is the true height of the individual.

Assume that the individual is walking in an oblique direction $\vec{d}$ relative to the camera plane, as shown by the solid black line in Fig. 2. Let the direction of the

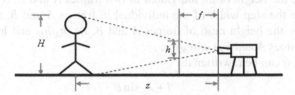

Fig. 1. Lateral view of camera and individual

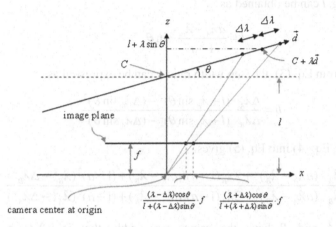

Fig. 2. Spatial relationship between camera and walking path

walking path relative to the camera plane be denoted by θ, where θ is measured in a counterclockwise direction from the camera plane. Furthermore, let C be the point of intersection between the optical axis of the camera and the walking path, and let l be the distance between C and the camera center. Thus, the position of the individual relative to the camera is given by $C + \lambda \vec{d}$, where λ is a scalar. The depth of field z is calculated as $l + \lambda \sin \theta$, and thus Eq. (1) can be rewritten as

$$h = \frac{f}{z} H = \frac{f}{l + \lambda \sin \theta} H . \tag{2}$$

Assume that the individual is walking in a straight line, the center of the head and the two legs lie on approximately the same spatial plane as that containing the walking path. Let $2\Delta\lambda$ be the distance between the two feet of the individual. Following the projection of the individual onto the camera plane, the distance between the two feet is given by

$$w = \frac{(\lambda + \Delta\lambda)\cos\theta}{l + (\lambda + \Delta\lambda)\sin\theta} \cdot f - \frac{(\lambda - \Delta\lambda)\cos\theta}{l + (\lambda - \Delta\lambda)\sin\theta} \cdot f \\ = \frac{2l\Delta\lambda\cos\theta}{(l + \lambda\sin\theta)^2 - (\Delta\lambda\sin\theta)^2} \cdot f \tag{3}$$

Let h_A and h_B be the height of an individual in two frames A and B, respectively, and let w_A and w_B be the step width of the individual in frames A and B, respectively. In addition, let a be the height ratio of frames A and B, i.e. h_A/h_B, and let b be the step width ratio of frames A and B, i.e. w_A/w_B.

From Eq. (2), a can be rewritten as

$$a = \frac{l + \lambda_B \sin\theta}{l + \lambda_A \sin\theta}.$$

Rearranging, l can be obtained as

$$l = \frac{a\lambda_A - \lambda_B}{1 - a} \cdot \sin\theta. \tag{4}$$

Similarly, from Eq. (3), the step width ratio b can be expressed as

$$b = \frac{\Delta\lambda_A}{\Delta\lambda_B} \cdot \frac{(l + \lambda_B \sin\theta)^2 - (\Delta\lambda_B \sin\theta)^2}{(l + \lambda_A \sin\theta)^2 - (\Delta\lambda_A \sin\theta)^2}. \tag{5}$$

Substituting Eq. (4) into Eq. (5) gives

$$b = \frac{\Delta\lambda_A}{\Delta\lambda_B} \cdot \frac{(a\lambda_A - \lambda_B)^2 + 2\lambda_B(1-a)(a\lambda_A - \lambda_B) + (1-a)^2(\lambda_B^2 - \Delta\lambda_B^2)}{(a\lambda_A - \lambda_B)^2 + 2\lambda_A(1-a)(a\lambda_A - \lambda_B) + (1-a)^2(\lambda_A^2 - \Delta\lambda_A^2)} \tag{6}$$

If frames A and B have the same step width, that is, $\Delta\lambda_A = \Delta\lambda_B = \Delta\lambda$ and $\lambda_B = \lambda_A + 2\Delta\lambda$, then Eq. (6) becomes

$$b = \frac{[(a-1)\lambda_A - 2\Delta\lambda]^2 + 2(\lambda_A + 2\Delta\lambda)(1-a)[(a-1)\lambda_A - 2\Delta\lambda] + (1-a)^2(\lambda_A^2 + 4\lambda_A\Delta\lambda + 3\Delta\lambda^2)}{[(a-1)\lambda_A - 2\Delta\lambda]^2 + 2\lambda_A(1-a)[(a-1)\lambda_A - 2\Delta\lambda] + (1-a)^2(\lambda_A^2 - \Delta\lambda^2)}$$

Rearranging, it can be shown that

$$[(b+3)(a-1)^2 + 8(a-1) + 4(1-b)]\Delta\lambda^2 = 0.$$

Since $\Delta\lambda = 0$ is a trivial condition, it follows that

$$(b+3)(a-1)^2 + 8(a-1) + 4(1-b) = 0. \tag{7}$$

Solving Eq. (7), the step width ratio in frames A and B is obtained as

$$b = \frac{1 - 3a}{a - 3} \tag{8}$$

In the canonical view, the height of the individual within each frame remains a constant due to the invariant depth of field of the camera. Thus, in the oblique view, a rectification procedure which tunes the height and width of the individual in each frame such that the height remains constant throughout the video sequence has the effect of converting the oblique images into approximate canonical equivalents.

Since the derivation of Eq. (8) is based on a constant step width, the use of this equation in performing the rectification process implies that the step width of the individual remains unchanged following the rectification procedure.

Assume that the height and step width of an individual in the oblique view are given by h and w, respectively. Equation 8 implies that if the height is to be adjusted to h', then the width should be adjusted to

$$w' = \frac{h - 3h'}{h' - 3h} \cdot w. \tag{9}$$

In general, the height of an individual in an image depends only on the position of the individual relative to the camera, i.e. the height is unaffected by the direction in which the individual walks. Thus, in the rectification process, the height of the individual in each frame is adjusted such that it is equal to the height value obtained from the frame in which the individual lies closest to the image center, i.e. the image in which the individual coincides with the optical axis (i.e. projection axis) of the camera. Let the corresponding height of the individual be denoted as h'. The aim of the rectification process is therefore to modify the width of the individual in each frame using Eq. (9) in such a way that the height of the individual is equal to h' in every case. Having rectified the width and height, the AR value in the equivalent canonical view is then computed by dividing the rectified width w' by the rectified height h'.

Figure 3(a) presents an illustrative AR wave for the case of an individual with no walking difficulties following an oblique path orientated at 20° to the camera plane. Note that the AR values are determined in accordance with Eqs. (2) and (3). Figure 3(b) presents the corresponding results obtained when applying the rectification procedure to the same image sequence. We can see that the algorithm performs considerably well on simulation.

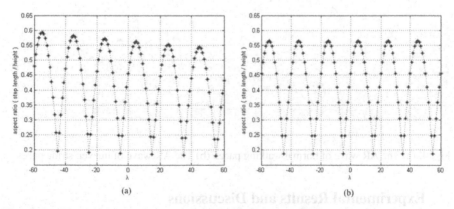

(a)   (b)

Fig. 3. (a) Unrectified AR wave for individual with no walking difficulties following path orientated at 20° to camera. (b) Rectified AR wave for individual with no walking difficulties following path orientated at 20° to camera plane.

3   Abnormal Stride Pattern Analysis from AR Wave

For each image frame, an aspect ratio (AR) can be obtained. Arranging this aspect ratio with time, a periodic wave, which will be called "AR wave", is obtained. Fig. 4(a) shows an AR wave of a normal walking pace. The peak areas in Fig. 4(a) represent the stance when the two legs are stretched far apart, while the valley areas represent the stance when the two legs are crossing. From Fig. 4(a), it can be seen that the valley areas is shaper than that in peak areas. This indicates that the movement on stance of crossing leg is faster than that spent on stance of striding out. This is due to the fact that on the stance of crossing leg, the center of body weight is close to the supporting foot and the person is in a more stable stance. As such he/she is free to make a faster movement. On the other hand, on the stance of striding out the weight center is far away from the supporting foot and the person in a less stable position. It would therefore take more efforts and time for the person to move the leg. From Fig. 4(a), we can also see that since a normal person walks relatively the same for both legs the shape and envelope of each cycle is similar.

As mentioned, in normal case each stride is almost of the same length, that is, all AR peaks have similar values. However, when a person limps, it is often that the length of one stride differs significantly from the next stride. The AR wave corresponds to this uneven stride case is shown in Fig. 4(b), from which we can see that the two consecutive strides have significantly different peak values, reflecting one stride having free movement and the other having difficulty of stretching out.

As mentioned, the stride characteristics of walking patterns are revealed in the AR wave, especially on the peaks and valleys positions. As such, the AR waves are segmented into the peaks and valleys. This is performed by firstly finding a separating axis which divides the AR waves into concave-up and concave-down arcs. Then features are extracted from the concave-up and concave-down arcs.

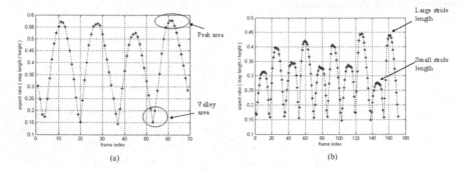

Fig. 4. (a) The AR wave of normal walking paces (b) The AR wave of uneven stride paces

4   Experimental Results and Discussions

To examine the performance of proposed rectifying algorithm, oblique-view videos of normal cases and uneven stride length cases are used for testing, and then rectified AR waves are used for analysis. Figure 5 presents three illustrative examples of the

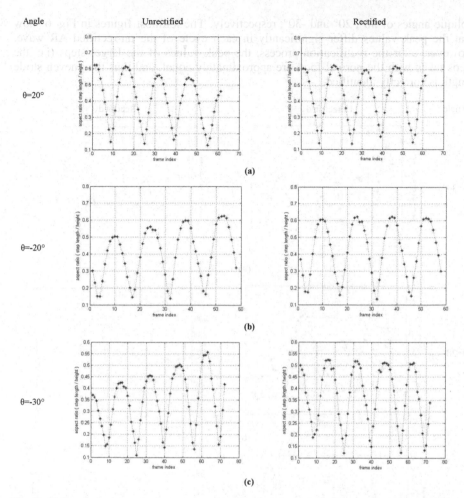

Fig. 5. Oblique-view AR waves before and after rectification for individuals with normal gait pattern

AR waves obtained before and after the rectification process for individuals with no walking difficulties following walking paths orientated at three different oblique angles. As discussed in Section 2, an oblique camera view angle causes the peak AR values to either increase or decrease as the individual passes in front of the camera. This tendency is clearly shown in the pre-rectification plots presented in the left column of Fig. 5. The drift in the peak AR value induces a large peak variance and therefore causes a misclassification of the gait pattern, i.e. the stride length is judged to be uneven whereas in practice it is actually even. However, the rightmost figures in Fig. 5 show that the proposed rectification mechanism results in a more uniform value of the peaks in each cycle of the AR wave. As a result, the peak variance is reduced, and thus the gait analysis system correctly recognizes a uniform stride length.

A second set of trials was performed in which an individual with an uneven stride length walked along walking paths orientated at various oblique angles to the camera plane. Figures 6(a)~(c) present the rectified and un-rectified AR waves obtained for oblique angles of 30°, 20° and -30°, respectively. The leftmost figures in Fig. 6 show that the peak values differ significantly in each cycle of the un-rectified AR wave. However, after the rectification process, the peak values of the larger steps (i.e. the steps taken with the normal foot) are approximately equal, and thus the uneven stride length is correctly identified.

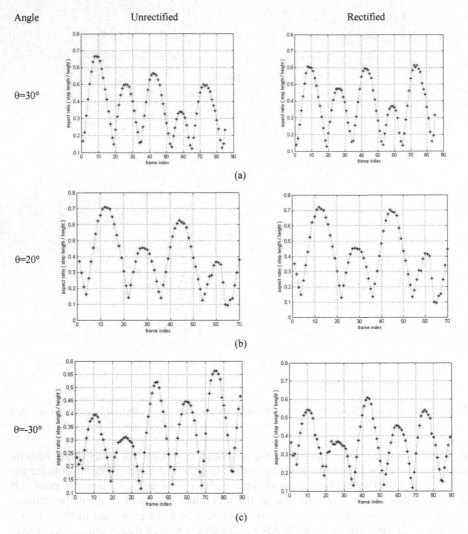

Fig. 6. Oblique-view AR waves before and after rectification for individual with an uneven stride length

5   Conclusion

We know that walking abnormality would reflect on walking stride patterns. Based on this consideration, this paper develops AR wave with view calibration for detecting abnormal walking patterns. With several practical examples, this paper demonstrated that walking with different walking difficulties generates different patterns in AR waves. Thus features can be extracted from the AR waves for the detection of walking abnormality. In addition, a rectification mechanism has been proposed for modifying the AR wave associated with an individual following an oblique walking path to an equivalent AR wave in the canonical view. The rectified AR waves are shown to present more accurate information in abnormal pattern detection.

Acknowledgment. This study was supported by the National Science Council of Taiwan, ROC, under Grant No. NSC 95-2221-E-6-374.

References

1. Sarkar, S., Phillips, P.J., Liu, Z., Vega, I.R., Grother, P., Bowyer, K.W.: The HumanID Gait Challenge Problem :   Data Sets, Performance, and Analysis. IEEE Trans. Pattern Analysis and Machine Intelligence 27(2) (February 2005)
2. Wang, L., Tan, T., Ning, H., Hu, W.: Silhouette Analysis-Based Gait Recognition for Human Identification. IEEE Trans. Pattern Analysis and Machine Intelligence 25(12), 1505–1518 (2003)
3. BenAbdelkader, C., Culter, R., Davis, L.: Stride and Cadence as a Bio-metric in Automatic Person Identification and Verification. In: Proc. Int. Conf. Automatic Face and Gesture Recognition (2002)
4. Cutler, R., BenAbdelkader, C., Davis, L.S.: Motion based recognition of people in eigengait space. In: Proc. IEEE Conf. Face and Gesture Recognition, pp. 267–272 (2002)
5. Lee, L., Grimson, W.E.L.: Gait analysis for recognition and classification. In: Proc. IEEE Conf. Face and Gesture Recognition, pp. 155–161 (2002)
6. Cunado, D., Nash, J.M., Nixon, M.S., Carter, J.N.: Gait extraction and de-scription by evidence-gathering. In: Proc. Int. Conf. Audio and Video Based Biometric Person Authentication, pp. 43–48 (1995)
7. Bauckhage, C., Tsotsos, J.K., Bunn, F.E.: Automatic detection of abnormal gait. Image and Vision Computing 27(1-2), 108–115 (2009)
8. Lin, K.W., Chung, P.C.: Vision-Based Gait Analysis. M. S. thesis, Institute of Computer and Communication, Cheng Kung University (2005)

Modeling and Recognizing Action Contexts in Persons Using Sparse Representation

Kai-Ting Chuang, Jun-Wei Hsieh*, and Yilin Yan

Depart. of Computer Science and Engineering, National Taiwan Ocean University,
No.2, Beining Rd., Keelung 202, Taiwan
shieh@mail.ntou.edu.tw

Abstract. This paper proposes a novel dynamic sparse representation-based classification scheme to treat the problem of interaction action analysis between persons using sparse representation. The occlusion problem and the difficulty to model complicated interactions are the major challenges in person-to-person action analysis. To address the occlusion problem, the proposed scheme represents an action sample in an over-complete dictionary whose base elements are the training samples themselves. This representation is naturally sparse and makes errors (caused by different environmental changes like lighting or occlusions) sparsely appear in the training library. Because of the sparsity, it is robust to occlusions and lighting changes. The difficulty of complicated action modeling can be tackled by adding more examples to the over-complete dictionary. Thus, even though the interaction relations are complicated, the proposed method still works successfully to recognize them and can be easily extended to analyze action events among multiple persons.

Keywords: Sparse Coding, Sparse Representation, Occlusions, daily events.

1 Introduction

Human action analysis [1]- [18]] is an important task in various application domains like video surveillance [1], video retrieval [18], human-computer interaction systems, and so on. Characterization of human action is equivalent to dealing with a sequence of video frames that contain both spatial and temporal information. The challenge in human action analysis is how to properly characterize spatial-temporal information and then facilitate subsequent comparison/recognition tasks. To treat this challenge, some approaches build various action syntactic primitives to represent and recognize events. For example, in [18], Park and Aggarwal used the "blob" concept to model and segment a human body into different body parts from which human events were analyzed using the dynamic Bayesian networks. Wang, Huang, and Tan [6] used the R transform to extract contour features from different frames and then proposed a HMM-based recognition scheme to analyze human behaviors. Some approaches decompose actions into sequences of key atomic action units which are referred to as atoms. For example, in [12], Gaidon, Harchaoui, Schmid proposed an atom

---

* Corresponding author.

J.-S. Pan et al. (Eds.): *Advances in Intelligent Systems & Applications*, SIST 21, pp. 531–541.
DOI: 10.1007/978-3-642-35473-1_53   © Springer-Verlag Berlin Heidelberg 2013

sequence model (ASM) to represent the temporal structure of actions and then recognize actions in videos using a sequence of "atoms" which are obtained by manual annotations. In addition to videos, humans can recognize activities based on only still images. However, the prerequisite that body parts or poses must be well estimated makes this scheme inappropriate for real-time analysis of human behaviors.

To avoid the difficulty of action primitive or body part extraction, some approaches extract feature points of interest and obtain their motion flows to represent and recognize actions. For example, Rosales and Sclaroff [13] proposed a trajectory-based recognition system to detect pedestrians in outdoor environments and then recognize their activities from multiple views using mixtures of Gaussian classifiers. In [9], Laptev *et al.* generated the concept of interest points (STIP) from images to flow volumes and then used it to extract various key frames to represent action events. The success of the above feature-flow methods strongly depends on a large set of well tracking points to describe action changes across frames.

In addition to event features, another key problem in event analysis is how to model the temporal and spatial dynamics of events. To treat this problem, Mahajan *et al.* [8] proposed a layer concept to divide the recognition task to three layers, *i.e.*, physical, logical, and event layers corresponding to feature extraction, action representation, and event analysis, respectively. Hidden Markov model (HMM) is another commonly used scheme to model event dynamics. For example, Messing, Pal, and Kautz [10] tracked a set of corners to obtain their velocity histories and then used HMM to learn activity event models. The challenges related to HMMs involve how to specify and learn the HMM model structure.

A particular action between two objects can vary significantly under different conditions such as camera views, person's clothing, object appearances, and so on. Thus, it is more challenging to analyze human events happening between two objects because of their complicated interaction relations. In addition, occlusions between objects often happen and lead to the failure of action recognition. In [17], Filipovych and Ribeiro used a probabilistic graphical model to recognize the primitive actor-object interaction events like "grasping" or "touching" a fork (or a spoon, a cup). In addition to videos, some previous works address joint modeling of human poses, objects and relations among them from still images. For example, in [15], Yao and Fei-Fei proposed a random field model to encode the mutual connections of components in the analyzed object, the human pose, and the body parts to recognize human-object interaction activities in still images. However, until now, reliable estimation of body configurations for persons in any poses remains a very challenging problem.

This paper addresses the problem of action analysis between persons (or human-object interactions) using sparse representation. Fig. 1 shows the flowchart of our system. As described before, the complicated interaction changes and the occlusion problem between two objects increase many challenges in action recognition. To the above problems, this paper proposes a novel dynamic sparse representation scheme to represent an event in an over-complete dictionary whose base elements are the training samples themselves. If sufficient training samples are collected from each action class, each test sample will be possibly represented as a linear combination of just the training sample from the same class. This representation is naturally sparse, involving only a small fraction of the overall training database. The sparse property also makes errors (caused by different environmental changes like lighting or occlusions) sparsely appear in the training library as a special case of training samples to be handled. The sparsity of error distribution increases the robustness of our scheme to occlusion. After that, a sparse reconstruction cost (SRC) is is proposed to classify action events to more categories. Even though the interaction relations are complicated, the proposed method still works successfully to recognize them and can be easily extended to analyze action events among multiple persons.

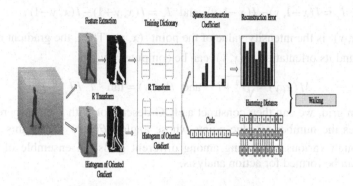

Fig. 1. Flowchart of the proposed system

2   Feature Extraction

This paper uses two features to represent an object shape, *i.e.*, R transform and HOG. Details of these two features are described as follows.

2.1   Radon Transform and R Transform

Radon transform in two dimensions is the integral transform consisting of the integral of an image over straight line.   Let $I(x, y)$ denote one input image and L be a straight line with the equation: $x\cos\theta + y\sin\theta = t$.   Then, the Radon transform of $I(x, y)$ along L is defined as follows:

$$Radon\{I\} = P(\theta, t) = \int_{-\infty}^{\infty}\int_{-\infty}^{\infty} I(x, y)\delta(x\cos\theta + y\sin\theta - t)dxdy , \qquad (1)$$

where $\theta \in [0, \pi]$ and $t \in (-\infty, \infty)$ 、 $\delta(x)$ is a Dirac delta function. In Eq.(1), the result of Radon Transform is a two dimensional signal.   It can be converted to 1D signal by the improved form of Radon transform [6]:

$$R(\theta) = \int_{-\infty}^{\infty} P^2(\theta, t)dt . \qquad (2)$$

Eq.(2) is the R transform of $I(x, y)$.   If $I(x, y)$ is normalized to a fixed size, the R transform is invariant under translation and scaling.

2.2   Histogram of Oriented Gradients

In addition to R transformation, this paper also uses the histogram of oriented gradients (HOG) [11] to describe a posture. Let I_x and I_y denote the central differences at point (x, y) are given by

$$I_x = I(x+1, y) - I(x-1, y) \text{ and } I_y = I(x, y+1) - I(x, y-1), \qquad (3)$$

where $I(x, y)$ is the intensity value of the point (x, y). Then, the gradient magnitude $M(x, y)$ and its orientation $\theta(x, y)$ can be computed by

$$M(x, y) = \sqrt{I_x^2 + I_y^2} \text{ and } \theta(x, y) = \tan^{-1} I_x / I_y. \qquad (4)$$

For a given grid, we can then construct a HOG descriptor with 8 bin, where each bin accumulates the number of edge points whose angles $\theta(x, y)$ fall in this angle bin. Then, through various combining among different grids, an ensemble of HOG descriptors can be formed for action analysis.

3   Sparse Representation

Sparse representation [3]-[5] is a technique to build an overcomplete dictionary to represent a target. In this paper, we utilize the sparse representation and dictionary learning techniques to design a novel framework to analyze action events happening between multiple persons.

3.1   Dictionary Initialization and Matching Pursuit

Let x denote an n-dimensional feature vector, i.e., $x \in R^n$. We say that it admits a sparse approximation over a *dictionary* D in $R^{n \times K}$, where each column vector is referred to as an *atom*. Consider a finite training set of signals $X = [x_1, x_2, ..., x_N] \in R^{n \times N}$. Then, one can find a linear combination of a "few" atoms from D that is "close" to the signal x; that is,

$$X \approx DA, \qquad (5)$$

where $A = [\alpha_1, ..., \alpha_N] \in R^{K \times N}$ is the set of combination coefficients in the sparse decomposition. Given X, the K-SVD algorithm maintains the best representation of each signal with strict sparsity constraints to learn the over complete dictionary D. It is an iterative scheme alternating between sparse coding of the training signals with respect to the current dictionary and an update process for the dictionary atoms so as to better fit the training signals. Then, the learning process can be formulated a joint optimization problem with respect to the dictionary D and A of the sparse decomposition as

$$\arg \min_{D,A} \| X - DA \|_2^2 \quad s.t. \ \forall i, \ \| \alpha_i \|_0 \leq T, \qquad (6)$$

where $D = [d_1, ..., d_K] \in R^{n \times K}, A = [\alpha_1, ..., \alpha_N] \in R^{K \times N}$, T is the most desired number of non-zero coefficients, and $\| \alpha_i \|_0$ is the l_0-norm which counts the number of nonzeros in a vector α_i. Eq.(6) can be formulated as another equivalent problem:

$$\arg\min_{D,\alpha_i} \| \alpha_i \|_0 \quad s.t. \| x_i - D\alpha_i \|_2^2 \le \varepsilon , \tag{7}$$

where ε is an error tolerance of reconstruction. One of common methods to solve α_i is the Orthogonal Matching Pursuit (OMP). OMP is a greedy method to iteratively solve the optimal α_i for each x_i while fixing D, i.e.,

$$\alpha_i = \min_A \| x_i - DA \|_2^2 \quad s.t. \| \alpha_i \|_0 \le T . \tag{8}$$

At each stage, the OMP selects an atom α_i from D that best resembles the residual.

3.2   Dictionary Learning via K-SVD

This paper applies the K-SVD to solve Eq.(8) through an iterative way with two stages, i.e., sparse coding stage and dictionary update stage. It optimizes **D** and **X** through a number of training iterations until convergence. Each iteration consists of a *sparse coding stage* that optimizes the coefficients in A and a *dictionary update stage* that improves the atoms in **D**. During the *sparse coding stage*, **D** is held while each α_i is optimized by solving Eq.(8) via the OMP scheme, and allowing ach coefficient vector to have no more than T nonzero elements. During the dictionary update stage, each column d_k in D is updated sequentially so that its coefficients can better represent X. The update process is the key inside of K-SVD which accelerates the optimization process of Eq.(6) while maintaining the sparsity requirement. Let d_k denote the k th column in D to be updated. In addition, we denote the coefficients that correspond to d_k, the k th row in A by α_k^{row}. Then, the cost function in Eq.(6) can be rewritten as follows:

$$\| X - DA \|_F^2 = \| X - \sum_{j=1}^{K} d_j \alpha_j^{row} \|_F^2 = \| X - \sum_{j \ne k} d_j \alpha_j^{row} - d_k \alpha_k^{row} \|_F^2$$

$$= \| E_k - d_k \alpha_k^{row} \|_F^2 .$$

The updated values of d_k and α_k^{row} can be obtained by solving

$$\arg\min_{d_k, \alpha_k^{row}} \| E_k - d_k \alpha_k^{row} \|_F^2 . \tag{9}$$

The K-SVD scheme suggests the use of the SVD to find alternative d_k and α_k^{row}. If the SVD of E_k is expressed as USV^T, d_k is updated by the first output basis vector u_1 and the non-zero values in α_k^{row} are adjusted to the product of the first singular value $S(1,1)$ and the first column of V.

4   Person-to-Person Action Recognition Using Sparse Representation

This section will deal with different challenges in human action analysis between persons, that is, how to properly characterize spatial-temporal information and how to perform the subsequent comparison/recognition tasks.

4.1   Action Event Representation

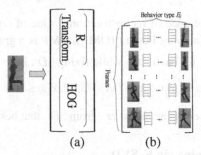

(a)   (b)

Fig. 2. Sparse representation for analyzing single-person action events. (a) Feature vector to represent an object at each frame. (b) Matrix to represent single person action.

To characterize spatial-temporal information of an action event, this paper uses the R-transform and HOG descriptors to describe each frame. For the R-transform, the angle ranges from $0°$ to $180°$ and is further sampled with $4°$. As to the HOG descriptor, the observed object is divided to $n_{grid} \times n_{grid}$ grids. Let h_R^X be a vector with 45 elements to represent the R-transform of an object X and h_{hog}^X denote another vector with $8n_{grid}^2$ elements to represent the HOG descriptor of X. Then, as shown in Fig.2 (a), a new feature vector $F^X = (h_R^X, h_{hog}^X)^T$ can be formed to represent X. Let $\oplus$ denote a vector concatenation operation between F^X and F^Y, i.e.,

$$F^X \oplus F^Y = (h_R^X, h_{hog}^X, h_R^Y, h_{hog}^Y)^T. \tag{10}$$

In addition, we use X_t to denote the version of X observed at the tth frame and the superscript k in X^k denote the kth object. If an action event A_X recorded with n_f frames is performed by X, a new feature vector can be formed to represent A through a vector concatenation operation. That is,

$$A_X = F^{X_1} \oplus ... \oplus F^{X_{n_f}} = (h_R^{X_1}, h_{hog}^{X_1}, ..., h_R^{X_{n_f}}, h_{hog}^{X_{n_f}})^T. \tag{11}$$

Let $n = n_f(45 + 8 * n_g^2)$. Then, $A_X \in R^n$ and denotes a normal feature. If an action type E_i is represented by K_i codes, we can use a matrix to represent E_i by the form:

$$E_i = (A_{X^1}, ..., A_{X^k}, ..., A_{X^{K_i}}), \tag{12}$$

where X^k denotes the kth object in E_i and $E_i \in R^{n \times K_i}$. Assume that there are L action types to be recognized and let $K = \sum_{i=1}^{L} K_i$. Then, the basis D in sparse representation for action recognition can be constructed by

$$D = (E_1, ..., E_L), \tag{13}$$

where $D \in R^{n \times K}$.

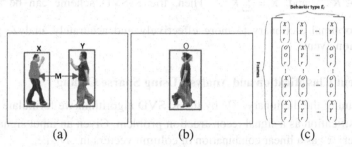

(a)   (b)   (c)

Fig. 3. Sparse representation for analyzing two-person action events. (a) Non-occluded frame. (b) Occluded frame. (c) Matrix to represent single person action event on the same type.

Different from the task of single-person event analysis, there will be different occlusion conditions happening when analyzing two-person interaction events. Assume the interaction events to be analyzed are performed by two persons X and Y. The visual descriptor to capture their spatial relations will change according to the condition whether X and Y are occluded. As Fig.3 (a), if X and Y are not occluded, a new feature descriptor F^{XY} is extracted for describing X and Y as follows:

$$F^{XY} = F^X \oplus F^Y \oplus (m) = (h_R^X, h_{hog}^X, h_R^Y, h_{hog}^Y, m)^T, \tag{14}$$

where m is the motion feature between X and Y. This paper scts m to the relative distance between X and Y. On the other hand, if X and Y are occluded together (see Fig.3 (b)), we replace X and Y with their occluded version O. Then, the descriptor to represent X and Y is constructed as follows:

$$F^{XY} = (h_R^O, h_{hog}^O, h_R^O, h_{hog}^O, m)^T, \tag{15}$$

where h_R^O is the R transform of O, h_{hog}^O is the HOG descriptor of O, and m is set to zero. Let A_{XY} denote the action event performed by two persons X and Y. If n_f frames are collected to represent A_{XY}, it can be constructed with the following sparse representation:

$$A_{XY} = F^{XY_1} \oplus ... \oplus F^{XY_t} \oplus ... \oplus F^{XY_{n_f}}. \tag{16}$$

As shown in Fig.3 (c), each column shows the structure of A_{XY}. In the two-person case, if an action type E_i is represented by K_i codes, we can use a matrix to represent E_i by the form:

$$E_i = (A_{XY^1}, ..., A_{XY^{K_i}}), \tag{17}$$

where X^k denotes the kth object, $E_i \in R^{n \times K_i}$. If there are L action types to be recognized, the library D in sparse representation for action recognition can be formed:

$$D = (E_1, ..., E_L),\qquad(18)$$

where $D \in R^{n \times K}$ and $K = \sum_{i=1}^{L} K_i$. Then, the K-SVD scheme can be applied to learn the optimal dictionary to more effectively and accurately analyze person-to-person action events.

4.2   Event Classification and Analysis Using Sparse Coding

After obtaining the dictionary D by the K-SVD algorithm, we formulate the action event classification as a signal reconstruction problem. Given an input signal $x \in R^n$, we consider x as a linear combination of column vectors in D, i.e.,

$$x = \alpha_1 d_1 + ... + ... + \alpha_K d_K,$$

where $d_k \in D$. Let $\alpha = (\alpha_1, ..., \alpha_K)$. The sparse solution α can be obtained by solving the following minimization problem:

$$\arg \min_{\alpha} \| \alpha \|_1 \quad s.t. \| x - D\alpha \|_2^2 \le \varepsilon.$$

This optimization problem can be efficiently solved via the second-order cone programming. Since there are L action types to be recognized, in Eq.(18), the library D is separated to L classes, i.e., $D = (E_1, ..., E_L)$. Let $\delta_i : R^n \to R^n$ be a function that selects the coefficients associated with the ith class. Then, using only the coefficients associated with class i, we compute the residual $r_i(x)$ between x and the approximated one:

$$r_i(x) = \| x - D\delta_i(\alpha) \|_2.\qquad(19)$$

We can classify x to its corresponding action type by assigning it to the event class that minimizes the residual, i.e.,

$$Type(x) = \arg \min_i r_i(x).\qquad(20)$$

5   Experimental Results

Fig. 4. Examples of real data for seven action types

To evaluate the performance of our proposed method, a real-time system to analyze different action events between two persons at different lighting conditions was implemented. There is no benchmarking database designed for evaluating algorithms to recognize two-person action events. Thus, two kinds of datasets were adopted in this paper for examining the effectiveness of our method, *i.e.*, synthetic and real videos. In this dataset, four kinds of action types were created, *i.e.*, waving, handshaking, running, and walking. For each action type, there were one hundred of action videos created for training and testing, where fifty videos were for training and another set of fifty videos were for testing. In addition to the four types, three extra action types were added in the real dataset for evaluating the effectiveness of our methods under real conditions. The three types are kinking, punching, and soccer-juggling, respectively. The dimension of video frame is 320×240 pixel elements.

Fig. 4 shows the examples of real data for the seven action types. Fig. 5 shows the results of two-person action recognition on the synthetic dataset. All the action types were correctly recognized. Table 1 shows the confusion matrix of two-person action recognition on the synthetic dataset using the SRC method. In this table, the "walking" action type is easily misclassified to the "running" type. The two action types are very similar except their speeds. The "handshaking" action type was sometimes misclassified to the "walking" type since their visual features are similar before handshaking. The average accuracy of the SRC method is about 86%.

Fig. 5. Results of action recognition on the synthetic dataset

Table 1. Confusion matrix of the SRC method on synthetic dataset

SRC				
Action Types	Handshaking	Greeting	Walking	Running
Handshaking	47/94%	0/0%	0/0%	3/6%
Greeting	0/0%	50/100%	0/0%	0/0%
Walking	8/16%	0/0%	40/80%	2/4%
Running	0/0%	0/0%	15/30%	35/70%

Table 2. Accuracy improvements of the SRC method after adding the speed feature

methods / Action Types	Handshaking	Greeting	Walking	Running	Average
SRC	94%	100%	86%	78%	90%

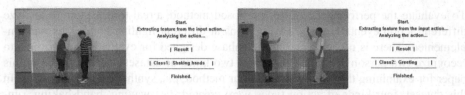

Fig. 6. Results of action recognition on real dataset in indoor environments

As to the real dataset, seven action types were recognized. The first six types focus on person-to-person action recognition. Fig. 5 shows the results of action type recognition in indoor environments. Fig. 6 shows the results of action type recognition in outdoor environments. Actually, our method can also be applied to recognize person-to-object action events. Thus, the last case is to recognize action events happening between an object (soccer) and a person. Fig. 7 shows the result of recognizing a person-to-object action event. Table 3 shows the confusion matrix of the SRC method on real data. The average accuracy of the SRC method is 80.54%. All the above experiments have proved that the proposed method is a robust, accurate, and powerful tool for action event analysis.

Fig. 7. Results of action recognition on real dataset in outdoor environments

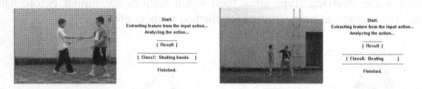

Fig. 8. Result of recognizing a person-to-object event

Table 3. Confusion matrix of the SRC method on real dataset

	SRC						
Action Types	Handshaking	Greeting	Walking	Running	Kicking	Punching	S-juggling
Handshaking	26/81.25%	0/0%	1/3.25%	0/0%	2/6.25%	3/9.25%	0/0%
Greeting	0/0%	30/93.75%	0/0%	0/0%	0/0%	2/6.25%	0/0%
Walking	2/6.25%	0/0%	25/78.125%	5/15.625	0/0%	0/0%	0/0%
Running	0/0%	0/0%	8/31.25%	24/75%	0/0%	0/0%	0/0%
Kicking	0/0%	0/0%	0/0%	0/0%	27/84.375%	5/15.625%	0/0%
Punching	2/6.25%	0/0%	0/0%	0/0%	4/12.5%	26/81.25%	0/0%
Soccer-juggling	1/3.4%	2/6.9%	2/6.9%	2/6.9%	1/3.4%	1/3.4%	20/69%

References

1. Weinland, D., Ronfard, R., Boyer, E.: A survey of vision-based methods for action representation, segmentation, and recognition. Computer Vision and Image Understanding 115(2), 224–241 (2011)
2. Poppe, R.: A survey on vision-based human action recognition. Image and Vision Computing 28(6), 976–990 (2010)
3. Aharon, M., Elad, M., Bruckstein, A.: K-svd: An algorithm for designing overcompletedictionries for sparse representation. IEEE Trans. on Signal Processing, 4311–4322 (2006)
4. Qiu, Q., Jiang, Z., Chellappa, R.: Sparse Dictionary-based Representation and Recognition of Action Attributes. In: IEEE Conference on Computer Vision (2011)
5. Mairal, J., Bach, F., Ponce, J., Sapiro, G.: Online learning for matrix factorization and sparse coding. J. Mach. Learn. Res. 11, 19–60 (2010)
6. Wang, Y., Huang, K., Tan, T.: Human activity recognition based on R transform. In: IEEE Conference on Computer Vision and Pattern Recognition (2007)
7. Kratz, L., Nishino, K.: Anomaly detection in extremely crowded scenes using spatio-temporal motion pattern models. In: IEEE Conference on Computer Vision and Pattern Recognition, pp. 1446–1453 (2009)
8. Mahajan, D., Kwatra, N., Jain, S., Kalra, P.: A framework for activity recognition and detection of unusual activities. In: International Conference on Graphic and Image Processing (2004)
9. Laptev, I., Perez, P.: Retrieving actions in movies. In: International Conference on Computer Vision (October 2007)
10. Messing, R., Pal, C., Kautz, H.: Activity recognition using the velocity histories of tracked keypoints. In: International Conference on Computer Vision (October 2009)
11. Dalal, N., Triggs, B.: Histograms of Oriented Gradients for Human Detection. In: IEEE Conference on Computer Vision and Pattern Recognition, vol. 1, pp. 886–893 (2005)
12. Gaidon, A., Harchaoui, Z., Schmid, C.: Actom Sequence Models for Efficient Action Detection. In: IEEE Conference on Computer Vision and Pattern Recognition (2011)
13. Rosales, R., Sclaroff, S.: 3D Trajectory Recovery for Tracking Multiple Objects and Trajectory Guided Recognition of Actions. In: Proc. of IEEE Conf. on Computer Vision and Pattern Recognition, vol. 2, pp. 117–123 (1999)
14. Nguyen, N.T., Bui, H.H., Venkatesh, S., West, G.: Recognition and monitoring high-level behaviours in complex spatial environments. In: IEEE International Conference on Computer Vision and Pattern Recognition, Madison, Wisconsin, USA, vol. 2, pp. 620–625 (June 2003)
15. Yao, B., Fei-Fei, L.: Recognizing Human-Object Interactions in Still Images by Modeling the Mutual Context of Objects and Human Poses. To appear in IEEE Transactions on Pattern Analysis and Machine Intelligence
16. Delaitre, V., Sivic, J., Laptev, I.: Learning person-object interactions for action recognition in still images. Advances in Neural Information Processing Systems (2011)
17. Filipovych, R., Ribeiro, E.: Recognizing Primitive Interactions by Exploring Actor-Object States. In: IEEE Conference on Computer Vision and Pattern Recognition, pp. 1–7 (June 2008)
18. Park, S., Park, J., Aggarwal, J.K.: Video Retrieval of Human Interactions Using Model-based Motion Tracking and Multi-Layer Finite State Automata. In: Bakker, E.M., Lew, M., Huang, T.S., Sebe, N., Zhou, X.S. (eds.) CIVR 2003. LNCS, vol. 2728, pp. 394–403. Springer, Heidelberg (2003)

References

1. Weinland, D., Ronfard, R., Boyer, E.: A survey of vision-based methods for action representation, segmentation and recognition. Computer Vision and Image Understanding 115(2), 224–241 (2011).

2. Poppe, R.: A survey on vision-based human action recognition. Image and Vision Computing 28(6), 976–990 (2010).

3. Alahari, M., Blake, M., Brookstein, A., Kaved: An algorithm for designing overcomplete dictionaries for sparse representation. IEEE Trans. on Signal Processing, 4311–4322(2000).

4. Qiu, Q., Jiang, Z., Chellappa, R.: Sparse Dictionary-based Representation and Recognition of Action Attributes. In: IEEE Conference on Computer Vision (2011).

5. Abdul, A., Pech, P., Rasse, A., Sapiro, G.: Online learning for matrix factorization and sparse coding. J. Mach. Learn. Res. 11, 19–60 (2010).

6. Wang, Y., Huang, K., Tan, T.: Human activity recognition based on R transform. In: IEEE Conference on Computer, Vision and Pattern Recognition (2007).

7. Kim, L., Nishino, K.: Automatic descriptor in extremely crowded scene using spatio-temporal motion pattern models. In: IEEE Conference on Computer, Vision and Pattern Recognition, pp. 1446–1453 (2009).

8. Marszałek, D., Kwasnica, N., Juhs, S., Kahn, P.: A framework for activity recognition and detection of unusual activities. In: International Conference on Graphic and Image Processing (2001).

9. Laptev, I., Pérez, P.: Retrieving actions in movies. In: International Conference on Computer Vision (October 2007).

10. Mocanu, R., Itti, C., Reid, H.: Action recognition using the velocity histories of tracked keypoints. In: International Conference on Computer Vision (October 2009).

11. Balal, N., Grings, H., Hiroshima, M.: Oriented Gradient as for Human Detection. In: IEEE Conference on Computer Vision and Pattern Recognition, vol. 1, pp. 886–893 (2005).

12. Gupta, A., Harehman, A., Schmid, C.: Action Sequence Models for Efficient Action Detection. In: IEEE Conference on Computer Vision and Pattern Recognition (2010).

13. Stauffer, W., Scharrat, S.: JRT: Real-Time Recognition, Tracking Multiple Objects and Trajectory-based Recognition of Actions. In: Proc. of IEEE Conf. on Computer, Vision and Pattern Recognition, vol. 2, pp. 117–175 (1999).

14. Nguyen, N.T., Bui, H.H., Venkatesh, S.L., West, G.: Recognition and monitoring high-level behaviors in complex spatial environment. In: IEEE International Conference on Computer, Vision and Pattern Recognition, Madison, Wisconsin, USA, vol. 2, pp. 620–625 (June 2003).

15. Yao, B., Fei-Fei, L.: Recognizing Human-Object Interactions in Still Images by Modeling the mutual context of Objects and Human Poses. In: IEEE Transactions on Pattern Analysis and Machine Intelligence.

16. Fillipovych, Z., Gibert, L., Izenman: person-object interactions for action recognition in still images. Advances in Neural Information Processing Systems (2009).

17. Filipovych, R., Ricci, E.: Recognizing Dynamic Human Actions by Modeling Actor-Object States. In: IEEE Conference on Computer, Vision and Pattern Recognition, pp. 1–7 (June 2008).

18. Park, S.J., Park, S., Aggarwal, J.K.: A hierarchical for Human Interactions Using Model-based Motion Tracking and Multi-layer Finite State Automata. In: Kalikar, K.M., Kew, M., Tuang, T.S., Zeng, W., Zhou, X.S. (eds.) CIVR 2003. LNCS, vol. 3758, pp. 394–403. Springer, Heidelberg (2003).

Efficient Parallel Knuth-Morris-Pratt Algorithm for Multi-GPUs with CUDA

Kuan-Ju Lin, Yi-Hsuan Huang, and Chun-Yuan Lin[*]

Department of Computer Science and Information Engineering,
Chang Gung University, Taoyuan 333, Taiwan, ROC
{g548462,rtetrtoo}@gmail.com, cyulin@mail.cgu.edu.tw

Abstract. String matching is an important technique among various applications. The traditional string matching algorithm needs the backtracking procedure and does the comparison repeatedly, thus these factors affect its efficiency. Knuth-Morris-Pratt (KMP) is one of well-known and efficient string matching algorithms. However, the computation time of KMP algorithm still is large for processing thousands of pattern strings. Current high-end graphics processing units (GPUs), contain up to hundreds cores per chip, are very popular in the high performance computing community. In this paper, we proposed an efficient parallel KMP algorithm, called KMP-GPU, for multi-GPUs with CUDA. The experimental results showed that the proposed KMP-GPU algorithm can achieve 97x speedups compared with the CPU-based KMP algorithm.

Keywords: String matching, Knuth-Morris-Pratt, CUDA, Graphics Processing Units, Parallel Processing.

1 Introduction

String matching is an important technique among various applications [10], such as the intrusion detection in the computer network, WWW search engine, information security [8], spelling checking, and biological sequence matching [6]. In the past, many string matching problems contain exact matching, fuzzy matching, and parallel matching have been proposed. A simple (exact) string matching problem can be defined as that there are two strings T (called the text string with a length of n) and P (called the pattern string with a length of m), represented as follows:

$$T=T_0\,T_1\,T_2\,T_3\,...T_n.$$
$$P=P_0\,P_1\,P_2\,P_3...P_m.$$

Assume that the strings T and P both are consisted of characters under the alphabet set (for example, the English letter: a through z) and the value of m is far less than that of n. The goal is to find whether the string P is appeared in the string T or not.

---

[*] Corresponding author.

J.-S. Pan et al. (Eds.): *Advances in Intelligent Systems & Applications*, SIST 21, pp. 543–552.
DOI: 10.1007/978-3-642-35473-1_54   © Springer-Verlag Berlin Heidelberg 2013

Several well-known string matching algorithms, such as Boyer-Moore (BM)[5] algorithm, RK algorithm, and Knuth-Morris-Pratt (KMP) algorithm, have been proposed to improve the performance of traditional algorithm (brute force algorithm). Rabin and Karp proposed the RK algorithm in 1987. The RK algorithm used the hashing technique to find pattern strings in a text string. For the text string with a length of n and p pattern strings with a combined length of m, the time complexities of its average and best cases both are $O(n+m)$, however, it is $O(nm)$ in the worst case. The KMP algorithm reduced the time complexity of RK Algorithm to $O(n)$ time complexity by using a pre-computation step to examine each character of text string only once [4]. In the KMP algorithm, the failure table of each pattern string is built before the matching procedure. Therefore, when detecting a mismatch, a shift action can be determined according to the failure table. By taking the advantage of this information to avoid the unnecessary comparisons, the KMP algorithm obtains the satisfied performance.

However, these methods still are very time-consuming for processing thousands of pattern strings under the modern CPU. Therefore, these methods were re-designed with a reconfigurable hardware device or software API in the past. Through parallel computation we can save lot of time in many string matching applications. More and more parallel algorithms for string matching have been published. For example, in 2011, Panwei *et al.* proposed the parallel KMP algorithm by using MPI [1]. And in recent time, the implementation of KMP algorithm based on MPI and OpenMP also be proposed [2].

Current high-end graphics processing units (GPUs), contain up to hundreds cores per chip, are very popular in the high performance computing community. GPU is a massively multi-threaded processor and expects the thousands of concurrent threads to fully utilize its computing power. Another method to get a good efficient is using OpenMP. By using OpenMP, we can control CPU multi-thread at the same time. Therefore, we can divide a task into several sub-tasks and each thread manages one of sub-tasks. The ease of access GPUs by using Compute Unified Device Architecture (CUDA) [12], as opposite to graphic APIs, has made the supercomputing available to the mass. CUDA uses a new computing architecture, named Single Instruction Multiple Threads (SIMT), and SIMT is different from the Flynn's classification [9]. The advantages of the computing power and memory bandwidth from modern GPUs have made porting applications on it become a very important issue. For example, in the computational biology, several algorithms or tools have been ported on GPUs with CUDA, such as MSA-CUDA, MUMmerGPU, CUDA-MEME, and CUDA-BLASTP.

Hence, in this paper, we use the GPU and OpenMP API to parallel executing KMP algorithm, called KMP-GPU, to get a good performance. The main idea is that, when the problem has j patterns, we create i processes to control i GPU cards, and then we distribute j/i patterns to each GPU card and swap the data from the host (CPU) to the device (GPU). After that we execute the KMP algorithm on each GPU. In GPU computation regard, we built each pattern string's failure table at first and then do the string matching bit by bit between the text and pattern strings. When detecting a mismatch, then we will look up the failure table to shift the pattern's position. When the comparison finished, swapping the matching result back to the host. Finally, we could solve the string matching problem by multi-GPUs. In the experimental tests, a machine NVIDIA C1060 was used to execute and evaluate our KMP-GPU algorithm in terms of the numbers of pattern strings and the length of text string. The speedup was computed by compare the execution time between the CPU-based and GPU-based computations. The results showed that the speedups increase when the numbers of pattern strings increase. This paper is organized as follows. In section 2, preliminary concepts are described concisely. Section 3 introduces the implementations of KMP-GPU algorithm by using CUDA and OpenMP. Experimental results were shown in section 4.

2   Preliminary Concepts

2.1   Brute Force Algorithm

The simplest algorithm for string matching problem is the brute force algorithm, where we compare the first element of the pattern string P with the first element of the text string T. If the first element of P string matches the first element of T string, then compare the second element of P string with the second element of T string, and so on until the entire P string is found. If a mismatch is found at any position of P string, we shift the pattern string over one character and repeat the comparison again. The comparison steps for an example can be represented as follows:

```
T: ABDAABAADX
P: ABA          (step a)
   ABA          (step b)
    ABA         (step c)
     ABA        (step d)
      ABA       (step e)
```

The brute force algorithm is easy to understand and implement, however, its performance is not suitable for processing many pattern string or long pattern string. Too many unnecessary comparisons were done in the brute force algorithm, for example, the step b is an unnecessary comparison due to the position 2 of P string is different to the position 1 of P string and this information is known before the comparison. Therefore, many string matching algorithms were proposed to skip the unnecessary comparisons by do the preprocessing step for the P string.

2.2   Knuth-Morris-Pratt Algorithm (KMP Algorithm)

The KMP algorithm was conceived in 1974 by Knuth and Pratt, and independently by James H. Morris; published jointly in 1977. The KMP matching algorithm is achieved by avoiding unnecessary comparisons with the elements of T string that have previously been involved in the comparison with some elements of P string to be matched. In order to achieve the goal, it should to do some additional steps to build a failure table for the P string, such as shown in Figure 1(a). The failure table of the P

Pos.	1	2	3	4	5	6	7
P	a	b	a	a	b	c	a
$F(P)$	0	0	1	1	2	0	1

(a)

Pos.	1	2	3
P	A	B	A
$F(P)$	0	0	1

(b)

Fig. 1. The examples of failure table for the pattern string. The first row is the position of P string; the second row is the P string; the third row is the failure table $F(P)$ for the P string.

string encapsulates the knowledge about how the P string matches against the shifts of itself. With the failure table, we can start to compare the P string with the T string. When detecting a mismatch, we look up the failure table of the P string to know how many characters should be shifted. Consider the following example, assume that the P and T strings are represented as follows:

T = ABDABADX
P = ABA

We can build the failure table for the P string as shown in Figure 1(b). Using the KMP algorithm, we only need to compare three times for entire T string to find all possible matches, see below. This example briefly shows that the performance of KMP algorithm is better than brute force algorithm.

    T: ABDAABADX
    P: ABA   (step a)
          ABA   (step b)
             ABA   (step c)

2.3 CUDA Programming Model

CUDA is an extension of C/C++ which users can write scalable multi-threaded programs for GPUs computing. The implementation of CUDA program is divided into two parts: host and device. The host mainly is executing by CPU and the device is mainly executing by GPU. The program which is executed on the device called a *kernel*. The *kernel* can invoke as a set of concurrently executing threads, and the *kernel* program will be executed by threads. These threads are represented in a hierarchical organization which can be combined into a thread block contains many threads. Threads within a thread block can communicate through a *per-block shared memory* (*PBSM*), whereas threads in different thread blocks cannot communicate or synchronize directly. In addition to *PBSM*, there are four kinds of memory types: *per-thread private local memory* (*LM*), *global memory* (*GM*) for data shared by all threads, *texture memory* (*TM*), and *constant memory* (*CM*). Among these memory types, *CM* and *TM* can be regarded as fast read only caches; the fastest memories are *registers* and *PBSM*.

The basic processing unit in the NVIDIA's GPU architecture is called the *Streaming Processor* (*SP*). There are many *SPs* which actually do the computation on GPU. A group of *SPs* can be combined into a *Stream Multiprocessor* (*SM*). While the program runs the *kernel* function, the GPU device schedules thread blocks for the execution on the *SM*. The threads running on the *SM* in a small group 32, called warps, is the *SIMT* scheme, every *SM* has a warp scheduler to execute warps. For example, NVIDIA GeForce GTX260, there is a 16KB of *PBSM* for each *SM* with 16,384 32-bit registers. The number of thread blocks assigned to the *SM* is affected by the *registers* and *PBSM* used in a thread block. *SM* can be assigned up to 8 thread blocks. The *GM*, *LM*, *TM*, and *CM* are all located on the GPU's memory. In addition to *PBSM* accessed by single thread block and *registers* only accessed by single thread, the other memory can be used by all the threads. The caches of *TM* and *CM* are limited to 8KB per *SM*. The best access cache strategy for *CM* is all threads read the same memory address. The texture cache is designed for threads to read between the proximity of address would be take a better reading efficiency.

2.4   OpenMP Programming Mode

The OpenMP Application Program Interface (API) supports multi-platform shared-memory parallel programming in C/C++ and FORTRAN on all architectures, including UNIX and Windows NT platforms. Jointly defined by a group of major computer hardware and software vendors, OpenMP is a portable, scalable model that gives programmers (for shared-memory parallel programming) a simple and flexible interface to develop parallel applications within platforms ranging from the desktop to the supercomputers.

3   Method

In this paper, a KMP-GPU algorithm was designed for multi-GPUs by using CUDA and OpenMP to get a good performance. Assume that there are one text string T with a length of n and j pattern strings P_j with the length of m. The question is whether the pattern P_j exists in text string T or not. If P_j exists in the text string T, than we return the position of text string T. If not, we return 0.

As the mentioned above, the designed KMP-GPU algorithm is divided into following steps:

Step 1: j pattern strings are combined into a one dimension *pattern array*.

Step 2: We uses the OpenMP API to create 'i' processes (CPU thread). Each process controls a GPU card. Then we divide the *pattern array* into 'i' parts. Each part of *pattern array* is processed by a GPU card (a CPU thread).

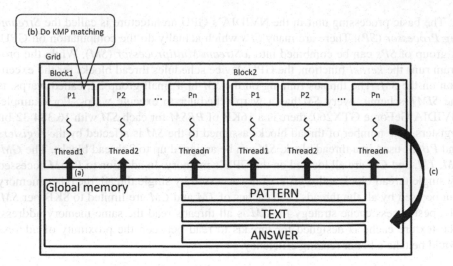

Fig. 2. The flowchart of KMP-GPU algorithm

Step 3: Allocate each GPU's *GM* space for the text string *T* and a part of pattern strings. After that we use each process moves one part of *pattern array* and the text string *T* from the host memory to the device memory

Step 4: For each GPU card, we create 'g' thread blocks; each thread block contains 'h' threads. Each thread has a unique ID (linearID) in a grid:
linearID = blockDim.x * blockIdx.x + threadIdx.x

Step 5: In a thread block, each thread processes a pattern string P_j to calculate its failure table and stored in a one-dimensional array.

Step 6: Run KMP algorithm in each thread to find the matching result and stored in a one-dimensional *result array*.

Step 7: Move the *result array* back to the host memory when finished all pattern strings.

The flowchart of KMP-GPU algorithm is shown in Figure 2. The pseudo-code of KMP-GPU is listed below.

GPU-Based KMP Algorithm (For each thread)

//Object: to find whether the pattern P_j exists in text string *T* or not. Return the position of text string *T* or 0.
//Load the text string *T* and *j* pattern strings P_j to the CPU memory and then transfers them to the *GM* of GPUs.

//blockDim.x is the built-in variable represents the size of block (number of threads in one thread block).

//blockIdx.x is the built-in variable represents the 1-D thread block index within the grid.

//threadIdx.x is the built-in variable represents the 1-D thread index within the thread block.

```
void KMP_kernel( char *P , char *T , int *answer)
{
    int linearId = blockDimx.x * blockIdx.x + threadIdx.x;
    char *position=P[linearID*patterSize];
    //To find the position of the pattern string Pj
        KMP( p , p.length , T , T.length , answer , linearId)
}
```

4   Performance Analysis

In this section, we will discuss the performance of CPU-based KMP algorithm and KMP-GPU algorithm. For each thread, it needs to calculate a failure table for assigned pattern string. The pseudo-code of failure table generation function is listed below.

Failure Table Generation Function (P)

```
1.      m ← Length[P]
2.      π[1] ← 0
3.      k ← 0
4.      for i ← 2 to m
5.          do while k>0 and P[k+1]≠P[i]
6.              do k ← π[k]
7.          if P[k+1] = P[i]
8.              then k ← k+1
9.          π[i] ← k
return π
```

The running time of failure table generation function is $O(m)$ by using the aggregate method of amortized analysis. The only tricky part is showing that **while** loop of lines 5-6 executes $O(m)$ times altogether. We shall show that it makes at most $m-1$ iterations. We start by making some observations about k. First, line 3 starts k at 0, and the only way that k increases is by the increment operation in line 8, which executes at most once per iteration of **for** loop of lines 4-9, Thus, the total increase in k is at most $m-1$. Second, since $k < i$ upon entering **for** loop and each iteration of loop increments i, we always have $k < i$. Therefore, the assignments in lines 2-9 ensure that $\pi[i] < i$ for all $i=1, 2, \ldots, m$, which means that each iteration of **while** decreases k. Third, k never becomes negative. Putting these facts together, we see that the total decrease in k from

while loop is bounded from above by the total increase in k over all iterations of **for** loop, which is m-1. Thus, **while** loop iterates at most m-1 times in all, and failure table generation function runs in time complexity $O(m)$ [11]. After the generation of failure table, each thread needs to process the assigned pattern string to find the matching result by the KMP algorithm. The pseudo-code of KMP matching function is listed below.

KMP Matching Function (T, P)

1. $n \leftarrow$ length[T]
2. $m \leftarrow$ length[P]
3. $\pi \leftarrow$ Failure table generation function (P)
4. $k \leftarrow 0$ //Number of characters matched.
5.   for $i \leftarrow 1$ to n //Scan the text string from left to right.
6.   do while $k > 0$ and $P[k + 1] \neq T[i]$
7.   do $k \leftarrow \pi[k]$ //Next character does not match.
8.   if $P[k + 1] = T[i]$
9.   then $k \leftarrow k + 1$ //Next character matches.
10.   if $k = m$ //Is all of T matched?
11.   then print the matching result
12. $k \leftarrow \pi[k]$ //Look for the next match.

From the similar arguments it follows that the KMP matching function requires $O(n)$ steps. The algorithm processes each character of the text string by doing a $O(1)$ computation and it will never go back. The only tricky part is showing that **for** loop in line 5 executes $O(n)$ times. Hence, the overall time complexity is $O(m + n)$.

Therefore, the time complexities for the CPU-based KMP algorithm and KMP-GPU algorithm are analyzed in the following, respectively.

CPU-Based KMP Algorithm

Assume that the numbers of pattern strings with equal length of m is p and the length of text string is n. If the CPU needs Tc seconds for doing an instruction, the time complexity of CPU-based KMP algorithm can be evaluated:

$$p \times O(m + n) \times Tc = O(pmTc + pnTc)$$

GPU-Based KMP Algorithm

Assume that the numbers of thread block is B and each thread block has T threads. Each thread in GPU takes Tg seconds for doing an instruction. Ideally, there are $B \times T$ threads can process the pattern strings concurrently. Therefore, the time complexity of KMP-GPU algorithm can be evaluated:

$$\frac{p}{B \times T} \times O(m + n) \times Tg$$

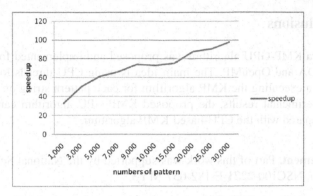

Fig. 3. The experimental speedups for KMP-GPU algorithm

Thus, the theoretical speedup can be determined:

$$\frac{p \times O(m+n) \times Tc}{\dfrac{p}{B \times T} \times O(m+n) \times Tg} = BT\frac{Tc}{Tg}$$

According to the above formula, the theoretical speedup increases when the number of threads increases. However, the number of threads is not infinite in practice. The number of threads in a thread block is bounded according to the memory usage and the number of SPs. Moreover, the number of concurrent thread blocks is bounded according to the number of SMs. Considering the Tc and Tg, in general, Tg is larger than Tc since the clock rate of CPU is faster than that of SP on GPU.

5   Experimental Results

The experimental tests were executed on an Intel Xeon CPU with a 2.27GHz clock rate and 4 GB of memory. The GPU cards used were the NVIDIA Tesla S1060 with four C1060 cards, each with 4GB of GM, 30 SMs and 240 SPs. The operating system used was Linux OS. The test data is a text string created randomly under a to z alphabet-set with a length of 100,000. The length of pattern string was set to 20 and more than 30,000 pattern strings are generated by randomly selected from the text string. The CUDA built-in variables gridDim.x (size of grid) and blockDim (size of block) were set to 256 and 128, respectively. The experimental results are shown in Figure 3 for various number of pattern strings. From Figure 3, we can see that the speedups increase when the numbers of pattern strings increase. When the number of pattern strings is small, the speedup is not good. The reason is that the number of threads (theoretical, 256×128) is larger than the number of pattern strings. Therefore, the most of computing power is wasted. When the number of pattern strings is larger than the number of threads, the KMP-GPU algorithm achieved 97x speedups compared with the CPU-based KMP algorithm.

6   Conclusions

In this paper, a KMP-GPU algorithm was proposed and implemented for multi-GPUs by using CUDA and OpenMP. The main idea is using CPU threads to control GPU cards, and then executing the KMP algorithm for each pattern string on GPU threads. From the experimental results, the proposed KMP-GPU algorithm can achieve 97x speedups compared with the CPU-based KMP algorithm.

Acknowledgement. Part of this work was supported by the National Science Council under the grant NSC100-2221-E-182-057-MY3.

References

1. Cao, P.: Parallel research on KMP algorithm. CECNet, 4253–4255 (2011)
2. Duan, G.: The implementation of KMP algorithm based on MPI and OpenMP. In: 9th International Conference on Fuzzy Systems and Knowledge Discovery, pp. 2511–2514 (2012)
3. Tumeo, A.: Accelerating DNA analysis applications on GPU clusters. In: 8th IEEE Symposium on Application Specific Processors, pp. 71–76 (2010)
4. Knuth, D.E., Morris, J.H., Pratt, V.R.: Fast pattern matching in strings. SIAM J. Comput. 6, 323–350 (1977)
5. Bayer, R.S., Moore, J.S.: A fast string searching algorithm. Communication of ACM, 762–772 (1977)
6. Cheng, L.L.: Approximate string matching in DNA sequences. In: 8th International Conference on Database Systems for Advanced Applications, pp. 303–310 (2003)
7. http://en.wikipedia.org/wiki/Rabin-Karp_string_search_algorithm
8. Peiravi, A.: Application of string matching in Internet security and Reliability. Journal of American Science, 25–3 (2010)
9. Flynn, M.: Some Computer Organizations and Their Effectiveness. IEEE Trans. Comput. C-21, 948 (1972)
10. SaiKrishna, V., Rasool, A., Khare, N.: String Matching and its Applications in Diversified Fields. International Journal of Computer Science Issues 9 (2012)
11. Cormen, T.H., Leiserson, C.E., Rivest, R.L., Stein, C.: Introduction To Algorithm. The MIT Press (2009)
12. Nickolls, J., Buck, I., Garland, M., Skandron, K.: Scalable parallel programming with CUDA. ACM Queue 6, 40–53 (2008)

Energy-Efficient Scheduling Based on Reducing Resource Contention for Multi-core Processors

Yan-Wei Chen, Mei-Ling Chiang, and Chieh-Jui Yang

Department of Information Management,
Nation Chi Nan University, Nantou, Taiwan, R.O.C.
{s98213522,joanna,s100213519}@ncnu.edu.tw

Abstract. Energy conservation is an important issue. In a multi-core and multi-processor system, some system resources are shared by processors, such as memory and processor's cache resources. When processors attempt to access shared resources at the same time, if processors cannot get the required shared resources immediately, it will result in the performance degradation and increased power consumption. In this paper, we have proposed a mechanism named energy-efficient scheduling which arranges tasks to run on processor cores in an appropriate way to avoid the access contention of shared resources and to save energy without sacrificing system performance too much.

The proposed energy-efficient scheduling is implemented in Linux kernel 2.6.33. The main work of implementation includes modification to the Linux kernel scheduler, the setting of the processor frequency, and the implementation of the system call for setting parameters of the system resources. Experimental results demonstrate that our energy-efficient scheduling can effectively save energy while avoiding the resource contention among processor cores to remain good performance.

Keywords: Linux, multi-core, scheduler, resource contention.

1 Introduction

Computer has become an indispensable tool for our daily life, and many jobs require a computer to improve operating speed in order to enhance the convenience of life. With the industrial and technological progress, the component of the processor becomes more and more sophisticated, and the clock rate is getting higher and higher. But with the higher clock rate, it will result in processor overheating and higher power consumption. The solution of handling overheating must take more cost of material, so manufacturers gradually focus on developing parallel processing with multi-core processor such as dual-core processor, quad-core processor and, eight-core processor. Manufacturers improve the processor performance and the ability of parallel processing by increasing the number of processors. However, the power consumption of such multi-core processor product is very high. Nowadays, the energy saving is an important issue, therefore how to achieve the higher processor performance indicator as much as possible with lesser electricity and energy has become a topic.

J.-S. Pan et al. (Eds.): *Advances in Intelligent Systems & Applications*, SIST 21, pp. 553–562.
DOI: 10.1007/978-3-642-35473-1_55   © Springer-Verlag Berlin Heidelberg 2013

In recent years, there are many researches about reducing power consumption, and they can be broadly divided into two types. The first type is that operating system bases on different task requirement to switch the suitable frequency of the processor to achieve energy saving. The newer processors usually have supported processor clock frequency switching, like Intel's Speedstep technology. Processor switching to lower clock frequency can reduce voltage, and operating system can work with lower power consumption. While system idling, switching to the lowest processor frequency is of a significant energy saving effect. However, the processor running at lower frequency may result in the longer task execution time. Therefore, switching to suitable processor frequency according to different type of tasks can achieve better ratio of performance to energy consumption. At present, latest version of Linux has implemented subsystem that can dynamically adjust the processor frequency depending on system loading. There are many user-level applications that also provide functions to dynamically adjust the processor frequency. The second type is that the operating system scheduler additionally considers processor resources (such as cache, execution units, etc.) while dispatching tasks to processors. Under multi-core processors architecture, operating systems regard each processor core as separate processor, but in fact some processor cores still have dependency because of using sharing resources such as cache, execution units, memory, etc.

In this paper, we discuss how to reduce system power consumption through the operating system scheduler. In the multi-core multi-processor system, according to the literature [11] study, each processor at the same physical package may access the same processor resource or memory. The system performance will decline as a result of resource contention. In this paper, we regard memory, processor cache resources as the critical resource that must be properly handled. We propose an energy-efficient scheduling mechanism which consists of processor frequency settings and operating system scheduling considerations of memory and processor cache resources. Through the modification of the scheduler in the Linux operating system, our energy-saving scheduling implementation can dispatch different types of the tasks to appropriate processors, and it also reduces the power consumption and negatives the impact incurred by the resource contention. It is expected to reduce power consumption without lowering the system performance as much as possible.

2   Background and Related Work

2.1   Energy Saving with Reduced Resource Contention

In the multi-core and multi-processor architecture, the processors would share some system resource. This may cause resource contention between processors if running tasks on different processors require the same resource, and it will not only extend task execution time but also cost unnecessary power consumption.

A. Merkel and F. Bellosa [10] proposed a mechanism based on Runqueue Sorting and Frequency Heuristic. The objective of Runqueue Sorting is to decrease memory contention between cores by combing complementary tasks such as memory-bound tasks and compute-bound tasks. When processor executes a task for a time slice, it will check memory intensity in this period. The scheduler will sort tasks in the runqueue according to their memory intensity. In the even-numbered processors, tasks are sorted with descending memory intensity in the runqueues of processor cores, while in the odd-numbered processors, tasks are sorted with ascending memory intensity like Figure 1. Runqueue Sorting mechanism ensures the task scheduler to properly schedule memory-bound tasks and compute-bound tasks, while avoiding memory contention.

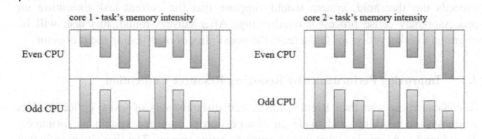

Fig. 1. Runqueue Sorting

Frequency Heuristic aims at properly switching the frequency of processor according to the different conditions. A memory threshold is set for determining task's memory intensity. If a task's memory intensity exceeds the threshold, it is classified as memory-bound task; otherwise it is compute-bound task. In the memory-bound task case, since processor spends most of its time slices to access memory instead of computing, so we can switch the processor frequency to a lower frequency. In the compute-bound task case, since it requires more processor computing ability, so we can switch processor frequency to a higher frequency. Through Runqueue Sorting and Frequency Heuristic, system reduces performance slightly whereas it can minimize energy consumption.

Besides, A. Merkel and F. Bellosa [11] proposed a load-balancing policy which migrates appropriate tasks among processors to balance resource utilization. System will examine Active Vector when performing load balancing. There are many columns in Active Vector, and each column represents a kind of sharing resource. The value of the column stands for the quantification which is required by the task. Every task has its own Active Vector, and processor also has its Active Vector which is the sum of all tasks' Active Vector in the runqueue. When system decides to perform load balancing, it will choose one destination processor and count the difference between destination's and current processor's Active Vector. Then system first will evaluate the difference between destination's and current processor's Active Vector after performing load balancing. If the difference of Active Vector would decrease after load balancing is performed, it is allowed to migrate tasks from current processor to destination processor. This load-balancing mechanism lets system resource be utilized in a more balanced manner and makes Runqueue Sorting and Frequency Heuristic mechanisms more efficient.

2.2   Thermal-Aware Scheduler

Processor's power dissipation is directly proportional to temperature. When processor executes under high frequency, its power dissipation and temperature remain increasing. While the processor temperature exceeds the critical threshold, the power dissipation would increase greatly. Liang Xia and Yongxin Zhu proposed a Thermal-Aware Scheduler[8] based on the round robin scheduling algorithm. System will determine every processor's temperature at every timer tick, when the temperature exceeds the threshold, system would suppose that the current task executing on processor may cause processor overheating. After current round, this task will be migrated to other processor to decrease the possibility of overheating the processor.

2.3   Improving Performance by Reducing Resource Contention

Rob Knauerhase and Paul Brett[9] has tested different sets of tasks, collected data while system working, and made an observation to improve system performance. They found two situations that would cause negative impact. The first situation is that processors compete with each other in accessing cache memory; the second, system does not consider the usage of shared resource among processors while performing load balancing. By this observation, this research schedules tasks in a way to avoid some unsuitable tasks sets running at the same time.

The other way is when the scheduler schedules tasks, it will consider other system resources such as bus transaction, stall cycle rate, and last level cache miss rate [12]. Each processor will choose tasks at the same time to form a better execution task set. Task set means those tasks in a set would be complementary to each other. A good task set will achieve better resource usage and improve system performance.

3   The Proposed Energy-Efficient Scheduling Mechanism

This section will describe the design and implementation of the proposed energy-efficient scheduling including assigning a ready-to-run task to the proper processor, sorting the tasks order in runqueue of the processor, modifying the load-balancing mechanism of Linux kernel, and setting the processor frequency.

3.1   Setting System Resource Usage in a Task

In multi-core multi-processor architecture, memory contention is of great impact on system performance, so we regard memory as an important resource in our work. We modify the *task_struct* data structure which maintains the information of a process in Linux kernel by adding a new variable called *mem* that stands for the degree of memory accessing. We also add a new system call to create a new task in the way similar to *fork()* system call, whereas, the task is given its *mem* value. The *mem* value ranges from 0 to 100. If the value is higher than 50, we regard this task as a memory-bound task, otherwise compute-bound task. Based on the *mem* value, the scheduler can determine whether the task should be assigned to odd-numbered or even-numbered processor, and sorts the task executing order in the runqueue.

3.2   System Hardware

In our study, we use ASUS TS500-E6 server which includes two Intel Xeon E5620 processors and 8G memory. Xeon E5620 is a quad-core processor and supports Hyper-threading technology [4]. There will be eight logical processors under Hyper-threading technology, and the system thinks each logical processor as an independent entity. Xeon E5620 also supports Intel SpeedStep technology which provides eight frequencies to switch, from high to low are 2401MHZ, 2400MHZ, 2267MHZ, 2133MHZ, 2000MHZ, 1867MHZ, 1733MHZ, and 1600MHz.

The cores in one physical package share L3 cache, and two logical processors in one physical processor will share L1, L2 cache and execution units. Hence our energy-efficient scheduling lets memory-bound tasks execute on even-numbered processor and compute-bound tasks execute on odd-numbered processor in order to avoid the situation that two logical processors in the same physical processor execute memory-bound tasks at the same time. Because executing memory-bound tasks at the same time may cause memory contention, the better way of task execution is to combine memory-bound tasks with compute-bound tasks. Running compute-bound task requires more execution units such as ALU, FLU, but less memory accessing. Processor executes complementary tasks could increase resource utilization.

3.3   Setting Processor Frequency

In our study, we use an ASUS TS500-E6 server of two quad-core processor with Hyper-threading technology. Therefore, there are 16 logical processors under system architecture. We set frequency of each processor according to its processor number. Basically, if the number is odd, this processor would execute memory-bound tasks, so we set lower frequency because memory-bound tasks do not need complicated computing. While the processor number is even, we set the higher frequency.

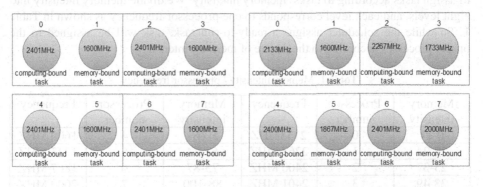

(a) Coarse-grained setting   (b) Fine-grained setting

Fig. 2. Two ways of setting processor frequency

We have implemented two ways of setting processor frequency as follows.

- Coarse-grained setting (i.e. only highest and lowest frequencies): Odd-numbered processor runs memory-bound tasks with lowest frequency, and even-numbered processor runs compute-bound tasks with highest frequency (2401MHz), as shown in Figure 2(a).
- Fine-grained setting (i.e. eight frequency levels): Odd-numbered processor runs with lower frequency, and even-numbered processor runs with higher frequency, as shown in Figure 2(b). Tasks are assigned to processors according to their memory intensity, such that tasks of higher memory intensity would be assigned to processors running with lower frequency.

3.4   Modification of Linux Kernel

In order to let memory-bound tasks and computing-bound tasks be assigned to odd--numbered and even-numbered processors, we modified some of the Linux kernel functions used by conventional tasks (i.e. non-real-time tasks) in *sched.c* and *sched_fair.c*. There are two settings proposed as we mentioned in Section 3.3. The way that both settings assign tasks and set processor frequency is different, but the way they sort tasks and perform load balancing is the same.

In the coarse-grained setting, we modify Linux kernel functions *find_idlest_group()* and *find_idlest_cpu()*. *Find_idlest_group()* starts to search idlest CPU group from the top of scheduling domain and returns the idlest CPU group. We devise two variables to separately record the sum of the loading of all odd-numbered processors and the sum of the loading of all even-numbered processors in the CPU group while calculating the loading of each CPU group. For example, if a compute-bound task is ready to run, we will compare all CPU groups' sum of the loading of all even-numbered processors and choose the CPU group whose sum of the loading is the lowest. Then, we find the idlest even-numbered processor in the CPU group through *find_idlest_cpu()* function.

In the fine-grained setting, we modify Linux kernel *select_task_rq_fair()* function to assign tasks according to tasks' memory intensity. We divide memory intensity into eight levels, and each level corresponds to one processor frequency as shown in Table 1. So while the scheduler assigns a ready-to-run task, task will be assigned to the proper processor according to the degree of memory intensity.

Table 1. Memory intensity / processor frequency

Memory intensity	Processor number	Frequency	Memory intensity	Processor number	Frequency
0-11	0	2133MHZ	50-61	4	1600 MHZ
12-24	1	2267 MHZ	62-74	5	1733 MHZ
25-37	2	2400 MHZ	75-87	6	1867 MHZ
38-49	3	2401 MHZ	88-100	7	2000 MHZ

3.5   Load Balancing of Energy-Efficient Scheduling

In Linux kernel, there are three situations to perform load balancing. The first is assigning a new task to the proper processor as mentioned in Section 3.4. The second one is when system performs load balancing periodically. The third one is migrating tasks to the idle processor. In our implementation, we assign compute-bound tasks to even-numbered processors and memory-bound tasks to odd- numbered processors. To avoid assigning tasks to unsuitable processors after performing load balancing, we modify load-balancing related functions of Linux kernel.

When timer interrupt occurs, Linux kernel will check if there is a CPU group with unbalanced loading in every scheduling domain [1]. If system is unbalanced, the kernel will invoke kernel functions *find_busiet_gruop()* and *find_busiest_cpu()* to find the busiest processor. Then, it will choose tasks from the busiest processor and migrate them to other processor.

In our load-balancing mechanism, we devise two variables to separately record the sum of the loading of all odd-numbered processors and the sum of the loading of all even-numbered processors in the CPU group while calculating the loading of each CPU group. When we want to balance the loading of processing memory-bound tasks, we compare the sum of the loading of all odd-numbered processors among CPU groups. The CPU group with the maximal sum of loading is the busiest CPU group. After finding the busiest CPU group, we start to determine every odd-numbered processor's loading in the busiest CPU group. The odd-numbered processor with the highest loading is the busiest processor in the group, so we will move its tasks to other odd-numbered processor of other CPU group. In other case, compute-bound tasks will be migrated to other even-numbered processor. The original processing flow of *find_busiest_group()* and *find_busiest_queue()* may also be used under this case when there are lots of memory-bound tasks executing on the odd-numbered processors and the even-numbered processor is idle because of no compute-bound tasks. In this case, we let memory-bound tasks be migrated to even-numbered processors.

4   Performance Evaluation

4.1   Experimental Environment

We use ASUS TS500-E6 server which consists of two E5620 quad-core processors supporting Hyper-threading technology and 8G memory, as described in Section 3.2. E5620 processor is based on Nehalem [5] and it's also NUMA [6] architecture.

4.2   Experimental Design

In order to evaluate the performance of our energy-efficient scheduling mechanism, we choose SPEC CPU2006 [3] as our benchmark. SPEC CPU2006 has 29 types of benchmarks. We design three test cases as follow. In case 1, the system runs 16 memory-bound tasks and 16 compute-bound tasks at the same time. In case 2, the system runs 12 memory-bound tasks sand 20 compute-bound tasks. In case 3, the system runs 20 memory-bound tasks and 12 compute-bound tasks.

While running benchmark, we use CM-9930R [7] to record the voltage usage of our server once per second and estimate the scaling of EDP (Energy Delay Product) [10]. We then compare the EDP among 3 test cases under the original Linux kernel 2.6.33 and the modified Linux kernel 2.6.33 running our energy-efficient scheduling.

4.3   Experimental Results

Figures 3 to 5 show the comparison of performance, EDP, and power consumption between the original Linux kernel 2.6.33 without energy-saving mechanism and the modified Linux kernel 2.6.33 running our energy-efficient scheduling mechanism. Four energy-saving settings for setting processor frequency are evaluated. Among them, "ondemand" stands for the frequency setting with ondemand configuration [2] which switches processor frequency dynamically according to the processor usage under CPUfreq subsystem, "conservation" stands for the frequency setting with conservative configuration [2] which switches processor frequency like "onedemand" does but more conservatively. "two-kind" stands for the proposed coarse-grained setting, and "eight-kind" stands for the proposed fine-grained setting as mentioned in Section 3.3. Performance is the time measured in executing benchmarks. Power consumption is the sum of power consumption for the period of executing tasks. The performance value multiplied by the sum of power consumption is EDP. If the ratio is less than 1, the system running with energy-efficient scheduling performs better; otherwise, it is worse than the original unmodified Linux kernel.

In case 1 that the amount of memory-bound tasks is equal to compute-bound tasks, the system running energy-efficient scheduling with the proposed fine-grained setting (i.e. eight-kind) performs best in performance data, EDP, and power consumption. It means execution time and power consumption are reduced at the same time.

In case 2 that there are more compute-bound tasks, the system running energy-efficient scheduling (i.e. two-kind or eight-kind) performs better than the original-Linux kernel running without energy-saving mechanism in EDP and power consumption. The execution time of benchmark is slightly longer because if there is an idle odd-numbered processor, both fine-grained setting and coarse-grained setting will assign compute-bound tasks to that idle processor in order to balance the load. Since odd-numbered processor remains running under lower frequency, so those migrated compute-bound tasks' finish times are extended. The fine-grained setting (i.e. eight-kind) performs better than coarse-grained setting (i.e. two-kind) in performance data. Because sometimes memory-bound tasks still need complicated computing, so let memory-bound tasks be executed under lower frequency instead of the lowest frequency will be better. When running under lower frequency, the execution time of memory-bound task is also extended. The system running energy-efficient scheduling with the proposed coarse-grained setting (i.e. two-kind) performs best in power consumption.

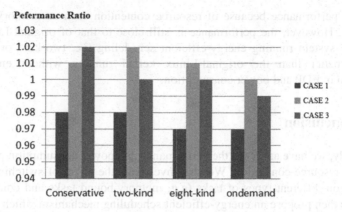

Fig. 3. Performance comparison relative to standard Linux scheduling

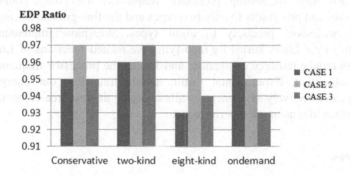

Fig. 4. Comparison of EDP relative to standard Linux scheduling

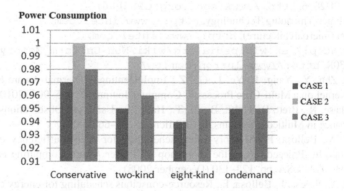

Fig. 5. Comparison of power consumption relative to standard Linux scheduling

In case 3, the amount of memory-bound tasks is more than compute-bound tasks. In fine-grained setting and coarse-grained setting, for load balancing, some memory-bound tasks are migrated to even-numbered processors and it will cause inefficient performance because of resource contention such as memory and cache contention. However, the performance is still close to that of original-Linux kernel. Again, the system running energy-efficient scheduling (i.e. two-kind or eight-kind) performs better than the original-Linux kernel running without energy-saving mechanism in EDP and power consumption.

5   Conclusion

In our study, we have analyzed the performance and power consumption phenomenon caused by resource contention. We also investigate the effect of switching processor frequency on different types of tasks (e.g. memory-bound tasks and compute-bound tasks). We then propose an energy-efficient scheduling mechanism which can classify tasks by their resource requirement and implement it in Linux kernel 2.6.33. We have designed two ways of setting processor frequency: the coarse-grained setting classifies tasks and processors to only two types and the fine-grained setting classifies tasks and processors precisely to eight types. We have implemented these mechanisms in the Linux kernel by modifying the related functions of Linux kernel scheduler to reduce resource contention, and setting the processor frequency through CPUfreq subsystem. Experimental results demonstrate that our energy-efficient scheduling can effectively save energy while avoiding the resource contention among processor cores to remain good performance.

References

1. Scheduling domains, http://lwn.net/Articles/80911/
2. Linux kernel CPUfreq subsystem, http://www.kernel.org
3. SPEC CPU2006, http://www.spec.org/CPU2006/
4. IntelR Hyper-Threading Technology, http://www.intel.com/
5. Nehalem (microarchitecture), http://www.intel.com/
6. NUMA, http://en.wikipedia.org/wiki/Non-Uniform_Memory_Access
7. CM-9930R, http://www.lutron.com.tw/
8. Xia, L., Zhu, Y., Yang, J., Ye, J., Gu, Z.: Implementing a Thermal-Aware Scheduler in Linux Kernel on a Multi-Core Processor. Computer Journal 53, 895–903 (2010)
9. Knauerhase, R., Brett, P., Hohlt, B., Li, T., Hahn, S.: Using OS Observations to Improve Performance in Multicore Systems. IEEE Micro 28, 54–66 (2008)
10. Merkel, A., Bellosa, F.: Memory-aware scheduling for energy efficiency on multicore processors. In: Proceedings of the Workshop on Power Aware Computing and Systems, San Diego, CA, USA, pp. 123–130 (December 2008)
11. Merkel, A., Stoess, J., Bellosa, F.: Resource-conscious scheduling for energy efficiency on multicore processors. In: Proceedings of the 5th European Conference on Computing Systems, Paris, France, pp. 153–166 (April 2010)
12. McGregor, R.L., Antonopoulos, C.D., Nikolopoulos, D.S.: Scheduling Algorithms for Effective Thread Pairing on Hybrid Multiprocessors. In: Proceedings of the 19th IEEE International Parallel and Distributed Processing Symposium (IPDPS 2005), Denver, CO, USA, pp. 47–58 (April 2005)

Effective Processor Allocation for Moldable Jobs with Application Speedup Model

Kuo-Chan Huang[1], Tse-Chi Huang[1], Yuan-Hsin Tung[2], Pin-Zei Shih[2]

[1] Department of Computer Science
National Taichung University of Education
No. 140, Min-Shen Road, Taichung, Taiwan
kchuang@mail.ntcu.edu.tw, rogevious@gmail.com
[2] Chunghwa Telecommunication Laboratories
12, Lane 551, Min-Tsu Road Sec. 5, Taoyuan, Taiwan 32617, R.O.C.
{yhdong,miashih}@cht.com.tw

Abstract. Traditionally, users who submit parallel jobs to supercomputing centers need to specify the amount of processors that each job requires. Job schedulers then allocate resources to each job according to the processor requirement. However, this kind of allocation has been shown leading to degraded system utilization and job turnaround time when mismatch between requirement and available resources occurs. System performance could be improved through the *moldable* property which most current parallel application programs have. With moldable property, parallel programs can exploit different parallelisms for execution at runtime. Previous research has shown potential performance improvement achieved by adaptive processor allocation based on the moldable property. This paper proposes effective processor allocation methods for moldable jobs, which can dynamically determine an appropriate number of processors for each job according to its speedup model and current workload situations. We evaluated the proposed approaches under three different usage scenarios through a series of simulation experiments. The experimental results indicate that our approaches outperform existing methods significantly, achieving up to 69%, 89%, and 98% performance improvement for the three usage scenarios, respectively.

Keywords: moldable property, adaptive processor allocation, application speedup model.

1 Introduction

Parallel job scheduling and allocation has long been an important research topic [1][3][4]. Users at traditional supercomputing centers usually need to specify the amount of processors required when submitting a parallel job. The workload management system will allocate computing resources to each job according to the specified processor requirement. If the specified amount of processors is larger than current available resources, the job would have to wait while the available resources are kept idle, resulting in degraded system utilization and job turnaround time.

J.-S. Pan et al. (Eds.): *Advances in Intelligent Systems & Applications*, SIST 21, pp. 563–572. .
DOI: 10.1007/978-3-642-35473-1_56   © Springer-Verlag Berlin Heidelberg 2013

However, the above situation is not necessary for most kinds of parallel applications. Only *rigid* jobs [16], which can only run with a specific number of processors, should strictly follow the above job submission procedure. On the other hand, system performance could be improved through the *moldable* property [16] which most current parallel application programs have. With moldable property, parallel programs can exploit different parallelisms for execution at runtime. Thus, users and job schedulers have the flexibility to select a suitable amount of processors for job execution.

Previous research [14] has shown potential performance improvement achieved by adaptive processor allocation based on the moldable property. The proposed adaptive processor allocation methods in [14] dynamically determine the number of processors to allocate just before job execution according to the amount of current available resources and job queue information. In this paper, we propose effective processor allocation methods for moldable jobs, which take into consideration of application speedup models [6] when making the decision of processor allocation. We adopt Downey's speedup models for parallel programs [6] in the proposed processors allocation methods. Downey's models have been shown capable of representing the workload characteristics of both real parallel applications and entire system workload [6][7], such as NAS benchmarks [17], SDSC workload [15], and CTC workload [15].

We conducted a series of simulation experiments to evaluate the proposed approaches under three different usage scenarios. In the first scenario, the application speedup model is used to automatically determine the most suitable amount of processors for a parallel program, relieving the burden of specifying a specific number of processors on job submission for users. In the second scenario, a user rents a virtual cluster from a cloud service provider for processing a large set of parallel computing jobs. The characteristic of this scenario is that all the jobs to be processed are ready from the very beginning, at time zero, and thus the system completion time is used to measure the overall system performance. In the third scenario, a cluster system is used to serve a lot of users who may submit different jobs at distinct time instants. Therefore, the cluster system has to deal with an online job scheduling and allocation problem. In this case, the performance metric is the average turnaround time of all jobs. The experimental results indicate that our approaches outperform existing methods significantly, achieving up to 69%, 89%, and 98% performance improvement for the three usage scenarios, respectively.

The remainder of this paper is organized as follows. Section 2 discusses the related work of application speedup models and moldable job scheduling. Section 3 presents the proposed processor allocation methods for moldable jobs. Section 4 evaluates the proposed methods with a series of simulation experiments. Section 5 concludes this paper.

2   Related Work

Upon job submission, if the specified amount of processors is larger than current available resources, the job would have to wait while the available resources are kept idle, resulting in degraded system utilization and job turnaround time. For rigid jobs

[16], such problem is usually dealt with by backfilling job scheduling approaches [2][5]. For moldable jobs [16], more flexible approaches are possible since the jobs can run with different parallelisms.

Cirne and Berman proposed an application-level scheduling approach for moldable jobs in [8,9], which is similar to our first usage scenario. Users provide a set of candidate requests with different processor requirements, and the application scheduler is used to adaptively select the most suitable request based on current system configuration and workload status. In this paper, our approaches do not require users to provide such candidate requests. On the other hand, the approaches automatically determine the most appropriate amount of processors to use based on the application speedup model.

In [10], Sabin et al. proposed an iterative algorithm which utilizes job efficiency information for scheduling moldable jobs. The proposed algorithm has higher computational complexity than our method since it is an iterative approach. In [11][12], Srinivasan et al. proposed a schedule-time aggressive fair-share strategy for moldable jobs, which adopts a profile-based allocation scheme. The strategy keeps track of the information about all the free-time blocks available in current schedule and scans all the blocks to find the most suitable one for a moldable job at each scheduling activity, considering the effects of partition size on the performance of the application. This strategy thus needs to have the knowledge of job execution time. On the other hand, our approaches do not require the information of job execution time. Only the parameters of application speedup models are needed to dynamically determine the appropriate number of processors, and the decision is made at job starting time instead of schedule time. Sun et al. proposed an adaptive scheduling approach for malleable jobs with periodic processor reallocations based on parallelism feedback of the jobs and allocation policy of the system in [13]. Malleable jobs [16] are similar to moldable jobs in that they both can run with different parallelisms in contrast to rigid jobs. However, malleable jobs are even more flexible in that they can change the amount of processors used dynamically during execution, while moldable jobs must determine the number of processors to use before execution and then fix the amount of processors throughout the entire execution period.

Huang proposed and evaluated four adaptive processors allocation heuristics for moldable jobs in [14]. The heuristics determine the amount of processors to use for each moldable job when it becomes the first job in the waiting queue. Only current available resources and job queue information are considered when making processor allocation decisions in the heuristics. In this paper, we extend the proposed heuristics to take into consideration of the application speedup model, aiming at improving the overall system performance further.

3   Effective Processor Allocation with Application Speedup Model

This section presents the effective processor allocation methods for moldable jobs taking advantage of application speedup models.

3.1   Speedup Model of Parallel Programs

Our processor allocation approaches adopt Downey's speedup model of parallel programs to take into consideration of both single job speedup and entire system performance. The model developed by Downey has been shown capable of representing the parallelism and speedup characteristics of real parallel applications [6][7].

The speedup of a job on n processors is defined as the ratio of the job's run time on a single processor to the job's run time on n processors:

$$S(n) = \frac{L}{T(n)} \tag{1}$$

Here, S is the speedup function, L is the effective sequential run time and $T(n)$ is the run time of the job on n processors. Downey's model is a non-linear function of two parameters [6]:

- σ *(sigma)* is an approximation of the coefficient of variance in parallelism within the job. It determines how close to linear the speedup is. A value of zero indicates linear speedup and higher values indicate greater deviation from the linear curve.
- A denotes the average parallelism of a job and is a measure of the maximum speedup that the job can achieve.

Downey proposed two speedup models with low and high variances, respectively in [6]. Figure 1 is a hypothetical parallelism profile for a program with low variance in degree of parallelism. The parallelism is equal to A, the average parallelism, for all but some fraction σ of the duration ($0 \leq \sigma \leq 1$). The remaining time is divided between a sequential component and a high-parallelism component (with parallelism chosen such that the average parallelism is A). The variance of parallelism is $V = \sigma (A - 1)^2$. The run time and speedup of a parallel program, as functions of cluster size, with the low-variance model are described in equations (2) and (3), respectively.

In the low variance profile have the following run time as functions of cluster size:

$$T(n) = \begin{cases} \dfrac{A - \sigma/2}{n} + \sigma/2 & 1 \leq n \leq A \\[2mm] \dfrac{\sigma(A - 1/2)}{n} + 1 - \sigma/2 & A \leq n \leq 2A - 1 \\[2mm] 1 & n \geq 2A - 1 \end{cases} \tag{2}$$

$$s(n) = \begin{cases} \dfrac{An}{A + \dfrac{\sigma}{2(n-1)}} & 1 \leq n \leq A \\[3mm] \dfrac{An}{\sigma\left(A - \dfrac{1}{2}\right) + n(1 - \sigma/2)} & A \leq n \leq 2A - 1 \\[3mm] A & n \geq 2A - 1 \end{cases} \tag{3}$$

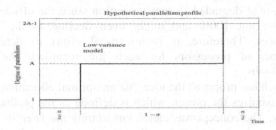

Fig. 1. The parallelism profile for low-variance speedup model

Figure 2 shows a hypothetical parallelism profile for a program with high variance in parallelism. The profile consists of a sequential component of duration σ and a parallel component of duration 1 and potential parallelism $A + A\sigma - \sigma$. A program with this profile would have the following run time and speedup as functions of cluster size, described in equations (4) and (5), respectively:

$$T(n) = \begin{cases} \sigma + \dfrac{A + A\sigma - \sigma}{n} & 1 \le n \le A + A\sigma - \sigma \\ \sigma + 1 & n \ge A + A\sigma - \sigma \end{cases} \quad (4)$$

$$S(n) = \begin{cases} \dfrac{nA(\sigma + 1)}{\sigma(n + A - 1) + A} & 1 \le n \le A + A\sigma - \sigma \\ A & n \ge A + A\sigma - \sigma \end{cases} \quad (5)$$

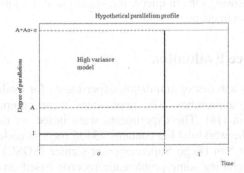

Fig. 2. The parallelism profile for high-variance speedup model

3.2   Effective Processor Allocation for Moldable Jobs

It is easy to speed up a single moldable job and usually can be achieved by giving the job more processors. However, processor allocation of moldable jobs often faces the dilemma of whether to increase a job's speedup as large as possible or not, since such speedup of a job might lead to enlarged turnaround time of another because the total number of processors in a system is usually fixed. Moreover, the speedup might be achieved at the cost of degraded system utilization since the efficiency of a parallel program is usually not 100% and might even decline as the number of used processors increases. Therefore, it is no trivial effort to determine the most appropriate number of processors for each job regarding the overall system performance of all jobs.

Previous research has proposed the idea that an optimal allocation for a parallel job is the one that maximizes the power, which is defined as the product of the speedup and the efficiency. The concept was called *calculating the knee* in [6]. The work in [14] proposed three adaptive processor allocation heuristics: *adaptive scaling down*, *adaptive scaling up and down*, *restricted scaling up and down*; considering only system workload and available resources. The experimental results in [14] show that the two heuristics with scaling up allocation in general outperform the adaptive scaling down heuristic. In this paper, we extend the two heuristics in [14] and adopt the above concept of knee in determining the number of processors to allocate each moldable job, as described in the following.

- *Adaptive scaling up and down protected.* When a parallel job becomes the first job in the waiting queue, if its originally specified number of processors is larger than the number of free processors, instead of keeping it waiting in queue, the system automatically scales the job down to use exactly the number of free processors. On the other hand, if the number of free processors is larger than the job's specified amount, the system automatically scales the job's amount of processors up to the minimum of total free processors and the optimal value determined by calculating the *knee* based on the job's speedup model.

- *Restricted scaling up and down protected.* This is a restricted version of the previous approach. In case that scaling up a job would in turn delay the start time of the following jobs in queue, the system scales a job up only if there are no jobs behind it in queue.

4   Performance Evaluation

This section presents a series of simulation experiments for evaluating the proposed processor allocation approaches. The approaches were compared with previous heuristics proposed in [14]. The experiments were based on the workload log on SDSC's SP2 [15]. The workload log contains 73496 records collected on a 128-node IBM SP2 machine at San Diego Supercomputer Center (SDSC) from May 1998 to April 2000. After excluding some problematic records based on the *completed* field [15] in the log, the simulations in this paper use 56490 job records as the input workload. The workload log was used to set up the job arrival time and job sizes. The two parameters, σ and A, for Downey's speedup models were generated randomly. The workload was assumed to be executed on a 128-processor cluster. In the following figures, *subtime* represents the submission times of jobs and log subtime indicates that the submission times of jobs were set according to the corresponding values in SDSC's SP2 log. The *load* is a parameter used to adjust the system loading by multiplying it and the original job runtime together.

Figure 3 shows the performance comparison for usage scenario I. In this scenario, the application speedup model is used to automatically determine the most suitable amount of processors for a parallel program, relieving the burden of specifying a specific number of processors on job submission for users. The results indicate that determining the amount of processors based on the knee calculation of the application speedup models, low variance speedup model and high variance speedup model in the figure, significantly outperforms the traditional approach where the amount of processors to used is decided arbitrarily by users, achieving up to 69% performance improvement in terms of average turnaround time. The performance improvement was calculated by dividing the performance difference between the original method and our method by the performance of the original method.

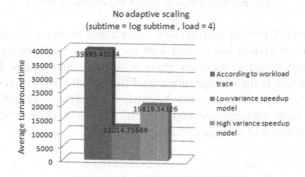

Fig. 3. Usage scenario I

Figures 4 and 5 present performance evaluation of the proposed approaches in usage scenario II. In this scenario, a user rents a virtual cluster from a cloud service provider for processing a large amount of parallel computing jobs. The characteristic of this scenario is that all the jobs to be processed are ready at time zero, and thus the system completion time is used to measure the overall system performance. The results demonstrate that our approaches outperform previous heuristics in [14] significantly for both low variance and high variance speedup models, achieving up to 89% performance improvement in terms of system completion time.

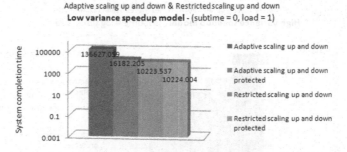

Fig. 4. Usage scenario II with low variance speedup model

Adaptive scaling up and down & Restricted scaling up and down
High variance speedup model - (subtime = 0, load = 1)
System completion time
100000
1000
10
0.1
0.001
133367.476
14651.539
6045.054
6033.478
Adaptive scaling up and down
Adaptive scaling up and down protected
Restricted scaling up and down
Restricted scaling up and down protected

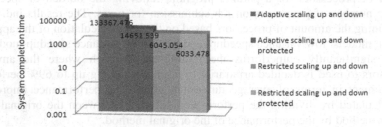

Fig. 5. Usage scenario II with high variance model

Figures 6 and 7 present performance comparison in usage scenario III. In this scenario, a cluster system has to deal with an online job scheduling and allocation problem where different jobs usually arrive at distinct time instants. In this case, the performance metric is the average turnaround time of all jobs. The results reveal that our approaches outperform previous heuristics in [14] significantly for both low variance and high variance speedup models, achieving up to 98% performance improvement in terms of average turnaround time.

Adaptive scaling up and down & Restricted scaling up and down
Low variance speedup model - (subtime = log time, load = 4)
Average turnaround time
100000
1000
10
0.1
0.001
18683.096
472.879
54.570969
31.33597
Adaptive scaling up and down
Adaptive scaling up and down protected
Restricted scaling up and down
Restricted scaling up and down protected

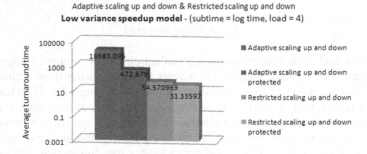

Fig. 6. Usage scenario III with low variance model

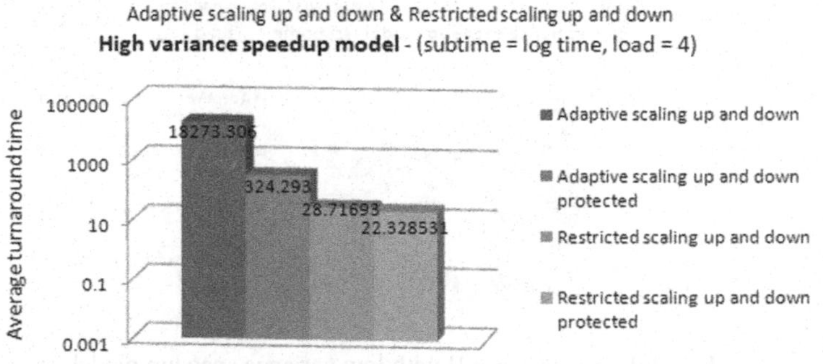

Adaptive scaling up and down & Restricted scaling up and down
High variance speedup model - (subtime = log time, load = 4)
Average turnaround time
100000
1000
10
0.1
0.001
18273.306
324.293
28.71693
22.328531
Adaptive scaling up and down
Adaptive scaling up and down protected
Restricted scaling up and down
Restricted scaling up and down protected

Fig. 7. Usage scenario III with high variance model

5   Conclusions

Traditionally, users at supercomputing centers usually need to specify the amount of processors required when submitting a parallel job. The system will allocate computing resources to each job according to the specified processor requirement. However, this kind of allocation has been shown leading to degraded system utilization and job turnaround time when mismatch between requirement and available resources occurs. Moreover, such kind of processor allocation is not necessary for most current parallel applications which usually have the moldable property and thus can run with different parallelisms. Taking advantage of the moldable property, the system has the potential to do a better job, determining the most appropriate amount of processors to use for each job automatically. This can relieve the burden of specifying a specific number of processors on job submission for users and improve the overall system performance further.

In this paper, we propose and evaluate effective processor allocation methods for moldable jobs, which take into consideration of application speedup models when making the decision of processor allocation. The proposed approaches were evaluated through a series of simulation experiments in three common system usage scenarios. The experimental results indicate that our approaches outperform existing methods significantly, achieving up to 69%, 89%, and 98% performance improvement for the three usage scenarios, respectively.

References

1. Feitelson, D.G.: A Survey of Scheduling in Multiprogrammed Parallel Systems. Research Report RC 19790 (87657), IBM T. J. Watson Research Center (October 1994)
2. Feitelson, D.G., Weil, A.M.: Utilization and Predictability in Scheduling the IBM SP2 with Backfilling. In: Proc. 12th Int'l Parallel Processing Symp., pp. 542–546 (April 1998)
3. Gibbons, R.: A Historical Application Profiler for Use by Parallel Schedulers. In: Proc. Job Scheduling Strategies for Parallel Processing, pp. 58–77. Springer (1997)
4. Lifka, D.: The ANL/IBM SP Scheduling System. In: Proc. Job Scheduling Strategies for Parallel Processing, pp. 295–303. Springer (1995)
5. Mu'alem, A.W., Feitelson, D.G.: Utilization, Predictability, Workloads, and User Runtime Estimate in Scheduling the IBM SP2 with Backfilling. IEEE Transactions on Parallel and Distributed Systems 12(6), 529–543 (2001)
6. Downey, A.B.: A Model for Speedup of Parallel Programs., UC Berkeley EECS Technical Report, No. UCB/CSD-97-933 (January 1997)
7. Downey, A.B.: A Parallel Workload Model and Its Implications for Processor Allocation. In: Proc. the 6th International Symposium on High Performance Distributed Computing (1997)
8. Cirne, W., Berman, F.: Using Moldability to Improve the Performance of Supercomputer Jobs. Journal of Parallel and Distributed Computing 62, 1571–1601 (2002)
9. Cirne, W., Berman, F.: Adaptive Selection of Partition Size for Supercomputer Requests. In: Feitelson, D.G., Rudolph, L. (eds.) IPDPS-WS 2000 and JSSPP 2000. LNCS, vol. 1911, pp. 187–207. Springer, Heidelberg (2000)

10. Sabin, G., Lang, M.: Moldable Parallel Job Scheduling Using Job Efficiency: An Iterative Approach. In: Proc. Job Scheduling Strategies for Parallel Processing, Saint Malo, France (June 2006)

11. Srinivasan, S., Krishnamoorthy, S., Sadayappan, P.: A Robust Scheduling Strategy for Moldable Scheduling of Parallel Jobs. In: Proc. 5th IEEE International Conference on Cluster Computing, pp. 92–99 (2003)

12. Srinivasan, S., Subramani, V., Kettimuthu, R., Holenarsipur, P., Sadayappan, P.: Effective Selection of Partition Sizes for Moldable Scheduling of Parallel Jobs. In: Sahni, S.K., Prasanna, V.K., Shukla, U. (eds.) HiPC 2002. LNCS, vol. 2552, pp. 174–183. Springer, Heidelberg (2002)

13. Sun, H., Cao, Y., Hsu, W.J.: Efficient Adaptive Scheduling of Multiprocessors with Stable Parallelism Feedback. IEEE Transactions on Parallel and Distributed System 22(4) (April 2011)

14. Huang, K.C.: Performance Evaluation of Adaptive Processor Allocation Policies for Moldable Parallel Batch Jobs. In: Proc. 3th Workshop on Grid Technologies and Applications (December 2006)

15. Parallel Workloads Archive,
    http://www.cs.huji.ac.il/labs/parallel/workload/

16. Feitelson, D.G., Rudolph, L., Schweigelshohn, U., Sevcik, K., Wong, P.: Theory and Practice in Parallel Job Scheduling. In: Feitelson, D.G., Rudolph, L. (eds.) IPPS-WS 1997 and JSSPP 1997. LNCS, vol. 1291, pp. 1–34. Springer, Heidelberg (1997)

17. NAS parallel benchmarks,
    http://www.nas.nasa.gov/publications/npb.html

Correctness of Self-stabilizing Algorithms under the Dolev Model When Adapted to Composite Atomicity Models

Chih-Yuan Chen[1], Cheng-Pin Wang[2], Tetz C. Huang[3,*], and Ji-Cherng Lin[3]

[1] Department of Computer Science and Information Engineering,
Taoyuan Innovation Institute of Technology, 414, Sec. 3, Chung-Shan East Road,
Chung-Li, Tao-Yuan 320, Taiwan
[2] General Education Center, Tzu Chi College of Technology,
880, Sec. 2, Chien-kuo Rd., Hualien, Taiwan
[3] Department of Computer Science and Engineering, Yuan-Ze University,
135 Yuan-Tung Road, Chung-Li, Tao-Yuan 320, Taiwan
cstetz@saturn.yzu.edu.tw

Abstract. In this paper, we first clarify that it is not a trivial matter whether or not a self-stabilizing algorithm under the Dolev model, when adapted to a composite atomicity model, is also self-stabilizing. Then we employ a particular "simulation" approach to show that if a self-stabilizing algorithm under the Dolev model has one of two certain forms, then it is also self-stabilizing when adapted to one of the composite atomicity models, the fair daemon model. Since most existing self-stabilizing algorithms under the Dolev model have the above-mentioned forms, our results imply that they are all self-stabilizing when adapted to the fair daemon model.

Keywords: Silent self-stabilizing algorithm, Composite atomicity, Read/write atomicity, Fair daemon model, Adaptation of algorithm.

1 Introduction

A *distributed system* consists of a set of loosely connected processors that do not share a global memory. It is usually modelled by a connected simple undirected graph $G = (V, E)$, with each node $x \in V$ representing a processor in the system and each edge $\{x, y\} \in E$ representing the link connecting processors x and y. Each processor has one or more *shared registers* and possibly some non-shared *local variables*, the contents of which specify the *local state* of the processor. Local states of all processors in the system at a certain time constitute the *global configuration* (or, simply, *configuration*) of the system at that time. The main restriction of the distributed system is that each processor in the system can only access the data (i.e., read the shared data) of its neighbors. Since a distributed algorithm is an algorithm that works in a distributed system, it

---

* Corresponding author.

J.-S. Pan et al. (Eds.): *Advances in Intelligent Systems & Applications*, SIST 21, pp. 573–586.
DOI: 10.1007/978-3-642-35473-1_57   © Springer-Verlag Berlin Heidelberg 2013

cannot violate this main restriction. In this paper, we adopt the point of view in Dolev et al. [6]. Thus, an *atomic step* is the "largest" step that is guaranteed to be executed uninterruptedly. A distributed algorithm uses *composite atomicity* if some atomic step contains (at least) a read operation and a write operation. A distributed algorithm uses *read/write atomicity* if each atomic step contains either a single read operation or a single write operation but not both.

1.1   Computational Models

Composite Atomicity Models. The *Dijkstra's central daemon model* (or, simply, *central daemon model*) was first introduced by Dijkstra [3] in 1974. Under this computational model, each processor is equipped with a *local algorithm* that consists of one or more rules of the form:

$$condition\ part \rightarrow action\ part.$$

The condition part (or *guard*) is a Boolean expression of registers of the processor and its neighbors, and the action part is an assignment of values to some registers of the processor. If the condition part of one or more rules in a processor is evaluated to be true, we say that the processor is *privileged* to execute the action part of any of these rules (or *privileged to make a move*). Under this computational model, if the system starts with a configuration in which no processor in the system is privileged, then the system is deadlocked. Otherwise, the *central daemon* in the system will randomly select exactly one privileged processor and exactly one executable rule in the processor's local algorithm, and let the selected processor execute the action part of the selected rule. The local state of the selected processor thus changes, which in the meantime results in the change of the global configuration of the system. The system will repeat the above process to change configurations as long as it does not encounter any deadlock situation. Thus the behavior of the system under the action of the algorithm can be described by *executions* defined as follows: an infinite sequence of configurations $\Gamma = (\gamma_1, \gamma_2, \dots)$ of a distributed system is called an *infinite execution* (*of the algorithm in the system*) under the central daemon model if for any $i \geq 1$, γ_{i+1} is obtained from γ_i after exactly one processor in the system makes a move in the ith *step* $\gamma_i \rightarrow \gamma_{i+1}$; a finite sequence of configurations $\Gamma = (\gamma_1, \gamma_2, \dots, \gamma_k)$ of a distributed system is called a *finite execution* (*of the algorithm in the system*) under the central daemon model if (1) $k = 1$, or for any $i = 1, 2, \dots, k - 1$, γ_{i+1} is obtained from γ_i after exactly one processor in the system makes a move in the ith step $\gamma_i \rightarrow \gamma_{i+1}$, and (2) no node is privileged in the last configuration γ_k.

The *distributed daemon model* was later considered by Burns [1] in 1987. The difference between the central daemon model and the distributed daemon model is the number of processors that make moves in a step of an execution of the algorithm. Under the central daemon model, exactly one privileged processor in the system is randomly selected by the central daemon to make a move in a step of an execution of the algorithm. Under the distributed daemon model, however, an arbitrary number of privileged processors are randomly selected by the *distributed daemon* to simultaneously make moves in a step. Thus, we can also define executions for the distributed daemon model as follows: an infinite sequence of configurations $\Gamma = (\gamma_1, \gamma_2, \ldots)$ of a distributed system is called an *infinite execution* (*of the algorithm in the system*) *under the distributed daemon model* if for any $i \geq 1$, γ_{i+1} is obtained from γ_i after a certain number of privileged processors selected by the distributed daemon simultaneously make moves in the ith *step* $\gamma_i \to \gamma_{i+1}$; a finite sequence of configurations $\Gamma = (\gamma_1, \gamma_2, \ldots, \gamma_k)$ of a distributed system is called a *finite execution* (*of the algorithm in the system*) *under the distributed daemon model* if (1) $k = 1$, or for any $i = 1, 2, \ldots, k - 1$, γ_{i+1} is obtained from γ_i after a certain number of privileged processors selected by the distributed daemon simultaneously make moves in the ith step $\gamma_i \to \gamma_{i+1}$, and (2) no node is privileged in the last configuration γ_k. As a consequence of the above definitions, the central daemon model is a restricted version of the distributed daemon model, i.e., the set of all distributed-daemon-model executions contains the set of all central-daemon-model executions.

The *weakly fair daemon model* (or, simply, *fair daemon model*) is a different restricted version of the distributed daemon model. Precisely, an execution Γ under the distributed daemon model is called an *execution under the fair daemon model* if for any suffix Γ' of Γ, no node can be privileged infinitely many times without making any move in Γ'. It follows immediately from this definition that every finite execution is a fair-daemon-model execution.

Dolev's Read/Write Atomicity Model. The above three computational models assume the composite atomicity. A single move (or atomic step) by a processor consists of reading registers of all its neighbors, making internal computations and then rewriting its own register (or registers). In 1993, Dolev et al., introduced a new type of computational model (we shall call it *Dolev's read/write atomicity model*, or simply, the *Dolev model*) in their famous paper [6] (cf. also [2], [4], [5]). Their model reflects more truthfully a real distributed system. Firstly, it assumes the more real read/write atomicity. Under such an assumption, each atomic step in the system consists of internal computations and either a single read operation or a single write operation. Secondly, under this model, it is assumed that each processor in the system runs its own program indefinitely and at its own pace; and the running of the program has to follow the order of the statements in the program. Therefore, algorithms operating under the Dolev model have different looks from algorithms operating under composite atomicity models (e.g., one can compare Algorithm DM with Algorithm ADV in Section 2). Under the Dolev model, the behavior of the system under the action of the algorithm can still be described by a sequence of configurations $\Gamma = (\gamma_1, \gamma_2, \ldots)$. As in Dijkstra's central daemon model, in any configuration γ_i ($i \geq 1$), a unique processor of the system is selected by a daemon to make a move in *the ith step* $\gamma_i \to \gamma_{i+1}$, which changes the system configuration from γ_i to γ_{i+1}. However, we should point out that due to the content of the algorithm and the way in which the algorithm is executed, the selection by this daemon is no longer random here under the Dolev model. In other words, any execution of the algorithm under the Dolev model has to obey certain restrictions.

(In Section 2, a precise definition of an *execution under the Dolev model*, given for a certain class of algorithms in which we are interested, will almost completely reflect such restrictions.)

1.2   Self-stabilization

A distributed algorithm and a subset of executions (of the algorithm under a certain computational model), called *legal executions*, are designed to solve a specific problem such as the shortest path problem, the bridge-finding problem, or the mutual exclusion problem, etc. Legal executions are designed in such a way that if the system is engaged in any legal execution, the solution of the problem can be easily seen. For instance, for the single-source shortest path problem, the legal executions are so designed that in every configuration of any legal execution, a shortest path between any processor and the source can be identified. A configuration γ is called a *safe configuration* or *legitimate configuration* (*with respect to an algorithm under a particular computational model*) if any execution (of the algorithm under that computational model) starting with γ is a legal execution.

An algorithm is *self-stabilizing* under a certain computational model (with respect to a particular problem) if any execution of the algorithm under that computational model contains a safe configuration. A self-stabilizing algorithm under a certain computational model is *silent* if any execution of the algorithm under that computational model contains a configuration γ such that in the suffix of the execution starting with γ, the values stored in all the local variables and all the shared registers of all the processors never change. Most of the existing self-stabilizing algorithms (e.g., [2], [7-22]) are silent.

1.3   Our Contributions

In Dolev et al. [5], the following was mentioned: "It should be noted that one step of the distributed daemon in which m processors move simultaneously can be simulated by our model: First let each of the m processors read all its neighbors' states, and then let them all move to their new states. Using this simulation, it is not hard to show that every mutual exclusion protocol that is self-stabilizing for read/write daemon is also self-stabilizing for distributed daemon". The above comment concerns the mutual exclusion problem in particular, and hence the following question naturally arises: Can the point of view and the logic of the above "simulation" approach be applied to any problem for which there exists a self-stabilizing algorithm under the Dolev model? In other words, for any such problem, can the above "simulation" approach be used to show that a self-stabilizing algorithm under the Dolev model, when adapted to the distributed daemon model, is also self-stabilizing?

Since we have found a counterexample (see Appendix) to show that a self-stabilizing algorithm under the Dolev model may not be self-stabilizing when adapted to the central daemon model and hence may not be self-stabilizing when adapted to the distributed daemon model, the answer to the above question is, to one's surprise, negative. This result shown by the counterexample also implies that applying the above simulation to an execution under the central daemon model may not result in an execution under the Dolev model. As a matter of fact, we are able to show that applying the above simulation to even an execution under the fair daemon model may not result in an execution under the Dolev model. Consequently, the above "simulation" approach cannot be used to show that a self-stabilizing algorithm under the Dolev model, when adapted to the fair daemon model, is also self-stabilizing.

By the above discussion, we have clarified that it is not a matter of course whether or not a self-stabilizing algorithm under the Dolev model, when adapted to a composite atomicity model, is also self-stabilizing. In this paper, we will show that if a self-stabilizing algorithm under the Dolev model has one of two particular forms, then it is also self-stabilizing when adapted to one of the composite atomicity models, the fair daemon model. To show this result, we need to develop a new simulation which is different from the above simulation suggested in Dolev et al. [5]. Since most existing self-stabilizing algorithms under the Dolev model (e.g., [2], [5], [6], [10], [16]) have the above-mentioned forms, our results imply that they are all self-stabilizing when adapted to the fair daemon model.

1.4 Organization of This Paper

The rest of this paper is organized as follows: In Section 2, an abstract form (Algorithm DM) of a certain class of silent self-stabilizing algorithms under the Dolev model is presented, and its adapted version (Algorithm ADV) to composite atomicity models is given. In Section 3, an example is exhibited to illustrate the adaptation in Section 2. Then in Section 4 it is verified that the adapted algorithm (Algorithm ADV) is self-stabilizing under the fair daemon model. In Section 5, some remarks conclude the whole discussion.

2   Adaptation from Algorithm DM into Algorithm ADV

As before, let $G = (V, E)$ denote a distributed system and for each $x \in V$, let $N(x)$ denote the set of all neighbors of x. Suppose Algorithm DM is a silent self-stabilizing algorithm under the Dolev model, with which G is equipped to solve a certain problem. Suppose Algorithm DM has the following abstract form:

Algorithm DM
{For any node x}
1 repeat forever
2   for each $y \in N(x)$ do
3   read $(r_{yx} := w_y)$
4   endfor
5   if $w_x \neq \mathcal{A}_x(r_{y_1 x}, r_{y_2 x}, \ldots, r_{y_{\delta(x)} x})$ then
6   write $(w_x := \mathcal{A}_x(r_{y_1 x}, r_{y_2 x}, \ldots, r_{y_{\delta(x)} x}))$
7   endif
8 endrepeat

where

1) w_x is a shared register of x, in which x writes and from which all neighbors of x read,
2) r_{yx} is a local variable of x, in which x stores the value that it reads from the shared register w_y of its neighbor y,
3) $\delta(x)$ is the number of neighbors of x, and
4) $y_1, y_2, \ldots, y_{\delta(x)}$ are all the neighbors of x.

Since Algorithm DM is a silent self-stabilizing algorithm under the Dolev model, a legal execution of Algorithm DM is defined to be an execution in any configuration of which the guard condition $[w_x \neq \mathcal{A}_x(r_{y_1x}, r_{y_2x}, \ldots, r_{y_{\delta(x)}x})]$ for each processor x is always false, and $r_{yx} = w_y$ always holds for each processor x and each neighbor y of x. Hence a safe configuration with respect to Algorithm DM is a configuration in which both conditions $[\forall x \in V, w_x = \mathcal{A}_x(r_{y_1x}, r_{y_2x}, \ldots, r_{y_{\delta(x)}x})]$ and $[\forall x \in V$ and $\forall y \in N(x), r_{yx} = w_y]$ hold.

As mentioned previously in Section 1, due to the content of the algorithm and the way in which the algorithm is executed, any execution of Algorithm DM under the Dolev model has to obey certain restrictions. For the rigor of our later proofs, we give here a precise definition to an execution of Algorithm DM under the Dolev model: A sequence $\Gamma = (\gamma_1, \gamma_2, \ldots)$ of configurations is called an *execution of Algorithm DM under the Dolev model* if:

(1) for each $i \geq 1$, γ_{i+1} is obtained from γ_i after a processor makes either a single read operation or a single write operation (the transition $\gamma_i \to \gamma_{i+1}$ from γ_i to γ_{i+1} is called the ith *step* of Γ),
(2) in Γ, each processor makes a read operation infinitely often,
(3) in Γ, once a processor makes a read operation, it must *complete a full round of reading all its neighbors* (i.e., it must complete a full execution of the loop from statement 2 to statement 4 in Algorithm DM), and
(4) in Γ, if the guard condition $[w_x \neq \mathcal{A}_x(r_{y_1x}, r_{y_2x}, \ldots, r_{y_{\delta(x)}x})]$ is evaluated to be true right after any processor x completes a full round of reading all its neighbors, then the write operation "write $(w_x := \mathcal{A}_x(r_{y_1x}, r_{y_2x}, \ldots, r_{y_{\delta(x)}x}))$" has to follow as the next move by x.

Next, we introduce Algorithm ADV, which is a naturally-adapted version of Algorithm DM to composite atomicity models, and operates also in G:

Algorithm ADV
{For any node x}
$R1 : w'_x \neq \mathcal{A}_x(w'_{y_1}, w'_{y_2}, \ldots, w'_{y_{\delta(x)}}) \to w'_x := \mathcal{A}_x(w'_{y_1}, w'_{y_2}, \ldots, w'_{y_{\delta(x)}})$

where

1) w'_x is the only shared register of x,
2) $\delta(x)$ is the number of neighbors of x, and
3) $y_1, y_2, \ldots, y_{\delta(x)}$ are all the neighbors of x.

Note that in order for Algorithm ADV to solve the same problem as Algorithm DM does, a legal execution of Algorithm ADV should be defined to be an execution in any configuration of which no node can be privileged. As a consequence, the legal executions of Algorithm ADV are precisely all the one-configuration executions, and the safe configurations are precisely all the configurations in which no node can be privileged. Thus, under any of the three composite atomicity models introduced in the introduction, Algorithm ADV is self-stabilizing if and only if all its executions are finite.

3   An Illustration

Suppose $G = (V, E)$ is a distributed system in which there is a unique special node r, called *root*. The following Algorithm DFS_DM is essentially the same as the DFS algorithm in Collin and Dolev [2]. Using the result in [2], one can easily convince oneself that Algorithm DFS_DM is a silent self-stabilizing algorithm under the Dolev model, with which G can be equipped to solve the Depth First Search (DFS) spanning tree problem.

Algorithm DFS_DM
{For the root r}
1 repeat forever
2   for each $y \in N(r)$ do
3   read ($read\_path_{yr} := write\_path_y$)
4   endfor
5   if $write\_path_r \neq \bot$ then
6   write ($write\_path_r := \bot$)
7   endif
8 endrepeat

{For any node $x \neq r$}
1 repeat forever
2   for each $y \in N(x)$ do
3   read ($read\_path_{yx} := write\_path_y$)
4   endfor
5   if $write\_path_x \neq \min_{\prec_{lex}} \{right_N(read\_path_{yx} \circ \alpha_y(x)) \mid y \in N(x)\}$ then
6   write ($write\_path_x := \min_{\prec_{lex}} \{right_N(read\_path_{yx} \circ \alpha_y(x)) \mid y \in N(x)\}$)
7   endif
8 endrepeat

In the above algorithm, for root or non-root x,

1) $write\_path_x$ is a shared register of x, in which x writes and from which all neighbors of x read,
2) $read\_path_{yx}$ is a local variable of x, in which x stores the value that it reads from the shared register $write\_path_y$ of its neighbor y,

3) α_x is a pre-given ordering of all the edges incident to x, and $\alpha_x(y)$ is the rank of edge $\{x, y\}$ in the ordering α_x,
4) the value of either $write\_path_x$ or $read\_path_{yx}$ is a sequence of at most N items, where $N \geq n$ (the number of nodes in G) is a constant positive integer, and each item in the sequence is either a positive integer or a special symbol $\bot$,
5) $\prec_{lex}$ is a lexicographical order over the set of all finite sequences of at most N items, and $\bot$ is the minimal character (for example, $(\bot, 1) \prec_{lex} (\bot, 1, 1) \prec_{lex} (\bot, 2) \prec_{lex} (1)$),
6) $read\_path_y \circ \alpha_y(x)$ is the concatenation of $read\_path_y$ with $\alpha_y(x)$ (for example, if $read\_path_y = (1, 2, 1)$ and $\alpha_y(x) = 3$, then $read\_path_y \circ \alpha_y(x) = (1, 2, 1, 3)$), and
7) the notation $right_N(w)$ refers to the sequence of the N least significant items of w (for example, $right_6(\bot, 3, 2, 1, 2, 1, 1) = (3, 2, 1, 2, 1, 1)$ and $right_6(2, 1, 2, 1, 1) = (2, 1, 2, 1, 1)$).

Since Algorithm DFS_DM is a silent self-stabilizing algorithm under the Dolev model, a legal execution of Algorithm DFS_DM is defined to be an execution in any configuration of which the guard conditions "$write\_path_r \neq \bot$" for the root r and "$write\_path_x \neq \min_{\prec_{lex}}\{right_N(read\_path_{yx} \circ \alpha_y(x)) \mid y \in N(x)\}$" for each processor $x \neq r$ are always false, and "$read\_path_{yx} = write\_path_y$" always holds for each processor x and each neighbor y of x. Hence a legal execution of Algorithm DFS_DM is an execution in any configuration of which both conditions $[write\_path_r = \bot$ and $\forall y \in N(r), read\_path_{yr} = write\_path_y]$ and $[\forall x \in V - \{r\}, write\_path_x = \min_{\prec_{lex}}\{right_N(read\_path_{yx} \circ \alpha_y(x)) \mid y \in N(x)\}$ and $\forall x \in V - \{r\}$ and $\forall y \in N(x), read\_path_{yx} = write\_path_y]$ hold.

According to the adaptation in Section 2, the naturally-adapted version of Algorithm DFS_DM to composite atomicity models is as follows:

Algorithm DFS_ADV
{For the root r}
$R0 : path_r \neq \bot \rightarrow path_r := \bot$

{For any node $x \neq r$}
$R1 : path_x \neq \min_{\prec_{lex}}\{right_N(path_y \circ \alpha_y(x)) \mid y \in N(x)\}$
$\rightarrow path_x := \min_{\prec_{lex}}\{right_N(path_y \circ \alpha_y(x)) \mid y \in N(x)\}$

Note that in the above algorithm, for root or non-root x,

1) $path_x$ is the only register of x,
2) the value of $path_x$ is a sequence of at most N items, where $N \geq n$ (the number of nodes in G) is a constant positive integer, and each item in the sequence is either a positive integer or a special symbol $\bot$, and
3) the notations α_x, $\alpha_x(y)$, $\prec_{lex}$, $right_N(\cdot)$ and $path_y \circ \alpha_y(x)$ have the same meanings as in Algorithm DFS_DM.

In order for Algorithm DFS_ADV to solve the DFS spanning tree problem as Algorithm DFS_DM does, a legal execution of Algorithm DFS_ADV should be defined to be an execution in any configuration of which no node can be privileged. As a consequence, the legal executions of Algorithm DFS_ADV are precisely all the one-configuration executions, and the safe configurations with respect to Algorithm DFS_ADV are precisely all the configurations in which no node can be privileged, i.e., the condition $[path_r = \perp$ and $\forall x \in V - \{r\}$, $path_x = \min_{\prec_{lex}}\{right_N(path_y \circ \alpha_y(x)) \mid y \in N(x)\}]$ holds. As mentioned previously, under any of the above three composite atomicity models introduced in the introduction, Algorithm DFS_ADV is self-stabilizing if and only if all its executions are finite.

4 Correctness of Algorithm ADV under the Fair Daemon Model

As mentioned previously, one of our main contributions in this paper is to show that Algorithm ADV is self-stabilizing under the fair daemon model. In this section, we will prove this by contradiction. So we first suppose Algorithm ADV is not self-stabilizing under the fair daemon model. Hence there exists an infinite execution $\Gamma' = (\gamma'_1, \gamma'_2, \ldots)$ of Algorithm ADV under the fair daemon model. By using the process below, we can simulate Γ' by an execution Γ of Algorithm DM under the Dolev model:

1. For each $i = 1, 2, \ldots$, let NP_i be the set of the nodes which are not privileged in γ'_i, and PM_i be the set of the nodes which are privileged in γ'_i and chosen by the fair daemon to make moves in the step $\gamma'_i \to \gamma'_{i+1}$.

2. Define γ_1 to be the configuration of the system G equipped with Algorithm DM such that for each $x \in V$, the value of w_x in γ_1 is equal to the value of w'_x in γ'_1; and for each $x \in V$ and each neighbor y of x, the value of r_{yx} in γ_1 is equal to the value of w'_y in γ'_1.

3. Starting with the configuration γ_1,
   (a) let nodes in $NP_1 \cup PM_1$ one by one make read operations (i.e., let them execute statements 2-4 of Algorithm DM), and then
   (b) let nodes in PM_1 one by one make write operations (i.e., let them execute statements 5-7 of Algorithm DM).

   (Note that the behavior of the system G under all these operations in (a) and (b) above can be represented by a finite sequence S_1 of configurations starting with γ_1. Let the last configuration of S_1 be γ_2, that is, $S_1 = (\gamma_1, \ldots, \gamma_2)$. One can easily see that for each node $x \in V$, the value of w_x in γ_2 is equal to the value of w'_x in γ'_2.)

4. Similarly, for any $i \geq 2$, starting with the configuration γ_i,
   (a) let nodes in $NP_i \cup PM_i$ one by one make read operations (i.e., let them execute statements 2-4 of Algorithm DM), and then
   (b) let nodes in PM_i one by one make write operations (i.e., let them execute statements 5-7 of Algorithm DM).

(Note that the behavior of the system G under all these operations in (a) and (b) above can be represented by a finite sequence S_i of configurations starting with γ_i. Let the last configuration of S_i be γ_{i+1}, that is, $S_i = (\gamma_i, \ldots, \gamma_{i+1})$. One can easily see that for each node $x \in V$, the value of w_x in γ_{i+1} is equal to the value of w'_x in γ'_{i+1}.)

5. From the above, we finally obtain an infinite sequence $\Gamma = (\gamma_1, \ldots, \gamma_2, \ldots, \gamma_3, \ldots)$ of configurations of G equipped with Algorithm DM.

Claim. Γ *is an execution of Algorithm DM under the Dolev model.*

Proof (Proof of claim).

(1) According to the above construction process of Γ, it is obvious that except for the initial configuration γ_1, any configuration in Γ is obtained from the preceding configuration after a processor makes either a single read operation or a single write operation.

(2) Suppose there exists a processor x that makes a read operation only a finite number of times in Γ. Then there exists a configuration γ_t in Γ such that x can not make any read operation after γ_t. According to the construction process of Γ, for all $l \geq t$, $x \notin NP_l \cup PM_l$. Thus, x is privileged, but is not selected by the fair daemon to make any move in the suffix $(\gamma'_t, \gamma'_{t+1}, \ldots)$ of Γ'. This contradicts the fairness of Γ'. Therefore, any processor makes a read operation infinitely often in Γ.

(3) According to the construction process of Γ, it is obvious that in Γ once a processor makes a read operation, it will complete a full round of reading all its neighbors.

(4) According to the construction process of Γ, if in Γ any processor x completes a full round of reading all its neighbors, then it must do so in S_i, for some $i = 1, 2, \ldots$. Thus, $x \in NP_i \cup PM_i$. If $x \in NP_i$, then the guard condition $[w_x \neq \mathcal{A}_x(r_{y_1 x}, r_{y_2 x}, \ldots, r_{y_{\delta(x)} x})]$ in statement 5 of Algorithm DM cannot be true right after x completes a full round of reading all its neighbors. If $x \in PM_i$, then the guard condition $[w_x \neq \mathcal{A}_x(r_{y_1 x}, r_{y_2 x}, \ldots, r_{y_{\delta(x)} x})]$ is evaluated to be true right after x completes a full round of reading all its neighbors. According to the construction process of Γ, the write operation "write $(w_x := \mathcal{A}_x(r_{y_1 x}, r_{y_2 x}, \ldots, r_{y_{\delta(x)} x}, w_x))$" will follow as the next move by x.

From all the above, we can conclude that Γ is an execution of Algorithm DM under the Dolev model.

Since Γ is an execution of Algorithm DM and Algorithm DM is a silent self-stabilizing algorithm under the Dolev model, there exists a safe configuration γ in Γ such that in the suffix $\overline{\Gamma}$ of Γ that starts with γ, the values stored in all the shared registers and local variables of all the nodes never change. Hence in $\overline{\Gamma}$ no node can make any write operation. In view of the construction process of Γ, $\gamma \in S_k$ for some $k = 1, 2, \ldots$. Hence in S_{k+1} no node can make any write operation. However, since $\Gamma' = (\gamma_1', \gamma_2', \ldots)$ is an infinite execution, there must be nodes selected by the fair (distributed) daemon to make moves in the step $\gamma_{k+1}' \to \gamma_{k+2}'$. In view of the construction process of Γ, these nodes must make write operations in S_{k+1}. A contradiction occurs. Therefore, our original supposition that Algorithm ADV is not self-stabilizing under the fair daemon model is false, and hence we obtain the following theorem.

Theorem 1. *Algorithm ADV is self-stabilizing under the fair daemon model.*

5 Concluding Remarks

In the introduction, we have clarified that it is not a matter of course whether or not a self-stabilizing algorithm under the Dolev model, when adapted to a composite atomicity model, is also self-stabilizing. Then, in the central part of this paper, we have shown that if Algorithm DM is a silent self-stabilizing algorithm under the Dolev model, then Algorithm ADV, a naturally adapted version of Algorithm DM, is a silent self-stabilizing algorithm under the fair daemon model. Note that in any silent self-stabilizing algorithm under the Dolev model that has the form of Algorithm DM, each processor only maintains a unique shared register, which can be read by all its neighbors. However, in the existing literature, there are also silent self-stabilizing algorithms under the Dolev model (e.g., [6], [10], [16]) that have the form of Algorithm DMM exhibited below, in which each processor maintains an individual register for each of its neighbors.

Algorithm DMM
{For any node x}
1 repeat forever
2   for each $y \in N(x)$ do
3   read $(r_{yx} := w_{yx})$
4   endfor
5   for each $y \in N(x)$ do
6   if $w_{xy} \neq \mathcal{B}_x(r_{y_1 x}, r_{y_2 x}, \ldots, r_{y_{\delta(x)} x})$ then
7   write $(w_{xy} := \mathcal{B}_x(r_{y_1 x}, r_{y_2 x}, \ldots, r_{y_{\delta(x)} x}))$
8   endif
9   endfor
10 endrepeat

where

1) w_{xy} is a shared register of x, in which x writes and from which its neighbor y reads,
2) r_{yx} is a local variable of x, in which x stores the value that it reads from the shared register w_{yx} of its neighbor y,
3) $\delta(x)$ is the number of neighbors of x, and
4) $y_1, y_2, \ldots, y_{\delta(x)}$ are all the neighbors of x.

Note that Algorithm DMM can be naturally adapted to composite atomicity models, and the resulting adapted version is as follows:

Algorithm ADV′
{For any node x}
$R1 : w'_x \neq \mathcal{B}_x(w'_{y_1}, w'_{y_2}, \ldots, w'_{y_{\delta(x)}}) \rightarrow w'_x := \mathcal{B}_x(w'_{y_1}, w'_{y_2}, \ldots, w'_{y_{\delta(x)}})$

where

1) w'_x is the only shared register of x,
2) $\delta(x)$ is the number of neighbors of x, and
3) $y_1, y_2, \ldots, y_{\delta(x)}$ are all the neighbors of x.

In view of the previous discussion in this paper, the following question comes out naturally: If Algorithm DMM is a silent self-stabilizing algorithm under the Dolev model, is Algorithm ADV′ a silent self-stabilizing algorithm under the fair daemon model? The answer to this question is actually also affirmative, and we will provide a more detailed explanation for it in the future version of this paper.

Appendix

In this appendix, we give a counterexample to show that a self-stabilizing algorithm under the Dolev model, when adapted to the central daemon model, may no longer be self-stabilizing. The counterexample which we are going to give is the Algorithm DFS_DM and its adapted version Algorithm DFS_ADV presented previously in Section 3.

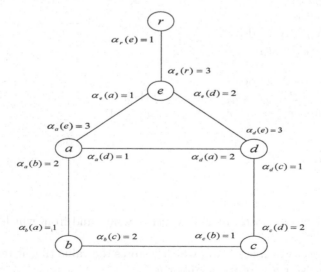

Fig. 1. A distributed system G labelled with edge ranks. (cf. the first line of page 6)

Table 1. A beginning portion of an infinite cyclic execution Γ of Algorithm DFS_ADV under the central daemon model for the system in Figure 1

	$path_r$	$path_e$	$path_a$	$path_b$	$path_c$	$path_d$
γ_1	⊥	(3)	(1,1,1,1,1,1)	(1,1,1,1,1,2)	(1,1,1,1,2,2)	(1,1,1,1,1,1)
γ_2	⊥	(3)	(1,1,1,1,1,2)	(1,1,1,1,1,2)	(1,1,1,1,2,2)	(1,1,1,1,1,1)
γ_3	⊥	(3)	(1,1,1,1,1,2)	(1,1,1,1,2,2)	(1,1,1,1,2,2)	(1,1,1,1,1,1)
γ_4	⊥	(3)	(1,1,1,1,1,2)	(1,1,1,1,2,2)	(1,1,1,1,1,1)	(1,1,1,1,1,1)
γ_5	⊥	(3)	(1,1,1,1,1,2)	(1,1,1,1,2,2)	(1,1,1,1,1,1)	(1,1,1,1,1,2)
γ_6	⊥	(3)	(1,1,1,1,2,2)	(1,1,1,1,2,2)	(1,1,1,1,1,1)	(1,1,1,1,1,2)
γ_7	⊥	(3)	(1,1,1,1,2,2)	(1,1,1,1,1,1)	(1,1,1,1,1,1)	(1,1,1,1,1,2)
γ_8	⊥	(3)	(1,1,1,1,2,2)	(1,1,1,1,1,1)	(1,1,1,1,1,2)	(1,1,1,1,1,2)
γ_9	⊥	(3)	(1,1,1,1,2,2)	(1,1,1,1,1,1)	(1,1,1,1,1,2)	(1,1,1,1,2,2)
γ_{10}	⊥	(3)	(1,1,1,1,1,1)	(1,1,1,1,1,1)	(1,1,1,1,1,2)	(1,1,1,1,2,2)
γ_{11}	⊥	(3)	(1,1,1,1,1,1)	(1,1,1,1,1,2)	(1,1,1,1,1,2)	(1,1,1,1,2,2)
γ_{12}	⊥	(3)	(1,1,1,1,1,1)	(1,1,1,1,1,2)	(1,1,1,1,2,2)	(1,1,1,1,2,2)
γ_{13}	⊥	(3)	(1,1,1,1,1,1)	(1,1,1,1,1,2)	(1,1,1,1,2,2)	(1,1,1,1,1,1)
γ_{14}	⊥	(3)	(1,1,1,1,1,2)	(1,1,1,1,1,2)	(1,1,1,1,2,2)	(1,1,1,1,1,1)
⋮	⋮	⋮	⋮	⋮	⋮	⋮

Consider now a distributed system G consisting of 6 nodes as illustrated in Figure 1. The number N in Algorithm DFS_ADV, with which G is equipped, is now set to 6. In Table 1 is exhibited a beginning portion of an infinite cyclic execution $\Gamma = (\gamma_1, \gamma_2, \ldots)$ of Algorithm DFS_ADV under the central daemon model, which has a period of 13 (note that $\gamma_1 = \gamma_{13}$). The shaded grid in each configuration of Γ in the table indicates the privileged node selected by the central daemon to make a move. As mentioned in the end of Section 3, Algorithm DFS_ADV is self-stabilizing under the central daemon model if and only if all its executions under the central daemon model are finite. Now that an infinite execution of Algorithm DFS_ADV under the central daemon model is found to exist, it is shown that Algorithm DFS_ADV is not a self-stabilizing algorithm under the central daemon model.

References

1. Burns, J.E.: Self-stabilizing Rings without Daemons. Technical report, Georgia Tech (1987)
2. Collin, Z., Dolev, S.: Self-stabilizing Depth-first Search. Inform. Process. Lett. 49, 297–301 (1994)
3. Dijkstra, E.W.: Self-stabilizing Systems in Spite of Distributed Control. Comm. ACM 17, 643–644 (1974)
4. Dolev, S.: Self-stabilization. MIT Press (2000)
5. Dolev, S., Israeli, A., Moran, S.: Self-stabilization of Dynamic Systems Assuming Only Read/Write Atomicity. In: 9th Annual ACM Symposium on Principles of Distributed Computing, Quebec, Canada, pp. 103–117 (1990)
6. Dolev, S., Israeli, A., Moran, S.: Self-stabilization of Dynamic Systems Assuming Only Read/Write Atomicity. Distrib. Comput. 7, 3–16 (1993)
7. Ghosh, S., Gupta, A.: An Exercise in Fault-containment: Self-stabilizing Leader Election. Inform. Process. Lett. 59, 281–288 (1996)

8. Ghosh, S., Gupta, A., Herman, T., Pemmaraju, S.V.: Fault-containing Self-stabilizing Algorithm. In: 15th ACM Symposium on Principles of Distributed Computing, Philadelphia, USA, pp. 45–54 (1996)
9. Ghosh, S., Gupta, A., Pemmaraju, S.V.: A Fault-containing Self-stabilizing Spanning Tree Algorithm. J. Comput. Inform. 2, 322–338 (1996)
10. Huang, T.C.: A Self-stabilizing Algorithm for the Shortest Path Problem Assuming Read/Write Atomicity. J. Comput. Syst. Sci. 71, 70–85 (2005)
11. Huang, S.T., Chen, N.S.: A Self-stabilizing Algorithm for Constructing Breadth-first Trees. Inform. Process. Lett. 41, 109–117 (1992)
12. Hsu, S.C., Huang, S.T.: A Self-stabilizing Algorithm for Maximal Matching. Inform. Process. Lett. 43, 77–81 (1992)
13. Huang, T.C., Lin, J.C.: A Self-stabilizing Algorithm for the Shortest Path Problem in a Distributed System. Comput. Math. Appl. 43, 103–109 (2002)
14. Huang, T.C., Lin, J.C., Chen, H.J.: A Self-stabilizing Algorithm which Finds a 2-center of a Tree. Comput. Math. Appl. 40, 607–624 (2000)
15. Huang, T.C., Lin, J.C., Chen, C.Y., Wang, C.P.: A Self-stabilizing Algorithm for Finding a Minimal 2-dominating Set Assuming the Distributed Demon Model. Comput. Math. Appl. 54, 350–356 (2007)
16. Huang, T.C., Lin, J.C., Mou, N.: A Self-stabilizing Algorithm for the Center-finding Problem Assuming Read/Write Atomicity. Comput. Math. Appl. 48, 1667–1676 (2004)
17. Ikeda, M., Kamei, S., Kakugawa, H.: A Space-optimal Self-stabilizing Algorithm for the Maximal Independent Set Problem. In: 3rd International Conference on Parallel and Distributed Computing, Applications and Technologies, Kanazawa, Japan, pp. 70–74 (2002)
18. Shukla, S., Rosenkrantz, D.J., Ravi, S.S.: Observations on Self-stabilizing Graph Algorithms on Anonymous Networks. In: 2nd Workshop on Self-stabilizing Systems, p. 7.1–7.15. Las Vegas, USA (1995)
19. Tsin, Y.H.: An Improved Self-stabilizing Algorithm for Biconnectivity and Bridge-connectivity. Inform. Process. Lett. 102, 27–34 (2007)
20. Turau, V.: Linear Self-stabilizing Algorithms for the Independent and Dominating Set Problems Using an Unfair Distributed Scheduler. Inform. Process. Lett. 103, 88–93 (2007)
21. Turau, V., Hauck, B.: A Self-stabilizing Algorithm for Constructing Weakly Connected Minimal Dominating Sets. Inform. Process. Lett. 109, 763–767 (2009)
22. Tzeng, C.H., Jiang, J.R., Huang, S.T.: A Self-stabilizing $(\Delta + 4)$-edge-coloring Algorithm for Planar Graphs in Anonymous Uniform Systems. Inform. Process. Lett. 101, 168–173 (2007)

Efficiently Extracting Change Data from Column Oriented NoSQL Databases

Yong Hu and Weiping Qu

University of Kaiserslautern, Germany
{hu,qu}@informatik.uni-kl.de

Abstract. This paper explores the appropriate change data capture approaches implemented by exploiting the MapReduce framework in the context of column-oriented NoSQL databases. Change data capture describes how the system acquires change data from data sources. It is the most significant preparatory task for the incremental re-computation. Based on analyzing the core features of column-oriented NoSQL databases, we define a notion of net-effect change data and propose three change data capture approaches, namely, Snapshot Differential, Log-based approach and Trigger-based approach (Change Tracking Table).

Keywords: Big data, MapReduce framework, Change data capture, Column Oriented NoSQL databases.

1 Introduction

"Big Data" is obtained more and more attentions in both research and industrial areas. The term *"Big"* indicates the size of data set is massive and the elapsed data processing time is not acceptable. Moreover, it depicts the size of data set is far beyond the current tried and true data processing approaches, e.g. exploiting SQL to query data stored in relational database system (RDBMS). Hence, new notable approaches need to be sophisticatedly taken into account to store and consume *Big Data*. The emergence of *"NoSQL Databases"* is to solve the problems caused by *Big Data*. The term *"NoSQL"* indicates a series of data storage systems for managing huge amounts of data without supporting SQL languages and multi-rows transaction. Such systems are usually conceived as, key-value store with support, elastic scalability and schema flexibility. The data in NoSQL databases are stored in the off-the-shelf commodity hardware connected with instable network. As yet, many companies have proposed their NoSQL database prototypes for handling *Big Data*. The most two famous prototypes are *"Big Table"* [9] which was proposed by Google in 2004 and *"Cassandra"* which was inspired by Amazon's *"Dynamo"* [6]. Both are classified into the column-oriented NoSQL database, as the data which belong to the same *"column"* are stored continuously in the disk.

Column-oriented NoSQL databases can play the role of data sink and data source. However, different from RDBMS, most NoSQL databases lack data processing engine. It denotes data query processing has to be implemented by programmers at the application layer. MapReduce [1] is the currently most popular parallel data

J.-S. Pan et al. (Eds.): *Advances in Intelligent Systems & Applications*, SIST 21, pp. 587–598.
DOI: 10.1007/978-3-642-35473-1_58   © Springer-Verlag Berlin Heidelberg 2013

processing framework utilized to cooperate with the column-oriented NoSQL databases. To extract information from column-oriented NoSQL database, programmers submit the desired user-defined Map and Reduce functions to the MapReduce framework and the framework takes care of parallel task scheduling and execution.

As yet, MapReduce jobs are always re-executed to incorporate with the changes of data source ("starting from scratch"). As the size of change data is usually much smaller than the size of the full data source, this approach is obviously inefficient. The better approach is the incremental re-computation, which only propagates the change data to the target data instead of re-computing the whole results. Currently, this approach is widely adopted by large companies, e.g. Google's *Percolator* [3], Yahoo's *continuous bulk processing* (CBP) [2] and Microsoft's *DryadInc* [6].

However, before these incremental re-computation approaches can be applied, the change data of data sources need to be extracted. This data processing step is usually called *change data capture* (CDC). In the context of column-oriented NoSQL databases, the design and implementation of CDC approaches are challenged by the application layer (MapReduce framework) and the peculiarities of column-oriented NoSQL databases which are introduced in Section 2.2.

In this paper, we first summarize and discuss the characteristics of column-oriented NoSQL databases, such as schema flexibility and key-value stores, data versioning, a new data model and a new CRUD (create, read, update and delete) operational model which could impact the design and implementation of the CDC approaches. Then, a notion of net-effect change data is defined to denote the desired data set of output for CDC approaches. Based on the predefined notion of net-effect change data, three CDC approaches i.e. Snapshot Differential, Log-based Approach and Trigger-based approach (Change Tracking Table) are deliberately designed and evaluated to adapt to the core features of column-oriented NoSQL databases.

The paper is organized as follows. In Section 2, the MapReduce framework and the characteristics of column-oriented NoSQL databases are introduced. Section 3 addresses the notion of net-effect change data. In Section 4, three change data capture approaches are described. Section 5 gives the performance for each CDC approach. The related work is discussed in Section 6 and Section 7 makes the conclusion.

2   MapReduce Framework and Column Oriented NoSQL Databases

The MapReduce framework is adopted as the application layer when extracting change data from column-oriented NoSQL databases. Hence, we will briefly describe its processing sequence and corresponding programming model. As extracted change data approaches are applied to column-oriented NoSQL databases, the characteristics of column-oriented NoSQL databases are summarized to conduct the design and implementation for CDC (change data capture) approaches.

2.1   MapReduce Framework

MapReduce [1] is a programming model and an associated implementation designed for parallel processing of massive data sets using distributed commodity hardware, which was proposed by Google in 2004. Programmers specify Map functions which take key/value pairs as input, perform various data manipulations and conversions, and output intermediate key/value pairs. The MapReduce framework then partitions and groups the intermediate results based on the same keys, and passes them to the Reduce functions which are also specified by the programmers. The Reduce function merges the intermediate values with the same intermediate keys. The MapReduce run-time system parallelizes the execution of these functions using data partitions and takes care of the tasks, such as distributed task scheduling, network communication and fault tolerance. The programming model of MapReduce framework is shown as follows:

$$map: \ (k_1, v_1) \rightarrow list(k_2, v_2)$$
$$reduce: \ (k_2, list(v_2)) \rightarrow list(v_2) \tag{1}$$

- Map Function. Map function takes an input key-value pair (k_1, v_1) and produces a set of intermediate key-value pairs (k_2, v_2).
- Reduce Function. Reduce function merges the intermediate key-value pairs $(k_2, list(v_2))$ together to produce a smaller set of values $list(v_2)$.

2.2   Characteristics of Column-Oriented NoSQL Databases

As change data derived from column-oriented NoSQL databases, the peculiarities of column-oriented NoSQL databases need to be first explored. The characteristics of column-oriented NoSQL databases are summarized based on HBase[1] [7] and Cassandra. The summarization mainly addresses the core properties of column-oriented NoSQL databases which could impact the CDC approaches.

1. New data model. In contrast to RDBMS, column-oriented NoSQL databases introduce a new concept called "*column-family*" besides the concepts of table, row and column. The column-family can be considered as a prefix of column names. The columns which belong to the same column-family are stored continuously in the disk.
2. Schema flexibility and key-value stores. The column-oriented NoSQL databases don't enforce a schema before data is loaded into database, e.g. in HBase, user can indicate the column-name when inserting a tuple. This property requires the change data must contain auxiliary metadata (e.g. column-family name and column name). Moreover, the actual data are organized as key-value pairs:
   (**key:***row_key+column-family+column+timestamp,* **value:***value*). Each key-value pair is stored as byte array in distributed file system. It indicates the CDC approaches must have the capability to precisely interpret the data type when accessing the data.
3. Data versioning. Under one column, each tuple stores multiple data versions which are sorted by their related timestamps (TSs). The value of TS indicates when a data version is generated. "Data versioning" implies the output of CDC approach may contain multiple change data versions for one data object.

---

[1]   HBase is an open-sourced implementation of Big Table. It's supported by Apache Software Fundation.

4. New CRUD (create, read, update and delete) operations. The column-oriented NoSQL databases support "Put" and "Delete" commands, when modifying the state of database. It does not distinguish between update and insert operations, as both will "put" a new data version into a table. To support the new data model of the column-oriented NoSQL databases, the operational granularity of deletion is classified into *data version*, *column* and *column-family*. "Delete" operation will not delete the data right way but to insert a marker called "tombstone" to mask the inserted values which TSs are equal to or less than the TS of tombstone. The TS of tombstone is assigned when deletion is issued by the user. Actual data deletion occurs at data compaction time. Figure 1 shows how tombstone is formatted as key-value pair w.r.t the individual deletion granularities in HBase. In Figure 1, each key part contains an auxiliary "operational type", such as *Put*, *Delete*, *DeleteColumn*, *DeleteFamily* to indicate the types of key-value pairs.

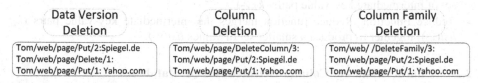

Data Version Deletion	Column Deletion	Column Family Deletion
Tom/web/page/Put/2:Spiegel.de Tom/web/page/Delete/1: Tom/web/page/Put/1: Yahoo.com	Tom/web/page/DeleteColumn/3: Tom/web/page/Put/2:Spiegel.de Tom/web/page/Put/1: Yahoo.com	Tom/web/ /DeleteFamily/3: Tom/web/page/Put/2:Spiegel.de Tom/web/page/Put/1: Yahoo.com

Fig. 1. Various deletion levels of column-oriented NoSQL databases

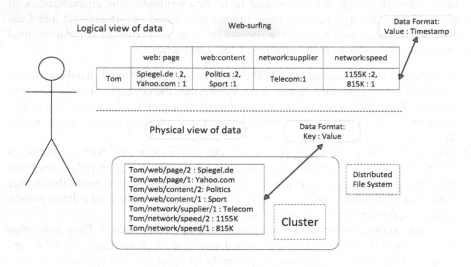

Fig. 2. Example of data stored in HBase

Figure 2 shows an example defined in HBase to illustrate some characteristics of column-oriented NoSQL databases. Suppose there is a table called Web-surfing which maintains the information when user browses internet. The Web-surfing table contains two column-families, i.e. web and network and each column family has two columns. For a given row, each column stores several data values (data versions) with their corresponding timestamps. The logical view of data represents the data structure which is seen by users. In contrast, the data persisted in distributed file system is formatted as key-value pairs depicted as physical view of data.

3   Net-Effect Change Data

Each change data object derived from column-oriented NoSQL databases is logically represented as a triplet: (*metadata, operation, value*). "Metadata" includes the information such as row_key, column-family, column name and timestamp (TS); "Operation" indicates whether the change data is put into or deleted from data source; "Value" denotes the value of data change. As MapReduce is adopted as the application layer for implementing the CDC approaches, it implies the output format of change data is key-value pair. For modeling the "put" change data, we use the original key-value pairs which are generated by the "Put" command. For modeling the deleted change data, we use the actual deleted data values rather than the tombstone. It means: 1) the deleted change data have only the data version granularity; 2) TS for each deleted data object is derived from the prior inserted one rather than the TS of tombstone.   The key-value representation of change data is depicted as:

- **Key:** *row_ key + column-family + column + TS + put or delete.*
- **Value:** *value*

The notion of *net-effect change data* is utilized to appraise the quality of change data set. It represents the change data has no "*noisy data*". "noisy data" denotes the change data set which has no dedication to the target data set. Hence, the change data propagation systems should avoid evaluating noisy data to promote the performance. We use I and D to denote put and deleted change data between time point t1 and t2 under a single column for a single row, respectively. The Venn diagram [14] shown in Figure 3 describes the relations between I and D.

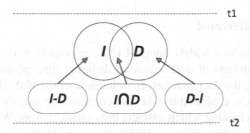

Fig. 3. Venn diagram to describe relations between I and D

In the picture, the change data set is divided into 3 portions, namely, I-D, $I \cap D$ and D-I. From the Venn diagram, following conditions hold:

- $\forall i \in (I - D), t1 < i.TS \leq t2;$
- $\forall n \in (I \cap D), t1 < n.TS \leq t2;$
- $\forall d \in (D - I), d.TS \leq t1;$

The aforementioned conditions and Venn diagram guide the notion of net-effect data, such that mere I-D and D-I could contribute to the target data. For the data values inside $I \cap D$, it's guaranteed the put operation occurs before deletion and the put and deleted change data emerge twice with equal timestamp. For example, suppose we first put a data value v_1 with TS t', t1<t'<t2, and then delete this value. Hence, (v_1, t') is an element of $I \cap D$. Obviously, (v_1, t') is not useful for the target data. In consequence, data elements which belong to $I \cap D$ are considered as noisy data. The notion of net-effect change data under a single column for a single row in time range (t1, t2) is defined as follows:

- $Put : \partial \in (I - D) \wedge !\exists \partial' \in (I - D), \partial.TS < \partial'.TS \leq t2.$
- $Delete : \partial \in (D - I) \wedge !\exists \partial' \in (D - I), \partial.TS < \partial'.TS \leq t1.$

In our definition, each net-effect change data needs to have the closest TS to t2 or t1and must not be an element of $I \cap D$. Moreover, it indicates the final results only contain the latest change data rather than multiple change data versions for one data object.

4   Extracting Change Data from Column Oriented NoSQL Database

In this section, three various CDC approaches, namely, Snapshot Differential, Log-based approach and Trigger-based approach (Change Tracking Table) are introduced to indicate how change data can be obtained from column-oriented NoSQL databases. The output of each CDC approach is restricted by the notion of net-effect change data to discard the noisy data.

4.1   Snapshot Differential

Snapshot Differential is a widely adopted CDC approach. It normally contains two steps: 1) taking snapshots of data source at different time points; 2) comparing the snapshots to acquire the data changes. In column-oriented NoSQL database, the MapReduce framework is exploited to take and compare the snapshots. Figure 4 depicts a MapReduce task chaining which extracts data changes from Web-surfing table between time t1 and t2. There are three MapReduce jobs in the figure. The first and the second MapReduce jobs only contain Map phase which is used to scan the Web-surfing table to gain the whole data versions for each tuple at time point t1 and t2, respectively. The third MapReduce job compares the snapshots to acquire the net-effect data changes.

We only give the pseudo code for the third MapReduce job as the first and the second MapReduce jobs are straightforward. "fam", "col", "val" and "TS" are used as the abbreviations of column-family, column, value and timestamp.

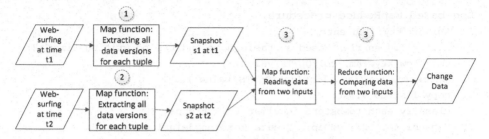

Fig. 4. MapReduce chaining of Snapshot Differential

```
Snapshot Differential MapReduce procedure
1.  Map (row_key/fam/col, TS/value){
2.    emit(row_key/fam/col, TS/value);}
3.  Reduce (row_key/fam/col, list(TS/value)){
4.    sort list based on TS in descending order;
5.    count occurrence occ for each TS/value in list;
6.    remove TS/value whose occ = 2;  // appear in both s1 and s2
7.    find the TS/value whose TS is nearest and ≤ t1;
8.    emit(row_key/fam/col/TS/delete, val);
9.    find the TS/value whose TS is nearest and ≤ t2;
10.   emit(row_key/fam/col/TS/put, val);}
```

Map function scans the two snapshots from the previous two MapReduce jobs. In Reduce function, if the occurrence of TS/value is equal to 2, it implies the same data version appears in both s1 and s2 (4-6). Hence, no data change happens. If a data version appears in s1 but not in s2, it indicates the data version is deleted. In contrast, if a data version appears in s2 but not in s1, it denotes the data version is generated. The output contains only the latest deleted and generated data version w.r.t the notion of net-effect change data (7-10).

4.2   Log-Based Approach

In the column-oriented NoSQL databases, each data node maintains a Write-ahead-log (WAL) which is utilized to recover the database when failure or crash emerges. On the other hand, the content of WAL can be exploited to extract the change data. WAL is a special type of redo log which denotes all the data modifications to a table must be applied after writing a log. A log entry will be recorded when user issues "Put" and "Delete" commands. Each log entry is organized as key-value format and stored in distributed file system. The key part contains the information such as log id, table name and log writing time. The value part includes the data modification.

Please note that, the log entries which record delete operations in WAL don't contain any data values but only tombstones. This property requires CDC to scan the whole WAL to detect the matching "Put" data regarding to the tombstones. The MapReduce procedure for analyzing WAL between time t1 and t2 is described as follows:

Log-based MapReduce procedure

```
1.  Map(log_key, log_entry){
2.  filter log_entries based on table_name and t1≤log.writing_time≤t2;
3.  emit(row_key/fam, col/Op/TS/value);}
4.  Reduce(row_key/fam, list(col/Op/TS/value)){
5.  copy list to vlist;
6.  classify each tombstone in vlist to 3 groups:
7.  Map<col,set<TS>> vdmap; // data version deletion
8.  Map<col,TS> cdmap;  // column deletion
9.  Long fd;  // column family deletion
10. remove tombstones in vlist;
11. foreach element e in vlist: // find out deleted data value
12.   if((vdmap contain e) or (cdmap.col.contain(e.col)&&
13.     e.TS ≤ cdmap.get(e.col)) or (e.TS ≤ fd))
14.    insert e into a cdelist based on column name;
15.   else
16.    insert e into an cinlist based on column name;
17. find element whose TS is nearest and ≤ t1 in cdelist;
18. emit(row_key/fam/col/TS/delete, value);
19. find element whose TS is nearest and ≤ t2 in cinlist and not in
    cdelist;
20. emit(row_key/fam/col/TS/put, value);}
```

The Map function iterates and filters the log entries according to desired table name and log's writing time (2). In Reduce function, each tomestone is classified based on its deleted granularity (5-9). Then, each "Put" key-value pair is detected by the predicate (12-16) to denote whether a key-value pair is a new generated value or a deleted value. In the predicate, four conditions are checked: 1) if the data is deleted by "Delete" tombstone (data version granularity); 2) if the data is deleted by "DeleteColumn" tombstone (column granularity); 3) if the data is deleted by "DeleteFamily" tombstone (column-family granularity); 4) if the data is new generated (15). Finally, the net-effect change data are acquired from cdelist and cinlist w.r.t the individual column (17-20).

4.3   Trigger-Based Approach (Change Tracking Table)

In the column-oriented NoSQL databases, the database state can be varied by two data operations, "Put" and "Delete". Consequently, if the column-oriented NoSQL databases support "*Trigger*", the data changes caused by these data modification events can be recorded in a predefined "Change Tracking Table". The change tracking table contains the same metadata as the original table and an auxiliary column-family called "Delete" which maintains the deleted data value. The content of a change tracking table can be logically viewed as a piece of log information for a specific table during a period of time. Figure 5 shows an example to illustrate the structure of the change tracking table. In the example, a change tracking table is created at time t1 to monitor the data modifications of Web-surfing table. Its initial content is empty. During the time point t1 and t2, five data operations are applied to Web-surfing table, i.e. putting CNN.com:5 and Bild.de:4 into web:page, putting Movie:4 into web:content, deleting CNN.com:5 in web:page and deleting 815K:1 in network:speed.

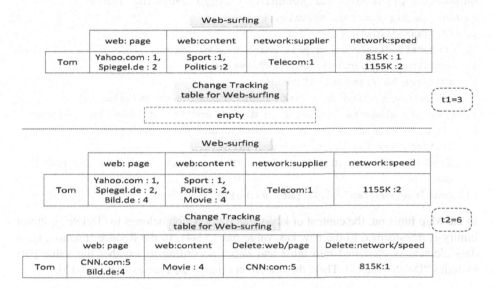

Fig. 5. An example of a change tracking table

For generating the change tracking table, two triggers: 1) for monitoring "Put" command; 2) for monitoring "Delete" command need to be defined:

```
Trigger Definition
1. AfterPut(Put put){
2.  extract metadata from put, i.e. row, fam, col;
3.  get put.TS and put.val from original table;
4.  insert (put.val, put.TS) into the change tracking table;}
5. BeforeDelete(Delete del){
6.  extract metadata from del, i.e. row, fam, col, Op;
7.  if(del.Op is data version level){
8.   search the data whose TS = del.TS and match del.fam/del.col;}
9.  else if(del.Op is column level){
10.  search the data whose TS ≤ del.TS and match del.fam/del.col;}
11. else if(del.lev is column family level){
12.  search the data whose TS≤ del.TS and match del.fam;}
13. insert matching data into tracking table;}
```

The trigger procedure monitors two types of data operations *put* and *deletion* at 3 different operational levels *data version*, *column* and *column-family*. Insertions to the change tracking table occurs in two occasions, "after put" and "before deletion".

As the content of change tracking table reveals the history of data modifications, scanning the change tracking table with certain time range can extract the desired net-effect change data. The following MapReduce pseudo code describes this data processing steps.

```
MapReduce procedure for Scanning Change Tracking Table
1. Map(row_key/fam/col, TS/value){
2.  if (fam=="Delete"){
3.    reconstruct key-value pair using column infor;
4.    emit(row_key/fam/col, Delete/TS/value);}
5.  emit(row_key/fam/col, TS/value);}
6. Reduce(row_key/fam/col, list(Delete/TS/value or TS/value)){
7.  classify elements in list into delist and inlist based on "Del"tag;
8.  find the element whose TS is nearest and ≤ t1 in delist;
9.  emit(row_key/fam/col/TS/delete, value);
10. find the element in inlist whose TS is nearest and ≤ t2 and not in
    delist;
11. emit(row_key/fam/col/TS/put, value);}
```

In the Map function, the content of key-value pairs which belongs to "Delete" column family is transformed to the new key-value pairs (3-4). In the Reduce function, each data element is classified into *inlist* and *delist* according to whether the value part includes "Delete" tag (7). Then, the net-effect change data are extracted (8-11).

5   Evaluation and Performance

We performed experiments on a six-node cluster (Xeon Quadcore CPU at 2.53GHz, 4GB RAM, 1TB SATA-II disk, Gigabit Ethernet) running Hadoop[2] (version 0.20.2-appending) and HBase (version 0.92.0), respectively. HBase is built on top of the HDFS (Hadoop distributed file system) [11]. The original data table is called "Base-ball". It includes 3 column-families, i.e. player, manager and team which have 28 columns, 10 columns and 48 columns, respectively[3]. Each column family contains maximum 3 data versions. The volume of each tuple is approximately 3.6 kilobytes. For keeping the deleted data value, the data compaction process is manually triggered rather than automatically executed by HBase. For implementing Log-based approach, WAL (Write-ahead Log) needs to be stored in extra disk to prevent log deletion. The performance of CDC approaches is displayed in Figure 6.

The Trigger-based (Change Tracking Table) approach shows the best performance, when the fraction of change data is low (approximately under 10%). However, for the middle and high percentages of data modifications, Log-based approach shows better performance than the other two approaches, even it needs to process more volume of data. The reason is that reading data from HDFS is approximately 5 times faster than

---

[2]   Hadoop is an open source implementation of MapReduce framework.
[3]   The test data is derived from http://www.baseball-databank.org/

reading the same amount of data from HBase when utilizing MapReduce framework. Hence, the Trigger-based approach shows better performance, only if the fraction of data modification is low. Snapshot approach shows the worst performance, as it needs to scan the whole table twice to take snapshots even only a small amount of data are modified. Please note we only test the case in which HBase only contains one data table (Baseball table). If more than one table exists in HBase, the performance chart may show different results as Log-based approach has to analyze more log entries.

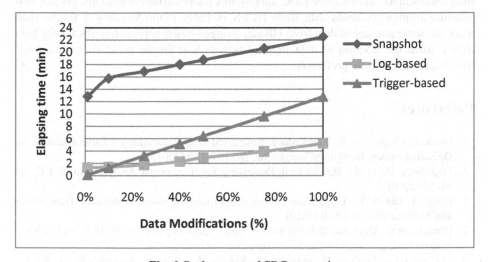

Fig. 6. Performance of CDC approaches

6   Related Work

The state of art of change data capture (CDC) in the context of column-oriented NoSQL databases is ambiguous. Few publications have mentioned the technologies about how to extract change data from column-oriented NoSQL databases. The possible analogous domain could be CDC approaches which are adopted by ETL tools [10, 12]. In ETL sphere, there are three popular CDC approaches, namely Audit Columns, Database Log Scraping and Snapshot Differential. "Audit columns" are appended at the end of each table to indicate the data modified date. "Database log scraping" analyzes transactional redo-log to obtain change data. "Snapshot differential" estimates snapshots taken at different time points to extract the data change. However, all of these approaches are designed under the restrictions and circumstances of RDBMS. In the context of column-oriented NoSQL databases, they cannot be directly utilized to extract change data because of the characteristics of column-oriented NoSQL databases, such as schema flexibility and key-value stores, no support of SQL and multi-rows transactions, data versioning, a new data model and a new CRUD operational model.

7   Conclusion

Based on analyzing the impact caused by the characteristics of column-oriented NoSQL databases, we proposed 3 change data capture (CDC) approaches, i.e. Snapshot Differential, Log-based approach and Trigger-based approach (Change Tracking Table) which are implemented by the MapReduce framework. The output of each approach coincides with the notion of net-effect change data. The performance chart shows the winner among three CDC approaches under certain conditions. As the MapReduce framework reads data from HDFS is faster (approximately 5 times) than reads the same amount of data from HBase, Trigger-based approach is always the best choice when the fraction of data modification is low. In the other situations, Log-based approach is more preferable.

References

1. Dean, J., Ghemawat, S.: MapReduce: Simplified Data Processing on Large Clusters. In: Operating System Design and Integration, pp. 137–150 (2004)
2. Logothetis, D., et al.: Stateful Bulk Processing for Incremental Analytics. In: SoCC, pp. 51–62 (2010)
3. Peng, D., Dabek, F.: Large-scale Incremental Processing Using Distributed Transactions and Notifications. In: OSDI (2010)
4. Popa, L., et al.: DryadInc: Reusing work in large-scalecomputations. In: HotCloud (2009)
5. Apache Hadoop, http://hadoop.apache.org
6. DeCandia, G., Hastorun, D., Jampani, M., Vogels, W.: Dynamo: Amazon's highly Available Key-value Store. In: SOSP 2007, Stevenson, Washington, USA, October 14-17 (2007)
7. Apache HBase, http://hbase.apache.org/
8. Apache cassandra, http://cassandra.apache.org/
9. Change, F., et al.: Bigtable: A Distributed Storage System for Structured Data. In: OSDI, pp. 205–218 (2006)
10. Kimball, R., Caserta, J.: The Data Warehouse ETL Toolkit: Practical Techniques for Extracting, Cleaning, Conforming, and Delivering Data. Wiley Publishing, Inc. (2004)
11. Apache HDFS, http://hadoop.apache.org/hdfs/
12. Jörg, T., Dessloch, S.: View Maintenance using Partial Deltas. In: BTW, pp. 287–306 (2011)
13. Jörg, T., Parvizi, R., Hu, Y., Dessloch, S.: Incremental Recompilations in MapReduce. In: CloudDB 2011, Glasgow, Scotland, U.K (2011)
14. Venn, J.: On the Diagrammatic and Mechanical Representation of Propositions and Reasonings. Philosophical Magazine and Journal of Science (July 1880)

Approaches for Data Synchronization on Mobile Peer-to-Peer Networks

Chuan-Chi Lai and Chuan-Ming Liu

Department of Computer Science and Information Engineering
National Taipei University of Technology
Taipei, Taiwan
{t100599007,cmliu}@ntut.edu.tw

Abstract. The energy and connectivity on the mobile devices in mobile peer-to-peer (MP2P) networks are limited. Since different mobile peers may store copies of shared information, the data synchronization on MP2P becomes crucial and challenging due to node mobility and limited resources. We propose effective approaches to solve the mentioned problem with a dynamic tree-like index structure and group-key agreement mechanism. The proposed approaches can effectively synchronize the data items and perform well in terms of the coverage of successfully synchronized nodes and the number of redundant messages. Last, we validate the proposed approaches through extensive simulation experiments and present our findings.

1 Introduction

The concept of Web 2.0 or social networks has been widely applied or realized in recent years [1]. This trend brings up some complicated and challenging issues about data sharing, publishing, and synchronizing. The applications on data sharing systems usually need the interaction between users and the data kept in different users should be consistent. Information newly published or updated should be synchronized correctly and promptly. Due to the node mobility, the data synchronization on a MP2P network becomes more interesting and crucial and has attracted much research attention in recent years [2–8].

The same data items on different nodes can be simply synchronized if the locations of the data items in the network are known. However, the information of the locations about the same data items is usually not easy to obtained in MP2P networks. Extra cost should be paid to have the data item's locations. The objective of this study is to design suitable data synchronization approaches for a data sharing system on an MP2P network, where the message flooding is avoided. The message cost is therefore reduced and more nodes can be updated successfully. We provide three different approaches on different architectures: Distributed Dynamic Tree-like Scheme (DDTS), DDTS with Group-Key agreement (DDTS-GK) mechanism [9] and DDTS-GK with a group information Server (DDTS-GK-S); and compare them with some existing works [3] carefully in the coverage of successfully synchronized nodes per update, the overhead of

J.-S. Pan et al. (Eds.): *Advances in Intelligent Systems & Applications*, SIST 21, pp. 599–608.
DOI: 10.1007/978-3-642-35473-1_59   © Springer-Verlag Berlin Heidelberg 2013

network load, and the amount of redundant messages. The effects of time period for an update, node speed, and node density are also discussed in this paper.

The rest of this paper is organized as follows. In section 2, we review the related work and introduce some background. The proposed solutions are given in Section 3, 4 and 5, respectively. Due to the constraint on the article length, we focus on the data synchronization process in the paper. Some other issues such as the election of supernodes and the process of node's join and departure will not be mentioned here. The simulation experiments are presented and discussed in Section 6. Last, we give the conclusion remarks in Section 7.

2   Related Work and Preliminaries

Some existing approaches for data data consistency problem divide the entire network topology into multiple geographical regions, groups or clusters for reducing the latency of search [2, 4, 5]. These approaches focus on the hit rate of cache, latency of query, or energy consumption, but do not address the problem of data consistency. Some related works [6] and [7] focus on replica data consistency and propose strategies for updating the replicated data items. [8] builds a collaborative application for ad-hoc wireless networks and can identify the need to replicate data across ad-hoc groups. Each peer has to maintain a global tree for the consistency of indices and data. In [3], a scheme for the consistency among cooperative caches, called DTCS, is proposed for mobile ad-hoc networks. DTCS uses Chord as the group management protocol and dynamically builds a data-updating tree from the Chord ring to maintain the consistency between data.

In this paper, the basic MP2P architecture is built upon a mobile ad hoc network. In our design, the connections between nodes in the physical network are considered. As a side effect, the topologies of overlay and physical networks are almost the same, thus eliminating the topology mismatching problem. We not only consider the basic architecture but also add a new mechanism of group maintenance in our design to deal with the complex MP2P networks.

In the experiments, three measurements will be measured. The first is the *network load* of system. In order to identify and analyze the advantages and disadvantages of each approach, we classify the messages into five types of messages: update query, topology control, data transmission, retransmission, and dropped messages. The second measurement is the *update coverage* of a data item D_i and defined as

$$Coverage(D_i) = \frac{c_t}{n} \times 100\%, \tag{1}$$

where c_t is the number of successfully updated nodes within time t and n is total number of nodes which contain data item D_i. The last measurement is the *message redundancy*. Given the total number of messages M and data item D_i, the message redundancy is defined as

$$Redundancy(D_i) = (1 - \frac{m_d + m_q}{M}) \times 100\%, \tag{2}$$

where m_d is the number of successful data propagation messages and m_q is the number of query messages. We also discuss the impacts of node speed and density on the update coverage and message redundancy in the experiments.

3   Distributed Dynamic Tree-Like Scheme

We now introduce the first approach, Distributed Dynamic Tree Scheme, on which the other two approaches are based. In DDTS, a new data index structure for managing nodes is proposed and the routing pattern for data synchronization is tree-like. Each node keeps two kinds of information: node description and the records of shared data items on that node. The node description includes *Node ID*: the identity of the node; *Appearance Time*: the time at which this node appears in the network; *Main Hash*: a hash value that characterizes the node ID and all the ID's of the data on this node; and *Neighbor List*: a list of 1-hop neighbors. The record of a shared data item on a node keep *Data ID*: the identity of the data item; *Timestamp*: the latest modified-time of the data item for data synchronization; and *Owner List*: a list of node's neighbors which also keep the data item.

Since nodes move over time, each node's neighbors and the neighbor list change frequently. In our approach, every data owner list will be updated as soon as the neighbor list changes. We adapt a gossip protocol for maintaining the node's status. Each node periodically sends messages to its new neighbors for notifying them what data it has. The new neighbors then reply to the node with what data items they have as well. Thus, the owner list of each data item can always be up-to-date and the routing pattern of messages is like a multicast tree. In short, the system can avoid broadcasting for data update and synchronization.

Synchronization Process of DDTS

When a node receives an update message or modifies a data item, this node will multicast an update message with the data item according the owner list. The node's neighbors which have the same data item will do the same update process after receiving the update message.

```
NODE_UPDATE_PROCESS()        // Update Process of a Node
    Input: Received new message m, shared data set D
1   d = PARSEMESSAGE(m)      // d, d_o are data items
2   d_o = FIND(d, D)
3   if d_o is not NIL
4       if d's timstamp > d_o's timstamp
5           UPDATE(d, d_o)
6   if d.TTL > 1
7       L = d's owner list
8       m = CREATEMESSAGE(d, d.TTL)
9       MULTICAST(m, L)
```

Although we use multicast instead of broadcast to reduce the number of redundant messages, it still could occur infinite-loop update. To avoid this, we make each node which has the latest copy of data to stop forwarding update messages if the new data's timestamp of the received update messages is older than the one it owns. The operation called NODE_UPDATE_PROCESS() is shown in following. The condition at 3th line checks whether the received data item is newer than the one it owns. The 9th line in UPDATE_PROCESS() forwards the update message. UPDATE_PROCESS() gives a TTL value to each message and then sends the message out according to the owner list. With the TTL, the number of messages for synchronization can be reduced.

Using the owner list, DDTS can avoid message flooding or broadcast when managing the data, thus reducing the amount of redundant messages. Hence, how to maintain the owner list becomes crucial. Due to node mobility, the owner list should be updated often since the neighbors change frequently. In DDTS, every data item has its owner list. If the all owner lists update at the same time or frequently, the computation cost will be large. We maintain the owner lists in the following two ways. The first one is that DDTS will periodically update the owner lists. The second one occurs when a node becomes a new neighbor. In this case, two nodes will identify each other to be a new neighbor, update the neighbor list, exchange the list of shared data items, and then update the owner lists. Both ways will call UPDATE_OWNER_LIST() to check whether each entity in the owner list is also in the neighbor list. The pseudocode is shown in following.

UPDATE_OWNER_LIST()   // Data d's owner list update process
   Input: Node's neighbor list $neighborlist$ and owner list $ownerlist$
1   A temporary list $list_{temp}$
2   **for** $i = 0$ **to** $ownerlist.length$
3   **if** $neighborlist$.contains($ownerlist[i]$)
4           $list_{temp}$.add($ownerlist[i]$)
5   $ownerlist = list_{temp}$

If an entity is not in the neighbor list, the owner list will remove this entity. The number of redundancy messages would be few because DDTS does not need too many extra operations or messages to control the synchronization process. The update process is also simple, lightweight, and suitable for mobile environments.

4   DDTS with Group-Key Agreement Mechanism

To effectively deal with the data management in mobile environments, the network is usually partitioned into several groups. As shown in Fig. 1, each group has at least one leader, called Group GateWay (GGW), which is responsible for gateway routing and group maintenance. In our second approach, the GGW is set to be the leader of a group. GGWs can help managing data more efficiently. The approach is based on DDTS so the node structure is kept. Besides, each GGW is appended with a new table to record the data IDs and hash keys of the sharing data items in its group.

Since nodes may migrate between groups, the same data items could exist in different groups. In this situation, how to synchronize the data items in different nodes and groups is challenging. Moreover, if many nodes modify the same data, how to verify the modifications is important. In our proposed approach, we use the group-key agreement mechanism to help the verification efficiently in terms of speed and message cost.

In the second approach, we adapt the group-key agreement mechanism [9] on GGW to verify and lock the data items in the group. Each group has at least one GGW. GGW is responsible to record indices of all data items in the group and generates a group lock with the hash values of the data items. For example, in Fig. 2, group G_1 generates the group lock, LOCK(G_1), using the hash values of data items D_1, D_3 and D_4. Then, the hash value $(Data\ ID)$ of each data item will be the key to unlock the group lock. A group lock can only be unlocked by the key of any data item in the group. So, LOCK(G_1) in Fig. 2 can be unlocked by Hash$_1$, Hash$_3$, or Hash$_4$.

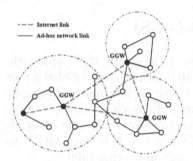

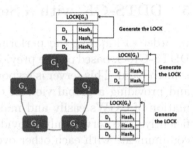

Fig. 1. An example of a partition on an MP2P network, where GGW denotes the Group GateWay

Fig. 2. An example for group key agreement, where $G_1, G_2, \ldots, G_5$ are groups, each group generates a lock, and only the group member can unlock it

Synchronization Process of DDTS-GK

In a group, the data synchronization is the same as DDTS. The data synchronization between groups will employ the group-key agreement mechanism. Each GGW may communicate with the others using the network links and this can help to improve the coverage of data synchronization between different groups.

```
GGW_UPDATE_PROCESS()      // Update Process of GGW
    Input: Received new message m and shared data set D
1   d = parseMessage(m)
2   d_o = FIND(d, D)
3   data_key = d_o.key   // Old data key
4   if d_o is not NULL
5       if d's timstamp > d_o's timstamp // Verify the data key
6           if UNLOCK(data_key) is TRUE
7               UPDATE(d, d_o)   // Update the data d_o
8               L = d's owner list
9               m = CREATEMESSAGE(d, d.TTL-1)
10              MULTICAST(m, L)
11              UPDATE_GROUP_KEY()    // Update the Group-Key
            // Flood update messages through the Internet links
12          UPDATEFORWARD(d.getData(), data_key, d.TTL-1)
```

The above pseudocode, GGW_UPDATE_PROCESS() shows the update process of GGW. When a GGW receives an update message for a data item, the GGW will check the message's type first, and then use the old key of the data item to unlock the group lock as the 6th line in GGW_UPDATE_PROCESS(). If the unlock function returns true value, it means that the old data items is in the group and need to be updated. Then, the GGW will send the new data item to its group members according to the owner list. Afterwards, a new lock for the group is generated and the GGW forwards the update message to the other GGWs. On the other hand, if the group lock can't be unlocked, it means that this group doesn't contain the data item which needs to be updated, then the GGW forwards the update message unless the value of TTL equals 0. With this mechanism, the number of redundant messages can be reduced and the update coverage can also be improved effectively.

5   DDTS-GK with a Server(DDTS-GK-S)

In order to improve the performance, we further propose the third approach, DDTS-GK-S, based on the previous approaches. In this approach, a global server is introduced. This server is responsible for maintaining the information of groups and providing a global view of the MP2P network. Each GGW can get information of GGWs easily and make Internet connections between each other efficiently. The server only provides the state of GGWs to help each GGW to communicate with each other over the Internet, such as routing table.

The update process of DDTS-GK-S is almost the same. Only one different part of the update process is that a GGW consults with this server about the information of all GGWs before transmitting update messages to all the other GGWs over the Internet. By this way, each GGW can connect and transmit the update messages to all the other GGWs precisely, and then the consistency of data items between different groups can be kept easily.

6   Simulation Results

This section presents the simulation results for the proposed approaches: DDTS is the basic scheme we proposed; DDTS-GK is combined with a group-key agreement mechanism; and DDTS-GK-S uses a server handle group management. For each scenario, we perform more than 10 runs to get the average results. We compare the performance of our proposed approaches, flooding strategies, and DTCS in [3] in terms of update coverage and message redundancy. For flooding strategies, we consider the fully flooding approach, Flooding, and a limited one, Flooding-TTL-3, where the flooding is limited within three hops. The impacts of node speed and density are also discussed. In the experiments, AODV [10] has been used as the underlying routing protocol. There are 1000 nodes in a 1000m×1000m network. The radius of a node's transmission range is from 20m to 40m and the retransmission interval is 1.5 sec. The node speed is uniformly generated within the range of [0, 20] m/s. Each node has at most 10 neighbors and 20 to 30 data items. The owner list is updated every 3 seconds. The total number of data items is 1000. We use Random WayPoint (RWP) mobility model, where a user may pause for up to 2 minutes to look for a destination. The bandwidth is set to 2Mb/s and the TTL is 3 hops. Furthermore, the document size is 512 KB and there are 10 queries issued per second. Initially, nodes are placed randomly in the area.

6.1 Message Redundancy

The results for message redundancy is presented in Fig. 3. All the proposed approaches are much better than the flooding-based strategies and the message redundancy of DDTS is close to 12% of total number of messages. DDTS-GK needs more communication not only between GGW and normal nodes but different groups, so its performance is little worse than DDTS. DDTS-GK-S has the least message redundancy since it uses a server to keep the routing information of all the groups. The GGWs can easily communicate with each other and avoid flooding messages between GGWs. DTCS only outperforms the flooding-based strategies since it still has to spend network resource to collect the latest information of network topology to build an updating multicast tree. This process causes excessive amount of topology control messages and dropped messages.

6.2 Update Coverage

We now consider the effectiveness of an update. After the simulation starts and all data items have been spread on the MP2P network, we randomly select a data item to modify, trigger the update process and then measure its coverage of successfully synchronized nodes. Nodes having the data items may be disconnected from the network frequently when moving. The probability for them to lose the update messages may be high. This will effect the update coverage. The results in Fig. 4 shows that the update coverage increases as time moves on. Although DDTS for update coverage is not better than flooding approaches, the update coverage still approaches 90% as time lasts. In DTCS, a global updating tree for updating data items can help. However, it takes time and high overhead to dynamically build the global updating tree. So it performs just a little better than the proposed DDTS. DDTS-GK and DDTS-GK-S both perform better than the flooding approaches. DDTS-GK can reach 95% update coverage because GGWs can maintain, spread, synchronize the data items between different groups efficiently. In fact, there is still much overhead of communications for data synchronization between groups. DDTS-GK-S has the best performance because it uses a server to provide each group's information and can avoid flooding messages between groups.

6.3 Network Load

We first discuss the network load for all the methods. There are ten nodes generating the new update queries per second in average. We observe the network load after the system has been established for a period of time. Fig. 5 show the average network loads of different approaches. Apparently, Flooding and Flooding-TTL-3 are the worst two approaches. Using flooding will causes the broadcast storm

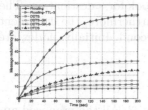 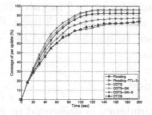 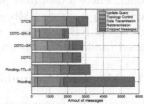

Fig. 3. Message Redundancy   **Fig. 4.** Update Coverage   **Fig. 5.** Network Load

problem and a great amount of dropped messages wastes the network resource. DTCS is just a little better than Flooding-TTL-3. DTCS builds a global updating multicast tree to update the data items. If the network scale becomes large or node's speed becomes faster, the data transmission would fail easily. To retransmit the message accurately, DTCS need to rebuild a new global updating multicast tree to complete the data retransmission. The more network topology changes, the higher the overhead of building global updating multicast tree is. So the number of messages for topology control and data retransmission becomes large. This show that DTCS may not be suitable to the network architecture we consider. In contrast, our three approaches can have better network load since our approaches can effectively control the amount of messages to be sent in the network. This also can be verified by the message redundancy in the following subsection.

6.4 Speed and Density

The last part in the simulation experiments, we discuss the impacts of different node speed and density. According to the previous results for update coverage and message redundancy, the curve of each approach doesn't increase obviously after 180 seconds. In other words, after 180 seconds, the update coverage will be steady. We thus use the results of steady update coverage to discuss.

As shown in Fig. 6, the update coverage of every approach decreases as node speed increases. In other words, no mater which approach the system uses, when more nodes move in high speed, the update coverage will be reduced. DDTS performs worst because it uses tree-like multicast for reducing the number of messages and has less information of data items to spread quickly. DTCS has more information about data items and node speed, so it can perform better than flooding-based approaches and DDTS. DDTS-GK has a better performance than DDTS and DTCS because it uses group management to reduce the time of synchronization process and allows different groups can communication with each other. DDTS-GK-S is the best one among these proposed approaches since different GGWs can communicate with each other more efficiently and correctly; thus, increasing the scale of update coverage in high-speed mobile environments.

Fig. 7 shows that the message redundancy increases when the node speed increases in all approaches. The curves of DDTS and DDTS-GK meet nearly

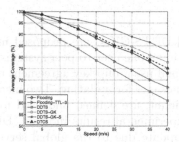
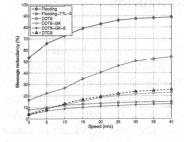

Fig. 6. Speed vs. Update Coverage

Fig. 7. Speed vs. Message Redundancy

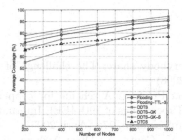
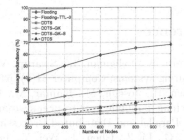

Fig. 8. Density vs. Update Coverage

Fig. 9. Density vs. Message Redundancy

at 5 m/s of node speed. When the node speed is less than 5 m/s, the network tends to be relatively static and DDTS-GK outperforms DDTS. In this case, the cost for managing the group is less and most of the updates can be done with less time using the communication between GGWs, compared to the update in DDTS using hop by hop communications. If the node speed is higher than 5 m/s, the network topology and connections change quickly. Node migrations between groups occur more frequently so that DDTS-GK needs more cost of messages to maintain the data consistency. Similar to DDTS-GK, DTCS needs to collect more information from the other nodes to build the updating data multicast tree. So, more redundant messages are generated than the proposed approaches due to the frequent changes on the network topology.

In Fig. 8 and Fig. 9, all the proposed approaches have better performance when network density(number of nodes) increases and DDTS-GK-S has the best performance with different node densities. As the density of nodes increases, the connectivity of nodes also increases. So, the failure rate of data synchronization and data transmissions could be reduced. Under such a circumstance, more nodes are active on the network and more message cost are necessary for managing the shared data and groups. As the number of nodes increase, DTCS not only has to collect more information from more the other nodes, but needs to build a larger updating data multicast tree for updating data items. Thus, more message redundancy occurs.

7   Conclusions

Three approaches, DDTS , DDTS-GK and DDTS-GK-s, are proposed in this paper for the data synchronization problem in mobile peer-to-peer systems. The approaches manage the nodes using groups and use group-key agreement mechanism to effectively control the number of messages for updating. Two measurements, update coverage and message redundant, are defined and used for the comparisons on the performance. As the experimental results indicate, DDTS-GK-S has the best coverage of successful update due to the global server. In general, when the node speed is high, the update coverage is reduced and the message redundancy arises. On the other hand, when the node density is high, the update coverage and message redundancy both increase.

References

1. Smith, K., Seligman, L., Swarup, V.: Everybody share: The challenge of data-sharing systems. Computer 41(9), 54–61 (2008)
2. Shen, H., Joseph, M., Kumar, M., Das, S.: Precinct: A scheme for cooperative caching in mobile peer-to-peer systems. In: Proceedings of 19th IEEE International on Parallel and Distributed Processing Symposium, p. 57a (April 2005)
3. Xe, G., Li, Z., Chen, J., Wei, Y., Issarny, V., Conte, A.: Dtcs: A dynamic tree-based consistency scheme of cooperative caching in mobile ad hoc networks. In: Proceedings of the 3rd IEEE International Conference on Wireless and Mobile Computing, Networking and Communications, pp. 48–56 (2007)
4. Chand, N., Joshi, R., Misra, M.: Cooperative caching strategy in mobile ad hoc networks based on clusters. Wireless Personal Communications 43, 41–63 (2007), doi:10.1007/s11277-006-9238-z
5. Chauhan, N., Awasthi, L.K., Chand, N., Joshi, R.C., Mishra, M.: A cooperative caching strategy in mobile ad hoc networks based on clusters. In: Proceedings of the 2011 International Conference on Communication, Computing & Security, ICCCS 2011, pp. 17–20. ACM, New York (2011)
6. Lu, H., Denko, M.: Replica update strategies in mobile ad hoc networks. In: Proceedings of the 2nd IFIP International Conference on Wireless and Optical Communications Networks, pp. 302–306 (2005)
7. Nawaf, M.M., Torbey, Z.: Replica update strategy in mobile ad hoc networks. In: Proceedings of the International Conference on Management of Emergent Digital EcoSystems, pp. 474–476 (2009)
8. Coatta, T., Hutchinson, N., Warfield, A., Won, J.: A data synchronization service for ad hoc groups. In: Proceedings of the 2004 IEEE Wireless Communications and Networking Conference, pp. 483–488 (2004)
9. Jeong, I.R., Lee, D.H.: Key agreement for key hypergraph. Computers & Security 26(7-8), 452–458 (2007)
10. Perkins, C., Royer, E.: Ad-hoc on-demand distance vector routing. In: Proceedings of Second IEEE Workshop on Mobile Computing Systems and Applications, WMCSA 1999, pp. 90–100 (February 1999)

On the Design of a Load Balancing Mechanism for ALE Middleware

Yi-Ting He, Yu-Chang Chen, and Chua-Huang Huang

Department of Information Engineering and Computer Science
Feng Chia University
Taichung, Taiwan
{yiting,ycchen,chh}@rfidlab.iecsfcu.edu.tw

Abstract. RFID middleware plays a very important role in the RFID applications, as whenever a large quantity of specifications need to be processed, or when the reader received a vast amount of data, the RFID middleware would be overloaded, and although traditional RFID middleware are usually duplicated completely among different machines and environment to balance the loads of the middleware, most of them do not conform with the characteristics of ALE. Hence, following the ALE characteristics, we propose our modularized middleware structure and partner it with our load balancing mechanism to achieve load balancing goal. Coupling these modules with the load balancing mechanism I designed, we would place each module at a node, and when a module node is overloaded, new nodes will be added to ease the load. On the other hand, when the load is of minimal, unneeded nodes can be shutdown.

In this paper, we propose a method to achieve load balancing in a distributed middleware that follows EPCglobal standard. The method according to predict the specifications for possible workload, and load status of each node to do the allocation of tasks and decide when to add and delete nodes to avoid inefficiency resources utilization.

Keywords: RFID, middleware, EPCglobal, ALE, load balancing.

1 Introduction

In recent years, radio frequency identification (RFID) technology has been widely used in various fields, such as warehouse management, logistics, retail, supply chain, access control, electronic ticket, and library application. Typically, an RFID application system is composed of tags, readers, application programs. With the expansion of the scale and scope of applications, a huge volume of tag data may be scanned by various readers that could bring heavy burdens to application programs. To deal with the problem of heavy data load and heterogeneous readers from different vendors, an RFID middleware is deployed between readers and application programs [1, 9]. The major functions of an RFID middleware are to interoperate with readers of various vendors, to filter out unnecessary tag data, and to aggregate tag data with given logical conditions.

J.-S. Pan et al. (Eds.): *Advances in Intelligent Systems & Applications*, SIST 21, pp. 609–618.
DOI: 10.1007/978-3-642-35473-1_60   © Springer-Verlag Berlin Heidelberg 2013

EPCglobal has released a collection of RFID standards for supply chain management [7]. One of them is Application Level Events (ALE) which is the middleware specification for filtering and aggregating tag data from various readers [6]. A number of RFID middleware systems are reported with variety of functions. Fosstrak open source RFID platform has developed Fosstrak ALE Middleware with LLRP Support according to EPCglobal standards [8]. Some other middleware systems even support application oriented functions. Dong, Wang, and Sheng present an RFID middleware based on complex event processing [4]. Dutta et al. propose an RFID middleware system with real time event handling [5].

Expansion of the scale of enterprise applications, much of the information may not be able to predict in advance, or in certain circumstances are required to deal with a lot of tag data. These may cause the RFID middleware in overloaded conditions, and adapt to the situation need to deploy multiple copies of middleware. However, this may cause the resource utilization problem. According to experiments, a study shows that 98% of processes only use 35% of CPU time and some 0.1% of the processes use upto 50% CPU time [13]. In order to effectively manage the resources of RFID middleware, load balancing is required. However, most distributed or parallel RFID middleware systems with load balancing usually reproduce middleware systems on different machines and environments.

Park et al. use the amount of tag data as load metrics and a connection pool mechanism to achieve load balancing [12]. This method collects all tags data read by a reader and place them in a connection pool to preprocess before filtering tag data. This method causes the problem when a large volume of tag data in the connection pool, the connection pool will be overloaded. Liu, Jie, and Hu also use connection pool in a distributed RFID middleware [10]. Their method fully reproduces middleware that may cause resource waste and it does not consider connection pool overloading problem. Park, Chae, and So present a centralized architecture which use an agent mechanism to perform load balancing [11]. They follow EPCglobal ALE specification that ECSpec is used as load metrics. This paper achieves load balancing through migration of ECSpec. However, it does not explain how migration is done and not consider the situations of larger number of ECSpecs and logical readers assigned with same physical readers. Cui and Chae also use agent and migration to solve load balancing problem [3]. They use readers as the load metrics and migrate readers. Chae et al. also propose five basic load balancing strategies [2]. Zou also present a load balancing mechanism using agent [14]. However, all of these works do not consider dependency between readers and ECSpecs that may cause larger burden of migration and result in data error.

In this paper, we propose a distributed module of RFID middleware that follows EPCglobal ALE specification. With the modular approach, one or more modules are executed on a processing node. A node is created only for an overloaded module. Both the number of ECSpec and node workload are used as metrics to achieve load balance. In addition, when the workload of a node is reduced under a given level, the node will be deleted when its processes are completed or are migrated to another node. This approach will utilize resources more efficiently and improve system scalability.

This paper is organized as follows. In Section 2, we give an overview of ALE standard and distributed ALE middleware in our RFID lab. In Section 3, we present the load balancing mechanism of the distributed ALE middleware, and in this section will elaborate on how the load balancing mechanism with our ALE middleware. Section 4 shows the experimental results of the load balancing mechanism. Section 5 is the conclusions and future work.

2   Overview of Distributed ALE Middleware

EPCglobal has released variety standards based on EPCglobal Architecture Framework [7]. In this section, we will briefly introduce ALE and our implementation of a distributed ALE middleware.

2.1   Application Level Events Middleware

Application Level Events (ALE) is a specification standard released by EPCglobal organization [6]. In Fig. 1, ALE plays the role of a middleware system between the applications and the readers. The major of ALE middleware is to filter and group tag data collected from readers, and to generate and forward resulting reports to back-end applications. The purpose is to reduce the load of back-end applications by sending only useful tag data to applications.

2.2   Distributed ALE Middleware

In this paper, we use distributed RFID middleware which according to ALE feature designed. Fig. 2 shows the architecture of the distributed RFID middleware. Following the ALE specification standard, the middleware system is divided into five modules: user interface module, event process module, data process module, report generator module, and reader control module.

1.   User interface module is used as interfaces that middleware clients can operate ALE through this entrance. For example, the client can define a event cycle or command cycle specification (ECSpec or CCSpec), define logical reader mapping, and set boundary cycle.

2.   Reader Control module is responsible for communication between the ALE middleware and readers. In this module, it contains LLRP reader module, customized reader module. LLRP module is implemented to support LLRP

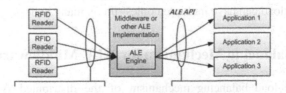

Fig. 1. ALE operations

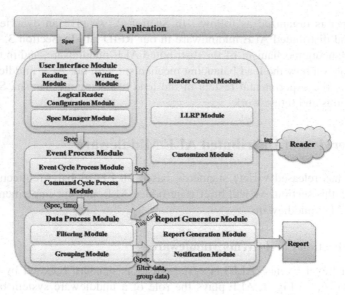

Fig. 2. Distributed ALE middleware

readers and customized reader module is implemented to support customized readers, including Alien ALR-9900, AWID MPR-1510, AWID MPR-2010BN, Impinj R1000, and our reader simulator.

3. Event process module will receives the processed information form user interface module. It is responsible for checking the ECSpec/CCSpec specifications, calculating the boundary cycle and notifying the reader control module. In this module, it contains event cycle process module, command cycle process module. We will focus on the event cycle process module which performs boundary cycle calculation.

4. Data process module is responsible for receiving data from event process module and reader control module and executes filtering and grouping operations of tag data according to event cycle specifications. This module consists of filtering module and grouping module.

5. Report generator module is responsible for receiving data from data process module and generates report according to ReportSpec. In this module, it contains report generation module and notification module. Report generation module will generate reports which are sent to ALE clients by notification module

These five modules can be executed on distributed processing nodes and a module with heave workload can have multiple copies run on many nodes.

3   Load Balancing Mechanism for ALE Middleware

We present the load balancing mechanism of the distributed ALE middleware. In addition to the five modules described in Section 2, we add two new modules:

specification manager module and load balancing module. Specification module coexists with user interface module to mange ECSpecs submitted from ALE clients and load balancing module is deployed in every module to monitor their workload to achieve load balance functions.

We define two cost functions of load balance metrics: one for calculating node load state and another for specification load:

$$L_{All(i)} = L_{cpu(i)} \times e_{cpu} + L_{mem(i)} \times e_{mem} + L_{que(i)} \times e_{que} \tag{1}$$

$$L_{S(i)} = L_{lr(i)} \times e_{lr} + L_{t(i)} \times e_t + L_{rs(i)} \times e_{rs} \tag{2}$$

In Equation 1, L_{All} is the loading cost of a distributed node that measures CPU usage, memory usage, and waiting job queue. The workload of CPU usage, memory usage and waiting job queue are denoted as L_{cpu}, L_{mem}, and L_{que}, respectively. L_{cpu} and L_{mem} are the value of usage percentages; L_{que} is the percentage of the waiting jobs and a specified maximum queue length. For these three workloads, each of them is assigned a weight e_{cpu}, e_{mem}, and e_{que}, respectively. The purpose of weights is to adjust the total loading cost to ensure $L_{All} \leq 1$. The weights e_{cpu}, e_{mem}, and e_{que} are assigned by system administrator and the best values can be determined by experiments. In Equation 2, L_s is the loading cost of a specification which is defined on the base of logical reader L_{lr}, boundary specification L_t, and report specification L_{rs}. L_{lr} is the number of physical readers designated in the logical readers of an ECSpec; L_t is the cycle frequencies and trigger specified in an ECSpec; L_{rs} is the number of for reports and filtering and grouping patterns in an ECSpec. Similarly, weights e_{lr}, e_t, and e_{rs} are the adjustment variables to make $L_s \leq 1$. The weights e_{lr}, e_t, and e_{rs} are assigned by system administrator and the best values can be determined by experiments.

Fig. 3 shows the functions of specification management module which is an addition to the user interface module. The specification management module contains specification load component and specification message delivery component. The specification load component analyzes logical readers, boundary conditions and report specification to determine specification cost. Logical reader definitions are used to calculate the number of physical readers in the ECSpec. For boundary conditions, the specification load component calculates event cycle frequency and the number of triggers. The specification load component also calculates the number report specification, filtering patterns and grouping patterns defined in an ECSpec. The specification message delivery component will send an ECSpec upon the requests of other nodes.

The load balancing module is deployed in all nodes. The operations of this module are performed in six steps as depicted in Figure 4:

Step 1: The five modules are executed as a pipeline as in Fig. 2. Event process module, reader control module, data process module, and report generator module will receive input message from its previous module. The message is sent by sender manager unit in message delivery component.

Step 2: Load information unit in the state information component will collect node workload and send the workload to message delivery component.

Fig. 3. Specification Manger module

Step 3: Load blackboard unit send the node workload to the state information component of its previously node. This step is to keep every node informed of the workloads of its succeeding nodes in the pipelined. Task dispatch and new node creation are decided by the workloads of the succeeding nodes.

Step 4: The workloads of succeeding nodes are forwarded to load balancing component for making load balancing decision.

Step 5: Balancing manager unit and balancing decision unit will make balancing decision and determine task dispatching policy.

Step 6: Workloads and task dispatch decisions are sent to location management component which does bookkeeping of task placement and node workloads.

The load balancing decision and task distribution mechanism are described in Fig. 5. With the loading cost and specification cost, balancing decision unit will first find whether a node has enough capability to accept the incoming task. If no acceptable node is found, a new node will be created. If only one node is acceptable, the task is dispatched to it; if many nodes are acceptable, a decision is made as the following. If a node with light load, i.e., workload less than 50%, is available, this node is chosen. If all nodes with medium load, i.e., workload between 50% and 80%, the specification load is considered whether to dispatch the task. When the specification load is much larger than the remanding node load or a node is heavily loaded, i.e., workload greater than 80%, a new node is created and the task is dispatched to the new node.

The load balancing mechanism analyzes both node workload and specification workload. The ALE middleware modules are distributed on different processing nodes and the processing nodes can be dynamically created and/or deleted according to execution status. There is no need to replicate entire ALE middleware to achieve load balance.

4   Experimental Results

We will show the experiment results of the distributed ALE middleware with load balancing mechanism. Using an ALE specification design tool developed by our

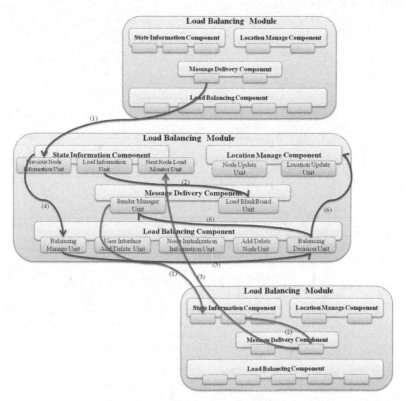

Fig. 4. Operation process with load balancing module

laboratory, we can set ALE specifications. We create 100 sets of EPCSpec as the testing data and run the experiments on different number of specifications. The experiments use five personal computers as the processing nodes each is a two-core CPU with four gigabytes memory. In the experiments, we first create a copy of each module in a node and then run the distributed ALE middleware with the load balancing mechanism with different number of nodes. The number used at the beginning is determined by the user and one or more than one modules can be allocated to the same node. Then, a given number of specifications are chosen as input data to the node running the user interface module. The experiment results of the distributed ALE middleware with load balancing mechanism is compared with the single node processing running a non-distributed ALE middleware.

Specification manager module is used to analyze and manage EPCSpec and is run on the node with the user interface module only. We test the user interface module with and without running specification manage module with different number of EPCSpec. In Fig. 6, we find the execution with specification management module is only 2% higher than not executing this module in all cases. It implies that analysis and management of EPCspec does not have significant impact.

In the experiments, the values of weights e_{cpu}, e_{mem}, and e_{que} in Equation 1 are set to 0.7, 0.2, and 0.1, respectively; the values of weights e_{lr}, e_t, and e_{rs} in Equation 2 are

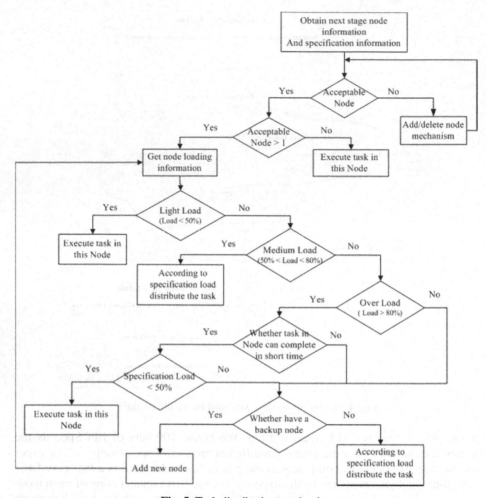

Fig. 5. Task distribution mechanism

set to 0.3, 0.4, and 0.3, respectively. In the experiments, the ECSpecs are submitted to the user interface module sequentially and each EPCSpec is terminated when its event cycle is repeated 50 times.   Fig. 7 shows the experimental results of the distributed ALE middleware with and without the load balancing mechanism. When the number of specifications increases to more than 50, distributed ALE with load balancing mechanism for three nodes and five nodes performs better than single-node ALE middleware. However, when the number of specification is less than 50, single-node non-distributed ALE middleware is still the choice. From the experimental results, we conclude that our load balancing mechanism for distributed ALE middleware can be benefit if only individual modules are duplicated when the workloads become heavy.

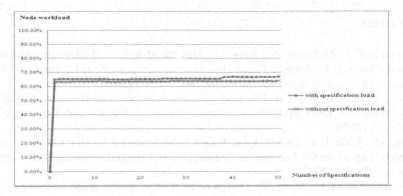

Fig. 6. Impact of specification management workload

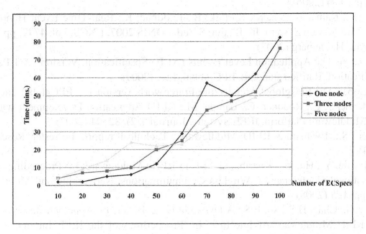

Fig. 7. Performance comparison of distributed ALE middleware

5   Conclusions and Future Work

In this paper, we present a distributed ALE middleware system with a modularized implementation following EPCglobal specification standard. A specification management module and a load balancing module are used to achieve load balancing. The load balancing mechanism considers both the costs of node load and specification load for dispatching modular tasks and creation/deletion of processing nodes. The experimental result shows that the load balancing mechanism can utilize computing resource more efficiently and improve system scalability.

The load balancing mechanism can be further improved by adjusting the weights of our load balancing mechanism dynamically based on different hardware environments and execution status. Also, specification dependency caused by logical reader assignment will be considered to avoid dispatching specifications using the same physical readers to different processing nodes. Boundary conditions can be further analyzed to improve the load balancing mechanism.

References

1. AI-Jaroodi, J., Mohamed, N., Jiang, H.: Distributed Systems Middleware Architecture from a Software Engineering Perspective. In: Proceedings of the IRI International Conference on Information Reuse and Integration, pp. 572–579 (2003)
2. Chae, H.S., Park, J.G., Cui, J.F., Lee, J.S.: An Adaptive Load Balancing Management Technique for RFID Middleware Systems. Software: Practice and Experience 40(6), 485–542 (2010)
3. Cui, J.F., Chae, H.S.: Mobile Agent Based Load Balancing for RFID Middlewares. In: Proceedings of the 9th International Conference on Advanced Communication Technology, pp. 973–978 (2007)
4. Dong, L., Wang, D., Sheng, H.: Design of RFID Middleware Based on Complex Event Processing. In: Proceedings of 2006 IEEE Conference on Cybernetics and Intelligent Systems, pp. 1–6 (2006)
5. Dutta, K., Ramamritham, K., Karthik, B., Laddhad, K.: Real-Time Event Handling in an RFID Middleware System. In: Bhalla, S. (ed.) DNIS 2007. LNCS, vol. 4777, pp. 232–251. Springer, Heidelberg (2007)
6. EPCglobal. The Application Level Events (ALE) Specification, Version 1.1.1 Part I: Core Specification. Ratified standard. EPCglobal, Inc. (2009)
7. EPCglobal. The EPCglobal Architecture Framework, version 1.4. EPCglobal, Inc. (2010)
8. Floerkemeier, C., Roduner, C., Lampe, M.: RFID Application Development with the Accada Middleware Platform. IEEE Systems Journal 1(2), 82–94 (2007)
9. Leaver, S.: Evaluating RFID Middleware. Technical report, Forrester Research, Inc. (2004)
10. Liu, F., Jie, Y., Hu, W.: Distributed ALE in RFID Middleware. In: Proceedings of the 4th International Conference on Wireless Communications, Networking and Mobile Computing, pp. 1–5 (2008)
11. Park, J.G., Chae, H.S., So, E.S.: A Dynamic Load Balancing Approach Based on the Standard RFID Middleware Architecture. In: Proceedings of the IEEE International Conference on e-Business Engineering, pp. 337–340 (2007)
12. Park, S.M., Song, J.H., Kim, C.S., Kim, J.J.: Load Balancing Method Using Connection Pool in RFID Middleware. In: Proceedings of the 5th ACIS International Conference on Software Engineering Research, Management & Applications, pp. 132–137 (2007)
13. Roush, E.T.: The Freeze Free Algorithm for Process Migration. Technical Report UIUCDCS-R-95-1924, University of Illinois at Urbana-Champaign. Dept. of Computer Science, Urbana, Illinois (1995)
14. Zou, X.: A Kind of RFID Middleware Load Balancing Algorithm Based on Agent. Advanced Materials Research 187, 282–286 (2011)

Platform-as-a-Service Architecture for Parallel Video Analysis in Clouds

Tse-Shih Chen, Tsiao-Wen Huang, Liang-Chun Yin,
Yi-Ling Chen, and Yi-Fu Ciou

Cloud Computing Research Center for Mobile Applications
Industrial Technology Research Institute, Hsinchu, Taiwan
tschen@itri.org.tw

Abstract. In this paper, we propose a Platform-as-a-Service (PaaS) architecture to facilitate the process of large scale video analysis. The proposed platform provides a unified framework to integrate and configure various video analysis engines, which can then work collaboratively to accomplish video analysis tasks with an user-specified workflow. In addition, users can adaptively adjust the amount of computation resources for speeding up jobs without affecting the dispatched jobs. Finally, we show a real video analysis application which is built on top of the proposed platform to demonstrate its effectiveness.

Keywords: PaaS, video analysis, workflow.

1 Introduction

In recent years, cloud computing becomes more and more popular and many new concepts have been proposed, such as Infrastructure-as-a-service (IaaS), Platform-as-a-service (PaaS) and Software-as-a-service (SaaS), which corresponds to different levels of services in cloud computing, respectively. Briefly speaking, IaaS aims to achieve consolidation of IT infrastructure and provides the methodology to share resources and can be easily scaled out on-demand. The Amazon EC2 service offering virtual machines for users is one typical example. PaaS [1,2,3], such as Google application engine, provides the environment that programmer can build and develop programs without worrying about the scalability of underlying hardware and software. Developers may concentrate on developing their novel cloud services more quickly and easily by adopting the services provided by PaaS systems. The services that offer end-user domain specific applications is SaaS. Some famous cloud applications, e.g. Gmail, Dropbox, are representative services that meet the different needs of customers. In a more systematic way, the building blocks from bottom up, IaaS is the fundamental layer of cloud computing and PaaS provides the software development tools for building SaaS.

Although a great diversity of PaaS for various purposes have been made publicly available to choose from, they can still be roughly categorized into two main classes, generic and specialized (domain-specific). Generic PaaS , such as, HEROKU [5], AWS Elastic Beanstalk [6], phpfog [7], are designed to allow for

J.-S. Pan et al. (Eds.): *Advances in Intelligent Systems & Applications*, SIST 21, pp. 619–626.
DOI: 10.1007/978-3-642-35473-1_61   © Springer-Verlag Berlin Heidelberg 2013

maximal flexibility to make and run any types of programs written in popular programming languages. Programmers may focus on building their applications on these platforms with the provided services without worrying about the complexity of allocating the necessary resources, e.g. CPU and storages. On the other hand, domain specific PaaS, such as [2,3], are designed to address specific purposes or needs when building specialized applications. In [2], a PaaS architecture focusing on real-time QoS processing is proposed. The use of hardware resource, such as CPU cycles, disk usage or network traffic, is well managed and optimized in order to achieve real-time performance required by application built on it. Although the applications built on such hardware-dependent PaaS may benefit from the highly controlled resources, they may also suffer from the difficulty of migrating to other platforms.

Traditionally, video analysis algorithms are based on processing the spatial/temporal relationships of pixels in consecutive frames or adjusting the variables in frequency domain of a sequence. The input video or extracted images are sequentially processed with some temporally related steps, e.g. to update background model or to track a moving object. To speed up processing, sometimes video analysis engines are treated as a video clip processor and the video clips may come from a subdivided lengthy video.

In this paper, we present a PaaS architecture for video analysis with configurable workflow. The video clips can be analytic by the unit in the PaaS and the analytic results are reported in general JSON format. Our PaaS architecture provides web service interface for SaaS developers that simplify the complexity of parallelizing video analysis tasks. The proposed interface adopting an easy-to-use workflow based scheduling procedure. The video processing workflow can be easily configured and result gathering is simple.

The rest of this paper is organized as follows. Section 2 presents the architecture of our PaaS platform. Section 3 describes the system workflow of our system. An exemplar application of video summarization is introduced in Section 4. Finally, we conclude this paper and discuss future directions in Section 5.

2   The Platform Architecture

The PaaS system consists of three major components. Being similar with MapReduce, the proposed PaaS system is backed by a computation server pool, which is formed by a scalable number of virtual machines that perform video analysis tasks. The job queue component plays the role of management and scheduling the assigned jobs in the middle. The front-end interface of our PaaS system provides the functionality of issuing jobs and gathering results. Fig. 1 illustrates the system architecture and the details of the main components and workflow are explained as follows:

- PaaS Controller (PC): PaaS Controller is the intermediate interface for getting requests from front-end applications and returning results. The SaaS applications send their video analytic jobs with relevant inputs, such as *SaaS*

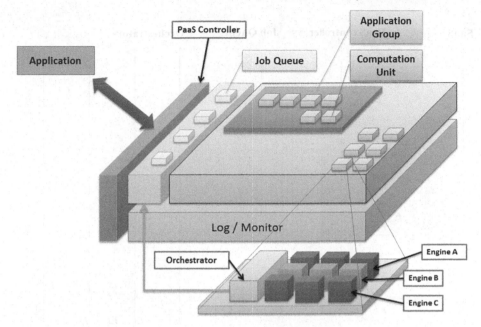

Fig. 1. PaaS architecture

ID, engine ID, engine workflow, engine parameters and *job priority* depending on the selected video analysis engines and related actions. For instance, in Fig. 1, a video analysis task may consist of three stages corresponding to different engines (A-B-C). The progress of the whole workflow is monitored and logged by the PaaS Controller. PaaS Controller also monitors job status and generates reports, which can then be queried by front-end applications as needed.

- Computation Unit (CU): Computation unit is a virtual machine where a variety of video analysis engines are installed and video analysis tasks are carried out. CUs may be of different types of operating systems and resources to host and meet the needs of different kinds of engines. *Orchestrator*, which is an administrative program running on each CU, is responsible for fetching jobs from job queues (described below), launching engines specified in the workflow, monitoring job progress, collecting results and error handling.
- Job Queue (JQ): The jobs received from applications with different configurations of engine workflows will be put into Job Queue after a validation process. Besides, in order to serve the jobs with different priorities, we designed a probability method for JQ such that higher priority jobs have a higher ratio to be picked up by the Computation Unit and vice versa. A job of higher priority will be put into the queue with higher process ratio and computation unit will pick up jobs by the order of the queues sorted by defined probability. Moreover, if the queue which has been picked by a computation unit is empty, then the next priority queue will be checked to avoid the starvation problem.

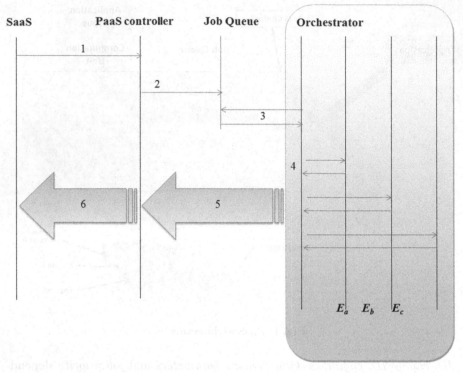

Fig. 2. PaaS workflow

– Video Analysis Engine (VAE) and other supporting engines: The proposed
PaaS system does not impose any restrictions on the types of supported engines as long as they can be launched through *command line interface* (CLI)
by orchestrator. A VAE may also communicate with orchestrator by writing
messages to standard output or error streams, which are compliant with a
pre-defined format. For example, a license plate recognition engine can notify orchestrator of the current progress or the recognized plate number by
printing messages to stdout. In addition to VAEs, a video analysis task may
also need other supporting engines, such as FTP client, to accomplish video
retrieval and fulfill the whole workflow.

3   System Workflow

From the above discussions, PaaS provides the analytic power for applications.
Video analysis jobs can be processed with an user-specified workflow. Result
collection and status report are also handled by PaaS system. In this section, we
explain the job processing flow and show our strategies in more details. Fig. 2
illustrates the workflow of the proposed system.

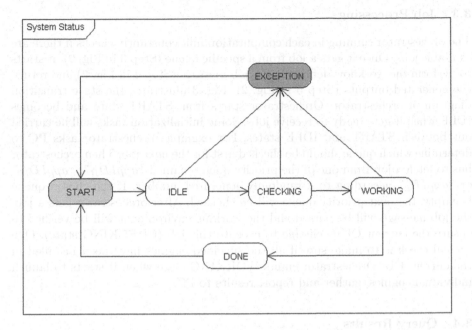

Fig. 3. Computation unit state transition diagram

3.1 Job Dispatching

When an application issues a job to the PC (Step 1 in Fig. 2), the submitted job will be inserted into a dedicated job queue named by *"appID"* (Step 2 in Fig. 2). Besides, the input parameters of each job will be created and formatted into a JSON string. The following example demonstrates a job message representation indicating a high priority job with a workflow consisting of three engines. Note that the job message also carries the parameters required to launch each engine in the workflow.

```
[{"appId":1},{"engineWorkflow":"1 2 3"},{"jobPriority":"HIGH"},
{"engineId":1,"param1":"value1", "param2":"value2"},
{"engineId":2,"param1":"value1", "param2":"value2"},
{"engineId":3,"param1":"value1", "param2":"value2"}]
```

3.2 Prioritized Job Queues

In order to enable the applications to differentiate jobs of various importance levels, we introduce priority queues for job scheduling in PaaS. The application can assign the job with parameter "jobPriority" to form jobs of High/Medium/Low priorities. The jobs with high priority will be put into a job queue with specified queue name *"appID_H"* and so on. Moreover, each job queue is associated with a different consuming rate in accordance with the corresponding priorities.

3.3 Job Processing

The orchestrator running in each computation unit constantly checks if there are available jobs. Once it gets a job from a specific queue (Step 3 in Fig. 2), it starts to perform the workflow defined in the job message (Step 4 in Fig. 2) and return the generated outputs (Step 5 in Fig. 2). Fig. 3 illustrates the state transition diagram of orchestrator. Orchestrator starts from START state and becomes IDLE when it gets ready to receive jobs. Some initialization tasks will be carried out between START and IDLE states. For example, orchestrator asks PC to determine which queue should be checked first for the next job. Then orchestrator has to fetch a job from one of the priority queues named "$appID_H$", "$appID_M$" or "$appID_L$" according to a randomly generated number. If the chosen queue is empty, the next priority queue will be checked. After orchestrator gets a job, the job message will be parsed and the working environment will be verified to ensure the current CU is eligible to execute this job (CHECKING state). One typical check is to make sure if all the specified engines have been installed in the current CU. Orchestrator enters WORKING state when it starts to launch individual engines, gather and report results to PC.

3.4 Query Results

The results, including execution status of CU, job progress and metadata generated by video analysis engines are collected and archived by orchestrator. PaaS Controller is responsible for responding to queries for results from front-end applications. Typically, applications can get their job results from PaaS by calling the provided APIs (Step 6 in Fig. 2). It is worth noting that the proposed PaaS system does not provide mass storage service. The uploaded or processed videos will not be persistently kept within PaaS. As a result, applications are responsible for retrieving their results before these files are purged.

4   Exemplar Application: Video Summarization

In this section, we demonstrate an exemplar application of video summarization which is built upon the proposed PaaS system. Nowadays, digital video surveillance systems are installed everywhere and continuously generate huge amount of data. However, mostly the recorded videos still rely on human inspection for threat detection. Video summarization aims to extract "meaningful" video segments by detecting *moving objects* in the scene and discarding still frames in the videos. This is accomplished by *background subtraction* techniques [4] Video summarization can drastically reduce the time needed to browse surveillance videos and will be very helpful for investigating security problems.

In this demo application, a video analysis job is composed of three stages, which corresponds to the engines listed in Table 1. A submitted job must contain the arguments required to launch each engine, such as FTP server IP and source file name, to form a valid job. Take video summarization as an example,

Table 1. Engines for Parallel Video Summarization

ID	Engine Name	Functions
1	FTP Client	Upload/Download files from FTP server.
2	Video Split/Merge	Split and Merge video files.
3	Video Summarization	Moving object detection with background subtraction.

Table 2. Performance of video summarization. Execution time is represented by seconds.

Video Info.	CU#	Splitting	Summarization	Merge	Total
	1	150	965	10	1125
157 MB (1280 × 1024, 537 sec)	10	153	101	9	263
	20	148	55	9	212
	1	361	2316	35	2712
537 MB (1280 × 1024, 1302 sec)	10	360	256	36	652
	20	365	126	37	528
	1	647	3660	255	4562
1 GB (1280 × 1024, 2388 sec)	10	643	405	257	1305
	20	645	205	254	1104

the implemented workflow starts from downloading a source video file to CU. The retrieved video is then subdivided into a number of video clips for parallel processing. A video summarization job is generated to be associated with a corresponding video clip. After all the processed video clips are collected and merged into one video file, the final result is uploaded back to FTP server to complete the whole workflow. To summarize, the workflow of video summarization process are composed of the following steps.

1. Issue the split job for a video file to split video into a number of clips (One CU).
2. Issue the summarization jobs for each video clip (N CU).
3. Issue the merge job to merge the processed clips from the last step (One CU).

Based on the proposed workflow above, we show the video summarization performance in Table 2. We processed three different sizes of videos (157 MB, 537 MB and 1 GB) with the same resolution (1280 × 1024) by various numbers of CUs, and show the corresponding processing time. In each row, the total processing time is reduced by using more CUs since more video clips can be processed simultaneously and the computation load is distributed to each CU. Note that in the above experiments the number of subdivided video clips is equal to that of CUs. The time required to split and merge different number of video clips does not vary very much. On the other hand, the overall video summarization time can be dramatically reduced by running more CUs.

As shown in Table 2, the overall processing time can be reduced by adopting more CUs. However, increasing the number of CUs may not consistently lead

Table 3. Performance of the summarization

Video Info.	Summarization time
1.90MB (60 sec)	90.0
1.05MB (30 sec)	44.2
0.42MB (10 sec)	15.8
0.26MB (5 sec)	8.5
0.11MB (1 sec)	1.9

to a linear speedup of the processing time due to more network traffic and disk I/O efforts. The optimal number between video/clip length and number of CUs is highly dependent on the network and hardware performance and each system may have different settings. Table 3 is the list of processing time for different video lengths under the same resolution 1280 × 1024. One can observe that the processing time of summarization is approximately 1.5 times of original video length. However, in the cases of shorter video lengths and more video clips, the overhead of network and system communication gradually becomes significant.

5   Conclusions and Future Directions

In this paper, we introduced a PaaS system for parallel video analysis with configurable workflow. Computationally intensive video analysis tasks can be partitioned into many independent subtasks in our system and processed simultaneously. A real video summarization application built upon the proposed PaaS platform is demonstrated to show its effectiveness. For future directions, we aim to do more system and application integration for more experiments, and improve the robustness and efficiency of our PaaS system.

References

1. Xu, M., Xie, F., Wang, H., Chen, Z.: A novel workflow based data processing platform as a service. In: Computing, Communications and Applications Conference, ComComAp (2012)
2. Boniface, M., Nasser, B., Papay, J., Phillips, S., Servin, A., Zlatev, Z., Yang, K.X., Katsaros, G., Konstanteli, K., Kousiouris, G., Menychtas, A., Kyriazis, D., Gogouvitis, S.: Platform-as-a-Service Architecture for Real-time Quality of Service Management in Clouds. In: Fifth International Conference on Internet and Web Applications and Services, ICIW 2010 (2010)
3. Rake-Revelant, J., Holschke, O., Offermann, P., Bub, U.: Platform-as-a-Service for business customers. In: 14th International Conference on Intelligence in Next Generation Networks, ICIN (2010)
4. Zivkovic, Z.: Improved adaptive Gaussian mixture model for background subtraction. In: Proceedings of IEEE International Conference on Pattern Recognition (ICPR 2004), pp. 28–31 (2004)
5. http://www.heroku.com/
6. http://aws.amazon.com/elasticbeanstalk/
7. https://www.phpfog.com/

A Translation Framework for Automatic Translation of Annotated LLVM IR into OpenCL Kernel Function

Chen-Ting Chang, Yu-Sheng Chen, I-Wei Wu, and Jyh-Jiun Shann

Dept. of Computer Science, National Chiao Tung University, Hsinchu, Taiwan
{deferplay,ansoncat}@gmail.com, {wuiw,jjshann}@cs.nctu.edu.tw

Abstract. Heterogeneous multi-core processor is proposed to accelerate applications using an application-specific hardware, such as graphics processing unit (GPU). However, heterogeneous multi-core processor is difficult to program. Therefore, OpenCL (Open Computing Language) standard recently has been proposed to reduce the difficulty. A program of OpenCL mainly consists of the host code (executed on CPU) and the device code (executed on GPU or other accelerators). LLVM (Low Level Virtual Machine) is a compiler infrastructure and supports a variety of front-ends into LLVM IR (Intermediate Representation). To help translate programs written by different programming languages of LLVM front-ends to OpenCL, this work defines some extensions of LLVM IR to represent the kernel function of OpenCL. Furthermore, a translation framework is designed and implemented to translate annotated LLVM IR to OpenCL kernel function.

Keywords: OpenCL, LLVM, heterogeneous multi-core.

1 Introduction

Heterogeneous multiprocessor platforms have much higher potential performance gain than homogeneous multiprocessor platforms [1]. According to the characteristics of a program, the program may be partitioned into different properties of tasks, and then the tasks will be scheduled on suitable processors, for example, a parallel code is suitable on GPU. Nevertheless, heterogeneous multiprocessor platforms are harder to program. To improve this issue, at present, there have been many parallel programming standards being proposed, one of them is OpenCL.

OpenCL is an open standard for parallel programming of the heterogeneous processors such as multi-core CPU, GPU, Cell/B.E., and DSP and so on[1]. The program of OpenCL mainly consists of the host code (executed on the CPU) and the device code (executed on the GPU or other accelerators). In OpenCL, the device code is called kernel function. To support different platforms, the kernel function usually exists in the format of source code and is compiled dynamically. However, a lot of programs written by many parallel programming frameworks cannot be executed on the platform of OpenCL. To help translate these programs to OpenCL, this work defines several extensions of LLVM IR to represent the kernel function of OpenCL. Furthermore, a translation framework called the <u>Annotated LLVM IR to OpenCL</u>

J.-S. Pan et al. (Eds.): *Advances in Intelligent Systems & Applications*, SIST 21, pp. 627–636.
DOI: 10.1007/978-3-642-35473-1_62   © Springer-Verlag Berlin Heidelberg 2013

<u>Kernel Function Translation Framework</u> is designed and implemented. The framework is derived from LLVM C back-end.

The rest of this work is structured as follows. Section 2 studies the background Section 3 then presents our translation framework. Next, Section 4 presents the experimental results. Conclusions are finally drawn in Section 5.

2   Background

This section introduces the parallel programming model of OpenCL and the compiler infrastructure of LLVM.

2.1   OpenCL

OpenCL is a standard for parallel programming of heterogeneous multiprocessors such as multi-core CPUs, GPUs, Cell/B.E., DSP and other [1]. OpenCL framework includes a language for writing kernel functions (executed on OpenCL devices), API, libraries and a runtime system to support software development. The core ideas of OpenCL specify four architecture models: platform model, execution model, memory model, and programming model [3].

The platform model is one host to connect one or more OpenCL devices. One or more compute units (CUs) compose an OpenCL device, and one or more processing elements (PEs) compose a compute unit. The processing elements within a device perform the computation. An OpenCL program is divided into two parts: (1) the host program which is executed on the host, and (2) the kernel which is executed on the OpenCL devices. OpenCL defines context and scheduling kernel to execute on OpenCL devices by the host program.

In OpenCL, all work-items executing a kernel have access to four distinct memory regions [3], namely global memory, constant memory, local memory and private memory. The global memory permits read/write access to all work-items in all work-groups. The constant memory is a region of global memory and only permits read access to all work-items. The local memory is shared by all work-items in the same work-group. The private memory is private to a work-item.

OpenCL supports data parallel programming model and task parallel programming model as well as the hybrids of these two models [3]. Data parallel programming model indicates that a series of instructions uses different element of memory objectives, scilicet each work item executes as same as program, inserts different data to execute work item by global ID or local ID. Task parallel programming model indicates that each work item of work space inside is absolutely independent executing kernel program then others item. In this model, each work item is equal to work in a single compute unit, which has only work item that executed only by it.

2.2   LLVM

LLVM (Low Level Virtual Machine) is a compiler infrastructure that is designed for compile-time, link-time, run-time, and "idle-time" optimization of programs written in arbitrary programming languages [4]. Figure 1 shows the LLVM framework, LLVM supports a variety of front-ends into LLVM IR, and LLVM provides analysis and optimization to transfer into optimum LLVM IR. Finally, the LLVM back-end generates high level languages or target assembly codes.

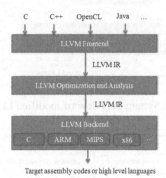

Fig. 1. LLVM framework

LLVM supports a language-independent instruction set and type system. Each instruction is in static single assignment form (SSA), i.e., each variable (called a typed register) is only assigned once [5]. SSA split operands value and storage location efficiently into a program, so that every definition gets its own version. Compiler optimization algorithms which are either enabled or strongly enhanced by the use of SSA, include constant propagation, value range propagation, sparse conditional constant propagation, dead code elimination, global value numbering, partial redundancy elimination, strength reduction, and register allocation.

The LLVM code representation has three different forms: as an in-memory compiler IR, as an on-disk bitcode representation (suitable for fast loading by a Just-In-Time compiler), and as a human readable assembly language representation [6]. This allows LLVM to provide a powerful intermediate representation for efficient compiler transformations and analysis, while providing a natural means to debug and visualize the transformations [6].

3   Design and Implementation of Kernel Function Translation

Our translation framework consists of three parts: LLVM front-end, LLVM optimization phase and LLVM back-end, as shown in Figure 2. The LLVM front-end translates program languages with OpenCL annotations into annotated LLVM IR. The LLVM optimization phase optimizes annotated LLVM IR by a series of LLVM optimization passes (e.g., dead code elimination). Kernel extraction extracts the kernel function from the annotated LLVM IR of the input program and inserts OpenCL specific extensions, such as address space qualifier, into LLVM IR. Finally, the LLVM back-end of OpenCL translates annotated LLVM IR into OpenCL kernel function.

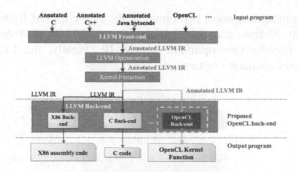

Fig. 2. System overview of modified LLVM

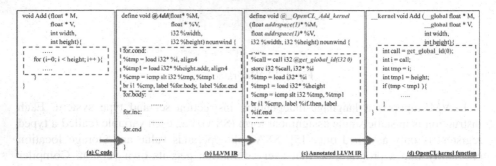

Fig. 3. Translation of the C code to OpenCL kernel function

Figure 3 illustrates how to translate C code to OpenCL kernel function through manually annotated LLVM IR. Firstly, the input program written by C is translated into LLVM IR by LLVM Clang C front-end as shown Figure 3 (b). The translated LLVM IR is manually inserted OpenCL annotations (e.g., kernel qualifier, address space qualifiers, and built-in functions of work-item). The OpenCL annotations are marked with red boldface in Figure 3 (c). The if statement in the annotated LLVM IR highlighted with blue dashed box is constructed from the for loop condition, and then loop's counter (e.g., increment and decrement) is replaced with built-in functions of work-item (e.g., get_globla_id(dim)). Finally, the annotated LLVM IR is translated into the OpenCL kernel function by our OpenCL back-end.

The OpenCL back-end consists of several steps as shown in Figure 4: initialization, function translation, and loop as well as basic block translation. The initialization declares all global variables, types and functions. In the function translation, the back-end translates the function prototype firstly. After translation, the back-end declares the local variable and then translate basic block within a function. If the back-end encounters a loop, it exits the basic block translation and enters loop translation. In the loop translation, the back-end constructs the loop body and then enters basic block translation again.

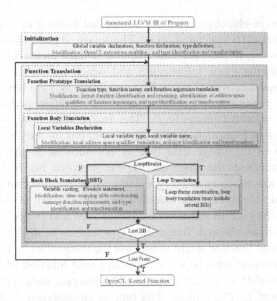

Fig. 4. Flow chart of the OpenCL back-end

The address space qualifier (i.e. location) of a variable in OpenCL program must be specified (such as global, constant, local or private); otherwise, the kernel function cannot be correctly executed because of the incorrect location of variable. To handle this problem, we add some extension in LLVM IR to indicate the type of and the address space qualifier of a variable. The mapping table between the address space qualifier of OpenCL and the address space attribute of LLVM IR is shown in Figure 5.

OpenCL		Annotated LLVM IR
Memory model	Address space qualifier	Address space attribute
Global memory	__global	1
Constant memory	__constant	3
Local memory	__local	2
Private memory	__private or none	no attribute

Fig. 5. The annotated LLVM IR mapping to the address space qualifiers of OpenCL

Some variable types used in OpenCL do not properly support in LLVM IR (e.g., half precision floating point, Image3D and Image2D). Thus, such variable types are difficult to translate back to C language from LLVM IR without any annotation. Half precision floating point (HPFP) data type is a 16 bit floating point format, and it must follow IEEE 754 -2008 half precision storage format. HPFP data type has one sign bit, five exponent bits, and ten mantissa bit. OpenCL supports HPFP data type and provides some operations of HPFP data type. Since LLVM IR did not support HPFP data type until LLVM 3.1 (Apr 2012), we also add some extension to support HPFP.

In OpenCL, the image object is a memory object that stores a two dimensional (2D) or three dimensional (3D) structured array. The image object consists of four parts: image data, dimensions of the image, description of each element in the image, and properties that describe usage information and which region to allocate from. In LLVM IR, the image data type of *image2d_t* and *image3d_t* are not supported. These image data types represent opaque structure type in annotated LLVM IR (e.g. struct._image2d_t or struct._image3d_t). The image data type is represented in LLVM IR as shown in Figure3-6. Due to the restrictions on the image data type, the variable operation of the image data type is not allowed in the kernel function. The OpenCL program uses the image data type of *image2d_t* or *image3d_t* based on the following restrictions: Image data type can only be the function argument, and it cannot be modified. Image data type cannot access the element of the image directly. OpenCL provides the built-in functions for image read and image write, pointers to image data type are not allowed. Image data type cannot be declared in a structure. The local variable and function return cannot use the image data type.

The sampler parameter must be set in the image read function when attempting to read the image variable in kernel function. The sampler data type is represented as *sampler_t* in OpenCL. The sampler data type is an unsigned 32 bits integer, and provides addressing mode, filter mode, normalized coordinates. The sampler data type has the following restrictions: it cannot be declared as a type of array or pointer. It cannot be defined without initializing local variables or as the return value in a function. It cannot be modified when the function argument is the sampler data type. In LLVM IR, sampler data type is not supported. The sampler data type represents i32 (integer) data type in annotated LLVM IR. Translation and identification of the sampler operation use the following steps:

I.   Identify the name of the sampler argument of each image read function call in the kernel function.
II. For each sampler argument, construct the data dependence graph between the sampler argument, the function arguments and local variables of the kernel function.
III. If the root of the data dependence graph is a function argument:
   i.   Change the data type of the function argument from i32 to sampler.
   ii. Replace the name of the sampler argument in the image read function call with the name of the function argument.
   iii. Eliminate all the local variables in the data flow graph, including the sampler argument.
IV. If the root of the data dependence graph is a local variable:
   i.   Change the data type of a local variable from i32 to sampler.
   ii. Initialize the declaration of the sampler argument to the initial value of the local variable.
   iii. Eliminate all the local variables in the data flow graph, excepting the sampler argument.

The proposed OpenCL back-end identifies a kernel function from the function names of the annotated LLVM IR. A function name with annotations __OpenCL and _kernel is a kernel function, e.g., void__OpenCL_Add_Kernel as shown in Figure 6 (a). The OpenCL backend will rename the function name in the translated OpenCL program by eliminating the __OpenCL and _kernel annotations, e.g., Add, as shown in Figure 6 (b). The proposed OpenCL back-end identifies the address space attribute from the function arguments of annotated LLVM IR. Refer to Figure 5 for the mapping between the address space attribute in annotated LLVM IR and the address space qualifiers in OpenCL. In Figure 6, the proposed OpenCL back-end translates addrspace(1) of annotation LLVM IR into __global of OpenCL kernel function.

define void @__OpenCL_Add_kernel (float addrspace(1)* %a) nounwind {

(a) Annotated LLVM IR

__kernel void Add { __global float *a) {

(b) OpenCL kernel function

Fig. 6. An example code segment for function prototype translation and identification

The memcpy function is used to copy a block of memory from the source location to the destination location. In OpenCL, the memcpy function has a restriction that its arguments cannot contain address space qualifiers (e.g., __global, __local, __constant). The syntax of memcpy function is represented as memcpy (destination, source, length) in OpenCL and memcpy (addrspace destination, addrspace source, length, align, isvolatile) in annotated LLVM IR.

Traditional C language cannot support vector type, while OpenCL and LLVM IR can. he Vector data type defined by OpenCL is defined with a type name (char, short, int, float, long and unsigned) followed by a literal value n that defines the number of elements in the vector, n = 2, 4, 8, 16. The vector data type defined by LLVM IR includes the number of elements and element type. In this work, all vector variables in the input kernel function are translated to the vector data type of LLVM IR. To correctly translate back to OpenCL kernel function, the OpenCL back-end must translate the vector data type (LLVM IR) to corresponding format in OpenCL. Figure 7 is an example of translating the vector data type between LLVM IR and OpenCL.

1	<4 x float>	
2	<16 x int>	LLVM IR
1	float4	
2	int16	OpenCL

Fig. 7. Example of translation between LLVM IR and OpenCL (vector data type)

The LLVM front-end would transform some local variables into global variables leading the program to have compiling and executing issue. To handle this isse, a mapping table is introduced as shown in Figure 8. For example, a variable of "__local int a" declared in OpenCL is considered as a global variable in LLVM IR.

OpenCL	Annotated LLVM IR	
Local variable declaration	Global variable declaration	Local variable declaration
__local int a;	*	
__local int a[1];	*	
__local struct a;	*	
__local int *a;		*
__local int4 a;		*
__local int *a[1];		*
__global struct a;		*
__global int *a;		*

Fig. 8. Function prototype identification and translation

Since the LLVM front-end will rename built-in function of OpenCL, the function calls of OpenCL program cannot call built-in function after translation. To overcome this problem, an alias mapping table is introduced as shown in Figure 9.

Function Class	OpenCL built-in function	Annotated LLVM IR
Math (C)	sqrt	sqrtf
Synchronization	Barrier (arg1)	Barrier (arg1 , arg2)
Math /native math	cos	_cos_f32
	Exp	_EXP_f32
Integer	abs	_abs_i32
Common	max	_max_2i32
Geometric	cross	_cross_4f32
3D image	imagef_image	__read_imagef_image3d4i32
2D image	imagef_image	__read_imagef_image2d2i32
Type conversion	convert_int4	__convert_int4_4u8
	convert_float4	__convert_int4_4f32
...		

Fig. 9. Alias mapping table

4   Experimental Results

This experiment is used to verify the correctness and performance of our OpenCL back-end. The experimental environment had a host with an Intel i7-920 and a device with NVidia GTX 460. The OpenCL back-end was developed based on LLVM 2.9. All benchmarks were selected from NVidia OpenCL SDK and translated to the annotated LLVM IR using the LLVM front-end developed by AMD. After performing optimization (-O2 in this study), the kernel function (LLVM IR format) is then translated to the C code format through OpenCL back-end. By executing the translated the kernel function on the device (i.e. NVidia GTX 460), the correctness and performance could be verified. Figure 10 shows the executed time of before and after translated kernel functions. Original OpenCL and translated OpenCL denotes the before and after translated kernel functions, respectively. According to the experimental results, all benchmarks could be executed correctly and almost no performance is loss.

NVidia SDK	Execution time of clock cycle		Ratio (T/O)
	Original OpenCL	Translated OpenCL	
MatrixMul	68,413,437	68,166,441	0.99639
ConvolutionSeparable	164,322,856	162,524,385	0.98905
BlackScholes	101,483,905	101,712,190	1.00224
Tridiagonal	282,236,392	283,328,057	1.00386
Transpose	585,751,696	585,751,175	0.99999
SortingNetworks	450,134,553	450,791,892	1.00146
SimpleMultiGPU	179,908,761	179,744,934	0.99873
Scan	4,764,150,345	4,764,139,172	0.99999
RadixSort	2,333,536	2,404,680	1.03048
QuasirandomGenerator	1,005,084	1,060,628	1.05526
DXTCompression	58,962,028	59,706,196	1.01262
DotProduct	45,546,896	43,581,255	0.93628
FDTD3d	1,121,176,436	1,136,779,335	1.01391
VectorAdd	191,660,748	190,942,509	0.99625
SobelFilter	15,052,436	15,098,284	1.00304
SimpleTexture3D	1,372,866	1,374,720	1.00135
SimpleGL	1,308,694	1,309,972	1.00097
RecursiveGaussian	24,529,491	24,326,601	0.99172
PostprocessGL	7,519,812	7,552,768	1.00438
MedianFilter	26,438,834	27,167,774	1.02757
DCT8x8	74,749,455	74,676,294	0.99902
MarchingCubes	1,579,408	1,583,066	1.00231
Nbody	8,043,914	8,015,520	0.99647
Particles	3,512,163	3,459,612	0.98503
MersenneTwister	857,732	840,019	0.97934
HiddenMarkovModel	369,065,620	371,369,797	1.00624
CopyComputeOverlap	10,221,842,360	10,210,928,766	0.99893
Histogram	90,223,175	95,434,984	1.05776
MatVecMul	1,044,741,185	1,056,151,563	1.01092
Average			1.00350

Fig. 10. Execution time of before and after translated kernel functions

5   Conclusion

In this work, we designed and implemented an OpenCL back-end in LLVM compiler infrastructure. According to the experimental results, the OpenCL back-end could correctly translate all benchmarks from NVidia SDK and almost no performance loss occurs. In this work, we only performed the default optimization (O2) of LLVM compiler framework on the kernel function. However, the program characteristic of the kernel function (with massive parallelism) is different from the sequential function or the function with limited parallelism. And, the target device architecture of OpenCL is significantly different from the host one. As consequence, most optimization passes in the default optimization does not perform well on the kernel function. To further increase the performance, designing the optimization pass specified for the kernel function would be of interest in the future. Furthermore, a new specification called SPIR (Standard Portable IR) [7] for OpenCL Kernel is released by Khronos Group Inc. at August 24, 2012. SPIR is a mapping from the OpenCL C programming language into LLVM IR. Obviously, the major objective of both our IR extension and SPIR is almost same. Therefore, modifying our framework to support SPIR would be another interesting topic.

References

1. Tsai, T.-C.: OMP2OCL Translator: A Translator for Automatic Translation of OpenMP Programs into OpenCL Programs. MS. Thesis, National Chiao Tung University (2010)
2. Hruska, J.: AMD Announces New GPGPU Programming Tools,
   http://hothardware.com/News/AMD-Announces-
   New-GPGPU-Programming-Tools
3. Khronos OpenCL Working Group, The OpenCL Specification 1.0 (2009)
4. Lattner, C., Adve, V.: LLVM: A compilation framework for lifelong program analysis & transformation. In: International Symposium on Code Generation and Optimization (2004)
5. Cytron, R., Ferrante, J., Rosen, B.K., Wegman, M.N., Zadeck, F.K.: Efficiently computing static single assignment form and the control dependence graph. In: ACM Transactions on Programming Languages and Systems (1991)
6. LLVM Developer Group, LLVM Language Reference Manual,
   http://llvm.org/docs/LangRef.html
7. Khronos Group, SPIR 1.0 Specification for OpenCL (2012)

Low Power Compiler Optimization
for Pipelining Scaling

Jen-Chieh Chang, Cheng-Yu Lee, Chia-Jung Chen, and Rong-Guey Chang

Department of Computer Science, National Chung Cheng University, Chia-Yi, Taiwan
{cjc99p,lcyu95m,ccj98p,rgchang}@cs.ccu.edu.tw

Abstract. Low power has played an increasingly important role for embedded systems. To save power, lowering voltage and frequency is very straightforward and effective; therefore dynamic voltage scaling (DVS) has become a prevalent low-power technique. However, DVS makes no effect on power saving when the voltage reaches a lower bound. Fortunately, a technique called dynamic pipeline scaling (DPS) can overcome this limitation by switching pipeline modes at low-voltage level. Approaches proposed in previous work on DPS were based on hardware support. From viewpoint of compiler, little has been addressed on this issue. This paper presents a DPS optimization technique at compiler time to reduce power dissipation. The useful information of an application is exploited to devise an analytical model to assess the cost of enabling DPS mechanism. As a consequence we can determine the switching timing between pipeline modes at compiler time without causing significant run-time overhead. The experimental result shows that our approach is effective in reducing energy consumption.

1 Introduction

Since most embedded systems are portable, reducing energy consumption to extend the lifetime of batteries has become a crucial issue. In recent years, many techniques have been proposed to address this issue. DVS is the famous one, which has been demonstrated by much work to be very effective [8,2,12]. It adjusts dynamically voltage and frequency to save power, as indicated in Equation 1.

$$E \propto f \times C \times V^2, \tag{1}$$

where A $\propto$ B means A is in direct ratio to B. However, DVS has no effect on energy saving when the voltage reaches its low bound because it becomes a constant [10]. Fortunately, with reference to Equation 2, energy is in direct ratio not only to the clock frequency and the square of the voltage, but also to instruction-per-cycle (IPC).

$$E \propto f \times V^2 \times t \propto f \times V^2 \times \frac{I_t}{f \times IPC} \propto \frac{V^2}{IPC} \tag{2}$$

Thus, we can reduce energy dissipation at low-voltage level based on IPC. Equation 1 and Equation 2 reveal that IPC is the key to power dissipation at low-voltage level. This fact shows that power will increase in the opposite direction of IPC and motivates our low-power idea to devise a DPS technique to evaluate the IPC and determine the

J.-S. Pan et al. (Eds.): *Advances in Intelligent Systems & Applications*, SIST 21, pp. 637–646.
DOI: 10.1007/978-3-642-35473-1_63   © Springer-Verlag Berlin Heidelberg 2013

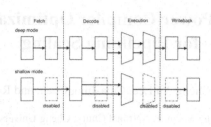

Fig. 1. Pipeline modes of DPS

switching timing between pipeline modes at compiler time. In DPS, the pipeline consists of deep mode and shallow mode, as shown in Figure 1. The deep mode is the default pipeline mode and the shallow mode is designed by dynamically merging adjacent pipeline stages of deep mode, where the latches between pipeline stages are made transparent and the corresponding feedback paths are disabled. In theory IPC is in inverse ratio to the pipeline depth, the IPC of deep mode may be smaller than that of shallow mode [6]. Therefore, executing applications in shallow mode will lead to the reduction of power dissipation. But this statement is not always true. In reality, many factors in deep pipeline mode will influence IPC [15,13]. Here are three examples.

(a) In deep mode, the branch penalty is about twice as large as that of shallow mode. It leads to the reduction of IPC, and then deep mode may consume more power than shallow mode.
(b) The deeper pipeline increases the execution latency. As a result, IPC will become smaller, but power consumption will be larger.
(c) In deep mode, the issue queue must be long and the number of loads of the reorder buffer must be large so that the reorder buffer can hold many in-flight instructions. In this case, this window pressure makes IPC smaller and raises the power dissipation of deep mode.

Hence, if we want to apply DPS to save power at low-voltage level, we must decide when the pipeline enter deep mode or shallow mode depending upon the IPC. Consider the voltage characteristic shown in Figure 2. In the first stage, DVS is applied to save power at high-voltage level. Although reducing voltage is very effective to low power, DVS fails in the second stage when the voltage reaches its lower bound. At the final stage, we can enable DPS to switch the pipeline modes based on IPC. Since IPC is affected by some factors, we can consider their impact on IPC to determine the switching timing between pipeline modes to save energy.

Previous work on DPS was proposed by architects with architectural support [10,7,4,14,3]. However, the research about how to solve this issue with compilation techniques remains open. In this paper, we present an optimization technique to enable DPS with respect to IPC at compile time. We first partition an application into many regions and then calculate the IPC of each region to determine the switching timing between pipeline modes based on our evaluating model. Since our work is performed at compiler time, the run-time overhead will be small and the hardware cost and complexity will be as minimal as possible. The experimental results prove that the energy reduction really benefit from our work.

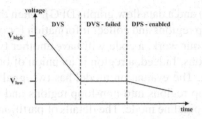

Fig. 2. Voltage characteristic

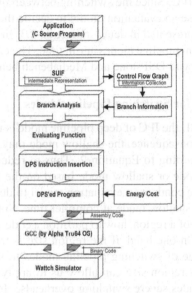

Fig. 3. The proposed DPS compilation system

The rest of this paper is organized as follows. Section 2 gives the overview of our work and then presents our approach in detail. The experimental results are shown in Section 3. Finally, we conclude our paper in Section 4.

2   The Proposed Approach

In this section, we focus on how our DPS approach is applied to applications to save energy at compiler time. We first introduce our basic idea in Section 2.1. In section 2.2, we depict the method to partition a code into regions and then present the evaluating function to decide the switching between pipeline modes. The mechanism to enable DPS is given in Section 2.3.

2.1   Basic Idea

Figure 3 shows our compilation system composed of SUIF framework [16], our proposed engine, and Wattch simulator. First, an application is compiled by SUIF as a control flow graph (CFG) and a data flow graph (DFG). Then the CFG and DFG are analyzed to identify the loop regions and collect information for our evaluating model and identify loop regions. In our work, a code will have another type of regions, non-loop regions, except loop regions. Indeed, a region is an union of basic blocks as a unit that our DPS can manipulate. The evaluation model has two goals: one is to partition the remaining part of the loop regions into non-loop regions and the other is enable each region to enter a suitable pipeline mode. The details of partitioning scheme is described in Section 2.2. To activate DPS, for each region we will insert the DPS-enable function in its entrance at compile time so that it can be executed in proper pipeline mode to save energy based on its IPC. Since the switching between pipeline modes best is very hard to decide, we propose an evaluation model to decide the timing during execution. The evaluation model is presented in detail in Section 2.3. In this way, the code will be switched between different pipeline modes at run time. The experiment is performed on the Wattch simulator [1] with DSPstone and Mediabench benchmark suites.

2.2   Evaluation Model for Switching Pipeline Modes

As mentioned in Section 1, the IPC of deep pipeline mode is not always larger than that of shallow mode. As a consequence, the shallow mode may have better power saving than the deep mode according to Equation 2. Thus to reduce power reduction, each region can enter deep mode or shallow mode based on the IPC during execution. To achieve this objective, we conduct an evaluation model to decide the switching timing between deep mode and shallow mode at compiler time. Since the calculation of IPC closely relates to the size of a region, how to partition a code into regions becomes very important to our work. On one hand, if the region size is too large, we may lose the chances to take advantage of switching pipeline modes to save energy. On the other hand, although the small region size can allow us to apply the DPS optimization to a code, it possibly generates severe switching overheads. However what the size of a region is the optimal solution for our approach is very hard to decide, thus we attempt to seek for a principle to guide our selection in this section. With our observations, since the loops usually dominate execution time and power consumption of a code, they are the key to our decision.

To use the loops to partition a code, they must be identified first and then be referred to divided the remaining part into non-loop regions. Below we present our partitioning approach and evaluation model. Given a code $G = (V, E)$, it is divided into two types of regions, Γ_1 and Γ_2, where $\Gamma_1, \Gamma_2 \subseteq V \times E, G = \Gamma_1 \bigcup \Gamma_2$, and $\Gamma_1 \bigcap \Gamma_2 = \emptyset$. Note $\emptyset$ represents the empty set; that is Γ_1 and Γ_2 are disjoint. We first define Γ_1 in case A and then use Γ_1 to define Γ_2 in case B.

Case A: Γ_1 is the set of regions, which are composed of loops. In other words, each region in Γ_1 only contains a loop.

After defining the loop regions, to classify the non-loop regions, we must present our evaluation model first. Then the evaluation model is applied to loop regions for categorizing the non-loop regions. Below we first give some assumptions and then present how to use them to divided the non-loop part of a code into non-loop regions and finally formalize our evaluation model.

For a code $G = (V, E)$, where $V = \{R_1, R_2, ..., R_n\}$ is the set of regions in G and $E = \{(u, v) \mid u, v \in V \text{ and } u \neq v\}$. That is, E is the set of edges between regions. For each region R_i, we assume:

· N_{R_i}: the number of instructions in region R_i, for $i = 1, 2, 3, \cdots$
· N_{b_i}: the number of branches in R_i
· B_{ij}: the jth branch in R_i
· $P_{B_{ij}}$: the probability that B_{ij} is taken
· C_i: the number of clock cycles that does not result from any branch in R_i
· CB_{ij}: the number of clock cycles that results from that B_{ij} is taken
· $\overline{CB}_{ij}$: the number of clock cycles that results from that B_{ij} is untaken

According to Equation 1 and Equation 2, IPC predominate the determination of switching pipeline modes during execution. With the information collected previously and the above terminologies, we can present our evaluating model as follows.

$$\Omega_{R_i} = \sum_{j=1}^{N_{b_i}} [P_{B_{ij}} \times CB_{ij} + (1 - P_{B_{ij}}) \times \overline{CB}_{ij}] + C_i \tag{3}$$

$$\Theta_{R_i} = N_{R_i} / \Omega_{R_i} \tag{4}$$

Equation 3 estimates the clock cycles required for each region of the target code, which is also applied to classify the non-loop regions. Equation 4 calculates the IPC for each region and is the guideline to enable DPS. Since the loop regions very likely dominates power dissipation of a code, we use the following parameter Λ with the aid of the evaluation function of regions in Γ_1 to partition the non-loop part of a code. Λ is defined as the maximum of all Ω_{R_i} in Γ_1. Formally, it can be described as follows.

$$\Lambda = \{\Omega_R \mid \exists R \in \Gamma_1 \text{ and } \Omega_R \geq \Omega_{R_i}, \text{for } i = 1, \cdots, n\} \tag{5}$$

Although the loops usually consume the majority of power dissipation for an application, using λ to partition the non-loop part can be furthermore improved. Instead, we adapt Λ as the new parameter by timing a α to it, where α is a real number. Thus Γ_2 can be defined on the basis of $\alpha\Lambda$ in the following case B.

The followings demonstrate how our partitioning approach works using the code segment selected from Matrix of DSPstone benchmark suite as an example. At first, there are two loops existing in this code; thus they are identified as two loop regions and $\Gamma_1 = \{R_{a_1}, R_{a_2}\}$. Then we compare $\Omega_{R_{a_1}}$ and $\Omega_{R_{a_2}}$ and define $\Lambda = \Omega_{R_{a_2}}$ since $\Omega_{R_{a_2}} > \Omega_{R_{a_1}}$. Afterwards, we let $\alpha = 1$ and thus $\alpha\Lambda = \Lambda$ is applied to categorize non-loop parts into regions of the second type. The evaluation value of the first non-loop code segment is smaller than Λ, so it is immediately identified as the first non-loop region, R_{b_1}. Similarly, the evaluation value of the second non-loop code segment is equivalent to Λ, so it is identified as another non-loop region, R_{b_2}. Finally, the evaluation value of the third non-loop code segment is larger than Λ, thus it is further classified into two regions of the second type R_{b_3} and R_{b_4} so that $\Omega_{R_{b_3}} = \Lambda$ and $\Omega_{R_{b_4}} < \Lambda$. As a consequence $\Gamma_2 = \{R_{b_1}, R_{b_2}, R_{b_3}, R_{b_4}\}$.

2.3 DPS Enabling

After the code partitioning has been done, to enable DPS, we insert a function DPS_enable () into its head of each region to make it executed in deep mode or shallow mode. The DPS_enable () is implemented as follows.

$$DPS\_enable() \begin{cases} Lda & \#SYSCALL\_DEEP2SHAW \\ Call\_Pal & \#131 \\ Lda & \#SYSCALL\_SHAW2DEEP \\ Call\_Pal & \#131 \end{cases}$$

DPS_enable() provides two functionalities to switch between pipeline modes with the system call of Alpha 21264 Call_Pal #131. #SYSCALL_DEEP2SHAW switches the pipeline from deep mode to shallow mode and #SYSCALL_SHAW2DEEP switches the pipeline from shallow mode to deep mode. In this way, we are able to determine the timing to switch pipeline modes. The Ω value of each region calculated by Equation 4 is used for DPS_enable() when the code is compiled by our system. Thus the code will dynamically enter the deep mode or shallow mode during execution after the DPS_enable() is inserted into it. Finally the optimized DPSed program is performed on the modified Wattch simulator.

3 Experimental Results

In Section 3.1, we introduce the system configuration of our work and present the experimental results in Section 3.2.

3.1 System Configuration

The underlying hardware is the Alpha 21264 processor, which contains one fetch buffer, four integer ALUs, two floating-point ALUs, one integer multiplier/divider, and one floating-point multiplier/divider, etc. In instruction window, RUU indicates register update unit and LSQ comprises load queue (LQ) and store queue (SQ). Its main features are summarized in Table 1. To perform our proposed approach, we extend the pipeline mode from one mode to two modes. We assume that the original pipelining mode is shallow mode and the new mode is deep mode by constructed by adding extra four stages to shallow pipeline. It is designed to dynamically disable one of each pair of stages by making the latches between pipeline stages transparent so that the processor can switch between these two pipeline modes. The software configuration is shown in Table 2. The SUIF compiler infrastructure is the front end of our system and generate CFG and branch information. The operating system is Tru64 UNIX for 64-bit instruction set architecture. The Wattch simulator is an architectural simulator that provides cycle-by-cycle simulation and detailed out-of-order issue with multi-level memory system. For keeping consistence with our DPS approach, it has been modified to support shallow pipeline mode and deep pipeline mode.

3.2 Experimental Results

In our experiment, the deep mode is the default pipeline mode and the shallow mode is chosen during execution if necessary. The energy reduction benefits by the switching between deep mode and shallow mode depending on the IPC of a region, which is

Table 1. Hardware configuration

Processor Core	
Pipeline length	4 cycles (shallow mode)
	8 cycles (deep mode)
Fetch buffer	8 entries
Functional units	4 Int ALU, 2 FP ALU, 1 Int mult/div,
	1 FP mult/div, 2 mem ports
Instruction win-dow	RUU=80, LSQ=40
Issue width	6 instructions per cycle: 4 Int, 2 FP
Memory Hierarchy	
L1 D-cache size	64KB, 2-way, 32B blocks,
L1 I-cache size	64KB, 2-way, 32B blocks,
L1 latency	1 cycle
L2	Unified, 2MB, 4-way LRU
	32B blocks, 11-cycle latency
Memory latency	100 cycles
TLB size	128-entry, fully-associative,
	30-cycle miss

Table 2. Software Configuration

OS and Software Configuration	
Profiler	SUIF
Compiler	MachSUIF
OS	Tru64 UNIX
Simulator	Wattch v.1.02 with DPS

calculated by equation 4. The experiment is performed on the Wattch simulator with DSPstone. For each program, the baseline is its original energy dissipation and the optimized energy is measured by performing our DPS approach. This benchmark is compiled by the Alpha compiler with default settings and linked with the intrinsic library on Tru64 UNIX operating system.

Figure 4 shows the energy reduction by comparing the baseline energy and the optimized energy for DSPstone. In this experiment, we let $\alpha = 0.25, 0.5, 1.0,$ and, 2.0 to measure the effects of various partitioning sizes of non-loop regions. The energy saving ranges from 2% to 35%, with a mean of reduction 17.8%. As the partitioning size of non-loop region λ becomes larger, the energy saving decreases slowly. With our observations, the large-size region eliminates some chances to switch the pipeline modes based on IPC and thus slightly increases energy consumption. Nine of these programs including adpcm, complex_multiply, complex_update, dot_product, fft, iir_biquad_one_sections, matrix, real_update, and startup, have better energy saving about from 22% to 35%. The reason is that there are many branches in them and thus our DPS approach can take advantage of them to save energy depending upon the contribution of branch penalty to IPC. With our experiences, our approach works better

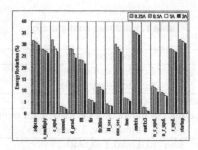

Fig. 4. Energy reduction of various λ of profile-based DPS for DSPstone

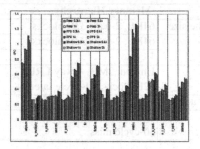

Fig. 5. Chang in IPC for three cases using DSPstone as a benchmark

for the codes with many loops and larger loops. Note that many programs have large outer loops such as event-driven programs or programs with GUI, which may include almost the entire programs. In this experiment, the typical example for above discussion is matrix testbench, and it has the best energy saving about 35%.

Figure 5 demonstrate the effect on IPC for three cases using DSPstone as the metric. Deep and Shallow represent deep and shallow pipeline modes; DPS indicates that applying our DPS approach to these benchmarks. They are still measured for various partitioning sizes of non-loop regions $0.25\lambda, 0.5\lambda, \lambda,$ and, 2λ. For DSPstone, the average IPCs of Deep case and Shallow case are 0.4 and 0.52. In DPS case, the average IPC is 0.45, which is between those of deep mode and shallow mode. In DSPstone, the IPCs of adpcm, fft, fir2dim, and matrix are larger. This is because since they are loop-intensive applications and the loops in them contribute a lot to the increase of IPC.

Figure 6 show the effect of our DPS approach on performance for DSPstone. The latency between pipelining stages is designed to be equivalent to increase performance and achieve resource sharing at each clock. In theory, the performance is in direct ratio to the number of pipelining stages and thus the longer pipeline will lead to the performance. Thus, the processor will result in slowdown when executing in shallow mode. The performance will be degraded if the pipelining stages are merged into shallow mode. In reality, the performance may be degraded due to many factors such as pipelining hazards, branch penalty, or switching overhead between pipeline modes. For each benchmark, the performance of deep mode with λ is the baseline to compare those of deep mode and our DPS method for various λs. For the performance of DSPstone in Figure 6, on average, our DPS approach leads to 6.23% degradation in performance.

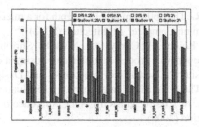

Fig. 6. Relative performance of various Λs for DSPstone

By contrast, the performance degradation of shallow pipeline mode is 61.62%, which is almost ten times larger than the above one. Although the DPS switches the pipelining modes based on the IPC to save energy, the switching slows down the processor compared to the high-speed execution in deep mode. In addition, larger λ has a better performance than smaller ones since it causes the pipeline to enter the shallow mode more infrequently.

4   Conclusions

DVS has been proven be very effective in low power optimizations, but it cannot further save energy when the voltage reaches its lower bound. Fortunately, DPS can overcome this limitation by adjusting pipeline modes based on IPC. Previous work resolved this issue with hardware techniques and thus increased hardware cost and design complexity. In this paper, we present a DPS technique to reduce power dissipation by proposing an evaluating model so that they can decide the timing of entering the proper pipeline mode. In contrast, our work can eliminate hardware overhead and reduce energy consumption according to the code behavior at compiler time. To investigate the effect of our approach, we perform the experiment with various criteria for DSPstone and Medieabench. In summary, the results show that smaller partitioning sizes of non-loop regions can create optimization space and loop-intensive applications provide more chances to optimize code to save energy.

References

1. Brooks, D., Tiwari, V., Martonosi, M.: Wattch: A framework for architectural level power analysis and optimizations. In: International Symposium on Computer Architecture (2000)
2. Burd, T., Brodersen, R.: Design issues for dynamic voltage scaling. In: International Symposium on Low Power Electronics and Design (2000)
3. Efthymiou, A., Garside, J.D.: Adaptive pipeline depth control for processor power-management. In: IEEE International Conference on Computer Design: VLSI in Computers and Processors (2002)
4. Ernst, D., Kim, N.S., Das, S., Pant, S., Rao, R., Pham, T., Ziesler, C., Blaauw, D., Austin, T., Flautner, K., Mudge, T.: Razor: A low-power pipeline based on circuit-level timing speculation. In: The 36th International Symposium on Microarchitecture (2003)
5. Gerndt, M.: Automatic parallelization for distributed-memory multiprocessing systems. Phd Thesis (1989)

6. Hartstein, A., Puzak, T.R.: The optimum pipeline depth for a microprocessor. In: ACM/IEEE International Symposium on Computer Architecture (2002)
7. Hiraki, M., Bajwa, R.S., Kojima, H., Corny, D.J., Nitta, K., Shridhar, A., Sasaki, K., Seki, K.: Stage-skip pipeline: A low power processor architecture using a decoded instruction buffer. In: International Symposium on Low Power Electronics and Design (1996)
8. Hsu, C.H., Kremer, U.: The design, implementation, and evaluation of a com-piler algorithm for cpu power reduction. In: The ACM SIGPLAN Conference on Programming Languages Design and Implementation (2003)
9. Kessler, R.E., McLellan, E.J., Webb, D.A.: The alpha 21264 microprocessor architecture. In: Intl Conf. Computer Design (1998)
10. Koppanalil, J., Ramrakhyani, P., Desai, S., Vaidyanathan, A., Rotenberg, E.: A case for dynamic pipeline scaling. In: International Conference on Compilers, Architecture, and Synthesis for Embedded Systems (2002)
11. Kornerup, J.: Mapping powerlists onto hypercubes. Masters Thesis (1994)
12. Krishna, C.M., Lee, Y.-H.: Voltage-clock-scaling adaptive scheduling techniques for low power in hard real-time systems. In: The 6th Real Time Technology and Applications Symposium (2000)
13. Lilia, D.J.: Reducing the branch penalty in pipelined processors. IEEE Computer 21(7), 47–55 (1988)
14. Manne, S., Grunwald, D., Klauser, A.: Pipeline gating: Speculation control for power reduction. In: ACM/IEEE International Symposium on Computer Architecture (1998)
15. Parikh, D., Skadron, K., Zhang, Y., Stan, M.: Power-aware branch prediction: Characterization and design. In: The IEEE Transactions on Computers (2004)
16. G. SUIF. Stanford University Intermediate Format, http://suif.stanford.edu

An Editing System Converting a UML State Diagram to a PLC Program

Yung-Liang Chang[1,2], Chin-Feng Fan[1], and Swu Yih[3]

[1] Dept. of Computer Science and Engineering
Yuan-Ze University, Chung-Li, Taiwan
csfanc@saturn.yzu.edu.tw
[2] Delta Electronics, Tao-Yuan, Taiwan
yl.chang@delta.com.tw
[3] Dept. of Computer Science and Information Engineering
Chien Hsin University of Science and Technology, Chung-Li, Taiwan
swuyih@uch.edu.tw

Abstract. UML is a modeling language commonly used in contemporary software or system development. Using UML models at the design stage is relatively simpler and better visualized than using one of the PLC languages specified in IEC 61131. This research developed an editing system, PSE (PLC State Diagram Editor), which can convert a UML state transition diagram into a PLC program to provide better visibility and quickly lead a non-professional PLC programmer into the field of PLC programming. Besides, PSE also support related application, such as model-based test cases generation.

Keywords: PLC, SFC (Sequential Flow Chart), IL (Instruction List), model-based test case generation.

1 Introduction

IEC 61131-3 [5] has standardized PLC programming languages into 5 different types; namely, Ladder (LD), Instruction List (IL), Sequential Flow Chart (SFC), Function Block Diagram (FBD), and Structure Text (ST). Therefore, a PLC programmer has different choices of programming languages based on his background and preferences. However, for a non-professional PLC programmer, the learning process may be long since PLC programs are different from other high-level programming languages. To expedite the learning process in PLC programming, this research developed an editing and translation system, PSE (PLC State diagram Editor), which supports the editing of a UML state transition diagram as well as the compilation of a state diagram into a PLC program. PSE is implemented as an extension to WPL Editor, a program-editing system made for the Delta DVP-PLC series; the first author is the major developer of the WPL Editor.

Section 2 provides related background. Section 3 presents our editing system and its implementation techniques. Section 4 shows a case study using this editor, followed by conclusion in Section 5.

J.-S. Pan et al. (Eds.): *Advances in Intelligent Systems & Applications*, SIST 21, pp. 647–655.
DOI: 10.1007/978-3-642-35473-1_64   © Springer-Verlag Berlin Heidelberg 2013

2   Related Background

2.1   WPL Editor

IEC 61131-3 [5] has standardized PLC programming languages into 5 different types; namely, Ladder (LD), Instruction List (IL), Sequential Flow Chart (SFC), Function Block Diagram (FBD), and Structure Text (ST). WPL is a PLC editing system developed for Delta Electronics, Inc. WPL supports three PLC languages, namely, LD, SFC and IL. WPL provides editing of programs written in these three languages and compilation of a LD and a SFC program to IL code, as well as other special functions, such as monitoring, uploading, downloading, and simulation of small DVP systems manufactured by Delta Electronics. This research extended the original WPL to include UML state transition diagrams as input for PLC programming along with associated enhancement and state-based applications, such as test case generation.

As to conversion from a UML state diagram to a PLC program, there are some other research activities such as Huang's [2]. However, Huang converted the input state diagram into a Ladder (LD) program. Our system extends WPL to provide a complete editing and a compilation environment for a state diagram and different PLC languages. An inputted state diagram is first converted into a SFC structure and then to IL code for simulation and testing. Moreover, other associated functions, such as XML representation of the input diagram, compilation of a ST program to IL code, and test case generation, are also provided.

2.2   Model-Based Test Case Generation

There are many different methods dealing with model-based test case generation [1,3,4]. We adapted the state-machine-based test case generation proposed by Kim, et al. [3]. They defined an EFSM(Extended finite state machines), on which they suggested to use a Breadth-First Search to generate test cases covering all paths, states, and transitions. One may assume <GStates, C0, GTrans> is an EFSM, GStates represents the set of states, C0 is the initial state, GTrans is the set of transition. If t $\in$ Gtrans, gs_i $\in$ Gstates. gti = (Ci, ei, gi, ai, C' i) $\in$ Gtrans, where Ci is a state, ei is an event, gi is a guard, ai is the action, C'i is the destination state. The sequece ((gs0, gt0), (gs1, gt1), (gs2, gt2)n..., (gsn, gtn)) represents a path. The following is their algorithm [3] to generate the test tree satisfying global state coverage.

> *make-state-coverage-testing-tree (**Node**)*
> > *begin*
> > *if all gs* $\in$ *GStates are visited **then return**;*
> > *for each **gt** = (**Ci, ei, gi, ai, C' i**)* $\in$ *GTrans **do begin***
> > > *if **Ci** = **Node** and **C'i** is not visited **then***
> > > *make **C'i** as a child of Node;*
> > *end*
> > *for each child **cNode** of Node **do begin***
> > > *make-state-coverage-testing-tree (**cNode**)*
> > *end*
> *end*

3   PLC State Diagram Editor (PSE)

This research aims to develop an editing system to convert a UML state transition diagram into a PLC program so as to support an non-professional user to construct PLC projects. The PSE is an extension to the WPL, the PLC editor by Delta Electronics. This new version is called as PSE (PLC State Diagram Editor). Fig. 1 shows the PSE execution steps. First, the input can be project specifications using a UML state diagram, or a Ladder (LD), or a Structure Text (ST) program. Then, the inputted state diagram will be converted into Sequential Flow Chart (SFC), which will in turn be compiled into Instruction List (IL) for further applications. The applications can be the original WPL functions, such as loading and simulation as well as the newly added test case generation.

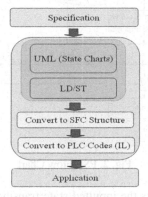

Fig. 1. PSE execution steps

The following sections focus on the state transition diagram related steps:

(1) Check the correctness of the state transition diagram

(2) Translate the input state diagram into a SFC program

(3) Compile the SFC to IL code

(4) Save the corresponding XML information for the inputted state diagram

(5) Generate test cases.

3.1   Step 1 and 2: Editing and Translate a State Diagram to a SFC Structure

Step 1 first checks the correctness of the inputted state diagram. The following requirements are checked:

1. Each state should be connected to some state. There is no isolated state.
2. Each edge, except the one from the start state and the one to the final state, has a source state and a destination state.

3.   The input diagram is restricted to have only one start and one final state. The
     start state and the final state will be checked for the correctness.

Step 2 converts the input state diagram into a corresponding SFC structure. The complete translation process on PSE is shown in Fig.2, in which the parts shown in yellow
are supported by the original WPL. The new editing system PSE converts an UML
state transition diagram or an inputted ST (Structure Text) program into a corresponding PLC IL program. The ST to IL compilation details are not described in this paper,
but can be found in [7].

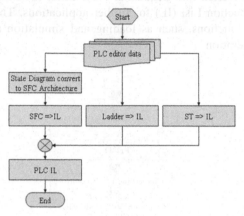

Fig. 2. PSE Flow Chart

In Step 2, the PSE converts the inputted state transition diagram into a SFC structure, including the following conversion patterns:

(a) A linear structure

(b) A loop structure

(c) A conditional fork-and-join structure

A simple linear structure in a state diagram can be translated straightforwardly to a
linear SFC, as shown in Fig.3. A loop structure in a state diagram can be a loop jumping back or forward to another state, or a loop going back to the source state, as
shown at the left side of Fig 4. Since SFC is in a data-flow style, the loop target is
only indicated with the state name, as shown at the right side of Fig. 4, no matter
whether the loop target is another state or the same state. We have augmented the
original state diagram with a fork-join (or concurrency) structure, similar to that in a
UML activity diagram, as shown at the left side of Fig. 5. It depicts the situation when
state S1 encountering the trigger event TRAN1, the execution will be carried out concurrently by state S2 and S3, which will later be merged and go to state S4 when the
trigger Tran2 occurs. The converted SFC structure is given at the right side of Fig 5.
A breadth-first-search algorithm is used to generate such translation.

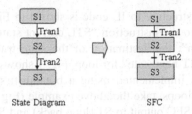

Fig. 3. A linear structure

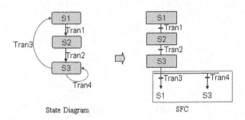

Fig. 4. A loop structure

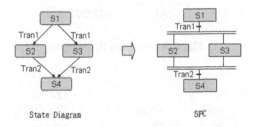

Fig. 5. A conditional fork and join

3.2 Step 3: Compiling a SFC Structure to IL Code

Step 3 uses the function provided by the WPL to convert the above SFC structure into IL code. IL (Instruction List) is the PLC language similar to assembly code. The following IL instructions are used in the target code of the compilation:

STL S* : STL represents the starting point from device S* for a SFC structure. The status of device S represents a state.

RET: RET ends a subroutine and returns to the general mode.

LD : LD sets the current result to value and save it to a register.

SET : SET sets the specified device point to be ON, which will remain ON until the RST (RESET) instruction is encountered. We use SET to describe the target of the next transition.

OUT : OUT outputs the result to a specified element. We use OUT instruction to represent loop structure.

Translation from a SFC structure to IL code is shown in Fig. 6. The start state of a transition is compiled into the instruction "STL current state"; the transition is compiled into "LD transition"; the destination of the state transition is translated into " SET next state" or "OUT next state" for loop. Fig. 7 shows a simple sample without loops. The translation is implemented using a breadth-first search order, which is depicted in Fig. 8. As to loops, take the above example (Fig. 4) for instance, loops in Fig. 9 are converted into SFC output to S1 (loop back) and S3 (loop to itself) through trigger event TRAN3 and TRAN4. Then the compiled IL code for these loops is:

        STL S3
        LD Tran3
        OUT S1
        LD Tran4
        OUT S3

Fig. 6. A converted IL template

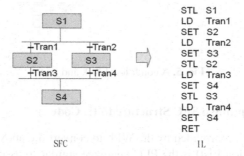

Fig. 7. A sample showing SFC to IL conversion

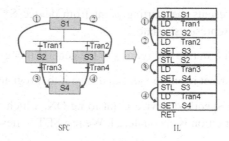

Fig. 8. Breadth-first order for converting SFC into IL

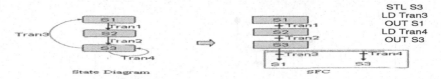

Fig. 9. Loops translation

3.3   Steps 4, 5: Associated Application

Besides, the PSE also provides associated applications, including converting inputted Structure Text (ST) code into IL, generating XML descriptions for the inputted state diagram, as well as generating model-based test cases from the inputted state diagram. The conversion from ST to IL is to make this extension to WPL complete. Details can be found in [7]. The XML files describing the inputted state diagram can be used for data exchange. The model-based testing case generation facilitates testing of the resulting IL code.

This research adapts the aforementioned Kim's algorithm [3] to generate model-based test cases to ensure the coverage of each state and each transition. Moreover, for the selection of parameter values, the equivalence portioning and boundary values approach [6] are used. Thus, for a tested variable with a specified legal range, the equivalence partitions are: (1) the legal part, (2) the part below the low bound, and (3) the part above the upper bound. Each of these partitions should be tested by at least one standard value. Multiple tests for a certain partition can also be performed if desired. The boundary testing tests the error prone boundary values; say, the low bound, low bound -1, low bound +1, upper bound, upper bound -1, and upper bound +1. Therefore, the PSE system not only outputs IL code for the original simulation purpose, it also generates model-based testing cases.

4   Case Study

We have applied the constructed tool on a case study: a Safety Injection System [8], which injects water into the vessel when a loss of coolant water accident occurs in a nuclear power plant. As the controller detects either the low water level or the high pressure below/over a predefined threshold, it will initiate water injection from Tank 1 into the vessel. When the water in Tank 1 is lower than a threshold, the injection from Tank 2 will be started. Fig. 10 is the state diagram. The following is a list of the states in the figure:

SI_Stop : Stop water injection
Low_Water_Mode : reactor vessel water level is low
  DW_HI_Press_Mode : pressure is high
Manual_Start_Mode : manually start the water injection
SI_Init : Safety injection from Tank 1.
SP_InWater : Safety injection from Tank 2

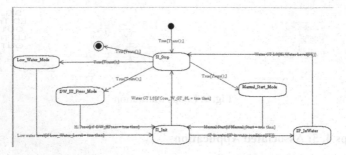

Fig. 10. State transition diagram for the Case study

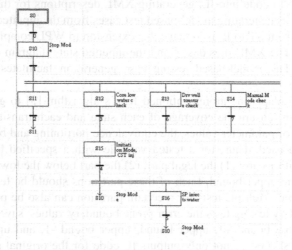

Fig. 11. Part of the converted SFC

000000	STL	S0
000001	LD	M1000
000002	SET	S10
000003	STL	S10
000004	LD	M1000
000005	SET	Y2
000006	LD	M1000
000007	SET	Y3
000008	LD	M1000
000009	RST	Y4
000012	LD	M1000
000013	RST	Y5
000016	LD	M1000
000017	RST	Y0
000020	LD	M1000
000021	RST	Y1
000024	LD	M1000
000025	RST	Y6
000028	LD	M1000
000029	RST	Y7
000032	LD	M1000
000033	SET	Y10
000034	LD	M1000
000035	SET	Y11
000036	LD	M1000
000037	SET	S11
000038	LD	M1000
000039	SET	S12
000040	LD	M1000
000041	SET	S13
000042	LD	M1000
000043	SET	S14

Fig. 12. Translated IL code

Fig. 13. Generated Test cases

Fig 11 shows part of the converted SFC. Fig 12 shows the compiled IL code. Fig 13 presents the generated test cases. Preliminary results show that the PSE editing system is valid and effective.

5   Conclusion

IT professionals are familiar with UML modeling; however, they may be not familiar with PLC programming. To expedite the PLC programming learning process for a layperson, this project has constructed an editing and translation system supporting inputs in a UML state diagram and translated it into a PLC program. This editing system, called as PSE, extended the WPL, the PLC editor provided by Delta Electronics. A UML state transition diagram is edited and checked for its correctness, and then it can be converted into a SFC program, which can be later compiled into IL code, an assembly level PLC language, for simulation and testing. Besides, test cases can be generated based on the inputted state model. The PSE can thus handle UML state diagrams as well as four out of the five PLC languages specified in IEC 61131-3, except Function Block Diagram (FBD). In the future, this system can be extended to handle FBD programs. This editing and translation tool supports a non-PLC programmer to perform PLC projects and facilitates the translation from a UML state diagram to different PLC languages. PLC programming can thus be done using software/system modeling.

References

1. Ferreira, R.D.F., Faria, J.P., Paiva, A.: C. R.: Test Coverage Analysis of UML State Machines. In: Third International Conference on Software Testing, Verification, and Validation Workshops (2010)
2. Huang, F.: State Diagrams: A New Visual Language for Programmable Logic Controllers. McMaster University master's thesis (2010)
3. Kim, Y.G., Hong, H.S., et al.: Test Cases Generation from UML State Diagrams. IEEE Proc.-Softw. 146(4) (1999)
4. Offutt, J., et al.: Generating Test Data from State-Based Specifications. The Journal of Software Testing, Verification and Reliability 13(1), 25–53 (2003)
5. Programmable controllers – Part 3: Programming languages, IEC61131-3, Second edn. (2003)
6. Pressman, R.: Software Engineering: A practitioner's Approach. McGraw-Hill (2005)
7. 張永良: 轉換狀態圖至PLC程式碼之編輯系統. 元 智 大 學資訊工程研究所碩士論文 (2012)
8. 廖本錦、游原昌等: ECCS/HPCF 數位控制邏輯分析報告. 能研究所技術報告 (2005)

Accurate Instruction-Level Alias Analysis for ARM Executable Code

Tat-Wai Chong and Peng-Sheng Chen*

Department of Computer Science and Information Engineering and
Advanced Institute of Manufacturing for High-tech Innovations,
National Chung Cheng University, Chia-Yi 621, Taiwan, ROC
funkwai@gmail.com, pschen@cs.ccu.edu.tw

Abstract. In this paper, we present an accurate, instruction-level, probabilistic alias analysis algorithm for ARM executable code, which reveals the probability of two registers holding the same value. The concept of memory regions (i.e., global, local, and heap variables) is borrowed from high-level programming languages and used to enhance the analytical accuracy. A memory information table (MIT) is proposed as a means of linking determinable stored values with the corresponding memory addresses in order to exploit unknown alias relationships. We also propose a simplified representation of probabilistic information, in order to enhance the efficiency of the proposed algorithm. The alias analysis algorithm was implemented using a post-link optimizer, Diablo. The experimental results show that the entire analytical procedure can be completed within a reasonable amount of time and memory. The proposed algorithm can provide alias information for about 80%-98% of memory reference relationships, and can estimate the probability that two registers refer to the same memory address in the 96.10% of the tested memory-disambiguation pairs, with an overall average accuracy of about 90%-100%.

Keywords: alias analysis, probability, assembly instruction.

1 Introduction

Alias analysis determines whether or not two or more memory references point to the same memory location. It is an important analysis technique, which allows software development tools to perform many types of aggressive optimization. For high-level programming languages, alias relationships are introduced by the features of the language (e.g., pointer variables, array indexing, and call-by-reference). For low-level assembly instructions, alias analysis provides information about the memory locations that a given register is associated with. The alias information of low-level instructions is usually used to analyze and superiorly optimize software programs whose source codes cannot be obtained. Consider the code fragment of ARM assembly instructions shown in Figure 1. If we determine that r2 and r4 do not refer to the same memory location, then the load instruction, S_3, can be moved above the store instruction, S_2, in order to hide the latency of memory access.

---

* Corresponding author.

J.-S. Pan et al. (Eds.): *Advances in Intelligent Systems & Applications*, SIST 21, pp. 657–667.
DOI: 10.1007/978-3-642-35473-1_65   © Springer-Verlag Berlin Heidelberg 2013

Alias analysis techniques classify aliases as *must* aliases (*definitely-points-to* relationships), which hold for all executions, or as *may* aliases (*possibly-points-to* relationships), which might hold for some executions, or as *probabilistic* aliases, which provide quantitative descriptions of how likely it is that the conditions will hold for the executions. A software development tool has few choices to handle superior optimizations if the *must* or *may* alias information is applied. With the increasing popularity of powerful and complex hardware/software systems, a smart software development tool needs more choices for further processing. The *probabilistic* alias analysis quantifies the alias information, potentially providing software development tools with many more choices from which to make intelligent decisions.

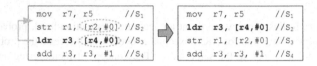

Fig. 1. Example of alias analysis

In this paper, we develop an accurate, instruction-level, probabilistic alias analysis algorithm for ARM executable code, which reveals the probability of two registers each holding the same value. The concept of memory regions (i.e., global, local, and heap variables), borrowed from high-level programming languages, is used to enhance the analytical accuracy. A memory information table (MIT) is proposed as a means to link determinable stored values with the corresponding memory addresses in order to uncover unknown alias relationships. We also propose a simplified representation of probabilistic information, in order to reduce the computational overheads. The alias analysis algorithm was implemented using Diablo [12], which is a retargetable, extensible, and reliable framework for static binary rewriting. Experimental results show that the entire analytical process can be completed within a reasonable time and with reasonable memory usage. The proposed algorithm can provide alias information for about 80%-98% of memory reference relationships, and can estimate the probability that registers refer to the same memory address in 96.10% of the tested memory-disambiguation pairs, with an overall average accuracy of about 90%-100%.

1.1 Contribution

This paper makes the following contributions.

- **Theoretical Algorithm.** It presents a flow-sensitive, context-insensitive algorithm for instruction-level alias analysis. The proposed algorithm integrates address abstraction, memory region, and MIT (i.e., linking determinable stored values with the corresponding memory addresses) techniques, in order to provide accurate alias information.
- **Simplified Representation of Probabilistic Information.** It presents a simplified representation of probabilistic information in order to improve analysis efficiency.

- Experimental Results. It presents experimental results that demonstrate the proposed algorithm can predict quantitative alias information and achieve good accuracy with acceptable resource consumption.

The remainder of this paper is organized as follows. Section 2 describes other works related to probabilistic alias analysis and instruction-level alias analysis. Section 3 presents the proposed algorithm for instruction-level alias analysis. Section 4 reports on the results obtained from the experimental evaluation of the proposed algorithm. Finally, we present conclusions in Section 5.

2   Related Works

Many methods have been developed for performing alias analysis in high-level programming languages [5,11,8]. In this paper, we propose an instruction-level alias analysis that provides accurate quantitative alias information. Accordingly, we now discuss two related areas; probabilistic alias analysis and instruction-level alias analysis.

2.1   Probabilistic Alias Analysis

Ju et al. [9] presented a general probabilistic memory disambiguation framework for estimating the probabilities of aliases among array references. Chen et al. [2] proposed a flow-sensitive, context-sensitive, interprocedural, probabilistic alias analysis algorithm. Their analysis method provides quantitative information on alias relationships. Both of these research works are focused on the alias analysis of high-level programming languages.

2.2   Instruction-Level Alias Analysis

Debray et al. [4] used an address descriptor (AD) to abstract memory addresses and developed an algorithm for including the effects of individual instructions on register values, in order to obtain instruction-level alias information. Fernandez and Espasa [6] proposed a speculative alias analysis method for executable code. They used a region-based alias analysis technique that introduced three separate memory regions (i.e., global, stack, and heap) for coarse-grain memory disambiguation. Guo et al. [7] presented a method for context-sensitive, partially flow-sensitive, low-level pointer analysis. The program memory was partitioned into several abstract structures (e.g., local variable space, incoming parameter space, and heap space). A special abstract structure, named the *unknown initial value*, was created in order to represent memory blocks whenever needed. The memory reference relationships were analyzed and recorded within a dataflow framework, by using the appropriate transfer functions. Their method simultaneously considered register values and the contents stored within the associated data objects , and produced accurate results. Balakrishnan and Reps [1] proposed a flowsensitive, context-insensitive, value-set analysis for accurately approximating the set of values that each data object holds at each program point. The global, local, and heap memory regions and *a-locs* were used to locate data objects.

Our work is most similar to that of Lu et al. [10], who proposed an instruction-level, probabilistic alias analysis for x86 executables. They used ADs to abstract memory objects, and integrated Chen's algorithm to compute the probabilities of alias relationships. The main difference between this previous work and the present study, is that we have added the concept of memory region, and use an MIT technique to model memory objects, in order to provide accurate alias information. Moreover, we have developed a simplified representation of probabilistic information in order to enhance the efficiency of the analysis.

3   Algorithm

This section describes the proposed algorithm. The method for instruction-level alias analysis that we developed, computes the probabilities for all given registers being associated with each of their possible values at every program point. A register r at program point B holds a value V with a probability $\mathcal{P}$, which is defined as follows:

$$\mathcal{P}((r, V), B) \overset{def}{=} \frac{E((r, V), B)}{E(B)}, \tag{1}$$

where $E((r, V), B)$ denotes the number of executions during which register r holds value V at program point B, and $E(B)$ is the number of times that the execution passes through B.

We first introduce the abstract representation of memory objects that will be used to represent register values, i.e., V in equation 1. We then describe the MIT technique and the simplified representation of probabilistic information. Finally, we describe the analysis algorithm.

3.1   Abstraction of Memory Object

An instruction-level alias analysis discloses the memory locations that a register is associated with. Due to the impossibility of enumerating memory locations, we are compelled to appropriately abstract them. The abstraction should use fewer system recourses (i.e., less memory) and have higher accuracy. Hence, we propose an enhanced address descriptor (EAD) for abstracting memory objects. An EAD consists of three elements: memory region, type, and AD. According to the memory layout of the software at runtime, the memory region may consist of a single global region, a separate region for each procedure, or a separate region for each heap-allocation statement. The type is used to indicate the nature of the register value. If we can determine that a register value is used for simple numeric computing (e.g., computation of loop iterations), then the type is *Numeric*. If, instead, a register value is used to calculate memory addresses, then the type is *Address*. In the proposed alias analysis, we assume that the type is *Numeric* if a register was assigned a value that falls into a particular range. In our implementation, if the immediate in a MOV instruction falls into the range 1 through 1024, then the type will be assigned as *Numeric*, otherwise it will be assigned as *Address*. For example, the type of register r0 will be assigned as *Numeric* after executing the instruction "MOV r0, #1". The chosen range begins at 1, because the NULL pointer is equal to 0.

The upper range limit of 1024 is a safe choice for programs running in Linux, since these have starting memory locations far larger than this number. The range can be adjusted according to the configuration of the target systems.

An AD is a pair $\langle I, M \rangle$, where I is either an instruction or one of the distinguished values $\{NONE, ANY\}$, and M is a set of *mod-k residues*. Given a program **P** and an instruction I at a program point p, let $val_p(I)$ denote the set of all values, w, such that, for some input to p, there is an execution path from the entry point of **P** to instruction I that causes I to move w into its destination register ($val_p(I) = \emptyset$ if I does not move a computed value into a register, or if control never reaches I). For any program **P**, $val_p(NONE) = 0$, while $val_p(ANY)$ is the set of all values.

3.2 Memory Information Table (MIT)

In our observation, for RISC processors, a register is usually assigned a value by load-related instructions. If an alias analysis does not keep track of memory content, the memory will become a black box, and load-related instructions will always load unknown data into registers. This situation leads to the imprecision during the alias analysis. Figure 2(a) shows an example where, without tracking stored values and the corresponding memory addresses, the value of r0 cannot be obtained after S_3. In order to solve this problem, we propose using the MIT technique to record the relationships between stored values and the corresponding memory addresses. Figure 2(b) illustrates the use of MIT. For ARM architectures, the memory address is represented by the expression: base-register $\pm$ offset. We appropriately record both the EAD of r2 and the corresponding memory address in the MIT. When analyzing the load instruction S_3, we search the MIT and assign the appropriate EAD (if found) to r0. Currently, in order to reduce the analysis overheads, we add an entry to the MIT only when the EAD of the current base register has been uniquely identified.

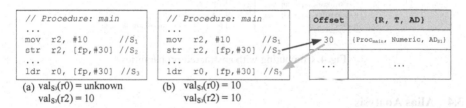

(a) $val_{S_3}(r0)$ = unknown
$val_{S_3}(r2) = 10$

(b) $val_{S_3}(r0) = 10$
$val_{S_3}(r2) = 10$

Fig. 2. Example of using MIT technique

3.3 Probabilistic Information

In order to improve the efficiency of the analysis, we propose a simplified representation of abstracted probabilistic information, and a heuristic approach to computing multiplications. We split the probability range into eight regions and three areas, as shown in Figure 3. The regions are used to denote the probabilities, and all probability calculations operate on these regions. In order to simplify these calculations, group the eight regions into three areas. The probability range is not uniformly distributed across the eight regions. Because values near the middle of the probability axis are not of use to software development tools, the ranges at the ends of the probability axis are more fine-grained.

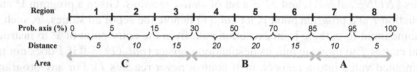

Fig. 3. Abstraction of probability

Figure 4(a) shows the method for computing probabilities. Assume we have two probabilities: P_1 and P_2, with $P_1 < P_2$. We can obtain $P_1 \times P_2$ from the areas that P_1 and P_2 belong to. Figure 4(b) shows how the eight ranges can be encoded by one byte (eight bits). This design can significantly reduce the memory required for data representation and eliminates the cost of floating-point computation. Figure 4(c) demonstrates how to compute probabilities within the proposed system.

(a)

P_1	P_2	$P_1 \times P_2$
A	A	P_1
B	A	P_1
B	B	$P_1 \gg 1\text{-bit}$
C	A	P_1
C	B	$P_1 \gg 1\text{-bit}$
C	C	$(00000001)_2$

(b)

Value	Probability range (%)
$(00000001)_2$	[0, 5]
$(00000010)_2$	(5, 15]
$(00000100)_2$	(15, 30]
$(00001000)_2$	(30, 50]
$(00010000)_2$	(50, 70]
$(00100000)_2$	(70, 85]
$(01000000)_2$	(85, 95]
$(10000000)_2$	(95, 100)
$(11111111)_2$	100

(c)

Example	Solution
$P_i = (00010000)_2$ in (50, 70], Area$_B$ $P_j = (00000010)_2$ in (5, 15], Area$_C$ $P_i \times P_j = ?$	$\because P_i > P_j$ $\therefore P_j$ is P_1, and P_i is P_2 $P_i \times P_j = P_i \gg 1 = (00001000)_2$ in (30, 50]

Fig. 4. Computing with abstracted probabilities

3.4   Alias Analysis

The proposed alias analysis is formulated as an iterative data-flow framework that includes both transfer functions and probability functions in order to formulate the effects of statements and program constructs. The main steps of the proposed alias analysis are divided into two phases. The first phase computes the set of alias relationships, and operates in the same way as do traditional alias analyses. The second phase computes, according to the proposed probability functions, the corresponding probabilities for the alias relationships obtained in the first phase. A flow-sensitive policy is adopted in the proposed analysis algorithm. Assume that IN_B and OUT_B respectively store the sets of alias relationships at the program points immediately before and after a basic block B, and that their corresponding sets of probabilities are IN_P_B and OUT_P_B. After initialization, for each basic block B, we first compute the IN_B and IN_P_B by using the appropriate meet operations of transfer and probability functions. In computing OUT_B, the effect of each instruction in the basic block B upon the alias relationships is estimated using the appropriate transfer and probability functions. The instructions whose execution will change a register value are considered to be effective. In our implementation, these instructions include ADC, ADD, AND, BIC, EOR, LDM, LDR, MLA, MOV, ORR, RSB, RSC, SBC, STM, STR, SUB, and SWP. The proposed algorithm iteratively re-computes all OUT_B and OUT_P_B until a steady state is attained. Finally, we can then obtain the probabilistic alias relationships at each program point. A detailed description of the data-flow analysis algorithm can be referred to in [10].

Table 1 gives an explicit formulation of the transfer function and probability function of an ADD instruction. For an ADD instruction, S, assume that the alias relationships $(r_{dst}, \langle R_{dst}, T_{dst}, I_{dst}, M_{dst} \rangle)$, $(r_1, \langle R_1, T_1, I_1, M_1 \rangle)$, and $(r_2, \langle R_2, T_2, I_2, M_2 \rangle)$ are in INS; with the corresponding probabilities being $\mathcal{P}_{dst}$, $\mathcal{P}_1$, and $\mathcal{P}_2$, respectively. The *mod-k residues* are used with $k = 2^m$, $m \in \mathbb{N}$. For an immediate operand, c, the EAD is $\langle Proc, Numeric, NONE, c \bmod k \rangle$. The formulations for other effective instructions can be derived by a similar approach.

Table 1. Transfer function and probability function for the ADD instruction

Instruction S	Transfer function F_S	Probability function $\mathcal{P}$
add r_{dst}, r_1, r_2	If $I_1 = NONE$, $I_2 \neq NONE$, and $R_1 = R_2$, then r_{dst} is changed to $\langle R_1, Address, I_2, (M_1 + M_2) \bmod k \rangle$.	$\mathcal{P}_1 \times \mathcal{P}_2$
	If $I_1 = NONE$, $I_2 \neq NONE$, and $R_1 \neq R_2$, then r_{dst} is changed to $\langle \perp, Address, I_1, (M_1 + M_2) \bmod k \rangle$.	$\mathcal{P}_1 \times \mathcal{P}_2$
	If $I_1 \neq NONE$, $I_2 = NONE$, and $R_1 = R_2$, then r_{dst} is changed to $\langle R_1, Address, I_1, (M_1 + M_2) \bmod k \rangle$.	$\mathcal{P}_1 \times \mathcal{P}_2$
	If $I_1 \neq NONE$, $I_2 = NONE$, and $R_1 \neq R_2$, then r_{dst} is changed to $\langle \perp, Address, I_1, (M_1 + M_2) \bmod k \rangle$.	$\mathcal{P}_1 \times \mathcal{P}_2$
	If $I_1 = I_2 = NONE$, then r_{dst} is changed to $\langle R_1, Address, NONE, (M_1 + M_2) \bmod k \rangle$.	$\mathcal{P}_1 \times \mathcal{P}_2$
	If $I_1 \neq NONE$ and $I_2 \neq NONE$, then the value of the destination register cannot be predicated. r_{dst} is conservatively defined as $\langle \perp, Address, S, 0 \rangle$.	1

4 Experiment

Our implementation of the proposed instruction-level alias analysis algorithm was based on the Diablo post-link optimizer, which is a retargetable link-time binary rewriting framework developed at Ghent University. An instrumentor based on FIT (a Flexible Instrumentation Toolkit) [3] was developed for collecting runtime information (i.e., the execution frequency of each basic block and the alias probability of registers of interest at runtime). The instrumented executable codes were executed, so as to generate profiling information, on a Create XScale-PXA270 EVB, with 32 MB of memory, running Linux kernel 2.6.15. The analyzer was executed on an Intel 3.0 GHz Pentium-D processor coupled to 3 GB of RAM and running Mandrake Linux 2.6.11-13. The experimental configuration used for the *mod-k residue* analysis was $k = 64$. The benchmark programs tested were MiBench, Olden, CommBench, and MediaBench2. Each benchmark program was compiled and statically linked using GCC 3.2.2 with the "-static" option, and then the static-linked executable code was fed into Diablo with default options for analyzing probabilistic alias information.

Table 2 lists the time and space requirements for each of the tested benchmark programs, and presents the statistics at static time for the number of functions, basic blocks, and instructions. For evaluating the proposed alias analysis algorithm, the memory reference instructions (i.e., load and store instructions) in a basic block are combined to check whether or not their referenced memory locations are disjoint. The term "Memory-dependent pairs" denotes the total number of these combinations appearing within a program at runtime.

Table 2. Analysis time (in seconds) and memory usage (in megabytes) for the tested benchmarks

Program	Funcs	Insns	BBLs (static)	BBLs (dynamic)	Memory-dependent pairs	Time	Memory
perimeter	784	71579	17743	813	1632	37	187
power	794	74746	18353	1067	2786	38	190
bh	842	79851	19421	1457	3145	39	200
bisort	784	71304	17667	735	1457	37	187
health	815	77026	18875	1025	2162	38	190
drr	783	72003	17912	1002	2108	37	187
rtr	803	73545	18302	1601	2332	37	184
basicmath	832	86852	19838	1701	3287	38	190
crc32	784	71067	17712	1033	2433	37	184
dijkstra	779	71023	17631	928	1929	36	181
fft	823	79788	18749	1381	1789	38	190
patricia	787	71513	17835	1484	2503	37	187
qsort	776	70763	17603	1260	2301	37	187
stringsearch	778	70847	17621	759	1429	37	187
susan	817	86752	19305	1174	163397	38	190
jpeg	972	82888	20320	1881	3792	41	209

Figure 5 shows the distribution of the alias relationship probabilities for each of the tested benchmark programs. With the alias information known, memory disambiguation can be decided according to the following steps. Consider two probabilistic alias relationships, $[(r1, \langle R_1, T_1, I_1, M_1 \rangle), \mathcal{P}_1]$ and $[(r2, \langle R_2, T_2, I_2, M_2 \rangle), \mathcal{P}_2]$. First, if $R_1 \neq R_2$, then we say that registers $r1$ and $r2$ refer to different memory locations with probability $\mathcal{P}_1 \times \mathcal{P}_2$. Otherwise, we check the relationship between T_1 and T_2. If $T_1 \neq T_2$, then registers $r1$ and $r2$ refer to different memory locations. If $T_1 = T_2$, then we need to check their I and M. If $I_1 = I_2$ and $M_1 = M_2$, then the probability that registers $r1$ and $r2$ refer to the same memory location is $\mathcal{P}_1 \times \mathcal{P}_2$. If $I_1 = I_2$ and $M_1 \neq M_2$, then $r1$ and $r2$ refer to different memory locations. For other cases, we cannot determine whether or not $r1$ and $r2$ refer to the same memory location. We define these relationships as "unknown". For all memory-dependent pairs in a basic block, we resolve the memory ambiguity by using alias information. Figure 6 shows the proportion of memory disambiguation that is accomplished by memory region, *mod-k residues*, or instruction inspection.

In order to evaluate the accuracy of quantitative alias information gleaned from the proposed probabilistic alias analysis, we compare the probabilities of alias relationships obtained from the proposed alias analysis with the real probabilities of alias relationships obtained from runtime profiling. Table 3 shows the distribution of precision for the tested benchmarks. The accuracy is computed by the formula: $1 - |\mathcal{P}_{computed} - \mathcal{P}_{runtime}|$. For 96.10% of the memory-dependent pairs, the proposed algorithm achieved 90–100% accuracy.

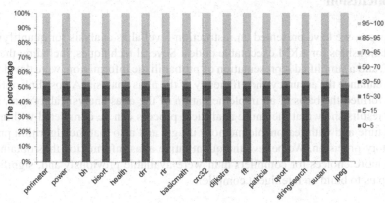

Fig. 5. Distribution of alias relationship probabilities

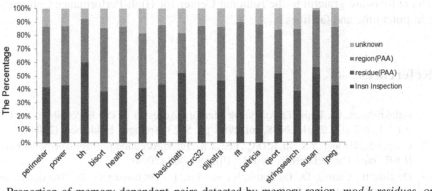

Fig. 6. Proportion of memory-dependent pairs detected by memory region, *mod-k residues*, or instruction inspection

Table 3. Evaluation of the accuracy of the proposed analysis

Program	90 — 100%	80 — 90%	70 — 80%	60 — 70%	50 — 60%	40 — 50%	30 — 40%	20 — 30%	10 — 20%	0 — 10%
perimeter	97.02%	0.21%	0.21%	0	0.43%	0	0.07%	0.07%	0	1.98%
power	96.03%	0	0	0.04%	0.41%	0	0.70%	0.17%	0	2.65%
bh	97.52%	0.03%	0.10%	0	0.55%	0	0.10%	0.03%	0	1.66%
bisort	96.87%	0	0.08%	0	0.24%	0	0.48%	0	0	2.33%
health	96.96%	0	0.11%	0.16%	0.27%	0	0.21%	0.05%	0.11%	2.14%
drr	97.44%	0.12%	0	0	0.17%	0.06%	0.29%	0	0	1.92%
rtr	96.03%	0.10%	0	0	0.54%	0	0.29%	0	0.05%	2.99%
basicmath	97.20%	0	0.22%	0.15%	0.30%	0	0.07%	0	0	2.06%
crc32	96.69%	0.05%	0.14%	0	0.52%	0	0.85%	0.05%	0	1.70%
dijkstra	97.05%	0.12%	0	0	0.42%	0	0.18%	0.06%	0	2.17%
fft	96.26%	0.06%	0.19%	0	0.31%	0.06%	0.37%	0.19%	0	2.56%
patricia	95.83%	0	0.09%	0.09%	0.91%	0	0.32%	0.05%	0	2.72%
qsort	94.00%	0	0.93%	0.16%	0.72%	0	1.35%	0	0.05%	2.79%
stringsearch	95.19%	0	0.25%	0	0.50%	0	0.25%	0.33%	0	3.49%
susan	99.73%	0	0.05%	0	0	0	0.01%	0.03%	0	0.18%
jpeg	87.78%	0.03%	0.28%	0	0.83%	0	2.52%	0.68%	0	7.89%
average	96.10%	0.05%	0.17%	0.04%	0.45%	0.01%	0.50%	0.11%	0.01%	2.58%

5   Conclusion

In this paper, we have presented an instruction-level alias analysis for quantifying the alias relationships for ARM executable codes. Several techniques, including the EAD, the MIT, and a simplified representation of probabilistic information, were developed in order to enhance the accuracy and efficiency of the alias analysis. A post-link optimizer, Diablo, formed the basis of an implementation of the alias analysis algorithm. Experimental results show that the entire analytical process can be completed within a reasonable time and with reasonable memory usage, and also that the algorithm provides satisfactory precision. We believe the quantitative alias information thus obtained can provide more choices for compilers than were available previously and, significantly, will help us to build an intelligent compiler.

Acknowledgements. This work was sponsored by the National Science Council of Taiwan under grants NSC-94-2218-E-194-014- and NSC-100-2221-E-194-034-MY2. The authors are grateful to the National Center for High-Performance Computing for computer time and facilities.

References

1. Balakrishnan, G., Reps, T.: Analyzing Memory Accesses in x86 Executables. In: Duesterwald, E. (ed.) CC 2004. LNCS, vol. 2985, pp. 5–23. Springer, Heidelberg (2004)
2. Chen, P.S., Hwang, Y.S., Ju, R.D.C., Lee, J.K.: Interprocedural probabilistic pointer analysis. IEEE Trans. Parallel Distrib. Syst. 15(10), 893–907 (2004)
3. De Bus, B., Chanet, D., De Sutter, B., Van Put, L., De Bosschere, K.: The design and implementation of fit: a flexible instrumentation toolkit. In: Proceedings of the 2004 ACM SIGPLAN-SIGSOFT Workshop on Program Analysis for Software Tools and Engineering, pp. 29–34. ACM Press, Washington (2004)
4. Debray, S., Muth, R., Weippert, M.: Alias analysis of executable code. In: POPL 1998: Proceedings of the 25th ACM SIGPLAN-SIGACT Symposium on Principles of Programming Languages, pp. 12–24. ACM, New York (1998)
5. Emami, M., Ghiya, R., Hendren, L.J.: Context-sensitive interprocedural Points-to analysis in the presence of function pointers. SIGPLAN Notices 29(6), 242–256 (1994); Proceedings of the ACM SIGPLAN 1994 Conference on Programming Language Design and Implementation
6. Fernandez, M., Espasa, R.: Speculative alias analysis for executable code. In: 2002 International Conference on Parallel Architectures and Compilation Techniques (September 2002)
7. Guo, B., Bridges, M.J., Triantafyllis, S., Ottoni, G., Raman, E., August, D.I.: Practical and accurate low-level pointer analysis. In: CGO, pp. 291–302. IEEE Computer Society (2005)
8. Hind, M.: Pointer analysis: haven't we solved this problem yet? In: PASTE 2001: Proceedings of the 2001 ACM SIGPLAN-SIGSOFT Workshop on Program Analysis for Software Tools and Engineering, pp. 54–61. ACM, New York (2001)
9. Ju, R.D.C., Collard, J.F., Oukbir, K.: Probabilistic memory disambiguation and its application to data speculation. SIGARCH Comput. Archit. News 27(1), 27–30 (1999)

10. Lu, Y.M., Chen, P.S.: Probabilistic alias analysis of executable code. International Journal of Parallel Programming 39(6), 663–693 (2011)
11. Ruf, E.: Context-insensitive alias analysis reconsidered. SIGPLAN Notices 30(6), 13–22 (1995); Proceedings of the ACM SIGPLAN 1995 Conference on Programming Language Design and Implementation
12. Van Put, L., Chanet, D., De Bus, B., De Sutter, B., De Bosschere, K.: Diablo: a reliable, retargetable and extensible link-time rewriting framework. In: Proceedings of the 2005 IEEE International Symposium On Signal Processing And Information Technology, pp. 7–12. IEEE, Athens (2005)

A Two-Leveled Web Service Path Re-planning Technique

Shih-Chien Chou and Chih-Yang Chiang

Department of Computer Science and Information Engineering
National Dong Hwa University, Taiwan
scchou@mail.ndhu.edu.tw

Abstract. During the execution of a web service path, violations of service level agreements (SLAs) will cause the path to be re-planed (healed). Existing re-planning techniques generally suffer from shortcomings of ignoring the effect of requirement change. This paper proposes a two-leveled path re-planning technique (TLPRP), which offers the following features: (a) TLPRP is composed of both meta and physical levels. The meta level re-planning senses environment parameters (e.g., requirement change and analysis/design errors) and re-plan the affected component paths produced by system design. (b) The physical level re-planning algorithm possesses the ability of web service path composition. (c) The re-planning algorithm embeds an access control policy that computes a successful possibility for every path to facilitate avoiding execution failure caused by web service access failure.

Keywords: Web service, re-planning, service level agreement (SLA), meta level agreement.

1   Introduction

When a requester intends to accomplish a function using web services and a composite of web services is selected, the invocation order forms a web service path. If problems occur during path execution, a re-planning process can heal the path. We first clarify the following concepts.

a. Web services are generally atomic (i.e., they offer few and cohesive functions) because: (a) a non-atomic web service is expected to be more expensive because of more functions offered and (b) requesters may only need parts of the functions in a non-atomic web service, which causes them to invoke atomic ones.

b. Customers (i.e., requesters) invoke web services to accomplish their requirements. A customer requirement is generally complicated that should be accomplished by a composite of web services instead of an atomic one.

Item a implies that a requirement is difficult to be accomplished by an atomic web service. Item b implies that most web service researches ignore software engineering process. In our opinion, software engineering process and web service application should cooperate. We think that web services cannot be applied earlier than system design because: (a) a requirement is generally complicated that cannot be accomplished by an atomic web service and (b) system design generally identifies atomic

J.-S. Pan et al. (Eds.): *Advances in Intelligent Systems & Applications*, SIST 21, pp. 669–679.
DOI: 10.1007/978-3-642-35473-1_66   © Springer-Verlag Berlin Heidelberg 2013

software components which are suitable to be accomplished by web services. According to the above description, we think that software engineering and web service application are dependent and the application of web services can follow the phase of system design.

Web service paths accomplish customer requirements. A path accomplishes one or more closely related requirements. During path execution, a web service may fail according to quality of service (QoS) or functionality dissatisfaction. In any case, the path should be healed (re-planned). Many re-planning techniques have been proposed. Some of them self-heal an ill path [2-4], some enhance the failure handing ability of BPEL for path healing [2, 5], and some use redundant web services [6-9]. Our survey reveals the shortcomings of the existing re-planning techniques as follows.

a. Most techniques operate on the web service level but ignore customers. Since web service paths accomplish customer requirements, changing customer requirements will cause the affected web service paths to become incorrect.
b. Most techniques ignore the correctness of functionality.
c. If re-planning is based on self-healing or web service replacement, unpredicted failures may invalidate the technique because no solution is offered for the failures.
d. If customers change requirements or an unpredicted failure occurs, the affected web service paths should be re-composed. Most techniques fail to take this point into consideration.
e. Web service access failures may cause path execution failure [2]. Most techniques fail to take access control into consideration.

We develop a new technique to overcome the shortcomings. The design philosophy of our re-planning technique is described in the following paragraphs.

As mentioned, system design identifies atomic software components. One or some closely related requirements can be accomplished by a composite of software components. The components should be executed following an order, which causes the components to form a component path. An element in a component path can be accomplished by a web service (complicated components can be decomposed). We call the activities before and after applying web services as the meta level activities and the physical level ones, respectively. After identifying component paths, our technique identifies web service paths for each component path following the procedure: (a) one or more web services are selected for each component in which the selected ones should fulfill the QoS criteria, (b) an access control policy is applied to filter out web services that cannot be invoked by the requester, (c) a composition algorithm is applied to compose multiple web service paths for the component path, (d) software engineers select one or more composed paths in which one is selected to execute and the others are spare paths, and (e) if violations of the service level agreement (SLAs) [10] are detected during path execution, the path is re-planned. The re-planning can be achieved using web services or sub-paths of the spare paths. This explains why we compose multiple paths. If spare paths cannot heal the ill path (e.g., the component path is changed because of requirement change), the re-planning algorithm is applied (the algorithm is in fact a path composition algorithm). According to the above procedure, our path re-planning technique offers the following features.

a. The re-planning technique is composed of both meta and physical levels. When meta level re-planning is required (e.g., when customers change requirements), the affected web service paths become incorrect and will be suspended immediately. After that, the affected component paths are re-designed and the corresponding web service paths are re-composed.
b. If the meta level re-planning does not occur, the physical level re-planning takes action when a violation of the physical level SLAs is detected. Since our path composition algorithm composes multiple paths, web services and even sub-paths of the spare paths can be selected to replace the failed web service. If the replacement cannot heal the path, the re-planning algorithm is initiated.
c. The re-planning algorithm embeds an access control policy to compute a successful possibility for every path. The possibilities facilitate avoiding path execution failure caused by web service access failure.

This paper proposes our path re-planning technique. It is composed of the meta and the physical levels. It is named TLPRP (two-leveled path re-planning).

2   Related Work

The technique in [5] uses dynamic description logic (DDL) to describe the preconditions and effects of web services. The DDL-described preconditions and effects are transferred into diagnosing processes embedded in a BPEL execution environment. When a web service is executing and a diagnosing process identifies mismatched predictions or effects, the web service will be isolated for path healing. The technique in [11] proposes a heuristic algorithm to evaluate whether the remaining resources (e.g., time and budget) are sufficient for the un-finished sub-path. If the answer is negative, the unfinished sub-path should be re-planned. The technique in [2] predicts seven types of web service failures and proposes solutions to recovery the failures. The recovery solutions are embedded in the BPEL execution environment. The technique in [12] uses semi-Markov model to predict the performance of an executing web service. If the predicted performance failed to fulfill the QoS criteria, the re-planning process is initiated. In this case, the re-planning process and the web service path are executed simultaneously, which reduce the time of re-planning. The technique in [3] proposes that the failure of an executing web service may cause other executing ones to fail because more than one close related web service of a path may be executed in parallel. It uses direct compensation dependency (DCD) and indirect compensation dependency (ICD) to identify the scope of web services affected by a failure. It then reselects web services to replaces those within the scope. The reselection should offer the least compensation. The technique in [4] belongs to the WS-DIAMOND project [13]. It proposes a QoS-driven, connector-based proactive healing architecture. Connectors can intercept the message sent to a web service and add QoS requirements to the message. After the execution of a web service, the QoS values are recorded in a log. The diagnostic engine periodically checks the log. When a possible violation of QoS criteria is detected, the technique initiates a re-configuration mechanism to self-heal the web service path. The technique in [14] compares the exact QoS values with the estimated ones. When a loop of web services is executing, the exact QoS values of the web services are used to evaluate whether the loop may violate the QoS-based SLAs. If the answer is positive, the technique uses an algorithm to identify the path slice (i.e. sub-path) that should be re-planned. The technique in [15] defines the scope of failed web service and embeds the scope in BPEL. When failure occurs, the BPEL engine initiates the healing process for the scope, which selects new web services to replace the failed ones. During the selection, the local and global business rules are used to filter out infeasible web services.

3   TLPRP

TLPRP is depicted in Figure 1. The dotted blocks are the meta and physical level re-planning mechanisms, respectively. Below we explain the figure.

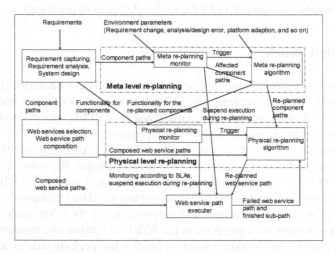

Fig. 1. TLPRP and the relatedactivitiesFunctionality

a. The top left block is the early phases of software engineering, including requirement engineering and system design. After system design, cohesive software components are identified. A composite of components solving one or some close related requirements constitute a component path.

b. The bottom left block is the selection and composition process. It selects web services for every component and composes web service paths for component paths. During selection and composition, an access control policy is applied.

c. The bottom block is a path executer.

d. The lower dotted block is the physical level re-planning mechanism. The physical re-planning monitor detects violations of the physical level SLAs. A violation will trigger the physical re-planning algorithm.
e. The upper dotted block is the meta level re-planning mechanism. The meta re-planning monitor detects the violates of the meta level SLAs related to requirement change, analysis/design errors, and platform adaption. A violation of the SLAs will trigger the meta re-planning algorithm.

3.1   Physical Level Re-planning

The physical level re-planning mechanism is composed of a physical re-planning monitor and an algorithm (see Figure 1) as described below.

Physical Re-planning Algorithm. We designed our physical re-planning algorithm as a path composition algorithm. Unlike most composition algorithms, our algorithm composes multiple web service paths for a component path. It is for possible replacement during re-planning. When composing web service paths, every path and web service should fulfill the QoS criteria. We suppose the value of every QoS criterion should be large. For a criterion such as cost that should be small, the requester should set a maximum value for the criterion to minus the criterion's value. To check QoS, every QoS criterion should be given a limitation. If a QoS criterion's value should be small, we let its value be its maximum value minus its original one. After that, every QoS criterion's value should be at least as large as its limitation. Details of the algorithm are shown in [16]. Below we use the component path in Figure 2 as an example to describe the algorithm.

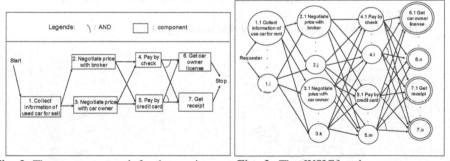

Fig. 2. The component path for the requirement "buy a used car"

Fig. 3. The WSIGfor the component path "Buy a used car"

When composing web service paths for the component path in Figure 2, web services for every component in the component path are first selected. Suppose the selection produces a web service invocation graph (WSIG) in Figure 3, in which one component in Figure 3 can be accomplished by multiple web services offering the same function. Double-circles in Figure 3 are the last web services that should be finished for a path. Logical relationships such as AND appearing in the outgoing arrows of a component will also appear in the outgoing arrows of the corresponding web services in a WSIG (please check Figures 2 and 3 to confirm this).

After selection, our two-leveled access control policy is applied. The upper level checks whether the requester possesses the attributes and credentials required by a web service to filter out web services that cannot be invoked by a requester. The lower level uses the credit level numbers of web services and security level numbers of arguments to evaluate the possibility of a requester that can invoke a web service using the formula below, in which: (a) POSIsucc is the possibility of a requester to successfully invoke a web service ws, (b) k is "wscln – maxsln", in which wscln is the credit level number of ws and maxsln is the maximum security level numbers of the arguments sent to ws, and (c) n is a threshold to facilitate computing POSIsucc.

$$
\begin{aligned}
& \text{POSIsucc} = 1, && \text{if } k \geq n \\
& POSI_{succ} = \frac{(k+n)}{2n}, && \text{if } k \text{ is between } -n \text{ and } (n\text{-}1) \\
& \text{POSIsucc} = 0, && \text{if } k < -n
\end{aligned}
\tag{1}
$$

Suppose Figure 4 is the WSIG after applying the upper level access control policy. To identify web service paths from the WSIG, the algorithm transfers it into a web service invocation tree (WSIT), because a leaf in a tree exactly belongs to one path. To transfer a WSIG into a WSIT, the operation TrTree listed below can be applied. After applying TrTree, Figure 4 will be transferred into Figure 5.

TrTree. Repeat

Identify a node NDi in WSIG with N incoming arrows in which N > 1

Duplicate N sub-WSIG rooted at NDi

Let every incoming arrow point to the root node of one duplicated subgraph

Until every node is pointed by only one incoming arrow

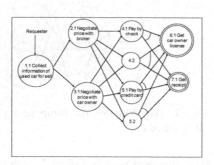 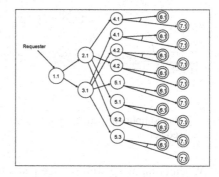

Fig. 4. The WSIG for the component path "buy a used car" after applying the upper level access control policy in which many web services have been filtered out

Fig. 5. The WSIT obtained from figure 4

After constructing a WSIT, the algorithm identifies web service paths by back-tracking the WSIT starting from its leaves. We use Figure 6 to explain the path identification process. Suppose node 1 is visited first. Then, the backtracking starts from node 1 up to 14. Since an AND covers two outgoing arrows of node 14, the backtracking rewinds into a forward depth first search process to identify node 10, 3, and 4. After that, the backtracking restarts from node 14 and backs to 17. The AND covering node 17 starts another depth first search. After that, the path containing nodes connected by dotted-lines in Figure 6 is identified.

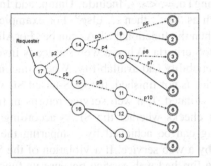

Fig. 6. Path identification example

Having identified all paths from a WSIT, our lower level access control policy evaluates the possibility of a requester that can invoke a web service using Formula 1. For example, p2 in Figure 6 is the possibility of a requester that can invoke web service 14. After the evaluation, the possibility of successfully finishing a path can be computed using $\prod_i P_i$. We call the possibility SUCCPTH. In addition to SUCCPTH, our algorithm also evaluates the value of QoS criteria of a path. We first compute the value of each QoS criterion for the path. For example, the cost of a path is the summation of the web services' costs in the path. After the value of every criterion has been computed, the expression $\sum_i (Qp_i * Wp_i)$ computes the QoS value QoSPTH for the path, in which Qpi is the ith QoS criterion's value, Wpi is the weight of Qpi, and $\sum_i Wp_i = 1$. After that, the overall value of a path PTHVAL is computed using the expression "SUCCPTH * w1 + QoSPTH * w2", in which w1 and w2 are respectively the weights of the two values. By referring to PTHVALs of the paths, the requester selects a path to execute and more or less others for spare paths.

Physical Re-planning Monitor. The physical re-planning monitor monitors web service path execution using physical level SLAs obtained from web service functionality and QoS criteria. Since we do not intend to design a QoS model, we only monitor the important QoS criteria of cost, time, reliability, and availability. A violation of the functional SLAs will cause the web service to be replaced. A violation of the QoS-based SLAs will cause the monitor to check whether the left resources are sufficient for the unfinished sub-path. The SLAs and the rules to monitor the execution of a web service path are described below.

a. The monitoring of web service function. It is impossible to know how a web service executes. We thus check the arguments and return data of a web service for the monitoring. The following operators are used for the checking: (a) logic operators such as AND, OR, NOT, ForAll, and ThereExist, (b) arithmetic operators such as +, -, *, and /, (d) comparison operators such as >, >=, =, <>, <=, and <, (d) set operators such as BelongTo, =, <>, -, Include, Union, and Intersection, and (e) condition operators such as "if ... then ... else". For example, if a web service sorts the data in array a, then the functional SLAs can be "ForAll i, a[i] >= a[i+1]".

b. The monitoring of QoS criteria. When a web service is invoked, a timeout counter is used to monitor reliability and availability. When timeout occurs and the execution of the web service is not finished, the QoS-based SLAs according to reliability or availability is violated. If a web service returns in time and the function is correct, the monitor checks whether the SLAs according to cost or time are violated. The checking can be achieved by comparing the actual cost/time spent with that promised by a web service. If a violation of the SLAs occurs, the monitor does not care the finished web service because its function is correct. Instead, the monitor checks whether the left budget/time is sufficient for the unfinished sub-path. If the checking passes, the sub-path executes normally. Otherwise, the physical re-planning algorithm will be triggered.

3.2   Meta Level Re-planning Mechanism

The meta level re-planning will be activated when environment parameters are detected, in which environment parameters include all factors that will cause the software to change. Example environment parameters are requirement change, analysis/design errors, and adaptation of platform. The physical level re-planning is enough if no environment parameter occurs. However, customers change requirements frequently, errors on requirement analysis or design are identified occasionally, and adaptation of platform is needed sometimes. Either of the cases will produce environment parameters. An environment parameter will change one or more component paths produced by system design, which will in turn change the corresponding web service paths. In this section, we describe the management of component path re-planning according to environment parameters. That is, this section describes the meta level re-planning.

The meta level re-planning mechanism is composed of a meta re-planning monitor and an algorithm (see Figure 1). The meta re-planning monitor senses environment parameters. As long as the monitor senses any environment parameter, it first identifies the component paths affected by the parameters. It then suspends the executing web service paths implementing the affected component paths. The suspension is necessary because the affected component paths as well as the corresponding web service paths should be changed (i.e., the affected paths become incorrect). After the suspension, the monitor triggers the meta re-planning algorithm to plan new component paths according the environment parameters sensed by the monitor. The above description reveals that the meta level SLAs used in the meta re-planning monitor include customer requirements, the results of requirement analysis and system design, the platform that executes the software, and so on. Any change of the above mentioned items results in a violation of the meta level SLAs (i.e., produces an environment parameter). Since the meta level monitor should sense environment parameters and identify the affected component paths, the monitor is difficult to automate. Therefore, the functions of the meta level monitor in TLPRP are performed by software engineers.

When a violation of the meta level SLAs occurs, the meta re-planning algorithm is triggered. The algorithm re-plans the affected component paths. The re-planning algorithm is actually a software maintenance process and performed by software engineers. During the re-planning, software engineers re-capture and re-analyze customer requirements, and re-design the affected component paths. The engineers will take the original component paths as a reference during the re-planning. Generally, the engineers should change the original component paths at least as possible to reduce the effort of the physical level re-planning (see the discussion in section 3.3).

According to the above description, the meta re-planning algorithm and the software engineering process can be the same. A typical software engineering process is composed of requirement capturing/analysis, system design, implementation, testing, and maintenance. The phases of implementation and testing in web service applications are replaced by web service selection, composition, and execution. Below we discuss the requirement and design phases in the meta level re-planning of TLPRP.

Requirement capturing and analysis identify what customers want. System design focuses on how to achieve customer requirements. In general, a requirement can be solved by more than one solution. For example, the requirement "travel around the world" can be solved using the solutions "take airplanes and stay in luxury hotels for the nights", "take ships and stay in the rooms of the ships for the nights", and "take ships or trains and stay in the rooms of the ships or in camps for the nights". Among the solutions of a requirement, which one will be selected? What dominate the selection? In fact, the identification of solutions for requirements and the selection of proper ones are classical problems. Many comprehensive researches can be identified. We use the existing research results to implement our meta re-planning algorithm.

The meta re-planning algorithm and the software engineering process can be the same. In the meta re-planning algorithm, we use existing techniques such as meetings or interacting with customers to identify customer requirements. We then use the use case diagram of UML [17] to represent the requirements. After that, we use existing system design process and metrics to design component paths. No matter what design process is used, our purpose is identifying components that meet design metrics such as high cohesion and low coupling. The design metrics, which are regarded as meta level QoS criteria in this paper can be identified from much published material. When selecting the criteria, we consider the following three factors: (a) localizing change effects, (b) fulfilling the budget and time requirements of customers, and (c) requiring that a component can be accomplished by a web service. From the first factor, we select the meta level QoS criteria "cohesion" and "coupling". From the second, we select "cost" and "time". From the third, we select "size" and "cohesion" (a cohesive component of proper size has more chance to be accomplished by a web service). Note that the physical level QoS criteria are generally offered by web service providers whereas the meta level ones are obtained from estimation.

With the meta level QoS criteria and the solutions identified by applying a system design process, the meta re-planning algorithm can select one optimal solution according to the meta level QoS criteria (here a solution is a component path). The re-planned component path is then sent to the physical re-planning algorithm for web service path re-planning. Remember that the physical re-planning algorithm composes multiple paths for possible replacement. On the contrary, the meta re-planning algorithm selects only one component path. The rationales are listed below:

a. In the physical level, web services for a component offers the same function. When a re-planning is required, using another one to replace the failed one is meaningful.
b. In the meta level, re-planning corresponds to changing solutions (i.e., changing component paths). In this situation, planning or re-planning multiple component paths for a requirement is meaningless.

4   Conclusion

This paper proposes a two-leveled path re-planning technique (TLPRP), which offers the following features.

a. TLPRP is composed of both meta and physical levels. The meta level re-planning senses environment parameters and re-plan the component paths.
b. The physical level re-planning algorithm is a composition algorithm. In other words, the algorithm offers the ability of path composition.
c. The re-planning algorithm embeds an access control policy to compute a successful possibility for every path.

References

1. Alves, A., Arkin, A., Askary, S., Barreto, C., Bloch, B.: Web Service Business Execution Language, v. 2.0, http://docs.oasis-open.org/wsbpel/2.0/OS/wsbpel-v2.0-OS.html
2. Subramanian, S., Thiran, P., Narendra, N.C., Mostefaoui, G.K., Maamir, Z.: On the Enhancement of BPEL Engines for Self-healing Composite Web Services. In: International Symposium on Applications and Internet, pp. 33–39 (2008)
3. Yin, Y., Zhang, B., Zhang, X., Zhao, Y.: A Self-healing Composite Web service Model. In: 2009 IEEE Asia-Pacific Service Computing Conference (IEEE APSCC), pp. 307–312 (2009)

4. Halima, R.B., Drira, K., Jmaiel, M.: A QoS-driven Reconfiguration management System Extending Web Services with Self-healing Properties. In: 16th IEEE International Workshops on Enabling Technologies: Infrastructure for Collaborative Enterprises, pp. 339–344 (2007)
5. Han, X., Shi, Z., Niu, W., Lin, F., Zhang, D.: An Approach for Diagnosing Unexpected Faults in Web Service Flows. In: 2009 International Conference on Grid and Cooperative Computing, pp. 61–66 (2009)
6. Jorge, S., Francisco, P., Marta, P., Ricardo, J.: Ws-replication: A Framework for Highly Available Web Services. In: 15th International Conference on World Wide Web, pp. 357–366 (2006)
7. Taher, Y., Benslimane, D., Fauvet, M.: Towards an Approach for Web Services Substitution. In: 10th International Database Engineering and Applications Symposium, pp. 166–173 (2006)
8. Yu, T., Lin, K.J.: Service Selection Algorithms for Web Services with End-to-end QoS Constraints. Journal of Information Systems and e-Business Management 3(2), 103–126 (2005)
9. Guo, H.: Optimal Configuration for High Available Service Composition. In: 15th International Conference on Web Services, pp. 280–287 (2007)
10. http://en.wikipedia.org/wiki/Service_level_agreement
11. Berbner, R., Spahn, M., Repp, N., Heckmann, O., Steinmetz, R.: Dynamic Replanning of Web Service Workflows. In: 2007 IEEE International Conference on Digital Ecosystems and Technologies, pp. 211–216 (2007)
12. Dai, Y., Zhu, Z., Li, D., Wang, H., Chen, Y., Zhang, B.: Low Cost Mechanism for QoS-aware Re-planning of Composite Service. In: 2009 International Symposium on Electronic Commerce and Security, pp. 287–291 (2009)
13. IST, The Web Services Diagnosability, Monitoring, and Diagnostic Project (2006), http://wsdiamond.di.uniti.it
14. Canfora, G., Penta, M.D., Esposito, R., Villani, M.L.: QoS-aware Replanning of Composite Web Services. In: 2005 IEEE International Conference on Web Services (2005)
15. Ren, W., Chen, G., Shen, H., Zhang, J.B., Low, C.P.: Dynamic Self-healing for Service Flows with Semitic Web Services. In: 2008 IEEE/WIC/ACM International Conference on Web Intelligence and Intelligent Technology, pp. 598–604 (2008)
16. Chou, S.-C., Jhu, J.-Y.: Access Control Policy Embedded Composition Algorithm for Web Services. In: 6th International Conference on Advanced Information Management and Service (2010)
17. http://www.uml.org/
18. OASIS, eXtensible Access Control Markup Language (XACML) Version 1.0. OASIS Standard 18 (2003)
19. W3C. Web Services Description Language (WSDL) 1.1, http://www.w3.org/TR/2001/NOTE-wsdl-20010315
20. http://www.w3.org/Submission/OWL-S
21. Hwang, S.-Y., Lim, E.-P., Lee, C.-H., Chen, C.-H.: Dynamic Web Service Selection for Reliable We Service Composition. IEEE Transactions on Services Computing 1(2), 104–116 (2008)
22. Chen, P.P.W., Lyu, M.R.: Dynamic Web Service Composition: A New Approach in Building Reliable Web Service. In: 22nd International Conference on Advanced Information Networking and Applications, pp. 20–25 (2008)

An Effective Flood Forecasting System Based on Web Services*

Ya-Hui Chang, Pei-Shan Wu, Yu-Te Liu, and Shang-Pin Ma

Department of Computer Science and Engineering, National Taiwan Ocean University
yahui@ntou.edu.tw

Abstract. A flood forecasting system usually needs to integrate many hydraulic modules, which may be legacy programs written in FORTRAN, running in heterogeneous environments, and differing in execution time. Besides, the data required for each module should be provided in real time, and the programs also need to be executed in a correct sequence. In this paper, we discuss how to build the flood forecasting system based on Web services to support module integration and data exchange. We also propose several workflow strategies for composing services, and perform experiments to determine which strategy is better.

1 Introduction

The disaster brought by heavy rain has become more and more serious in Taiwan for the past few years due to global warming. In certain areas such as the Linbien city, which is surrounded by the South China Ocean and passed by two rivers, *i.e.*, the Linbien River and the Lili River, the situation is even worse due to the complex interaction of rain, rivers, waves and tides. It is well known that the Morakot typhoon in 2009 seriously damaged that area. Therefore, it is very important to build an effective flood forecasting system for early warning so that the authorities can evacuate residents in time.

A few experts in the hydraulic discipline have developed several program modules with different functionality for that specific area. Particularly, the Hydrol module is constructed to predict the river overbank, the WaveTide module is used to estimate the future overtopping discharge of the ocean, and the TwoFD module is designed to estimate the water level of the whole area. A complete flood forecasting system needs to integrate these modules to provide comprehensive warning messages, but it is not a trivial task based on the following reasons:

- These legacy hydraulic modules are written in the FORTRAN programming language. Moreover, some modules only function well in the LINUX environment, while others are designed to run in the Windows operating system. The flood forecasting system should be able to invoke modules operated in different environments.

---

* This work was partially supported by the National Science Council under Contract No. NSC 101-2625-M-019-004-.

J.-S. Pan et al. (Eds.): *Advances in Intelligent Systems & Applications*, SIST 21, pp. 681–690.
DOI: 10.1007/978-3-642-35473-1_67   © Springer-Verlag Berlin Heidelberg 2013

- There exists complex data dependency relationship between these modules. For example, the output of the WaveTide module is needed by both the Hydrol module and the TwoFD module for real-time computation. Therefore, the integrated system should ensure that these modules run in a correct order, and each module gets the data it requires.
- The Hydrol Module takes a few minutes to compute, while the WaveTide and TwoFD Modules may take up to an hour. Since the running time of modules differs a lot, we wish to design a good workflow strategy so that the throughput of the whole system can increase.

There have been many flood forecasting systems seen in literature. They are built based on different architectures to meet their own needs. For example, the European Flood Alert System (FEAS) [10], which intends to provide flood forecasting in trans-national river basins, is a large system and applies the grid architecture to integrate several national hydrological and meteorological services. The Web-based Flood Forecasting System (WFFS) [8] and the Flood Early Warning System (FEWS) [7], which are small to medium sized systems, are component-based and apply the Java technique. On the other hand, the concepts of *services*, particularly *Web services*, attract a lot of attention recently due to their flexibility and reusability. For example, the researchers in [5,6] use services to allow others accessing their water resources represented in data warehouses or databases. The researchers in [4] identify the functions which are frequently required by a flood forecasting system, such as formatting data, and represent them as Web Services, while the researchers in [3] intend to represent the water resource models as Web services.

In this research, we also propose to build the flood forecasting system using Web services, since Web services can be invoked through networks without concerning about the differences of the underlying environment. Besides, there exist softwares to assist in composing Web services into different execution sequence. These properties reduce the cost of maintaining the system. The contributions of this paper are summarized as follows:

- We wrap several legacy hydraulic modules written in FORTRAN as Web services, so that these modules can be easily composed and reused as the basis of a complete flood forecasting system.
- We discuss different workflow strategies to compose these services and design several experiments to examine their performance.
- We implement the whole system based on the .Net solution [2], particularly the Windows Communication Foundation (WCF) for Web services and the Windows Workflow Foundation (WF) for flow control. It is shown that this system can function effectively.

The rest of this paper is organized as follows. In Section 2, we introduce the functionality and the output of the legacy hydraulic modules. In Section 3, we describe the whole architecture for the flood forecasting system. In Section 4, we describe the Web services designed for hydraulic modules, and propose several different workflow strategies. Finally, experimental results and the system prototype are shown in Section 5, and conclusions are given in Section 6.

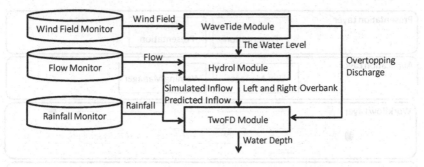

Fig. 1. Data Exchange between Hydraulics Modules

2   Preliminaries

We first introduce the functionality of the existing hydraulic modules. Recall that the flood forecasting system discussed in this paper is designed especially for the Linbien city, which is surrounded by the ocean and rivers.

During a typhoon period, the strong wind usually causes storm surge and swell, and there might also exist tide. *The WaveTide module* first applies the data got from the *wind wield monitor* to calculate the overtopping discharge of the coastal structure, and simulates the water level of the downstream. Second, *the Hydrol module* then evaluates how the heavy rain affects the river flow. It will apply the data got from the *rainfall monitor* and output simulated inflow of the upstream, where there are no flow monitors, and also output predicted inflow for the next three hours. Moreover, it will base on the calculated or observed flow information and the water level of the downstream outputted from the WaveTide module, to estimate the left and right overbank of each river section. Finally, *the TwoFD module* divides the area into around 40,000 small grids. It uses the technique of parallel computing and bases on the output of the other two modules to calculate the water level of each grid, and to predict which area should be warned.

Figure 1 summarizes how these modules exchange data, and their outputs are represented in plain text files as listed in Table 1. For example, the file "HydrolInput.dat" outputted by the WaveTide module provides the water level at a particular location and a particular time. The piece of sample data indicates

Table 1. Output Data of Hydraulic Modules

Module	File name	Sample Data
WaveTide	HydrolInput.dat	(12:00:00, 198738.643, 2480062.259, 0.0098)
	TwoFDInput.dat	(12:00:00, 200588.092, 2478948.852, 6.8, 1)
TwoFD	Output.txt	(12:00:00, 201944.681, 2485393.236, 38.5)
Hydrol	K_RealXX.dat	(12:00:00, 8.63, 2.1)
	K_PredXX.dat	
	SecData_LinbienXX.dat	(12:00:00, 0.098, 0.012, 729.971, 109.793)
	SecData_LiliXX.dat	

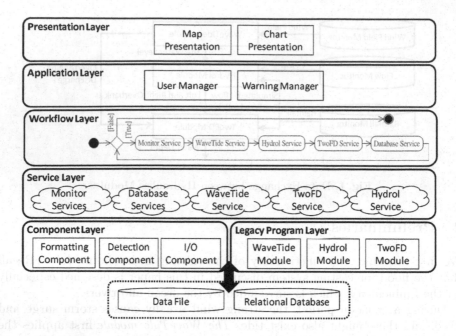

Fig. 2. The Architecture of the Flood Forecasting System

that the water level is 0.0098 m at the coordinates (198738.643, 2480062.259) at the time 12:00:00. The contents of other files are similar and we omit the detailed discussion due to space limitation. One thing to note is that the symbol "XX" represents an identifer for a certain river or a river section.

3   Architecture

We now describe the architecture for the flood forecasting system, which is designed to correctly coordinate those legacy hydraulic modules running in heterogeneous environments, and to output the warning message in real-time.

As shown in the bottom of Figure 2, this system operates two types of data. The first one is the data files outputted by existing hydraulic modules, as described in Section 2. The second one is a relational database additionally maintained to assist in efficient data retrieval, display, and backup. Next, the *Legacy System Layer* consists of the existing hydraulic modules as discussed before, and the *Component Layer* consists of several utilities programs. Particularly, the *formatting components* are used to transform the data acquired from the monitor to the proper format required by the existing hydraulic modules. The *detection components* are used to detect if the monitored data have been transmitted to the local directory. The *I/O components* are responsible for the interaction with the relational database software, including storing and retrieving data.

Table 2. Interfaces of Hydraulic Services

Service	Input Parameter	Output Parameter
WaveTide	NowWind	HydroInput, TwoFDInput
Hydrol	NowFlow, NowRainFall HydrolInput	K_Real, K_Pred, SecData_LiLi, SecData_LinPien
TwoFD	NowRainFall, TwoFDInput, K_Real, K_Pred, SecData_LiLi, SecData_LinPien	TwoFDResult

On top of the Legacy System Layer and the Component Layer, we implement several Web services. The *WaveTide service*, the *TwoFD service*, and the *Hydrol service*, correspond to the three existing hydraulic modules. By wrapping as Web services, these modules can be invoked from different operating systems, and provide information upon request. The *Monitor services* actually correspond to several services, each of which is designed to get the data from a particular monitor station and provides data in the proper format. The *Database services* refer to those services which involve the interaction with the relational database software, which can be further classified into ToRDB services and FromRDB services. The former type of services is responsible for automatically exporting data calculated by hydraulic modules to relational databases, and the latter type of services is used by the upper *Application Layer*. Note that there is a *Workflow Layer* above the Service Layer. It is used to compose those services in a correct order, and those composite services can be created based on users' demand.

Finally, the *Application Layer* and the *Presentation Layer* complete the construction of the flood forecasting system as a Web-based system. Particularly, *the User Manager* classifies users as administrators or ordinary users and allow them to browse different Web pages. The *Warning Manager* displays warning messages on the Web pages by maps or charts.

4   Hydraulic Services

In this section, we first discuss how to wrap the three existing hydraulic modules as Web services, which are called collectively as *hydraulic services*. We then propose several possible ways of composing those services.

4.1   Interfaces of Hydraulic Services

As described in Section 2, each existing hydraulic module is written in FORTRAN, which operates on and outputs plain text files. For easy correspondence, we let the output parameters of each hydraulic Web service directly correspond to the original output files as listed in Table 1, and design the input parameters based on their data dependency relationship as depicted in Figure 1. The interfaces of the three hydraulic services are summarized in Table 2. In the table, the three parameters *NowWind*, *NowRainFall*, and *NowFlow* represent the data got from the wind field monitor, the rainfall monitor, and the flow monitor, respectively, while the parameters *HydrolInput*, *TwoFDInput*, and *TwoFDResult*

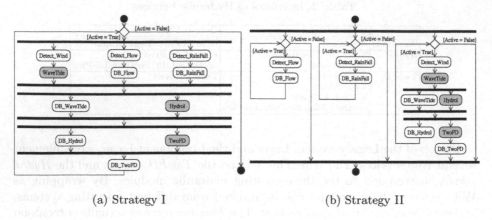

(a) Strategy I   (b) Strategy II

Fig. 3. Workflow Strategies I and II

represent the contents of the files "HydrolInput.dat", "TwoFDInput.dat", and "Output.txt", respectively. An exception is the output parameter of the Hydrol service, since the associated module outputs more than 70 files. If we let each file correspond to a parameter, there will be too many parameters to process. Therefore, we let the same type of data represented by the same parameter, and there are only four output parameters for that service as shown in Table 2.

Based on the discussion above, it is obvious that we need the transformation between plain text files and parameters. Therefore, we wrap the following three parts of program fragments in sequence into a hydraulic Web service:

- transforming the input parameters into input text files
- invoking the legacy hydraulic module, which operates on the input text files, and outputs the calculated results as text files
- transforming the output text files into output parameters

4.2 Composing Web Services

As depicted in Figure 1, the hydraulic services need to be invoked in a correct order for continuous calculation. Moreover, the output of each hydraulic service also conveys useful information, which should be represented in the relational database for further retrieval. Since there exist obvious time difference when executing the three hydraulic services individually, we propose several possible strategies of composing these services and discuss the underlying rationale. The process considered starts from gathering the input data till representing the output data in the relational databases, and covers the following services: (1) the three hydraulic services, *i.e.*, *WaveTide*, *Hydrol*, and *TwoFD*, (2) the monitor services, *i.e.*, *Detect_Wind*, *Detect_Flow*, and *Detect_RainFlow*, and (3) the ToRDB services, *i.e.*, *DB_RainFlow*, *DB_Flow*, *DB_WaveTide*, *DB_Hydrol*, and *DB_TwoFD*.

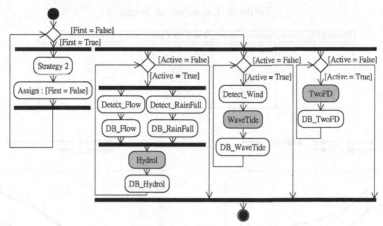

Fig. 4. WorkFlow Strategy III

Each workflow strategy is depicted as a UML activity diagram. In particular, each rounded square represents a Web Service. An exception is the rounded square denoted by the word "Assign", which represents the value assignment to a particular variable. We also let each Web Service associated with one or more variables corresponding to its output parameters, which are omitted in the figure for simplicity. In addition, a diamond represents a conditional statement, a pair of parallel thick dark lines represent parallel processing, and an arrow represents sequential processing.

First, Figure 3(a) shows the baseline strategy. That is, a hydraulic service is invoked as soon as all its dependent services have produced new values. In addition, those services which are not related, such as *DB_Flow* and *WaveTide*, is scheduled to run in parallel. A special boolean variable *Active* is used to indicate if the process is currently active or has been terminated by the administrator.

The second strategy, as depicted in Figure 3(b), considers the large time differences between services, and wishes to represent as many as possible data in the relational database. Recall that the Hydrol service only takes minutes to compute, but the WaveTide service and the TwoFD service may take up to an hour. Besides, the flow monitor and the rainflow monitor transmit monitored data every ten minutes. Therefore, we let the monitor services continuously execute without waiting. In addition, after the Hydrol service has executed and transmitted data to the relational database, we immediately start a new execution instance while the TwoFD service is still running. By doing this, we can shorten the time of waiting the slow service to finish.

The final strategy further extends this idea, and makes each service run independently purely based on the current available values of the input parameters. The corresponding activity diagram is depicted in Figure 4. It is divided into two parts. The left part represents the first execution, where there exist no values for the input parameters. It will first execute based on the second strategy, and then assign the variable *First* the value *False*, so the remaining executions will apply the right part of the workflow, where each hydraulic module runs independently.

Table 3. Experiment Setup

WaveTide (min)	Hydrol (min)	TwoFD (min)	Max Difference (min)
1	6	11	10
1	11	21	20
1	16	31	30
1	21	41	40
1	26	51	50
1	31	61	60

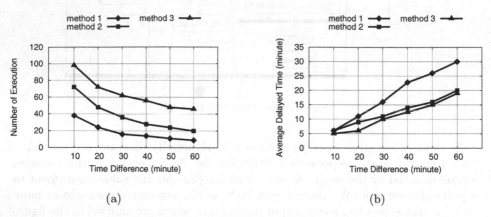

Fig. 5. Evaluations of Different Workflow Strategies

5   Evaluation

In this section, we perform several experiments to evaluate the different workflow strategies, and demonstrate the constructed flood forecasting system.

5.1   Evaluation of Workflow Strategies

In this section, we evaluate the proposed workflow strategies by controlling the execution time of the three hydraulic services. As listed in Table 3, we fix the execution time of the WaveTide service, and increase the execution time of the other services. The maximum time difference of these hydraulic services are denoted in the last column.

In the first experiment, we compare the total number of execution for all hydraulic services within five hours. Since each execution outputs useful information, we prefer the workflow strategy with the highest number. As illustrated in Figure 5(a), we can see that the numbers decrease along with the values of time differences, but strategy III is always the best one.

In the next experiment, we consider the TwoFD service, which requires the outputs of the other two hydraulic services. We calculate the *average delayed time* for this service as follows, where the subscript i represents the i_{th} execution, *start* represents the time when the TwoFD service starts execution, while *input* represents the time associated with its input parameters:

Fig. 6. The Homepage of the Flood Forecasting System

$$\frac{\sum_{i=1}^{n}(start_i - input_i)}{n}$$

If this value is small, the TwoFD service can perform its calculation based on more recent data, and its outputs will be more convincible. As illustrated in Figure 5(b), the average delayed time of the three workflow strategies all increase along with the maximum time difference, but strategy III is still the best.

5.2   System Prototype

The flood forecasting system is built as a Web-based system. Those hydraulic services discussed in Section 4 are implemented in the back end and their computed results are stored in the relational database. FromRDB services are then applied to retrieve required information from the database for display. The whole system is implemented by using the programming languages ASP.net and Javascript. The Web server is IIS and the relational database server is MS SQL Server 2008. As shown in Figure 6, the homepage of the system is mainly the map of the Linbien city annotated with warning messages. Particularly, each river section is noted with a short line, with arrows to indicate river overbank. The arrows depicted by the ocean, as seen in the lower left corner of the map, represent wave overtop. The set of small gray circles show the areas which are inundated. This shows the effectiveness of our approach, and we omit the detailed discussion due to space limitation.

6   Conclusion

In this paper, we demonstrate how to wrap several legacy hydraulic programs as Web services to build a flood forecasting system. We have successfully implemented this system, and currently test it under the real environment. In the future, we will also investigate the open standard such as OpenMI [1,9], so that the proposed services can be more reusable for other systems.

References

1. Open modeling interface, http://www.openmi.org
2. Windows .net solution, http://msdn.microsoft.com
3. Goodall, J.L., Robinson, B.F., Castronova, A.M.: Modeling water resource systems using a service-oriented computing paradigm. Environmental Modelling & Software 26(5), 573–582 (2011)
4. Diaz, L., Granell, C., Gould, M.: Service-oriented applications for environmental models: Reusable geospatial services. Environmental Modelling & Software 25(2), 182–198 (2010)
5. Dzemydiene, D., Maskeliunas, S., Jacobsen, K.: Sustainable management of water resources based on web services and distributed data warehouses. Technological and Economic Development of Economy 14(1), 38–50 (2008)
6. Goodall, J.L., Horsburgh, J.S., Whiteaker, T.L., Maidment, D.R., Zaslavskye, I.: A first approach to web services for the national water information system. Environmental Modelling & Software 23(4), 404–411 (2008)
7. Hydraulics, D.: Delft-fews, an open shell flood forecasting system, http://www.wldelft.nl/soft/fews/int/index.html
8. Li, X.-Y., Chau, K.W., Cheng, C., Li, Y.S.: A web-based flood forecasting system for shuangpai region. Advances in Engineering Software 37(3), 146–158 (2006)
9. Liao, Y.-P., Lin, S.-S., Chou, H.-S.: Integration of urban runoff and storm sewer models by using openmi framework. Journal of Hydroinformatics 14(3) (2012)
10. Thielen, J., Bartholmes, J., Ramos, M.-H., de Roo, A.: The european flood alert system-part 1: Concept and development. Hydrology and Earth System Sciences 13(2) (2009)

A Simulation Environment for Studying the Interaction Process between a Human and an Embedded Control System

Chin-Feng Fan[1], Cheng-Tao Chiang[1], and Albert Yih[2]

[1] Dept. of Computer Science and Engineering
Yuan-Ze University, Taiwan
[2] Dept. of Comptuer Science & Information Engineering
National Normal University, Taiwan
csfanc@saturn.yzu.edu.tw

Abstract. The operator in a computer-controlled system plays an important role to handle different abnormal situations, and behaves as the last layer of defense for the system. Study of the human-machine interaction process can be conducted in a simulation environment. Ptolemy II is a widely-used modeling and design environment for embedded control systems. However, it does not provide an operator computing model. This research extends Ptolemy II to include an operator domain and then constructs a simulation and testing environment with 2D animation for a safety injection system. We obtained valuable experience in modeling a complex heterogeneous system using the extended Ptolemy II.

Keywords: Embedded control systems, Ptolemy II, operator domain, interaction process.

1   Introduction

Embedded computer-controls have been increasingly incorporated into many safety-critical systems in such areas as transportation, aviation, military, and nuclear power plants. Simulation and testing of such embedded systems are particularly critical to ensure safe operation of these systems. These computer-controlled systems are typically heterogeneous and require several different computing models, including continuous and discrete models, to model their physical and logical subsystems. Ptolemy II is a popular open-source software framework supporting design and simulation of concurrent, real-time, embedded systems [7, 8, 9]. Ptolemy II provides several different models of computation, so called "domains". Thus, Ptolemy II is a good choice to model a heterogeneous embedded control system.

On the other hand, for a computer-controlled system, a human or an operator is needed to monitor the system to ensure its correct operation, and also to perform manual operation in case of hardware or software failures. Therefore, the operator's mental model needs also to be simulated in order to analyze the human-machine

J.-S. Pan et al. (Eds.): *Advances in Intelligent Systems & Applications*, SIST 21, pp. 691–699.
DOI: 10.1007/978-3-642-35473-1_68   © Springer-Verlag Berlin Heidelberg 2013

interaction process. Note that the term *"interaction process"* used in this paper refers to the *sequences* of interaction between a human and a computer controlled system in a safety-critical domain. Study of such an interaction process aims to observe whether the interaction sequence leads to an accident or a recovered state. However, Ptolemy II does not provide an operator computing model. In order to study human-machine interaction, this research first added an operator domain to Ptolemy II, and then constructed a simulation environment for a safety-injection system using the extended Ptolemy II. Human-machine interaction at both normal and abnormal situations can then be tested and analyzed using this simulation environment. This paper reports our experience in applying the extended Ptolemy II to modeling human interaction with a heterogeneous and embedded control system.

Section 2 provides related background. Section 3 presents our approach and simulator. Section 4 shows the simulation results, followed by conclusion in Section 5.

2   Related Background

2.1   Ptolemy II

Ptolemy II is a java-based open source framework, developed by EECS Department of UC Berkeley, aiming to support modeling, simulation, and design of concurrent, real-time, embedded systems [7,8,9]. Ptolemy II provides a hierarchical structure for heterogeneous computation in an actor-oriented fashion. The components in Ptolemy II are called as "actors" drawn using Vergil, the graphic-user interface provided by Ptolemy II. Each actor has input and output ports to communicate with other actors. Tokens are used for message passing between actors. The "director" coordinates the scheduling and execution of these actors. Both directors and actors can be modeled in a hierarchical fashion. Ptolemy II interprets the execution and interaction of its actor components through so called "domains", which are models of computation. A model of computation behaves as "laws of physics" to governs the interaction of actors in the model. Ptolemy II has defined many domains, such as CSP (communicating sequential processes), CT (continuous time), DE(discrete event), FSM(Finite State machines), etc. However, Ptolemy II has not yet specified an operator domain.

2.2   Case Study : A Safety Injection System

The safety injection system in a chemical or nuclear power plant is frequently used as a case study in formal specification research [3]. In this paper, we use a complex and pragmatic safety injection system as a case study. We assume that there are three tanks in the system, namely, the Vessel and two supplier tanks: Tank 1 (CST: Condensate Storage Tank) and Tank 2 (SP: Suppression Pool). The safety injection system, will function when a leaking incident (LOCA :loss of coolant) leads to a low (Vessel) water level (less than 1.5). The leaked vessel water goes to Tank 2. Then, the safety injection controller starts the water Pump and turns on the Valve of Tank 1 first; however, when the water level of Tank 1 is lower than a set point(say, 2.6 meter) or that of Tank 2 is higher than another set point (say, 7.1 meter), the controller turns on Tank 2's valve for injection. The injection will be stopped when the Vessel water level reaches Level 8. The operator routinely monitors the system. When abnormality arises, due to software errors or hardware failures, the operator should manually control the injection system.

3   Extending an Operator Domain in Ptolemy II

We first add an "operator" domain into Ptolemy II. In the current prototype, the following actors are currently added into the domain:
  (a) Plan actor
  (b) Observer Actor
  (c) Decision Actor
  (d) Action Actor
  (e) OperatorComposite actor

The "OperatorComposite" actor facilitates potential hierarchical construction of actors. The "plan" actor can be further extended to support complex operator computation. The steps to add a new domain in Ptolemy II are described below:

1. We first added the Root of the Operator domain (say, "package Ptolemy.domains.operator.kernel") to the Ptolemy II source code. Then, the java program implementing "Operator" Director was added under this Root. The Operator Director was currently designed based on DE (Discrete Event) director with modifications in the Receiver part.
2. The java code implementing the above actors was added in Ptolemy II. The "Observer" Actor collects system information every second, and sends it to the "Decision" Actor, which will make judgment. The "Action" Actor was implemented as a Modal Model. Results from the Decision Actor may trigger different state transitions and operator actions.
3. We then designed the icons of Operator Director and Actors by adding an XML file "operator.xml" under "Ptolemy\domains\operator". Thus, the user can select and drag these icons in Vergil, the editor in Ptolemy II. The added operator domain in Vergil is shown in Fig.1.

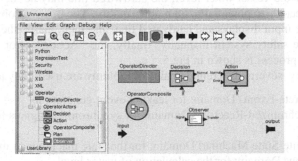

Fig. 1. The Operator domain in Vergil

4   A Simulation and Testing Environment for a Safety Injection System

A simulation and testing environment for a safety injection system has been constructed using Ptolemy II. The environment includes three parts: a gate-level testing tool, a simulator for the safety injection system, and a simulator of the human operator.

Our toolset consists of three major parts (Fig. 2):

(1) Logic-gate Testing Tool
(2) Operator control simulator
(3) Safety injection system simulator

The "Logic-gate Testing Tool" supports gate-level testing and mutation tests. A safety-critical system is usually designed with multiple duplicate sets of input signals or computing logic to ensure the correctness and consistency of the input or logic. We assume that the controller of the safety injection system gets input from 4 sets of water level sensors, 4 sets of bypass signals, and 4 sets of testing signals; the controller then makes a judgment based on 2-out-of-4 logic, that is, at least two signals out of the four input signals are identical, then that input value will be accepted. The logic is implemented as that shown in Fig.3. For testing purpose, our implementation allows each gate in this logic structure to be changed by using a pick list to simulate implementation errors. For example, an AND gate can be erroneously omitted or implemented as an OR gate. Thus, gate-level data flow testing schemes, such as MC/DC [4] or CCC [6], can be tested using this tool.

The "safety injection system" simulator simulates hardware devices, such as pumps and valves, and the software controller. Physical properties of each hardware device are modeled. The three tanks are modeled in a similar fashion as that in the Two-Tank model example given in Ptolemy [1]. Take Tank 2 (SP) for example, its physical logic is shown in Fig.4. The change of water amount depends on the current water amount, input flow, and output flow; and then, the height of water in the tank will be converted into output water level ("SP_Level").The controller is modeled using a state machine.

The "Operator control" simulator models the Operator role in the system. The newly added operator domain is used. Inputs include various displays as well as a hardwired line for the vessel water level; the hardwired line may be the last resort for the operator to recognize potential system abnormality. The operator observes the current state from the input displays and hardwired line, and then, makes the decision whether to control the safety system manually; finally, the "Action" actor may perform the actions. This process is shown in Fig.5.

In summary, several different domains in Ptolemy are used; they are:

- DE(Discrete-Event) Domain: for leaking event generation
- SR (Synchronized-Reactive) Domain: for synchronized signals for gate level testing.
- FSM(Finite-State Machine) Domain: for the state machines of most components. Continuous Domain: for the calculation of water level.

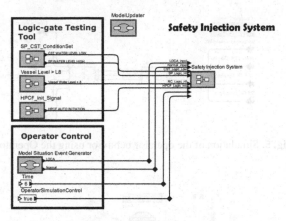

Fig. 2. The Simulation and Testing Environment

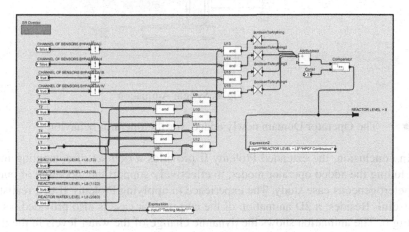

Fig. 3. The Gate-level Testing Tool

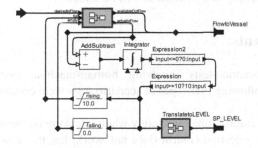

Fig. 4. The Physical Logic of Tank 2

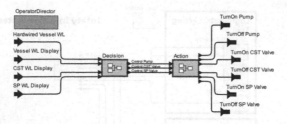

Fig. 5. Simulation of the operator behavior using the Operator domain

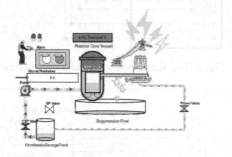

Fig. 6. System Animation

- The Operator Domain newly added: for the operator behavior.

In conclusion, the extended Ptolemy II provides a rich set of computing models, including the added operator model, to effectively support the modeling of our complex heterogeneous case study. The experience in applying Ptolemy II to a realistic case is useful. Besides, a 2D animation of the current test case is also provided, as shown in Fig.6. The animation shows the dynamic change of the water levels in the three tanks as well as the operator's reaction, if any. However, the current implementation of the operator domain is still preliminary. In the next version, in-depth details and more generic actors, including fuzzy decision, will be added.

5   Experiments

We have run scenario tests involving human-machine interaction using this enviroment. The following four types of scenario tests were conducted:

(1) The computer control functions safely when a leaking incident occurs.
(2) The automatic injection control does not work, but the abnormality is detected and handled properly by the operator.
(3) The automatic injection system does not work and the abnormality is undetected by the operator.
(4) The automatic injection system does not work and the abnormality is detected by the operator with a time delay.

The first case is designed to verify the correctness of our model and design. The result of the first case is shown in Fig.7. The upper part of the figure shows the water changes in the Vessel and supplier Tank 1. The lower part of the figure shows the water changes in the Vessel and supplier Tank 2. When a water leaking incident occurs, the vessel water level is descreasing, and then the controller turns on the pump and the valve of the supply Tank 1 (CST) around time unit 11. The controller switched to inject water from the supply Tank 2 (SP) once the Tank 1 water level lower than a set point at time unit 12. Since the leaking situation does not change during the entire simulation. Therefore, Fig. 9 shows that there are three cycles that vessel water level changed from Level 1.5 to Level 8 with water injection, and that there are three times that the Tank 2 (SP) valve was turned on. This test case verifies that our model and design are correct since the computer control works as expected.

The second case analyzes the operator's role in an abnormal situation. We assume that either the computer control did not work or the hardware fails, and the water level display falsely indicated a previously safe level of the vessel water. This is so called "false indication" failure, which is the most dangerous failure situation. We assume that the operator directly saw the hardwired water level line and noticed the abnormality, and then the operator manually turned on the water pump and the CST/SP valve. The resulting figure, shown in Fig.8, is similar to that of the first case except that the devices are all turned on manually. This case proves that the newly added operator domain works satisfactorily.

The third case is similar to the above false indication case, yet in this case the operator was misled by the false displays and failed to be aware of the dangerous stituation; then disasters occur (Fig.9). This case and the above case demonstrate the importance of the operator's role in a computer-controlled system. Without an alert operator, any hardware/software failure may lead to a catastraphic accident.

Similarly, in Case 4, the alertness of the operator with a time delay was tested; that is, the operator did not notice the abnormality at first, and a few minutes later, he was finally alert or aware of this problem. The results depend on the delayed response time; if the delay is short, it is possible to prevent the accident; otherwise, the result is hazardous. Accurate formulae for water changes and various physical phenomena can be calculated, and then the tolerable delay of the operator's reaction can be estimated. This figure of the tolerable delay is critical to a safety system design. Once this number is estimated, proper redesign of the system or prevention mechanisms, such as assertions or multiple independent displays, can then be implemented to enhance system safety. It was reported that a 6-minute delay is the unbound to prevent such an accident [5].

Further detailed variations of the above scenario tests can be performed for safety analysis in human-machine interaction. All the experiments showed that our toolset provides a good testing environment to study of the interaction process between a human and an embedded control system. The gate-level mutation tests using this toolset were also examined and found to be satisfactory to test MC/DC test cases [4] or mutation test (fault injection) cases [11]. This environment can also effectively support our other research activities in false indication [10] and human-machine interaciton [2].

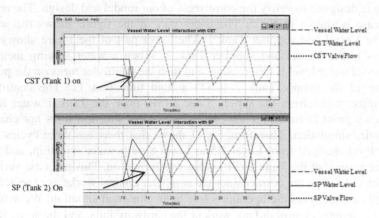

CST (Tank 1) on

SP (Tank 2) On

Fig. 7. Case 1: Computer control works

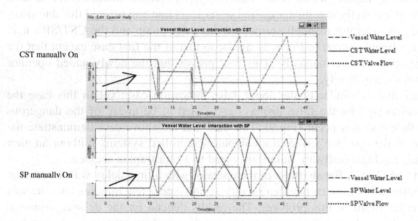

CST manually On

SP manually On

Fig. 8. Case 2: Handled by the operator

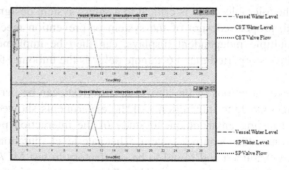

Fig. 9. Case 3: An abnormal situation unaware of by the operator

6   Conclusion

Embedded control systems have been prevailing in various industries. The operator plays a critical role in monitoring the operation of such systems. The operator also functions as the last layer of defense. Human interaction with the embedded control system can be analyzed in a simulation environment. Ptolemy II is a popular modeling and design tool for real-time, concurrent, computer control and embedded systems. However, Ptolemy II does not explicitly provide the operator modeling schemes. This research extends Ptolemy II by adding the operator domain. This research has constructed a simulation and testing environment for a safety injection system in Ptolemy II to study human interaction with the embedded control system. We gained useful experience in applying the extended Ptolemy II to modeling a complex heterogeneous system. Preliminary results show that the constructed toolset provides a good environment for analyzing human-machine interaction scenarios. In the next version, we will further extend the operator domain by adding more actors such as fuzzy decisions so that it can be generic enough to apply to other domains. We will also apply for permission to officially add the operator domain to the Ptolemy II software for public use.

References

1. Derler, P., Lee, E.A.: FuelSystem2Tanks (2012),
   `http://ptolemy.eecs.berkeley.edu/ptolemyII/ptII8.1/jnlp-modelingCPS/ptolemy/demo/FuelSystem/FuelSystem2Tanks.htm`
2. Fan, C., Tsai, S., Tseng, W.: Development of safety analysis and constraint detection techniques for process interaction errors. Annals of Nuclear Energy 38, 547–557 (2011)
3. Heitmeyer, C., et al.: Automated Consistency Checking of Requirements Specifications. ACM Transaction on Software Engineering and Methodology 5(3), 231–261 (1996)
4. Hayhurst, K., Veerhusen, D., et al.: A Practical Tutorial on Modified Condition/Decision Coverage. NASA / TM-2001-210876 (2001)
5. Huang, H.: Study of Nuclear Power Plant Safety Effect by Interactions between Operator and Digital Instrumentation and Control System. Ph.D. dissertation, ESS Dept., National Tsing Hua University, Taiwan (2007)
6. Jee, E., et al.: A data flow-based structural testing technique for FBD programs. Information and Software Technology 51, 1131–1139 (2009)
7. Lee, E.A.: Disciplined Heterogeneous Modeling. In: Petriu, D.C., Rouquette, N., Haugen, Ø. (eds.) MODELS 2010, Part II. LNCS, vol. 6395, pp. 273–287. Springer, Heidelberg (2010)
8. Derler, P., Lee, E., Sangiovanni-Vincentelli, A.: Modeling Cyber-Physical Systems. Proceedings of the IEEE (Special Issue on CPS) 100(1), 13–28 (2012)
9. Ptolemy II Project, EECS Dept., UC Berkeley (2012),
   `http://ptolemy.eecs.berkeley.edu/ptolemyII/`
10. Tseng, W., Fan, C.: Scenario analysis of false indication in computer-control systems. To appear in Annals of Nuclear Energy (2012)
11. Voas, J.: Fault Injection for the Masses. Computer 30, 129–130 (1997)

A Flexible and Re-configurable Service Platform for Multi-user Mobile Games

Yu-Sheng Cheng, Chun-Feng Liao, and Don-Lin Yang

Department of Information Engineering and Computer Science
Feng Chia University, Taichung, Taiwan
{m0005691,cfliao,dlyang}@fcu.edu.tw

Abstract. Recent trends suggest that the multi-user mobile game is becoming increasingly popular due to the success of social network applications. Multi-user mobile games differ from traditional multi-user on-line games in that the network connections are transient and that the bandwidths are limited. Moreover, there are few supports for constructing this type of applications, causing game developers have to do a lot of redundant works in dealing recurring issues such as connection and presence management. This paper aims to provide a flexible framework and a re-configurable game server that simplifies the development of multi-user mobile games. Preliminary performance evaluation shows that the proposed platform is more efficient compared to the traditional one. Besides, two mobile games that are constructed based on the proposed platform are also introduced to validate the feasibility.

Keywords: Mobile Games, Application Framework, Game Server.

1 Introduction

The computing capability of mobile devices has been improved rapidly in the last few years. The landscape of mobile computing is also changed significantly because of the introduction of the "App" business model. An App is a dynamically deployable software component running on a component container of a mobile device. Among various types of Apps, the multi-user mobile game has been one of the most popular categories recently. Similar to traditional multi-user on-line games, a multi-user mobile game is essentially distributed. Hence, the App has to deal with low-level tasks such as presence and communication management, which, if not carefully designed, will significantly affect the usability of the App. Although there are many commercial or open source libraries for developing games, most of them focus on stand-alone mobile games. This paper aims to propose a service platform for multi-user mobile games, which takes care of low-level and complex issues for game developers so that they can focus on the design of game contents and user experience. The contribution of this paper is therefore two folds: 1) an easy-to-use and flexible application framework that encapsulates low-level tasks of multi-user mobile games to relief the burden of game developers. The framework is design by considering the

J.-S. Pan et al. (Eds.): *Advances in Intelligent Systems & Applications*, SIST 21, pp. 701–710.
DOI: 10.1007/978-3-642-35473-1_69　　　　　© Springer-Verlag Berlin Heidelberg 2013

challenges in a distributed mobile environment such as limited bandwidths and transient connections. 2) A re-configurable game server that collaborates with the aforementioned framework to facilitate efficient communication and presence management is also proposed. Besides, the implementation of the proposed platform is available as an open source project hosted the Software Creative Society Service Platform of Information Technology Software Academy (ITSA)[2].

2   Related Work

Numerous platforms have been proposed to deal with the issues of large-scale multi-user on-line games. Most of them focusing on the latency, consistency, and scalability [5]. However, few of them deal with the challenges in the mobile environments in which the network connections are unstable and bandwidths are restricted. The client framework proposed by Kao et al. [6], is one of the few attempts that deals with the issues multi-user games in mobile environments. Nevertheless, the proposed platform focuses on protocol design and does not provides an application framework. In addition, it does not support presence management, which is a key feature of a multi-user game platform. GASP [9] is an open source and easy-to-use platform for mobile games, but is tightly coupled on the J2ME platform.

Recently, many game engines have been proposed to speed-up the development of mobile games. To name a few, Peker and Can [8] proposed a systematical way for developing mobile game engine based on design patterns [3]. As reported by Xie, most of mobile engines focus on scene management, screen rendering, or 3D-related effects [11] instead of distributed issues. The objective of the paper differs from the previous works in that we aim to devise a platform for multi-user mobile game that deal with the communication and presence management issues. The design of flows and protocols proposed in this work are also encapsulated as an open source game server and an application framework.

3   Requirements

This section discusses the needs of multi-user games for the mobile devices. In such type of games, the players form parties dynamically either to accomplish some goals or to be opponents in a contest. In both cases, there must be a logically centralized place for players to meet and then to team up. From technical points' of view, a game server is apparently required for client registration and team formation. In addition, the joining or leaving of clients are also important events so that these events have to be delivered to the clients. The issues mentioned above is called presence management.

When the games are conducted, there are a great deal of communicating messages that must be multiplexed and to delivered to appropriate clients. Moreover, according to the investigation report of OECD [7], the average bandwidths in Taiwan for download and upload are only 1.26 Mbps and 0.17 Mbps, respectively. Thus, the size and amount of the communicating message have to be kept as small as possible since the bandwidths of mobile games are limited.

Besides, since most of the libraries or application frameworks provided by mobile platforms only provide primitive supports for networking so that the the the developers of multi-user mobile games has to handle a lot of low-level issues repeatedly, including the design and implementation of game server, protocol, and message multiplexing. Consequently, the needs of a common service platform for multi-user mobile games can be summarized below.

Functional Requirements. The game server has to implement the functionality of managing user profile, users' presence, and team forming. The developers are able to access or to manipulate these data through APIs in the application.

Non-functional Requirements. A well-designed framework is required to shorten the development time and relief the developers from the burden of networking issues so that it has to be flexible, easy to learn and customizable. In addition, the consumption of bandwidths has to be kept as small as possible so that the overall communication is reasonably efficient.

4   Platform Design and Implementation

To fulfill the requirements mentioned in the previous section, we propose a service platform that aims to be flexible, easy to learn, customizable, and reasonably efficient to support game developers. The overall architecture of proposed platform is shown in Fig. 1, which can be divided into two parts: the game server and the Apps deployed on the mobile devices.

The game server provides infrastructural services such as connection management, event notification, activity management, configuration management, and storing user profiles. In addition, the server is equipped with a set of communication channels and is responsible for multiplexing and routing messages among clients. The communication channels are constructed based on clustered MOMs (Message-Oriented Middleware) so that the channels are loosely coupled and scalable. In other words, the use of MOM enhances the flexibility and efficiency of mobile games. Besides, the server supports a set of presence management API. Presence management API defines a process including login, finding waiting players, creating a party, joining a party, and starting a game. Game developers is able to use the API to help players forming the teams or contests.

The client-side application framework is part of an App, which aims to support the networking details and presence management and interacting with the game server. The application framework encapsulates frequently used multi-user game communication logic, such as the routine for processing wiring format and for routing messages, into high-level and easy to use APIs so that even developers that are not familiar with multi-user game programming more easily.

The wiring format for client-server communication is designed based on JSON (Java Script Object Notation) [4], which is a popular and cross-platform message format. As a result, the platform is interoperable and highly extensible. For instance, both an iOS-based mobile App and an Android-based App is able to interact to the game server via JSON message format.

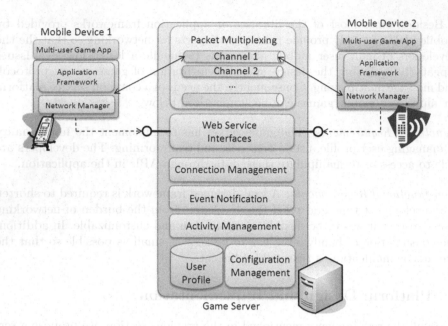

Fig. 1. System architecture

The details of the presence and communication management are discussed in detail in the following sub-sections.

4.1 Presence Management

Presence management, a common functional requirement in multi-user games, refers the the capability of a system to be aware of users' presences and to group users based on certain criteria. Almost all multi-user games require presence management. It can be observed that most of presence management tasks follow the similar process, as shown in Fig. 2.

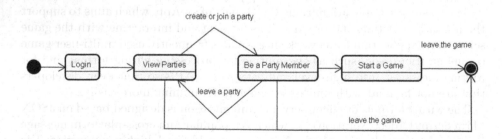

Fig. 2. General flow for presence management

When a game starts, the program flow enters the "Login" state so that the system is able to be aware of user presences. After correct name and password is entered, the flow enters "party forming" state in which an user either creates a new party or joins an existing party created by other ones. The user who creates the party is called the host of the party. Users are also freed to create, leave and join different parties repeatedly at this stage. Finally, if the users are satisfied with the members of current party, the game will be initiated based on the hosts' commands.

From a technical's points of view, the presence management process is realized by coordinating the game server and client-side applications. The game server keeps track of important presence events such as creating a new party or joining/leaving an existing party. If the connections are lost accidentally, then the presence information can be inconsistent. Therefore, the game server has to periodically check the presence of client based on the timestamp reported regularly by the clients. Whenever a client enters a new state, the server notifies party members through the communication channel so that the client can react appropriately. The client-side application framework supports presence management based on the State pattern [3].

4.2 Communication Management

In the proposed platform, all running processes in a multi-user mobile game are abstracted as two basic types: the *Producer* and the *Consumer*. Assuming that process A sends a message to process B, then A is a *Producer* and B is a *Consumer*. For example, if a player invokes an action in the game such as move or chat, then these events are reified as messages which transmitted by a *Producer*. These messages are then multiplexed by the game server and then are delivered to appropriate *Receivers*. However, if the framework send a message whenever the user takes an action, then messages will be sent too frequently, causing poor network performance. Hence, we also design a message pool to serve as the cache and the buffer on the client side so that the flooding problem can be prevented. In this design, the framework regularly checks the pool and attempts to consolidates messages. When the pool is full, *Producer* cleans up the pool and then flush messages to the game server.

Consumer and *Producer* also play the role of the gateway to the MOM. The design is based on the Adapter pattern [3]. Table 1 shows MOM-related operations of the *Consumer* and *Producer* interfaces.

The communications of multi-user mobile games can be divided into two styles. One is to request resources such as looking up information about a user, talking to a user, or buying an item used in games. Such style of communication is similar to the traditional RPC (Remote Procedure Call). Hence, the framework provides an interface *Callable* and an abstract class *Caller* for this purpose.

Since each game has different logics for resource management, the *Callable* has to be realized and to be deployed on the game server. Figure 3 depicts the overall design of the framework. The *Callable* interface implements *DurableConsumer*

Table 1. MOM related operations in the *Consumer* and *Producer* interfaces

Operation	Description
setURL	Specify the address of game server.
setTopic	Specify the name of publish-subscribe communication channel.
setQueue	Specify the name of point-to-point communication channel.
isPersistent	When the server is crashed or is restarted, the messages are still contain in the server.
isDurable	If the message can not be delivered due to client fails, the message will be resent when the client is recovered.

so the messages can be processed even when the game server restarts. On the client side, the developers have to realize the abstract class *Caller*, which is able to call *Callable* services deployed on the serve and to receive the responses at the same time.

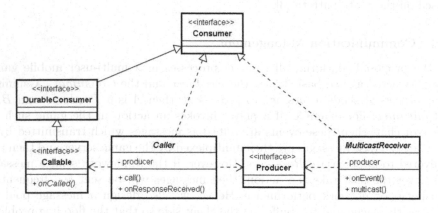

Fig. 3. The class diagram of the application framework

A client may want to publish a message to a specific group of clients. Such style is implemented as *MulticastReceiver* in the framework. When a process initiates an instance of *MulticastReceiver*, it is registered with a session and a key that identifies the session. Therefore, the clients having the same key is able to share information. As shown in Fig. 3, the event handling logic is realized in the *onEvent* method, whereas the *multicast* method sends the messages to a group associating to the session key.

To ensure the interoperability while maintaining reasonable network efficiency, in the transport layer, we implement the wiring formats among game server and clients based a platform independent lightweight message-oriented protocol called STOMP (Streaming Text Orientated Messaging Protocol)[10]. On top of STOMP, the communicating messages is encoded in JSON (JavaScript Object

Notation)[4] format. JSON is easy for machines to parse and to generate. Also, it can be decoded efficiently. Compared with XML-based encoding format, JSON is more compact and therefore more efficient.

5   Evaluation

This section reports the evaluations on the proposed platform. First, we briefly introduce and discuss two multi-user mobile games constructed based on the proposed platform. After that, we conducted experiments to evaluate the performance of the game server with proposed message pooling and lightweight STOMP-based protocol against the one that is not using these techniques.

5.1   Feasibility

We verify the feasibility by implementing two multi-user mobile games based on the proposed platform. The Chinese Chess game is a multi-user mobile game that allows multiple pairs of players playing on-line at the same time (see Fig. 4a). The Art of Bomb shown in Fig. 4b is a game that allows up to four people as a contest group. In each group, the users are able to attack to one another based on different kinds of weapons. Similar to the Chinese Chess game, several contest groups are conducted simultaneously on the game server.

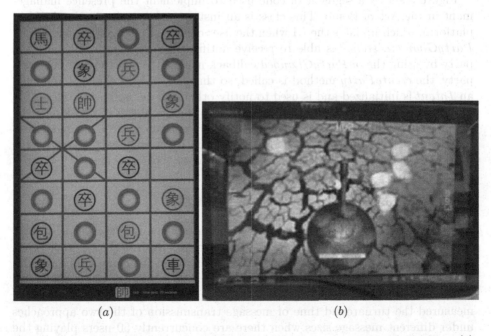

(a)   (b)

Fig. 4. Multi-user mobile games implemented based on the proposed platform (a) Chinese Chess; (b) Art of Bomb

```
public class Gallery extends Activity implements   PartyChangeListener {
    public void createParty () {
        state.nextState(myParty, me, false);
        Intent intent = new Intent();
        intent.setClass(GalleryActivity.this, CreatePartyActivity.class);
    }

    public void onPartyChanged(List<Party> parties) {
        imageAdapter.setData(parties);
        Message msg = uiHandler.obtainMessage();
        msg.what = GalleryActivity.UPDATE_GALLERY;
        uiHandler.sendMessage(msg);
    }
    ...
```

Fig. 5. Code segment for presence management

Figure 5 shows a segment of code used to implement the presence management in the Art of Bomb. This class is an instance of Activity in the Android platform, which updates the UI when the user creates or changes the party. The *PartyChangeListener* is able to receive notifications upon user changing the party by using the *onPartyChanged* callback method. When the user creates a party, the *createParty* method is called, so that the state is changed and then an *Intent* is initialized and is used to notify other *PartyChangeListener*s.

It is also worthy to mention that based on our experiences in developing the games, it takes about 40 work days for a 2-person team to design a moderate-scale multi-user mobile game (the time to design UI is excluded). Based on the proposed platform, it can be reduced to 18-20 work days in average.

5.2   Performance

In the experiments we evaluated the performance of the game server with message pooling and lightweight STOMP-based protocol against the one that is not using these techniques. The communication channels of the game server are constructed based on ActiveMQ [1], an open source MOM server, where the MOM server is modified to facilitate the message pooling and protocol handling. We measured the turnaround time of message transmission of the two approaches under different message sizes when there are concurrently 50 users playing the game. In the experiments, the simulated user actions are triggered repeatedly every 500 milliseconds. The results is shown in Fig. 6, where the X axis is the

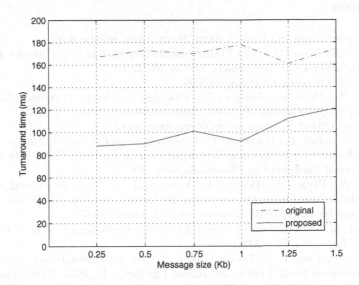

Fig. 6. Performance of the proposed platform

size of messages and Y axis is the turnaround time in milliseconds. It can be observed that when the size of messages increases, both approaches are able to scale linearly. However, the proposed approach outperforms the default protocol.

6   Conclusion

As multi-user mobile games becoming popular, time to market is one of the most important success factors for mobile games. In addition, the performance of Apps also has great impacts on user experience. This paper addresses these needs by proposing a service platform for multi-user mobile games. The proposed platform consists of a reconfigurable game server which enables efficient communications and presence management. In addition, the platform also provides an application framework that can significantly reduce the overall development time and thus makes the developed mobile games time-to-market. We are currently designing more cross-cutting functionalities such as billing and loading balancing on the game server and then supports these functionalities using easy-to-use APIs on the client-side. Moreover, the client-side application framework is limited to Android-based mobile devices. Thus, we are currently working on an iOS version in order to support more clients.

Acknowledgements. This work is sponsored by the Ministry of Education, Taiwan, under National Education Program on Information and Software Technologies, Value-Added Software Application Development Project, Category-B and by National Science Council under grant 101-2218-E-035-008 and 101-2221-E-035-082.

References

1. The Apache ActiveMQ Project, http://activemq.apache.org/
2. Cheng, Y.S., Liao, C.F.: The Flexible Multi-user Mobile Game Service Platform Project. Software Creative Society Service Platform, http://of.itsa.org.tw/projects/506
3. Gamma, E., Helm, R., Johnson, R., Vlissides, J.: Design Patterns: Elements of Reusable Object-Oriented Software. Addison-Wesley (1994)
4. JavaScript Object Notation, http://www.json.org/
5. Jiang, X., Safaei, F., Boustead, P.: Latency and Scalability: A Survey of Issues and Techniques for Supporting Networked Games. In: Proc. IEEE 13th Malaysia International Conference on Communication (2005)
6. Kao, Y.W., Peng, P.Y., Hsieh, S.L., Yuan, S.M.: A Client Framework for Massively Multiplayer Online Games on Mobile Devices. In: Proc. IEEE International Conference on Convergence Information Technology (2007)
7. Lin, T., et al.: Performance analysis of 3G wireless networking in Taiwan. Consumer Report Magazine (375) (2008)
8. Peker, A.G., Can, T.: A Design Goal and Design Pattern Based Approach for Development of Game Engines for Mobile Platforms. In: Proc. International Conference on Computer Games (2011)
9. Pellerin, R., Delpiano, F., Duclos, F., Gressier-Soudan, E., Simatic, M.: GASP: an open source gaming service middleware dedicated to multiplayer games for J2ME based mobile phones. In: Proc. International Conference on Computer Games (2005)
10. The Streaming Text Orientated Messaging Protocol Specification, version 1.1 (March 2011), http://stomp.github.com/stomp-specification-1.1.html
11. Xie, J.: The Research on Mobile Game Engine. In: International Conference on Image Analysis and Signal Processing (2011)

High-Performance 128-Bit Comparator Based on Conditional Carry-Select Scheme

Shun-Wen Cheng, Jhen-Yuan Li, and Wei-Chi Chen

Dept. of Electronic Engineering, Far East University, Hsin-Shih 74448, Tainan, Taiwan
swcheng@ieee.org

Abstract. The comparator is a basic arithmetic component of digital systems. An individual, high-performance, 128-bit hardware comparator core plays an important role on many computing applications of nowadays. The study proposes a high-speed comparator design. Based on modified 1's complement principle and conditional carry select scheme, the proposed design has small transistor count and short propagation delay. Post-layout simulations based on TSMC 0.18μm CMOS process has completed. It shown the three-output, 128-bit CMOS comparator of the proposed architecture with 18-inverter buffers only needs 3,138 transistors and spends 1.16ns.

Keywords: magnitude comparator, digital comparator, 1's complement, conditional carry adder, CMOS, digital IC and VLSI.

1   Introduction

Comparator is a basic and useful arithmetic component in the VLSI design, which detects whether one input is equal (EQ) to or greater than (GT) or less than (LT) the other input [1]. A 64-bit microprocessor requires a 40-bit tag comparator due to the 50-bit physical address, which has been increasing every generation [2]. Recently, tree-based comparators have been proposed in [3] and [4] where dynamic Manchester structures are used to facilitate the comparison process. An individual, high-performance, 128-bit hardware comparator core plays an important role on many computing applications of nowadays. For example, each equipment on the Internet, such as a computer or cell phone, must be allocated an IP address in order to communicate with other equipments. IPv6 (Internet Protocol version 6) is a revision of the IPv4 (IP version 6); IPv6 was developed to deal with the long-anticipated problem of IPv4 running out of addresses. IPv6 implements a new addressing system that allows for far more addresses to be assigned than with IPv4. IPv6 uses 128-bit addresses, allowing for 2^{128}, or approximately 3.4×10^{38} addresses — more than 7.9×10^{28} times as many as IPv4, which uses 32-bit addresses. Therefore, a high-speed 128-bit comparator plays an important role.

Moreover, compare and swap elements of data are vital for sorting [5]. In conventional computer systems, instruction COMPARE and instruction SUBTRACT often share the hardware. This can reduce layout and power cost.

J.-S. Pan et al. (Eds.): *Advances in Intelligent Systems & Applications*, SIST 21, pp. 711–720.
DOI: 10.1007/978-3-642-35473-1_70   © Springer-Verlag Berlin Heidelberg 2013

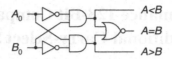

Fig. 1. 1-bit magnitude comparator

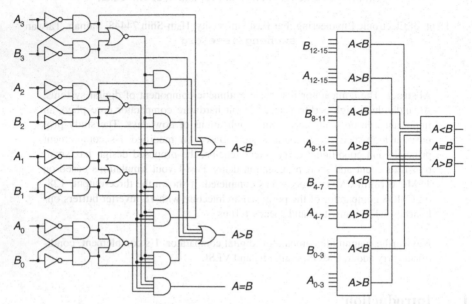

Fig. 2. 4-bit magnitude comparator

Fig. 3. 16-bit magnitude comparator

Classical digital comparators are displayed in Fig. 1 [1], [6]. The circuit compares two binary digit A and B, and produces three output: A>B, A=B, A<B. Due to the fanout limitation of CMOS logic [6], 4-bit comparator in Fig. 2 is the basic constructive unit. Fig. 3 indicates that a 16-bit comparator needs five 4-bit comparators. The example reveals circuit cost/complexity of a (2k)-bit comparator are often not only twice than a k-bit comparator. If someone needs to process long digit integer comparison, then directly design a corresponding hardware comparator, the circuit cost will become very large. Therefore a compact, high-performance comparator core is very important. This paper is organized as follows. In Section 2, it presnts the feasibility study of modified 1's complement for three-output comparator design. Then the proposed conditional carry-select comparator architecture is demostarted in Section 3. The final section draws conclusions.

2   Modified 1's Complement Rule for Three-Output Comparator Design

In one's complement representation, subtraction is performed by addition of a negative integer. This eliminates the need for a separate subtraction unit. As demonstrated in Fig. 4, when $A > B$ ($A=X$ and $B=Y$), the carry-out bit $Cout = GT = 1$. When $A=B$ ($A=B=X$), the Carry-out bit $Cout = GT = 0$, when $A<B$ ($A=Y$ and $B=X$), the Carry-out bit $Cout = GT = 0$. Additionally, if the scheme always adds a fixed carry, then if $A \geq B$, bit $Comp = 1$. If $A < B$, bit $Comp = 0$ [7]. At this time, a common two-output magnitude comparator is obtained, however, it is still not fit for the purpose of tag identification applications. Thus we let LT is the inverted signal of bit $Comp$, and then get the Equal signal $EQ = \overline{GT + LT}$, as shown in Table 1. Thus the three-output (GT, LT, and EQ) magnitude comparator is achieved.

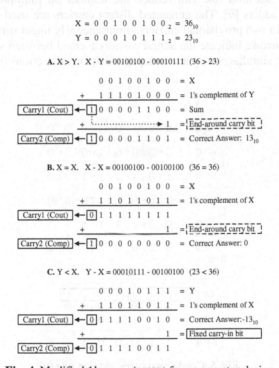

Fig. 4. Modified 1'c complement for comparator design

Table 1. Modified 1'c complement for the three-output comparator design

Status bit	$GT = $ Carry1(Cout)	Carry2($Comp$)	$LT = \overline{Carry2(Comp)}$	$EQ = \overline{GT + LT}$
$X > Y$	1	1	0	0
$X = X$	0	1	0	1
$Y < X$	0	0	1	0

3   Proposed Conditional Carry-Select Comparator Scheme

3.1   Conditional Carry Addition Rule

In the conditional sum adder, many multiplexers are used to select the sum output data [8]. This makes the multiplexer tree large and irregular. However, the sum outputs are useless for the carry selection procedure. It is clear that only the carriers are used in this network reduction procedure. Fig. 5 shows the proposed conditional carry addition rule, which is an improvement of the conditional-sum addition rule. The conditional-carry addition rule can reduce the number on multiplexer gates in the conditional sum adders [9]. The generated distant carriers are used to select the true carry outputs from two provisional carriers simultaneously under different carry input conditions. The arrows indicate the actual carriers formed between sections. The carries are generated simultaneously and independently on all sections [9].

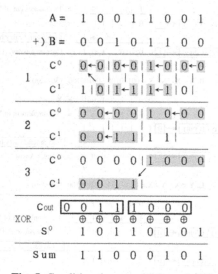

Fig. 5. Conditional carry addition rule [9]

3.2   Conditional Carry-Select Comparator Scheme

For achieve the three-output comparator demand, the proposed architecture is modified from the conditional carry adder (CCA) in Fig. 6 [9]. $Carry = AB + AC + BC = AB + (A + B) C$. Now if $C=0$, $Carry = AB$. If $C=1$, $Carry = AB + (A+B) = A + B$. All shaded multiplexers are removed from the CCA in order to select carry bits C_2, C_4, C_5, and C_6. Originally there is no carry-in in the addition of A_0 and B_0. For the fixed carry-in bit, additional three 2-to-1 multiplexers is necessary.

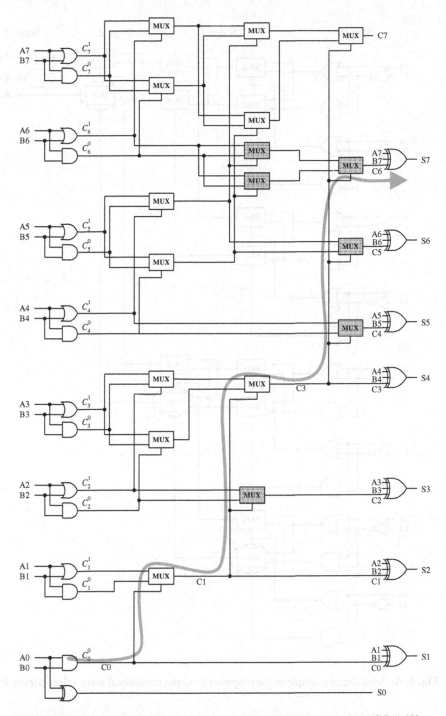

Fig. 6. Circuit and critical delay path of the 8-bit conditional carry adder (CCA) [9]

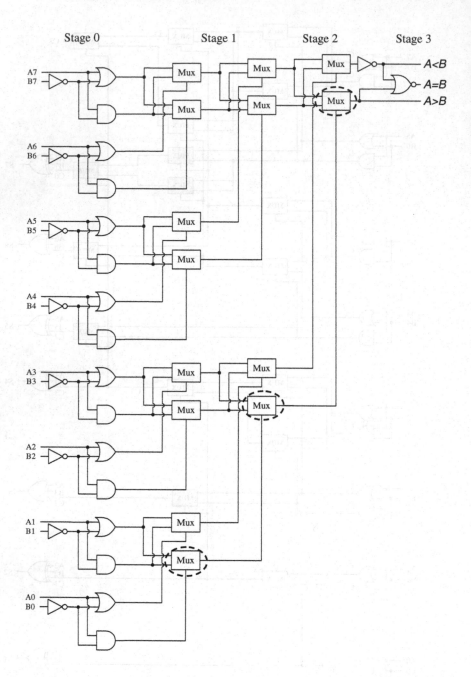

Fig. 7. An 8-bit circuit example of the proposed 3-output conditional carry-select comparator

Table 2. Multiplexer count of the proposed comparator scheme

Bit Number of Comparator	Total Multiplexer Count Method 1	Total Multiplexer Count Method 2
2^1	2	$2^1(=2^2-2)$	2
2^2	4	$2^2+2^1(=2^3-2)$	2+4=6
2^3	8	$2^3+2^2+2^1$ $(=2^4-2)$	**6+8=14**
2^4	16	$2^4+2^3+2^2+2^1$ $(=2^5-2)$	14+16=30
2^5	32	$2^5+2^4+2^3+2^2+2^1$ $(=2^6-2)$	30+32=62
2^6	64	$2^6+2^5+2^4+2^3+2^2+2^1$ $(=2^7-2)$	62+64=126
2^7	128	$2^7+2^6+2^5+2^4+2^3+2^2+2^1$ $(=2^8-2)$	126+128=254

Follow the deduction in Table 2, the total of MUX gates of N-bit comparator is,

$$\sum_{k=1}^{M} 2^k = 2^{k+1} - 2, \quad \text{where } M = \log_2 N.$$

Figure 7 shows an 8-bit comparator example of the proposed comparator architecture. The 8-bit comparator needs 14 (=2+4+8) 2-to-1 multiplexers. The schematic need eight inverters to generate complementary values of input B. The total transistor count is one two-input NOR gates, nine inverters, eight two-input OR gates, and eight two-input AND gates, and fourteen 2-to-1 multiplexers. The static CMOS AND gate and OR gate internally generate their complementary signals [9]. They can provide for the use of stage-1 multiplexers, so this reduces the requirements of inverter.

The total gate count of 3-output N-bit comparator is,

$$NOR2 \times 1 + INV \times (N+1) + AND2 \times N + OR2 \times N + Mux2to1_{Stage-1} \times N +$$

$$Mux2to1_{Higher-Stages} \times \left\{ [\sum_{k=1}^{M} (2^{k+1} - 2)] - N) \right\},$$

where $M = \log_2 N$.

So the total transistor count of the 8-bit static CMOS comparator is (2p+2n)×1+ (1p+1n)×9 + (3p+3n)×8 + (3p+3n)×8 + (2p+2n)×8 + (3p+3n)×(14-8) = 93p + 93n. The total transistor count of N-bit static CMOS comparator is,

$$(6p + 6n)N + (3p + 3n)(N+1) + (3p + 3n)\left\{ [\sum_{k=1}^{M} (2^{k+1} - 2)] - N) \right\},$$

where $M = \log_2 N$. If $N = 64$ bit, and ten buffers are used for increasing driving capability, thus the total transistor count is $765 \times 2 + 10 \times 4 = 1,570$. If $N = 128$ bit, and eighteen buffers are used for increasing driving capability, thus the total transistor count is $1,533 \times 2 + 18 \times 4 = 3,138$.

Use pass-transistor logic to construct the proposed comparator architecture, will have the fewest transistor count. The total transistor count of the 8-bit pass-transistor logic comparator is $(2p+2n) \times 1 + (1p+1n) \times 3 + (4p+4n) \times 8 + (1p+1n) \times 8 + (2p+2n) \times (14-8) = 57p + 57n$. The total transistor count of N-bit static CMOS comparator is,

$$(5p+5n)(N+1)+(2p+2n)\left\{[\sum_{k=1}^{M}(2^{k+1}-2)]-N)\right\},$$

where $M = \log_2 N$. If $N = 64$ bit, and ten buffers are used for increasing driving capability, thus the total transistor count is $449 \times 2 + 10 \times 4 = 938$. If $N = 128$ bit, and eighteen buffers are used for increasing driving capability, thus the total transistor count is $897 \times 2 + 18 \times 4 = 1,866$. The proposed comparator circuit also easily partitions to several stage pipelines for increasing the data throughput. And the conditional carry-select comparator can combine with conditional carry adder to increase and hardware sharing.

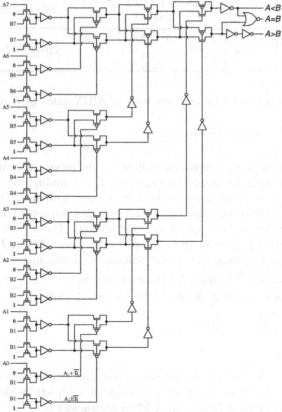

Fig. 8. Proposed 8-bit conditional carry-select comparator with pass-transistor logic needs 114 transistors (57P+57N) only. And a 64-bit comparator by this way only needs 938 transistors.

4   Conclusion

The circuit complexity information of the proposed conditional carry-select comparator is listed in Table 3. The author found the transistor count of the new design is less than that required in the conventional design, while the transistor count of the new design with static CMOS is only approximately half of the conventional design. The comparisons of comparator design are based upon TSMC 0.18μm 1P6M CMOS Process Technology. As displayed in Table 4, compare with other 64-bit comparator designs, the proposed comparator scheme is very competitive. The transistor count of the 128-bit comparator using static CMOS needs 3,138 transistors and the worst propagation delay spends 1.16ns.

Table 3. Transistor count comparison of various bit-length comparators

Bit Number of Comparator	Classic design with static CMOS [1], [6]	The proposed design with static CMOS	The proposed design with pass-transistor logic
2	35 p + 35 n	21 p + 21 n	15 p + 15 n
4	79 p + 79 n	45 p + 45 n	29 p + 29 n
8	182 p + 182 n	93 p + 93 n	57 p + 57n
16	370 p + 370 n	189 p + 189n	113 p + 113n
32	758 p + 758 n	381 p + 381 n	225 p + 225 n
64	1,522 p + 1,522 n	765p + 765 n	449p + 449 n
128	3,790 p + 3,790 n	1,533 p + 1,533 n	897 p + 897 n

Table 4. Transistor count and delay comparison of various 64-bit comparators

64-bit Comparator Design	Number of transistor	Delay on 0.18μm process
Huang and Wang [10] 2003	1,640	752 ps
Lam and Tsui [11] 2007	3,386	453 ps
Kim and Yoo [12] 2007	964	1005 ps
Frustaci, Lanuzza, and Corsonello [13] 2010	1,365	633 ps
Chuang, Li, and Sachdev [14] 2012	1,206	642 ps
The proposed conditional carry-select comparator design with static CMOS	1,570	660 ps
The proposed conditional carry-select design with pass-transistor logic	938	710 ps

Acknowledgement. This study was supported in part by the National Science Council (NSC) of Taiwan, under 2012 Grant NSC 101-2815-C-269-012-E. The author would like to thank the TSMC and the National Chip Implementation Center (CIC) of the National Applied Research Laboratories (NARL) of Taiwan, for supporting process technology and service. The author would like to thank the editor and anonymous reviewers for their helpful comments in improving the quality of this paper.

References

1. Hwang, K.: Computer Arithmetic–Principles, Architecture and Design. John Wiley & Sons (1979)
2. Suzuki, H., Kim, C., Roy, K.: Fast Tag Comparator Using Diode Partitioned Domino for 64-bit Microprocessors. IEEE Trans. on Circuits and Systems I: Regular Papers 54(2), 322–328 (2007)
3. Perri, S., Corsonello, P.: Fast Low-cost Implementation of Single-clockcycle Binary Comparator. IEEE Trans. Circuits Syst. II, Exp. Briefs 55(12), 1239–1243 (2008)
4. Frustaci, F., Perri, S., Lanuzza, M., Corsonello, P.: A New Low-power High-speed Single-clock-cycle Binary Comparator. In: IEEE Int'l Symp. Circuit and System, ISCAS 2010, pp. 317–320 (2010)
5. Knuth, D.E.: Sorting and Searching. Addison-Wesley (1973)
6. Weste, N.H.E., Eshraghian, K.: CMOS VLSI Design — A Circuits and Systems Perspective, 3rd edn. Addison–Wesley (2005)
7. Cheng, S.-W.: High-speed Magnitude Comparator with Small Transistor Count. In: 2003 IEEE Int'l Conference on Electronics, Circuits and Systems, ICECS 2003, vol. 3, pp. 1168–1171 (2003)
8. Sklansky, J.: Conditional-Sum Addition Logic. IRE Transactions on Electronic Computers EC 9(2), 226–231 (1960)
9. Cheng, K.-H., Cheng, S.-W.: 64-bit High-Performance Power-Aware Conditional Carry Adder Design. IEICE Trans. on Electronics E 88-C(6), 1322–1331 (2005)
10. Huang, C.-H., Wang, J.-S.: High-Performance and Power-Efficient CMOS Comparators. IEEE J. Solid-State Circuits 38, 254–262 (2003)
11. Lam, H.-M., Tsui, C.-Y.: A Mux-based High-Performance Single-Cycle CMOS Comparator. IEEE Trans. on Circuits and Systems II: Express Briefs 54(7), 591–595 (2007)
12. Kim, J.-Y., Yoo, H.-J.: Bitwise Competition Logic for Compact Digital Comparator. In: 2007 IEEE Asian Solid-State Circuits Conference, ASSCC 2007, pp. 59–62 (2007)
13. Perri, S., Corsonello, P.: Fast Low-Cost Implementation of Single-Clock-Cycle Binary Comparator. IEEE Trans. on Circuits and Systems II: Express Briefs 55(12), 1239–1243 (2008)
14. Chuang, P., Li, D., Sachdev, M.: A Low-Power High-Performance Single-Cycle Tree-Based 64-Bit Binary Comparator. IEEE Trans. on Circuits and Systems II: Express Briefs 59(2), 108–112 (2012)

A Multiplier-Free Noise Trapped Touch Algorithm for Low Cost 4x4 Matrix Panel Design

Yu-Hsaing Yu, Qi-Wen Wang, and Tsung-Ying Sun, Member, IEEE

Department of Electrical Engineering, National Dong Hwa University, Hualien, Taiwan, R.O.C
{d9823010,u9823036}@ems.ndhu.edu.tw, sunty@mail.ndhu.edu.tw

Abstract. A low cost 4x4 matrix touch panel which based on a multiplier-free noise trapped touch algorithm is discussed in this paper. We focus on the method that filters out the noise without long delay to make a fast response and the implement detail about how to put the method into ATMEL tinyAVR, an 8-bit, low-cost microcontroller. By scheduling the measure task, process task and display task, each touch key switch is allowed to have their own independent LED indicator that shares the limited I/O with the sensing matrix. The touch panel is connected to the host controller through a 4-wire telecommunication cable and also powered by the cable. As a result, the touch panel encounters the noise coupled by the cable other than the 50/60Hz noise picked up by the user.

Keywords: Touch algorithm, FIR notch filter, multiplier-free, noise trape.

1   Introduction

In the local control of a building management system, it is needed to replace traditional power switch with a small signal pushbutton that generate edge signal to trigger the on/off event running by the microcontroller to enable both local and remote control. In general, a standard small signal pushbutton cannot stand permanently. Because its contact may oxidize quickly in a wet environment and the mechanism will be worn out after a long time usage. Besides, a standard one doesn't equip LED which helps user to find the button easily in the dark. For a good one that has LED and gold plated contact is too expensive to deploy control panels in a building massively. A control panel with capacitive touch technology is more applicable to a building management system. It is wear free, expandable and a LED can be add-on easily. Through the layout in the PCB and resistor that control the charge current, a touch algorithm can sense which touch pad is pressed by user and thus accomplish the function of traditional pushbutton. As the touch pad increased, only the resistor is required for a new pad. As for the design that using pushbutton, a whole pushbutton is needed which is more expensive than a resistor.

2   Problems and Related Works

Though the conevtient feature of touch panel is better than a traditional keypad panel, it is based on the assumption that touch panel would not fail under the operating

J.-S. Pan et al. (Eds.): *Advances in Intelligent Systems & Applications*, SIST 21, pp. 721–730.
DOI: 10.1007/978-3-642-35473-1_71   © Springer-Verlag Berlin Heidelberg 2013

environment. Actually, there are many noises even in our living spaces. For example, a household appliance can generate impulse noise when starting-up and periodic noise while it try to control output power of heater or light by switching related method. These noises may impact other electronic device directly in the same loop or is emitted over the printed circuit board (PCB) and pick up by the human boy. Human bodies pick up many noises that would challenge the touch design as if we were an antenna. Once the sensing pad has been touched, not only the capacitance is changed but also the noises are induced to disturb the measuring procedure. To detect the touch event without disturbance, it is necessary to obtain the very low frequency signal caused by the touch event and filter out the higher one. In [1], an IIR low-pass filter which cost almost no memory and can operate without multiplication is employed to remove unwanted noise. The design in [2] utilizes the cascade architecture combined by IIR and FIR structure, this ideal is quite good only that the order is high and cannot notch the dedicated noise effectively. There is also a filter design without multiplication for communication [3] can be used to filter out the dedicated noise, however, a cost-effective way is needed for the low-cost microprocessor.

3   The Proposed Touch Algorithm

In this work, the touch panel sends key number that touched by user to room controller under the noisy environment. It is a hard work because it is also hard to image the cable that panel needs to connect to the room controller is routed together with sprawled AC power cords to the room controller. Although there are many methods to wipe out the noise via hardware or software approaches, under the limited budget, the choice is limited. Due to the consideration of cost and system integration, ATTiny2313V from ATMEL Corporation is chosen to host the measurement of capacitance, displaying the key state and the transmission of data. ATTiny2313V has one UART interface and just enough I/O for designing the 4x4 matrix touch panel. Moreover, it is cheap, register-rich and equipped with 128-byte EEPROM for storing parameters and coefficients that needed for the algorithm. It's also capable of generating an interrupt caused by a pin change event and has a 16-bit timer/counter to get the charging time in the pin change interrupt service routine (ISR) to fulfill the RC based capacitance measurement method used in our design. However, the lack of multiplier of ATTiny2313V makes it difficult to deal with the signal and noises. Accordingly, we propose a multiplier-free noise trapped touch algorithm for this dilemma.

3.1   Touch Algorithm

The schema of the proposed touch algorithm is depicted as Fig. 1. There are three major parts for processing the measured data and outcome the triggered key number:

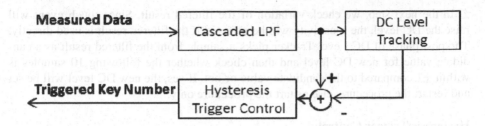

Fig. 1. Schema of the proposed algorithm

Cascaded LPF

The low-pass filter (LPF) in our design is constructed of 1st-order IIR LPF following by a 6th-order FIR LPF. The IIR filter from [1] that used to track DC value is suitable for sweeping out most of the undesirable noise while keeping touch signal nearly unchanged. The difference equation of the 1st-order IIR LPF is as follow:

$$y[n] = y[n-1] + k \cdot (x[n] - y[n-1]) \tag{1}$$

To make the response faster with respect to the short time touch event, we choose the feedback gain (k) of 0.125, which is twice as big as the gain in [1]. However, this is a tradeoff between response time and attenuation. As a result, the strong 60 Hz noise does not disappeared after the 1st-order IIR LPF. Therefore, we cascade the second FIR LPF to take over the mission to eliminate the 60 Hz noise. The post filter is 6th-order FIR LPF and the coefficients are friendly to multiplierless microcontroller. The following equation describes the impulse response of the filter:

$$h[n] = \begin{cases} \{1,3,5,5,3,1\}, & 0 \le n \le 5, \\ 0, & \text{otherwise.} \end{cases} \tag{2}$$

With the coefficients, there are zeros locating at 60 Hz and 120 Hz when the sampling frequency is 180 Hz. The zeros are very useful to notch the dedicated periodic noise such as well-known 50/60 Hz noise with a simple FIR structure.

DC Level Tracking

Obviously, the measured data is composed of the static value of capacitance, the dynamic value of capacitance and noise. A precise DC level (static value of capacitance) is required to ensure the sensitivity. In the previous work, the noise component is decayed by the cascaded LPF but is not clean enough because the consideration toward response time. We use the IIR DC tracker here again to remove the residual noise. It also comes to the tradeoff between response time and attenuation again, but quite easy to make a decision because the accuracy of DC level is more important than the rate of its adjustment. Feedback gain of 0.03125 is picked for this part.

In the next step, we check variation of the filtered result. Since touch event will raise the DC level, the sensitivity will be affected if the filtered result is used directly. The post phase of DC Level Tracker picks a sample from the filtered result as a candidate value for new DC level and then check whether the following 10 samples is within ±1 compared to the candidate value or not. If yes, the new DC level will be set and restart the procedure, else restart the procedure only.

Hysteresis Trigger Control

Either touch event or noise fluctuates the input data, a single threshold cannot fulfill the trigger control. With the high threshold for touched state and low threshold for untouched state, the touch flag is set to 1 if input data is greater than the high threshold and clear to 0 while lower than the low threshold. And then the trigger flag is set to 1 when touch flag goes from 0 to 1 and clear to 0 after a run. This mechanism for single pad case is quite stable. For two dimension trigger, the first thing is to check which row and column is triggered. If only one row or column is triggered, then touch flag take the place of the trigger flag of another column or row. In our design, the 4x4 matrix sensing pad is allowed to perform sliding control by the method to turn on four lights in a flash.

3.2   Scheduling

With the tick concept from OS, the sampling time for multiple measurements can be controlled. Also, the displaying of LED matrix can be inserted into the schedule and share the same I/O that is used for sensing the capacitance without affect the measurement. Fig. 2 shows the schedule of the touch panel that the proposed algorithm is run on it.

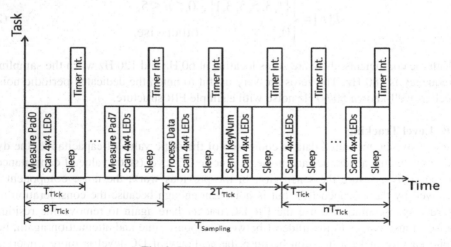

Fig. 2. Schedule of the touch panel

3.3   Resource Arrangement

Due to the resource of a low cost microcontroller is low as well, it is important to arrange the available resources efficiently to make the design possible. The proposed algorithm is allowed to share I/Os with other purposes and its low memory footprint reduce the resource consumption. With the rich register file of ATTiny2313V, the register overlapping between background and ISR can be reduced and thus reduce the stack requirement needs by context switching. As a result, the mostly remained SRAM space could be used to store data without stack overflow problem. The following three tables show the resources used by the proposed algorithm.

Table 1. Summary of resource usage

Item	Available in ATTiny2313V	Usage
I/O	18	18
SRAM (Data)	128 bytes	86 bytes
FLASH (Code)	2048 bytes	918 bytes

Table 2. I/O usage of the touch panel

Pin No.	Function
1	Reset/Debug
4,5	Oscillator
2,3	UART
6	LED Action Indicator
7,8,9,11	LED Column Sources
12,13,14,15	Column Sensing Pads/LED Row Sinkers Select
16,17,18,19	Row Sensing Pads

Table 3. SRAM usage of the proposed algorithm for a 4x4 matrix sensing

Item	Format	Size (byte)
Filtered result of IIR low-pass filter	8x2 bytes	16
Filtered result of FIR low-pass notch filter	8x2 bytes	16
Filtered result of IIR DC tracker	8x2 bytes	16
DC level	8x2 bytes	16
Candidate for DC level	8x2 bytes	16
Stability counter for the candidates	8x4 bits	4
Touch flag	8x1 bit	1
Trigger flag	8x1 bit	1
Total		86

4   Design of the Filters

There are quite a lot of constraints in a low cost microcontroller when implementing a digital filter in it. Typically, there is no multiplier and the memory space for filter to store data is not so rich. While encounter this adverse circumstance, the choice of filter is not so much and the performance of a filter may be degraded. Besides, the undesirable characteristic of the filter is confronted by using the substituted one.

4.1   IIR DC Tracker

Step Response
In addition to frequency response, step response should be taken into consideration as well. In [4], the good and bad waveforms are listed. The 1st-order IIR DC Tracker in [1] has long rise time with a small feedback gain.   As Fig. 3 shown, the attenuation is good but it takes long time to reach 90% of the final value when a step signal is inputted. To make a clear response to a fast and the slight touch event, the step response cannot be ignored. In our design, we decide to pay more attention to step response and compensate the attenuation by cascade a 6th-order FIR LPF. The magnitude response and step response of IIR DC tracker in our design are as Fig. 4 depicted, which makes fast response to a step liked signal such as a touch event.

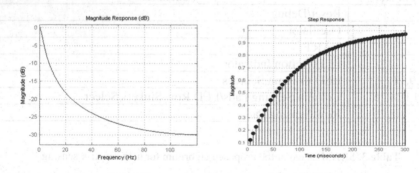

Fig. 3. Responses of Eq. (1) with $k = 1/16$

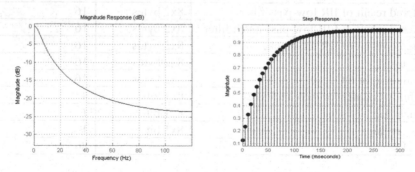

Fig. 4. Responses of Eq. (1) with $k = 1/8$

Tracking Error

Use shift operator as a makeshift for the multiplier is a very common technique. However, tracking error that cannot be neglected is occurring because the shifting operation cuts off the least significant part of the difference between tracking result and target to be tracked. Fig. 5 shows the DC level tracking result processed by the normal method that without considerations for numerical problems after power-on. It is clear to see the tracking error stuck at -30 for a while. But for the touch event occurred at about the 950th sample served as a disturbance, the tracking error is hard to converge due to the small gain that equivalent to do right shift five times would remove the difference lower than 32.

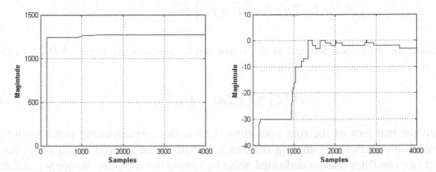

Fig. 5. DC level tracking and its error without care for numerical issue

To deal with numerical problem, the data path is reviewed and we decided to move the post-scaling function from cascade LPF block to the front of the Hysteresis Trigger Control block. In cascade LPF block, the integral coefficients cause the network of the LPF has a maximum magnitude of 18 at DC. The purpose of post-scaling function is to keep the DC gain close to 0 dB by shifting the filtered result four times to the right side. If the operation can wait until the DC level is computed, the least significant part of the difference can be feedback to DC level and thus reduce the error. With the method is applied, the result is demonstrated by Fig 6.

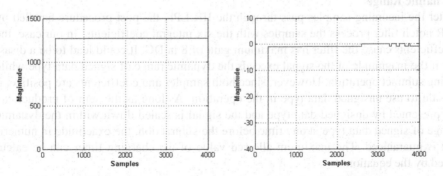

Fig. 6. DC level tracking and its error with care for numerical issue

4.2   FIR Low-Pass Notch Filter

Multiplierless

To filter out the dedicated periodic noise by using an 8-bit, multiplierless microcontroller, the most possible way is to put the zeros of the filter at the frequency belong to the noise. For a 2nd order FIR LPF that has two zeros located at left half of unity circle on z-plane and therefore we have two roots $z = -a \pm jb$, the network transfer function can be represented as follow:

$$H(z) = \left[1-(-a-jb)z^{-1}\right]\left[1-(-a+jb)z^{-1}\right]$$
$$= 1 + 2az^{-1} + \left(a^2 + b^2\right)z^{-2} \tag{3}$$

Because the zeros are located at the unity circle. By substituting $a^2 + b^2 = 1$, the equation can be rewritten as follows:

$$H(z) = 1 + 2az^{-1} + z^{-2} \tag{4}$$

When the real part of the root goes from 0 to 1, the corresponding notch center is moved from quarter of sampling frequency to half of the sampling frequency. As a result, we can filter out the dedicated noise by tuning the sampling frequency and the real part of the root. The following equation describes the operation in detail:

$$f_{center} = \frac{\cos^{-1}(-a)}{2\pi} \cdot f_{sampling} \tag{5}$$

With help from Eq. (4) and Eq. (5), it is easy to notch a dedicated noise within limited times of shift-add/subtract procedure. By cascading or taking convolution of multiple coefficient sets, the filter can notch more than single frequency or pay more attention to the same frequency.

Dynamic Range

After the incoming samples pass through the IIR LPF, the post procedure handled by FIR notch filter process the samples with the six integral coefficients. In our case, the coefficients cause the filter has maximum gain of 8 in DC. It could lead to be a disaster if the magnitude of the signal exceeds the dynamic range of signed data type while being subtract operation. However, since both samples and coefficients are positive, it is safe to use unsigned data type in the operation. As long as the sum of product can be presented by unsigned data type and the signal is scaled down within the dynamic range of signed data type at any time before the subtraction, the exactitude in numerical is guaranteed. The maximum allowed value of the charging timer can be calculated by the equation:

$$Timer_{max} = \frac{2^{bit\ depth}}{\sum_{i=0}^{N-1} a_i} \quad (4)$$

Where $N-1$ is the order of FIR filter and a_i is i-th coefficient of FIR filter. According to the maximum allowed value, the clock prescaler can be selected to acquire preferred sensibility and the compare match register can be set to ensure that time count is within the maximum allowed value.

5   Experiment

In the experiment, the original raw data is sampled by the touch panel and send back to the host PC via room controller for demonstration.   Fig. 7 shows the raw data and its filtered result processed by the proposed cascaded filter. In the scene, four quick and slight touches happened after the panel starts sampling for five second, then the finger holds on the panel for a long time. The noise induced by finger maybe not so clear when just a quick touch but very obvious if the finger stay onto the panel. After filtering, the noise is removed and the quick and slight touches haven't been eliminated because we take the step response of our filter seriously.

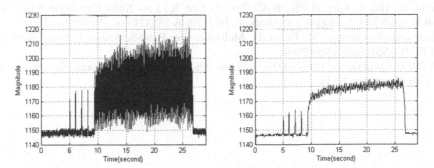

Fig. 7. Original raw data (left) and its filtered result (right)

To quantify the effect of the filter, the following equations from [4] are used to calculate the SNR for a capacitive sensor:

$$SNR(dB) = 20\log\left(\frac{TouchStrength}{NoiseTouched_{RMS100}}\right) \quad (6)$$

$$TouchStrength = SignalTouched_{AVG100} - SignalUntouched_{AVG100} \quad (7)$$

$$NoiseTouched_{RMS100} = \sqrt{\frac{\sum_{n=0}^{99}\left(SignalTouched[n]-SignalTouched_{AVG100}\right)^2}{100}} \qquad (8)$$

For the data samples represented in Fig. 7, the SNR is improved from 5.8075 dB to 26.2131 dB.

6   Conclusion

The proposed algorithm can filter out the dedicated heavy noise while keeps the weak signal caused by quick and slight touch event just by shift-add/subtract operation. For the cost sensitive project that using low cost microcontroller without multiplier and the resource is limited, the proposed touch algorithm would be a help. Moreover, the LED scanning task can be scheduled into the software procedure and share I/O pins with sensing pads without any effect.

References

1. Texas Instruments Corporation, MSP430 Capacitive Single-Touch Sensor Design Guide (January 2008)
2. Shin, H., Lee, J., Jang, H., So, B.-C., Yun, I., Lee, K.: Low Noise Capacitive Sensor for Multi-touch Mobile handset's applications. In: IEEE A-SSCC Conf. (November 2010)
3. Santraine, A., Leprince, S., Taylor, F.: Multiplier-Free Band-Selectable Digital Filters. In: IEEE ICASSP Conf. (May 2004)
4. ATMEL Corporation, QTAN0079: Buttons, Sliders and Wheels Touch Sensor Design Guide (November 2011)

Design of a Dynamic Parallel Execution Architecture for Multi-core Systems

Shiang Huang, Jer-Min Jou, Cheng-Hung Hsieh, and Ding-Yuan Lin

VLSI/CAD Group
Department of Electrical Engineering National Cheng Kung University Tainan, Taiwan
jjmjjmjjm3@gmail.com

Abstract. In this paper, a new dynamic Parallel Execution Architecture, DEAL, which can execute the dynamic parallel threads and handle the dynamically incorrect data access caused by parallel execution threads efficiently, is proposed. DEAL combines the concept of multi-thread speculation with the transactional memory into a new model and then threads can be efficiently executed in parallel in it. DEAL can detect incorrect parallel data access immediately and resolve them to keep data consistent among threads and ensure the threads do not violate the data dependences during parallel execution dynamically. Based on experimental results, we find that the performance of parallel applications running in DEAL can be significantly faster, 1.4 speed-ups at least, than those running in sequential, which demonstrated that the DEAL can execute parallel threads and manage data and resolve incorrect data access among them efficiently.

Keywords: parallel execution, complexity, multi-core system.

1 Introduction

Current proposals rely on changing traditional cache architecture to handle version management or conflict detection to support and resolve these two challenges. Traditional approaches result in either unavoidable design complexity, or high performance costs. This study proposes dynamic execution architecture with log (DEAL), a useful solution which follows a new design approach that fully decouples [16] transaction processing from the traditional cache system in order to resolve these two challenges. Using additional hardware supports version management and conflict detection.

The most important functions of data management in the dynamic parallel execution architecture are version management and conflict detection, which are divided into two categories: eager and lazy.

TCC [5] uses both lazy version management and lazy conflict detection, and Eazy HTM [13] uses lazy version management and eager conflict detection. In the proposed design, DEAL uses eager version management and eager conflict detection. A log base and log pointer are used to store old values, as in LogTM, while additional units, such as the read/write signature, are used to record the data's state. Conflict detection must be performed before each data access since the data can only be accessed if there is no conflict.

J.-S. Pan et al. (Eds.): *Advances in Intelligent Systems & Applications*, SIST 21, pp. 731–740.
DOI: 10.1007/978-3-642-35473-1_72   © Springer-Verlag Berlin Heidelberg 2013

Ideally, DEAL should use eager version management and eager conflict detection, because:

- Eager version management puts new values "in place," making commits faster than aborts. This makes sense when commits are much more common than aborts, which is what we generally find.
- Eager conflict detection finds conflicts early, thereby reducing wasted work by conflicting transactions. This is less complex than lazy detection, since standard coherence makes implementing eager conflict detection more efficient.

On the other hand, Speculation also has many features similar to transactional memory. Thread-Level Speculation (TLS) allows the compiler to automatically parallelize portions of code in the presence of statically ambiguous data dependences, thus extracting parallelism between whatever dynamic dependences actually exist at run-time. Therefore, in speculative execution, each data access needs to check other speculative execution threads, using the order to decide on the resolution. Based on these characteristics, the new design of the system includes some additional hardware to support transactional memory and speculation.

The remainder of this paper is as follows. Section 2 details the decoupled parallel execution architecture design approach. Section 3 gives the mechanism of version management and conflict detection. DEAL's evaluation results are presented in Section 4. The concluding remarks are given in Section 5.

2   DEAL : Dynamic Parallel Execution Architecture with Log

The DEAL design decouples transaction processing from the traditional cache system. It does not rely on the data cache system, which is contrary to the contemporary HTM design approach, which usually requires tagging transactional states on cache blocks, or needs to bring back conflict detection information onto cache coherence protocol messages. The traditional architecture is not altered, but additional hardware for conflict detection and version management is included. Because lazy version management uses a buffer to store new data, the buffer will overflow when the transaction is too large. Eager version management and eager conflict detection are therefore adopted, referring to LogTM and including additional hardware to support conflict detection and version management in the design.

2.1   The Design of the Function Unit in DEAL

In the proposed design, the architecture includes: processors, a run-time dynamic data manager (RDM), a shared L2 cache and an L2 transaction signature cache (L2 TSC). Included are not only an original memory management unit (MMU) and L1 cache within each processor, but also an additional L1 transaction signature cache (L1 TSC). Also, control processor for task management and assignment is added.

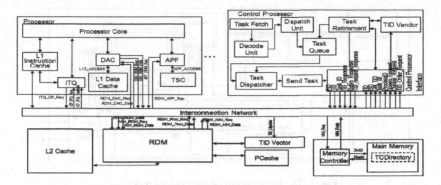

Fig. 1. The design of the architecture's overview

2.2   The Design of the Function Unit in the Control Processor

In order to handle task management, TID allocation and the order of the speculative threads a control processor (CP) is designed. The control processor consists of the TID Vendor, TID Order Table, FID Table, Task Queue and Processors Map, as shown in Figure 2. The control processor counts the workload for every processor, and records the order of the speculative thread and the TID current usage and status. The control processor differs from other processers in that it does not need to execute; it just performs management jobs. The control processor (master) assigns tasks to other processors (slave), and the slave processors execute the task and send a message back to the control processor when the task is complete.

The following is the design of the control processor:

1. Task Fetch reads the TaskID into the control processor.
2. The Decode unit decodes the TaskID.
3. The Dispatch Unit sends the TaskID into the Task Queue, and waits to assign.
4. The Task Queue manages all the tasks in this system, assuming each processor can contain eight tasks, and there are four processors. Eight tasks are pre-fetched, so the Task Queue has forty entries to store TaskIDs.
5. The Task Dispatcher selects the executable ProcessorID (PID) and assigns the waiting task in the Task Queue.
6. The Processor Map records the current workload of every processor.
7. The Send Task sends the TaskID to the selected processor.
8. The TID manager records the usage and states of every TID.
9. The FID Table is used in speculative mode to record the FamilyID (FID); the program has a different TID, but the same FID in speculative mode.
10. The TID Order Table is used in speculative mode; it uses the "linked-list" method to record the complete program order in speculative mode.
11. Task Retirement handles the messages from the processors. The processors will send messages to the control processor when the task is complete, and then the Task Retirement checks the task. If the task can be committed, Task Retirement will send messages to clean up the Task Queue and TID Manager.

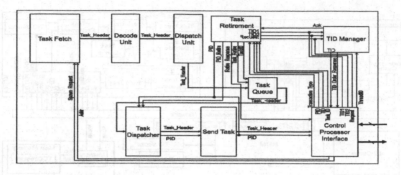

Fig. 2. The input/output of the control processor

2.3   The Design of Other New Units

Additional hardware is included in order to support transactional memory and specu-lation, including: a Control Processor, L1 TSC, L2 TSC, RDM, TID Vector and TCDirectory. The following explains the design concept and capability.

This paper refers to LogTM for version management; Log Base, Log Pointer and Log Filter [2] are joined in each processor. When the thread begins in transactional mode or speculative mode, it stores the new values "in place," and stores the old value in the log. Before new data replaces old data in the transaction, the old data must be stored in the log. However, some addresses are accessed often, and will be stored in the log many times. When the transaction is aborted and rolled back, the old data will be restored. A log filter is used to reduce the overhead on storing and restoring the data at the same address. When a thread stores to a block not found in its log filter, DEAL logs the block and adds its address to the log filter. Stores to addresses in the log filter are not logged again; this not only reduces the backup cost, but also accele-rates the rollback time.

To increase transactional check speed, a small cache, named L1 transaction signa-ture cache (L1 TSC), is used in each core to buffer the transactional states for recently accessed blocks. L1 TSC's space requirement is small because it does not contain the requested block's data. L1 TSC buffers transaction coherence information for recently accessed blocks, through which subsequent transactional check requests can be quick-ly acknowledged if the request matches the cached state (without sending the transac-tional check request out of the chip). Only when the request misses L1 TSC or it does not match the cached state will this request be sent to the run-time dynamic data man-ager (RDM) where it can eventually be arbitrated. L2 TSC holds recently accessed TCDirectory entries for transactional check requests. The structure of the TCDirecto-ry entry buffered in L2 TSC is slightly different from its in-memory structure. When loaded from the main memory to L2 TSC, the TCDirectory entry is extended with two entries for every sharer transaction: a P flag to indicate whether or not this sharer information is currently buffered in some processor's L1 TSC, whose processor id is indicated by the PID entry. When the check request is missed in L1 TSC, this request will be sent to the RDM and L2 TSC will be checked. If L2 TSC misses again, the RDM will obtain the information from the TCDirectory for conflict checking.

A TID Vector is used to accelerate conflict detection. The TID Vector records the current TID usage, and when a conflict is detected, the RDM will check the TID usage. The TID Vector is also responsible for answering TID query requests sent by the RDM to test whether or not a certain TID is currently active. When the thread begins in transactional or speculative mode, it will send a request to the control processor for a TID. When the TID Manager assigns a TID back to the processor, it will also send the TID to the TID Vector. The TID Vector is like the TID Manager's cache, and the RDM can check the TID usage faster by sending a query to the TID Vector.

In this design, each processor has its own In-Node Task Queue (ITQ). When a processor receives a task from the control processor, it puts the task into the ITQ and waits for execution. Furthermore, when the task is waiting for execution in the ITQ, it can pre-fetch some instructions into the ITQ, improving system performance.

2.4 The Design of the Execution Mode in DEAL

The architecture is designed so that it can be divided into three modes when executed: Non-Transactional Mode, Transactional Mode and Speculative Transaction Mode.

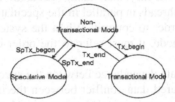

Fig. 3. The mode switching graph

The thread generally executes in the non-transactional mode. When the thread begins executing transaction or speculation, it will change the mode. When the transaction or speculation is finished, the thread will return to non-transactional mode.

(1) In non-transactional mode, it is a general sequential execution. The control processor fetches the TID, and assigns it to a processor for execution. The processor will send a commit request to the control processor for commit when the task is complete. The Task Retirement will commit and retire directly in non-transactional mode.
(2) In the transactional mode, a transaction will execute in parallel with other transactions. Additional hardware is used to detect the conflict between the threads. The processor will send a request to the control processor's TID Manager in order to get a TID when the thread begins the transactional mode. The processor will send a commit request to the control processor for commit when the task is finished. The Task Retirement will commit and retire directly in the transactional mode.
(3) In the speculative mode, the program will spawn many threads for speculative execution. The data dependence between the speculative threads must be detected correctly; additional hardware is therefore used to detect conflicts.

The speculative mode is similar to the transactional mode. The processor will send a request to the control processor's TID Manager in order to get a TID when the thread begins in speculative mode. In addition, The TID Manager will send the TID to the TID Order Table in order to record the program order. When there is a conflict between the speculative threads, the processor must send a query to the TID Order Table in order to get the order between the speculative threads. The system uses this order to decide how to resolve the conflict (abort or stall). Moreover, the processor will send a commit request to the control processor for commit when the task is finished. Task Retirement must send the TID to the TID Order Table to check whether or not the TID is the first in the list, because speculation must be completed in order. Therefore, only the first TID in the list can be retired.

2.5   Programming Model

The following is the design of the programming model in DEAL. When the parallel program accesses the shared data, it will begin the transactional mode in order to execute. In the transactional mode it will check all of the data access for conflicts. There is no order between the transactions in transactional mode; therefore, the transactions can commit out of order. On the other hand, speculation divides the program and speculatively executes the threads in parallel in the speculative mode. So these threads must follow the original order to commit. When the system detects the data dependence in the speculative mode, it must refer to the order of the program in order to preserve the data accuracy.

In this design, there is data dependence between the speculative threads in speculative mode, and there is shared data conflict between the transactions in the transactional mode. Because the speculative and transactional modes have different data space, there is no problem with data dependence and shared data conflict.

3   Version Management and Conflict Detection

The proposed design implements version management as in LogTM. The new values are stored "in place," and the old values are stored in a log using log base and log pointer. In the following, the TID Vector, TCDirectory, L1 TSC and L2 TSC entries, as shown in Table 1 are described. The TCDirectory tracks a sharer transaction's access information for a specific granularity of the memory block. The TCDirectory includes tag, state, v and TID. Tag is the data address. State is the data state, of which there are three types: state R means that there is one or several transactions to read this data, state W means that there is one transaction to write this data, state U means that there is no transaction to access this data. L2 TSC's structure is almost the same as that of the TCDirectory. L2 TSC joins P and PID, and the P and PID field is used to temporarily bind a transactional state with the processor whose L1 TSC has buffered this block's access permission. L2 TSC holds recently accessed TCDirectory entries for all of the processors' transactional check requests. Finally, L1 TSC is the cache of L2 TSC, which holds recently accessed L2 TSC.

Table 1. The TID Vector, TCDirectory, L1 TSC and L2 TSC entries

Tag	State	V	TID	V	TID	......	V	TID

(R/W/U | Sharer Transaction List)

TCDirectory entry structure : to track sharer transaction's access information for a specific granularity of the memory block.

Tag	State	TID

L1 TSC (L1 Transaction Signature Cache) block structure : to buffer the transactional states for recently accessed blocks in each core.

Tag	State	V	TID	P	PID	......	V	TID	P	PID

(R/W/U | Sharer Transaction List)

L2 TSC (L1 Transaction Signature Cache) entry structure : to hold recently accessed TCDirectory entries for transactional check requests.

V	TID

TID Vector : to record the current usage of TID usage, when a conflict is detected that RDM will check the TID usage.

Conflict requests are made according to the address and state for conflict detection, as in the following cases. First, when there is a reading request at address A in transactional or speculative mode, and when address A's state is R or W in L1 TSC, the check request is given permission. Second, when there is a writing request at address A in the transactional or speculative mode, and when address A's state is W in L1 TSC, the check request is given permission. Third, when there is a reading request at address A in the non-transactional mode, and when address A's state is R or U in L1 TSC, the check request is given permission. Fourth, when there is writing request at address A in the non-transactional mode, and when address A's state is U in L1 TSC, the check request is given permission.

In order to ensure that all conflicts can be detected, L1 TSC cannot modify the state itself when it misses. The processor will instead send a request to the RDM, and the RDM uses L2 TSC or the TCDirectory's state to detect the conflict. For this reason, L1 TSC is responsible for storing recently accessed data states to provide rapid conflict detection.

The following are two examples of conflict resolution, as shown in Figure 4. The mechanism "write has higher priority than read, first write first win" Is used. In Figure 4(a), the thread PID0, TID2 sends a write request to L2 TSC, and finding the address has been read by PID0, TID3, the RDM sends a query to the TID Vector to check whether or not the TID is active, and the RDM gets the information that TID3 is active. Because of the mechanism "write has higher priority," processor0 will send a message to processor1 to roll back and restart TID3.

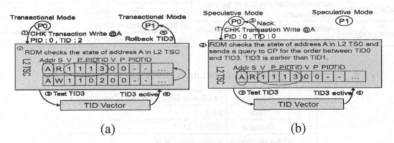

Fig. 4. Conflict resolution

In the speculative mode, as shown in Figure 4(b), version management and conflict detection are the same as in the transactional mode, but different in regard to conflict detection. The control processor will record the order between the speculative threads in the speculative mode. Every time a conflict is detected, the RDM must send a query to the TID Order Table to check the order between the speculative threads. The RDM then decides on the conflict resolution according to the thread order. In Figure 4(b), the thread PID0, TID0 sends a writing request to L2 TSC, and finds that the address has been read by PID1, TID3. The RDM sends a query to the TID Vector to check whether or not the TID is active, and the RDM gets the information that TID3 is active. Because TID0 and TID3 are working in speculative mode, the RDM must send a query to the TID Order Table to check the order between TID0 and TID3, and the RDM gets the information that TID3 is in front of TID0. Therefore, TID0 is writing the data too early, and it must be rejected.

4   Evaluation

In order to simulate the structure of the proposed design, it was necessary to modify QEMU's source code. The code added is used to simulate the behavior, version management and conflict detection in the proposed architecture. Included in the additional code are the unit that we are joining, check point, log base, log pointer, log filter, TID manager and some additional instructions. The instructions and functions added in QEMU for DEAL are shown in Table 2. Table 3 shows the system model parameters used.

Table 2. New instructions and functions

Instruction	Action
TM_system_startup	The system starts transaction execution, and shows current system information:CPU number, Log size, TID size, and to initiates the variable:Log, LogPointer, TID, TIDOrder, TIDVector, and Abort counter.
TM_system_shutdown	When the system finishes transaction execution in parallel, it shows the system information:usage of Log, TID usage, and Abort count, and deletes the system information:Log, LogPointer, TID, TIDOrder, TIDVector, and Abort counter.
Transaction_begin	Begin transactional mode and get the TID and PID. Back up the CPSR before execution.
Transaction_end	Finish transactional mode and release the TID. Clean up the Log and LogPointer, etc.
Speculative_transaction_begin	Begin speculative mode and get the TID and PID. Back up the CPSR before execution.
Speculative_transaction_commit	Finish speculative mode and release the TID. Clean up the Log and LogPointer, etc.

Instruction	Action
Transaction_load	The load instruction in transactional & speculative mode, it will check the conflicts before load.
Transaction_store	The store instruction in transactional & speculative mode, it will check the conflicts before store.
tm_trd_chk	The system must detect conflicts before loading the data in transactional mode.
tm_twt_chk	The system must detect conflicts before storing the data in transactional mode.
conflict_detection	Check for whether there are conflicts or not.
add_log	Back up old data into Log.
add_TCD_sharer	Modify the TCDirectory.
abort_sharers	Abort the transaction.

Table 3. Simulation environment

HOST	Processor	Intel Core-i5 760 2.8GHz Quad-Core Processor
	RAM	2 GB
	Simulation	QEMU 1.0.1
	Operating System	Ubuntu 10.04 - 64 bit
	Linux Kernel	2.6.32
Guest	System	Versatile Express Development Board
	Processor	ARM Cotex-a9 Quad-Core Processor
	RAM	1 GB
	Operating System	Linux 2.6.38 Embedded
	Tool Chain	gcc-4.5.2 for ARM

Once the instructions are defined and the codes are added, the simulation and the evaluation of the architecture can begin. Ten examples are run, including: Linked-list, Histogram - no conflict, Histogram - low conflict, Histogram - high conflict, Labyrinth - 32*32*3*64, Labyrinth - 64*64*3*64, Intruder - a10 l4 n2048, Intruder - a10 l8 n4096, Ssca2 - s13 i1.0 u1.0 l3 p3 and Ssca2 - s13 i1.0 u1.0 l9 p9. These programs are the examples in STAMP [10, 11].

Execution mode can be divided into single thread (single core), two threads (dual cores) and four threads (quad cores). The following are the execution time and the performance improvement. Figure 5 shows the improved performance speed between the programs. It is clear that most programs will show improved performance using the proposed architecture, since there are few conflicts in these programs. Although aborts take time to execute in parallel, the benefits of parallelism still improve the overall performance. To demonstrate this, a program with many conflicts in the histogram is written; if it has high conflict, it must require more time to execute in parallel for abort. In speculation, there are three test patterns: for loop speculative unrolling, RGB to YUV and histogram. A for loop is unrolled, and these threads are speculatively executed, divided into no dependency, low dependency and high dependency. The dependency is the data dependence between every loop. The RGB to YUV program reads a picture and translates the pixels. The picture is divided into four parts for execution in parallel. Finally, the histogram is also divided into four parts for execution in parallel.

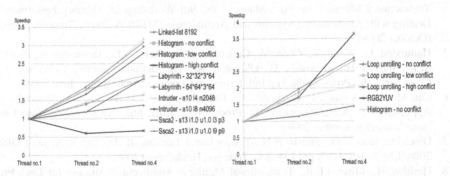

Fig. 5. Improved speed of the programs, and the improved speed of the speculative programs

5   Conclusion

We presented a Dynamic Parallel Execution Architecture, DEAL, for Multi-Core Systems, which executes threads in parallel efficiently and dynamically detects and resolves parallel execution conflicts using the combined concepts of transaction memory and speculation. With the decoupled and distributed design, the DEAL is able to have better scalability and modularity when scaling system architecture and maintaining performance. Furthermore, we have integrated the concept of speculation with the transactional memory into a new model, speculative transactional memory. The DEAL can provide the execution of speculative transactions and normal transactions in parallel. Also, the RDM and TID vector units in DEAL can alternate the detection mechanism according the execution mode. In the speculative mode, they turn on the violation detection to detect the data dependence violations between speculative transactions. In the normal mode, they perform the conflict detection to detect the conflicts between transactions instead of violations. DEAL is designed and simulated on the QEMU emulator, and several parallel applications are used to accomplish the design evaluation. The experimental results show that our DEAL improves parallel performance in 1.4x speedups at least.

References

1. Moore, K.E., Bobba, J., Moravan, M.J., Hill, M.D., Wood, D.A.: LogTM: Log-based Transactional Memory. In: The proceedings of the 12th Annual International Symposium on High Performance Computer Architecture (HPCA 12), Austin, TX, February 11-15 (2006)
2. Yen, L., Bobba, J., Marty, M.R., Moore, K.E., Volos, H., Hill, M.D., Swift, M.M., Wood, D.A.: LogTM-SE: Decoupling Hardware Transactional Memory from Caches. In: The Proceedings of the 13th Annual International Symposium on High Performance Computer Architecture (HPCA 13), Phoenix, AZ, February 10-14 (2007)
3. Bobba, J.: Hardware Support For Efficient Transactional and Supervised Memory Systems. Ph.D. Thesis, The University of Wisconsin - Madison (2010)
4. Lupon, M., Magklis, G., González, A.: Version Management Alternatives for Hardware Transactional Memory. In: Proceedings of the 9th Workshop on Memory Performance: Dealing with Applications, Systems and Architecture (MEDEA 2008), Toronto, Canada (October 2008)
5. Hammond, L., Wong, V., Chen, M., Carlstrom, B.D., Davis, J.D., Hertzberg, B., Prabhu, M.K., Wijaya, H., Kozyrakis, C., Olukotun, K.: Transactional Memory Coherence and Consistency. In: Procs. of the 31st Intl. Symp. on Computer Architecture (June 2004)
6. Cao Minh, C., Chung, J., Kozyrakis, C., Olukotun, K.: STAMP: Stanford Transactional Applications for Multi-Processing. In: Procs. of the IEEE Intl. Symp. on Workload Characterization (September 2008)
7. Dice, D., Shalev, O., Shavit, N.N.: Transactional Locking II. In: Dolev, S. (ed.) DISC 2006. LNCS, vol. 4167, pp. 194–208. Springer, Heidelberg (2006)
8. Herlihy, M., Moss, J.E.B.: Transactional Memory: Architectural Support for Lock-Free Data Structures. In: Procs. of the 20th Intl. Symp. on Computer Architecture (May 1993)
9. Bobba, J., Moore, K.E., Yen, L., Volos, H., Hill, M.D., Swift, M.M., Wood, D.A.: Performance Pathologies in Hardware Transactional Memory. In: Procs. of the 34th Intl. Symp. on Computer Architecture (June 2007)
10. Sohi, G.S., Roth, A.: Speculative multithreaded processors. Computer 34(4), 66–73 (2001)
11. Shaogang, W., Weixia, X., et al.: DTM: Decoupled Hardware Transactional Memory to Support Unbounded Transaction and Operating System. In: International Conference on Parallel Processing, ICPP 2009 (2009)
12. Bellard, F.: QEMU, a Fast and Portable Dynamic Translator. In: Proceeding of USENIX Annual Technical Conference, pp. 41–46 (2005)
13. Tomic, S., Perfumo, C., et al.: EazyHTM: EAger-LaZY hardware Transactional Memory. In: IEEE/ACM International Symposium on Microarchitecture, MICRO 42 (2009)

A Distributed Run-Time Dynamic Data Manager for Multi-core System Parallel Execution

Wen-Hsien Chang, Jer-Min Jou, Cheng-Hung Hsieh, and Ding-Yuan Lin

Department of Electrical Engineering
National Cheng Kung University, Tainan, Taiwan, R.O.C
jjmjjmjjm3@gmail.com

Abstract. In this paper, we propose a new Distributed Run-Time Dynamic Data Manager (DRDM) to manage the dynamic data among parallel threads and to handle the dynamically incorrect data access caused by parallel execution threads efficiently. Also, we combine the concept of multi-thread speculation with the transactional memory into a new model and all the dynamic data can be managed by the DRDM. The DRDM can detect incorrect data access immediately and resolve them to keep data consistent among threads and ensure the threads do not violate the data dependences during execution dynamically. We have demonstrated that the performance of parallel applications running with the DRDM can be at least 1.4 times faster than those running in sequential and thus the DRDM can manage data and resolve incorrect data access among threads efficiently.

Keywords: Multi-Core System, Dynamically Parallel Execution, Thread-level Speculation, Transactional Memory.

1 Introduction

There are many parallel execution models of multi-core systems used to improve the performance of programs. A popular one of them is the transactional memory (TM) [1] which replaces the parallel parts in conventional parallel program with transactions in which they may compete for shared data. The speculative parallel execution [2, 3] is another approach to parallelize programs, in which a sequential program is parallelized into speculative threads and they are executed in parallel despite uncertainty as to whether those threads are actually independent. To achieve multi-core parallelism more flexibility and efficiency, we should integrate them here instead of separate designs.

Moreover, some of the implementations for those parallel execution models usually spend a lot of storage structures for storing the information of the dynamic data usage among parallel threads [4, 5]. Some of them usually require modifying the original cache system, such as the L1 cache in [6]. As a result, they all suffer from low scalability and high complexity when scaling more processors due to the increasing hardware structures, and more complicated data dependencies and conflicts between threads. Moreover, they are not able to provide run-time support for both speculative parallel execution and transactional memory so far.

J.-S. Pan et al. (Eds.): *Advances in Intelligent Systems & Applications*, SIST 21, pp. 741–750.
DOI: 10.1007/978-3-642-35473-1_73   © Springer-Verlag Berlin Heidelberg 2013

To address these issues, we not only present a new model, speculative transactional memory, which is a combination of speculation and transactional memory, but also design a Distributed Run-Time Dynamic Data Manager (DRDM) to manage data among parallel threads executing in this new model efficiently and to detect and resolve incorrect data access which is generated dynamically due to competing for data shared and violating data dependences. The speculative transactional memory has two modes, the speculative mode (SP mode) and the normal mode. In the SP mode, the sequential part of a program is partitioned into several threads and the code section for computation of each thread is wrapped in some speculative transactions. Although these speculative transactions may have data dependencies, they still run in parallel but must commit in order. In the normal mode, a program is parallelized into non-speculative transactions and they runs in parallel.

The remainder of this paper is organized as follows. Section 2 describes the design concept of the DRDM, the data consistence protocol, data conflict detection mechanism, data version management and the execution model of the speculative transactional memory. Section 3 describes the design of the DRDM. Section 4 then evaluates the performance, and concludes in Section 5.

2   Design Concept of Distributed Run-Time Dynamic Data Manager

The scalability and efficiency are the first consideration when designing the Distributed Run-Time Dynamic Data Manager (DRDM) including a global part and several local parts. To avoid hardware overhead and maintain scalability when scaling processors, the storage structures for storing the information of the data usage among parallel threads are extracted from the processors and centralized in the particular storage structure. Also, we design a global part of the DRDM shared for all processors which manages the particular storage structure. The global part of the DRDM tracks all the common data among parallel threads and stores this data consistence information to the particular storage structure. The data consistence information includes the sharer threads of the data and also indicates the data is shared by one or more sharer threads, or occupied by the owner thread.

Although the basic concept of parallel execution in the speculative parallel and the transactional memory are different, in fact the mechanism for managing data among parallel threads is similar. Therefore, it is possible to implement the data management and data violation and conflict detection of them all in the DRDM and to combine them together. The new model, called the speculative transactional memory, is introduced in this paper. The proposed speculative transactional memory provides two modes, the normal mode and the speculative (SP) mode. A program running based on the non-speculative transactional memory model is called in the normal mode. In this mode, the parallel threads with transactions are called the transaction threads. When executing transaction threads dynamically in parallel, they may cause data inconsistent due to competing shared data concurrently, as known as the data conflicts. The global part of the DRDM must not only manage all the data competed by the transaction threads but also detect the data conflicts among them. If the transaction threads completes without any data conflicts, they perform commit to make the new values of the data modified by them visible to others. In the SP mode, a program is fully parallelized into a couple of speculative transaction threads (SP-TX thread). Because the SP-TX threads are ordered in the original program, they probably violate the data dependence due to the early read and write by a SP-TX thread with later order during parallel execution. In this mode, the global part of the DRDM performs the data dependence violation detection to detect the violations among SP-TX threads instead of the data conflict detection. Also, the SP-TX threads must commit in order to ensure the SP-TX threads with later order can use the proper data and thus keep all the data consistent. All the data in both the two modes are managed properly and correctly by using a dedicated data consistence protocol and this protocol is implemented in the global part of the DRDM.

When a transaction thread wants to access data, the processor executing it will send a data access check request to the global part of the DRDM first. Then the global part of the DRDM use the data consistence information to detect there are any data conflicts or any data dependence violations caused by the requestor or not. If not, the transaction thread is able to get the access permission of data and then begins to read/write data actually. Sometimes, other transaction threads will be aborted since a certain transaction thread wins the write permission or a certain SP-TX thread with higher order reads/writes the data. Moreover, if we only use the single global part of the DRDM to manage all the data among parallel transaction threads and to detect incorrect data access in both the two modes, the global part will become the bottleneck of the multi-core systems. To reduces the overhead of the global part of the DRDM and improves overall performance, we have designed a local part of the DRDM to buffer recently gotten access permission. In this way, the transaction threads and SP-TX threads send the transaction thread read/write check request to the local part of the DRDM at the first instead of the global part of it. The execution flow of the dynamically parallel data access check request is summarized in Fig. 1.

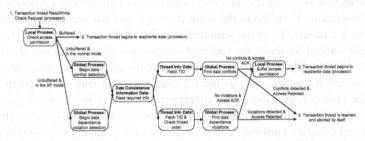

Fig. 1. The distributed execution flow of a dynamically parallel data access request

The data consistence protocol used in the new speculative transactional memory model is shown in Fig. 2, which manages all the data among parallel threads and maintain the data consistency by the proper state transition. This protocol is implemented in the global part of the DRDM. Every data can be in either one of the three states, U (Uncached), R (Read) or W (Written). The state U indicates that the data is never used by any transaction threads. When a transaction thread reads or writes the data in state U, the state of the data will transit to state R by a read or transit to state W by a write. The state R represents that the data is shared by one or more transaction threads. The state W indicates that the data is occupied by only one transaction thread, called the owner.

Based on the concepts described above, we shall describe the whole design of DRDM in detail in the next section.

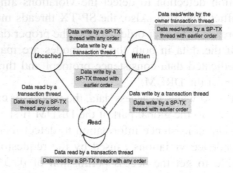

Fig. 2. The proposed data consistence protocol for the new model

3   Design of Distributed Run-Time Dynamic Data Manager

To increase efficiency and reduce the hardware overhead, the Distributed Run-Time Dynamic Data Manager (DRDM) is designed in a distributed and decoupled manner. The DRDM is composed of a global part shared for all processors and some local parts for each processor in a multi-core system. The global part of the DRDM is called the Global Run-Time Dynamic Data Manager (GRDM) which manages and records all the dynamic data access operations among parallel execution threads, those information is called the data consistence information. The complete data consistence information is stored in the main memory, and the recently used part of it is buffered in a virtualized storage structure in GRDM. When the GRDM requires the information without buffered, the required information will be returned from the main memory to GRDM. The GRDM also will manage the other cache structures shared by all the transaction threads, such as the L2 Cache. In this way, we combine the data consistence protocol with the cache coherence by sharing most of common components (e.g. decoder, arbiter...), likes the idea of Smart Memory [7], rather than design dedicated and complex hardware components for each of them. The GRDM processes requests from processors and main memory. The request processing flow of the GRDM is shown in Fig. 3.

When a processor executing a transaction thread sends a transaction thread read/write check request to the GRDM, the GRDM arbitrates requests first since it receives requests from many sources, such as the processors and the main memory.

Fig. 3. The processing flow of requests of the GRDM

Then, the request selected by a very efficient arbiter [12] will be decoded. The decoded requests are then buffered in a queue. If the request is a cache coherence type, such as the L2 cache read/write or data from main memory, it will be scheduled to the unit for managing the L2 cache. If the request is a transaction thread data access check request (TRD/TWT CHK Req), it will be sent to the unit which tracks all the data among threads and use this data consistence information to detect there is a data conflict/data dependence violation or not. If the required data or the data consistence information is not in the GRDM, the GRDM will generate a read request to the main memory to acquire data. If the request is accomplished by the GRDM, it will return "ACK" or "Reject" if the request is a check request, sometimes there will be many conflicting threads to be aborted by sent abort request. The "ACK" reply indicates the requestor transaction thread sent the check request has gotten the access permission. If the request is the "TRD CHK Req", the "TRD ACK" represents it gets the read permission and then the transaction thread is able to read the requested data. If the request is the "TWT CHK Req", the "TWT ACK" means it gets the read and write permission since it becomes the owner of the requested data, and then the transaction thread is able to read and write the requested data. If the request is a L2 cache read, the data will be return to the processor.

To reduce the need for requesting the GRDM frequently and improve the overall performance, we design the per-processor Local Run-Time Dynamic Data Manager (LRDM) as the local part of the DRDM to store recently acquired access permission. As the result, the transaction thread read/write check request (TRD/TWT CHK Req) will be sent to the LRDM first instead of the GRDM directly (refer to Fig. 1). If the LRDM has buffered the required access permission, the LRDM replies the "ACK" directly and it is no need to send the same request to the GRDM again. If the LRDM has no required access permission, the TRD/TWT CHK Req will be sent to the GRDM. Then, the GRDM performs the data conflict detection and data dependence violation detection and returns the result to the LRDM. The LRDM buffers the access permission when the GRDM replies the "TRD ACK" or "TWT ACK".

3.1   Global Run-Time Dynamic Data Manger Design

Based on the behaviors described above, the architecture of the GRDM is designed and presented in Fig. 4. The GRDM consists of seven major units. The very fast Request Arbiter (RA) [8] arbitrates requests from the Request Queue (RQ) which buffers requests from the processors or the main memory. The Request Decoder (RD) decodes the selected request. The Miss Status Holding Register (MSHR) provides storage for requests which cause a cache miss. The RD would not decode requests with the address which has been recorded in the MSHR, and if so, the RD stores it in the MSHR. The requests in the MSHR would not be processed until the required data or required data consistence information is returned from the main memory. The Request Scheduler (RS) dispatches decoded requests to the Conflict Detection Unit (CDU) or the State Management Unit (SMU) according to the request type. A transaction read/write check requests sent by a non-speculative or a speculative transaction will be processed by the CDU. A data access request will be processed by the SMU and the Data Access Unit (DAU). The SMU and the DAU manage the Tag Array and the Data Array of the L2 Cache respectively. The SMU implements the protocol of the L2 cache coherence. The DAU reads or updates data in the L2 cache. If the L2 cache misses, the request will be bypassed to the Message Generation Unit (MGU). The CDU implements the data consistence protocol described in section 3.1 and thus it manages all the data consistence information and stores them in the Data Consistence Directory (DCD).

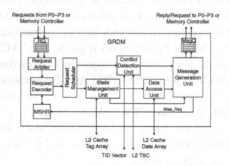

Fig. 4. The architecture of the GRDM

The CDU also manages the virtualized storage structure, called Level 2 Thread Signature Cache (L2 TSC), which buffers the partial data consistence information in the DCD. When processing the check requests, the CDU will access the TID Vector and the L2 TSC first and then performs the violation detection for speculative transactions or the conflict detection for non-speculative transactions. The result of the CDU and the SMU/DAU will be sent to the MGU. The MGU generates reply to processors or memory controller if the request is completed. If the L2 TSC misses or the L2 cache misses, that is the request could not be processed at once, it requires more operations. The MGU will generates a memory read request if the L2 cache misses and a DCD entry read request if a L2 TSC misses to main memory. The MGU also stores these requests to the MSHR.

3.2   Local Run-Time Dynamic Data Manager Design

Fig. 5 shows the architecture of the LRDM. The Decode Unit (DU) decodes the read/write check requests from processor or replies from the GRDM. The MSHR is as same as the one in the GRDM which buffers the requests with an L1 TSC cache miss. The Access permission Filtering Unit (APFU) manages the L1 TSC and process check requests by examining the L1 TSC. If the L1 TSC has the necessary access permission corresponding to the requested data, the APFU will acknowledge the Message Generation Unit (MGU) to generate "ACK" to processor; otherwise, when the L1 TSC misses, the MGU would pass the same request to the GRDM.

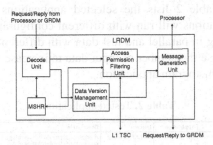

Fig. 5. The architecture of the LRDM

When a thread in both of the SP mode and the normal mode is aborted by the GRDM, all the data modified by them is not permitted and need to be discarded. Therefore, we should manage the old values and the new values of all the data properly for each thread and this would be designed in the local part, the LRDM.

4   Evaluation

This section evaluates our DRDM. To measure the performance of the normal mode and the SP mode, we construct a simulation environment and select six transactional applications to evaluate the DRDM.

4.1   Experimental Environment

We use QEMU [9] as a baseline full system simulator on which our DRDM is built. The baseline processor is the ARM Cortext-A9 quad-core processor and some modifications are required, including new instruction extension for speculative parallel execution model and transactional memory model and additional structures for version management (Log Filter, Log Base and Log Pointer). Table 1 lists the simulation environment of our experiment.

To measure the performance and efficiency of DRDM, we design two applications, histogram and RGB to YUV transformation, for the speculative mode. For the normal

Table 1. Simulation Parameters

Host	OS	Ubuntu 10.04 64bit
	Linux Kernel	version 2.6.32
	QEMU	version 1.0.1
	CPU	Intel Core-i5 760 2.8GHz x86/x64 Quad-Core Processor
	RAM	2GB
Guest	System	Versatile Express Development Board
	CPU	ARM Cortex-A9 Quad-Core Processor
	RAM	1GB
	OS	Linux 2.6.38 Embedded
	Tool Chain	gcc-4.5.2 for ARM

mode, we modify the histogram application and select three transactional applications from STAMP [10]. Table 2 lists the selected applications used to evaluate our DRDM. All the applications will run with different configurations, such as the number of threads executing in parallel and test data with different contention levels. For example, we can increase the number of input data to increase the number of conflicts between threads in the histogram application.

Table 2. Tested applications

Mode	Application	Domain	Description
Speculation/ Normal	histogram	Image Processing	A graphical representation showing a visual impression of the distribution of data
Speculation	RGB to YUV Transformation	Image Processing	A transformation from RGB value to YUV value for pixels in an image
Normal	linked list	General	Construction of Linked list
	labyrinth	Engineering	Routes paths in maze
	intruder	Security	Detects network intrusions
	ssca2	scientific	Creates efficient graph representation

4.2   Experimental Results

First, we evaluate and summaries experiment is to evaluate the performance of applications running in normal mode. Fig. 6 shows the speedup of the five transactional applications and all the results are compared with their sequential version. The post fix of each application represents the configuration: the "no" indicates the application runs in sequential, the "t4" indicates the number of threads executing in parallel is four and the "low" indicates low data contention. Most of the applications have significant performance improvement except histogram with high data contention. The histogram with high data contention results in a lot of conflicts because the GRDM takes the eager conflict detection, so that the re-executions have made too much overhead to afford. The granularity of transaction in the histogram and the linked list are smaller than others from STAMP. When they run with lower data contention, the performance improves significantly. The performance are also affected by the characteristics of the application, such as the mentioned granularity of transaction, the number of transactions to be executed, the degree of data contention and the number of data accessed by each transaction.

The second experiment is to evaluate the performance of the applications executing in the SP mode. Fig. 7 shows the speedup of the two speculative parallel applications and all the results are compared with their sequential version. The post fix of each

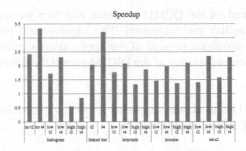

Fig. 6. The speedup of tested applications in the normal mode

application represents the configuration: the "sp" indicates the application runs in sp mode and the "t4" indicates the number of threads executing in parallel is four. The RGB2YUV can be highly parallelized than histogram. These two applications have only one speculative transaction per thread. In other words, we divide the application into two threads when running in two parallel threads and the code section of each thread is wrapped in a speculative transaction. Then, the two threads will run in SP mode and commit in order. Here, we only evaluate the performance of the applications without violations between threads due to the lacking support for context-switch in QEMU.

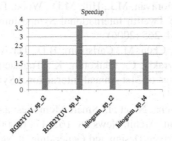

Fig. 7. The speedup of tested speculative parallel applications

5   Conclusions

With the decoupled and distributed design, the Distributed Run-Time Dynamic Data Manager (DRDM) is able to have better scalability and modularity when scaling processors and thus maintain performance. Furthermore, we have integrated the concept of speculation with the transactional memory into a new model, speculative transactional memory. The DRDM can provide run-time support for the execution of speculative transactions and normal transactions. Also, the CDU of the GRDM can alternate the detection mechanism according the execution mode. In the SP mode, the CDU turns on the violation detection to detect the data dependence violations between speculative transactions. In the normal mode, the CDU performs the conflict detection to detect the conflicts between transactions instead of violations. Finally, the DRDM is designed and simulated on the QEMU emulator and then we select several parallel applications to accomplish the evaluation. The experimental results show that our DRDM improves performance in most of the applications significantly (at least 1.4x speedup) due to the well efficiency of the DRDM except the histogram with high data contention.

Acknowledgement. This work was supported in part by the National Science Council under the Grants NSC100-2221-E-006- 179.

References

1. Herlihy, M., Eliot, J., Moss, B.: Transactional Memory: Architectural Support For Lock-free Data Structures. In: Proc. of 20th Internat. Sympos. on Computer Architecture, pp. 289–300 (1993)
2. Hammond, L.S.: Hydra: a chip multiprocessor with support for speculative thread-level parallelization. Stanford Univ. (2002)
3. Steffan, J.G.: Hardware support for thread-level speculation. Carnegie Mellon University (2003)
4. Baek, W., Bronson, N., Kozyrakis, C., Olukotun, K.: Making nested parallel transactions practical using lightweight hardware support. In: Proc. of 24th ACM Internat. Confer. on Supercomputing, Tsukuba, Ibaraki, Japan (2010)
5. Moore, K.E., Bobba, J., Moravan, M.J., Hill, M.D., Wood, D.A.: LogTM: log-based transactional memory. In: The Twelfth International Symposium on High-Performance Computer Architecture, pp. 254–265 (2006)
6. Hammond, L., Wong, V., Chen, M., Carlstrom, B.D., Davis, J.D., Hertzberg, B., Prabhu, M.K., Honggo, W., Kozyrakis, C., Olukotun, K.: Transactional memory coherence and consistency. In: Proceedings of 31st Annual International Symposium on Computer Architecture, pp. 102–113 (2004)
7. Mai, K., Paaske, T., Jayasena, N., et al.: Smart Memories: a modular reconfigurable architecture. SIGARCH Comput. Archit. News 28, 161–171 (2000)
8. Jou, J.M., et al.: Model-Driven Design and Generation of New Multi-Facet Arbiters: From the Design Model to the Hardware Synthesis. IEEE Transactions on CAD of Integrated Circuits and Systems 30, 1184–1196 (2011)
9. Bellard, F.: QEMU, a fast and portable dynamic translator. Presented at the Proceedings of the Annual Conference on USENIX Annual Technical Conference, Anaheim, CA (2005)
10. Chi Cao, M., JaeWoong, C., Kozyrakis, C., Olukotun, K.: STAMP: Stanford Transactional Applications for Multi-Processing. In: IEEE International Symposium on Workload Characterization, IISWC 2008, pp. 35–46 (2008)

A Novel Defragmemtable Memory Allocating Schema for MMU-Less Embedded System

Yu-Hsaing Yu, Jing-Zhong Wang, and Tsung-Ying Sun, Member, IEEE

Department of Electrical Engineering, National Dong Hwa University, Hualien,
Taiwan, R.O.C.
{d9823010,610123001}@ems.ndhu.edu.tw, sunty@mail.ndhu.edu.tw

Abstract. A new approach of memory allocating that allowed to perform memory defragment without the help of the memory management unit (MMU) is proposed in this paper. With the proposed allocating schema, memory allocation can be more precise without suffering from internal and external fragmentation and hence the storage cost is reduced as physical memory space can fit the actual requirement in an MMU-less embedded system. With the ability to maximizing the utilization of memory, the schema is needed by an allocator which delivers the high speed but low volume memory to the task for time-critical operation. As a result, the overall cost could be reduced by the increasing of efficiency.

Keywords: Defragmentation, MMU-less, memory allocation.

1 Introduction

Dynamic memory allocation brings the highest efficient way to distribute memory space to the application to store data or even if a program without waste. This method shows its capability in most of undetermined difficulties due to unknown inputs, such as the variation of reign of interesting (ROI) or the other unpredictable things that need to be stored by memory. However, the intrinsic to the dynamic allocation also bring back some side effects to memory. The biggest problem that is really hard to avoid is fragmentation, or to be more precise, named external fragmentation and internal fragmentation. The first one that we pay more attention in this paper is happening due to multiple allocating and freeing with different size. The typical example is a computer system running OS, where there are many tasks acquiring memory dynamically. If the system has tendency to make memory space fragmented, the memory will be used up as time goes by even if the total available space is sufficient large to fulfill the requirement. For a system equipped with memory management unit (MMU), the fragmentation may not be a critical problem, just mapping the discrete fragments into a continuous space to serve the need of application. However, it is impossible to re-map the memory map for solving the fragment problem in a MMU-less computer system such as DSP and most of microcontrollers. Due to avoid the fragment problem in a MMU-less system, the memory size should be much larger than the dedicated size for accepting the incoming memory allocation request.

J.-S. Pan et al. (Eds.): *Advances in Intelligent Systems & Applications*, SIST 21, pp. 751–757.
DOI: 10.1007/978-3-642-35473-1_74   © Springer-Verlag Berlin Heidelberg 2013

2   Related Works

A MMU-equipped embedded system can remap the fragmented memory spaces into a whole large block or even perform defragment operation to merge those pieces into continuous physical block. Therefore, fragmentation problem only causes performance issue in a MMU-equipped embedded system. For a MMU-less embedded system, only countermeasures that try to prevent fragmentation as possible as they could be taken to increase the utilization of memory. A simplest allocator formats the memory space into the chunks with uniform size. Since the chunks are in one size, it is not easy to be fragmented with a larger size. However, the slack of allocation makes it inefficient to fulfill the variance of the requirements. In order to use the memory efficiently, Buddy systems [1] splits the large block into two half pieces that close to the requirement. Its advanced version also take care the efficiency issue and increase the performance [2], [3]. As for the slab allocator [4] the memory space has been formatted into chunks with different size, and the closest matched memory space will be assigned to the pointer. There is also a research follow up the slab allocator [5]. CAMA [6] splits and merge the blocks according to the need to avoid fragmentation.

Those methods do decrease the fragment spaces but more or less produce fragmentation, which doesn't happen when using a dynamic allocation method that never make slack allocation. Now that both external fragmentation and internal fragmentation are hard to avoid, it is badly in need of defragment method in a MMU-less embedded system. In this paper, we propose a preliminary study of the approach to make defragmentation possible for the MMU-less embedded system.

3   The Proposed Allocation Method

The fragmentation lead to high memory device cost because this phenomenon lower the utilization of memory. As a result, designers are forced to use a higher volume memory device than it was planned. There are two approaches to increase the efficiency of using memory. The first one is preventing memory from being fragmented. Nevertheless, fragmentation is still hard to exterminate, so the second one show its importance to solve the problem by defragment the memory. Therefore, we design a mechanism to allow defragmentation available in MMU-less embedded system and packages it into an allocator by adding a simple allocation algorithm with size-cared descriptor table.

3.1   MMU-Less Defragmentable Allocation Schema

Due to enable the defragment in a MMU-less embedded system, the allocated memory block should be moveable. The operation is really easy in an embedded system which equipped with MMU since the MMU is capable of redirecting the address holding by the pointer to the new space without inform the pointer about the change. However, it brings a brainteaser to designer who wants to implement the behavior in another platform without the help of MMU. With the limitation of original allocation style, reallocating the allocated memory space seems to be the only way to move the given space filling with the data. We can trace the following code to understand how to find the breakthrough to allow the MMU-less embedded system defragment the memory.

```
void *malloc(size_t size);
```

This function prototype is the memory allocation function provided by the stand C library. The function simply returns the address of the memory space that fulfills the given size if it can provide. Once the address is handed over to the pointer successfully, the task can utilize the given space to store data for processing. The changes caused by a memory allocation function are described as Fig. 1. The pointer is uninitialized at first. After calling the memory allocation function, the pointer is now point to 0x80020000 and ready for use.

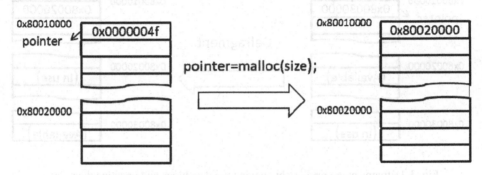

Fig. 1. Initializing a pointer by allocating new memory space to it

The value of the pointer can only be affect when assign a value to it manually. If the address of the pointer can pass to the allocator while calling the memory allocation function, the location of the data obtained by the pointer can change automatically as the allocator wish. It would be very convenient to defragment the memory if the allocated memory spaces can be moved without assign the changes to the pointer. To achieve this requirement, the function prototype needs to be rewritten as follow:

```
void malloc(void *ptr_addr, size_t size);
```

The operation of modified memory allocation method is shown as Fig. 2. The behavior of the modified method is almost identical to the original method, only that the address of pointer leaves a record in the allocator. Once the allocator keeps the address of the pointer, the allocator can move the location of the data block that is in use except the data block is under processing or being reference. The scene that a memory space which is in use in Fig. 2 is released in Fig. 3, and the allocator performs the defragment process by moving the data block obtained by the pointer.   After all of the data is transferred to the new memory space, the allocator redirects the pointer from old location to the new one through indirect addressing using the address of the pointer that hold by the allocator.

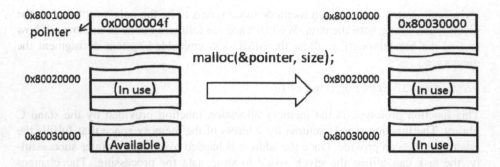

Fig. 2. Initializing a pointer by allocating new memory space to it with the proposed schema

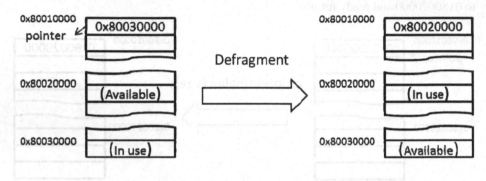

Fig. 3. Defragment the memory by moving the data block and updating the pointer

Accordingly, the fragmented memory spaces are able to be merged into a large and continuous space without the help of MMU.

3.2   Applying to an Allocator

The proposed allocating schema need more space to store the extra information in the descriptor. We give an example here to explain the basic requirement to construct the descriptor. For the free space list, the basic structure is as following:

```
typedef struct freelist_T {
  unsigned int start_address;
  unsigned int size;
  struct freelist_T *next;
}freelist;
```

As for the allocated space list, the basic structure has an extra attribute compare to the free space list to store the address of the pointer. The basic structure of the allocated space lists as following:

```
typedef struct allocatedlist_T {
  void *pointer_address;
  unsigned int start_address;
  unsigned int size;
  struct allocatedlist_T *next;
}allocatedlist;
```

It is clear to see the proposed schema has the only difference at the attribute to store the address of the pointer, which is exactly the key to make defragmentation possible in a MMU-less embedded platform. As long as the address of the pointer contains the data structure of the descriptor, the structure can be modified arbitrarily.

Then it comes to the operation combined with a defragmenter. The following steps are the basic procedures that we suggest when moving a data block:

1. Calculate the transporting time for the data block.
2. Obtain the idle time from OS.
3. Abort if the time is not enough to move the data block, else go to next step.
4. Transport the data block to the destination with direct memory access (DMA).
5. Update the address of the pointer.

Note that we suggest using DMA for transporting. Because defragmentation costs the resource to merge the scattered memory spaces, the extra cost should be minimized in an embedded system.

4   Limitation and Workaround

However, the proposed allocation method does not really remap the memory space as MMU functions so it cannot replace the role of MMU. Although is it possible to re-map the memory via software approach, the index method not so direct to be implemented and hence its efficiency is too low to make it acceptable. Since the allocator can change the value of the pointer because it holds the address of the pointer, the proposed allocation method is sensitive to the synchronization issue between the real address that owning to the pointer currently and the address registered in the allocator when the last time the memory allocation function was called. Here we list the two main problems with the proposed method and its workaround.

4.1   Copying of the Pointer

It is common to obtain a copy of the pointer. For example, to sending an array to a function, the most popular way is passing the beginning address of the array to the function as an input argument rather than packages the array into a structure. During the above procedure, the value of the pointer is duplicated. It could be a disaster if defragment operation is started before the function is done because the pointer will be redirected by the allocator after the operation is finished and thus cause the address that hold by each other are asynchronous. Even though the defragment operation can only be executed in the idle state between the tasks, the scenario still possible to happen due to any possible combinations of code. Due to fix the problem, the designer has to change to the attribution of the descriptor in the allocator by calling the following function or the identical one:

```
void IsBlockMovable(pointer, false);
```

Once the function is called, it is safe to enter critical code section that has potential to induce synchronization issue. Due to make the data block movable again, just call the function and set the second argument to true to re-enable the given data block to be transported during the defragment operation.

4.2   Pointer in the Allocated Memory

Sometimes the pointer is placed inside an allocated memory space pointed by another pointer. For instance, a linked list is one of the types. It is dangerous to handle a data block that contains any pointer. As soon as the data block is moved, the allocator will lose the address of the pointer. In order to move the data block safely, the related address of pointers recording in the allocator needs to be added up with the comparative offset caused by the moving of the data block to make the address synchronized. But it could be very inefficient to travel through the entire descriptor list every time before perform defragmentation to find the address of pointer that is within the range of data block to be moved. And the fact that not all of the data blocks contain a pointer. Therefore, the lowest requirement way is adding a counter to the structure of the descriptor to count involved pointers. As long as the counter reaches its top, the allocator stops searching and start to transport the data block.

5   Experiment

In the experiment, we demonstrate the proposed schema by compare to the same allocator without defragment. The allocator is a memory pool construct with 16384 4-byte chunks. After the random freeing and allocating by three pointers with bounded random size request, the allocator without defragment finally failed to allocate 14652 bytes to pointer1 and the free space is fragmented shown as Fig. 4. It is absolutely foreseeable that with the help of the proposed schema, the fragmented spaces could be merged into a whole large space with maximum defragment effort. Once the continuous free space is capable of satisfying the request, the experiment could be continued.

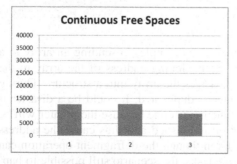 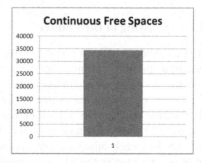

Fig. 4. Without defragment (left) and defragmented result (right)

6   Conclusion

The proposed defragmentable allocating mechanism has proved its capability to solve the fragmentation problem in MMU-less embedded system by cooperating with a simple allocator. With the ability to defragment the fragment memory, the allocator is allowed to assign the fittest memory space to place the data without worrying about the fragmentation problem. Such the high efficient allocating method would be popular when apply to the management of low volume memory device. For example, the internal SRAM will benefit from this method. As a result, the designer is free from management the internal SRAM manually while optimizing the code. Though it is not so perfect comparing to the real MMU that implemented in hardware, it does useful to improve the efficiency of using memory by solving the fragmentation problem when MMU is not available in the given chip.

References

1. Peterson, J.L., Norman, T.A.: Buddy Systems. Communications of the ACM 20(6), 421–431 (1977)
2. Barkley, R.E., Lee, T.P.: A lazy buddy system bounded by two coalescing delays. In: SOSP 1989 Proceedings of the Twelfth ACM Symposium on Operating Systems Principles, pp. 167–176. ACM Press, New York (1989)
3. Defoe, D.C., Cholleti, S.R., Cytron, R.K.: Upper Bound for Defragmenting Buddy Heaps. In: LCTES 2005 Proceedings of the 2005 ACM SIGPLAN/SIGBED Conference on Languages, Compilers, and Tools for Embedded Systems, pp. 222–229. ACM Press, New York (2005)
4. Bonwick, J.: The slab allocator: An object-caching kernel memory allocator. In: USENIX Summer, pp. 87–98 (1994)
5. Dongwoo, L., Junghoon, K., Ungmo, K., Young Ik, E., Hyung Kook, J., Won Tae, K.: A fast lock-free user memory space allocator for embedded systems. In: International Conference on Computational Science and Its Applications (ICCSA), pp. 20–23. IEEE Press, New York (2011)
6. Herter, J., Backes, P., Haupenthal, F., Reineke, J.: CAMA: A Predictable Cache-Aware Memory Allocator. In: 23rd Euromicro Conference on Real-Time Systems (ECRTS), pp. 23–32. IEEE Press, New York (2011)

Hardware Acceleration Design for Embedded Operating System Scheduling

Jian-He Liao, Jer-Min Jou, Cheng-Hung Hsieh, and Ding-Yuan Lin

Department of Electrical Engineering, National Cheng Kung University, Tainan 701, Taiwan
jjmjjmjjm3@gmail.com

Abstract. This study examines the scheduling hardware for an embedded operating system (OS). This scheduler, which implements task sorting and choosing, is deployed when a new task enters in the system. The scheduler always limits the performance of an embedded operating system, so we consider designing the scheduler within the hardware to accelerate the performance of the OS. Therefore, hardware is used which involves an inserting and removing task in the red-black tree [9] and a checking of the red-black tree with regards to whether or not its rules are being followed. Additionally, the software communicates with the hardware by sending task data and is designed specifically to build the red/black tree into the hardware. Finally, the scheduler chooses a task from the red-black tree and tells the software to execute the program. In experiments, the performance of the embedded operating system scheduling hardware improves beyond the existing software by 13%.

Keywords: Scheduling, Hardware, Operating system, ARM, Red black tree.

1 Introduction

The embedded system contains a microprocessor, operating system, root file system, driver and applications. The OS handles communication between the software and hardware.

The OS must quickly allocate hardware to deal with many applications. In this way, the OS can increase performance. As the scheduler is a key component in the hardware allocation, decreasing its execution time is the major focus of this study.

In [8], the OS implements some functions in the hardware [7] and can significantly improve performance within the software. Even though the elasticity of hardware is poorer than the software, the algorithm of the operating system is static, so the elasticity problem can be ignored.

After analyzing the OS execution process, it can be seen that scheduling can be used in the OS, this study chooses scheduling to accelerate the hardware design in order to effectively enhance performance. There, the Linux kernel 2.6.28 [16] is used, and its scheduler is the denoted as the completely fair scheduler (CFS).

The remainder of this paper is organized as follows: Section 2 introduces actions pertaining to the scheduler, such as the main algorithm. Section 3 describes the

J.-S. Pan et al. (Eds.): *Advances in Intelligent Systems & Applications*, SIST 21, pp. 759–767.
DOI: 10.1007/978-3-642-35473-1_75   © Springer-Verlag Berlin Heidelberg 2013

architecture of the scheduling process. Section 4 describes the experimental results and performance analysis. Section 5 offers some conclusions.

2   Scheduler Behavior

2.1   Basic Scheduler Introduction

The scheduling of the operating system allocates the executive order of the task nodes. Then, the OS uses the CFS. In order to implement the CFS, every task describes the amount of time which calls upon virtual runtime to represent the CPU wait time. When the virtual runtime of the task is smaller than the waiting time, the processer's time is shorter. Then, the CFS uses the red-black tree to sort task nodes.

There are four main functions in the CFS: (1) deleting current task information, (2) inserting task node in the red-black tree, (3) removing the task node from the red-black tree and (4) implementing the context switch. In the transmission, large amounts of data will occupy the bus bandwidth; this is not in the hardware design.

Implementation of the OS scheduler involves the following: the first case is when a task state transition time, such as when scheduling a termination, schedules a sleep and also includes a task creation (fork); the second is when the virtual runtime of the current task is greater than the ideal execution time; the third bypasses the CPU and directly calls the scheduler; the last one is such that when scheduling returns from interruptions, exceptions and system calls, it will check the need for scheduling.

2.1.1   Scheduling Algorithm
This repeats the action to insert and remove nodes from the red-black tree in the scheduler and holds the red-black tree to the above rules. Figure 1 shows the scheduling algorithm. The upper right block is inserting the task node in the red-black tree; the lower right block is removing the task node from the red-black tree.

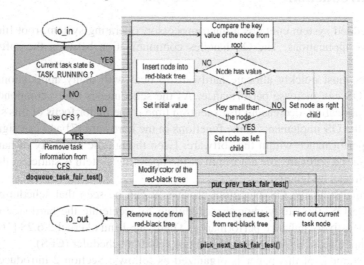

Fig. 1. Scheduling algorithm

2.1.2 Inserting Node

Whenever a task is created or task state transition is triggered for the implementation of the inserted task node, the inserting action of the red-black tree and the binary tree are similar. Compare the key value of the task from the root of the red-black tree: if the key value is less than the target task's, then search the left side child node; if the key value is greater than the target task's, then search the right side child node. Eventually, this process will find the inserting point. The red-black tree coupled with color properties can maintain the longest path of the tree no more than the shortest path twice.

The inserting node is necessary with regard to following the above five rules. When the parent node of the inserted node is black, then there is no violation of any rule. Otherwise, a modification of the color is required. There are two important functions, left rotation and right rotation, to modify color.

2.1.3 Removing Node

Whenever the task termination or a task execution time is greater, then the ideal execution time is triggered to remove the node. Removing the node requires a similar approach. First, determine its parent and child nodes and connect it. Then, confirm whether the red-black tree conforms to the rules or not. When the removing node is black will violate the rules and must be fixed.

3 Scheduling Architecture Design

This section describes the analysis of the proposed scheduling framework, and development platform based on ARM [10], then analyzes the system architecture, hardware and software systems, implementation processes and introduces some software and hardware developments.

3.1 System Architecture of the Hardware/Software Scheduling

Hardware modules achieve the scheduling circuit, and scheduling circuit and bus communicate through scheduling wrapper control. Scheduling wrapper control contains an input buffer, output buffer, control circuit and scheduling circuit. The input buffer and output buffer use the Xilinx BRAM (Block RAM) component library to generate. Then, the control circuit contains the state of the data read/write machine, the BRAM controller and scheduling circuit controller. The scheduling circuit is achieved by the red-black tree algorithm. The software modules include a node conversion, data transmission and other programs. Then, the data transmission of the standard character device accesses mmap(), write() and read().

3.1.1 Scheduling Data Structure

In order to communicate with the buffer of the hardware, the software is designed with the following two storage areas. First, is the runqueue, each CPU has a runqueue which stores the order of the tasks, a context switch count, the next task to be run and so on. Second, is the structure of the task node; each task has its own structure stored task information.

3.2   Interface of the Scheduling Software and Hardware

Between software and hardware, communication regarding received and sent data must exist [1]. Therefore, all of the scheduling interfaces are shown as in Figure 2. The software sends a writing data signal and after the hardware receives a signed notification, it will receive data from the software until the software sends the stop signal. At this time, the hardware will begin scheduling until finishing the scheduling algorithm and will notify the software to receive data. When the software receives all of the data, the hardware will enter an initializing state.

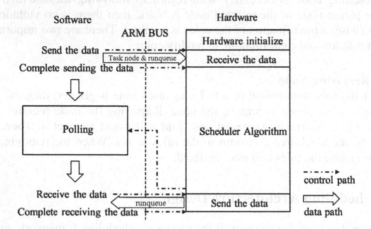

Fig. 2. Interface of the scheduling software and hardware

3.2.1   Transmission Data and Format

Transmission data requires bus communication between software and hardware; the software side transmission data uses iowrite32() and ioread32() [14]. The iowrite32() is 32bits of data sent to a specified memory location, and the ioread32() is 32bits read from the specified memory location. The hardware can only judge from the location of the memory read or write. When the signal is writing, the next hardware clock cycle will receive the information from the ARM bus. If the signal is reading, the next hardware clock cycle will send the information to the ARM bus.

Transmission data involves sending and receiving and, as such, the software sends the task data and order of the red-black tree to the hardware, but software receives only the task order of the red-black tree from the hardware.

3.3   Scheduling Hardware Design

The proposed scheduling wrapper internal contains a scheduler, a read and write controller, a data selection counter and output selection, shown in Figure 3.

After the scheduling circuit receives the information, roughly four hundred clock cycles are spent in the output data, this part is the simple scheduling algorithm.

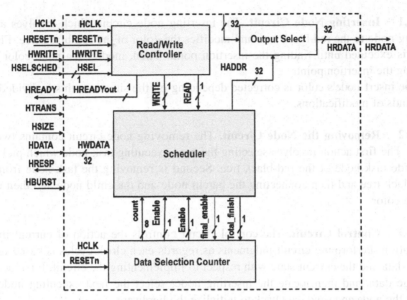

Fig. 3. Scheduling wrapper

3.3.1 Read/Write Controller

The read and write controller is used to control the AHB read and write signals. The AMBA protocol [11] allows the scheduling wrapper to correctly access the data through the state machine. The AHB bus sends read and write enable signals to the hardware buffer and the hardware buffer outputs an HREADY signal which notifies the AHB that data is ready and complete written or not. HWRITE represents read and write operations and HSEL represents access to the memory address of the Logic Tile.

AHB FSM state diagram, which is divided into three states: ST_IDLE, ST_BURST_READ and ST_BURST_ERITE.

3.3.2 Data Selection Counter

A data selection counter is used to control the input and output data. The input data control receives the input signal and the input counter increases until the input data control receives the stop signal. Then, the output data is the same, when the output data control receives the output signal, the output counter increases until the output data control receives the stop signal. After the output data control receives the stop signal, the scheduling hardware will initialize.

3.3.3 Scheduling Hardware

The scheduling hardware contains three parts: inserting node circuit, removing node circuit and control circuit. The inserting and removing node circuits are designed according to the scheduling algorithm and use control/data flow graph to optimize data. The control circuit controls all of the hardware when receiving data, sending data and initializing hardware.

3.3.3.1   Inserting Node Circuit. The inserting node circuit which involves an inserting node to the red-black tree and modifies the color of the red-black tree. First, a loop is executed until finding the insertion point. Then, modify the node color after finding the insertion point.

The insert node's color is corrected depending on the situation; this is divided into six kinds of modifications.

3.3.3.2   Removing the Node Circuit. The removing node circuit contains two actions. The first action involves selecting the next executing task node which picks the left side task node of the red-black tree. Second is removing the task node from the red-black tree and then connecting the parent node and the child node and then modifying color.

3.3.3.3   Control Circuit. The control circuit controls the action of current implementation. Performing circuit judgments as regards each clock cycles is based on the input data and the current state with respect to implementing the circuit. First, acquire storage data, and then locate the inserting node, select the next executing node, remove the node and send data back to initialize the hardware.

When the scheduling hardware finishes removing the node, the signal of finishing scheduling will be sent, notifying software to receive data from the hardware.

4   Experimental Results and Performance Analysis

This section describes experimental results and discusses the performance of the cases in the proposed platform architecture.

4.1   Development Environment

This study achieves system architecture on the ARM9 development platform and modifies the Embedded Linux OS ported to the platform board to develop its application. The system development environment can be divided into hardware and software development environments. The following paragraphs will introduce the hardware and software development environments and processes.

The proposed development platform is a Versatile Platform Board with FPGA [12], and is connected to the PC Host by a USB Debugger Port which is equivalent to the ARM Multi-ICE (or Real View ICE) with the functionality of the hardware debugger. A UART Port RS232 connects with the PC Host and a Linux text message transmits through the serial port to the super-terminals giving commands in a text interface window. The Network Port is connected with an RJ45 and the development board uses an NFS (Network File System) connected with a Linux Host to accelerate the speed of the software development.

4.2   Hardware and Software Measurement Results

The experimental test uses the scheduling hardware to implement tasks and compute the time costs, comparing performance against the scheduler of pure software. The experimental test project is divided into three parts: one contains the inserting node action, another removes the node and the last focuses on the time consumed by the transmission of information.

The software measurement comprises a system call function to measure the execution time of each block, and the system function gettimeofday(), whose resolution close to nanoseconds.

4.3   Experimental Results and Analysis

First, a performance analysis of the scheduling hardware is undertaken. Then, the clock cycles of the each part are calculated. There are five aspects to the scheduling hardware: finding out an insertion point, the color modification of the insertion node, selecting the next executing node, discerning both the parent node and the child node and the color modification of the removing node. Each aspect will test when the red-black tree has 0, 1, 2, 3, 4, 5, 10, 20, 100, 200, 500 or 1000 nodes consumption with regard to the clocks, as shown in Table 1.

Table 1. Time consumption of the hardware

Stratum	Test items / The number of nodes in the tree	Inserting node time — Find inserting point	Inserting node time — Color modifying	Removing node time — Selecting next node	Removing node time — Find parent & child node	Removing node time — Color modifying	Total (no inserting node)
0	0	2	0	3	0	0	5(3)
1	1	2	0	6	4	11	23(22)
2	2	9	3	6	4	8	30(18)
2	3	12	3	6	4	8	33(18)
3	4	12	3	6	4	8	33(18)
3	5	12	10	8	4	18	52(30)
4	10	15	17	8	4	25	69(36)
5	20	18	24	10	4	32	88(42)
6	50	21	31	12	4	39	107(55)
7	100	24	31	14	4	44	117(62)
8	200	27	38	16	4	51	136(71)
9	500	30	38	18	4	52	142(74)
10	1000	33	38	20	4	51	146(75)

Measurement hardware and software performance are calibrated in several ways. Here, the scheduler is set to 1500 implementation instances, it has different number of nodes on the tree and it records the execution time.

In the hardware, the frequency of the implementation is 35MHz, so a clock cycle is about 28.57ns. A 32bit data packet received from the software side takes 15 clock cycles. For the hardware scheduling, transmission data needs about 24 documents, so the scheduling hardware takes up 15ms in the transmission. The average taken from the above data describes a common situation in all cases, as shown in Table 2.

Table 2. Inserting and removing the node statistics

The number of nodes in the tree (Test items)	The consuming time of the software scheduler (ms)	The consuming time of the hardware scheduler(ms)		Speed up
		Executing time	Transmission time	
1	15.030	0.556		0.91
2	15.371	1.091		0.90
3	15.944	1.091		0.94
4	16.593	1.091		0.97
5	17.514	1.498		1.00
10	18.859	2.119	15.944	1.04
20	20.614	2.675		1.11
50	22.586	3.274		1.18
100	24.628	3.681		1.25
200	26.239	4.237		1.30
500	28.007	4.558		1.37
1000	30.419	4.708		1.47
平均	20.984	2.548	15.944	1.13

When the number of nodes is greater than four, the performance of the hardware will be better than the software; The general average exhibits more than a 13% increase in performance. With more nodes, the time consumed by the software scheduling will approach a constant, but also match O(logN) (N is the node number of the tree).

5   Conclusion

This study proposes a design for an embedded scheduling hardware and a software OS with the scheduling hardware aimed at accelerating the efficiency of scheduling in the OS. As such, hardware technology is used to increase its performance. The scheduling achieves the red-black tree algorithm, which contains an inserting node, removing node, color modification and other algorithms.

Notably, hardware is used to implement a scheduling algorithm and reduce transmissions. In order to achieve the effect of improved performance, a new structure is proposed to be used in transmission, which can read data from the hardware and write back to the original structure quickly. Finally, the performance of scheduling is enhanced by 13% using only a small amount of hardware.

As regards system bottlenecks is the transmission of software and hardware, the majority of the execution time of the scheduling hardware is spent on sending and receiving data, which does not significantly improve performance unless it can reduce the transmission of information or increase the transmission speed.

References

1. Nakanot, T., Utamaz, A., Itabashis, M., Shiomiz, A., Imai, M.: Hardware Implementation of a Real-time Operating System. In: TRON 1995 Proceedings of the 12th TRON Project International Symposium, p. 34 (1995)
2. Kohout, P., Ganesh, B., Jacob, B.: Hardware support for real-time operating systems. In: Proceedings of the 1st IEEE/ACM/IFIP International Conference on Hardware/Software Codesign and System Synthesis (CODES+ISSS 2003), pp. 45–51 (2003)
3. Morton, A., Loucks, W.M.: A hardware/software kernel for system on chip designs. In: Proceedings of the 2004 ACM Symposium on Applied Computing, SAC 2004, pp. 869–875 (2004)
4. Lee, J., Mooney III, V.J., Daleby, A., Ingstrom, K., Klevin, T., Lindh, L.: A comparison of the RTU hardware RTOS with a hardware/software RTOS. In: Proceedings of the 2003 Asia and South Pacific Design Automation Conference (ASP-DAC 2003), pp. 683–688 (2003)
5. Nácul, A.C., Regazzoni, F., Lajolo, M.: Hardware scheduling support in SMP architectures. In: Proceedings of the Conference on Design, Automation and Test in Europe (DATE 2007), pp. 642–647 (2007)
6. Park, S., Hong, D.-S., Chae, S.-I.: A hardware operating system kernel for multi-processor systems. IEICE Electronics Express 5(9), 296–302 (2008)
7. Kuacharoen, P., Shalan, M.A., Mooney III, V.J.: A Configurable Hardware Scheduler for Real-Time Systems. In: Proceedings of the International Conference on Engineering of Reconfigurable Systems and Algorithms, pp. 96–101 (2003)
8. Castrillon, J., Zhang, D., Kempf, T., Vanthournout, B., Leupers, R., Ascheid, G.: Task management in MPSoCs: an ASIP approach. In: Proceedings of the 2009 International Conference on Computer-Aided Design (ICCAD 2009), pp. 587–594 (2009)
9. Weisstein, E.W.: Red-Black Tree. From MathWorld–A Wolfram Web Resource, http://mathworld.wolfram.com/Red-BlackTree.html
10. ARM Staff, RealView Platform Baseboard for ARM926EJ-S User Guide (2003-2007)
11. ARM Staff, AMBA™ Specification, 2000-2003 Rev. 2.0 (1999)
12. ARM Staff, RealView LT-XC4VLX100+ Logic Tile User Guide (2006-2007)
13. Bovet, D.P., Cesati, M.: Understanding the Linux Kernel, 3rd edn. O'Reilly (2006)
14. Corbet, J., Rubini, A., Hartman, G.K.: Linux Device Drivers, 3rd edn. O'Reilly (2005)
15. ARM, http://www.arm.com/
16. Linux Source, http://www.kernel.org/

Asynchronous Ring Network Mechanism with a Fair Arbitration Strategy for Network on Chip

Jih-Ching Chiu, Kai-Ming Yang, and Chen-Ang Wong

Department of Electrical Engineering, National Sun Yat-Sen University, Kaohsiung, Taiwan
chiujihc@ee.nsysu.edu.tw, {d953010024,onondessuu037}@gmail.com

Abstract. The multi-core systems are implemented on homogeneous or heterogeneous cores, in order to design the better NOC. It must consider the performance, scalability, simplifies hardware design and arbitration strategy. The routers are designed with asynchronous circuits and it's no queuing. Ring topology with multi-transaction bus architecture. It could make multiple packets to access the bus at the same time. When the component increases, the central arbiter circuit become more complexity, this thesis presents SAP (self-adjusting priority) scheme that can fairly adjust priorities of each components at distributed arbiter. When contentions occur for the same network resource, a winner will hand over its priority to the loser path. This principle can guarantee that the opportunity of winners will be decreased priority in the next connection. In opposition, these losers can obtain the higher priority than before. The proposed scheme not only offers fair strategy, but also simplifies hardware design.

Keywords: Index Terms: Communication architecture, on-chip communication, system-on-chip.

1 Introduction

COMMUNICATION architecture constitutes the infrastructure of system-on-chip (SoC) and multiprocessor systems, more components can exist on a single chip. Therefore, communication architectures have created major challenges of scalability, speed, and efficiency of arbitration management. On-chip interconnection system represents one of the major elements which has to be optimized in designing a complex digital system. However, when the number of components increases, the communication between each component is more frequent. In the traditional single-transaction bus architectures[9] limits the performance of SoC, multi-transaction communication can perform multiple accesses in one bus cycle by a multiple transaction[1], The concept of the multi-transaction bus has been widely used in current products such as the IBM Cell BE [2][3]. Asynchronous ring topology with multi-transaction bus architecture, it will be applied in torus topology and mesh topology to improve the transmission performance. In the multi-core processors, each component must transmit the packet header to NoC[10], and therefore needs to control the packet how to transfer. The arbiter is divided into centralized and distributed arbiter, when the components increases, the central arbiter circuit become very complex and

J.-S. Pan et al. (Eds.): *Advances in Intelligent Systems & Applications*, SIST 21, pp. 769–777.
DOI: 10.1007/978-3-642-35473-1_76   © Springer-Verlag Berlin Heidelberg 2013

increased latency, when the component increases the distributed arbiter will not make the router additional burden. Therefore distributed arbitration has better scalability, simple circuit. An efficient strategy should support fair management because the starvation problems of unfair strategies cause poor performance. The main focus of this work is arbitration strategy for fairness of link bandwidth. The contributions of proposed strategy include the following: (1) we introduce a novel arbitration, called Self-Adjusting Priority (SAP) schedule, swapping dynamical priorities. (2) Low complexity: SAP scheme can provide fairness by simple operations and be implemented easily. (3)Based on ring network [8], SAP scheme can also be extended to other topologies.

The remainder of this paper is organized as follows: In Section II, we briefly review the background and four previous arbitration strategies. In Section III, the proposed SAP scheme is presented. In Section IV, we simulate and analysis performance of SAP and others strategies. Finally, the conclusion is shown in Section V.

2   Background and Previous Works

SAP arbitration strategy is extendable to other topology based on ring network such as 2D mesh and tours. In this section, novel bus architecture for ring network. Circuit switching[4] is the implementation of asynchronous communication[11][12]. It is easy to implement because data and packet header can be separated [5]. Each packet is serialized into a sequence of flow control units (flits). When a head flit does not arrive at the destination router yet, this routing path is named "pending path" in this paper. When the head flit of a packet arrives at a destination router, this routing path is named "successful connection". Considering the efficient bandwidth for ring network, the multi-transaction bus architecture is proposed [1]. This bus architecture can perform multiple accesses in one bus cycle by multiple transactions [1]. This concept of multi-transaction bus can improve efficiently the bandwidth and has been widely used in the current products, such as IBM Cell BE [2][3].

2.1   Arbitration Strategy

As traffic flits through network, arbitration strategy in each router must fairly determine injected head flit and traveling head flits in the network when traveling and injected flits are routed via the same output port. Head flits owning the lower priority will loss bandwidth. Only one of head flits owning the highest priority can further travel to the next router most approaches adopt four strategies (Fixed-priority, remaining-path, distance-based and age-based) as follow:

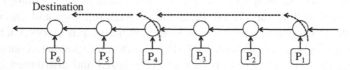

Fig. 1. The principle for SAP Scheme

(1) Fixed-Priority Strategy (FP)[7]:
Fixed-priority strategy is simplest but un-fairest arbitration. Fig. 1 shows an example of communication pattern for an 6-router network, in which router 1 and 4 are sending head flits to node 6. Assume P_4 owning higher priority than P_1. P_1 will always is blocked by P_4 when contention occurs.
(2)Remaining-Path Strategy (RP):
A head flit closest to destination will get the highest priority. For an example in Fig. 1, as there are two competing pending path, P_4 will get the available bandwidth, because it is closer to the destination. Specially, a routing path has fewer hops away from source to destination. It means also the head flit of this path closer to destination. In other words, routing paths with shorter distance likely are arbitrated as winner paths.
(3)Distance-Based Strategy(DB)[7]:
Distance-based strategy gives long-distance communication higher priorities. For an example in Fig. 1, as there are two competing pending path, P_1 will get the available bandwidth, because pending path from P_4 with fewer hops away lose arbitration. Therefore, some routing paths with short distance cannot arrive at destination for a very long period because of several routing paths with longer distance in network.
(4)Age-Based strategy (AB)[6]:
With age-based strategy, a routing path with the oldest age wins the arbitration when two or more routing paths arbitrate for a shared resource. Although age-based strategy is known to provide global fairness, it becomes complex to count pending path time in network.

3   SAP Scheme

3.1   The Basic Principle of SAP Scheme

The basic principle of SAP scheme is to decrease opportunity of successful connection for winners and to increase opportunity of successful connection for losers in the next injection. When two pending paths must travel through the same output at the same router, the pending path winning arbitration will swap its priority with losers as a new priority in the next injection. It means that this winner path router must decrease the opportunity of successful connection in the next injection. In opposition, losers can obtain the higher priority from the winner so as to attach higher priority at the next injection. It means that this loser path router will increase the opportunity of successful connection. From the point of view of global fairness,

Each router performing SAP scheme must maintain the uniqueness. (Two and more sharing the same priorities are not allowed to exist in one network system.) Implementation of SAP scheme operates two simple steps to maintain the uniqueness of priorities. 1: Comparing the priorities. When contention occurs, priorities of two pending paths will be compared to arbitrate a winner. In step2, SAP performs swapping operation to facilitate the fair arbitration. However, by swapping priorities, the priority of head flit become weakens gradually, during the process of this adjusting priority. As the example of Fig. 2 (a), a pending path from P_1 breaks others with lower priority router (P_3 and P_5). This attached priority on the pending path from P_1 will becomes weakens gradually. Because swapped priorities are used for the next injection, this work defines another Consistent Priority (CP) for the current injection. CP indicates the priority for the current injection. It can be also viewed as the strength of pending paths for the current connection. The swapped priority is called Adjusting Priority (AP).

As shown Fig. 2(b), the head flit of pending paths is attached with CP and AP. Values of CP and AP are the priority of source router in the initial state. The CP will be not exchanged as the head flits travel through intermediate routers. Therefore, when SAP performs comparing operations, it will be used to arbitrate a winner. On the other hand, when SAP performs swapping operations, it will be used to exchange with losers. This value represents the strength of connections for the next injection. After the successful connection, the exchanged AP becomes the new priority of the source router of pending path.

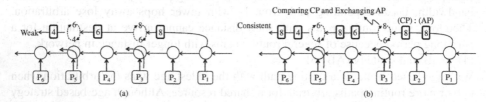

Fig. 2. The Principle of SAP Scheme

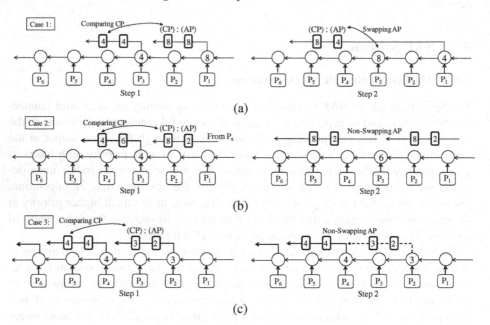

Fig. 3. Connection patterns for SAP scheme

3.2   Mechanism for SAP Scheme

As a pending path travels through the network, it may encounter another pending path with lower CP and *break* that with lower CP. Based on each other's AP, *breaking* condition will distinguish case 1 and case 2 as shown Fig.3 (a) and Fig.3 (b). On the other hand, a pending path may also encounter another pending path with higher CP. In this condition, this pending path must be *blocked*. Likewise, based on each other's AP, *blocked* condition will distinguish case 3 and case 4 as shown Fig.3 (c) and Fig.3 (d). These cases must follow SAP principle: 1. A winner must obtain lower AP than current that when it breaking a pending path. 2. A loser should avoid obtaining lower AP than current when it is broken or blocked. The detailed SAP operations are shown as follow:

Case 1: One pending path with a higher AP breaks another. As shown the step 1 in Fig. 3 (a), one pending path has higher CP and AP P_1 than another from P_3. After comparing CPs, P_1 will break P_3. In step 2, for SAP principle, AP of P_1 will be swapped. The pending path from P_1 will obtain new AP (Priority 4). The router of loser path (P_3) must be interrupted but obtains the higher priority from the P_1. For the next injection, P_3 can attach the higher priority (Priority 8 in this example) so as to become the stronger connection than before. From another point of view, CP represents the strength of connections of traveling head flits.

Case 2: In Fig. 3(b), one pending path encounters contention with another owning a higher CP, but the AP is lower. Under SAP principle, winner path can travel continuously to destination. However, the router of loser path will not swap AP to obtain a lower priority than itself. Therefore, loser path will not swap AP as shown step2 in Fig. 3(b).

Case 3: As shown step 1 in Fig. 3(c), when a head flit encounters another pending path with low CP from the direction of the source node. The pending path with low CP (loser path) must be blocked because it cannot win arbitration. In this case, AP will be not exchanged as shown step 2 in Fig. 3(c).

3.3   SAP Scheme for 4-Degree Network Topologies

As shown Fig. 4, the SAP mechanism for the ring network can be extended to torus and mesh topologies. To illustrate the proposed arbitration strategy for mesh and torus network topologies, we describe examples of routing and flow control on these different network topologies in Fig. 4. The higher CP indicates the winner path On the opposite is the loser paths in Fig.4 (a), there are three pending path is transition, gray routers are means collided. As shown in Fig.4 (b), when there are two pending paths, one is attaches CP 5 toward R3, the other is pending path with CP 2, pending path will return AP via the same channel. Pending path with higher CP will handing over its higher AP to the loser path and then forward toward the destination router, As shown in Fig.4 (c), SAP scheme has to arbitrate one, the winner must observe the SAP scheme handing over its AP to the loser path and then forward toward the destination router. As shown in Fig.4 (d), the loser path (with CP 2) will be canceled, and waiting until available to transfer these paths does not affect each other. In short, when two pending paths go though the same output at a router, the SAP can be used to arbitrate a winner path for the most topologies.

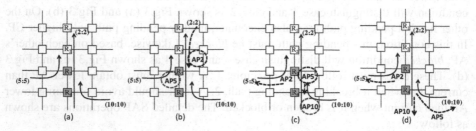

Fig. 4. The routing rule of pending paths

4   Experimental ResultsIn

This section using NS2 [13], we will present experimental results about impacts of performance. Fix priority, Distance base (Db), Renaming path (Rp), aging-based and SAP strategies will be compared.

4.1   Dual-Ring Topology

Due to the concept in Hyperscalar computing, more often communicate with devices on neighboring places, so the transmission distance poisson distribution of an average of 2 to determine the receiving destination, The transmission of information in the case of heavy load, we analyze in the injection rate 90% and the packet number of poisson distribution by an average of 64, The total amount of data is 160000. Because it is a short distance transmission, SAP strategy can exchange priority to increase opportunity of successful connection.

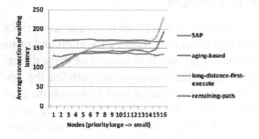

Fig. 5. Average latency

Table 1. Waiting latency per connection

Node16 Dual ring Packet 64 (poisson mean)	SAP	Aging based	Long distance first-execute	Renaming path
The maximum waiting latency of connection	1254	1010	5530	17306
WFC	104.7873	109.9973	152.2084	138.8079
Average connection of the maximum waiting latency	9.3035	5.94	36.46	124.676

WFC: Waiting cycle per connection

As shown Fig. 5, SAP strategy is fairness at each node, and the waiting latency is less than aging-based. But aging-based is the fairest as shown table1, average connection of the maximum waiting latency is means the lower number has the higher fairness.

4.2   Torus Topology

At 64 Node Torus topology, the transmission distance is average distribution to determine the receiving destination, the transmission of information in the case of heavy load, we analyze in the injection rate 90% and the packet number of poisson distribution by an average of 16, The total amount of data is 640000. As shown Fig. 6, Aging-based and SAP strategy can have a fair strategy. The number of packet collisions is decreased in the transmission path, so that the performance of Distance-path and remaining-path is better than SAP.

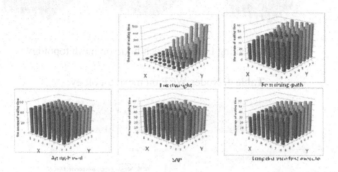

Fig. 6. The waiting latency at each node on torus topology

Table 2. 64 nodes WCP

Node64 Torus Packet 16 (poisson mean)	SAP	Aging-based	Long-distance-first-execute	Remaining-path
WPC	46.337	49.212	41.570	44.723
Normalized Waiting latency	1	1.06205	0.89712	0.96517

WCP : Waiting cycle per connection

Table 3. 576 nodes WCP

Torus Node576 Packet 16 (poisson mean)	SAP	Aging-based	Long-distance-first-execute	Remaining-path
WPC	200.3532	209.9437	201.1096	204.9217
Normalized Waiting latency	1	1.04786	1.00377	1.0228

WPC : Waiting cycle per connection

As shown table 2, long-Distance first execute has the better performance at 64 node, we increase the distance to 576 node, because the distance based has long latency at short distance, as shown table. 3, SAP strategy become the best performance.

4.3   Mesh Topology

As shown Fig.7, due to restriction on the location of the mesh topology, the transmission in the topology edge has the poor performance. This paper presents the weight of the merger SD (SAP + Distance) strategy to solve the restrictions on the location of the Mesh. The transmission distance is average distribution to determine the receiving destination, the transmission of information in the case of heavy load, we analyze in the injection rate 90% and the packet number of poisson distribution by an average of 16, The total amount of data is 640000.

Mesh topology will be the impact on the location and cause weight exchange not balance. We use the weight of the merger SD (SAP + Distance) strategy to solve the Mesh topology restrictions on the location, to achieve fairness transmission. Transmission speed is less than the Aging-based strategy, but very close.

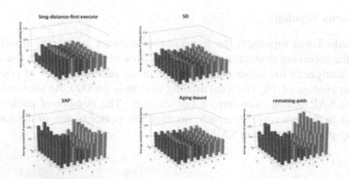

Fig. 7. The waiting latency at each node on mesh topology

Table 4. Nodes WCP on mesh topology

Mesh Node64 Packet 16 (poisson mesn)	SAP	SD(SAP +Distance)	Aging-based	Long-distance-first-execute	Remaining-path
WPC	75.02405	60.50607	60.34797	61.81031	73.22472
Normalized Waiting latency	1	0.80649	0.79974	0.82387	0.97601

WPC : Waiting cycle per connection

5   Conclusion

In the simulation, SAP arbitration mechanism, under the heavy load of high-density collision to exchange priority, the pending path winning arbitration will swap its priority with losers as a new priority in the next injection. it's also to improve the packet connection opportunities, so the SAP strategy at short distance will reduce the waiting latency. In the multi-core systems will communicate frequently the component on the neighboring places, SAP strategy can also reach each component has a fair transition, but on the Mesh topology makes the exchange of the priority is not easy, therefore this thesis presents the weight of the merger SD (SAP + Distance) strategy to solve the restrictions on the location of the Mesh to reach a fair transmission opportunity and improve the transmission performance.

References

1. Lu, R., Cao, A., Koh, C.: SAMBA-Bus: A High Performance Bus Architecture for System-on-Chips. IEEE Transactions on VLSI Systems 15(1), 69–79 (2007)
2. Handbook, Cell Broadband Engine Programming Handbook, Version 1.1. IBM (published April 24, 2007)
3. Gschwind, M., Peter Hofstee, H., Flachs, B., Hopkins, M., Watanabe, Y., Yamazaki, T.: Synergistic processing in cell's multicore architecture. IEEE Computer Society
4. Rantala, V., Lehtonen, T., Plosila, J.: Network on Chip Routing Algorithms, TUCS Technical Report, No. 779 (August 2006)

5. Wolkotte, P.T., Smit, G.J., Rauwerda, G.K., Smit, L.T.: An energy efficient reconfigurable circuit-switched network-on-chip. In: Proc. Int. Parallel Distrib. Process. Symp., p. 155a (2005)
6. Abts, D., Weisser, D.: Age-based packet arbitration in large-radix k-ary n-cubes. In: SC (2007)
7. Lee, M.M., Kim, J., Abts, D., Marty, M., Lee, J.W.: Probabilistic Distance-Based Arbitration: Providing Equality of Service for Many-Core CMPs. IEEE/ACM Micro, 509–519 (December 2010)
8. Liljeberg, P., Plosila, J., Isoaho, J.: Self-timed ring architecture for SoCpplications. In: Proc. SOC Conference, pp. 359–362 (September 2003)
9. AMBA Specification. ARM Ltd. Hall
10. Wiklund, D., Liu, D.: SoCBUS: Switched Network on Chip for Hard Real Time Embedded Systems. In: Proc. Int'l Parallel and Distributed Processing Symp. 2003, pp. 78–85 (2003)
11. Thonnart, Y., Vivet, P., Clermidy, F.: A fully-asynchronous low-power framework for GALS NoC integration. In: Proc. Conf. DATE, pp. 33–38 (2010)
12. Sheibanyrad, A., Panades, I.M., Greiner, A.: Systematic comparison between the asynchronous and the multi-synchronous implementations of a network-on-chip architecture. In: Proc. Des., Autom. Test Eur. Conf., p. 1090 (2007)
13. Ali, M., Welzl, M., Adnan, A., Nadeem, F.: Using the NS-2 network simulator for evaluating network-on-chips (NoC). In: Proc. IEEE 2nd Int. Conf. Emerging Technol., p. 506 (2006)

Energy-Aware Compiler Optimization for VLIW-DSP Cores

Yung-Cheng Ma, Tse-An Liu, and Wen-Shih Chao

Department of Computer Science and Information Engineering,
Chang-Gung University
yungcheng.ma@gmail.com

Abstract. VLIW-DSP processor cores are widely used in embedded
SoCs. Improving the energy efficiency becomes one of the key issues in
designing a VLIW-DSP core. This paper proposes compiler optimization algorithms to reduce the register file power in a VLIW-DSP processor. The optimization is targeted to VLIW processors in which each
execution slot is associated with a low-powered local register file. Instruction scheduling and register allocation algorithms are proposed to
direct operand accesses to the local register files. We propose energy-
aware list scheduling algorithm to reduce cross-slot data dependencies
without affecting the program execution time. Constrained by the instruction scheduling result, energy-aware register allocation is performed
through weighted graph coloring. Evaluation with MiBench benchmark
suite shows that our approach reduces over 50% of data transfer energy with low hardware cost. This research shows a cost-effective way to
design an energy-efficient VLIW- DSP processor.

Keywords: energy-aware instruction scheduling, weighted graph coloring, register allocation, VLIW-DSP processor.

1 Introduction

In recent years, VLIW DSP processors are widely used in SoC (system-on-chip)
platforms. Many SoC platforms (such as [1], [2], and [3]) contain DSP (digital
signal processing) processor cores to provide high performance processing over
streamed data. VLIW (very-long-instruction-word) architecture is the major architecture for the DSP cores (such as [4] and [5]). Energy efficiency is one of the
key issues to design embedded VLIW-DSP cores.

Register file design plays the key role in designing an energy-efficient VLIW-
DSP core. The register file power grows as a square function of access ports ([6]).
The major approach is to design some means of partitioned register file architecture [4][5][7][8]. The ELM (Efficient Low-power Microprocessor) architecture [9]
is the extremal approach to reduce the register file power: having each execution
slot associated with a small local register file. While it is believed that such an
extremal architecture has the potential to save data-supply power, the required
compiler support is not mentioned.

J.-S. Pan et al. (Eds.): *Advances in Intelligent Systems & Applications*, SIST 21, pp. 779–788.
DOI: 10.1007/978-3-642-35473-1_77   © Springer-Verlag Berlin Heidelberg 2013

This paper proposes compiler optimization algorithms to exploit energy-efficiency of a VLIW-DSP core with ELM-like architecture. Energy efficiency is exploited through instruction scheduling and register allocation. Evaluation over MiBench ([10]) benchmark suite shows the effectiveness of our approach.

2 Related Work

Modern VLIW-DSP cores are some means of partitioned register file architecture [4,5,7]. The registers set is partitioned into clusters. Each cluster has a register file connected to limited amount of functional units. The bandwidth for inter-cluster data transfer is quite small. The Texas Instruments C6000-series DSP processor ([4]) is an 8-issue VLIW processor with two homogeneous clusters. The Freescale Starcore DSP processor ([5]) adopts the heterogeneous clustered architecture design: the address arithmetic cluster has two address generation units (AGU) and the data arithmetic cluster has four execution slots for arithmetic operations. Lin et. al [7] proposes ping-pong register file for inter-cluster data exchange. Terechko et. al [11] give a classification on the clustered architectures and propose compiler scheduling algorithms for such clustered architectures.

Our approach to reduce register file power is motivated by the ELM (Efficient Low-power Microprocessor) architecture [9]. The key idea is to associate a low-powered local register file with each execution slot. The success lies on the compiler optimization to direct most of operand accesses to local register files.

3 The Energy-Aware Code Optimization Problem

Figure 1 shows the target architecture — the ELM-like VLIW data path. In this architecture, execution slots exchange data through a **shared register file** (SRF). Besides SRF, each execution slot has a **local register file** (LRF) connected only to functional units within the execution slot.

The architecture lies on the compiler optimization to reduce the dynamic energy on accessing register operands. Zyuban et. al [6] states that the dynamic energy per access to a register file is proportional to the amount of access ports. The SRF has lots of access ports and hence each access has higher energy consumption. On the other hand, accessing a LRF is cheap on energy cost. It is the task of the program allocation in a compiler to direct most of operand accesses to local register files.

The program allocation problem is outlined as follows. We are given the intermediate representation (IR) of a code fragment without branch operations in SSA-form. The code fragment is represented as a series of machine-level operations $\{I_0, I_1, I_2, ...\}$ accessing a set of temporaries (variables) $TR = \{t_0, t_1, t_2, ...\}$. The **program allocation** allocates operations and temporaries onto execution slots and register files in the target architecture, aimed at minimizing the energy of accessing register operands without affecting the program execution time.

Instruction scheduling sets the constraints to register allocation and plays the central role to improve energy efficiency. Instruction scheduling divides the set of temporaries into two classes: shared and local temporaries. A **shared temporary**

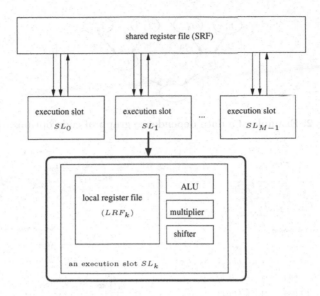

Fig. 1. The target architecture — ELM-like VLIW data path

is a temporary accessed by multiple execution slots and has to be allocated in the power-hungry SRF. On the other hand, a **local temporary** is only accessed by one execution slot and is preferred to be in the low-powered LRF. The optimization is to reduce cross-slot dependencies in an ASAP (as-soon-as-possible) schedule.

Consider the following code fragment as an example.

```
I0: t1 = t0*c1
I1: t3 = t2+c2
I2: t4 = t1*t3
I3: t5 = t5+1
I4: t7 = t6*c3
I5: t8 = t5*t7
I6: t9 = t4+t8
```

Figure 2 shows the data dependence graph (DDG) with dependent temporaries marked with edges. We assume that the latency for each operation is of 1 cycle. Figure 3 illustrates the optimization guideline. We draw a schedule as a DDG layout. The preferred schedule has only one cross-slot edge (I_5, I_6) and the temporary t_8 is the only shared temporary that has to be in SRF. Remaining temporaries are local temporaries and are allocated in LRFs. On the other hand, with the non-preferred schedule and resulted register allocation, the program is executed with higher amount of accesses through the power-hungry SRF.

4 Energy-Aware List Scheduling

We propose energy-aware list scheduling following the optimization guideline. The algorithm assigns each operation to an empty position in the schedule chart following some topological order of the DDG. Design issues are as follows:

- Issue 1: the **sequencing policy** to establish the topological order.
- Issue 2: the **operation placement policy** to place each operation on a position of the schedule chart.

We start from discussion the operation placement policy.

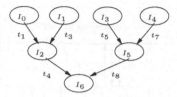

Fig. 2. Example of a data dependence graph of code optimization

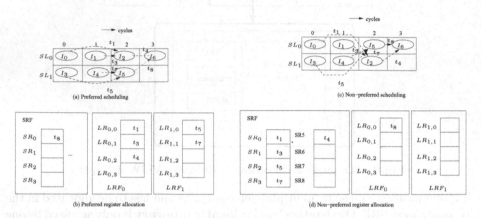

Fig. 3. Example of the optimization guideline

4.1 Operation Placement Policy

We define the *affinity* to guide the position selection. Consider to place operation I_j on a partial schedule S. The affinity of I_j to an execution slot SL_k, denoted $AF(SL_k, I_j)$, is the predicted amount of edges connecting I_j to operations at SL_k. The concept is formalized through *direct* and *indirect* affinity. Two operations I_0 and I_1 have **direct affinity** if there is an edge in G connecting the two vertices. The measure, denoted $DAF(I_0, I_1)$, is as follows.

$$DAF(I_0, I_1) = \begin{cases} 1 \text{ if}(I_0, I_1) \in E(G) \\ 0 \text{ if}(I_0, I_1) \notin E(G) \end{cases}$$

Two operations I_0 and I_1 have **indirect affinity** if they have some common descendant. The measure, denoted $IAF(I_0, I_1)$, is

$$IAF(I_0, I_1) = 2^{-d},$$

where d is the shortest distance in DDG to a common descendant of I_0 and I_1. The **affinity** between two operations is the sum of direct and indirect affinity.

$$AF(I_i, I_j) = DAF(I_i, I_j) + IAF(I_i, I_j)$$

Let $A_{op}^{-1}(SL_k)$ be the set of operations already in SL_k. The **affinity** from I_j to SL_k is the total affinity over all operations already scheduled at SL_k.

$$AF(SL_k, I_j) = \sum_{I_i \in A_{op}^{-1}(SL_k)} AF(I_i, I_j)$$

Figure 4 shows examples of the affinity measure. Consider to schedule the DDG in Figure 4(a) with the operation index as the scheduling sequence. Figure 4(b) shows the first snapshot, which extends a partial schedule containing I_0. Affinity measures from I_1 to slots SL_0 and SL_1 are as follows.

$$AF(SL_0, I_1) = AF(I_0, I_1) = 2^{-1}$$

$$AF(SL_1, I_1) = AF(SL_2, I_1) = 0$$

Indirect affinity leads I_1 to be scheduled at SL_0 and makes the chance to save data transfer energy among operations I_0, I_1, and I_4. The second snapshot in Figure 4(c) is to place I_5. Affinity measures are as follows.

$$AF(SL_0, I_5) = AF(I_0, I_5) + AF(I_1, I_5) + AF(I_4, I_5) = 2^{-1} + 2^{-1} + 2^{-1} = \frac{3}{2}$$

$$AF(SL_1, I_5) = AF(I_2, I_5) + AF(I_3, I5) = 1 + 1 = 2$$

Affinity from I_5 to SL_0 comes from indirect affinity while the affinity to SL_1 comes from direct affinity. By making direct affinity higher than indirect affinity, the affinity measure leads I_5 to be placed at the preferred slot SL_1.

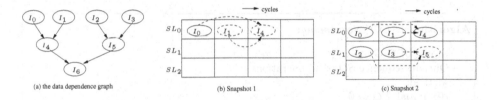

(a) the data dependence graph   (b) Snapshot 1   (c) Snapshot 2

Fig. 4. Snapshots of scheduling by affinity

4.2   Establishing the Scheduling Sequence

The sequencing guideline is: trying to build a topological order that interleaves operations with immediate common ancestors and descendants. Scheduling two operations with immediate common ancestors or descendants in parallel generates un-avoidable cross-slot dependencies. We try to select a set of parallel operations in which no two operations having immediate common ancestors and descendants. Figure 5 shows the effect. For a set of parallel operations shown in Figure 5(a), it is preferred to execute $\{I_0, I_3\}$ in cycle 0. With the selection, we have chance to reduce cross-slot edges even with ASAP placement. In Figure 5(b), both I_1 and I_4 are scheduled in their earliest available time. The affinity measure leads I_1 and I_4 to slot SL_0 and avoids all cross-slot data transfer.

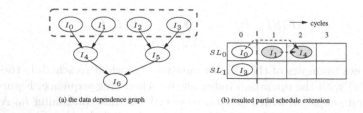

(a) the data dependence graph   (b) resulted partial schedule extension

Fig. 5. Guideline to select parallel operations

The sequencing guideline is realized by finding independent sets in an *operand sharing graph* (OSG). An **operand sharing graph** is defined over a set of parallel operations. A vertex of OSG stands for an operation and an edge connects two operations having immediate common ancestors and descendants. An independent set in a graph H is a set of vertices in which no two vertices are adjacent by an edge ([12]) in H. An independent set of OSG thus stands for a set of operations having no immediate common ancestors and descendants.

Figure 6 shows the IS-LTF algorithm to establish the scheduling sequence. The algorithm is featured by (1) finding independent sets (IS) from operand sharing graphs, and (2) following the longest-tail-first (LTF) policy to establish the sequence. We use the greedy algorithm from [12] to extract an independent set $IS(u)$ containing vertex u in OSG.

Algorithm *ISLTF_Sequencing*:

- Input: DDG G
- Output: sequence SQ of all operations
- Method:
    (0) Initial: $SQ \leftarrow \emptyset$
    (1) leveling operations in $V(G)$ by $Level(I_j)$
        * $Level(I_j)$ is the longest path from the source vertex of G to I_j
    (2) **for** each level i **do**
        (2.1) establish operand sharing graph G_i over vertices V_i in level i
        (2.2) **while** G_i is not empty **do**
            (2.2.1) pick $u \in V(G_i)$: the vertex stands for the operation with the longest tail in DDG
            (2.2.2) establish $IS(u)$: the maximum independent set containing u in G_i
            (2.2.3) append vertices in $IS(u)$ to sequence SQ in non-increasing order of tail length
            (2.2.4) remove all vertices in $IS(u)$ from G_i

Fig. 6. Framework of the IS-LTF sequencing algorithm

4.3 Summary: The EALS Algorithm

We summarize previous discussions to form the scheduling algorithm in Figure 7. Each operation I_i is scheduled at its earliest available time, $\tau = EAT(I_i)$, which is the earliest cycle to satisfy all dependencies in DDG. When there are multiple empty positions at cycle $EAT(I_i)$, the affinity measure leads I_i to the slot SL_k that is predicted to generate highest amount of intra-slot DDG edges.

Algorithm $EALS$:

Input: (G, ES)

Output: schedule S

Method:

(1) find operation sequencing SQ by Algorithm $ISLTF\_Sequencing$
(2) **for** each operation I_i in the order of SQ **do**
  (2.1) select cycle τ:

$$\tau = EAT(I_i)$$

  (2.2) select slot SL_k:

$$SL_k = \mathrm{argmax}\{AF(SL_m, I_i)|SL_m \in ES \text{ and } (SL_m, \tau) \text{ is empty}\}$$

  (2.3) place I_i at position (SL_k, τ) of the schedule chart S

Fig. 7. The energy-aware list scheduling algorithm

5 Energy-Aware Register Allocation with Weighted Graph Coloring

We now turn to energy-aware register allocation problem. Assume that the shared register file is capable of storing all register operands. We focus on selecting local temporaries for local register files to fit the capacity constraints.

Allocation for LRFs is optimized through weighted graph coloring. Modified from traditional graph coloring problem, the **maximum weighted K-colorable sub-graph** (MWKS) problem is defined for the energy-aware allocation. The modification is that each vertex is associated a weight to indicate energy contribution of a temporary. The optimization problem is to find the sub-graph with maximum total weight that is K-colorable: vertices adjacent by an edge are not painted the same color.

Figure 8 shows an example of the MWKS problem. It is impossible to paint the graph with $K = 3$ colors due to the clique $\{v_0, v_1, v_2, v_3\}$. The painting in Figure 8 shows the optimal solution for the instance of MWKS problem.

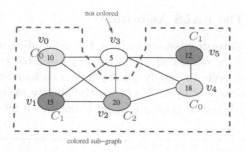

colored sub–graph

Fig. 8. Example of MWKS problem

Energy contribution of a temporary is estimated from the access count. For a temporary t_j allocated in the shared register file, the energy contribution is:

$$\hat{e}(t_j) = \hat{e}_{wr} + \hat{e}_{rd} * |COP(t_j)|,$$

where $\hat{e}_{wr}$ and $\hat{e}_{rd}$ is the energy per write and read to the SRF, respectively.

To allocate temporaries onto the local register file LRF_k, the interference graph $IG_k = (V, E)$ is defined as follows. The vertex set is the set of all local temporaries of SL_k. Similar to traditional register allocation, we put an edge $e = (t_i, t_j)$ in E if the two temporaries interferes on the live range. The weight of a vertex representing t_i is $\hat{e}(t_i)$, the energy contribution of t_i if t_i is allocated in SRF. By setting the color set as the set of registers in LRF_k, an algorithm for MWKS problem tries to select the set of local temporaries with maximum energy contribution to fit into LRF_k.

An example of weighted interference graph for local register file allocation is shown in Figure 9. This example assumes that each read and write consumes the same amount of energy on a register file.

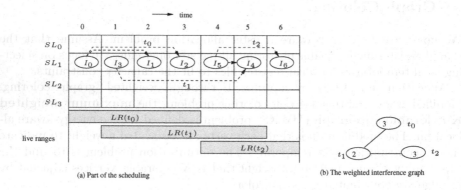

(a) Part of the scheduling   (b) The weighted interference graph

Fig. 9. Example of weighted interference graph

We follow the Chaitin's algorithm framework ([13]) to devise a heuristic for the MWKS problem. Our modification is the policy to select a victim not to be assigned a color. We select the minimum weighted vertex neighboring to the set of lowest degree vertices as the victim. The degree of a graph is the minimum degree among all vertices. With this policy, removing a victim reduces the graph degree by 1 and helps other vertices to be colorable.

6   Evaluation of the Architecture and Program Allocation

We take MiBench ([10]) as the benchmark programs for the evaluation. The proposed algorithms are implemented to obtain program allocation results. Energy consumption of register files is obtained using CACTI power model [14] for 65nm process. The evaluation is for a 4-issue VLIW architecture.

Figure 10 shows the evaluation results. Figure 10(a) shows the access ratio to SRF. Figure 10(b) shows the energy saving ratio:

$$ESR = \frac{\text{Energy with the proposed approach}}{\text{Energy if all operands are in SRF}}.$$

The result indicates that, associating each execution slot with 4 local registers, we will obtain more than 50% reduction on dynamic data-supply energy.

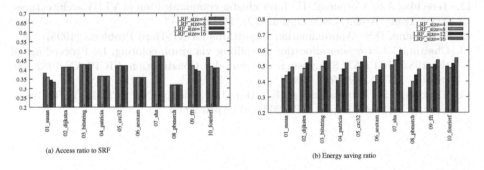

(a) Access ratio to SRF

(b) Energy saving ratio

Fig. 10. Energy efficiency of the proposed approach

7   Conclusions

The high power consumption contributed by register files has been bothering VLIW-DSP designers for more than one decade. The result of this research shows an optimistic way to attack the data-supply energy: associating a small local register file with each execution slot. The contributed compiler optimization algorithms successfully exploit the energy efficiency of the architecture. The most attractive point of this approach is the low hardware cost: each execution slot requires only 4 local registers to obtain more than 50% energy saving. This result turns the research direction back to traditional VLIW philosophy: keeping the hardware simple and left the difficult part to the compiler.

References

1. Texas Instruments, OMAP 5 mobile applications platform (2012)
2. Philips: Philips nexperiahighly integrated programmable system-on-chip (2012)
3. St. Nomadik: St nomadik multimedia processor (2012)
4. Texas Instruments, Tms320c6455 fixed-point digital signal processor (2008)
5. Freescale Semiconductor, Tuning C code for StarCore-based digital signal processors (2008)
6. Zyuban, V., Kogge, P.: The energy complexity of register files. In: Proceedings of the 1998 International Symposium on Low Power Electronics and Design, ISLPED 1998, pp. 305–310. ACM, New York (1998)
7. Lin, Y.-C., Lu, C.H., Wu, C.-J., Tang, C.-L., You, Y.-P., Moo, Y.-C., Lee, J.-K.: Effective code generation for distributed and ping-pong register files: A case study on pac vliw dsp cores. J. Signal Process. Syst. 51, 269–288 (2008)
8. Nagpal, R., Srikant, Y.: Compiler-assisted power optimization for clustered VLIW architectures. Parallel Computing 37(1), 42–59 (2011)
9. Dally, W., Balfour, J., Black-Shaffer, D., Chen, J., Harting, R., Parikh, V., Park, J., Sheffield, D.: Efficient embedded computing. Computer 41(7), 27–32 (2008)
10. Guthaus, M., Ringenberg, J., Ernst, D., Austin, T., Mudge, T., Brown, R.: Mibench: A free, commercially representative embedded benchmark suite. In: 2001 IEEE International Workshop on Workload Characterization, WWC 4, pp. 3–14 (December 2001)
11. Terechko, A.S., Corporaal, H.: Inter-cluster communication in VLIW architectures. ACM Trans. Archit. Code Optim. 4(2), 11 (2007)
12. Hochbaum, D.S.: Approximation Algorithms for NP-Hard Problems (1995)
13. Chaitin, G.J.: Register allocation & spilling via graph coloring. In: Proceedings of the 1982 SIGPLAN Symposium on Compiler Construction, SIGPLAN 1982, pp. 98–105. ACM, New York (1982)
14. Thoziyoor, S., Muralimanohar, N., Ahn, J.H., Jouppi, N.P.: Cacti 5.1. HP Laboratories Technical Report HPL-2008-20 (April 2008)

On the Variants of Tagged Geometric History Length Branch Predictors

Yeong-Chang Maa and Mao-Hsu Yen

Department of Computer Science and Engineering
National Taiwan Ocean University
Keelung, Taiwan
ycmaa@mail.ntou.edu.tw

Abstract. With the incessant pursuit for high performance, cost effective and power efficient processor design in recent years, how to provide performance with affordable hardware and power consumption has become an important issue. In this paper, we study and evaluate several variants on the TAgged GEometric history length (TAGE) branch predictors for better power, cost and performance portfolio, including fast-TAGE (f-TAGE), Fixed-Interleaving-TAGE (FI-TAGE), Non-Fixed-Interleaving TAGE (NFI-TAGE) and Bit-flipping-Interleaving TAGE (BI-TAGE). We analyze and empirically study our proposed scheme along with the original TAGE with respect to branch prediction accuracy, critical path delay, hardware overhead and power consumption. It is shown, among the proposed variants that f-TAGE fares best, reducing critical path delay by over 20% while preserving prediction accuracy at affordable hardware and power requirements.

Keywords: f-TAGE, FI-TAGE, NFI-TAGE, BI-TAGE, critical path delay, branch prediction, variable length history predictor, processor architecture.

1   Introduction

For the past two decades, branch prediction has still been very important to improve processor performance. Branch prediction resolves control dependencies exposed in program instructions and improve instruction level parallelism, thus enabling deeper processor pipeline (and higher clock rate) or wider instruction issue for better performance results.

To achieve better processor pipeline utilization, advanced branch prediction techniques have been employed. It has been reported that variable length history branch predictors (VLHBP), such as FPB, PLB, TAGE and O-GEHL give higher prediction accuracy than traditional counterparts, such as gshare [11][12][15][17]. This paper empirically studies several variants of the exemplar TAGE predictor of the above VLHBPs: f-TAGE, FI-TAGE, NFI-TAGE and BI-TAGE... Among them, the proposed f-TAGE not only preserves branch prediction accuracy but also reduces critical path delay (by over 20%) at affordable hardware and power requirements.

The rest of this paper is organized as follows. Section 2 will review branch prediction and related background. Section 3 introduces variable length history branch

J.-S. Pan et al. (Eds.): *Advances in Intelligent Systems & Applications*, SIST 21, pp. 789–807.
DOI: 10.1007/978-3-642-35473-1_78   © Springer-Verlag Berlin Heidelberg 2013

predictor operation and architecture of the proposed TAGE variants. Section 4 presents simulation-based evaluation results and discussions. Finally section 5 gives the conclusion of the paper.

2   Background

This section reviews the branch prediction techniques and architecture for branch predictor (or branch prediction unit), including the branch direction predictor and branch target buffer.

Branch prediction technique and branch target buffer were first introduced by Lee and Smith in 1984 [1]. Static prediction techniques use plain guesses of global branch behaviors or semantics based dichotomy of branch and jump instructions to forecast branch directions, while dynamic counterparts rely on hardware automata and branch histories to predict the outcomes and target addresses for branch instructions. Although branch prediction can be carried out by static or dynamic approaches, aggressive pipelining predominantly favors the dynamic approach for its higher accuracy and less performance penalty from pipeline stalls. Hence this section focuses on the review of dynamic prediction techniques and related hardware design, i.e., the direction predictor, the BTB and the neural perceptron predictors.

2.1   Branch Prediction Unit (BPU)

The branch prediction unit (or branch predictor) is set in the instruction fetch stage in the processor pipeline. When (branch) instruction address enters BPU, it will concurrently access the direction predictor (DP) and the branch target buffer (BTB), as shown in Figure 1.

Since high performance processors adopt pipelined architecture, branch prediction/handling takes place across the instruction fetch (IF) stage, the instruction decode (ID) stage and the execution stage (EX). At IF stage, if the instruction address (branch PC) is hit in BTB, then target address will be used as the next instruction address. If the address is missed in BTB, the branch instruction will be recognized at ID stage. If the instruction is branch, then the branch target address will be written into BTB at EX stage.

2.2   Direction Predictor (DP)

The major goal for branch prediction is to decide whether a branch will be Taken (T) or Not Taken (NT). Branch prediction can be made statically or dynamically. The focus of this study, dynamic prediction, counts on hardware aids to predict branch direction on-the-fly. Different dynamic direction predictors were proposed by various researchers: bimodal predictor, Yeh's two-level adaptive predictor [2], McFarling's gshare, gselect and combined predictors [3], the Skewed predictor [4], Agree predictor [5], Bi-Mode predictor [6], YAGS predictor [7], and Filter [8] predictor. Lately neural network techniques such as perceptron have been applied for dynamic branch prediction [9] [10] and also inspired the introduction of variable length history branch predictors.

Figure 2 shows the most famous gshare branch predictor. It uses an XOR hashing function of branch address and branch history to provide index to the pattern history table (PHT), which in turn provides branch prediction result. The hashing function avoids aliasing problem inside PHT, which leads to higher prediction accuracy and thus better processor performance.

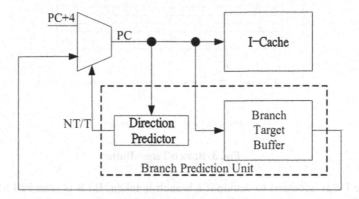

Fig. 1. Branch Prediction Unit

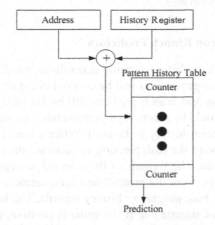

Fig. 2. Gshare branch predictor diagram

2.3   Branch Target Buffer (BTB)

BTB, a special purpose cache, is part of the branch prediction unit of processor pipeline's IF stage. Each cache entry has two fields: the branch instruction address and the branch target address, as shown in Figure 3.

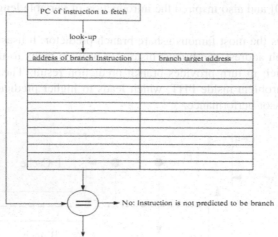

PC of instruction to fetch

look-up

address of branch Instruction	branch target address

No: Instruction is not predicted to be branch

Yes: then instruction branch and the target address is outputtec

Fig. 3. Branch Target Buffer

While DP is accessed to decide if a branch is taken, BTB is searched to get its target address, without awaiting address computation result at EX stage. Thus, early resolution of target address eases the pipeline inefficiency due to branches and greatly improves processor performance.

2.4   Neural Perceptron Branch Predictors

Perceptron based branch predictor has many generations, we discuss here the first and second generations. Later generations will be covered in the next section.

The first generation neural branch predictor [9] by Jimènez used perceptron to determine if the branch should be taken. A perceptron has a number of weights, depending on the length of branch history to be used. When a branch arrives, predictor will input the branch address to the hash function to generate the index to the perceptron table to pick a perceptron. The weights of the selected perceptron then are used with the branch histories to perform dot product and accumulate results in y, as shown in Figure 4 (y: output, w_i: bias weight, x_i: history record). The branch prediction is determined by the value of summation: if the value is positive, then the branch is predicted Taken, otherwise Not Taken, as shown in Figure 5. Finally the final branch outcomes will update the perceptron weights. If the branch outcome agrees with the prediction, then weights will be incremented, otherwise decremented accordingly.

The second generation neural predictor [10] improves operation latency by replacing the dot product with addition and subtraction, as exemplified in Figure 6. If the recorded branch history is taken, then an addition is performed. Otherwise a subtraction is performed. The rest of the operation is the same as the first generation (as shown in Figure 5).

$$y = w_0 + \sum_{i=1}^{n} x_i w_i.$$

Fig. 4. The dot product of weights and history record

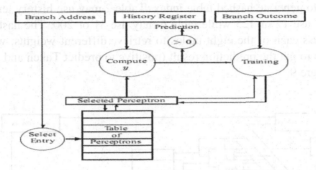

Fig. 5. The Neural perceptron branch predictor structure

The neural branch predictor counts on training algorithm to adjust the weights of perceptron for branch prediction, as shown in Figure 7. The final outcome of branch prediction, correct or incorrect, will be fed back to update the weights of the perceptron via the training mechanism.

Fast Path-Based (FPB) branch predictor [11] and Piecewise Linear Branch (PLB) predictor [12] followed as further improved variants of neural predictor, but they are omitted due to space limit. Readers can refer to [11] [12] [27] for more details.

bit	0	1	2	3	Bias
result	NT	T	T	NT	Input
Branch history	−1	1	1	−1	1
Weights	1	30	−2	−20	10
Prediction	−1 + 30 − 2 + 20 + 10 = 57 ≥ 0 → Predict Taken				

Fig. 6. The second generation of Neural Perceptron Branch Predictor

if *prediction* ≠ *outcome*

$$W[i,0] := W[i,0] + \begin{cases} 1 & \text{if } outcome = taken \\ -1 & \text{if } outcome = not\_taken \end{cases}$$

for j **in** $1..h$ **in parallel do**

$$W[i,j] := W[i,j] + \begin{cases} 1 & \text{if } outcome = G[j] \\ -1 & \text{if } outcome \neq G[j] \end{cases}$$

end for
end if
$G := (G << 1) \text{ or } outcome$

Fig. 7. Training algorithm

3 Variable Length History Predictors

3.1 Optimized GEometric History Length Branch Predictor, O-GEHL

O-GEHL predictor [15][16] is a way of using multiple tables to compute outputs for branch forecast, as illustrated in Figure 13. Eight fixed table for branch histories are used. Each table has independent hash function for access. For the 8KB configuration, Table 0 has 2K entries; each entry is 5-bit weight. Table 1 has 1K entries; each entry is 5-bit weight. Table 3 through Table 7 has 2K entries; each entry is 4-bit weight. To avoid table access conflicts, table 2, table 4 and table 6 may use branch history of variable length to generate hashed table index. Table 2 may use history length of 5 or 79; table 4 may use 12 or 125 while table 6 may use 31 or 200. The hashed indexes are used to access each of the eight tables to retrieve different weights, which are in turn summed up to get the prediction result (positive to predict Taken and vice versa), as shown in Figure 9.

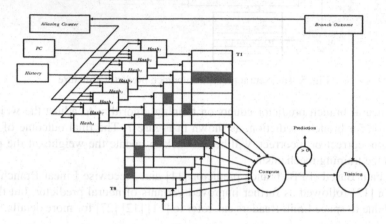

Fig. 8. Optimized GEometric History Length branch predictor structure

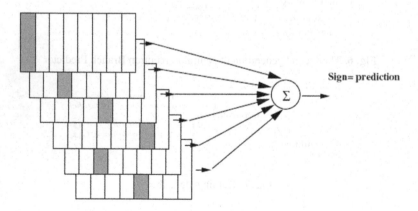

Fig. 9. O-GEHL prediction operation

3.2   TAgged GEometric History Length Branch Predictor, TAGE

TAGE predictor [17] [18] combines a default predictor (such as gshare) table, T0, with some tables of variable history length for branch forecast. Each table has independent hash function using different historical length; usually the latter table (T2, T3, etc.) tends to use longer history length. Every table entry has three fields: predicted (*pred*), label (*tag*) and the effective bits (*u*).   When the program counter sends in a branch address, the hash function for each table will produce an individual index to access the corresponding table. The tag field of each selected table entry will be matched against the tag field of incoming hashed index. If there is a match in a higher table, the resulting *pred* field will be selected and override results from lower tables (e.g., T3 overrides T2, T2 overrides T1, and T1 overrides T0). If there is no match among higher tables, the prediction from the default predictor will be used. The operation of (5 tables, T0 plus T1~T4) TAGE is shown in Figure 10. TAGE predictor may use as many as 12 tables (T0 plus T1~T12).

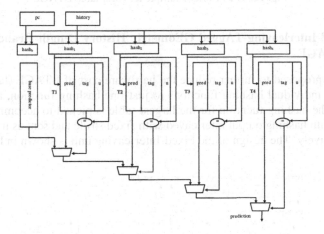

Fig. 10. TAgged GEometric history length Branch Predictor structure

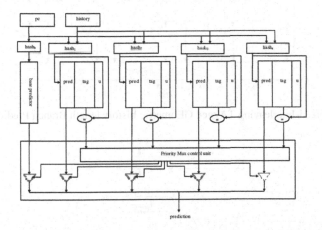

Fig. 11. Fast TAgged GEometric history length Branch Predictor structure

3.3   Fast-TAgged GEometric History Length Branch Predictor, f-TAGE

Because the serial multiplexer chain takes up consider-able amount of in the critical path, f-TAGE (fast TAGE) is proposed. f-TAGE is a variant from TAGE, with the serial multiplexer chain replaced by the priority multiplexer as depicted in Figure 11. The design of the priority multiplexer for 5-table f-TAGE is shown in Figure 12.

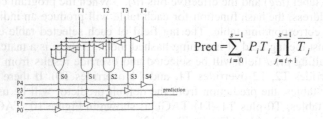

$$Pred = \sum_{i=0}^{s-1} P_i T_i \prod_{j=i+1}^{s-1} \overline{T_j}$$

Fig. 12. The Priority Multiplexer for 5-table f-TAGE

3.4   Fixed-Interleaving-TAgged GEometric History Length Branch Predictor, FI-TAGE

FI-TAGE is proposed to further reduce the hardware cost of f-TAGE (fast TAGE) by merging the individual hashing functions to just one hashing function, as depicted in Figure 13. The hashing fucntion for the largest table is picked to accommodate for all the tables, with hashing output interleaved with fixed stride and sent as indices to each table respectively. The design of the Fixed-Interleaving unit is shown in Figure 14.

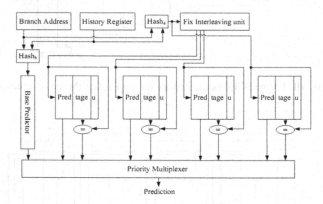

Fig. 13. Fixed-Interleaving TAgged GEometric history length Branch Predictor structure

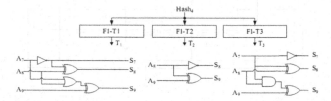

Fig. 14. The fixed-interleaving unit of FI-TAGE predictor

3.5   Non-Fixed-Interleaving-TAgged GEometric History Length Branch Predictor, NFI-TAGE

NFI-TAGE is another proposal similar to FI-TAGE to further reduce the hardware cost of f-TAGE (fast TAGE) by merging the individual hashing functions to just one hashing function, as depicted in Figure 15. The hashing fucntion for the largest table is picked to accommodate for all the tables, with hashing output interleaved with different strides and sent as indices to each table respectively. The design of the Non-Fixed-Interleaving unit is shown in Figure 16.

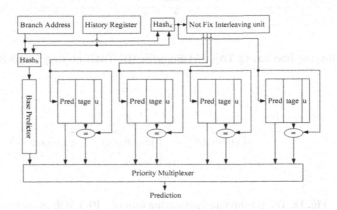

Fig. 15. Non-fixed-interleaving TAgged GEometric history length Branch Predictor structure

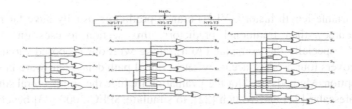

Fig. 16. The Non-fixed-interleaving unit for NFI-TAGE predictor

3.6   Bit-flipping-Interleaving-TAgged GEometric History Length Branch Predictor, BI-TAGE

BI-TAGE is a third proposal similar to FI-TAGE to further reduce the hardware cost of f-TAGE (fast TAGE) by merging the individual hashing functions to just one hashing function, as depicted in Figure 17. The hashing fucntion for the largest table is picked to accommodate for all the tables, with hashing output bits flipped and sent as indices to each table respectively. The design of the Non-Fixed-Interleaving unit is shown in Figure 18.

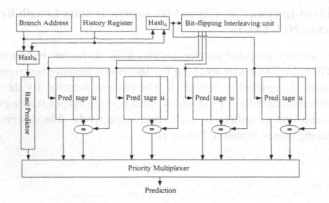

Fig. 17. Bit-flipping-Interleaving TAgged GEometric (BI-TAGE) history length Branch Predictor structure

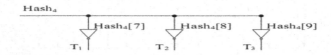

Fig. 18. The Bit-flipping-Interleaving unit of BI-TAGE predictor

4   Evaluation and Discussion

Since variable length history (VLH) branch predictors usually have far better prediction accuracy than traditional predictors, this section focuses on evaluations of VLHBP, especially the exemplar TAGE, along with our proposed variants. The evaluation covers four aspects: accuracy rate, critical path delay, hardware cost and power consumption. Methods used for the evaluation include the architectural simulators such as Simplescalar [21] and Wattch [22], to simulate SPEC 2000 [23] benchmark execution on the Alpha-21264 [24] processor; the CACTI tool [25] along with the logic effort [26] for delay estimation, and the PrimeTime PX [28] for power estimation.

4.1   Accuracy Rate

We apply Simplescalar/Alpha-21264 simulator along with SPEC 2000 benchmarks to evaluate the accuracy rate of these VLH branch predictors. The simulation settings are given in Table 1.

Table 1. Configurations for Simplescalar/Wattch simulation

Processor Core	
Instruction Window Issue width Function Units	RUU=64, LSQ=32 4 instructions per cycle 4 Int ALU, 1 Int mult/div 2 FP ALU, 1 FP mult/div
Memory Hierarchy	
L1 D-cache size L1 I-cache size L1 latency L2 Memory latency TLB Size	64KB, 4-way, 32B block 64KB, 2-way, 32B block 1 cycle Unified, 1MB, 4-way, LRU 11 cycle latency 200 cycle 64-entry, 30 cycle miss penalty
Branch Predictor	
Branch target buffer Return-address-stack	64-entry, 8-way 32-entry
Prediction Strategy	FPB (1KB / 2KB / 4KB / 8KB / 16KB / 32KB / 64KB / 128KB / 256KB) PLB (4KB / 8KB / 16KB / 32KB / 64KB / 128KB / 256KB) O-GEHL (8KB / 16KB /32KB) TAGE (8KB / 16KB / 32KB) TAGE_PM (8KB / 16KB / 32KB)

Since TAGE is reported to fare best among the VLH branch predictors [29], we focus on the results for TAGE and the variants thereof. The accuracy metric is misprediction rate. We take the misprediction rate of all SPEC2000 integer and floating point benchmarks to represent the accuracy rate of to compare the baseline TAGE predictor along with the variants. Figure 19 shows the accuracy result for the FI-TAGE predictor with table size from 8 through 256 KB.

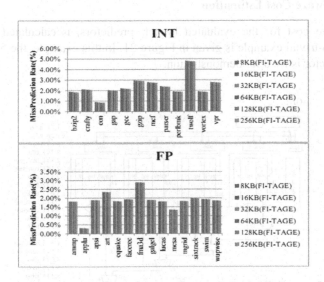

Fig. 19. The accuracy rate of FI-TAGE branch predictors of different configuration

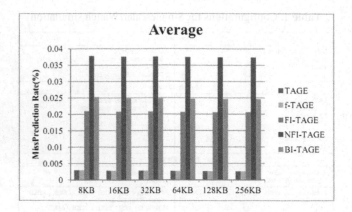

Fig. 20. Misprediction rate of TAGE, f-TAGE, FI-TAGE, NFI-TAGE and BI-TAGE

Figure 20 shows the accuracy result for TAGE, f-TAGE and the other three variants, FI-TAGE, NFI-TAGE and BI-TAGE. Since f-TAGE does not change the hashing arrangement, it shows the same result as the TAGE predictor. To save some hardware cost, the other three predictors tried to employ one hashing function for different history-length tables, yet they all suffer from higher misprediction rates. TAGE and f-TAGE can have misprediction rate less than 1%, in contrast the other variants have rates around 2~3.5%.

4.2   Hardware Cost Estimation

The hardware cost for the evaluated branch predictors. is calculated in transistor count. A non-trivial example is given in Figure 21. In this example, the 5-table TAGE branch predictor is used for demonstration.

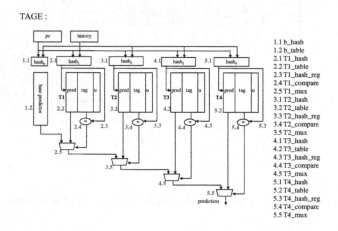

Fig. 21. The hardware cost example of 5-table TAGE branch predictor

Example : (8KB)
N_{bank} : bank number
N_{entry} : entry number
N_{pred} : 3-bit counter
N_{tag} : tag bit number
N_u : use-bit number
(each bit use SRAM tailor calculation, six transistors)

```
Base : N_bank : 1    N_entry : 10240    decoder_14 : 196612
T1 : N_bank : 1    N_entry : 512    N_pred : 3    N_tag : 7    N_u : 2    decoder_9 : 6138
T2 : N_bank : 1    N_entry : 512    N_pred : 3    N_tag : 7    N_u : 2    decoder_9 : 6138
T3 : N_bank : 1    N_entry : 1024    N_pred : 3    N_tag : 8    N_u : 2    decoder_10 : 12284
T4 : N_bank : 1    N_entry : 1024    N_pred : 3    N_tag : 8    N_u : 2    decoder_10 : 12284
```

```
1.1 b_hash : N_XOR2 * 14
1.2 b_table : N_bank * N_entry * 6 + decoder_14 (每個位元皆以SRAM計算故乘6)
Base : 1.1 + 1.2
12 * 14 + 1 * 10240 * 6 + 589852 = 712900
```

```
2.1 T1_hash : N_XOR2 * 4
2.2 T1_table : N_bank * N_entry * ( N_pred + N_tag + N_u ) * 6 + decoder_9
2.3 T1_hash_reg : N_tag * N_Dff
2.4 T1_compare : N_tag * N_XOR2
2.5 T1_mux : N_mux
T1 : 2.1 + 2.2 + 2.3 + 2.4 + 2.5
12 * 4 + 1 * 512 * ( 3 + 7 + 2 ) * 6 + 12306 + 7 * 16 + 7 * 12 + 6 = 49428
```

```
3.1 T2_hash : N_XOR2 * 6
3.2 T2_table : N_bank * N_entry * ( N_pred + N_tag + N_u ) * 6 + decoder_9
3.3 T2_hash_reg : N_tag * N_Dff
3.4 T2_compare : N_tag * N_XOR2
3.5 T2_mux : N_mux
T2 : 3.1 + 3.2 + 3.3 + 3.4
12 * 6 + 1 * 512 * ( 3 + 7 + 2 ) * 6 + 12306 + 7 * 16 + 7 * 12 + 6 = 49452
```

```
4.1 T3_hash : N_XOR2 * 10
4.2 T3_table : N_bank * N_entry * ( N_pred + N_tag + N_u ) * 6 + decoder_10
4.3 T3_hash_reg : N_tag * N_Dff
4.4 T3_compare : N_tag * N_XOR2
4.5 T3_mux : N_mux
T3 : 4.1 + 4.2 + 4.3 + 4.4
12 * 10 + 1 * 1024 * ( 3 + 8 + 2 ) * 6 + 26644 + 8 * 16 + 8 * 12 + 6 = 106876
```

```
5.1 T4_hash : N_XOR3 * 6 + N_XOR2 * 4
5.2 T4_table : N_bank * N_entry * ( N_pred + N_tag + N_u ) * 6 + decoder_10
5.3 T4_hash_reg : N_tag * N_Dff
5.4 T4_compare : N_tag * N_XOR2
5.5 T4_mux : N_mux
T4 : 5.1 + 5.2 + 5.3 + 5.4
26 * 6 + 12 * 4 + 1 * 1024 * ( 3 + 8 + 2 ) * 6 + 26644 + 8 * 16 + 8 * 12 + 6 = 106960
```

```
Total : Base + T1 + T2 + T3 + T4
712900 + 49428 + 49452 + 106876 + 106960 = 1025640
```

Fig. 21. *(continued)*

Because TAGE is a multi-table structure, we separately calculate the cost for each table. We use gshare predictor for the table T0 in our evaluation. For T1 to T12, they have their own decoder, hash function and TAG length. For SRAM cell, we assume a 6 transistor circuit is used.

We compared the hardware cost (table storage plus overhead) for TAGE and all proposed variants, including f-TAGE, FI-TAGE, NFI-TAGE and BI-TAGE.as shown in Figure 22 and Table 2. It can be observed that the cost for the f-TAGE is slightly higher (0.1~0.2%) while the cost for the other three variants are somewhat lower (around 0.5%).

Hardware Cost

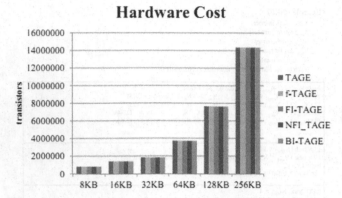

Fig. 22. The total hardware cost of branch predictors

Hardware Cost	TAGE	f-TAGE	FI-TAGE	NFI-TAGE	BI-TAGE
8KB	100.00%	100.01%	99.99%	100.01%	99.98%
16KB	100.00%	100.01%	99.94%	99.95%	99.93%
32KB	100.00%	100.02%	99.41%	99.42%	99.41%
64KB	100.00%	100.01%	99.72%	99.72%	99.72%
128KB	100.00%	100.00%	99.86%	99.87%	99.86%
256KB	100.00%	100.00%	99.93%	99.93%	99.93%

Fig. 23. The Relative Comparison of total hardware cost

4.3   CriticalPath Delay Estimation

Figure 23 depicts the Instruction Fetch stage pipeline circuit paths for branch predictors. Path 1 is for accessing instruction cache memory. Path 2 is the path for traversing the direction predictor. Figure 24 gives critical path delay calculation breakdown for f-TAGE predictor with varying table sizes.

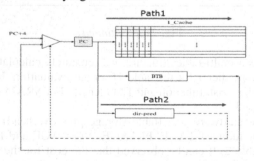

Fig. 24. The Instruction Fetch stage pipeline circuit paths for the branch predictors scheme

Figure 25 and Table 3 show the critical path comparison results for TAGE predictor and the other four variants. Except for the case of 8KB/16KB table size, f-TAGE and the rest excel TAGE by up to 20%. Among them, f-TAGE and BI-TAGE seem to perform better the rest two variants, FI-TAGE and NFI-TAGE.

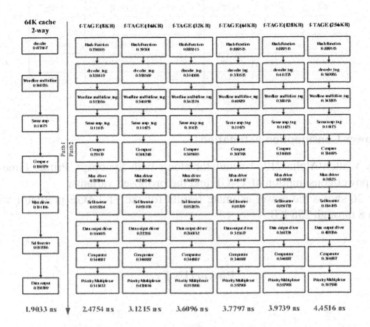

Fig. 25. Path Delay of f-TAGE schemes with 2-way 64-entry BTB

Critial Path Delay

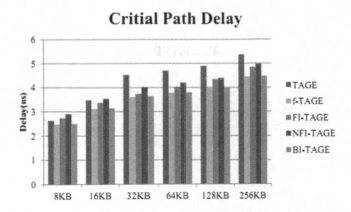

Fig. 26. Critical Path Delay Comparison of TAGE, f-TAGE schemes and the other three variants

Table 2. The Relative Comparison of Critical Path Delay

delay	TAGE	f-TAGE	FI-TAGE	NFI-TAGE	BI-TAGE
8KB	100.00%	94.30%	103.97%	110.19%	95.06%
16KB	100.00%	89.75%	97.05%	101.75%	90.33%
32KB	100.00%	79.76%	82.57%	88.42%	80.20%
64KB	100.00%	80.49%	85.89%	89.37%	80.91%
128KB	100.00%	81.26%	88.80%	89.83%	81.67%
256KB	100.00%	82.93%	90.23%	92.84%	83.30%

4.4   Power Consumption Estimation

We apply Wattch/Alpha-21264 simulator along with Synopsys Prime Time to estimate the power consumption of TAGE and the proposed variants. .

Figure 26 shows the power estimate results for the FI-TAGE predictor with table
size from 8 through 256 KB.

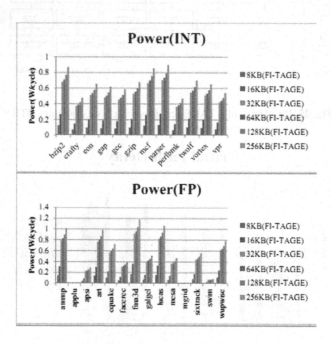

Fig. 27. Power consumption estimates of FI-TAGE branch predictors of different configuration

Figure 27 and Table 4 show the power estimate comparison results for TAGE predictor and the other four variants. As shown from the table, the FI-TAGE, NFI-TAGE and BI-TAGE predictors perform slightly better in power consumption by up to 0.5% .

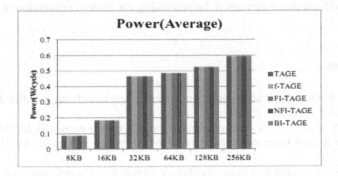

Fig. 28. Power Consumption Comparison of TAGE, f-TAGE schemes and the other three variants

Table 3. The Relative Comparison of Power Consumption

Power	TAGE	f-TAGE	FI-TAGE	NFI-TAGE	BI-TAGE
8KB	100.00%	100.00%	99.99%	100.03%	100.00%
16KB	100.00%	100.00%	99.96%	99.98%	99.97%
32KB	100.00%	100.00%	99.43%	99.47%	99.44%
64KB	100.00%	100.00%	99.74%	99.74%	99.76%
128KB	100.00%	100.00%	99.88%	99.91%	99.89%
256KB	100.00%	100.00%	99.95%	99.97%	99.95%

From above results, it can be observed that f-TAGE branch predictor has the highest prediction accuracy, and can cut the critical path delay by as high as 20%. The other three TAGE variants, FI-TAGE, NFI-TAGE and BI-TAGE can somewhat reduced the path delay and reduce the hardware overhead and power consumption by less than 1%, yet the incurred higher misprediction seem to suggest that it does not pay to go for further variation on the TAGE predictor design.

5   Conclusion

In this paper, we propose improved variants for the highly accurate TAGE predictor: f-TAGE by applying the Priority Multiplexer to reduce multi-level multiplexer gate delay and hashing function arrangement to save hardware and power overhead.

We analyze and empirically study these proposed schemes with respect to prediction accuracy, critical path delay, hardware requirement and power consumption. Although three variants, FI-TAGE, NFI-TAGE and BI-TAGE do not fare so well, the proposed f-TAGE can improve critical path delay by up to 21% and hardware overhead increase less than 1% as compared to the original TAGE predictor. This proposed f-TAGE predictor appears to be promising for future exploitation or implementation.

References

1. Lee, J., Smith, A.J.: Branch prediction strategies and branch target buffer design. IEEE Computer, 6–22 (January 1984)
2. Yeh, T.Y., Patt, Y.N.: Two-level adaptive training branch prediction. In: Proceedings of the 24th Annual International Symposium on Microarchitecture, pp. 51–61 (November 1991)
3. McFarling, S.: Combining branch predictors. Digital Equipment Corporation, WRL Tech. Note TN-36 (1993)
4. Michaud, P., Seznec, A., Uhlig, R.: Trading conflict and capacity aliasing in conditional branch predictors. In: Proceedings of the 24th Annual International Symposium on Computer Architecture, pp. 292–303 (May 1997)
5. Sprangle, E.E., Chappell, R., Alsup, M., Patt, Y.: The Agree predictor: A mechanism for reducing negative branch history interference. In: Proceedings of the 24th Annual International Symposium on Computer Architecture (May 1997)
6. Lee, C.-C., Chen, I.-C.K., Mudge, T.N.: The bimode branch predictor. In: Proceedings of the 30th Annual ACM/IEEE International Symposium on Microarchitecture, pp. 4–13 (December 1997)
7. Eden, A., Mudge, T.: The YAGS Branch Prediction Scheme. In: Proc. 31st International Symposium on Micro-architecture, pp. 69–77 (November 1998)
8. Chang, P., Evers, M., Patt, Y.: Improving branch prediction accuracy by reducing pattern history table interference. In: Proceedings of the International Conference Parallel Architecture and Compilation Techniques, pp. 48–57 (October 1995)
9. Jiménez, D.A., Lin, C.: Dynamic branch prediction with perceptrons. In: The Seventh International Symposium on High-Performance Computer Architecture, HPCA, pp. 197–206 (2001)
10. Jiménez, D.A., Lin, C.: Neural Methods for Dynamic Branch Prediction. ACM Transactions on Computer Systems 20(4), 369–397 (2002)
11. Jiménez, D.A.: Fast path-based neural branch prediction. In: Proceedings of 36th Annual IEEE/ACM International Symposium on Microarchitecture, MICRO 36, December 3-5, pp. 243–252 (2003)
12. Jiménez, D.A.: Piecewise linear branch prediction. In: Proceedings of 32nd International Symposium on Computer Architecture, ISCA 2005, June 4-8, pp. 382–393 (2005)
13. Jiménez, D.A.: Improved Latency and Accuracy for Neural Branch Prediction. ACM Transactions on Computer Systems 23(2), 197–218 (2005)
14. St. Amant, R., Jiménez, D.A., Burger, D.: Low-power, high-performance analog neural branch prediction. In: 2008 41st IEEE/ACM International Symposium on Microarchitecture, MICRO 41, November 8-12, pp. 447–458 (2008)

15. Seznec, A.: Analysis of the O-GEometric history length branch predictor. In: Proceedings of 32nd International Symposium on Computer Architecture, ISCA 2005, June 4-8, pp. 394–405 (2005)
16. Seznec, A.: Genesis of the O-GEHL branch predictor. Journal of Instruction Level Parallelism (April 2005)
17. Seznec, A., Michaud, P.: A case for (partially)-tagged geometric history length predictors. Journal of Instruction Level Parallelism (April 2006)
18. Seznec, A.: The L-TAGE predictor. Journal of Instruction Level Parallelism 9 (April 2007)
19. Jimènez, D.A., Keckler, S.W., Lin, C.: The impact of delay on the design of branch predictors. In: Proceedings of 33rd Annual IEEE/ACM International Symposium on Microarchitecture, MICRO 33, pp. 67–76 (2000)
20. Stark, J., Evers, M., Patt, Y.N.: Variable length path branch prediction. SIGOPS Oper. Syst. Rev. 32, 5 (1998)
21. Todd, M., et al.: SimpleScalar 3.0, http://www.simple-scalar.com/
22. Brooks, D.: Wattch Version 1.02, http://www.eecs.harvard.edu/~dbrooks/wattch-form.html
23. Standard Performance Evaluation Corporation, http://www.specbench.org/osg/cpu2000/
24. Kessler, R.: The Alpha 21264 microprocessor. IEEE Micro 19(2), 24–36 (1999)
25. http://www.hpl.hp.com/research/cacti/
26. Sutherland, I., et al.: Logic Effort: Designing Fast CMOS Circuits. Morgan Kaufmann series
27. Wang, Y.: Study and Analysis of Variable Length History Predictors. Master Thesis, Dept. of Computer Science and Engineering, National Taiwan Ocean University, Keelung, Taiwan (January 2011)
28. SYNOPSYS, http://www.synopsys.com/home.aspx, (online) http://www.synopsys.com/Tools/Implementation/SignOff/Pages/PrimeTime.aspx
29. Maa, Y.-C., et al.: Evaluating and Improving Variable Length History Branch Predictors. In: Proceedings of 2010 International Computer Symposium, Tainan, Taiwan (December 2010)

Author Index

Printed in the United States
By Bookmasters